Lecture Notes in Artificial Intelligence

Edited by R. Goebel, J. Siekmann, and W. Wahlster

Subseries of Lecture Notes in Computer Science

W9-CYX-764

Guoyin Wang   Tianrui Li
Jerzy W. Grzymala-Busse   Duoqian Miao
Andrzej Skowron   Yiyu Yao (Eds.)

Rough Sets and Knowledge Technology

Third International Conference, RSKT 2008
Chengdu, China, May 17-19, 2008
Proceedings

 Springer

Volume Editors

Guoyin Wang
Chongqing University of Posts and Telecommunications, P.R. China
E-mail: wanggy@cqupt.edu.cn

Tianrui Li
Southwest Jiaotong University, Chengdu, P.R. China, E-mail: trli@swjtu.edu.cn

Jerzy W. Grzymala-Busse
University of Kansas, Lawrence, KS, USA, E-mail: jerzy@ku.edu

Duoqian Miao
Tongji University, Shanghai, P.R. China, E-mail: miaoduoqian@163.com

Andrzej Skowron
Warsaw University, Poland, E-mail: skowron@mimuw.edu.pl

Yiyu Yao
University of Regina, Saskatchewan, Canada, E-mail: yyao@cs.uregina.ca

Library of Congress Control Number: 2008926413

CR Subject Classification (1998): I.2, H.2.4, H.3, F.4.1, F.1, I.5, H.4

LNCS Sublibrary: SL 7 – Artificial Intelligence

ISSN   0302-9743
ISBN-10   3-540-79720-3 Springer Berlin Heidelberg New York
ISBN-13   978-3-540-79720-3 Springer Berlin Heidelberg New York

Springer is a part of Springer Science+Business Media

springer.com

© Springer-Verlag Berlin Heidelberg 2008
Printed in Germany

Typesetting: Camera-ready by author, data conversion by Scientific Publishing Services, Chennai, India
Printed on acid-free paper   SPIN: 12265817   06/3180   5 4 3 2 1 0

Preface

This volume contains the papers selected for presentation at the Third International Conference on Rough Sets and Knowledge Technology (RSKT 2008) held in Chengdu, P.R. China, May 16–19, 2008.

The RSKT conferences were initiated in 2006 in Chongqing, P.R. China. RSKT 2007 was held in Toronto, Canada, together with RSFDGrC 2007, as JRS 2007. The RSKT conferences aim to present state-of-the-art scientific results, encourage academic and industrial interaction, and promote collaborative research in rough sets and knowledge technology worldwide. They place emphasis on exploring synergies between rough sets and knowledge discovery, knowledge management, data mining, granular and soft computing as well as emerging application areas such as bioinformatics, cognitive informatics, and Web intelligence, both at the level of theoretical foundations and real-life applications.

RSKT 2008 focused on five major research fields: computing theory and paradigms, knowledge technology, intelligent information processing, intelligent control, and applications. This was achieved by including in the conference program sessions on rough and soft computing, rough mereology with applications, dominance-based rough set approach, fuzzy-rough hybridization, granular computing, logical and mathematical foundations, formal concept analysis, data mining, machine learning, intelligent information processing, bioinformatics and cognitive informatics, Web intelligence, pattern recognition, and real-life applications of knowledge technology. A very strict quality control policy was adopted in the paper review process of RSKT 2008. Firstly, the PC Chairs reviewed all submissions. Some submissions, not meeting the quality standards of the conference, were rejected in this step. Then, 184 papers were subjected to a double-blind review process. In this step, every paper was examined by at least two reviewers. In all, 105 papers were initially selected by the PC Chairs according to reviewers' comments. Among these initially selected papers, some were conditionally approved, subjected to revision, and then additionally evaluated. Finally, 91 papers were accepted for RSKT 2008.

This volume contains 100 papers, including 3 invited keynote papers, 6 invited tutorial papers, and 91 contributed papers.

We are grateful to our Honorary Chairs, Ruqian Lu and Lotfi A. Zadeh, for their support and visionary leadership. We also acknowledge the scientists who kindly agreed to give the keynote and tutorial lectures: Ruqian Lu, Witold Pedrycz, Feiyue Wang, Andrzej Skowron, Hung Son Nguyen, Roman Slowinski, Salvatore Greco, Guoyin Wang, Yiyu Yao, and Mihir Kumar Chakraborty. We also wish to express our deep appreciation to all Conference Chairs, Organizing Chairs, Special Session Chairs, Industry Chairs, Publicity Chairs, Steering Committee Chairs, Steering Committee members, Program Committee members, and all reviewers.

We wish to thank all the authors who submitted high-quality papers and all conference participants.

We greatly appreciate the co-operation, support, and sponsorship of various institutions, companies, and organizations, including: Southwest Jiaotong University (SWJTU), Beijing Institute of Technology (BIT), Chongqing University of Posts and Telecommunications (CQUPT), University of Regina (UofR), International Rough Set Society (IRSS), Rough Sets and Soft Computation Society of the Chinese Association for Artificial Intelligence (CRSSC), and IEEE Chengdu Section.

Last but not least, we are thankful to Alfred Hofmann of Springer for support and co-operation during preparation of this volume.

May 2008

Guoyin Wang
Tianrui Li
Jerzy Grzymala-Busse
Duoqian Miao
Andrzej Skowron
Yiyu Yao

RSKT 2008 Conference Committee

Honorary Chairs	Ruqian Lu, Lotfi Zadeh
Conference Chairs	Pingzhi Fan, Roman Slowinski, Shusaku Tsumoto, Wojciech Ziarko
Program Chairs	Jerzy Grzymala-Busse, Duoqian Miao, Andrzej Skowron, Guoyin Wang, Yiyu Yao
Organizing Chairs	Yunde Jia, Tianrui Li
Special Session Chairs	Salvatore Greco, Weizhi Wu
Industry Chair	Dominik Slezak
Publicity Chairs	Jianchao Han, Jingtao Yao, Zhimin Gu
Steering Committee Chairs	Qing Liu, James F. Peters, Lech Polkowski
Conference Secretaries	Haibo Cheng, Jing Dai, Feng Hu, Hengchao Li, Xiaoping Qiu

Steering Committee

Aijun An
Malcolm Beynon
Hans-Dieter Burkhard
Cory Butz
Gianpiero Cattaneo
Nicholas Cercone
Mihir K. Chakraborty
Juan-Carlos Cubero
Didier Dubois
Ivo Duentsch
Aboul E. Hassanien
Masahiro Inuiguchi
Etienne Kerre
Yuefeng Li

Jiye Liang
Tsau Young Lin
Pawan Lingras
Jiming Liu
Jie Lu
Victor Marek
Ernestina
   Menasalvas-Ruiz
Sadaaki Miyamoto
Masoud Nikravesh
Setsuo Ohsuga
Ewa Orlowska
Sankar K. Pal
Witold Pedrycz

Vijay V. Raghavan
Sheela Ramanna
Da Ruan
Wladyslaw Skarbek
Jerzy Stefanowski
Zbigniew Suraj
Julio V. Valdes
Hui Wang
S.K. Michael Wong
Huanglin Zeng
Justin Zhan
Bo Zhang
Wen-Xiu Zhang
Ning Zhong

Program Committee

Rakesh Agrawal
Qiusheng An
Mohua Banerjee
Haider Banka
Jan Bazan
Theresa Beaubouef
Rajan Bhatt

Maciej Borkowski
Tom Burns
Cornelis Chris
Chien-Chung Chan
Davide Ciucci
Jianhua Dai
Martine De Cock

Jitender Deogun
Patrick Doherty
Jiali Feng
Maria C.
   Fernandez-Baizan
Philippe Fortemps
Bernhard Ganter

Non-committee Reviewers

Yang Gao
Syamal Kumar Samanta
Min-Ling Zhang

Sponsoring Institutions

Southwest Jiaotong University
Beijing Institute of Technology
Chongqing University of Posts and Telecommunications
University of Regina
International Rough Set Society
Rough Set and Soft Computation Society of Chinese Association for Artificial
    Intelligence
IEEE Chengdu Section

Table of Contents

Rough Mereology with Applications

Dominance-Based Rough Set Approach

Formal Concept Analysis

Data Mining

Machine Learning

Intelligent Information Processing

Bioinformatics and Cognitive Informatics

Web Intelligence

Pattern Recognition

Real-Life Applications of Knowledge Technology

Knowware: A Commodity Form of Knowledge

Ruqian Lu

Academy of Mathematics and System Science,
The Chinese Academy of Sciences, Beijing 100080, P.R. China

Abstract. Knowledge and Intelligence have a much closed relationship. Knowledge is both the crystallization and source of intelligence. Knowledge embodies intelligence, and intelligence emerges from knowledge. Every ICAX system (Intelligent Computer Aided X, where X may mean any domain, such as education, design or manufacturing, etc.), such as ICAI (I = Instruction), ICAD (D = Design), ICAM (M = Manufacturing), etc., has its intelligence based on a content rich knowledge base. In this sense, we may have the formula: ICAX = CAX + X knowledge base. Using this formula, we have developed a methodology of generating knowledge based system automatically. The core idea is to develop a domain-oriented pseudo-natural language (PNL for short), where PNL means a normalized subset of some natural language, which can be easily parsed by computer. Each domain expert may use this language to write down his knowledge and experience. A PNL compiler then compiles 'program's written in this PNL to form a domain knowledge base. Combined with a preexisting system shell, a prototype of the knowledge based system is automatically generated. We have applied this idea to automatic generation of ICAI and ICASE (SE = Software Engineering) systems. The following problem is how to generalize this idea. Can the development of knowledge base and system shell be done by different people or groups? Can the knowledge base be easily renewed or even become an independent commodity? Finally, we have got an answer to this problem. The commodity form of such knowledge base is knowware. In general, knowware is a commodity form of knowledge.

More precisely, knowware is a commercialized knowledge module with documentation and intellectual property, which is computer operable, but free of any built-in control mechanism, meeting some industrial standards and embeddable in software/hardware. The process of development, application and management of knowware is called knowware engineering. Three different knowware life cycle models are discussed: the furnace model, the crystallization model and the spiral model. Software/knowware co-engineering is a mixed process involving both software engineering and knowware engineering issues. It involves three parallel lines of developing system components of different types. The key issues of this process are how to guarantee the correctness and appropriateness of system composition and decomposition. The ladder principle, which is a modification of the waterfall model, and the tower principle, which is a modification of the fountain model, are proposed.

G. Wang et al. (Eds.): RSKT 2008, LNAI 5009, pp. 1–2, 2008.

References

1. Lu, R.Q., Jin, Z., Wan, R.L.: Requirement Specification in Pseudo-natural Language in PROMIS. In: Proceeding of 19th International Computer Software and Applications Conference (COMPSAC), pp. 96–101 (1995)
2. Lu, R.Q., Cao, C.G., Chen, Y.H., Mao, W.J., Han, Z.G.: The PLNU Approach to Automatic Generation of ICAI Systems. Science in China Series A 38(Supplement), 1–11 (1995)
3. Lu, R.Q.: Text Mining by Pseudo-Natural Language Understanding. In: Wang, J. (ed.) Encyclopedia of Data Warehousing and Mining, 2nd edn., Idea Group Reference (to appear, 2008)
4. Lu, R.Q.: From Hardware to Software to Knowware: IT's third liberation? IEEE Intelligent systems 20(2), 82–85 (2005)
5. Lu, R.Q., Jin, Z.: Beyond Knowledge Engineering. Journal of Computer Science and Technology 21(5), 790–799 (2006)
6. Lu, R.Q.: Towards a Software/Knowware Co-engineering. In: Lang, J., Lin, F., Wang, J. (eds.) KSEM 2006. LNCS (LNAI), vol. 4092, pp. 23–32. Springer, Heidelberg (2006)
7. Lu, R.Q.: Knowware, Knowware Engineering and Knowware/Software Co-engineering. In: ICCS 2007, Invited Talk, Beijing, China (2007)
8. Lu, R.Q.: Knowware: the Third Star after Hardware and Software, Polimetrica, Italy (2007)

Granular Computing in Multi-agent Systems

Witold Pedrycz

Department of Electrical & Computer Engineering
University of Alberta, Edmonton AB T6R 2G7 Canada and
System Research Institute, Polish Academy of Sciences
Warsaw, Poland
pedrycz@ee.ualberta.ca

Abstract. In recent years multi-agent systems have emerged as one of the interesting architectures facilitating distributed collaboration and distributed problem solving. Each node (agent) of the network might pursue its own agenda, exploit its environment, develop its own problem solving strategy and establish required communication strategies. Within each node of the network, one could encounter a diversity of problem-solving approaches. Quite commonly the agents can realize their processing at the level of information granules that is the most suitable from their local points of view. Information granules can come at various levels of granularity. Each agent could exploit a certain formalism of information granulation engaging a machinery of fuzzy sets, interval analysis, rough sets, just to name a few dominant technologies of granular computing. Having this in mind, arises a fundamental issue of forming effective interaction linkages between the agents so that they fully broadcast their findings and benefit from interacting with others.

Keywords: Granular computing, Fuzzy sets, Rough sets, Multi-agent systems, Communication schemes, Granularity of information.

1 Introduction

There has been a growing interest in agent systems and their collaborative structures of multi-agent topologies. There is a great deal of methodological and algorithmic pursuits as well a wave of application-oriented developments cf. [1][4][6][10][20][21] Given the nature of the problem tackled by such systems where we commonly encounter nodes (agents) operating quite independently at various levels of specificity, it is very likely that the effectiveness of the overall system depends directly upon a way in which the agents collaborate and exchange their findings.

In this study, we are interested in the development of schemes of interaction (communication) in multi-agent systems where exchange of findings obtained locally (at the level of individual agents) are represented as information granules [12][13][14][15][17][18][19][22] rather than plain numeric entities (which might not be feasible or very much limited in terms of knowledge representation). There are a number of important and practically relevant issues dealing with various ways of expressing incoming evidence available to an individual agent which expresses findings in the format available to all other agents in the network.

G. Wang et al. (Eds.): RSKT 2008, LNAI 5009, pp. 3–17, 2008.

Each agent could exploit its own formalism of information granules be it fuzzy sets [2][7][8][17], rough sets [12][13][14][15], rough fuzzy sets or fuzzy rough sets [7], type-2 fuzzy sets [6] and others including hybrid models. What becomes quite apparent in the realization of the communication schemes is an increase of the sophistication of the information granules which are exchanged between agents and need to be translated into an internal format acceptable to the particular agent. An example of this tendency is an emergence of type-2 fuzzy sets (in cases when originally we have dealt with fuzzy sets). We demonstrate that the concept of rough sets becomes central to our investigations as a conceptual and algorithmic realization of the communication mechanisms.

The study is structured in a top-down manner. Some generic architectural and functional insights into granular agents are presented in Section 2. Sections 3-5 present the conceptual underpinnings of communication realized between the agents and a way of dealing with multiple evidence and related aggregation issues. The consecutive sections concentrate on the detailed algorithmic developments of the concepts.

2 Granular Agents and Multi-agent Systems: Architectural and Functional Insights

In a nutshell, by a granular agent we mean a processing module which realizes processing carried out at the level of information granules (no matter what formalism of information granulation is being used there). The module comes with substantial processing capabilities, is able to carry out some learning and enhancements on a basis of locally available experimental evidence. It communicates its findings to other agents and engages into some collaborative pursuits. Each agent operates at its own level of information granularity

A general scheme of a multi-agent system can be schematically outlined in Figure 1. Note that the communication layer of each agent (which comes in the form of a certain stratum) plays a pivotal role in establishing a sound and effective

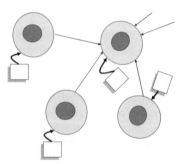

Fig. 1. An overview of a multi-agent system; each agent develops and evolves on a basis of experimental data that are locally available and communicates with other agents. Distinguished are the computing core (dark center) and the communication layer of the nodes (surrounding light color region) present in the multi-agent systems.

collaborative interaction which becomes essential when building distributed models, forming distributed control strategies and constructing distributed classification architectures, just to name the most representative categories of tasks.

The communication scheme in which an agent accepts some result of processing offered by some other agent has to deal with an issue of representation of the incoming evidence in the setting of the information granules.

3   Agent Communication: Internal Representation of Incoming Evidence

Agent accepts findings coming from other agents and expresses them in the format which is pertinent to its own processing. We can view this process as translating an input evidence X with the aid of a vocabulary of information granules {A1, A2, ..., Ac} pertinent to the agent under discussion. Both X and Ai could exhibit a significant diversity in terms of their underlying formalism of information granulation. In spite of this possible diversity, some general representation guidelines can be envisioned. First, we can describe X by considering an extent to which X and Ai overlap considering that this concept is reflective of the notion of closeness (resemblance) between these two information granules. Anticipating that such a quantification might not be able to capture the entire matching process, we consider X and Ai in the context of an extent to which X is included in Ai. The predicate of inclusion itself could be gradual viz. returning a certain numeric quantification with values confined to the unit interval. Denote the results of this representation by λ_i and μ_i, respectively

$$\lambda_i = \tau(X \cap A_i) \tag{1}$$

$$\mu_i = \tau(X \subset A_i) \tag{2}$$

where the operation τ is used here to schematically capture the realization of the operations of overlap and inclusion. Overall, the scheme realized above isgraphically represented in Figure 2.

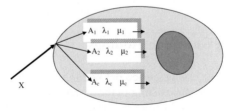

Fig. 2. The representation of incoming evidence through the operations of overlap and inclusion

It is interesting to underline that the general operations (1) – (2) exhibit a close resemblance with the very nature of rough sets. The quantification conveyed by the values of λ_i and μ_i relate directly with the lower and upper approximations forming the holy grail of rough sets. Note however that in the formulation above we have not confined ourselves to any specific formal representation of information granules. This representation underlines the fundamental feature of the concept of rough sets which stretches beyond the original concept introduced by Pawlak. It is needless to say the implementation details depend upon the character of the information granules. The quality of the incoming evidence X is assessed vis-à-vis existing information granules A_i by plotting the values of (1) and (2). There are two interesting boundary cases:

Numeric information. It leads to the equality $\mu_i = \lambda_i$
Totally *unknown* piece of evidence. In this case $\lambda_i=1$ for all i=1, 2, ..., c and $\mu_i=0$.

The plots of these two boundaries are shown in Figure 3. There are, obviously, a lot of other scenarios whose visualization manifests in the form of points positioned over the main diagonal. The closer the point is to the main diagonal, the tighter the bounds are and the higher the quality of the input evidence (assessed with regard to A_i).

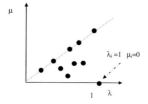

Fig. 3. Visualization of λ_i and μ_i for various nature of the incoming evidence

Referring to the characterization of X in terms of A_is, it is worth noting that by default we considered that the specificity of X is higher than the information granules A_i and therefore it becomes legitimate to talk about the relationship of inclusion. The notion of granularity (which is intuitively appealing) requires here further quantification which depends upon the formalism of information granules. The simplest one would envision is to count the number of elements in the information granule or summing up the degrees of membership (which is more suitable when dealing with fuzzy sets).

4   Communicating Granular Findings

The results of granular processing carried out within the bounds of a certain agent are next broadcasted to other agents existing in the system. To do so, the agent realizes its findings in the form of a certain information granule. Typically, for the agent we encounter a collection of information granules in some input space (say, some receptor space) and a family of information granules in the output space (e.g., a space of actions, decisions, etc.). There could be a fairly advanced web of connections between them which could be either "hardwired" or it may exhibit some level of plasticity which is essential in supporting learning capabilities. Rule-based architectures such as e.g., fuzzy rule-based models are examples of such granular architectures. The result of processing are expressed via degrees of overlap and inclusion pertaining to the individual information granules in the output space.

Referring to the way in which the input evidence has been captured, the internal processing realized by the agent returns a vector of degrees of overlap γ ($=[\gamma_1 \ \gamma_2 \ \ldots \ \gamma_m]$) and degrees of inclusion η($=[\eta_1 \ \eta_2 \ \ldots \ \eta_m]$). Those need to be translated into some information granule where in this construct we engage the corresponding information granules B_j. Being more formal, we are concerned with the following inverse problem:

-for given vectors γ and η information granules B_i and a family of constraints

$$\gamma_i = \tau(B \cap B_i) \quad \eta_i = \tau(B \subset B_i) \quad i=1, 2, \ldots, m$$

determine B. The graphical visualization of the underlying problem is illustrated in Figure 4.

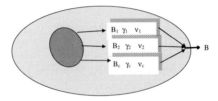

Fig. 4. The essence of communicating granular findings

5   Acceptance of Multiple Input Evidence and Its Representation

Evidence coming from different agents is expressed in terms of Ai producing the results conveyed in the format (1) – (2). As we encounter several sources of information which might be in some interaction, they need to be reconciled or aggregated [3]. Schematically we display this situation as included in Figure 5.

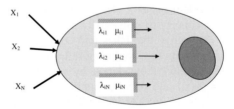

Fig. 5. Multiple source evidence and its reconciliation prior to further processing by the computing core of the agent

The crux of the construct is to reflect upon the nature of reconciled evidence which has to be taken into account when proceeding with further processing realized by the agent. Intuitively, any sound aggregation would return some quantification at the same level of granularity as the originally available evidence. For instance, from the statistical perspective, we could contemplate using average, modal or median as a meaningful descriptor of the available evidence. A more suitable approach would be the one in which we convey not only the single numeric quantity but an information granule whose role is to quantify the diversity of available sources of evidence.

In what follows, we discuss several detailed computing realizations which support the implementation of the individual communication mechanisms presented so far.

6 Realization of Acceptance of Input Evidence through the Use of Logically Inclined Logic Operators

In the realization of (1) – (2), we quantify an overlap and inclusion holding between X and A_i. With this regard the following formulas are commonly encountered

$$\lambda_i = \sup_{x \in \mathbf{X}}[X(x)tA_i(x)] \tag{3}$$

and

$$\mu_i = \inf_{x \in \mathbf{X}}[(1 - X(x))sA_i(x)] \tag{4}$$

In fuzzy sets (3) is referred as possibility measure while (4) comes is a necessity measure. The notation of "t" and "s" is used to denote t-norms and t-conorms (s-norms) which are commonly encountered as realizations of logic operators of *and* and *or*. The above formulas apply directly to information granules formalized as sets and fuzzy sets. Considering even a fairly specific case when dealing with finite space $\mathbf{X}$, the above formulas may exhibit shortcomings. First, the max (sup) and min (inf) operations are non-interactive so they tend to lead to the results that are getting close to 1 (in case of (3)) and 0 (for (4)). Second, the result is entirely dependant upon the extreme entry in the whole set of results $X(x)tA_i(x)$ for the possibility measure and $(1-X(x))sA_i(x)$ for the necessity measure. These two measures could have quite evident drawbacks by leading to the computing machinery which produces results where the bounds used in the description of X are getting quite loose converging to 1 and 0, respectively. To alleviate this problem, the underlying concept is to incorporate some knowledge about the statistics of the partial results of computing t- norms and t-conorms for the individual elements of $\mathbf{X}$. This leads us to statistically-grounded logic operators. We introduce a concept of statistically augmented (directed) logic connectives by constructing a connective that takes into consideration a statistically driven aggregation with some weighting function being reflective of the nature of the underlying logic operation. Furthermore let us denote by z_j the result of t-norm aggregation of $X(x_j)$ and $A_i(x_j)$, $z_i = X(x_j)$ and $A_i(x_j)$ assuming that the space $\mathbf{X}$ is finite, card $(\mathbf{X})$ =N.

6.1 SOR Logic Connectives

The (SOR) connective is defined as follows. Denote by $w(x)$ a monotonically non-decreasing weight function from [0,1] to [0,1] with the boundary condition $w(1) = 1$. The result of the aggregation of $\mathbf{z} = [z_1, z_2, \ldots, z_N]$, denoted by $SOR(\mathbf{z}; w)$, is obtained from the minimization of the following expression (performance index) Q

$$Q = \sum_{j=1}^{N} w(z_j) \,|\, z_i - y \,| \qquad\qquad \text{Min}_y \, Q \qquad\qquad (5)$$

where the value of "y" minimizing the above expression is taken as the result of the operation SOR(z, w) = y. Put it differently, we have SOR(z, w) = arg $\min_{y\in[0,1]} \sum_{k=1}^{N} w(z_k) \,|z_k - y\,|$ The weight function "w" is used to model a contribution of different membership grades to the result of the aggregation. Several models of the relationships "w" are of particular interest; all of them are reflective of the *or* type of aggregation

(a) w(z) assumes a form of a certain step function

$$w(z) = \begin{cases} 1 \text{ if } z \geq z_{max} \\ 0, \text{otherwise} \end{cases} \qquad\qquad (6)$$

where z_{max} is the maximal value reported in z. This weight function effectively eliminates all the membership grades but the largest one. For this form of the weight function, we effectively end up with the maximum operator, SOR(z, w) =max $(z_1, z_2, ..., z_N)$

(b) w(z) is equal identically to 1, w(z) =1. It becomes clear that the result of the minimization of the following expression

$$\sum_{j=1}^{N} |z_j - y| \qquad\qquad (7)$$

is a median of z, median(z). Subsequently SOR(z, w) = median(z). Interestingly, the result of the aggregation is a robust statistics of the membership grades involved in this operation.

We can consider different forms of weight functions. In particular, one could think of an identity function w(z) = z. There is an interesting and logically justified alternative which links the weight functions with the logic operator standing behind the logic operations. In essence, the weight function can be induced by various t-conorms (s-norms) by defining w(z) to be in the form w(z) = zsz. In particular, for the maximum operator, we obtain the identity weight function w(z) =max(z,z) = z. For the probabilistic sum, we obtain w(z) = (z+z-z*z) = 2z(1-z). For the Lukasiewicz *or* connective, the weight function comes in the form of some piecewise linear relationship with some saturation region, that is w(z) = max(1, z+z) = max (1, 2z). The plots of these three weight functions are included in Figure 6.

In general, the weight functions (which are monotonically non-decreasing and satisfy the condition w(1) = 1) occupy the region of the unit square as portrayed in Figure 7. Obviously the weight functions induced by t-conorms are subsumed by the weight functions included in Figure 7.

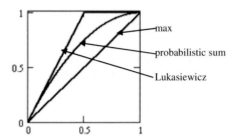

Fig. 6. Plots of selected weight functions induced by selected t-conorms

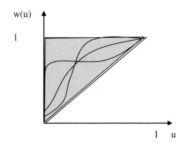

Fig. 7. Examples of weight functions generating SORs logic operators induced by t-conorms; all of them are localized in the shaded region of the unit square

For all these weight functions implied by t-conorms, the following inequality holds median($\mathbf{z}$) $\leq$ SOR($\mathbf{z}$, w) $\leq$ max($\mathbf{z}$).

6.2 SAND Logic Connectives

The statistically grounded AND (SAND) logic connective is defined in an analogous way as it was proposed in the development of the SOR. Here w(x) denotes a monotonically non-increasing weight function from [0,1] to [0,1] with the boundary condition w(0)=1. The result of the aggregation of $\mathbf{z} = [z_1, z_2, ..., z_N]$, denoted by SAND($\mathbf{z}$; w), is obtained from the minimization of the same expression (3) as introduced before. Thus we produce the logic operator SAND($\mathbf{z}$, w) = y with "y" being the solution to the corresponding minimization problem.

As before, we can envision several models of the weight function; all of them are reflective of the *and* type of aggregation

(a) w(z) assumes a form of some step function

$$w(z) = \begin{cases} 1 \text{ if } z \leq z_{min} \\ 0, \text{otherwise} \end{cases} \tag{8}$$

where z_{min} is the minimal value in $\mathbf{z}$. This weight function eliminates all the membership grades but the smallest one. For this form of the weight function, we effectively end up with the maximum operator, SAND($\mathbf{z}$, w) =min $(z_1, z_2, ..., z_N)$

(b) for w(z) being equal identically to 1, w(z) =1, SAND becomes a median, namely SAND(**z**, w) = median(**z**).

(c) more generally, the weight function is defined on a basis of some t-norm as follows, w(z) =1- ztz. Depending upon the specific t-norm, we arrive at different forms of the mapping. For the minimum operator, w(z) =1- min(z,z) =1-z which is a complement of "z". The use of the product operation leads to the expression w(z) =1- z^2. In the case of the Lukasiewicz *and* connective, one has w(z)=1-max(0, z+z-1) =1-max(0, 2z-1).

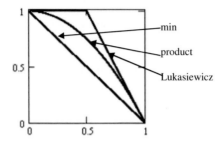

Fig. 8. Examples of weight functions used in the construction of the SAND operation

If we confine ourselves to monotonically non-increasing functions of [0,1] with the boundary condition of w(0) =1, they can be illustrated as shown in Figure 9. Note that the general inequality relationship holds min(**z**) $\leq$ SAND(**z**, w) $\leq$ median(**z**).

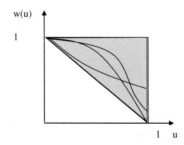

Fig. 9. Localization of weight functions induced by t-norms generating SANDs logic operators

Investigating the fundamental properties of the logic connectives, we note that the commutativity and monotonicity properties hold. The boundary condition does not hold when being considered with respect to a single membership grade (which is completely understood given the fact that the operation is expressed by taking into consideration a collection of membership grades). Assuming the t-norm and t-conorm driven format of the weight function (where we have w(1) =1 and w(0) =0 for *or* operators and w(0)=1 and w(1)=1 for *and* operators) we have SOR(**1**, w) =0, SAND(**0**, w) = 0. The property of associativity does not hold. This is fully justified given that the proposed operators are inherently associated with the processing of all membership grades not just individual membership values.

A brief numeric example serves as an illustration of the concepts of the statistically grounded logic operators. The collection of membership grades to be aggregated consists of 13 values located in the unit interval

$$\{ 0.4\ 0.1\ 0.8\ 0.6\ 0.5\ 0.4\ 0.35\ 0.9\ 1.0\ 0.55\ 0.22\ 0.1\ 0.7\}$$

The median of these membership grades is 0.50. The optimization of the SOR and SAND operators leads to the following aggregation results for selected t-norms and t-conorms:

SOR operator   max: 0.70,   Lukasiewicz: 0.55, probabilistic sum: 0.60
SAND operator min: 0.35   Lukasiewicz: 0.35 product: 0.40

We note that all SOR values are located above the median while the SAND operators generate aggregations with the values below the median. For the given data set, the specific values of the aggregation depend on the character of the weight function. The monotonicity in the weight functions induced by the corresponding t- and t-conorms is fully reflected in the order of the aggregation results. In other words, as the weight function implied by the max function is shifted to the right in comparison with the one induced by the probabilistic sum, then the result of SOR for the min is higher than the one for the SOR formed with the aid of the probabilistic sum.

Clearly the introduced SAND and SOR operators can also help deal with the lack of discriminatory capabilities of the possibility and necessity measures.

If we consider that the membership grades to be aggregated are governed by some probability density function $p(x)$ then the optimization problem can be expressed in the following format

$$Q= \int_0^1 w(x) \mid x - m \mid p(x)dx = \int_0^{m_0} w(x)(x - m_0)p(x)dx + \int_{m_0}^1 w(x)(m_0 - x)p(x)dx \qquad (9)$$

7 Reconstruction of Information Granules

The problem of communication of granular computing realized by the agent is to construct the information granule which leads to the determination of B given B_is and the results of processing conveyed by the two vectors γ and η. A careful inspection of these relationships reveals that when the intersection and inclusion operations are realized with t-norms and t-conorms, we end up with a system of fuzzy relational equations and the determination of B comes as a result of the solution to inverse problem.

Let us recall that we have

$$\gamma_i = \sup_{x \in X}[B(x)tB_i(x)] \qquad (10)$$

$$\eta_i = \inf_{x \in X}[(1 - B(x))sB_i(x)] \qquad (11)$$

One can look at (10) and (11) and treat these relationships as a system of equations to be solved with respect to B. There is no unique solution to neither the first nor the second set of equations. There are, however extreme solutions that are unique to the problem at hand. Their construction is supported by the theory of fuzzy relational equations (as a matter of fact, (10) as a sup-t composition of B and B_i). In light of the fundamental results available in the theory, the membership function of this maximal fuzzy set (mapping) induced by the B_i reads as

$$\hat{B}_i(x) = B_i(x) \rightarrow \gamma_i = \begin{cases} 1 & \text{if } B_i(x) \le \gamma_i \\ \gamma_i & \text{otherwise} \end{cases} \tag{12}$$

The above formula applies to the t-norm implemented as a minimum operator. In general, for any continuous t-norm (12) reads in the form

$$\hat{B}_i(x) = B_i(x) \rightarrow \gamma_i = \sup[a \in [0,1] \mid atB_i(x) \le \gamma_i] \tag{13}$$

When using the entire family of B_is (that leads to the intersection of the partial results presented above

$$\hat{B} = \bigcap_{i=1}^{c} \hat{B}_i \tag{14}$$

The theory of fuzzy relational equations [5][16] plays the same dominant role in the case of the necessity computations. It is worth noting that we are faced with so-called dual fuzzy relational equations. Let us rewrite (11) as follows

$$\eta_i = \inf_{x \in \mathbf{X}} [B^*(x) s B_i(x)] \tag{15}$$

where $B^*(x) = 1 - B(x)$. Here the minimal solution to (15) for B_i and η_i given reads in the form

$$\tilde{B}_i^*(x) = B_i(x) \varepsilon \, \eta_i = \begin{cases} \eta_i, & \text{if } B_i(x) < \eta_i \\ 0, & \text{otherwise} \end{cases} \tag{16}$$

Again the above formula applies to the maximum realization of the t-conorm. More generally, we have

$$\tilde{B}_i^*(x) = B_i(x) \varepsilon \eta_i = \inf\{a \in [0,1] \mid a s B_i(x) \ge \eta_i\} \tag{17}$$

The partial results constructed in this manner are combined by taking their union

$$\tilde{B}^* = \bigcup_{i=1}^{c} \tilde{B}_i^* \tag{18}$$

In conclusion, (14) and (18) become a description of the information granule expressed in terms of bounds of membership values that is $[\min(1 - \tilde{B}^*(x), \hat{B}(x)), \max(1 - \tilde{B}^*(x), \hat{B}(x))]$ for each x.

An illustrative example of the discussed concept is shown in Figure 10 when the family of B_is involves Gaussian membership functions with modal values of 1.9, 3.4 and 5 and equal spreads of 0.8. Here we consider several vectors of numeric values of γ and η. The resulting information granule comes in the form of the type-2 fuzzy set, and interval-valued fuzzy set.

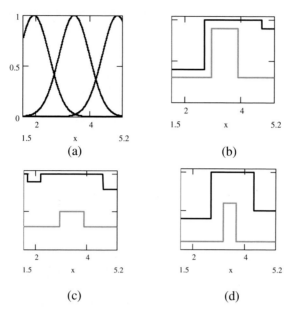

Fig. 10. A collection of Gaussian membership functions (a) and resulting information granules for $\gamma = [0.4\ 1.0\ 0.9]$ and $\eta = [0.1\ 0.7\ 0.1]$ (b) $\gamma = [0.9\ 1.0\ 0.8]$ $\eta = [0.5\ 0.7\ 0.1]$ (c), and $\gamma = [0.4\ 1.0\ 0.5]$ $\eta = [0.3\ 0.9\ 0.4]$ (d)

8   Aggregating and Representing Multiple Evidence: A Principle of Justifiable Granularity

A single agent accepts evidence from other agents. Denote these incoming entities by $X_1, X_2, ..., X_N$. Their characterization in terms of A_i returns the values of $\lambda_{i1}, \lambda_{i2}, ... , \lambda_{iN}$ and $\mu_{i1}, \mu_{i2}, ... , \mu_{iN}$. These levels of overlap or inclusion could be different as being reflective of the diversified evidence provided by the entities available to this agent. To quantify the results in some synthetic manner, we proceed with a certain unified construct formed on the basis of the numeric values of λ_{i1} and λ_{i1}, respectively. This construct offers an interesting conceptual alternative which leads to the emergence of granular constructs. In particular this may lead to type-2 fuzzy sets. From the algorithmic perspective, we transform numeric quantities into a single granular representative. Its design is guided by the principle of *justifiable* granularity. To illustrate its essence, we consider a finite collection of numeric entities, say $\{\lambda_{ij}\}$ or $\{\mu_{ij}\}$. To unify the notation to e used, let denote these numeric evidence by $\{w_{ij}\}$, $j=1, 2, ..., N$. The underlying idea of the principle of *justifiable granularity* [11] is to represent the numeric entries w_{ij} in the form of a certain fuzzy set Ω which "spans" over the data in such a way that it "represents" the data to the highest extent while being maintained quite specific (detailed). The modal value of the fuzzy set constructed in this way is taken as the median of $\{w_{ij}\}$. The left-hand and the right –hand side of the support of the membership function is constructed independently on the basis of the available data. Consider those elements w_{ij} whose values are lower than w, that is $w_{ij} < w$ where "w" is the median of the set $\{w_{ij}\}$. Denote by "a" the lowest value in [0,1] the left-hand side of the support of Ω.

There are two requirements guiding the determination of the "optimal" value of "a":

> (a) maximize the experimental evidence of the fuzzy set; this implies that we tend to "cover" as many numeric data as possible, viz. the coverage has to be made as high as possible. Graphically, in the optimization (maximization) of this requirement, we move "a" down to zero as much as possible.
>
> The sum of the membership grades $W(w_{ij})$, $\sum_i W(w_{ij})$ has to be maximized which reflects that the experimental evidence is high enough.
>
> (b) Simultaneously, we want to make the fuzzy set as specific as possible so that is comes with some well defined semantics. This requirement is met by making the support of Ω as small as possible, that is $\min_a |w - a|$

To accommodate these two conflicting requirements, we combine (a) and (b) into a form of a single scalar index which in turn has to be maximized. The following expression is one of the viable alternatives which captures these two requirements at the same time

$$\max_{a \neq u} \frac{\sum_i W(w_{ij})}{|w - a|} \tag{19}$$

The same construct is realized for the upper bound of the support of the fuzzy set. The construct does not restrict itself to any specific form of the membership function. In particular, we could consider triangular or parabolic membership functions; they need to be symmetric, though. The crux of the above construct is illustrated in Figure 11.

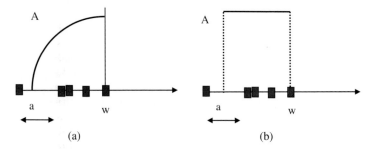

(a)   (b)

Fig. 11. The development of information granule formed on a basis of the principle of justifiable granularity: (a) forming the membership function, and (b) construction of the characteristic function

The above construct equally well applies to the optimization of the characteristic function in case we consider sets as information granules. In this case, the sum is simply the cardinality of the experimental data "covered" contained in the set as illustrated in Figure 11(b).

The information granule formed in this manner reflects the diversity of the evidence available to the agent. Note that the granularity has emerged because of this phenomenon of diversity the agent was exposed to.

9 Conclusions

In this study, we elaborated on the role of effective communication mechanisms in multi-agent and showed that given various perspectives and mechanisms of computing supported individual agents there is a need to develop schemes of interaction at the level of information granules. We have formulated the main communication mechanisms by starting with conceptual aspects and offering detailed algorithmic developments.

There are several observations of a general nature that are worth spelling out:

- The communication is quantified by describing relationships between information granules in terms of their overlap and inclusion. This description emphasizes the relevance of this principle which associates well to rough sets
- Any exposure to multiple sources of evidence leads to the emergence of information granules; the principle of justifiable granularity is a compelling illustration of the way in which information granules are constructed
- The quantification of interaction between agents gives rise to information granules of higher type (say, type-2 fuzzy sets).

References

1. Acampora, G., Loia, V.: A Proposal of Ubiquitous Fuzzy Computing for Ambient Intelligence. Information Sciences 178(3), 631–646 (2008)
2. Bezdek, J.C.: Pattern Recognition with Fuzzy Objective Function Algorithms. Plenum Press, N. York (1981)
3. Bouchon-Meunier, B.: Aggregation and Fusion of Imperfect Information. Physica-Verlag, Heidelberg (1998)
4. Cheng, C.B., Chan, C.C.H., Lin, K.C.: Intelligent Agents for E-marketplace: Negotiation with Issue Trade-offs by Fuzzy Inference Systems. Decision Support Systems 42(2), 626–638 (2006)
5. Di Nola, A., Pedrycz, W., Sessa, S.: Fuzzy Relational Structures: The State of Art. Fuzzy Sets & Systems 75, 241–262 (1995)
6. Doctor, F., Hagras, H., Callaghan, V.: A Type-2 Fuzzy Embedded Agent to Realise Ambient Intelligence in Ubiquitous Computing Environments. Information Sciences 171(4), 309–334 (2005)
7. Dubois, D., Prade, H.: Rough–Fuzzy Sets and Fuzzy–Rough Sets. Int. J. General Systems. 17(2–3), 191–209 (1990)
8. Nguyen, H., Walker, E.: A First Course in Fuzzy Logic. Chapman Hall, CRC Press, Boca Raton (1999)
9. Hoppner, F., et al.: Fuzzy Cluster Analysis. J. Wiley, Chichester, England (1999)
10. Kwon, O., Im, G.P., Lee, K.C.: MACE-SCM: A Multi-Agent and Case-Based Reasoning Collaboration Mechanism for Supply Chain Management under Supply and Demand Uncertainties. Expert Systems with Applications 33(3), 690–705 (2007)

11. Pedrycz, W., Vukovich, G.: Clustering inThe Framework of Collaborative Agents. In: Proc. 2002 IEEE Int. Conference on Fuzzy Systems(1), pp. 134–138 (2002)
12. Pawlak, Z.: Rough sets. Int. J. Comput. Inform. Sci. 11, 341–356 (1982)
13. Pawlak, Z.: Rough Sets. Theoretical Aspects of Reasoning About Data. Kluwer Academic Publishers, Dordrecht (1991)
14. Pawlak, Z., Busse, J.G., Slowinski, R., Ziarko, R.W.: Rough sets. Commun. ACM. 38(11), 89–95 (1995)
15. Pawlak, Z., Skowron, A.: Rudiments of Rough Sets. Information Sciences 177(1), 3–27 (2007)
16. Pedrycz, W.: Fuzzy Relational Equations: Bridging Theory, Methodology and Practice. Int. J. General Systems. 29, 529–554 (2000)
17. Pedrycz, W.: Knowledge-Based Clustering. J. Wiley, Hoboken (2005)
18. Yao, Y.Y.: Two Views of the Theory of Rough Sets in Finite Universes. Int. J. Approximate Reasoning. 15, 291–317 (1996)
19. Yao, Y.Y.: Probabilistic Approaches to Rough Sets. Expert Systems 20(5), 287–297 (2003)
20. Yu, R., Iung, B., Panetto, H.: A Multi-Agents Based E-maintenance System with Case-based Reasoning Decision Support. Engineering Applications of Artificial Intelligence 16(4), 321–333 (2003)
21. Wang, T.W., Tadisina, S.K.: Simulating Internet-based Collaboration: A Cost-benefit Case Study using a Multi-agent Model. Decision Support Systems 43(2), 645–662 (2007)
22. Zadeh, L.A.: Toward a Generalized Theory of Uncertainty (GTU)-an Outline. Information Sciences 172, 1–40 (2005)

Linguistic Dynamic Systems for Computing with Words and Granular Computing

Fei-Yue Wang

Institute of Automation,
The Chinese Academy of Sciences, Beijing 100080, P.R. China

Abstract. The term of Linguistic Dynamic Systems (LDS) was originally coined in earlier 1990s to reformulate logic programming for computing with words and granular computing dynamically and numerically. In the earlier stage of its development, fuzzy sets were used as the logic foundation for its analytical formulation and cell-to-cell mappings were applied in its computational framework. Therefore, many concepts and methods developed in ODE-based conventional dynamic systems can be used directly for computing with words and granular computing in LDS. Actually, cell mappings lead analysis in LDS to search problems in cell spaces, thus various search methods and techniques in Artificial Intelligence can be utilized for LDS. However, the procedure of transformation from fuzzy logic in hyper-cubes to cell-to-cell mappings in cell spaces is quite tedious and involving ad hoc steps in the process.

Both rough sets and type-2 fuzzy sets can be very useful in the improvement or even reformulation of LDS. Rough sets can lead to a connection between LDS and data mining, as well as granular computing, especially in dynamic construction and computing of value, variable, concept, and ontology granulation. There is a natural connection between LDS and type-2 fuzzy sets. As a matter of fact, a cell in cellular structured hyper-cubes is a specialized type-2 fuzzy set. New concepts and methods developed in the emerging type-2 fuzzy sets could be used in LDS for better design and improved computational efficiency in analysis of rule-based linguistic control systems. In this presentation, we will discuss and investigate the issues related to the relationship among LDS, computing with words, granular computing, and other methods.

G. Wang et al. (Eds.): RSKT 2008, LNAI 5009, p. 18, 2008.
© Springer-Verlag Berlin Heidelberg 2008

Rough Set Approach to KDD
(Extended Abstract)

Hung Son Nguyen and Andrzej Skowron

Institute of Mathematics, Warsaw University
Banacha 2, 02-097 Warsaw, Poland
son@mimuw.edu.pl, skowron@mimuw.edu.pl

This tutorial is a survey on rough set theory and some of its applications in Knowledge Discovery from Databases (KDD). It will also cover the practice guide to analysis of different real life problems using rough set methods as well as the presentation of Rough Set Exploration System (RSES) what can be treated as a preliminary material for the main conference and associated workshops.

Rough Set theory was introduced by Zdzisław Pawlak in the early 80's and has currently reached a level of high visibility and maturity [3,2,4,5,6]. Originally, rough sets, whose main philosophy is based simply on indiscernibility and discernibility of objects, were presented as an approach to concept approximation under uncertainty. This brilliantly simple idea has been successively expanded in the last twenty years. Many effective methods for data analysis have been developed on the basis of rough set theory.

In recent years, a growth of interest in rough set theory and its applications can be seen in the number of research papers submitted to international workshops, conferences, journals and edited books, including two main biannual conferences on rough sets and the special sub-line of LNCS series. A large number of efficient applications of rough sets in Knowledge Discovery for various types of databases have been developed. Rough sets are applied in many domains, such as medicine, finance, marketing, telecommunication, conflict resolution, text mining, intelligent agents, image analysis, pattern recognition, bioinformatics (e.g., see, [4,5,6] and the bibliography in these papers).

This tutorial is intended to fulfill the needs of many researchers to understand the rough set methodologies for mining of standard and nonstandard data. The methodology based on rough sets can serve as a useful tool to complement capabilities of other data mining methods. The tutorial should help the audience to find out if some of the presented methods may support their own KDD and data mining (DM) research.

The tutorial is intended to occupy four slots of 45 minutes each and to cover the following topics:

- Fundamentals of rough set theory;
- Rough set approach to KDD;
- Examples of rough set based methods for data reduction, rule extraction, discretization, decomposition, hierarchical learning;
- Practical guide for Rough Set Exploration System (RSES);

G. Wang et al. (Eds.): RSKT 2008, LNAI 5009, pp. 19–20, 2008.

- Some exemplary applications of RSES (with exercises);
- Emergent challenging problems.

The second part of this tutorial is targeted to young researchers who want to apply the rough set approach to real-life problems in KDD.

For more readings the readers are referred to the survey papers [4,5,6] and the bibliography in these papers, in particular to [1], as well as to the bibliography accessible from http://rsds.univ.rzeszow.pl/.

References

1. Nguyen, H.S.: Approximate boolean reasoning: Foundations and applications in data mining. In: Peters, J.F., Skowron, A. (eds.) Transactions on Rough Sets V. LNCS, vol. 4100, pp. 334–506. Springer, Heidelberg (2006)
2. Pawlak, Z.: Rough sets. International Journal of Computer and Information Sciences 11, 341–356 (1982)
3. Pawlak, Z.: Rough Sets: Theoretical Aspects of Reasoning about Data. System Theory, Knowledge Engineering and Problem Solving, vol. 9. Kluwer Academic Publishers, Dordrecht, The Netherlands (1991)
4. Pawlak, Z., Skowron, A.: Rudiments of rough sets. Information Sciences 177(1), 3–27 (2007)
5. Pawlak, Z., Skowron, A.: Rough sets: Some extensions. Information Sciences 177(1), 28–40 (2007)
6. Pawlak, Z., Skowron, A.: Rough sets and boolean reasoning. Information Sciences 177(1), 41–73 (2007)

Dominance-Based Rough Set Approach to Reasoning about Ordinal Data - A Tutorial

Roman Słowiński

Institute of Computing Science, Poznań University of Technology,
60-965 Poznań, and Systems Research Institute,
Polish Academy of Sciences, 01-447 Warsaw, Poland
roman.slowinski@cs.put.poznan.pl

This tutorial lecture intends to introduce the D*ominance-based Rough Set Approach* (DRSA) methodology for reasoning about *ordinal data*. DRSA, proposed by Greco, Matarazzo and Słowiński (see e.g. [2,6]), extends the classical Rough Set approach by handling background knowledge about ordinal evaluations of objects and about monotonic relationships between these evaluations. In DRSA, the indiscernibility or tolerance relation among objects, which is used in the classical Rough Set approach, has been replaced by the dominance relation – the only relation uncontested in multiattribute pairwise comparisons when attribute scales are ordered. The lecture starts with principles of DRSA and goes through the application of DRSA to fuzzy-rough hybridization [1], to end with DRSA to case-based reasoning [3], which builds on this hybridization. This tutorial prepares the ground for a second tutorial lecture on applications of DRSA to decision analysis.

There are various reasons for taking into account the *order* in data analysis. The ordering of data describing decision situations is naturally related to preferences on considered condition and decision attributes. For example, in credit rating, the degree of regularity in paying previous debts by a consumer is a condition attribute with a value set (scale) clearly ordered, e.g., unstable, acceptable, very stable; on the other hand, the decision attribute evaluating the potential risk of lending money to a consumer has also a clearly ordered value set (scale), e.g., high-risk, medium-risk, low-risk; these two scales are ordinal scales, typical for *evaluation criteria*; moreover, there exists a natural monotonic relationship between the two attributes: the more stable the payment of the debt, the less risky is the new credit. Both these aspects: ordering of values sets, and monotonic relationship between condition attributes (criteria) and the decision attribute, are taken into account in DRSA, which is new in machine learning.

While the above example is typical for data describing *multiple criteria classification* problems (called also *ordinal classification*), DRSA is also relevant in case where preferences are not considered but a kind of monotonicity relating ordered attribute values is meaningful for the analysis of data at hand. In general, monotonicity concerns relationship between different aspects of a phenomenon described by data, e.g., "the larger the house, the higher its price" or "the more a tomato is red, the more it is ripe". The qualifiers, like "large house", "high

G. Wang et al. (Eds.): RSKT 2008, LNAI 5009, pp. 21–22, 2008.

price", "red" and "ripe", may be expressed either in terms of some measurement units, or in terms of degrees of membership to some fuzzy sets.

In recent years, one can observe a growing interest in ordinal classification within *machine learning*. The proposed approaches are based on statistical analysis of data, such as support vector machines, decision trees, loss function and ordinal regression approaches. DRSA has been proposed well before. It extends the classical Indiscernibility-based Rough Set Approach (IRSA) on ordinal classification, while maintaining all good properties of IRSA [4]. Being proposed on the ground of rough sets, DRSA escapes the statistical approach giving a non-invasive methodology for reasoning about ordinal data.

Looking at DRSA from *granular computing* perspective, we can say that DRSA permits to deal with ordered data by considering a specific type of information granules defined by means of dominance-based constraints having a syntax of the type: "x is at least R" or "x is at most R", where R is a qualifier from a properly ordered scale. In condition and decision space, such granules are *dominance cones*.

Decision rules induced from dominance-based rough approximations put in evidence the monotonic relationship between conditions and decision. For example, in credit rating, the decision rule could be:

"if the customer has the bank account since at least 3 years, his income is at least 2000\$/month, and his previous payment of the debt was at least acceptable, then the new credit is acceptable with care or better".

The DRSA rules involve partial profiles which are compared with evaluation of objects using a simple dominance relation.

The DRSA decision rules can be characterized by the usual confidence and coverage factors, however, as shown in [5], some Bayesian confirmation measures account in a more meaningful way for the attractiveness of these rules.

References

1. Greco, S., Inuiguchi, M., Słowiński, R.: Fuzzy rough sets and multiple-premise gradual decision rules, Internat. J. of Approximate Reasoning. 41, 179–211 (2006)
2. Greco, S., Matarazzo, B., Słowiński, R.: Rough sets theory for multicriteria decision analysis. European J. of Operational Research 129, 1–47 (2001)
3. Greco, S., Matarazzo, B., Słowiński, R.: Dominance-based Rough Set Approach to Case-Based Reasoning, in V. In: Torra, V., Narukawa, Y., Valls, A., Domingo-Ferrer, J. (eds.) MDAI 2006. LNCS (LNAI), vol. 3885, pp. 7–18. Springer, Heidelberg (2006)
4. Greco, S., Matarazzo, B., Słowiński, R.: Dominance-based Rough Set Approach as a proper way of handling graduality in rough set theory. In: Peters, J.F., Skowron, A., Marek, V.W., Orłowska, E., Słowiński, R., Ziarko, W. (eds.) Transactions on Rough Sets VII. LNCS, vol. 4400, pp. 36–52. Springer, Berlin (2007)
5. Greco, S., Pawlak, Z., Słowiński, R.: Can Bayesian confirmation measures be useful for rough set decision rules? Engineering Applications of Artificial Intelligence 17, 345–361 (2004)
6. Słowiński, R., Greco, S., Matarazzo, B.: Rough set based decision support, ch. 16. In: Burke, E.K., Kendall, G. (eds.) Search Methodologies: Introductory Tutorials in Optimization and Decision Support Techniques, pp. 475–527. Springer, New York (2005)

Dominance-Based Rough Set Approach for Decision Analysis - A Tutorial

Salvatore Greco

Faculty of Economics, University of Catania,
Corso Italia, 55, 95129 Catania, Italy

Scientific analysis of decision problems aims at giving the decision maker (DM) a recommendation concerning a set of objects (called also alternatives, solutions, acts, actions, cases, candidates) evaluated wiht respect to a plurality of their characteristics considered relevant for the problem at hand, and called attributes. For example, a decision can regard:

1) diagnosis of pathologies for a set of patients, being the objects of the decision, and the attributes are symptoms and results of medical examinations,
2) assignment to classes of risk for a set of firms, being the objects of the decision, and the attributes are ratio indices and other economic indicators such as the market structure, the technology used by the enterprises, the quality of the management and so on,
3) selection of a car to be bought from a given set of cars, being the objects of the decision, and the attributes are maximum speed, acceleration, price, fuel consumption and so on,
4) ordering of students applying for a scholarship, being the objects of the decision, and the attributes are scores in different disciplines.

The following three most frequent decision problems are typically distinguished:

- classification, when the decision aims at assigning objects to predefined classes,
- choice, when the decision aims at selecting the best objects,
- ranking, when the decision aims at ordering objects from the best to the worst.

Looking at the above examples, we can say that 1) and 2) are classification problems, 3) is a choice problem and 4) is a ranking problem. The above categorization can be refined with respect to classification problems by distinguishing between

- nominal classification, called also taxonomy, when neither the value sets of attributes nor the predefined classes are preference ordered,
- ordinal classification, called also sorting, when both the value sets of attributes and the predefined classes are preference ordered [1] (even if in this case it is also possible to take into consideration attributes with non ordered value sets [2]).

G. Wang et al. (Eds.): RSKT 2008, LNAI 5009, pp. 23–24, 2008.

Looking at the above examples, one can say that 1) is a nominal classification problem, while 2) is an ordinal classification problem. Classical Rough Set Approach (CRSA) proposed by Pawlak [5] cannot deal with preference order in the value sets of attributes. Thus, among all listed above decision problems, CRSA supports nominal classification only. For ordinal classification, choice and ranking it is necessary to generalize the CRSA, so as to take into account preference orders. This generalization, called Dominance-based Rough Set Approach (DRSA), has been proposed by Greco, Matarazzo and Slowinski [1],[3]. Moreover, it has been proved that CRSA is a specific case of DRSA, and, therefore, any application of rough set approach to decision problems can be effectively dealt with using DRSA [4]). In this tutorial we present applications of DRSA to many real life decision problems, emphasizing the advantages of this approach with respect to competitive approaches.

References

1. Greco, S., Matarazzo, B., Słowiński, R.: Rough set theory for multicriteria decision analysis. European Journal of Operational Research 129, 1–47 (2001)
2. Greco, S., Matarazzo, B., Słowiński, R.: Rough sets methodology for sorting problems in presence of multiple attributes and criteria. European Journal of Operational Research 138, 247–259 (2002)
3. Greco, S., Matarazzo, B., Słowiński, R.: Decision rule approach. In: Figueira, J., Greco, S., Erghott, M. (eds.) Multiple Criteria Decision Analysis: State of the Art Surveys, pp. 507–563. Springer, Berlin (2005)
4. Greco, S., Matarazzo, B., Słowiński, R.: Dominance-based Rough Set approach as a Proper Way of Handling Graduality in Rough Set Theory. In: Peters, J.F., Skowron, A., Marek, V.W., Orłowska, E., Słowiński, R., Ziarko, W. (eds.) Transactions on Rough Sets VII. LNCS, vol. 4400, pp. 36–52. Springer, Heidelberg (2007)
5. Pawlak, Z.: Rough Sets. Kluwer, Dordrecht (1991)

Introduction to 3DM: Domain-Oriented Data-Driven Data Mining*

Guoyin Wang[1,2]

[1] Institute of Computer Science and Technology,
Chongqing University of Posts and Telecommunications,
Chongqing 400065, P.R. China
[2] School of Information Science and Technology,
Southwest Jiaotong University,
Chengdu 600031, P.R. China
wanggy@ieee.org

Abstract. Recent advances in computing, communications, digital storage technologies, and high-throughput data-acquisition technologies, make it possible to gather and store incredible volumes of data. It creates unprecedented opportunities for large-scale knowledge discovery from huge database. Data mining (DM) technology has emerged as a means of performing this discovery. There are countless researchers working on designing efficient data mining techniques, methods, and algorithms. Many data mining methods and algorithms have been developed and applied in a lot of application fields [1]. Unfortunately, most data mining researchers pay much attention to technique problems for developing data mining models and methods, while little to basic issues of data mining [2].

In this talk, some basic issues of data mining are addressed. What is data mining? What is the product of a data mining process? What are we doing in a data mining process? What is the rule we should obey in a data mining process? Through analyzing existing data mining methods, and domain-driven (or user-driven) data mining models [3-5], we find that we should take a data mining process as a process of knowledge transformation. Based on this understanding of data mining, a conceptual data mining model of domain-oriented data-driven data mining (3DM) is proposed [2]. The relationship between traditional domain-driven (or user-driven) data mining models and our proposed 3DM model is also analyzed. Some domain-oriented data-driven data mining algorithms for mining such knowledge as default rule [6], decision tree [7], and concept lattice [8] from database are proposed. The experiment results for these algorithms are also shown to illustrate the efficiency and performance of the knowledge acquired by our 3DM data mining algorithms.

Keywords: Data mining, machine learning, rough set, data driven, domain driven, domain oriented.

---

* This work is partially supported by National Natural Science Foundation of P. R. China under Grants No.60573068 and No.60773113, Program for New Century Excellent Talents in University (NCET), Natural Science Foundation of Chongqing, and Science & Technology Research Program of the Municipal Education Committee of Chongqing of China (No.060517).

References

1. Wu, X.D., Kumar, V., Quinlan, J.R., Ghosh, J., Yang, Q., Motoda, H., McLachlan, G.J., Ng, A., Liu, B., Yu, P.S., Zhou, Z.H., Steinbach, M., Hand, D.J., Steinberg, D.: Top 10 Algorithms in Data Mining. Knowledge and Information Systems 14(1), 1–37 (2008)
2. Wang, G.Y.: Domain-Oriented Data-Driven Data Mining (3DM): Simulation of Human Knowledge Understanding. In: Zhong, N., Liu, J., Yao, Y., Wu, J., Lu, S., Li, K. (eds.) Web Intelligence Meets Brain Informatics. LNCS (LNAI), vol. 4845, pp. 278–290. Springer, Heidelberg (2007)
3. Zhao, Y., Yao, Y.Y.: Interactive Classification Using a Granule Network. In: Proc. of the 4th IEEE Int. Conf. on Cognitive Informatics, Irvine, USA, pp. 250–259 (2005)
4. Zhang, C., Cao, L.: Domain-Driven Data Mining: Methodologies and Applications. In: Li, Y.F., Looi, M., Zhong, N. (eds.) Advances in Intelligent IT - Active Media Technology 2006, pp. 13–16. IOS Press, Amsterdam (2006)
5. Cao, L., Lin, L., Zhang, C.: Domain-driven In-depth Pattern Discovery: A practical methodology. [Research Report], Faculty of Information Technology, University of Technology, Sydney, Australia (2005)
6. Wang, G.Y., He, X.: A Self-Learning Model under Uncertain Condition. Journal of Software 14(6), 1096–1102 (2003)
7. Yin, D.S., Wang, G.Y., Wu, Y.: Data-Driven Decision Tree Learning Algorithm Based on Rough Set Theory. In: Tarumi, H., Li, Y., Yoshida, T. (eds.) Proc. of the 2005 International Conference on Active Media Technology (AMT 2005), Takamatsu, Kagawa, Japan, pp. 579–584 (2005)
8. Wang, Y., Wang, G.Y., Deng, W.B.: Concept Lattice Based Data-Driven Uncertain Knowledge Acquisition. Pattern Recognition and Artificial Intelligence 20(5), 636–642 (2007)

Granular Computing: Past, Present, and Future

Yiyu (Y.Y.) Yao

Department of Computer Science, University of Regina, Canada

Granular computing offers a new paradigm of computation that exploits varying sized data, information, knowledge, and wisdom granules. It may be interpreted in terms of nature-inspired computing, human-centered computing, and
machine-oriented computing. In this tutorial, I look at the past and the present,
and probe the future, of granular computing based on the triarchic theory/model
of granular computing.[1]

Emergence of Granular Computing

Granular computing may be viewed as both old and new. It is old in the sense
that its basic ideas, principles, strategies, methodologies, and tools have indeed
appeared and reappeared in many branches of science and various fields of computer science under different names and with different notations. It is new by
its distinctive goals and scopes. Based on existing studies, granular computing
is growing into a school of thought and a full-scale theory, focusing on problem-
solving and information processing in the abstract, in the brain, and in machines.

Investigations of concrete models, including rough sets and fuzzy sets, are a
major driving force behind granular computing. In a wider context, motivations
for the emerging study of granular computing are much similar to the emergence
of the study of systems theory a few decades earlier. A general systems theory
may be viewed as an interdisciplinary study of complex systems in nature, society, and science. It investigates the principles common to all complex systems
and abstract models of such diverse systems. Granular computing may be viewed
as an interdiscriplary study of computations in nature, society, and science and
abstract models of such computations, with an underlying notion of multiple
levels of granularity. It extracts and studies principles, strategies, and heuristics
common to all types of problem solving and information processing.

The Triarchic Theory of Granular Computing

Granular computing is an interdisciplinary and a multidisciplinary study, emerging from many disciplines. Results from cognitive science and psychology on
human guessing, knowing, thinking and languages provide evidence to support
the philosophical view of granular computing that humans perceive, understand,

---

[1] More detailed information about the triarchic model can be found in my papers
available at `http://www.cs.uregina.ca/~yyao/grc_paper`. The reference section of
those papers contains an extensive literature that, unfortunately, cannot be included
here. I would like to thank Professors Guoyin Wang and Tianrui Li for the support
and encouragement when preparing the tutorial.

G. Wang et al. (Eds.): RSKT 2008, LNAI 5009, pp. 27–28, 2008.

and represent the real world in multiple levels of granularity and abstraction. Results from the general systems theory, regarding efficient evolution of complex systems and organized complexity, suggest that multilevel hierarchical granular structures used in granular computing are suitable for describing complex real-world problems. The effective methodology of structured programming, characterized by top-down design and step-wise refinement, is generally applicable to other types of problem solving, and hence may provide a methodological foundation for granular computing. The systematic study of human problem solving, strategies and heuristics, knowledge represent! ation and search, in artificial intelligence may provide the necessary models, methods, and tools required by granular computing. Advances in information processing and related systems may help in establishing a computational basis of granular computing.

The Triarchic Theory of granular computing reflects the above integration. It consists of the philosophy, methodology and computation perspectives, based on multilevel and multiview hierarchical granular structures. While a single hierarchy offers one representation and understanding with many levels of granularity, multiple hierarchies afford a complete understanding from many views. Briefly, the three perspectives are:

- **Philosophy: a way of structured thinking**. It combines analytical thinking for decomposing a whole into parts and synthetic thinking for integrating parts into a whole.
- **Methodology: a method of structured problem solving**. It promotes systematic approaches, effective principles, and practical heuristics and strategies used effectively by humans in solving real-world problems. Three tasks are considered: constructing granular structures, working within a particular level of the structure, and switching between levels.
- **Computation: a paradigm of structured information processing**. Two related basic issues are representations and processes. Representation covers the formal and precise description of granules and granular structures. Processes may be broadly divided into the two classes: granulation and computation with granules.

Future of Granular Computing

Falmagne insightfully pointed out that a "scientific field in its infancy ... requires loving care in the form of a steady diet of fresh ideas and results." A steady diet for granular computing may be provided in several ways. First, we need to emphasize on a balanced approach by considering all the three perspectives. Second, we need to stress for diversity, in order to arrive at a unity. Third, we need to cross the boundaries of different disciplines, in order to observe their commonalities and to use them for granular computing. Forth, we need to shift from machine-centered approaches to human-centered approaches. Fifth, we need to move beyond the current rough sets and fuzzy sets dominated research agenda. Finally, we need to learn to appreciate both the ideas and concrete results. After all, what makes granular computing attractive is its ideas. It is exciting to investigate how to realize those ideas.

Rough Logics with Possible Applications to Approximate Reasoning

Mihir Kr. Chakraborty

Department of Pure Mathematics, University of Calcutta
35, Ballygunge Circular Road, Calcutta 700 019, India
mihirc99@vsnl.com

Representations of the lower and upper approximations of a set in the context of an approximation space as modal operators in the first order language of modal logics, are quite natural and widely familiar now to the rough-set community. According to the perception of an observer, objects of a universe (of discourse) are clustered. These are the basic information granules (or quanta). With respect to the information available, objects belonging to the same cluster are indistinguishable. It may not always be the case that the clusters are mutually disjoint.

Now, an indiscernibility relation I may be defined such that for any objects a, b of the universe, aIb holds if (and only if) a and b belong to the same cluster. It is quite reasonable to assume that I is at least reflexive and symmetric.

The relation I may be formally interpreted as the (general) accessibility relation of the Kripke-models of modal logic-systems. Although the accessibility relation in the models of modal logics need not be symmetric (and even reflexive), if I is taken to be indiscernibility there should not be any valid reason of its being non-symmetric. The corresponding modal system turns out to be the system B. In Pawlaks rough set systems since I is defined in terms of information tables with respect to attribute-value systems, the indiscernibility relation I turns out to be transitive too. Thus, I being an equivalence relation, the corresponding modal logic-system has to be $S5$.

In the present lecture, however, we shall concentrate on two basic types of clustering of the universe, viz. the covering based clustering and partition based clustering of the objects of the universe – the first giving rise to relations which are reflexive and symmetric, i.e. tolerances, while the second giving rise to equivalences (reflexive, symmetric and transitive).

Now, the main support of the inference machine (methodology) in a logic or broadly speaking in reasoning, rests on the rule Modus Ponens (M.P.) which says that "If from a premise set X, the sentences A and $A \rightarrow B$ (If A then B) are both derivable then to infer B from X".

In the context of approximate reasoning, this rule of inference is usually relaxed in various ways. For example in the fuzzy logic literature one gets fuzzy modus ponens rule like "If A' and $A \rightarrow B$ are derivable from X where A' is similar to or in some sense related to A, then to infer B' from X".

A typical example of the above extended MP rule that is noticed in fuzzy literature is

G. Wang et al. (Eds.): RSKT 2008, LNAI 5009, pp. 29–30, 2008.
© Springer-Verlag Berlin Heidelberg 2008

" If mango is red then mango is ripe.
This mango is very red.
Hence this mango is very ripe."

A similar approach shall be taken here in the rough-context where A' , the first premise, is not exactly the same as the antecedent of the second premise but similar to it.

The MP rule, however, is equivalent to the following rule that we call MP (modified). "If $A, B \rightarrow C$ both follow from X and $A \rightarrow B$ is universally true, then to infer C from X." By "universally true" we mean that under all states of affairs, the wff $A \rightarrow B$ is true. This, in turn reduces to the claim that, the states of affairs making A true makes B true also, i.e. $Int(A) \subseteq Int(B)$, where $Int(A)$ is the set of the states of affairs making A true.

Now, in rough-set context one can define various notions of inclusion, viz. rough lower/ upper inclusion, rough inclusion and others. In the case of Pawlak rough sets, we get the following ultimate categories of inclusion

$$\overline{A} \subseteq \underline{B}, A \subseteq B, \underline{A} \subseteq \underline{B} \wedge \overline{A} \subseteq \overline{B}, \underline{A} \subseteq \underline{B}, \overline{A} \subseteq \overline{B}, \underline{A} \subseteq \overline{B}.$$

Thus $Int(A)$ and $Int(B)$ may have now all the above five various inclusion relations of which one (viz. the second one) gives the classical inclusion.

Rough logic based on various "Rough MP" rules may now be defined by

$$\frac{A, B \rightarrow C \text{ roughly follow from } X}{Int(A) \text{ rough-included (Rough Inc) in } Int(B)}$$
$$C \text{ roughly follows from } X$$

where "Rough Inc" is any of the rough inclusions mentioned above. In fact, a hierarchy of rough-logics shall be obtained.

Besides these rough MP rules there have been some other variants of MP rules in rough set context.

In the lectures we shall focus on

- definitions of various rough MP rules
- rough logics developed with the help of these rules
- the interrelations between them
- the semantics of these logics
- suggested applications of rough MP rules.

References

1. Banerjee, M.: Logic for rough truth. Fund. Informaticae 71(2-3), 139–151 (2006)
2. Banerjee, M., Chakraborty, M.K.: Rough consequence and rough algebra. In: Ziarko, W.P. (ed.) Rough Sets, Fuzzy Sets and Knowledge Discovery, Proc. RSKD 1993, pp. 196–207. Springer, London (1994)
3. Bunder, M.W.: Rough consequence and Jaśkowski's D2 logics. Technical Report 2/04. School of Mathematics and Applied Statistics, University of Wollongong, Australia (2004)
4. Bunder, M.W., Banerjee, M., Chakraborty, M.K.: Some rough consequence logics and their interrelations. Transactions on Rough Sets (to appear)

A Comparison of Six Approaches to Discretization—A Rough Set Perspective

Piotr Blajdo[1], Jerzy W. Grzymala-Busse[2], Zdzislaw S. Hippe[1],
Maksymilian Knap[1], Teresa Mroczek[3], and Lukasz Piatek[3]

[1] Department of Expert Systems and Artificial Intelligence,
University of Information Technology and Management, 35-225 Rzeszow, Poland
{pblajdo,zhippe,mknap}@wsiz.rzeszow.pl
[2] Department of Electrical Engineering and Computer Science,
University of Kansas, Lawrence, KS 66045, USA
and
Institute of Computer Science,
Polish Academy of Sciences, 01-237 Warsaw, Poland
jerzy@ku.edu
[3] Department of Distributed Systems,
University of Information Technology and Management, 35-225 Rzeszow, Poland
{tmroczek,lpiatek}@wsiz.rzeszow.pl

Abstract. We present results of extensive experiments performed on
nine data sets with numerical attributes using six promising discretization methods. For every method and every data set 30 experiments of
ten-fold cross validation were conducted and then means and sample
standard deviations were computed. Our results show that for a specific
data set it is essential to choose an appropriate discretization method
since performance of discretization methods differ significantly. However,
in general, among all of these discretization methods there is no statistically significant worst or best method. Thus, in practice, for a given
data set the best discretization method should be selected individually.

Keywords: Rough sets, Discretization, Cluster analysis, Merging intervals, Ten-fold cross validation, Test on the difference between means,
F-test.

1 Introduction

Many real-life data contain numerical attributes whose values are integers or real
numbers. Mining such data sets requires special techniques, taking into account
that input data sets are numerical. Most frequently, numerical attributes are
converted into symbolic ones during a special process, called discretization [9],
before the main process of knowledge acquisition. In some data mining systems
both processes, discretization and knowledge acquisition, are executed at the
same time. Examples of such systems are C4.5 [14], CART [1], and MLEM2 [10].

Our objective was to compare, for the first time, the most promising discretization techniques [4,9,11,15] through extensive experiments on real-life data sets

G. Wang et al. (Eds.): RSKT 2008, LNAI 5009, pp. 31–38, 2008.

and using ten-fold cross validation, standard statistical test on the difference between means and F-test. The oldest method, among our six approaches to discretization, is based on conditional entropy and was presented in [7]. The remaining five approaches are implemented in the data mining system LERS (Learning from Examples based on Rough Sets) [8]. One of them is based on a hierarchical method of cluster analysis, two on a divisive method of cluster analysis, and remaining two on merging intervals. Our results show that there is no best or worst method. Additionally, for a specific data set, difference in performance between different discretization techniques is significant and the best discretization method should be selected individually, trying as many techniques as possible.

2   Discretization Methods

Knowledge acquisition, such as rule induction or decision tree generation, from data with numerical attributes requires converting numerical values of an attribute into intervals. The process of converting numerical values into intervals, called *discretization*, is usually done as a preprocessing, before the main process of knowledge acquisition. In some data mining systems, such as C4.5, CART, and MLEM2, both processes: discretization and knowledge acquisition are conducted simultaneously. In this paper we will discuss discretization as a preprocessing.

For a numerical attribute a with an interval $[a, b]$ as a range, a partition of the range into n intervals

$$\{[a_0, a_1), [a_1, a_2), ..., [a_{n-2}, a_{n-1}), [a_{n-1}, a_n]\},$$

where $a_0 = a$, $a_n = b$, and $a_i < a_{i+1}$ for $i = 0, 1, ..., n - 1$, defines discretization of a. The numbers a_1, a_2,..., a_{n-1} are called *cut-points*.

Discretization methods in which attributes are processed one at a time are called *local* [4,9] (or *static* [5]). On the other hand, if all attributes are considered in selection of the best cut-point, the method is called *global* [4,9] (or *dynamic* [5]). Additionally, if information about the expert's classification of cases is taken into account during the process of discretization, the method is called *supervised* [5].

Many discretization methods [9] are used in data mining. In this paper we will use three approaches to discretization based on cluster analysis, two additional methods that will use similar principles, and, for comparison, a well-known discretization method based on minimal conditional entropy. All of these methods are global and supervised.

The simplest discretization methods are local methods called Equal Interval Width and Equal Frequency per Interval [4,9]. Another local discretization method [7] is called a Minimal Class Entropy. The conditional entropy, defined by a cut-point q that splits the set U of all cases into two sets, S_1 and S_2 is defined as follows

$$E(q, U) = \frac{|S_1|}{|U|} E(S_1) + \frac{|S_2|}{|U|} E(S_2),$$

where $E(S)$ is the entropy of S and $|X|$ denotes the cardinality of the set X. The cut-point q for which the conditional entropy $E(q, U)$ has the smallest value is selected as the best cut-point. If k intervals are required, the procedure is applied recursively $k-1$ times. Let q_1 and q_2 be the best cut-points for sets S_1 and S_2, respectively. If $E(q_1, S_1) > E(q_2, S_2)$ we select q_1 as the next cut-point, if not, we select q_2.

2.1 Globalization of Local Discretization Methods

We will present an approach to convert local discretization methods to global [4]. First, we discretize all attributes, one at a time, selecting the best cut-point for all attributes. If the level of consistency is sufficient, the process is completed. If not, we further discretize, selecting an attribute a for which the following expression has the largest value

$$M_a = \frac{\sum_{B \in \{a\}^*} \frac{|B|}{|U|} E(B)}{|\{a\}^*|}.$$

In all six discretization methods discussed in this paper, the stopping condition was the level of consistency [4], based on rough set theory introduced by Z. Pawlak in [12]. Let U denote the set of all cases of the data set. Let P denote a nonempty subset of the set of all variables, i.e., attributes and a decision. Obviously, set P defines an equivalence relation $\wp$ on U, where two cases x and y from U belong to the same equivalence class of $\wp$ if and only if both x and y are characterized by the same values of each variable from P. The set of all equivalence classes of $\wp$, i.e., a partition on U, will be denoted by P^*.

Equivalence classes of $\wp$ are called *elementary sets* of P. Any finite union of elementary sets of P is called a *definable set* in P. Let X be any subset of U. In general, X is not a definable set in P. However, set X may be approximated by two definable sets in P, the first one is called a *lower approximation of X in P*, denoted by $\underline{P}X$ and defined as follows

$$\bigcup \{Y \in P^* | Y \subseteq X\}.$$

The second set is called an *upper approximation of X in P*, denoted by $\overline{P}X$ and defined as follows

$$\bigcup \{Y \in P^* | Y \cap X \neq \emptyset\}.$$

The lower approximation of X in P is the greatest definable set in P, contained in X. The upper approximation of X in P is the least definable set in P containing X. A *rough set of X* is the family of all subsets of U having the same lower and the same upper approximations of X.

A *level of consistency* [4], denoted L_c, is defined as follows

$$L_c = \frac{\sum_{X \in \{d\}^*} |\underline{A}X|}{|U|}.$$

Practically, the requested level of consistency for discretization is 100%, i.e., we want the discretized data set to be *consistent*.

2.2   Discretization Based on Cluster Analysis and Interval Merging

The data mining system LERS uses two methods of cluster analysis, agglomerative (bottom-up) [4] and divisive (top-down) [13], for discretization. In agglomerative techniques, initially each case is a single cluster, then clusters are fused together, forming larger and larger clusters. In divisive techniques, initially all cases are grouped in one cluster, then this cluster is gradually divided into smaller and smaller clusters. In both methods, during the first step of discretization, *cluster formation*, cases that exhibit the most similarity are fused into clusters. Once this process is completed, clusters are projected on all attributes to determine initial intervals on the domains of the numerical attributes. During the second step (*merging*) adjacent intervals are merged together.

Initially all attributes were categorized into numerical and symbolic. During clustering, symbolic attributes were used only for clustering stopping condition. All numerical attributes were normalized [6] (attribute values were divided by the attribute standard deviation).

In agglomerative discretization method initial clusters were single cases. Then the distance matrix of all Euclidean distances between pairs of cases was computed. The closest two cases, a and b, compose a new cluster $\{a, b\}$. The distance from $\{a, b\}$ to any remaining case c was computed using the Median Cluster Analysis formula [6]:

$$\frac{1}{2}d_{ca} + \frac{1}{2}d_{cb} - \frac{1}{4}d_{ab},$$

where d_{xy} is the Euclidean distance between x and y. The closest two cases compose a new cluster, etc.

At any step of clustering process, the clusters form a partition π on the set of all cases. All symbolic attributes define another partition τ on the set of all cases. The process of forming new clusters was continued as long as $\pi \cdot \tau \leq \{d\}^*$.

In divisive discretization method, initially all cases were placed in one cluster C_1. Next, for every case the average distance from all other cases was computed. The case with the largest average distance was identified, removed from C_1, and placed in a new cluster C_2. For all remaining cases from C_1 a case c with the largest average distance d_1 from all other cases in C_1 was selected and the average distance d_2 from c to all cases in C_2 was computed. If $d_1 - d_2 > 0$, c was removed from C_1 and put to C_2. Then the next case c with the largest average distance in C_1 was chosen and the same procedure was repeated. The process was terminated when $d_1 - d_2 \leq 0$. The partition defined by C_1 and C_2 was checked whether all cases from C_1 were labeled by the same decision value and, similarly, if all cases from C_2 were labeled by the same decision value (though the label for C_1 might be different than the label for C_2). The process of forming new clusters was continued until $\pi \cdot \tau \leq \{d\}^*$.

Final clusters were projected into all numerical attributes, defining this way a set of intervals. The next step of discretization was merging these intervals to reduce the number of intervals and, at the same time, preserve consistency. Merging of intervals begins from *safe merging*, where, for each attribute, neighboring intervals labeled by the same decision value are replaced by their union.

Table 1. Data sets

| Data set | | Number of | |
	cases	attributes	concepts
Australian	690	14	2
Bank	66	5	2
Bupa	345	6	2
German	1000	24	2
Glass	214	9	6
Iris	150	4	3
Segmentation	210	19	7
Wine	178	13	3
Wisconsin	625	9	9

The next step of merging intervals was based on checking every pair of neighboring intervals whether their merging will result in preserving consistency. If so, intervals are merged permanently. If not, they are marked as un-mergeable. Obviously, the order in which pairs of intervals are selected affects the final outcome. In our experiments we started either from an attribute with the most intervals first or from an attribute with the largest conditional entropy.

3   Experiments

Our experiments were conducted on nine data sets, summarized in Table 1. All of these data sets, with the exception of *bank*, are available on the University of California at Irvine *Machine Learning Repository*. The bank data set is a well-known data set used by E. Altman to predict a bankruptcy of companies.

The following six discretization methods were used in our experiments:

- Cluster analysis divisive method with merging intervals with preference for attributes with most intervals, coded as 00,
- Cluster analysis divisive method with merging intervals with preference for attributes with largest conditional entropy, coded as 01,
- Merging intervals with preference for attributes with most intervals, coded as 10,
- Merging intervals with preference for attributes with largest conditional entropy, coded as 11,
- Globalized minimal class entropy method, coded as 13,
- Cluster analysis hierarchical method, coded as 14.

Every discretization method was applied to every data set, with the level of consistency equal to 100%. For any discretized data set, the ten-fold cross

Table 2. Error Rates—Means

| Data set | Methods of discretization | | | | | |
	00	01	10	11	13	14
Australian	16.01	14.62	16.32	14.81	15.51	15.88
Bank	4.70	4.35	3.69	4.29	4.50	3.03
Bupa	36.81	36.82	36.72	37.38	42.73	35.90
German	31.11	31.32	31.12	31.34	29.99	29.30
Glass	31.21	28.54	28.65	28.78	42.63	31.21
Iris	3.26	3.29	3.29	3.31	8.71	4.02
Segmentation	14.67	15.00	16.51	12.87	49.24	17.05
Wine	7.32	7.27	7.21	7.42	2.40	6.26
Wisconsin	21.03	19.45	20.76	19.06	20.87	20.05

validation experiment for determining an error rate was repeated 30 times, with different re-ordering and partitioning the set U of all cases into 10 subsets, where rule sets were induced using the LEM2 algorithm [3,8]. The mean and standard deviation were computed for every sequence of 30 experiments.

Then we used the standard statistical test about the difference between two means, see, e.g., [2]. With the level of significance at 0.05, the decision: reject H_0 if $Z \geq 1.96$ or $Z \leq -1.96$, where H_0 is the hypothesis that the performance of two respective methods do not differ. For example, for *Australian* data set, the value of Z for methods 00 and 01 is 5.55, hence they do differ—as follows from Table 2, method 01 is better (the error rate is smaller). In general, for *Australian* data set, methods 01 and 11 are significantly better than all remaining methods while methods 01 and 11 do not differ significantly. For remaining methods situation is more complicated: methods 13 and 14 do not differ significantly, but method 13 is obviously worse than methods 01 and 11 and is better than methods 00 and 10. Method 14, though worse than 01 and 11, does not differ significantly from methods 00 and 10. On the other hand, methods 00 and 01 do not differ significantly between each other and method 14. A similar analysis was conducted for every data set, the details are skipped because of the page limit for this paper.

As follows from Tables 2–3, performance of discretization methods varies with the change of the data set. The question is if there exists a universally best or worst method. The appropriate test here is the F-test, based on the analysis of variance. The variance s_n^2 of sample means is equal to 6.07. The mean s_d^2 of all sample variances is equal to 178.0. Thus the test statistics F, where

$$F = \frac{s_n^2}{s_d^2}(n)$$

Table 3. Error Rates—Standard Deviations

Data set	Methods of discretization					
	00	01	10	11	13	14
Australian	0.88	1.05	1.02	0.88	0.88	0.78
Bank	1.28	1.24	1.10	1.95	0.63	0.00
Bupa	1.79	1.41	1.43	1.41	2.06	1.83
German	1.15	0.80	0.81	0.90	0.87	0.84
Glass	1.98	2.20	2.10	2.06	2.36	1.42
Iris	0.20	0.17	0.46	0.33	1.05	0.62
Segmentation	1.42	1.35	1.31	1.06	2.22	1.31
Wine	0.98	1.11	1.09	1.13	0.83	1.26
Wisconsin	0.56	0.43	0.55	0.64	0.74	0.41

and $n = 9$ (the number of data sets), is equal to 0.307. Since F is less than 1, we do not need to look to the F-table to know that these discretization methods do not show a statistically significant variation in performance.

4   Conclusions

Our paper presents results of experiments in which six promising discretization methods were used on nine data sets with numerical attributes. All six methods were global and supervised. Results of all six methods, the discretized input data, were used for rule induction using the same LEM2 rule induction algorithm. The performance of discretization, for every method and every data set, was evaluated using 30 experiments of ten-fold cross validation. As a result, we conclude that

- for a specific data set, difference in performance between different discretization methods is significant,
- there is no universally best or worst discretization method. In different words, difference in performance for our six discretization methods, evaluated on all nine data sets, is not significant.

Thus, for a specific data set the best discretization method should be selected individually.

References

1. Breiman, L., Friedman, J.H., Olshen, R.A., Stone, C.J.: Classification and Regression Trees, Wadsworth & Brooks, Monterey, CA (1984)
2. Chao, L.L.: Introduction to Statistics. Brooks Cole Publishing Co., Monterey (1980)

3. Chan, C.C., Grzymala-Busse, J.W.: On the attribute redundancy and the learning programs ID3, PRISM, and LEM2. Department of Computer Science, University of Kansas, TR-91-14 (1991)
4. Chmielewski, M.R., Grzymala-Busse, J.W.: Global discretization of continuous attributes as preprocessing for machine learning. Int. Journal of Approximate Reasoning 15, 319–331 (1996)
5. Dougherty, J., Kohavi, R., Sahami, M.: Supervised and unsupervised discretization of continuous features. In: 12-th International Conference on Machine Learning, pp. 194–202. Morgan Kaufmann, San Francisco (1995)
6. Everitt, B.: Cluster Analysis, 2nd edn. Heinmann Educational Books, London (1980)
7. Fayyad, U.M., Irani, K.B.: On the handling of continuous-valued attributes in decision tree generation. Machine Learning 8, 87–102 (1992)
8. Grzymala-Busse, J.W.: A new version of the rule induction system LERS. Fundamenta Informaticae 31, 27–39 (1997)
9. Grzymala-Busse, J.W.: Discretization of numerical attributes. In: Klösgen, W., Zytkow, J. (eds.) Handbook of Data Mining and Knowledge Discovery, pp. 218–225. Oxford University Press, New York (2002)
10. Grzymala-Busse, J.W.: MLEM2: A new algorithm for rule induction from imperfect data. In: 9th International Conference on Information Processing and Management of Uncertainty in Knowledge-Based Systems ESIA Annecy, France, pp. 243–250 (2002)
11. Grzymala-Busse, J.W., Stefanowski, J.: Three discretization methods for rule induction. Int. Journal of Intelligent Systems. 16, 29–38 (2001)
12. Pawlak, Z.: Rough Sets. International Journal of Computer and Information Sciences 11, 341–356 (1982)
13. Peterson, N.: Discretization using divisive cluster analysis and selected postprocessing techniques, University of Kansas, Internal Report. Department of Computer Science (1993)
14. Quinlan, J.R.: C4.5: Programs for Machine Learning. Morgan Kaufmann Publishers, San Mateo (1993)
15. Stefanowski, J.: Handling continuous attributes in discovery of strong decision rules. In: Polkowski, L., Skowron, A. (eds.) RSCTC 1998. LNCS (LNAI), vol. 1424, pp. 394–401. Springer, Heidelberg (1998)

Adaptive Classification with Jumping Emerging Patterns*

Pawel Terlecki and Krzysztof Walczak

Institute of Computer Science, Warsaw University of Technology,
Nowowiejska 15/19, 00-665 Warsaw, Poland
P.Terlecki, K.Walczak@ii.pw.edu.pl

Abstract. In this paper a generic adaptive classification scheme based on a classifier with reject option is proposed. A testing set is considered iteratively, accepted, semi-labeled cases are used to modify the underlying hypothesis and improve its accuracy for rejected ones. We apply our approach to classification with jumping emerging patterns (JEPs). Two adaptive versions of JEP-Classifier, by support adjustment and by border recomputation, are discussed. An adaptation condition is formulated after distance and ambiguity rejection strategies for probabilistic classifiers. The behavior of the method is tested against real-life datasets.

Keywords: jumping emerging pattern, adaptive classification, classification with reject option, transaction database, local reduct, rough set.

1 Introduction

Classification belongs to major knowledge discovery tasks. It has been approached with a wide spectrum of tools, like basic statistical methods, support vector machines, neural networks or pattern/rule-based solutions. Also, real-life applications can be found in various fields, like medicine, business, industry etc. Our attention has been drawn by a problem of reliable classification under a small training set. In fact, many solutions in this area employ adaptive mechanisms. In supervised learning, they are present in widely known approaches like AdaBoost and evolutionary algorithms. They also enhance semi-supervised strategies, e.g. a self-training EM algorithm for EEG signal classification ([1]) or an adaptive algorithm for high-dimensional data ([2]).

Here, we seek accuracy improvement by modifying a model during a testing phase. An adaptive generic scheme that employs a classifier with reject option is proposed. Initially, one performs supervised learning with labeled training cases. Then, for a testing set, labels are predicted in an iterative manner. Accepted cases are assumed semi-labeled and used to modify the classifier, while the rest becomes an input for the next iteration. This way harder cases can be approached with an enriched model. When no further improvement is possible, the remaining objects are handled with the version of a final classifier without reject option.

---

* The research has been partially supported by grant No 3 T11C 002 29 received from Polish Ministry of Education and Science.

G. Wang et al. (Eds.): RSKT 2008, LNAI 5009, pp. 39–46, 2008.

Classification with reject option is a classical problem approached most often with the statistical framework. In practice, it remains a common choice when a cost of incorrect classification is high, like in medical diagnosis or email filtering ([3]). The uncertain cases are rejected and usually undergo separate processing. In our case, adaptations with incorrectly labeled objects are harmful for the underlying model and should be prevented by rejections.

Our generic scheme is used to improve classification with jumping emerging patterns (JEPs, [4]) in transaction databases. This relatively new concept refers to patterns that are present in one class and absent from others and, thus, highly discriminative ([5]). Recent works show a direct relation between JEPs and the rough set theory ([6]). We discuss and experimentally compare two adaptation strategies: support adjustment and border recomputation. Also, an adaptation condition is proposed basing on distance and outlier rejection approaches ([7]).

The paper is organized as follows. Section 2 introduces formal background for classification problem in transaction databases. In Sect. 3 a general adaptation scheme employing a classifier with reject option is proposed. Fundamentals on emerging patterns and their use in classification are given in Sect. 4. Section 5 covers adaptive versions of JEP-Classifier and discusses an adaptation condition. In Sect. 6 experimental results are presented. The paper is summarized in Sect. 7.

2   Classification Problem

This section formulates the classification problem considered in the paper. Since our scheme is concretized for classifiers based on emerging patterns, we introduce definitions relevant to transaction databases. The convention follows [6].

Let a transaction system be a pair $(\mathcal{D}, \mathcal{I})$, where $\mathcal{D}$ is a finite sequence of transactions $(T_1, .., T_n)$ (database) such as $T_i \subseteq \mathcal{I}$ for $i = 1, .., n$ and $\mathcal{I}$ is a non-empty set of items (itemspace). Let a decision transaction system be a tuple $(\mathcal{D}, \mathcal{I}, \mathcal{I}_d)$, where $(\mathcal{D}, \mathcal{I} \cup \mathcal{I}_d)$ is a transaction system and $\forall_{T \in \mathcal{D}} |T \cap \mathcal{I}_d| = 1$. Elements of $\mathcal{I}$ and $\mathcal{I}_d$ are called condition and decision items, respectively. For each $c \in \mathcal{I}_d$, we define a decision class sequence $C_c = (T_i)_{i \in K}$, where $K = \{k \in \{1, .., n\} : c \in T_k\}$. The notations C_c and $C_{\{c\}}$ are used interchangeably.

Given a system $(\mathcal{D}, \mathcal{I}, \mathcal{I}_d)$, we define a classifier as any function $f : \{T - \mathcal{I}_d : T \in \mathcal{D}\} \mapsto \mathcal{I}_d$, i.e. it is able to predict a single class for any transaction from the considered domain. A classifier with reject option is defined as any function $f_R : \{T - \mathcal{I}_d : T \in \mathcal{D}\} \mapsto \mathcal{I}_d \cup R$, where R refers to a rejection class (bucket). If a classifier is not confident about a decision for a given transaction it gets assigned to a rejection class and called *rejected* (*accepted*, otherwise).

In the supervised learning, a classifier is constructed basing on information from a training sequence and assessed with a testing one. In our case, the testing sequence also affects the way the model evolves. The whole sequence is assumed to be known upfront, which makes the procedure independent a transaction order. In practice, one may consider adaptation on separate testing subsequences arriving at different points of time. In consequence, earlier cases cannot benefit from adaptation performed with the bundles considered later.

3   Adaptive Classification

Let us consider a decision transaction system $(\mathcal{D}, \mathcal{I}, \mathcal{I}_d)$. For $D \in \mathcal{D}$, we introduce the notation $MSEQ(D)$ as a set of possible sequences of elements from D. F (F_R) denote a set of possible classifiers (with reject option). Now, let us describe adaptation of a classifier (with reject option) by a function: $F \times MSEQ(\mathcal{D}_{adapt}) \mapsto F$ $(F_R \times MSEQ(\mathcal{D}_{adapt}) \mapsto F_R)$, where $\mathcal{D}_{adapt} = (T_{adapt,k})$ and $T_{adapt,|\mathcal{I}_d|*(i-1)+j-1} = T_i - \mathcal{I}_d \cup \{I_j\}$, for $i \in \{1, .., |\mathcal{D}|\}$, $j \in \{1, .., |\mathcal{I}_d|\}$. Note that $\mathcal{D}_{adapt}$ is a sequence of combinations of all transactions from $\mathcal{D}$ with all possible classes from $\mathcal{I}_d$. In other words, adaptation uses a sequence of transactions, called further an adaptation sequence, to modify a classifier.

In the paper, we propose an iterative self-training scheme that employs an internal classifier with reject option $f \in F_R$. We assume that this option can be disabled. In each iteration, unclassified transactions are passed through the internal classifier f. Its answers for accepted transactions become answers for the whole scheme. We treat them as confident and use for adaptation. If any transactions in a particular step were accepted, the model was possibly enriched and can be used to classify transactions rejected so far. In this case, these transactions become an input for the next iteration. When no adaptation happens anymore, depending on the type of classification, one may reject remaining transactions or classify them with a final version of f without reject option.

Algorithm 1 takes an internal classifier f and a testing sequence D_{test}, and produces a sequence of classes C. The flag $reject$ enables classification with reject option. Two functions has to be defined to obtain an instance of the whole classifier. The first one, $classify(indices, seq\text{-}of\text{-}transactions, classes, classifier, reject)$ uses an internal $classifier$ with/without reject option basing on a parameter $reject$ to classify transactions from $seq\text{-}of\text{-}transactions$ of specified $indices$. It returns a sequence of classes, where only answers referring to $indices$ are affected and the rest is copied from a sequence $classes$. The second one, $adapt(indices, seq\text{-}of\text{-}transactions, classes, classifier)$ performs adaptation of a given $classifier$ by transactions and classes indicated by $indices$. In order to concretize this general concept, one has to make three choices: an internal classifier, rejection condition and adaptation method.

4   Emerging Patterns

Our meta-scheme is concretized for classifiers based on emerging patterns, a concept introduced to capture highly discriminating features between two classes of transactions ([4]). Historically, first proposition is known as CAEP ([8]) and works by aggregating ρ-EPs. It was followed, in particular, by JEP/CEP/SEP-Classifier and DeEP ([5,9]). The choice of a particular type does not affect the general idea of our approach.

Let us consider a decision transaction system $(\mathcal{D}, \mathcal{I}, \mathcal{I}_d)$. For a database $D = (T_i)_{i \in K \subseteq \{1,...,n\}} \subseteq \mathcal{D}$, we define a complementary database $D' = (T_i)_{i \in \{1,...,n\}-K}$. The support of an itemset $X \subseteq \mathcal{I} \cup \mathcal{I}_d$ in a sequence $D = (T_i)_{i \in K} \subseteq \mathcal{D}$ is

---

Algorithm 1. Input: $D_{test}, f, reject$, **Output:** C

1: $C = (R)_{k \in \{1,..,|D_{test}|\}}$, $K_{unclassified} \Leftarrow \{1,..,|D_{test}|\}$, $f_{old} \Leftarrow f$, $f_{new} \Leftarrow NULL$
2: **while** $f_{old} \neq f_{new}$ **do**
3:     $K_{accepted} \Leftarrow \{k \in \{1,..,|D_{test}|\} : C_k \neq R\}$
4:     $C \Leftarrow classify(K_{unclassified}, D_{test}, C, f_{new}, \textbf{true})$
5:     $K_{unclassified} \Leftarrow \{k \in \{1,..,|D_{test}|\} : C_k = R\}$
6:     $f_{old} \Leftarrow f_{new}$
7:     $f_{new} \Leftarrow adapt(K_{adapt}, D_{test}, C, f_{new})$,
       where $K_{adapt} = \{k \in \{1,..,|D_{test}|\} : C_k \neq R\} - K_{accepted}$
8: **end while**
9: **if** $\neg reject$ **then**
10:     $C \Leftarrow classify(K_{unclassified}, D_{test}, C, f_{new}, \textbf{false})$
11: **end if**
12: **return** C

---

defined as $supp_D(X) = \frac{|\{i \in K : X \subseteq T_i\}|}{|K|}$, where $K \subseteq \{1,..,n\}$. Given two databases $D_1, D_2 \subseteq \mathcal{D}$, the growth-rate of an itemset $X \in \mathcal{P}$ from D_1 to D_2 is defined as $GR_{D_1 \to D_2}(X) = GR(supp_{D_1}(X), supp_{D_2}(X))$, where, for further convenience, for $m_1, m_2 \in \mathbb{R}_+ \cup \{0\}$: $GR(m_1, m_2) = 0$, for $m_1, m_2 = 0$; $= \infty$, for $m_1 = 0$ and $m_2 \neq 0$; $= \frac{m_2}{m_1}$, otherwise. A ρ-emerging pattern is defined as an itemset $X \subseteq \mathcal{I}$, for which $GR(X) \geq \rho$. Similarly, we define a jumping emerging pattern (JEP) from D_1 to D_2 as an itemset $X \subseteq \mathcal{I}$ with an infinite growth-rate, $GR_{D_1 \to D_2}(X) = +\infty$. The set of all JEPs from D_1 to D_2 is called a JEP space and denoted by $JEP(D_1, D_2)$. JEP spaces can be described concisely by means of borders ([4]). For $c \in \mathcal{I}_d$, we use a border $< \mathcal{L}_c, \mathcal{R}_c >$ to represent a JEP space $JEP(C'_c, C_c)$. $\mathcal{L}$ and $\mathcal{R}$ are called a left and a right bound, respectively.

Let us consider a decision transaction system $(\mathcal{D}, \mathcal{I}, \mathcal{I}_d)$ and define $D = \{T - \mathcal{I}_d : T \in \mathcal{D}\}$. In general, each of these classifiers can be defined according to the following template: (1) class-wise collections, $\{\mathcal{P}_k\}_{k \in \mathcal{I}_d}$, where $\mathcal{P} = \bigcup_{k \in \mathcal{I}_d} \mathcal{P}_k \subseteq D$; (2) interestingness measures, $m_{k,i} : \mathcal{P}_k \mapsto \mathbb{R}$, for $k \in \mathcal{I}_d$ and $i \in \{1..M\}$, where M is the number of measures; (3) scoring function, $s : \mathbb{R}^M \mapsto \mathbb{R}$. Class membership is expressed by a labeling function, $(L_k)_{k \in \mathcal{I}_d} = L : D \mapsto \mathbb{R}^{|\mathcal{I}_d|}$, where $L_k(X) = \sum_{P \in \mathcal{P}_k \wedge P \subseteq X} s(m_{k,1}(P),.., m_{k,M}(P))$ for a given transaction $X \in D$. This is the case of pattern-based classifiers that such labels do not have precise probabilistic interpretation. Nevertheless, one is usually interested in a single class and this labeling is hardened: $argmax_{k \in \mathcal{I}_d} L_k(X)$.

5   Adaptive JEP-Classifier

Our generic self-training scheme can be concretized for EP-based solutions. We describe how adaptation can be incorporated in JEP-Classifier ([5]).

Throughout this section we consider a decision transaction system $(\mathcal{D}, \mathcal{I}, \mathcal{I}_d)$, an internal classifier $f : \{T - \mathcal{I}_d : T \in \mathcal{D}\} \mapsto \mathcal{I}_d$ and a training sequence D_{train}. In order to clearly present adaptation process, let us consider a sequence of classifiers obtained by subsequent modifications. For a given step $i \in \mathbb{N} \cup \{0\}$, we have a classifier $f^i \in F$ and an associated sequence $D^i_{train} = D_{train} + \sum_{j \in \{1,...,i\}} D^j_{adapt}$, where D^j_{adapt} for $j \in \{1,..,i\}$ are successive adaptation sequences. For brevity, we represent concatenation of sequences by addition. We also define respective decision classes $C^i_{train,k}$, for $i \in \mathbb{N} \cup \{0\}$ and $C^j_{adapt,k}$, for $j \in \{1,..,i\}$, where $k \in \mathcal{I}_d$. At the beginning, one has an initial classifier f^0 trained with a sequence $D^0_{train} = D_{train}$. When a respective sequence of classifiers is consider, superscripts are added only to the parameters that are affected by adaptation.

5.1 Support Adjustment

A common strategy to distinguish importance of patterns is to assign to them specific weights. Following this idea, we put forward an approach placed among existing propositions for emerging patterns. It employs a modified JEP-Classifier. In i-th step, f^i can be characterized as follows: (1) patterns: $\{\mathcal{P}_k\}_{k \in \mathcal{I}_d} = \{\mathcal{L}_k\}_{k \in \mathcal{I}_d}$ for $JEP(C'_{train,k}, C_{train,k})$; (2) measures: $m^i_{k,1}(P) = supp_{C^i_{train,k}}(P)$, $m^i_{k,2}(P) = supp_{C^i_{train,k}'}(P)$, $m^i_{k,3}(\cdot) = |C^i_{train,k}|$, $m^i_{k,4}(\cdot) = |C^i_{train,k}'|$, in $(D^i_{train}, \mathcal{I}, \mathcal{I}_d)$; (3) pattern scoring function $s(m_1, m_2) = \frac{GR(m_1,m_2)}{GR(m_1,m_2)+1} * m_1$.

Note that, once this classifier is trained, we have $m_{k,2}(P) = \infty$ for $P \in \mathcal{P}_k$ and $k \in \mathcal{I}_d$. Therefore, it behaves exactly like regular JEP-Classifier. The reason for the extension of the scoring function is that we want to modify the importance of patterns through adaptation. In fact, patterns are treated as emerging patterns with growth rate equal to $GR(m_{k,1}(P), m_{k,2}(P))$. Scoring is performed like in CAEP. We also decided not to perform normalization of scores ([8]), since this approach is not present in all EP-based classifiers ([5]).

It is advantageous that a training sequence is not necessary to compute supports for each adaptation. For each transaction in adaptation sequence, it is sufficient to increment by one respective supports of the patterns it contains. The proof is omitted due to space limitations and analogical for $m_{k,2}, m_{k,4}$.

Theorem 1. $\forall_{i \in \mathbb{N}} \forall_{k \in \mathcal{I}_d} \forall_{P \in \mathcal{P}_k} m^i_{k,1}(P) = \frac{m^{i-1}_{k,1}(P) * m^{i-1}_{k,3}(P) + |\{j : P \subseteq T_j \wedge T_j \in C^i_{adapt,k}\}|}{|m^i_{k,3}(P)|} \wedge$ $m^i_{k,3}(P) = m^{i-1}_{k,3}(P) + |C^i_{adapt,k}|$.

5.2 Border Recomputation

Although support adjustment enriches our knowledge on pattern importance, one is deprived of discovering new patterns. A more invasive approach is to assume that newly accepted transactions are correctly labeled, add them to a training sequence and recompute pattern collections and supports.

In i-th step, a classifier f^i requires collections $\{\mathcal{P}^i_k\}_{k \in \mathcal{I}_d} = \{\mathcal{L}^i_k\}_{k \in \mathcal{I}_d}$ for $JEP(C^i_{train,k}', C^i_{train,k})$ and supports in each class for all the patterns, $m^i_{k,1}(P) = supp_{C^i_{train,k}}(P)$, for $P \in \mathcal{P}_k$ and $k \in \mathcal{I}_d$, all computed in $(D^i_{train}, \mathcal{I}, \mathcal{I}_d)$. A matching pattern is scored with $s(m) = m$, for $m \in \mathbb{R}$.

There are several important drawbacks of this approach. First of all each adaptation requires maintenance of borders for each class ([4]). In consequence, not only minimal, but also maximal patterns have to be stored. Secondly, although negative supports, equal to 0, are not present, in order to recompute positive supports a sequence D_{train}^i is stored. Instead, one may use another maintainable structure that provides supports for patterns, e.g. a hash tree. Regardless, the size of the classifier grows with the number of adaptations.

5.3   Adaptation Condition

In general, a testing transaction should be rejected when a classifier is not certain of its decision. One may look at the training sequence and assess how representative is the neighborhood of this transaction ([10]). In another approach, a condition can be based on labels predicted by a classifier ([7]). For probabilistic classifiers, the problem of minimizing risk given certain costs of possible decisions is solved ([11]). Following this path, one considers two types of rejection: outlier/distance and ambiguity. The first one expresses how certain a classifier is about each label, while the other one - decisiveness of a class choice. We formulate analogical criteria for labeling functions of EP-based classifiers. As pointed out, these functions do not have probabilistic interpretation, which makes a formal discussion much harder. Instead, an experimental verification is presented.

Let us perform a class-wise normalization with the largest value for training transactions in a respective class of i-th step, namely: $L_k'(X) = \frac{L_k(X)}{M_k^i}$ for $M_k^i = max_{T \in C_{train,k}^i} L_k(T)$, $k \in \mathcal{I}_d$ and $X \in D = \{T - \mathcal{I}_d : T \in \mathcal{D}\}$. The normalizing values are stored with a classifier and maintained. Note that this normalization does not prevent labels from going beyond 1 for certain unseen transactions. If a transaction is accepted, a respective normalizing value is updated.

Now, let us move to the actual adaptation criteria. A given transaction $X \in D$ is accepted with a classifier's decision $c \in \mathcal{I}_d$, only if the following criteria are met: $L_c'(X) > t_{distance}$ and $\frac{L_c'(X)}{L_k'(X)} > t_{ambiguity}$ for all $k \in \mathcal{I}_d$. The values $t_{distance} \in [0,1]$, $t_{ambiguity} \in \mathbb{R}_+$ are certain thresholds. They can be estimated experimentally on a training sequence as a tipping point of accuracy.

6   Experimental Results

Our tests investigated behavior of JEP-Classifier (JEPC), Adaptive JEP-Classifier with support adjustment (JEPC-Adjust) and with border recomputation (JEPC-Recompute). Accuracy, a ratio of adaptation (Adapt) and percentage of correct adaptations (AdAcc) were collected. The datasets originated from UCI repository. Due to their relational form, JEP spaces used in classifiers were obtained with RedApriori ([6]). Border maintenance was performed by incremental routines ([4]). Classification statistics were assessed by means of a modified k-fold cross validation scheme, which uses one fold for training and the rest for testing.

According to Tab.1, JEPC was outperformed by both adaptive approaches in all the investigated cases except *wine* and JEPC-Recompute is not a winner only for *tic-tac-toe*. The biggest improvement towards JEPC of > 8 p.p. occurred for

Table 1. Accuracy, adaptation ratio and adaptation accuracy for Adaptive JEP-Classifier in a modified 5-fold cross validation, $t_{distance} = 20\%$, $t_{ambiguity} = 2$

Dataset	Trans	Attrs	Items	JEPC Accu	JEPC-Adjust			JEPC-Recompute		
					Accu	Adapt	AdAcc	Accu	Adapt	AdAcc
balance	625	4	20	71.32	73.40	60.96	79.00	79.76	63.32	80.10
breast-wisc	699	9	29	96.17	96.21	92.45	98.30	96.92	94.13	98.33
car	1728	6	21	83.17	87.30	75.27	93.62	90.35	84.23	92.31
election	202	16	32	70.67	71.16	70.92	83.25	72.52	73.89	83.42
geo	402	10	78	49.07	57.59	63.74	62.24	60.45	90.98	66.23
heart	270	13	22	78.70	79.91	79.26	87.03	81.57	82.22	87.39
house	435	16	48	90.17	91.09	88.51	94.61	92.24	92.64	94.29
lymn	148	18	59	71.28	80.57	78.89	86.94	81.08	93.75	86.31
monks-1	556	6	17	84.71	94.15	90.51	93.89	94.24	99.82	94.28
monks-2	601	6	17	64.89	67.18	60.11	71.21	70.84	57.99	73.24
monks-3	554	6	17	94.68	96.80	95.31	97.16	96.84	96.16	97.04
tic-tac-toe	958	9	27	81.86	90.08	82.88	90.81	89.77	96.76	91.07
wine	178	13	35	90.59	90.31	89.33	95.60	93.96	98.46	95.44
zoo	101	16	35	76.73	78.96	79.46	90.34	81.19	92.57	86.90

balance, geo, monks-1, lymn. Although each initial classifier was created with a small training sequence, JEPC-Adjust was able to achieve a significant gain only by modifying importance of patterns. In fact, average accuracy difference over all the datasets for JEPC-Recompute—JEPC is equal to 3.29 p.p. When JEPC-Recompute, we gain another 1.70 p.p. At the same time, the choice of the adaptation condition is confirmed by high adaptation ratio and accuracy. On the whole, the results are promising. Even a simple strategy used in JEPC-Adjust gives already much improvement, whereas high performance of JEPC-Recompute is traded off expensive model modifications.

7   Conclusions

In this paper, we have proposed an adaptive classification scheme and investigated its impact on overall accuracy. The procedure is iterative and employs a classifier with reject option. In each step a certain part of a testing sequence is classified and used to modify the classifier. The process is repeated for a rejected part as an input until no new transactions are accepted. To the best of our knowledge, the fact that a testing sequence is analyzed as a whole and each transaction may be classified multiple times with presumably more exact versions of the model distinguish our approach from existing adaptive algorithms.

Our concept has been used to improve classification with jumping emerging patterns (JEPs). We have considered two approaches combined with the same distance/ambiguity reject strategy. The first one, support adjustment, employs an internal JEP-based classifier with reject option and the labeling function from CAEP. Adaptation does not change the original set of patterns and models pattern confidence to be JEPs in the domain by modifying positive and negative support. The second option, border recomputation, uses an internal JEP-Classifier with reject option and adapts to accepted cases by recomputing class-wise borders and pattern supports. Although maintenance of borders can be efficiently done in incremental manner, support calculations require a complete transaction sequence and the structure grows with subsequent adaptations.

Our experiments have assumed that relatively small training sequences are available. Both solutions outperform classical JEP-Classifier in terms of accuracy. The difference is especially visible when an initial model behaves poorly. In almost all cases, border recomputation is a winner. However, the price is a much longer classification time and growing classifier's complexity.

References

1. Li, Y., Guan, C.: An extended em algorithm for joint feature extraction and classification in brain-computer interfaces. Neural Comput. 18(11), 2730–2761 (2006)
2. Zhang, Q., Landgrebe, J.D.: Design of an adaptive classification procedure for the analysis of highdimensional data with limited training samples. In: Monostori, L., Váncza, J., Ali, M. (eds.) IEA/AIE 2001. LNCS (LNAI), vol. 2070, p. 476. Springer, Heidelberg (2001)
3. Delany, S.J., Cunningham, P., Doyle, D., Zamolotskikh, A.: Generating estimates of classification confidence for a case-based spam filter. In: Muñoz-Ávila, H., Ricci, F. (eds.) ICCBR 2005. LNCS (LNAI), vol. 3620, pp. 177–190. Springer, Heidelberg (2005)
4. Dong, G., Li, J.: Mining border descriptions of emerging patterns from dataset pairs. Knowl. Inf. Syst. 8(2), 178–202 (2005)
5. Li, J., Dong, G., Ramamohanarao, K.: Making use of the most expressive jumping emerging patterns for classification. Knowl. Inf. Syst. 3(2), 1–29 (2001)
6. Terlecki, P., Walczak, K.: Jumping emerging pattern induction by means of graph coloring and local reducts in transaction databases. In: An, A., Stefanowski, J., Ramanna, S., Butz, C.J., Pedrycz, W., Wang, G. (eds.) RSFDGrC 2007. LNCS (LNAI), vol. 4482, pp. 363–370. Springer, Heidelberg (2007)
7. Mascarilla, L., Frélicot, C.: Reject strategies driven combination of pattern classifiers. Pattern Analysis and Applications 5(2), 234–243 (2002)
8. Dong, G., Zhang, X., Wong, L., Li, J.: Caep: Classification by aggregating emerging patterns. In: Int. Conf. on Disc. Sc., pp. 30–42 (1999)
9. Fan, H., Ramamohanarao, K.: Efficiently mining interesting emerging patterns. In: Dong, G., Tang, C.-j., Wang, W. (eds.) WAIM 2003. LNCS, vol. 2762, pp. 189–201. Springer, Heidelberg (2003)
10. Fumera, G., Pillai, I., Roli, F.: Classification with reject option in text categorisation systems. In: ICIAP, p. 582. IEEE Computer Society, Los Alamitos (2003)
11. Chow, C.: On optimum recognition error and reject tradeoff. IEEE Transactions on Information Theory 16, 41–46 (2003)

Two-Phase Rule Induction from Incomplete Data

Huaxiong Li[1,2], Yiyu Yao[2], Xianzhong Zhou[1], and Bing Huang[1]

[1] School of Management and Engineering, Nanjing University,
Nanjing, Jiangsu, 210093, P.R. China
huaxiongli@gmail.com, zhouxz@nju.edu.cn, hbhuangbing@126.com
[2] Department of Computer Science, University of Regina,
Regina, Saskatchewan, S4S 0A2, Canada
yyao@cs.uregina.ca

Abstract. A framework of learning a new form of rules from incomplete data is introduced so that a user can easily identify attributes with or without missing values in a rule. Two levels of measurement are assigned to a rule. An algorithm for two-phase rule induction is presented. Instead of filling in missing attribute values before or during the process of rule induction, we divide rule induction into two phases. In the first phase, rules and partial rules are induced based on non-missing values. In the second phase, partial rules are modified and refined by filling in some missing values. Such rules truthfully reflect the knowledge embedded in the incomplete data. The study not only presents a new view of rule induction from incomplete data, but also provides a practical solution.

Keywords: Missing attribute values, Filled-in values, Two-phase rule induction.

1 Introduction

A major focus of machine learning and data mining research is to extract useful knowledge from a large amount of data. For such a purpose, the integrality of data is very important. However, real-world data sets frequently contain missing values, i.e., attribute values of objects may be unknown or missing [14]. To deal with missing values, many methods have been proposed [2,3,4,5,6,8,1,10]. They may be broadly classified into three categories. The first category mainly focuses on transforming incomplete data into complete data by filling in the missing values. Rules are induced from the completed data. The second category fills in the missing values during the process of rule induction. The third category considers tolerance relations or similarity relations defined based on missing values [5,6,10].

The third category may be considered as a special case of the first category. In fact, one may first fill in the missing values and then derive a similarity relation. One disadvantage of filling in attribute values before the learning process is that the learning goals and algorithms are not directly considered. Another

G. Wang et al. (Eds.): RSKT 2008, LNAI 5009, pp. 47–54, 2008.

disadvantage is that all missing values are filled in although some of them are not necessary. Since rules normally contain only a subset of all (*attribute, value*) pairs, we do not need to fill in all missing values. To avoid those problems, one can combine the processes of filling in the missing values and learning together. Algorithms like C4.5 [1] fill in missing attribute values according to some special learning goals. In addition, missing values are filled in only when the demand arises in the learning process.

Two fundamental important problems still remain in the existing algorithms for inducing rules from incomplete data. One is the use of the filled-in values. Any method of filling in missing values is based on certain assumption about the data, which may not be valid. However, rule learning algorithms treat filled-in values as if they are the original values. This may result in rules having more number of filled-in values and less number of the original values. That is, we may obtain rules that more fit the filled-in values. Although these rules may have good statistical characteristics, they are in fact not reliable. The other issue is the use of induced rules by users. Without a clear distinction between the filled-in values and the original values, a user may find it difficult to interpret and apply rules. In fact, a user may misuse a rule by putting more weights on filled-in values. Solutions for those two problems require new ideas and methodologies.

The objective of this paper is to propose a new framework of rule induction by separating filled-in and the original values in both the learning process and the induced rules. A two-phase model is suggested. The first phase induces partial rules based only on the original values. Rules induced in the first phase will be associated with a quantitative measures such as confidence, coverage and generality [11,13]. In the second phase of rule induction, filled-in values are taken into account so that the performance of the rules may improve. A new form of rules is introduced, in which known attribute values and filled-in attribute values are used. A user can easily identify attributes with or without missing values in rules.

2   A Framework of Two-Phase Rule Induction

The two-phase framework of rule induction from incomplete data uses a new form of rule involving both the known attribute values and filled-in values. In the first phase of rule induction, rules is induced only based on known attribute values. Many approaches of machine learning and data mining methods, such as concept learning, decision tree and rough set-based learning, can be used. One may associate certain quantitative measures to express the strength of a rule. In the second phase of rule induction, missing attribute values are filled in according to a certain method, and rules are induced based on the repaired data. The performance of the new rule induced in the second phase should be superior (higher) than that of rule induced in the first phase.

The main ideas of two-phase rule induction can be illustrated by a simple example. Suppose r is a rule induced in the first phase based on known attribute values with a measure α_1:

$$r : \phi_1 \wedge \phi_2 \rightarrow \psi \quad (m = \alpha_1),$$

where ϕ_1 and ϕ_2 are concepts or conditions based on known attribute values, and ψ is a decision concept, and m is a measure of the rule. By filling in missing attribute values, a new concept ϕ_3 may appear and a new rule r^* may be induced as follows:

$$r^* : \phi_1 \wedge \phi_2 \wedge \phi_3^* \rightarrow \psi \quad (m = (\alpha_1, \alpha_2)),$$

where "*" indicates that ϕ_3 is a concept based on a filled-in value while other two concepts ϕ_1 and ϕ_2 are based on the original values, and α_1 is the measure of rule based on known attribute values and α_2 is the measure of rule based on both known attribute value and filled-in values. A user can easily identify attributes with or without missing values in this form of rules. Suppose $m = (\alpha_1, \alpha_2)$ represents the confidence of a rule. It may be necessary to require that $\alpha_2 > \alpha_1$, otherwise there is no need to fill in the missing values. Thus, the conditions for filling in missing values can be analyzed in the second phase. It is not necessary to filling in all missing values.

A high level outline of the two-phase rule induction method is:

Phase 1: $R = \emptyset$;
　　　Repeat the following operations
　　　　　Add rule to R;
　　　　　Remove a rule from R;
　　　　　Replace rule in R by one or more rules;
　　　Until R satisfies a termination condition

Phase 2: $R^* = R$;
　　　Repeat the following operations
　　　　　Replace a rule in R^* by one or more rules
　　　　　with filled-in values;
　　　Until R^* satisfies a termination condition

In the Phase 1, the condition ϕ of each rule $\phi \rightarrow \psi$ in R must be expressed using only existing attribute values. Furthermore, each rule is associated with the value of a quantitative measure. The termination condition is typically related to the measure. In the phase 2, one may fill in missing values and add extra conditions that contain only filled-in values. At the end of Phase 2, a rule is of the form:

$$\phi \wedge \phi^* \rightarrow \psi,$$

where $\phi \rightarrow \psi$ is a rule in R. In addition, the rule is associated with another value based on the measure. In general, if unknown value is good enough to induce a reasonable rule, a rule may not necessarily use filled-in value. That is, ϕ^* is optional.

3   An Algorithm of Two-Phase Rule Induction

The high-level schematic algorithm presented in the last section offers a wide class of algorithms for two-phase rule induction. In the first phase, we can adopt and adapt any existing rule induction algorithm. In the second phase, we can use any of the existing approaches for filling in missing values. As an illustration, we present a concrete algorithm of two-phase rule induction, as shown in Figure 1. In this algorithm, we assume that the performance of a rule is measured by its confidence. The details of the algorithm are explained below.

3.1   Induction of Partial Rules Based on Known Value

In the first phase of rule induction, we present an algorithm using a concept learning method. The process of rule induction can be described as finding a class of conjunctive concepts [12], each covering a subset of a given decision concept, and the union of these conjunction concepts can cover all instances.

As mentioned in [12], extension of a k-conjunction concept formed by adding a new atomic formula to the $(k-1)$-conjunction concept is a subset of that of $(k-1)$-conjunction concept. The most general 1-conjunction concepts can cover the most number of objects in universe. We search the 1-conjunction concept first. Suppose ϕ_1 is a 1-conjunction concept, if the extension of a 1-conjunction concept $m(\phi_1)$ is the subset of the extension of a decision concept ψ, a certain rule can be extracted, i.e., $\phi_1 \rightarrow \psi$. Otherwise, we search other 1-conjunction concept that can be appropriately classified into a decision concept. If all 1-conjunction concepts have been checked, we start with 2-conjunction concepts, and we may induce a rule as: $\phi_1 \wedge \phi_2 \rightarrow \psi$. This process will go on until all objects are covered or a certain condition is met.

In general we can not expect to find a set of certain rules or high performance rules that covers all objects, especially for incomplete data and inconsistent data. We introduce measures such as confidence, coverage and generality to quantify the performance of a rule. In the algorithm of Figure 1, we introduce two thresholds for the confidence measure.

3.2   Refinement of Rules Based on Filled-In Missing Values

At the end of the first phase of rule induction, rules and partial rules have been induced. In a special case, all rules are not partial rules, and all objects in universe have been covered, then it is not necessary to fill in missing values, and the second phase of rule induction is not required. Otherwise, it is necessary to fill in the missing values so that some new rules with better performance may be induced.

There are many methods to deal with filling in the missing values. Grzymala-Busse and Hu discuss and compare nine different approaches to fill in missing attribute values. For example, two of them are known as "most common attribute value" and "concept most common attribute value"[2]. A main objective of two-phase rule induction is to introduce a new pattern of rules. Any approach of

Input: a training set of examples S,
           a pair thresholds (α_1, α_2) of a performance measure,
           a conjunction order threshold K,
           a partition of the training set $\Pi = \{\pi_1, \cdots, \pi_N\}$
               corresponding to decision concepts set $\Psi = \{\psi_1, \cdots, \psi_N\}$
Output: a set of rules R^*
Phase 1:
           Set $R = \emptyset$;
           Set $k = 1$;
           While S is not empty and $k \leq K$
               For each k-conjunction ϕ
                   Select a π_i with the maximum value of $measure(\phi \rightarrow \psi_i)$;
                   If $measure(\phi \rightarrow \psi_i) \geq \alpha_2$
                       Add rule $\phi \rightarrow \psi_i$ to R;
                       $S = S - S \cap m(\phi)$;
               Add all k-conjunction rules satisfying the following conditions:
                   (a). can not be further specialized,
                   (b). performance $\geq \alpha_1$,
               into R;
               Delete instances correctly covered by those rules from S;
               $k = k + 1$;
Phase 2:
           Set $R^* = R$;
           While (there exists a $r_i = \phi_i \rightarrow \psi_i \in R^*$
                   satisfying $measure(r_i) < \alpha_2$ and
                   the number of attributes in it $< K$)
               Select an attribute c_j that has missing values
                       Fill in missing values of c_j;
                       Set $\Phi^* = \{$all c_j related concepts$\}$;
                       Select a set of rules $\{r_i^* = \phi_i \wedge \phi^* \rightarrow \psi_i\}$
                           with the largest measures and
                           covering most of instances of rule r_i;
                       Replace r_i by r_i^*;
           Delete those rules that do not satisfy required conditions from R^*;
           Return R^*

Fig. 1. A Two-phase Rule Induction Algorithm

filling in missing value may be used. When filling in the missing value, the filled-in values must help to increase the performance of rules. In the algorithm of Figure 1, we use the simplest "most common attribute value" method to fill in missing values.

3.3   An Example

We use a simple example of cars given in Table 1, taken from   [6], to illustrate the main idea of two-phase rule induction from incomplete data. We adopt the confidence as a performance measure of rules. The set of all objects in the table, i.e., object 1-6, is partitioned into three classes according to the decision attribute d:

$$(d, good), \quad (d, excellent), \quad (d, poor),$$

which form three decision concepts and the corresponding extensions are:

$$m(d, good) = \{1, 2, 4, 6\}, \quad m(d, excellent) = \{5\}, \quad m(d, poor) = \{3\}.$$

Assume further that $K = 3$, $\alpha_1 = 0.5$ and $\alpha_2 = 0.9$.

Table 1. An Incomplete Information Table

Car	Price	Mileage	Size	Max-Speed	d
1	high	low	full	low	good
2	low	*	full	low	good
3	*	*	compact	low	poor
4	high	*	full	high	good
5	*	*	full	high	excellent
6	low	high	full	*	good

Phase 1: In the first phase, rules are induced only from known values. The most general 1-conjunction concepts are considered and they are given by:

$$(Price, high), \quad (Price, low), \quad (Mileage, high), \quad (Mileage, low)$$
$$(Size, full), \quad (Size, compact), \quad (Max\text{-}Speed, high), \quad (Max\text{-}Speed, low).$$

Their extensions are:

$$m(Price, high) = \{1, 4\}, \qquad m(Price, low) = \{2, 6\},$$
$$m(Mileage, high) = \{6\}, \qquad m(Mileage, low) = \{1\},$$
$$m(Size, full) = \{1, 2, 4, 5, 6\}, \qquad m(Size, compact) = \{3\},$$
$$m(Max\text{-}Speed, high) = \{4, 5\}, \qquad m(Max\text{-}Speed, low) = \{1, 2, 3\}.$$

According to the thresholds $\alpha_1 = 0.5$ and $\alpha_2 = 0.9$, we have the following 1-conjunction rules:

$$r_1 : (Price, high) \rightarrow (d, good) \quad (m = 1),$$
$$r_2 : (Price, low) \rightarrow (d, good) \quad (m = 1),$$
$$r_3 : (Mileage, high) \rightarrow (d, good) \quad (m = 1),$$
$$r_4 : (Mileage, low) \rightarrow (d, good) \quad (m = 1),$$
$$r_5 : (Size, compact) \rightarrow (d, poor) \quad (m = 1).$$

They do not cover the entire universe. We thus search for 2-conjunction rules. In this case, only one rule can be found:

$$r_6 : (Size, full) \wedge (Max\text{-}Speed, high) \rightarrow (d, excellent) \quad (m = 0.5).$$

All objects in the table are now covered by these rules, and the first phase of rule induction terminates.

Phase 2: All rules, with the exception of r_6, satisfy the conditions as specified by threshold values. For r_6, we need to add attributes with missing values. Since r_6 only covers instance 5, we only need to fill in missing values for object 5. Based on the idea of "most common attribute values," one may fill in either *low* or *high* for attribute *Mileage*. Similarly, for attribute *Price* we can fill in either *low* or *high*. Suppose we fill in the value *high* for attribute *Mileage*, we obtain the following rule:

$$r_6^* : \ (Size, full) \wedge (Max\text{-}Speed, high) \wedge (Mileage, high)^* \rightarrow (d, excellent)$$
$$(m = (0.5, 1)).$$

On the other hand, filling-in the value *low* for attribute *Price* produces the rule:

$$r_7^* : \ (Size, full) \wedge (Max\text{-}Speed, high) \wedge (Price, low)^* \rightarrow (d, excellent)$$
$$(m = (0.5, 1)).$$

Filling in other values would not produce a satisfactory rule. Both rules with filled-in values seem to be reasonable, although rule r_6^* seems more plausible than rule r_7^*. Typically, a car with low price may not be full size and have high max-speed.

Unlike existing methods, the majority of rules produced in the first phase is satisfactory. We only need to fill in one missing value in this example. It demonstrates the advantage of the two-phase rule induction approach.

4   Conclusion

This paper proposes a framework of two-phase rule induction from incomplete data. The main idea is to divide the process of rule induction into two phases. The aim of the first phase is to induce rules only from known attributes values. If rules induced in the first phase are good enough, it may not be necessary to fill in missing values. If needed, the second phase fills in missing values and refine rules obtained in the first phase. A new form of rules is introduced by explicitly denoting the known values and the filled-in values and associating with two levels of performance. As future research, we plan to study systematically combinations of learning algorithms in the first phase and filling in missing values algorithms in the second phase.

Acknowledgments

This research is partially supported by the National Natural Science Foundation of China under grant No.70571032, the scholarship awarded by China State Scholarship Fund under grant No.2007100699, and a Discovery Grant from NSERC Canada.

References

1. Greco, S., Matarazzo, B., Slowinski, R.: Handling Missing Values in Rough Set-Analysis of Multi-attribute and Multi-criteria Decision Problems. In: Zhong, N., Skowron, A., Ohsuga, S. (eds.) RSFDGrC 1999. LNCS (LNAI), vol. 1711, pp. 146–157. Springer, Heidelberg (1999)
2. Grzymala-Busse, J.W., Hu, M.: A Comparison of Several Approches to Missing Attribute Values in Data Mining. In: Ziarko, W., Yao, Y. (eds.) RSCTC 2000. LNCS (LNAI), vol. 2005, pp. 378–385. Springer, Heidelberg (2001)
3. Grzymala-Busse, J.W., Grzymala-Busse, W.J.: An Experimental Comparison of Three Rough Set Approaches to Missing Attribute Values. In: Peters, J.F., Skowron, A., Düntsch, I., Grzymała-Busse, J.W., Orłowska, E., Polkowski, L. (eds.) Transactions on Rough Sets VI. LNCS, vol. 4374, pp. 31–50. Springer, Heidelberg (2007)
4. Grzymala-Busse, J.W.: Data with Missing Attribute Values: Generalization of Indiscernibility Relation and Rule Induction. In: Peters, J.F., Skowron, A., Grzymała-Busse, J.W., Kostek, B.z., Świniarski, R.W., Szczuka, M. (eds.) Transactions on Rough Sets I. LNCS, vol. 3100, pp. 78–95. Springer, Heidelberg (2004)
5. Kryszkiewicz, M.: Rough Set Approach to Incomplete Information Systems. Information Sciences 112, 39–49 (1998)
6. Kryszkiewicz, M.: Rules in Incomplete Information Systems. Information Sciences 113, 271–292 (1999)
7. Mitchell, T.M.: Generalization as Search. Artificial Intelligence 18, 203–226 (1982)
8. Polkowski, L., Artiemjew, P.: Granular Classifiers and Missing Values. In: Proc. of ICCI 2007, pp. 186–194 (2007)
9. Quinlan, J.R.: C4.5: Programs for Machine Learning. Morgan Kaufmann, San Mateo (1993)
10. Stefamowski, J., Tsoukeas, A.: On the Extension of Rough Sets under Incomplete Information. Int. J. of Intel. Sys. 16, 29–38 (1999)
11. Yao, J.T., Yao, Y.Y.: Induction of Classification Rules by Granular Computing. In: Alpigini, J.J., Peters, J.F., Skowron, A., Zhong, N. (eds.) RSCTC 2002. LNCS (LNAI), vol. 2475, pp. 331–338. Springer, Heidelberg (2002)
12. Yao, Y.Y.: Concept Formation and Learning: A Cognitive Informatics Perspective. In: Proc. of the ICCI 2004, pp. 42–51 (2004)
13. Yao, Y.Y., Zhong, N.: An Analysis of Quantitative Measures Associated with Rules. In: Kapoor, S., Prasad, S. (eds.) FST TCS 2000. LNCS, vol. 1974, pp. 479–488. Springer, Heidelberg (2000)
14. Zhang, S.C., Qin, Z.X., Ling, C.X., Sheng, S.L.: Missing is Useful: Missing Values in Cost-sensitive Decision Trees. IEEE Trans. on Know. and Data Eng. 17, 1689–1693 (2005)

Comparison of Lazy Classification Algorithms Based on Deterministic and Inhibitory Decision Rules

Paweł Delimata[1], Mikhail Moshkov[2], Andrzej Skowron[3], and Zbigniew Suraj[4]

[1] Chair of Computer Science, University of Rzeszów
Rejtana 16A, 35-310 Rzeszów, Poland
pdelimata@wp.pl
[2] Institute of Computer Science, University of Silesia
Będzińska 39, 41-200 Sosnowiec, Poland
moshkov@us.edu.pl
[3] Institute of Mathematics, Warsaw University
Banacha 2, 02-097 Warsaw, Poland
skowron@mimuw.edu.pl
[4] Chair of Computer Science, University of Rzeszów
Rejtana 16A, 35-310 Rzeszów, Poland
zsuraj@univ.rzeszow.pl

Abstract. In the paper, two lazy classification algorithms of polynomial time complexity are considered. These algorithms are based on deterministic and inhibitory decision rules, but the direct generation of rules is not required. Instead of this, for any new object the considered algorithms extract from a given decision table efficiently some information about the set of rules. Next, this information is used by a decision-making procedure. The reported results of experiments show that the algorithms based on inhibitory decision rules are often better than those based on deterministic decision rules.

Keywords: Rough sets, Decision tables, Deterministic decision rules, Inhibitory decision rules.

1 Introduction

In the paper, the following classification problem is considered: for a given decision table T [11,12] and a new object v generate a value of the decision attribute on v using values of conditional attributes on v.

We compare two lazy [1,9] classification algorithms based on deterministic and inhibitory decision rules of the forms

$$a_1(x) = b_1 \wedge \ldots \wedge a_t(x) = b_t \Rightarrow d(x) = b,$$

$$a_1(x) = b_1 \wedge \ldots \wedge a_t(x) = b_t \Rightarrow d(x) \neq b,$$

respectively, where $a_1, \ldots, a_t$ are conditional attributes, $b_1, \ldots, b_t$ are values of these attributes, d is the decision attribute and b is a value of d. By $V_d(T)$ we denote the set of values of the decision attribute d.

G. Wang et al. (Eds.): RSKT 2008, LNAI 5009, pp. 55–62, 2008.

The first algorithm (D-algorithm) was proposed and studied by J.G. Bazan [2,3,4]. This algorithm is based on the deterministic decision rules. For any new object v and each decision $b \in V_d(T)$ we find (using polynomial-time algorithm) the number $D(T, b, v)$ of objects u from the decision table T such that there exists a deterministic decision rule r satisfying the following conditions: (i) r is true for the decision table T, (ii) r is realizable for u and v, and (iii) r has the equality $d(x) = b$ on the right hand side. For the new object v we choose a decision $b \in V_d(T)$ for which the value $D(T, b, v)$ is maximal. Note that this approach was generalized by J.G. Bazan [2,3,4] to the case of approximate decision rules, and by A. Wojna [16] to the case of decision tables with not only nominal but also numerical attributes.

The second algorithm (I-algorithm) is based on the inhibitory decision rules. For any new object v and each decision $b \in V_d(T)$ using a polynomial-time algorithm it is computed the number $I(T, b, v)$ of objects u from the decision table T such that there exists an inhibitory decision rule r satisfying the following conditions: (i) r is true for the decision table T, (ii) r is realizable for u and v, and (iii) r has the relation $d(x) \neq b$ on the right hand side. For the new object v we choose a decision $b \in V_d(T)$ for which the value $I(T, b, v)$ is minimal. Hence, for v we vote, in a sense, for the decision b for which there are weakest arguments "against". Note that in [10] the dissimilarity measures are used for obtaining arguments against classification of a given handwritten digit to some decision classes. These arguments can be interpreted as inhibitory rules.

Results of experiments show that the algorithm based on inhibitory decision rules is, often, better than the algorithm based on deterministic decision rules.

This work was inspired by results of comparison of deterministic and inhibitory rules for information system $S = (U, A)$ [11,12], where U is a finite set of objects and A is a finite set of attributes (functions defined on U). The considered rules are of the following form:

$$a_1(x) = b_1 \wedge \ldots \wedge a_t(x) = b_t \Rightarrow a_{t+1}(x) = b_{t+1},$$
$$a_1(x) = b_1 \wedge \ldots \wedge a_t(x) = b_t \Rightarrow a_{t+1}(x) \neq b_{t+1},$$

where $a_1, \ldots, a_{t+1}$ are attributes from A and $b_1, \ldots, b_{t+1}$ are values of these attributes. We consider only true and realizable rules. *True* means that the rule is true for any object from U. *Realizable* means that the left hand side of the rule is true for at least one object from U. We identify objects from U and tuples of values of attributes from A on these objects. Let V be the set of all tuples of known values of attributes from A. We say that the set U can be described by deterministic (inhibitory) rules if there exists a set Q of true and realizable deterministic (inhibitory) rules such that the set of objects from V, for which all rules from Q are true, is equal to U.

In [14,15] it was shown that there exist information systems $S = (U, A)$ such that the set U can not be described by deterministic rules. In [7] it was shown that for any information system $S = (U, A)$ the set U can be described by inhibitory rules. It means that the inhibitory rules can express essentially more information encoded in information systems than the deterministic rules. This fact is a motivation for a wider use of inhibitory rules, in particular, in classification algorithms and in algorithms for synthesis of concurrent systems [14]. To compare experimentally the classification quality based on inhibitory and deterministic rules, we create two similar families of lazy classification algorithms based on deterministic and inhibitory rules, respectively [6]. Results of experiments show that the algorithms based on inhibitory rules are noticeably better than those based on deterministic rules.

The paper consists of seven sections. In Sect. 2, we recall the notion of decision table. In Sects. 3 and 4, we describe notions of deterministic and inhibitory decision rules. Sect. 5 contains definitions of two lazy classification algorithms. Results of experiments are discussed in Sect. 6.

2   Decision Tables

Let $T = (U, A, d)$ be a *decision table*, where $U = \{u_1, \ldots, u_n\}$ is a finite nonempty set of *objects*, $A = \{a_1, \ldots, a_m\}$ is a finite nonempty set of *conditional attributes* (functions defined on U), and d is the *decision attribute* (function defined on U). We assume that for each $u_i \in U$ and each $a_j \in A$ the value $a_j(u_i)$ and the value $d(u_i)$ belong to ω, where $\omega = \{0, 1, 2, \ldots\}$ is the set of nonnegative integers. By $V_d(T)$ we denote the set of values of the decision attribute d on objects from U.

Besides objects from U we consider also objects from $\mathcal{U}(T) = \omega^m$. The set $\mathcal{U}(T)$ is called the *universe* for the decision table T. For any object (tuple) $v \in \mathcal{U}(T)$ and any attribute $a_j \in A$ the value $a_j(v)$ is equal to j-th integer component of v.

3   Deterministic Decision Rules

Let us consider a rule

$$a_{j_1}(x) = b_1 \wedge \ldots \wedge a_{j_t}(x) = b_t \Rightarrow d(x) = b, \tag{1}$$

where $t \geq 0$, $a_{j_1}, \ldots, a_{j_t} \in A$, $b_1, \ldots, b_t \in \omega$, $b \in V_d(T)$ and numbers $j_1, \ldots, j_t$ are pairwise different. Such rules are called *deterministic decision* rules. The rule (1) is called *realizable for an object* $u \in \mathcal{U}(T)$ if $a_{j_1}(u) = b_1, \ldots, a_{j_t}(u) = b_t$ or $t = 0$. The rule (1) is called *true for an object* $u_i \in U$ if $d(u_i) = b$ or (1) is not realizable for u. The rule (1) is called *true for* T if it is true for any object from U. The rule (1) is called *realizable for* T if it is realizable for at least one object from U. By $Det(T)$ we denote the set of all deterministic decision rules which are true for T and realizable for T.

Our aim is to recognize, for given objects $u_i \in U$ and $v \in \mathcal{U}(T)$, and given value $b \in V_d(T)$ if there exists a rule from $Det(T)$ which is realizable for u_i and v and has $d(x) = b$ on the right hand side. Such a rule "supports" the assignment of the decision b to the new object v.

Let $M(u_i, v) = \{a_j : a_j \in A, a_j(u_i) = a_j(v)\}$ and $P(u_i, v) = \{d(u) : u \in U, a_j(u) = a_j(v)$ for any $a_j \in M(u_i, v)\}$. Note that if $M(u_i, v) = \emptyset$, then $P(u_i, v) = \{d(u) : u \in U\} = V_d(T)$.

Proposition 1. *Let $T = (U, A, d)$ be a decision table, $u_i \in U$, $v \in \mathcal{U}(T)$, and $b \in V_d(T)$. Then in $Det(T)$ there exists a rule, which is realizable for u_i and v and has $d(x) = b$ in the right hand side, if and only if $P(u_i, v) = \{b\}$.*

Proof. Let $P(u_i, v) = \{b\}$. In this case, the rule

$$\bigwedge_{a_j \in M(u_i, v)} a_j(x) = a_j(v) \Rightarrow d(x) = b \tag{2}$$

belongs to $Det(T)$, is realizable for u_i and v, and has $d(x) = b$ in the right hand side.

Let us assume that there exists a rule (1) from $Det(T)$, which is realizable for u_i and v, and has $d(x) = b$ in the right hand side. Since (1) is realizable for u_i and v, we have $a_{j_1}, \ldots, a_{j_t} \in M(u_i, v)$. Since (1) is true for T, the rule (2) is true for T. Therefore, $P(u_i, v) = \{b\}$.   □

From Proposition 1 it follows that there exists a polynomial algorithm recognizing, for a given decision table $T = (U, A, d)$, given objects $u_i \in U$ and $v \in \mathcal{U}(T)$, and a given value $b \in V_d(T)$, if there exists a rule from $Det(T)$, which is realizable for u_i and v, and has $d(x) = b$ in the right hand side. This algorithm constructs the set $M(u_i, v)$ and the set $P(u_i, v)$. The considered rule exists if and only if $P(u_i, v) = \{b\}$.

4   Inhibitory Decision Rules

Let us consider a rule

$$a_{j_1}(x) = b_1 \wedge \ldots \wedge a_{j_t}(x) = b_t \Rightarrow d(x) \neq b, \tag{3}$$

where $t \geq 0$, $a_{j_1}, \ldots, a_{j_t} \in A$, $b_1, \ldots, b_t \in \omega$, $b \in V_d(T)$, and numbers $j_1, \ldots, j_t$ are pairwise different. Such rules are called *inhibitory decision* rules. The rule (3) is called *realizable for an object* $u \in \mathcal{U}(T)$ if $a_{j_1}(u) = b_1, \ldots, a_{j_t}(u) = b_t$ or $t = 0$. The rule (3) is called *true for an object* $u_i \in U$ if $d(u_i) \neq b$ or (3) is not realizable for u_i. The rule (3) is called *true for T* if it is true for any object from U. The rule (3) is called *realizable for T* if it is realizable for at least one object from U. By $Inh(T)$ we denote the set of all inhibitory decision rules which are true for T and realizable for T.

Our aim is to recognize for given objects $u_i \in U$ and $v \in \mathcal{U}(T)$, and given value $b \in V_d(T)$ if there exists a rule from $Inh(T)$, which is realizable for u_i and v, and has $d(x) \neq b$ in the right hand side. Such a rule "contradicts" the assignment of the decision b to the new object v.

Proposition 2. *Let $T = (U, A, d)$ be a decision table, $u_i \in U$, $v \in \mathcal{U}(T)$, and $b \in V_d(T)$. Then in $Inh(T)$ there exists a rule, which is realizable for u_i and v, and has $d(x) \neq b$ in the right hand side, if and only if $b \notin P(u_i, v)$.*

Proof. Let $b \notin P(u_i, v)$. In this case, the rule

$$\bigwedge_{a_j \in M(u_i, v)} a_j(x) = a_j(v) \Rightarrow d(x) \neq b \tag{4}$$

belongs to $Inh(T)$, is realizable for u_i and v, and has $d(x) \neq b$ in the right hand side.

Let us assume that there exists a rule (3) from $Inh(T)$, which is realizable for u_i and v, and has $d(x) \neq b$ in the right hand side. Since (3) is realizable for u_i and v, we have $a_{j_1}, \dots, a_{j_t} \in M(u_i, v)$. Since (3) is true for T, the rule (4) is true for T. Therefore, $b \notin P(u_i, v)$. □

From Proposition 2 it follows that there exists a polynomial algorithm recognizing for a given decision table $T = (U, A, d)$, given objects $u_i \in U$ and $v \in \mathcal{U}(T)$, and a given value $b \in V_d(T)$ if there exists a rule from $Inh(T)$, which is realizable for u_i and v, and has $d(x) \neq b$ in the right hand side. This algorithm constructs the set $M(u_i, v)$ and the set $P(u_i, v)$. The considered rule exists if and only if $b \notin P(u_i, v)$.

5   Classification Algorithms

Let $T = (U, A, d)$ be a decision table. We consider the following classification problem: for an object $v \in \mathcal{U}(T)$ predict the value of the decision attribute d on v using only values of attributes from A on v. To this end, we use the D-classification algorithm (D-algorithm) and the I-classification algorithm (I-algorithm).

D-algorithm is based on the use of the parameter $D(T, b, v)$, $b \in V_d(T)$. This parameter is equal to the number of objects $u_i \in U$ for which there exists a rule from $Det(T)$, that is realizable for u_i and v, and has $d(x) = b$ in the right hand side. From Proposition 1 it follows that there exists a polynomial algorithm which for a given decision table $T = (U, A, d)$, a given object $v \in \mathcal{U}(T)$ and a given value $b \in V_d(T)$ computes the value $D(T, b, v) = |\{u_i : u_i \in U, P(u_i, v) = \{b\}\}|$.

D-algorithm: For given object v and each $b \in V_d(T)$ we find the value of the parameter $D(T, b, v)$. As the value of the decision attribute for v we choose $b \in V_d(T)$ such that $D(T, b, v)$ has the maximal value. If more than one such b exists then we choose the minimal b for which $D(T, b, v)$ has the maximal value.

I-algorithm is based on the use of the parameter $I(T, b, v)$, $b \in V_d(T)$. This parameter is equal to the number of objects $u_i \in U$ for which there exists a rule from $Inh(T)$, that is realizable for u_i and v, and has $d(x) \neq b$ in the right hand side. From Proposition 2 it follows that there exists a polynomial algorithm which for a given decision table $T = (U, A, d)$, a given object $v \in \mathcal{U}(T)$ and a given value $b \in V_d(T)$ computes the value $I(T, b, v) = |\{u_i : u_i \in U, b \notin P(u_i, v)\}|$.

I-algorithm: For given object v and each $b \in V_d(T)$ we find the value of the parameter $I(T, b, v)$. As the value of the decision attribute for v we choose $b \in V_d(T)$ such that $I(T, b, v)$ has the minimal value. If more than one such b exists then we choose the minimal b for which $I(T, b, v)$ has the minimal value.

6   Results of Experiments

We have performed experiments with D-algorithm and I-algorithm and decision tables from [8] using DMES system [5]. Some attributes in tables are discretized, and missing values are filled by algorithms from RSES2 [13]. We removed attributes of the kind "name" that are distinct for each instance. To evaluate the accuracy of an algorithm on a decision table (the percent of correctly classified objects) we use either train-and-test method or cross-validation method.

Let us, for some algorithm and some table, use n-fold cross-validation method for the estimation of the accuracy. Then for this table we obtain n accuracies $x_1, \ldots, x_n$. As the final accuracy we use the value $\bar{x} = \frac{1}{n} \sum_{i=1}^{n} x_i$ which is the arithmetic mean of $x_1, \ldots, x_n$. *Maximal relative deviation* for $x_1, \ldots, x_n$ is equal to $\max \left\{ \frac{|x_i - \bar{x}|}{\bar{x}} : i = 1, \ldots, n \right\}$. This value characterizes algorithm stability.

Table 1 contains results of experiments for D-algorithm and I-algorithm and initial decision tables from [8]. Columns "D ac" and "I ac" contain accuracy of D-algorithm and I-algorithm. Columns "D mrd" and "I mrd" contain maximal relative deviations of accuracies in the case when cross-validation method is used.

Table 1. Results of experiments with initial decision tables

Decision table	D ac	I ac	D mrd	I mrd	Decision table	D ac	I ac	D mrd	I mrd
monk1	89.8	89.8			lenses	71.6	76.6	0.534	0.565
monk2	80.0	80.0			soybean-small	57.5	57.5	0.652	0.652
monk3	93.9	93.9			soybean-large	85.6	85.9		
lymphography	78.5	79.2	0.272	0.279	zoo	85.1	94.0	0.177	0.063
diabetes	75.2	75.2	0.168	0.168	post-operative	65.5	64.4	0.152	0.310
breast-cancer	76.4	76.4	0.121	0.121	hayes-roth	92.8	85.7		
primary-tumor	35.7	37.5	0.322	0.290	lung-cancer	40.0	40.0	0.665	0.665
balance-scale	78.7	76.6	0.099	0.097	solar-flare	97.8	97.8		

For 3 decision tables the accuracy of D-algorithm is greater than the accuracy of I-algorithm, for 5 decision tables the accuracy of I-algorithm is greater than the accuracy of D-algorithm, and for 8 decision tables D-algorithm and I-algorithm have the same accuracy. The considered algorithms are not stable.

Table 2 contains results of experiments for D-algorithm and I-algorithm and modified decision tables from [8]. For each initial table from [8] we choose a number of attributes different from the decision attribute, and consider each such attribute as new decision attribute. As the result we obtain the same number of new decision tables as the number of chosen attributes (this number can be found in the column "New"). The column "D opt" contains the number of new tables for which the accuracy of D-algorithm is greater than the accuracy of I-algorithm. The column "I opt" contains the number of new tables for which the accuracy of I-algorithm is greater than the accuracy of D-algorithm. The columns "D aac" and "I aac" contain the average accuracy of D-algorithm and I-algorithm for new tables.

Table 2. Results of experiments with modified decision tables

Decision table	New	D opt	I opt	D aac	I aac
lymphography	9	4	3	54.9	55.3
primary-tumor	3	1	2	64.8	65.7
balance-scale	4	0	4	19.5	23.9
soybean-large	5	0	3	86.2	86.4
zoo	1	0	1	73.0	74.0
post-operative	8	1	4	56.4	57.4
hayes-roth	4	1	3	45.5	49.9
lung-cancer	7	0	1	58.7	59.1
solar-flare	7	1	3	67.0	67.8

For 8 new decision tables the accuracy of D-algorithm is greater than the accuracy of I-algorithm, for 24 new decision tables the accuracy of I-algorithm is greater than the accuracy of D-algorithm, and for 16 new decision tables D-algorithm and I-algorithm have the same accuracy. Note also that for each of the considered initial tables the average accuracy of I-algorithm for new tables corresponding to the initial one is greater than the average accuracy of D-algorithm.

7 Conclusions

Results of experiments show that the algorithm based on inhibitory decision rules is, often, better than the algorithm based on deterministic decision rules. It means that inhibitory decision rules are as relevant to classification algorithms as deterministic decision rules. There is an additional (intuitive) motivation for the use of inhibitory decision rules in classification algorithms: the inhibitory decision rules have much more chance to have larger support than the deterministic ones.

In our future investigations we are planning to continue the comparison of classification algorithms based on deterministic and nondeterministic (contained on the right hand side more than one value of decision attribute) decision rules. Moreover, to improve the stability of classification, we plan to combine approximate decision rules (using them as arguments "for" decisions specified by the right hand sides of such rules) with generalized inhibitory rules (used as arguments "against" decisions defined by the right hand sides of such rules).

References

1. Aha, D.W. (ed.): Lazy Learning. Kluwer Academic Publishers, Dordrecht (1997)
2. Bazan, J.G.: Discovery of Decision Rules by Matching New Objects Against Data Tables. In: Polkowski, L., Skowron, A. (eds.) RSCTC 1998. LNCS (LNAI), vol. 1424, pp. 521–528. Springer, Heidelberg (1998)
3. Bazan, J.G.: A Comparison of Dynamic and Non-Dynamic Rough Set Methods for Extracting Laws from Decision Table. In: Polkowski, L., Skowron, A. (eds.) Rough Sets in Knowledge Discovery, pp. 321–365. Physica-Verlag, Heidelberg (1998)

4. Bazan, J.G.: Methods of Approximate Reasoning for Synthesis of Decision Algorithms. Ph.D. Thesis. Warsaw University (in Polish) (1998)
5. Data Mining Exploration System (Software), http://www.univ.rzeszow.pl/rspn
6. Delimata, P., Moshkov, M., Skowron, A., Suraj, Z.: Two Families of Classification Algorithms. In: An, A., Stefanowski, J., Ramanna, S., Butz, C.J., Pedrycz, W., Wang, G. (eds.) RSFDGrC 2007. LNCS (LNAI), vol. 4482, pp. 297–304. Springer, Heidelberg (2007)
7. Moshkov, M., Skowron, A., Suraj, Z.: On Maximal Consistent Extensions of Information Systems. In: Conference Decision Support Systems (Zakopane, Poland, December 2006), University of Silesia, Katowice, vol. 1, pp. 199–206 (2007)
8. UCI Repository of Machine Learning Databases, University of California, Irvine http://www.ics.uci.edu/~mlearn/MLRepository.html
9. Nguyen, H.S.: Scalable Classification Method Based on Rough Sets. In: Alpigini, J.J., Peters, J.F., Skowron, A., Zhong, N. (eds.) RSCTC 2002. LNCS (LNAI), vol. 2475, pp. 433–440. Springer, Heidelberg (2002)
10. Nguyen, T.T.: Handwritten Digit Recognition Using Adaptive Classifier Construction Techniques. In: Pal, S.K., Polkowski, L., Skowron, A. (eds.) Rough-Neural Computing: Techniques for Computing with Words, pp. 573–585. Springer, Heidelberg (2003)
11. Pawlak, Z.: Rough Sets – Theoretical Aspects of Reasoning about Data. Kluwer Academic Publishers, Dordrecht (1991)
12. Pawlak, Z., Skowron, A.: Rudiments of Rough Sets. Information Sciences 177, 3–27 (2007); Rough Sets: Some Extensions. Information Sciences 177, 28–40 (2007); Rough Sets and Boolean Reasoning. Information Sciences 177, 41–73 (2007)
13. Rough Set Exploration System, http://logic.mimuw.edu.pl/~rses
14. Skowron, A., Suraj, Z.: Rough Sets and Concurrency. Bulletin of the Polish Academy of Sciences 41, 237–254 (1993)
15. Suraj, Z.: Some Remarks on Extensions and Restrictions of Information Systems. In: Ziarko, W., Yao, Y. (eds.) RSCTC 2000. LNCS (LNAI), vol. 2005, pp. 204–211. Springer, Heidelberg (2001)
16. Wojna, A.: Analogy-Based Reasoning in Classifier Construction. (Ph.D. Thesis). In: Peters, J.F., Skowron, A. (eds.) Transactions on Rough Sets IV. LNCS, vol. 3700, pp. 277–374. Springer, Heidelberg (2005)

Consistency and Fuzziness in Ordered Decision Tables

Yuhua Qian, Jiye Liang, Wei Wei, and Feng Wang

Key Laboratory of Computational Intelligence and Chinese Information Processing of
Ministry of Education, Taiyuan, 030006, Shanxi, China
jinchengqyh@126.com, ljy@sxu.edu.cn, mecdang@cityu.edu.hk

Abstract. In this paper, we focus on how to measure the consistency of an ordered decision table and the fuzziness of an ordered rough set and an ordered rough classification in the context of ordered information systems. The membership function of an object is defined through using the dominance class including itself. Based on the membership function, we introduce a consistency measure to assess the consistency of an ordered decision table and define two fuzziness measures to compute the fuzziness of an ordered rough set and an ordered rough classification. Several examples are employed to illustrate their mechanisms as well. These results will be helpful for understanding the uncertainty in ordered information systems and ordered decision tables.

Keywords: Ordered decision table, Consistency, Fuzziness.

1 Introduction

Rough set theory, introduced by Pawlak [1, 2], has been conceived as a tool to conceptualize and analyze various types of data. It can be used in the attribute-value representation model to describe the dependencies among attributes and evaluate the significance of attributes and derive decision rules. It has important applications to intelligence decision and cognitive sciences, as a tool for dealing with vagueness and uncertainty of facts, and in classification [3-8]. Rough-set-based data analysis starts from a data table, called information systems. The information systems contains data about objects of interest, characterized by a finite set of attributes [9-14].

The original rough sets theory does not consider attributes with preference-ordered domains, that is, criteria. However, in many real situations, we are often faced with the problems in which the ordering of properties of the considered attributes plays a crucial role. One such type of problem is the ordering of objects. For this reason, Greco, Matarazzo, and Slowinski [15, 16] proposed an extension of rough set theory, called the dominance-based rough sets approach (DRSA) to take into account the ordering properties of criteria. This innovation is mainly based on substitution of the indiscernibility relation by a dominance relation.

G. Wang et al. (Eds.): RSKT 2008, LNAI 5009, pp. 63–71, 2008.
© Springer-Verlag Berlin Heidelberg 2008

Because the notion of consistency degree [1] is defined for a decision table, in some sense, it could be regarded as measures for evaluating the decision performance of a decision table [17, 18]. Nevertheless, the consistency degree has some limitations. For instance, the consistency of a decision table could not be well depicted by the consistency degree when its value achieve zero. As we know, the fact that consistency degree is equal to zero only implies that there is no decision rule with the certainty of one in the decision table. Hence, the consistency degree of a decision table cannot give elaborate depictions of the consistency for a given decision table. Therefore, we introduced three new measures to assess the entire decision performance of a decision-rule set extracted from a complete/incomplete decision table [18, 19]. So far, however, how to assess the consistency of an ordered decision table has not been reported. In addition, like classical rough set theory, there exist some fuzziness of an ordered rough set and an ordered rough classification in the dominance-based rough sets approach.

The rest of the paper is organized as follows. Some basic concepts of ordered information systems and ordered decision tables are briefly reviewed in Section 2. In Section 3, how to measure the consistencies of a set and an ordered decision table are investigated. In Section 4, we propose fuzziness measures of an ordered rough set and an ordered rough classification in an ordered decision table. Section 5 concludes this paper with some remarks.

2   Preliminaries

In this section, we recall some basic concepts of ordered information systems and ordered decision tables.

An information system (IS) is an quadruple $S = (U, AT, V, f)$, where U is a finite nonempty set of objects and AT is a finite nonempty set of attributes, $V = \bigcup_{a \in AT} V_a$ and V_a is a domain of attribute a, $f : U \times AT \rightarrow V$ is a total function such that $f(x, a) \in V_a$ for every $a \in AT$, $x \in U$, called an information function. A decision table is a special case of an information system in which, among the attributes, we distinguish one called a decision attribute. The other attributes are called condition attributes. Therefore, $S = (U, C \cup d, V, f)$ and $C \cap d = \emptyset$, where the set C is called the condition attributes and d is called the decision attribute.

If the domain (scale) of a condition attribute is ordered according to a decreasing or increasing preference, then the attribute is a criterion.

Definition 1.[20] A decision table is called an ordered decision table (ODT) if all condition attributes are criterions.

It is assumed that the domain of a criterion $a \in AT$ is completely pre-ordered by an outranking relation $\succeq_a$; $x \succeq_a y$ means that x is at least as good as (outranks) y with respect to criterion a. In the following, without any loss of generality, we consider a condition criterion having a numerical domain, that is, $V_a \subseteq \mathbf{R}$ ($\mathbf{R}$ denotes the set of real numbers) and being of type gain, that is,

$x \succeq_a y \Leftrightarrow f(x,a) \geq f(y,a)$ (according to increasing preference) or $x \succeq_a y \Leftrightarrow f(x,a) \leq f(y,a)$ (according to decreasing preference), where $a \in AT$, $x, y \in U$. For a subset of attributes $B \subseteq C$, we define $x \succeq_B y \Leftrightarrow \forall a \in B, f(x,a) \geq f(y,a)$. In other words, x is at least as good as y with respect to all attributes in B. In general, the domain of the condition criterion may be also discrete, but the preference order between its values has to be provided.

In a given ordered information system, we say that x dominates y with respect to $B \subseteq C$ if $x \succeq_B y$, and denoted by $xR_B^\geq y$. That is $R_B^\geq = \{(y,x) \in U \times U \mid y \succeq_B x\}$. Obviously, if $(y,x) \in R_B^\geq$, then y dominates x with respect to B.

Let B_1 be attributes set according to increasing preference and B_2 attributes set according to decreasing preference, hence $B = B_1 \cup B_2$. The granules of knowledge induced by the dominance relation $R_B^\geq$ are the set of objects dominating x, that is

$$[x]_B^\geq = \{y \mid f(y,a_1) \geq f(x,a_1)(\forall a_1 \in B_1) \text{ and } f(y,a_2) \leq f(x,a_2)(\forall a_2 \in B_2)\}$$
$$= \{y \in U \mid (y,x) \in R_B^\geq\}$$

and the set of objects dominated by x,

$$[x]_B^\leq = \{y \mid f(y,a_1) \leq f(x,a_1)(\forall a_1 \in B_1) \text{ and } f(y,a_2) \geq f(x,a_2)(\forall a_2 \in B_2)\}$$
$$= \{y \in U \mid (x,y) \in R_B^\geq\},$$

which are called the B-dominating set and the B-dominated set with respect to $x \in U$, respectively.

Let $U/R_B^\geq$ denote classification on the universe, which is the family set $\{[x]_B^\geq \mid x \in U\}$. Any element from $U/R_B^\geq$ will be called a dominance class with respect to B. Dominance classes in $U/R_B^\geq$ do not constitute a partition of U in general. They constitute a covering of U.

3   Consistency of an Ordered Decision Table

In this section, we deal with how to measure the consistency of an ordered decision table.

Let $S = (U, AT)$ be an ordered information system, $P, Q \subseteq A$, $U/R_P^\geq = \{[x_1]_P^\geq, [x_2]_P^\geq, \cdots, [x_{|U|}]_P^\geq\}$ and $U/R_Q^\geq = \{[x_1]_Q^\geq, [x_2]_Q^\geq, \cdots, [x_{|U|}]_Q^\geq\}$. We define a partial relation $\preceq$ as follows: $P \preceq Q \Leftrightarrow [x_i]_P^\geq \subseteq [x_i]_Q^\geq$ for any $x_i \in U$, where $[x_i]_P^\geq \in U/R_P^\geq$ and $[x_i]_Q^\geq \in U/R_Q^\geq$. If $P \preceq Q$, we say that Q is coarser than P (or P is finer than Q).

Let $S = (U, C \cup d)$ be an ordered decision table, $U/R_C^\geq = \{[x_1]_C^\geq, [x_2]_C^\geq, \cdots, [x_{|U|}]_C^\geq\}$ and $U/R_d^\geq = \{[x_1]_d^\geq, [x_2]_d^\geq, \cdots, [x_{|U|}]_d^\geq\}$. If $C \preceq d$, then S is said to be a consistent ordered decision table; otherwise, S is said to be inconsistent.

Firstly, we investigate the consistency of the dominance class $[x_i]_C^\geq$ ($i \in \{1, 2, \cdots, |U|\}$) with respect to d in an ordered decision table.

Let $S = (U, C \cup d)$ be an ordered decision table, $[x_i]_C^\geq \in U/R_C^\geq$ a dominance class and $U/R_d^\geq = \{[x_i]_d^\geq : x_i \in U\}$. For any object $x \in U$, the membership function of x in the dominance class $[x_i]_C^\geq$ is defined as

$$\delta_{[x_i]_{\bar{C}}^{\geq}}(x) = \begin{cases} \frac{|[x_i]_{\bar{C}}^{\geq} \cap [x]_{\bar{d}}^{\geq}|}{|[x_i]_{\bar{C}}^{\geq}|}, & if \ x = x_i; \\ 0, & if \ x \neq x_i. \end{cases} \tag{1}$$

Where $\delta_{[x_i]_{\bar{C}}^{\geq}}(x)$ denotes a fuzzy concept.

If $\delta_{[x_i]_{\bar{C}}^{\geq}}(x) = 1$, then the dominance class $[x_i]_{\bar{C}}^{\geq}$ can be said to be consistent with respect to d. In other words, if $[x_i]_{\bar{C}}^{\geq}$ is a consistent set with respect to d, then $[x_i]_{\bar{C}}^{\geq} \subseteq [x_i]_{\bar{d}}^{\geq}$. This generates a fuzzy set $F_{[x_i]_{\bar{C}}^{\geq}}^d = \{(x, \delta_{[x_i]_{\bar{C}}^{\geq}}(x)) \mid x \in U\}$ on the universe U.

Definition 2. Let $S = (U, C \cup d)$ be an ordered decision table, $[x_i]_{\bar{C}}^{\geq} \in U/R_{\bar{C}}^{\geq}$ a dominance class and $U/R_{\bar{d}}^{\geq} = \{[x_1]_{\bar{d}}^{\geq}, [x_2]_{\bar{d}}^{\geq}, \cdots, [x_{|U|}]_{\bar{d}}^{\geq}\}$. A consistency measure of $[x_i]_{\bar{C}}^{\geq}$ with respect to d is defined as

$$C([x_i]_{\bar{C}}^{\geq}, d) = \sum_{x \in U} \delta_{[x_i]_{\bar{C}}^{\geq}}(x), \tag{2}$$

where $0 \leq C([x_i]_{\bar{C}}^{\geq}, d) \leq 1$.

Proposition 1. The consistency measure of a consistent dominance class in an ordered decision table is one.

In the following, based on the above discussion, we research the consistency between the condition part and the decision part in an ordered decision table.

Definition 3. Let $S = (U, C \cup d)$ be an ordered decision table, $U/R_{\bar{C}}^{\geq} = \{[x_1]_{\bar{C}}^{\geq}, [x_2]_{\bar{C}}^{\geq}, \cdots, [x_{|U|}]_{\bar{C}}^{\geq}\}$ and $U/R_{\bar{d}}^{\geq} = \{[x_1]_{\bar{d}}^{\geq}, [x_2]_{\bar{d}}^{\geq}, \cdots, [x_{|U|}]_{\bar{d}}^{\geq}\}$. A consistency measure of C with respect to d is defined as

$$C(C, d) = \frac{1}{|U|} \sum_{i=1}^{|U|} \sum_{x \in U} \delta_{[x_i]_{\bar{C}}^{\geq}}(x), \tag{3}$$

where $0 \leq C(C, d) \leq 1$ and $\delta_{[x_i]_{\bar{C}}^{\geq}}(x)$ is the membership function of $x \in U$ in the dominance class $[x_i]_{\bar{C}}^{\geq}$.

Example 1. An ODT is presented in Table 1, where $U = \{x_1, x_2, x_3, x_4, x_5, x_6\}$ and $C = \{a_1, a_2, a_3\}$.

Table 1. An ordered decision table

U	a_1	a_2	a_3	d
x_1	1	2	1	1
x_2	3	2	2	2
x_3	1	1	2	1
x_4	2	1	3	2
x_5	3	3	2	1
x_6	3	2	3	2

In this table, from the definition of dominance classes, one can obtain that the dominance classes determined by C are

$[x_1]_C^{\geq} = \{x_1, x_2, x_5, x_6\}$, $[x_2]_C^{\geq} = \{x_2, x_5, x_6\}$, $[x_3]_C^{\geq} = \{x_2, x_3, x_4, x_5, x_6\}$, $[x_4]_C^{\geq} = \{x_4, x_6\}$, $[x_5]_C^{\geq} = \{x_5\}$, $[x_6]_C^{\geq} = \{x_6\}$;

and the dominance classes determined by d are

$[x_1]_d^{\geq} = [x_3]_d^{\geq} = [x_5]_d^{\geq} = \{x_1, x_2, x_3, x_4, x_5, x_6\}$, $[x_2]_d^{\geq} = [x_4]_d^{\geq} = [x_6]_d^{\geq} = \{x_2, x_4, x_6\}$.

From formula (1), one has that

$C([x_1]_C^{\geq}, d) = 1$, $C([x_2]_C^{\geq}, d) = \frac{2}{3}$, $C([x_3]_C^{\geq}, d) = 1$, $C([x_4]_C^{\geq}, d) = 1$, $C([x_5]_C^{\geq}, d) = 1$ and $C([x_6]_C^{\geq}, d) = 1$. Therefore,

$C(C, d) = \frac{1}{|U|} \sum_{i=1}^{|U|} \sum_{x \in U} \delta_{[x_i]_C^{\geq}}(x) = \frac{1}{6}(1 + \frac{2}{3} + 1 + 1 + 1 + 1) = \frac{17}{18}$.

Proposition 2. The consistency measure of a consistent ordered decision table is one.

Proof. Let $S = (U, C \cup d)$ be an ordered decision table, $U/R_C^{\geq} = \{[x_1]_C^{\geq}, [x_2]_C^{\geq}, \cdots, [x_{|U|}]_C^{\geq}\}$ and $U/R_d^{\geq} = \{[x_1]_d^{\geq}, [x_2]_d^{\geq}, \cdots, [x_{|U|}]_d^{\geq}\}$. If S is consistent, then, for any $x_i \in U$, one has $[x_i]_C^{\geq} \subseteq [x_i]_d^{\geq}$. Hence, when $x = x_i$, we have $\delta_{[x_i]_C^{\geq}}(x) = \frac{|[x_i]_C^{\geq} \cap [x_i]_d^{\geq}|}{|[x_i]_C^{\geq}|} = \frac{|[x_i]_C^{\geq}|}{|[x_i]_C^{\geq}|} = 1$; otherwise, $\delta_{[x_i]_C^{\geq}}(x) = 0$. Therefore,

$$C(C, d) = \frac{1}{|U|} \sum_{i=1}^{|U|} \sum_{x \in U} \delta_{[x_i]_C^{\geq}}(x) = \frac{1}{|U|} \sum_{i=1}^{|U|} (1 \cdot 1 + (|U| - 1) \cdot 0) = 1.$$

Thus, the consistency measure of a consistent ordered decision table is one.

4 Fuzziness of an Ordered Rough Set and an Ordered Rough Classification

In this section, we present fuzziness measures of an ordered rough set and an ordered rough classification in an ordered decision table.

In the literature, Greco et al. [15, 16] proposed the rough set theory for multi-criteria decision analysis. For any $X \subseteq U$ and $B \subseteq C$, the lower and upper approximation of X with respect to the dominance relation $R_B^{\geq}$ are defined as $\underline{R_B^{\geq}}(X) = \{x \in U \mid [x]_B^{\geq} \subseteq X\}$ and $\overline{R_B^{\geq}}(X) = \{[x]_B^{\geq} \mid [x]_B^{\geq} \cap X \neq \emptyset\}$. Unlike classical rough set theory, one can easily notice the properties $\underline{R_B^{\geq}}(X) = \{[x]_B^{\geq} \mid [x]_B^{\geq} \subseteq X\}$ and $\overline{R_B^{\geq}}(X) = \{[x]_B^{\geq} \mid [x]_B^{\geq} \cap X \neq \emptyset\}$ do not hold.

Let $S = (U, AT)$ be an ordered information system and $X \subseteq U$. For any object $x \in U$, the membership function of x in X is defined as

$$\mu_X(x) = \frac{|[x]_{AT}^{\geq} \cap X|}{|[x]_{AT}^{\geq}|} \tag{4}$$

where $\mu_X(u)$ $(0 \leq \mu_X(u) \leq 1)$ represents a fuzzy concept. It can generate a fuzzy set $F_X^{AT} = \{(x, \mu_X(x)) \mid x \in U\}$ on the universe U. Based on this membership function, one can define a fuzzy measure of a given rough set induced by the attribute set AT as follows.

Definition 4. Let $S = (U, A)$ be an ordered information system and $X \subseteq U$. A fuzziness measure of the rough set X is defined as

$$E(F_X^{AT}) = \sum_{i=1}^{|U|} \mu_X(x_i)(1 - \mu_X(x_i)). \qquad (5)$$

Proposition 3. The fuzziness measure of a crisp set equals zero in an ordered information system.

Proof. Let X be a crisp set in the ordered information system $S = (U, AT)$, then $\underline{R_{AT}^\geq}(X) = \overline{R_{AT}^\geq}(X)$. Hence, for any $x \in U$, one can get that if $x \in \underline{R_{AT}^\geq}(X)$, then $[x]_{AT}^\geq \subseteq X$, thus $\mu_X(x) = 1$; and if $x \notin \overline{R_{AT}^\geq}(X)$, then $x \notin \overline{R_{AT}^\geq}(X)$, i.e., $[x]_{AT}^\geq \cap X = \emptyset$, thus $\mu_X(x) = 0$. Therefore, one has that $\mu_X(x)(1 - \mu_X(x)) = 0$, that is $E(F_X^{AT}) = 0$. This completes the proof.

Proposition 4. The fuzziness measure of a rough set is the same as that of its complement set in an ordered information system.

Proof. Let X be a rough set in the ordered information system $S = (U, AT)$ and X^c is its complement set on the universe U, i.e., $X^c = U - X$. For any $x \in U$, one has that

$$\mu_X(x) + \mu_{X^c}(x) = \frac{|X \cap [x]_{AT}^\geq|}{|[x]_{AT}^\geq|} + \frac{|X^c \cap [x]_{AT}^\geq|}{|[x]_{AT}^\geq|} = \frac{|[x]_{AT}^\geq|}{|[x]_{AT}^\geq|} = 1,$$

i.e., $\mu_{X^c}(x) = 1 - \mu_X(x)$. Thus, for any $x \in U$, one can obtain that $\mu_X(x)(1 - \mu_X(x)) = \mu_{X^c}(x)(1 - \mu_{X^c}(x))$, i.e., $E(F_X^{AT}) = E(F_{X^c}^{AT})$.

Assume that the decision attribute d makes a partition of U into a finite number of classes; let $\mathbf{D} = \{D_1, D_2, \cdots, D_r\}$ be a set of these classes that are ordered, that is, for all $i, j \leq r$ if $i \geq j$, then the objects from D_i are preferred to the objects from D_j. The sets to be approximated are an upward union and a downward union of classes, which are defined as $D_i^\geq = \bigcup_{j \geq i} D_j$, $D_i^\leq = \bigcup_{j \leq i} D_j$, $(i \leq r)$ [15, 16]. The statement $x \in D_i^\geq$ means "x belongs to at least class D_i", whereas $x \in D_i^\leq$ means "x belongs to at most class D_i". In the following, we review the definitions of the lower and upper approximations of $D_i^\geq$ $(i \leq r)$ with respect to the dominance relation $R_C^\geq$ in an ODT [20].

Definition 5. [15, 16] Let $S = (U, C \cup d)$ be an ODT, $A \subseteq C$ and $\mathbf{D} = \{D_1, D_2, \cdots, D_r\}$ the decision induced by d. Lower and upper approximations of $D_i^\geq$ $(i \leq r)$ with respect to the dominance relation $R_C^\geq$ are defined as

$$\underline{R_C^\geq}(D_i^\geq) = \{x \in U \mid [x]_C^\geq \subseteq D_i^\geq\}, \quad \overline{R_C^\geq}(D_i^\geq) = \bigcup_{x \in D_i^\geq} [x]_C^\geq.$$

Denoted by $\underline{R_C^\geq}(\mathbf{D}) = (\underline{R_C^\geq}(D_1^\geq), \underline{R_C^\geq}(D_2^\geq), \cdots, \underline{R_C^\geq}(D_r^\geq))$, $\overline{R_C^\geq}(\mathbf{D}) = (\overline{R_C^\geq}(D_1^\geq), \overline{R_C^\geq}(D_2^\geq), \cdots, \overline{R_C^\geq}(D_r^\geq))$. $(\underline{R_C^\geq}(\mathbf{D}), \overline{R_C^\geq}(\mathbf{D}))$ are called the rough decision induced by C. For any object $x \in U$, the membership function of x in $\mathbf{D}$ is defined as

$$\mu_{\mathbf{D}}(x) = \frac{|[x]_{\bar{C}}^{\geq} \cap D_j^{\geq}|}{|[x]_{\bar{C}}^{\geq}|} \quad (u \in D_j), \tag{6}$$

where $\mu_{\mathbf{D}}(x)$ $(0 \leq \mu_{\mathbf{D}}(x) \leq 1)$ represents a fuzzy concept. It can generate a fuzzy set $F_{\mathbf{D}}^{C} = \{(x, \mu_{\mathbf{D}}(x)) \mid x \in U\}$ on the universe U.

Definition 6. Let $S = (U, C \cup d)$ be an ordered information system and $\mathbf{D} = \{D_1, D_2, \cdots, D_r\}$. A fuzziness measure of a rough decision is defined as

$$E(F_{\mathbf{D}}^{C}) = \sum_{i=1}^{|U|} \mu_{\mathbf{D}}(x_i)(1 - \mu_{\mathbf{D}}(x_i)), \tag{7}$$

where $\mu_{\mathbf{D}}(x_i)$ denotes the membership function of $x_i \in U$ in the decision $\mathbf{D}$.

Example 2. (Continued from Example 1.) Suppose that $D_1 = \{x_2, x_4, x_6\}$ and $D_2 = \{x_1, x_3, x_5\}$. From formula (6), we have that $\mu_{\mathbf{D}}(x_1) = \frac{|[x_1]_{\bar{C}}^{\geq} \cap D_2^{\geq}|}{|[x_1]_{\bar{C}}^{\geq}|} = 1$, $\mu_{\mathbf{D}}(x_2) = \frac{|[x_2]_{\bar{C}}^{\geq} \cap D_1^{\geq}|}{|[x_2]_{\bar{C}}^{\geq}|} = \frac{2}{3}$, $\mu_{\mathbf{D}}(x_3) = \frac{|[x_3]_{\bar{C}}^{\geq} \cap D_2^{\geq}|}{|[x_3]_{\bar{C}}^{\geq}|} = 1$, $\mu_{\mathbf{D}}(x_4) = \frac{|[x_4]_{\bar{C}}^{\geq} \cap D_1^{\geq}|}{|[x_4]_{\bar{C}}^{\geq}|} = 1$, $\mu_{\mathbf{D}}(x_5) = \frac{|[x_5]_{\bar{C}}^{\geq} \cap D_2^{\geq}|}{|[x_5]_{\bar{C}}^{\geq}|} = 1$, $\mu_{\mathbf{D}}(x_6) = \frac{|[x_6]_{\bar{C}}^{\geq} \cap D_1^{\geq}|}{|[x_6]_{\bar{C}}^{\geq}|} = 1$. Therefore, $E(F_{\mathbf{D}}^{C}) = \sum_{i=1}^{6} \mu_{\mathbf{D}}(x_i)(1 - \mu_{\mathbf{D}}(x_i)) = 1 \times (1 - 1) \times 5 + \frac{2}{3} \times \frac{1}{3} = \frac{2}{9}$.

Proposition 5. In an ordered decision table $S = (U, C \cup d)$, the fuzziness measure of a crisp decision equals zero.

Proof. Let $\mathbf{D} = \{D_1, D_2, \cdots, D_r\}$ be a crisp decision in the ordered decision table, i.e., $\underline{R_{\bar{C}}^{\geq}}(D_j^{\geq}) = \overline{R_{\bar{C}}^{\geq}}(D_j^{\geq})$, $j = \{1, 2, \cdots, r\}$. Hence, for any one has that $[x]_{\bar{C}}^{\geq} \subseteq D_j^{\geq}$. Thus, $\mu_{\mathbf{D}}(x) = \frac{|[x]_{\bar{C}}^{\geq} \cap D_j^{\geq}|}{|[x]_{\bar{C}}^{\geq}|} = \frac{|[x]_{\bar{C}}^{\geq}|}{|[x]_{\bar{C}}^{\geq}|} = 1$, $\forall x \in U$. Therefore, $\mu_{\mathbf{D}}(x_i)(1 - \mu_{\mathbf{D}}(x_i)) = 0$, $i \leq |U|$, i.e., $E(F_{\mathbf{D}}^{C}) = 0$. This completes the proof.

5   Conclusions

In this study, we have constructed the membership function of an object through using the dominance class including itself. Based on the membership function, we have introduced a consistency measure to calculate the consistency of an ordered decision table and fuzziness measures to compute the fuzziness of an ordered rough set and an ordered rough classification in the context of ordered information systems. Their mechanisms and validity have been shown by several illustrative examples. These results will be helpful for understanding the uncertainty in ordered decision tables.

Acknowledgements. This work was supported by the national high technology research and development program (No. 2007AA01Z165), the national natural science foundation of China (No. 70471003, No. 60573074), the foundation of doctoral program research of the ministry of education of China (No. 20050108004) and key project of science and technology research of the ministry of education of China (No. 206017).

References

1. Pawlak, Z.: Rough Sets: Theoretical Aspects of Reasoning about Data, System Theory, Knowledge Engineering and Problem Solving. Kluwer, Dordrecht (1991)
2. Pawlak, Z., Skowron, A.: Rudiments of Rough Sets. Information Sciences 177, 3–27 (2007)
3. Dubois, D., Prade, H.: Rough Fuzzy Sets and Fuzzy Rough Sets. International Journal of General Systems 17, 191–209 (1990)
4. Düntsch, I., Gediga, G.: Uncertainty Measures of Rough Set Prrediction. Artificial Intelligence 106, 109–137 (1998)
5. Gediga, G., Düntsch, I.: Rough Approximation Quality Revisited. Artificial Intelligence 132, 219–234 (2001)
6. Jensen, R., Shen, Q.: Fuzzy-rough Sets assisted Attribute Selection. IEEE Transactions on Fuzzy Systems 15(1), 73–89 (2007)
7. Liang, J.Y., Dang, C.Y., Chin, K.S., Yam Richard, C.M.: A New Method for Measuring Uncertainty and Fuzziness in Rough Set Theory. International Journal of General Systems 31(4), 331–342 (2002)
8. Qian, Y.H., Liang, J.Y.: Rough Set Method Based on Multi-granulations. In: 5th IEEE Conference on Cognitive Informatics I, pp. 297–304 (2006)
9. Guan, J.W., Bell, D.A.: Rough Computational Methods for Information Systems. Artificial Intelligence 105, 77–103 (1998)
10. Jeon, G., Kim, D., Jeong, J.: Rough Sets Attributes Reduction Based Expert System in Interlaced Video. IEEE Transactions on Consumer Electronics 52(4), 1348–1355 (2006)
11. Kryszkiewicz, M.: Rough Set Approach to Incomplete Information Systems. Information Sciences 112, 39–49 (1998)
12. Liang, J.Y., Li, D.Y.: Uncertainty and Knowledge Acquisition in Information Systems. Science Press, Beijing (2005)
13. Liang, J.Y., Qian, Y.H.: Axiomatic Approach of Knowledge Granulation in Information Systems. In: Sattar, A., Kang, B.-h. (eds.) AI 2006. LNCS (LNAI), vol. 4304, pp. 1074–1078. Springer, Heidelberg (2006)
14. Qian, Y.H., Liang, J.Y.: Combination Entropy and Combination Granulation in Incomplete Information Systems. In: Wang, G.-Y., Peters, J.F., Skowron, A., Yao, Y. (eds.) RSKT 2006. LNCS (LNAI), vol. 4062, pp. 184–190. Springer, Heidelberg (2006)
15. Greco, S., Matarazzo, B., Slowinski, R.: Rough Sets Theory for Multicriteria Decision Analysis. European Journal of Operational Research 129, 11–47 (2001)
16. Greco, S., Matarazzo, B., Slowinski, R.: Rough sets Methodology for Sorting Problems in Presense of Multiple Attributes and Criteria. European Journal of Operational Research 138, 247–259 (2002)
17. Qian, Y.H., Liang, J.Y.: Evaluation Method for Decision Rule Sets. In: An, A., Stefanowski, J., Ramanna, S., Butz, C.J., Pedrycz, W., Wang, G. (eds.) RSFDGrC 2007. LNCS (LNAI), vol. 4482, pp. 272–279. Springer, Heidelberg (2007)

18. Qian, Y.H., Liang, J.Y., Li, D.Y., Zhang, H.Y., Dang, C.Y.: Measures for Evaluating the Decision Performance of a Decision Table in Rough Set Theory. Information Sciences 178(1), 181–202 (2008)
19. Qian, Y.H., Liang, J.Y., Dang, C.Y., Zhang, H.Y., Ma, J.M.: On the Evaluation of the Decision Performance of an Incomplete Decision Table. Data and Knowledge Engineering. doi:10.1016/j.datak.2007.12.002.
20. Shao, M.W., Zhang, W.X.: Dominance Relation and Rules in an Incomplete Ordered Information System. International Journal of Intelligent Systems 20, 13–27 (2005)

Fast Knowledge Reduction Algorithms Based on Quick Sort[*]

Feng Hu[1,2], Guoyin Wang[1,2], and Lin Feng[1,2]

[1] [1] School of Information Science and Technology,
Southwest Jiaotong University,
Chengdu 600031, P.R. China
[2] Institute of Computer Science and Technology,
Chongqing University of Posts and Telecommunications,
Chongqing 400065, P.R. China
{hufeng, wanggy}@cqupt.edu.cn

Abstract. Many researchers are working on developing fast data mining methods for processing huge data sets efficiently. In this paper, we develop some efficient algorithms for knowledge reduction based on rough sets. In these algorithms we use the fact that the average time complexity for the quick sort algorithm for a two dimensions table with n rows and m columns is just $n \times (m + logn)$ (not $m \times n \times logn$). Experiment results also show the efficiency of these algorithms.

Keywords: Huge data, Knowledge reduction, Rough set, Quick sort, Divide and conquer.

1   Introduction

Knowledge reduction is one of the most important contributions of rough set theory [1] to data mining. Usually, there are several steps in knowledge reduction based on rough set theory. That is, data preparation, data discretization, attribute reduction, value reduction. Attribute reduction is the key for knowledge reduction. Many researchers proposed some algorithms for attribution reduction [2-9]. These reduction algorithms can be classified into two categories: reduction without attribute order and reduction with attribute order. The former does not take the domain experts' priori knowledge into consideration in the knowledge reduction process. It generates knowledge from data only. The later considers the domain experts' priori knowledge (attribute order) in the knowledge reduction process. Knowledge generated from data in this kind of methods also depends on domain experts' prior knowledge. However, the efficiencies of these reduction

---

[*] This paper is partially supported by National Natural Science Foundation of China under Grants No.60773113 and No.60573068, Program for New Century Excellent Talents in University (NCET), Natural Science Foundation of Chongqing under Grant No.2005BA2003, Science & Technology Research Program of Chongqing Education Commission under Grant No.KJ060517.

algorithms in dealing with huge data sets are not high enough. There are two reasons: one is the time complexity, and the other is the space complexity.

Quick sort for a two dimensions table is an important operation in data mining. In huge database processing based on rough set theory, it is a basic operation to divide a decision table into indiscernible classes. Quick sort could be used to deal with this problem. Some researchers found that reduction algorithms will become quicker when database management system and quick sort are used for reduction generation [10]. If the data of a two dimensions table has uniform distribution, then the average time complexity of quick sort for a two dimensions table with m attributes and n records was usually considered to be $O(n \times logn \times m)$ [4,5,6]. However, we find in [11] that this time is just $O(n \times (logn + m))$. Based on this finding, we may further revise knowledge reduction algorithms and improve their performance.

In this paper, we will reanalyze the time complexity of quick sort for a two dimensions table briefly, propose an algorithm for computing attribute core based on divide and conquer method, and develop some improved knowledge reduction methods based on the idea of quick sort. Some simulation experiment results will show the performance of our improved methods.

2   Time Complexity of Quick Sort

The process of sorting a two dimensions table with m attributes $(k_1, k_2, ..., k_m)$ and n records is as follows.

Step 1, according to the first attribute k_1, the quick sort algorithm for one dimension table is used to sort the decision table and generate an ordered partition on k_1.

Step 2, the quick sort method for one dimension table is used to sort each partition part respectively and generate an ordered partition of the original table on both k_1 and k_2.

Step 3, repeat the sort process of step 2 using the other attributes $k_3, k_4, ..., k_m$ until all attributes have been used to sort the decision table. A partition part containing only one record needs not to be further sorted using remaining attributes in this process.

In order to discuss about the quick sort algorithm for a two dimensions table, a procedure and a function are used.

Procedure 1: $TwoDimension\_QuickSort(int\ r, int\ low, int\ high)$
// r is the number of keywords(attributes), low and $high$ are the pointers pointing to the first record and the last records respectively.
{
   IF $(r > m)$ THEN return;
   IF $(low == high)$ THEN return;
   $CanBePartition = false$;
   FOR $j = low + 1$ TO $high$ DO
   {   IF $(S[L[j][r]] \neq S[L[low][r]])$ THEN
      { $CanBePartition = true$;   BREAK;}

```
}
IF (CanBePartition == true) THEN
{   average = (S[L[low]][r] + S[L[(low + high)/2]][r] + S[L[high]][r])/3;
    mid=CALL Partition(r, low, high, average);
    CALL TwoDimension_QuickSort(r, low, mid);
    CALL TwoDimension_QuickSort(r, mid + 1, high);
}
IF (CanBePartition == false) THEN
{   CALL TwoDimension_QuickSort(r + 1, low, high); }
}
```

Function 1: *Partition(int r, int low, int high, int average)*

```
{
    i = low; j = high;
    WHILE (i < j) DO
    {   WHILE (S[L[j]][r] > average)&&(i < j) DO j − −;
        Swapping L[i] and L[j];
        WHILE (S[L[i]][r] ≤ average)&&(i < j) DO i + +;
        Swapping L[i] and L[j];
    }
    IF (S[L[i]][r] > average) THEN RETURN i − 1;
    ELSE RETURN i;
}
```

Algorithm 1: Quick Sort Algorithm for a Two Dimensions Table

Input: A two dimensions table $S[1..n][1..m]$

Output: An ordered two dimensions table $S[1..n][1..m]$

Step 1: FOR $i = 1$ TO n DO $L[i] = i$;

Step 2: CALL $TwoDimension\_QuickSort(1, 1, n)$;

Step 3: RETURN S.

When the *Partition* function is invoked once, the number of comparison is $(high - low + 1)$. The time complexity of Algorithm 1 mainly depends on the *TwoDimension_QuickSort* procedure . Now, let's analyze the running time of the *TwoDimension_QuickSort* procedure. Suppose $T(r, n)$ be the time complexity of the *TwoDimension_QuickSort* procedure(where, $n = high - low + 1$). $T(r, n)$ can be approximated by the following recursive equation [11]:

$$T(r, n) = \begin{cases} (a) \ 1 & if \ n = 1 \\ (b) \ 1 & if \ r = m + 1 \\ (c) \ n + T(r + 1, n) & if \ n > 1, 1 \leq r \leq m, and \\ \quad the \ values \ of \ n \ records \ key \ k_r \ are \ unique) \\ (d) \ 2n + T(r, n - k) + T(r, k) & if \ 1 \leq r \leq m, 1 \leq k < n, n > 1, \\ \quad and \ the \ values \ of \ n \ records \ on \ key \ k_r \ are \ not \ unique) \end{cases} \quad (1)$$

Suppose that T_0 represents the total time cost of invoking (a) and (b), T_1 represents the total time cost of invoking (c), and T_2 represents the total time cost of invoking (d). The average time complexity of *TwoDimension_QuickSort* procedure would be: $T(r, n) = T_0 + T_1 + T_2$, where, the average and worst time cost of T_1 is $O(m \times n)$, $T_0 + T_2$ can be approximate by the following recursive equation [11]:

$$T_0 + T_2 = \begin{cases} 1 & if\ n = 1 \\ 2n + T(n-k) + T(k) & if\ 1 \le k < n,\ n > 1 \end{cases} \tag{2}$$

It is the same as the time cost of quick sort for a one dimension array. That is, $T_0 + T_2 = T_1(n) = O(n \times logn)$. Thus, $T(r, n) = T_0 + T_1 + T_2 = T_1 + T_1(n) = O(n \times m) + O(n \times logn) = O(n \times (m + logn))$.

Hence [11], $T(r, n) = O(n \times (m + logn))$.

That is, the average time complexity of Algorithm 1 is $O(n \times (m + logn))$.

Since Algorithm 1 is a recursive algorithm, the recursive stack space is considered in the computation of space complexity. In $TwoDimension\_QuickSort$ procedure, when $TwoDimension\_QuickSort$ is invoked recursively, the **three** parameters: $(r, low, high)$ are needed to be stored into the stack. So, the recursive stack space complexity $S(r, n)$ of quick sort for a two dimensions table can be approximated by the following recursive equation [11]:

$$S(r, n) = \begin{cases} 0 & if\ n \le 1 \\ 0 & if\ r > m \\ 0 & if\ n > 1, 1 \le r \le m, and \\ & the\ values\ of\ n\ records\ key\ k_r\ are\ unique \\ 3 + S(r, n/2) & if\ n > 1, 1 \le r \le m, 1 \le k \le n - 1, and \\ & the\ values\ of\ n\ records\ on\ key\ k_r\ are\ not\ unique \end{cases} \tag{3}$$

Hence [11], $S(r, n) = O(logn)$.

Because array $L[1..n]$ needs $O(n)$ space, the space complexity of Algorithm 1 is $O(n) + O(logn) = O(n)$.

3   Algorithm for Computing Attribute Core

Theorem 1. Given a decision table $S = <U, C \cup D, V, f>$. For any $c(c \in C)$, according to $U/\{c\}$, S is divided into $k(k = |IND(U/\{c\})|)$ sub-decision tables $S_1, S_2,..., S_k$, where, $S_k = <U_k, (C - \{c\}) \cup D, V_k, f_k>$, and $\forall_{x \in U_i} \forall_{y \in U_i} c(x) = c(y)(1 \le i \le k)$ and $\forall_{x \in U_i} \forall_{z \in U_j} c(x) \ne c(z)(1 \le i < j \le k)$. Suppose $Core_i(1 \le i \le k)$ be the core of the sub decision table S_i, and $Core$ be the attribute core of the decision table S. Then, $\bigcup_{1 \le i \le k} Core_i \subseteq Core \subseteq \{c\} \cup \bigcup_{1 \le i \le k} Core_i$.(If c is a core attribute, $Core = \{c\} \cup \bigcup_{1 \le i \le k} Core_i$)

According to the Theorem 1, an efficient algorithm for computing the core of a decision table based on divide and conquer can be developed.

Algorithm 2: Computation of Attribute Core Based on Divide and Conquer Method
Input: A decision table $S = <U, A = C \cup D, V, f>$
Output: Attribute core $(Core)$ of S
Step 1: $Core = \phi$;
Step 2: IF $|C| == 1$ THEN
{   IF $Pos_C(D) \ne Pos_{C-\{c_1\}}(D)$ THEN $Core = C$;
    GOTO Step 3;

```
}
ELSE   //|C| > 1
{   Choose the first attribute c₁ in C;
```

Choose the first attribute c_1 in C;

Calculate $Pos_C(D)$ and $Pos_{C-\{c_1\}}(D)$ using divide and conquer method;

IF $Pos_C(D) \neq Pos_{C-\{c_1\}}(D)$ THEN $Core = Core \cup \{c_1\}$;

According to U/c_1, divide S into $|U/c_1|$ sub decision tables $S_1^1, S_2^1, ...,$ $S_{|U/c_1|}^1$;

$C = C - \{c_1\}$;

Calculate all the attribute cores of the sub decisions tables $Core_1, Core_2,$ $..., Core_{|U/c_1|}$ recursively;

}

Step 3: RETURN $Core = Core \cup Core_1 \cup Core_2 \cup ... \cup Core_{|U/c_1|}$;

Suppose $n = |U|$, $m = |C|$. The time complexity of the Algorithm 2 is $O(n \times m^2)$ [12]. Suppose $n = |U|, m = |C|, p = max(|V_i|)(1 \leq i \leq |C|)$. Then, the space complexity of the Algorithm 2 is $O(n + p \times m)$ [12].

4   Fast Knowledge Reduction Based on Attribute Order

If the idea of divide and conquer is adopted in the attribute reduction of rough set theory, some more efficient knowledge reduction algorithms may be developed. In this section, a quick attribute reduction algorithm based on the divide and conquer method is proposed.

4.1   Attribute Order

An attribute reduction algorithm based on attribute order was proposed by Wang in 2001 [7]. For the convenience of illustration, some basic notions about attribute order are introduced here.

Given a decision table $S = < U, C \cup D, V, f >$, we could define an attribute order relation over $C(SO : c_1 \prec c_2 \prec ... \prec c_{|C|})$. Suppose M is the discernibility matrix of S [2]. For any $\delta \in M$, the attributes of δ inherit the order relation of SO from left to right. The first element of δ is called the label of δ.

For c_j, we define a set $[c_j] = \{\delta | \delta = c_j B, \delta$ inherit the order relation of SO from left to right, $\delta \in M\}$. Hence, M can be divided into equivalent classes defined by attribute labels defining a partition $\{[c_1], [c_2], ..., [c_{|C|}]\}$ of M denoted by $M/L(SO)$. be the partition. The partition can be also expressed by $M/L(SO) = \{[1], [2], ..., [|C|]\}$ [7,8].

In order to compute the attribute reduction of a decision table, we should calculate its non-empty label attribute set at first. Using the divide and conquer method, the following algorithm for computing non-empty label attribute set is developed.

Algorithm 3: Computation of Non-empty Label Attribute Set $L(SO)$

Input: A decision table $S = < U, A = C \cup D, V, f >$ and a given attribute order $SO : c_1 \prec c_2 \prec ... \prec c_{|C|}$

Output: The non-empty label attribute set R_1 of S

Step 1: $R_1 = \phi; r = 1; OSet_1^1 = U;$

FOR $j = 1$ TO $|C|$ DO $NonEmptyLabel[j] = 0;$

Step 2: $NonEmptyLabelAttr(1, OSet_1^1);$

Step 3: FOR $j = 1$ TO $|C|$ DO

IF $NonEmptyLabel[j] == 1$ THEN $R_1 = R_1 \cup \{c_j\};$

Step 4: RETURN R_1 .

Function 2: $NonEmptyLabelAttr(int\ r, ObjectSet\ OSet)$

$\{$ IF $r < (|C| + 1)$ THEN

$\{$ IF $|OSet| == 1$ THEN RETURN;

ELSE IF $Pos_C(D) == \phi$ THEN RETURN;

ELSE IF $|V(c_r)| == 1$ THEN

$\{$ $r = r + 1;\ NonEmptyLabelAttr(r, OSet);$

$\}$

ELSE

$\{$ $NonEmptyLabel[r] = 1;$

Divide $OSet$ into $|V(c_r)|$ parts: $OSet_1, OSet_2, ..., OSet_{|V(c_r)|};$

$r = r + 1;$

FOR $i = 1$ TO $|V(c_r - 1)| + 1$ DO

$NonEmptyLabelAttr(r, OSet_i^r);$

$\}$

$\}$

$\}$

4.2   Attribute Reduction Based on Attribute Order and Divide and Conquer Method

Given an attribute order of the condition attributes in a decision table, using the Algorithm 3 and divide and conquer method, an efficient attribute reduction algorithm is developed.

Algorithm 4: Attribute Reduction Based on Divide and Conquer Method

Input: A decision table $S = <U, A = C \cup D, V, f>$ and a given attribute order
$SO:\ c_1 \prec c_2 \prec ... \prec c_{|C|}$

Output: Attribute reduction R of S

Step 1: $U_1^1 = U;\ R = \phi.$

Step 2: Compute the positive region $Pos_C(D)$ of S.

Step 3: Compute the non-empty label attribute set R_1 by Algorithm 3.

Step 4: //Suppose $c_{N'}$ be the maximum label attribute of R_1.

$R = R \cup \{c_{N'}\};\ R_1 = R_1 - R;\ C = R \cup R_1;$

IF $R_1 == \phi$ THEN RETURN R;

ELSE

$\{$ Generate a new attribute order: $c_1^1 \prec c_2^1 \prec ... \prec c_{|R|}^1 \prec c_1^2 \prec c_2^2 \prec ... \prec$

$c_{|R_1|}^2 (c_i^1 \in R, c_j^2 \in R_1);$

Compute a new non-empty label attribute set R_1 by Algorithm 3;

GOTO Step 4.

}

Suppose $n = |U|$ and $m = |C|$, the average time complexity of the Algorithm 4 is $O(m \times n \times (m + logn))$. Its space complexity is $(m + n)$.

Algorithm 4 depends on attribute order. However, it is provided by domain experts since in some domain it is easy for domain experts to provide the order [7,8].

5   Experiment Results

In order to test the efficiency of Algorithm 4, we have done an experiment. In the experiment, the KDDCUP99 data set is used. $10\%, 20\%, 30\%, ...$, and 100% records of the KDDCUP99 data set are selected randomly to create a testing data set each time respectively. The Equal Frequency Discretization Algorithm [13] is used to discretize these 10 data sets. The domains of all attributes are [0, 255]. The parameters of the computer used here are P4 2.6G CPU, 512M RAM, and Windows XP. The experiment results are shown in Table 1.

Table 1. Results of the test for Algorithm 4

Rate of records	Num of records	Num of total attributes	Running time(Sec)	Num of attributes in the reduction	Memery usage(KB)
10%	489843	41	61.232417	29	198804
20%	979686	41	135.023143	30	246972
30%	1469529	41	218.743598	31	303232
40%	1959372	41	283.406260	30	344984
50%	2449216	41	384.723173	32	395144
60%	2939059	41	469.103032	32	444576
70%	3428902	41	602.920099	34	493672
80%	3918745	41	661.663162	33	543172
90%	4408588	41	753.895816	33	592372
100%	4898432	41	13337.205877	34	641896

From Table 1, we can find that the processing speed of the Algorithm 4 is very fast. However, when the number of records reaches to 4898432, the processing speed slows down greatly. The reason is that the Algorithm 4 will use too much memory when the number of records reaches to 4898432. The using rate of CPU is below 10% in this case. Therefore, the space complexity of an algorithm plays a very important role in dealing with huge data sets.

6   Conclusions and Future Works

In this paper, the complexities of quick sort method for a two dimensions table with m attributes and n records are analyzed. Its average time and space complexities are find to be $O(n \times (m + logn))$ and $O(n)$ respectively. Moreover, an algorithm for computing attribute core is developed. Its time complexity is $O(n \times m^2)$. Besides, given an attribute order of a decision table, a quick attribute reduction algorithm based on divide and conquer method is developed. Its time complexity is $O(m \times n \times (m + logn))$, and space complexity is $O(m + n)$.

References

1. Pawlak, Z.: Rough sets. International Journal of Computer and Information Sciences 11, 341–356 (1982)
2. Skowron, A., Rauszer, C.: The Discernibility Functions Matrics and Functions in Information Systems. In: Slowinski, R. (ed.) Intelligent Decision Support - Handbook of Applications and Advances of the Rough Sets Theory, pp. 331–362. Kluwer Academic Publisher, Dordrecht (1992)
3. Hu, X.H., Cercone, N.: Learning in Relational Database: A Rough Set Approach. International Journal of Computional Intelligence 11(2), 323–338 (1995)
4. Nguyen, H.S., Nguyen, S.H.: Some Efficient Algorithms for Rough Set Methods. In: The Sixth International Conference, Information Procesing and Management of Uncertainty in Knowledge-Based Systems (IPMU 1996), Granada, Spain, July 1-5, 1996, vol. 2, pp. 1451–1456 (1996)
5. Wang, G.Y., Yu, H., Yang, D.C.: Decision Table Reduction Based on Conditional Information Entropy. Chinese Journal of computers 25(7), 759–766 (2002)
6. Liu, S.H., Cheng, Q.J., Shi, Z.Z.: A New Method for Fast Computing Positve Region. Journal of Computer Research and Development 40(5), 637–642 (2003)
7. Wang, J., Wang, J.: Reduction Algorithms Based on Discernibility Matrix: the Ordered Attributed Method. Journal of Computer Science and Technology 11(6), 489–504 (2001)
8. Zhao, M., Wang, J.: The Data Description Based on Reduct. PhD Thesis, Institute of Automation, Chinese Academy of Sciences. Beijing, China (in Chinese) (2004)
9. Mikhail, J.M., Marcin, P., Beata, Z.: On Partial Covers, Reducts and Decision Rules with Weights. In: Peters, J.F., Skowron, A., Düntsch, I., Grzymała-Busse, J.W., Orłowska, E., Polkowski, L. (eds.) Transactions on Rough Sets VI. LNCS, vol. 4374, pp. 211–246. Springer, Heidelberg (2007)
10. Qin, Z.R., Wu, Y., Wang, G.Y.: A Partition Algorithm for Huge Data Sets Based on Rough Set. Pattern Recognition and Artificial Intelligence 19(2), 249–256 (2006)
11. Hu, F., Wang, G.Y.: Analysis of the Complexity of Quick Sort for Two Dimension Table. Chinese Journal of Computers 30(6), 963–968 (2007) (in Chinese)
12. Hu, F., Wang, G.Y., Xia, Y.: Attribute Core Computation Based on Divide and Conquer Method. In: Kryszkiewicz, M., Peters, J.F., Rybinski, H., Skowron, A. (eds.) RSEISP 2007. LNCS (LNAI), vol. 4585, pp. 310–319. Springer, Heidelberg (2007)
13. Wang, G.Y.: Rough Set Theory and Knowledge Acquisition. Xi'an Jiaotong University Press, Xi'an (2001) (in Chinese)

Multiple-Source Approximation Systems: Membership Functions and Indiscernibility

Md. Aquil Khan and Mohua Banerjee

Department of Mathematics and Statistics,
Indian Institute of Technology,
Kanpur 208 016, India
mdaquil,mohua@iitk.ac.in

Abstract. The work presents an investigation of multiple-source approximation systems, which are collections of Pawlak approximation spaces over the same domain. We particularly look at notions of definability of sets in such a collection μ. Some possibilities for membership functions in μ are explored. Finally, a relation that reflects the degree to which objects are (in)discernible in μ is also presented.

Keywords: Approximation spaces, Rough sets, Rough membership functions.

1 Introduction

In this article, we are interested in situations where information is obtained from different sources about the same set of objects, giving rise to a sequence of Pawlak approximation spaces [5,8] over the same domain. This has been discussed, for instance, in [13,3]. In a more general setting, multi-agent systems have also been studied (cf. [1], [4], [10], [9]). Our focus, however, is on the following.

Definition 1. *A* multiple-source approximation system (MAS) *is a tuple* $(U, \{R_i\}_{i \in N})$, *where* U *is a non-empty set,* N *an initial segment of the set of positive integers, and* $R_i, i \in N$, *is an equivalence relation on the domain* U.

We present the basic notions in Section 2. Different kinds of definability of sets that one may have in MAS's, are given in Section 3, and some ensuing properties discussed.

For a Pawlak approximation space (U, R), the *rough membership function* f^X [8,7] quantifies the degree of relative overlap between a set $X (\subseteq U)$ and the equivalence class $[x]$ for an object x of U. If $f^X(x) > f^X(y)$, y has a greater possibility to be an element of X compared to x. When we consider an MAS μ, the situation is clearly different. In Section 4, two simple instances of membership function for μ are defined. Further, we consider the approximation space (U, ind_μ) obtained from a given MAS $\mu := (U, \{R_i\}_{1 \leq i \leq n})$, by taking $ind_\mu := \cap_{i=1}^{n} R_i$. This has been considered, for instance, in [13] while discussing common knowledge of a group of agents. A relation that reflects the degree to which objects are (in)discernible

G. Wang et al. (Eds.): RSKT 2008, LNAI 5009, pp. 80–87, 2008.

in μ is presented, and related to the previously defined notions. Furthermore, some observations about the rough membership function in (U, ind_μ) are made. For lack of space, we must omit many propositions on the notions defined in the paper. Section 5 concludes the article.

2 Basic Notions in Multiple-source Approximation Systems

Let us consider the following example.

Example 1. Suppose we have information regarding the attribute set {transport facilities(Tra), law and order(LO), literacy(Li)} for the cities Calcutta(Cal), Mumbai(Mum), Delhi(Del), Chennai(Chen), Bangalore(Ban) and Kanpur(Kan) from four different agencies M_1, M_2, M_3 and M_4:

	M_1			M_2			M_3			M_4		
	Tra	LO	Li	Tra	LO	Li	Tra	LO	Li	Tra	LO	Li
Cal	a	a	g	a	a	a	a	a	g	g	g	a
Mum	g	g	a	a	a	g	a	g	p	p	a	g
Del	g	g	a	a	a	a	a	g	p	a	a	g
Chen	g	p	g	a	a	g	a	p	g	p	p	g
Ban	a	a	g	a	a	g	a	g	p	p	a	g
Kan	p	g	g	p	g	p	a	p	g	p	p	g

Here g, a, p stand for *good, average* and *poor*. Let $U := \{Cal, Mum, Del, Chen, Ban, Kan\}$. We then have an MAS $(U, \{R_i\}_{1 \leq i \leq 4})$, where
$R_1 := \{\{Cal, Ban\}, \{Mum, Del\}, \{Chen\}, \{Kan\}\};$
$R_2 := \{\{Cal, Del\}, \{Mum, Chen, Ban\}, \{Kan\}\};$
$R_3 := \{\{Cal\}, \{Mum, Del, Ban\}, \{Chen, Kan\}\};$
$R_4 := \{\{Cal\}, \{Mum, Ban\}, \{Del\}, \{Chen, Kan\}\}.$

Let us consider a subset $X := \{Cal, Mum, Del\}$ of U and suppose we wish to define this set in terms of the attributes (Tra), (LO), (Li). We obtain different lower/upper approximations of X, e.g. $\underline{X}_{R_1} = \{Mum, Del\}$ and $\underline{X}_{R_2} = \{Cal, Del\}$. Moreover, in terms of types of elements, we shall have more possibilities to be considered. For instance, there may be an object which is a possible element of X in every (U, R_i), but may or may not be a positive element of X there; or, it may be a possible element of X in some (U, R_i), but a negative element of X in the rest. We give the following definitions.

Let $\mu := (U, \{R_i\}_{i \in N})$ be an MAS, and $X \subseteq U$.

Definition 2. *The* strong lower approximation $\underline{X}_{s_\mu}$, weak lower approximation $\underline{X}_{w_\mu}$, strong upper approximation $\overline{X}_{s_\mu}$, *and* weak upper approximation $\overline{X}_{w_\mu}$ *of* X, *respectively, are defined as follows.*

$$\underline{X}_{s_\mu} := \bigcap \underline{X}_{R_i}; \quad \underline{X}_{w_\mu} := \bigcup \underline{X}_{R_i}; \quad \overline{X}_{s_\mu} := \bigcap \overline{X}_{R_i}; \quad \overline{X}_{w_\mu} := \bigcup \overline{X}_{R_i}.$$

If there is no confusion, we shall omit μ as the subscript in the above definition. The relationship between the defined sets is:

$$\underline{X}_s \subseteq \underline{X}_w \subseteq X \subseteq \overline{X}_s \subseteq \overline{X}_w.$$

Note that if $\mu := (U, \{R\})$ then $\underline{X}_s = \underline{X}_w = \underline{X}_R$, and $\overline{X}_s = \overline{X}_w = \overline{X}_R$. So in the special case of a single approximation space, the weak/strong lower and upper approximations are just the standard lower and upper approximations respectively.

We thus obtain a partition of the domain U by the five disjoint sets $\underline{X}_s$, $\underline{X}_w \setminus \underline{X}_s$, $\overline{X}_s \setminus \underline{X}_w$, $\overline{X}_w \setminus \overline{X}_s$, and $(\overline{X}_w)^c$. Note that the possibility of an element $x \in U$ to belong to X on the basis of information provided by μ, reduces as we go from $\underline{X}_s$ to $(\overline{X}_w)^c$.

Definition 3. $x \in U$ *is said to be a*

certain positive element *of* X, *if* $x \in \underline{X}_s$,
possible positive element *of* X, *if* $x \in \underline{X}_w \setminus \underline{X}_s$,
certain negative element *of* X, *if* $x \in (\overline{X}_w)^c$,
possible negative element *of* X, *if* $x \in \overline{X}_w \setminus \overline{X}_s$, *and*
certain boundary element *of* X, *if* $x \in \overline{X}_s \setminus \underline{X}_w$.

The following can be shown.

Proposition 1

1. $\underline{X \cap Y}_s = \underline{X}_s \cap \underline{Y}_s$; $\overline{X \cup Y}_w = \overline{X}_w \cup \overline{Y}_w$.
2. $\overline{X \cap Y}_s \subseteq \overline{X}_s \cap \overline{Y}_s$; $\underline{X \cup Y}_w \supseteq \underline{X}_w \cup \underline{Y}_w$.
3. $\underline{X^c}_s = (\overline{X}_w)^c$; $\underline{X^c}_w = (\overline{X}_s)^c$; $\overline{X^c}_s = (\underline{X}_w)^c$; $\overline{X^c}_w = (\underline{X}_s)^c$.
4. *If* $X \subseteq Y$ *then* $\underline{X}_s \subseteq \underline{Y}_s$, $\underline{X}_w \subseteq \underline{Y}_w$, $\overline{X}_s \subseteq \overline{Y}_s$ *and* $\overline{X}_w \subseteq \overline{Y}_w$.
5. $\underline{X}_w = (\underline{X}_w)_w$; $\overline{X}_s = (\overline{X}_s)_s$; $\overline{X}_w = (\overline{X}_w)_w = (\overline{X}_s)_w$; $(\underline{X}_s)_w \subseteq \underline{X}_w$.

Remark 1. We observe from Proposition 1 (3) that $\overline{X}_w$ is the dual of $\underline{X}_s$, while $\overline{X}_s$ is the dual of $\underline{X}_w$.

3   Different Notions of Definability

Let $\mu := (U, \{R_i\}_{i \in N})$, and $X \subseteq U$.

Definition 4. X *is said to be*

lower definable *in* μ, *if* $\underline{X}_w = \underline{X}_s$,
upper definable *in* μ, *if* $\overline{X}_w = \overline{X}_s$,
strong definable *in* μ, *if* $\overline{X}_w = \underline{X}_s$, i.e. every element of U is either certain positive or certain negative,
weak definable *in* μ, *if* $\overline{X}_s = \underline{X}_w$, i.e. X does not have any certain boundary element.

It is then not difficult to obtain the following for lower and upper definable sets.

Proposition 2

1. X is lower definable in μ, if and only if $\underline{X}_{R_i} = \underline{X}_{R_j}$, for each $i, j \in N$, i.e. the sets of positive elements in all the approximation spaces of μ are identical. Similarly, X is upper definable in μ, if and only if the sets of negative elements in all the approximation spaces of μ are identical.
2. X is both lower and upper definable in μ, if and only if the sets of boundary elements in all the approximation spaces of μ are the same.
3. X is upper definable, if and only if X^c is lower definable.
4. If X and Y are upper definable then so are $\overline{X}_s, \underline{X^c}_w, \underline{X^c}_s, \overline{X}_s \cup \overline{Y}_s, \overline{X}_s \cap \overline{Y}_s$.
5. Arbitrary union (intersection) of upper (lower) definable sets is also upper (lower) definable.

Using Proposition 2 (4) and Remark 1, one would get corresponding properties for lower definable sets. We also observe that the collection of upper (lower) definable sets is not closed under intersection (union) – cf. Example 2 below.

Example 2. Let us consider an MAS $\mu := (U, \{R_1, R_2\})$ where $U := \{a, b, c, d\}$, $U/R_1 := \{\{a, c\}, \{b\}, \{d\}\}$ and $U/R_2 := \{\{a, b\}, \{c, d\}\}$. The subsets $Y_1 := \{a\}$, $Y_2 := \{c\}$ of U are lower definable, but their union, i.e. the set $Y_1 \cup Y_2 = \{a, c\}$, is not lower definable. Similarly, the subsets $Z_1 := \{a, b, d\}$, $Z_2 := \{b, c, d\}$ are upper definable, but the set $Z_1 \cap Z_2 = \{b, d\}$ is not upper definable.

For strong definable sets, we have the following.

Proposition 3

1. $\emptyset, U$ are both strong definable.
2. If $X (\subseteq U)$ is strong definable then X^c is also strong definable.
3. Arbitrary union and intersection of strong definable subsets of U are also strong definable.
4. The following are equivalent.
   (i) X is strong definable in μ.
   (ii) There is a collapse of the regions given in (*) following Definition 2:
   $$\underline{X}_s = \underline{X}_w = X = \overline{X}_s = \overline{X}_w.$$
   (iii) X is both lower and upper definable in μ, and X is definable in some approximation space of μ.
   (iv) X is definable in each approximation space of μ.

Remark 2. From this Proposition, it may be concluded that the collection of all strong definable sets forms a complete field of sets.

It is clear that, if X is definable in an approximation space of μ, it is weak definable in μ. Not much more can be said about the collection of these sets, as we see, for instance, from the following.

Example 3. Consider the MAS of Example 2. The subsets $Y_1 := \{a, b\}$ and $Y_2 := \{a, c\}$ of U are weak definable, but the set $Y_1 \cap Y_2 = \{a\}$ is not weak definable. Similarly, the subsets $Z_1 := \{b\}$, $Z_2 := \{c, d\}$ are weak definable, but the set $Z_1 \cup Z_2 = \{b, c, d\}$ is not weak definable.

4   Membership Functions and Degree of Indiscernibility

We only consider MASs with finite cardinality.

Let $\mu := (U, \{R_i\}_{1 \leq i \leq n})$ be an MAS and $X \subseteq U$. Given two objects $x, y \in U$, one may ask which one has a greater possibility to be an element of X. In other words, one may be interested in a function which will determine the relative possibilities of objects to be an element of a given set, based on the information provided by μ. A natural and simple candidate for such a membership function would be the mean of the rough membership functions corresponding to the constituent approximation spaces.

Given μ and $X \subseteq U$, the *membership function* $M_\mu^X(x) : U \to [0, 1]$ is defined as

$$M_\mu^X(x) := \frac{\sum_{i=1}^n f_i^X(x)}{n},$$

where $f_i^X(x) := \frac{|[x]_i \cap X|}{|[x]_i|}$, i.e. f_i^X is the rough membership function for X in the ith approximation space of μ.

Using $M_\mu^X(x)$, we can define lower and upper approximation of X with some arbitrary level of precision $\pi \in (1/2, 1]$ as follows [15]:

$$\underline{X}_\mu^\pi := \{x : M_\mu^X(x) \geq \pi\},, \text{ and } \overline{X}_\mu^\pi := \{x : M_\mu^X(x) > 1 - \pi\}.$$

Thus if we take $\pi = 1$, then $\underline{X}_\mu^\pi = \underline{X}_s$ and $\overline{X}_\mu^\pi = \overline{X}_w$. From this it follows that a set $X \subseteq U$ is strong definable, if and only if $\underline{X}_\mu^\pi = \overline{X}_\mu^\pi$ for $\pi = 1$.

In another approach, one can determine the relative possibility of an object of U to be an element of a subset X, by considering the number of instances when the object is a positive/negative element of X in the constituent approximation spaces of μ. We note that if an object is a positive (negative) element of X in even one approximation space, it cannot be a negative (positive) element of X in any approximation space. Thus we have three situations: $x, y \in \underline{X}_w$; $x, y \notin \underline{X}_w$; and $x \in \underline{X}_w$, but $y \notin \underline{X}_w$.

A mathematical formulation of this idea may be given by considering a membership function G_μ^X which is the mean of the three valued functions μ_i^X, $i = 1, 2, \ldots n$, defined as follows.

$$\mu_i^X(x) := \begin{cases} 1 & \text{if } x \in \underline{X}_i \\ \frac{1}{2} & \text{if } x \in \overline{X}^i \setminus \underline{X}_i \\ 0 & \text{if } x \in (\overline{X}^i)^c \end{cases}$$

i.e. $G_\mu^X(x) := \frac{\sum_{i=1}^n \mu_i^X(x)}{n}$. An easy observation leads to the following.

Proposition 4

1. $G_\mu^X(x) = 1$, if and only if $x \in \underline{X}_s$.
2. $G_\mu^X(x) = 0$, if and only if $x \in (\overline{X}_w)^c$.
3. $1/2 < G_\mu^X(x) < 1$, if and only if $x \in \underline{X}_w \setminus \underline{X}_s$.
4. $G_\mu^X(x) = 1/2$, if and only if $x \in \overline{X}_s \setminus \underline{X}_w$.
5. $0 < G_\mu^X(x) < 1/2$, if and only if $x \in \overline{X}_w \setminus \overline{X}_s$.

It is not difficult to see that G_μ^X and M_μ^X match on $\underline{X}_s$ and $(\overline{X}_w)^c$. But we may have $x \in \underline{X}_w$ and $y \in \overline{X}_s \setminus \underline{X}_w$ such that $M_\mu^X(y) > M_\mu^X(x)$, although $G_\mu^X(y) < G_\mu^X(x)$.

4.1 Degree of Indiscernibility of Objects in μ

Let us now consider indiscernibility in μ, by combining the information available from each source. An equivalence relation that would reflect this combination is the intersection of all the indiscernibility relations in μ. Let us denote the intersection by ind_μ, i.e. $ind_\mu := \cap_{i=1}^n R_i$.

We observe that

$$\underline{X}_s \subseteq \underline{X}_w \subseteq \underline{X}_{ind_\mu} \subseteq X \subseteq \overline{X}_{ind_\mu} \subseteq \overline{X}_s \subseteq \overline{X}_w.$$

It is easy to see that every weak definable set in μ is definable in the approximation space (U, ind_μ). Thus the number of definable sets in (U, ind_μ) is larger than the number of weak (or strong) definable sets in μ.

In a more general scenario, one may also wish to consider the number of approximation spaces in which these objects are indistinguishable. This motivates us to consider a fuzzy relation based on MASs.

Definition 5. *The* degree of indiscernibility *of objects* $x, y \in U$ *in* μ, *is given by a fuzzy relation* R *on* $U \times U$ *defined as:*

$$R(x, y) = \frac{|rel(x, y)|}{n}, \quad x, y \in U,$$

where $rel(x, y) := \{R_i : xR_iy\}$ *(so that* $0 \leq |rel(x, y)| \leq n$*).*

In a standard way [3], fuzzy lower and upper approximations of a set $X \subseteq U$, $\tilde{L}_\mu^X, \tilde{U}_\mu^X : U \to [0, 1]$ *may be defined.*

$$\tilde{L}_\mu^X(x) := \min\{1 - R(x, y) : y \notin X\}, \text{ and}$$
$$\tilde{U}_\mu^X(x) := \max\{R(x, y) : y \in X\}.$$

R is clearly a (fuzzy) reflexive and symmetric relation.

Proposition 5 (Properties of $\tilde{L}_\mu^X$, $\tilde{U}_\mu^X$)

1. $x \in \underline{X}_{ind_\mu}$, *if and only if* $\tilde{L}_\mu^X(x) > 0$.
2. $x \in \overline{X}_{ind_\mu}$, *if and only if* $\tilde{U}_\mu^X(x) = 1$.
3. $\tilde{L}_\mu^X(x) = 0 \Rightarrow x \notin \underline{X}_w$; $\tilde{L}_\mu^X(x) = 1 \Leftrightarrow x \in \underline{X}_s$.
4. $\tilde{U}_\mu^X(x) = 1 \Rightarrow x \in \overline{X}_s$; $\tilde{U}_\mu^X(x) = 0 \Leftrightarrow x \notin \overline{X}_w$.
5. $\tilde{L}_\mu^X(x) = 0$ *and* $\tilde{U}_\mu^X(x) = 1 \Rightarrow x \in \overline{X}_s \setminus \underline{X}_w$.
6. $\tilde{L}_\mu^{X \cup Y}(x) \geq max\{\tilde{L}_\mu^X(x), \tilde{L}_\mu^Y(x)\}$; $\tilde{U}_\mu^{X \cup Y}(x) = max\{\tilde{U}_\mu^X(x), \tilde{U}_\mu^Y(x)\}$.
7. $\tilde{L}_\mu^{X \cap Y}(x) = min\{\tilde{L}_\mu^X(x), \tilde{L}_\mu^Y(x)\}$; $\tilde{U}_\mu^{X \cap Y}(x) \leq min\{\tilde{U}_\mu^X(x), \tilde{U}_\mu^Y(x)\}$.
8. $\tilde{L}_\mu^{X^c}(x) = 1 - \tilde{U}_\mu^X(x)$; $\tilde{U}_\mu^{X^c}(x) = 1 - \tilde{L}_\mu^X(x)$.
9. *If* $X \subseteq Y$, *then* $\tilde{L}_\mu^X \leq \tilde{L}_\mu^Y$ *and* $\tilde{U}_\mu^X \leq \tilde{U}_\mu^Y$.

4.2   Another Membership Function for μ

We now consider the rough membership function with respect to the indiscernibility ind_μ, i.e. $f^X_{ind_\mu} = \frac{|[x]_{ind_\mu} \cap X|}{|[x]_{ind_\mu}|}$.

Proposition 6

1. If $x \in \underline{X}_w$ then $f^X_{ind_\mu}(x) = 1$.
2. If $x \notin \overline{X}_s$ then $f^X_{ind_\mu}(x) = 0$.
3. If $x \in \overline{X}_s$ then $0 \le f^X_{ind_\mu}(x) \le 1$.
4. If x, y are such that at least one of them is not in $\underline{X}_w$,
   $f^X_{ind_\mu}(x) \ge f^X_{ind_\mu}(y) \Rightarrow G^X_\mu(x) \ge G^X_\mu(y)$.
   The same holds if at least one of them is not in $(\overline{X}_s)^c$.
5. If x, y are such that at least one of them is not in $\overline{X}_s \setminus \underline{X}_w$,
   $G^X_\mu(x) \ge G^X_\mu(y) \Rightarrow f^X_{ind_\mu}(x) \ge f^X_{ind_\mu}(y)$.

We note that for $x, y \in \underline{X}_w$, or $x, y \in (\overline{X}_s)^c$, we have $f^X_{ind_\mu}(x) = f^X_{ind_\mu}(y) = 1$ but $G^X_\mu(x) >=< G^X_\mu(y)$. Similarly, for $x, y \in \overline{X}_s \setminus \underline{X}_w$, we have $G^X_\mu(x) = G^X_\mu(y)$ but $f^X_{ind_\mu}(x) >=< f^X_{ind_\mu}(y)$. Also note that for $x \in \underline{X}_w \setminus X_w$ and $y \in (\overline{X}_s)^c$, we always have $G^X_\mu(x) > G^X_\mu(y)$ and $f^X_{ind_\mu}(x) > f^X_{ind_\mu}(y)$ but we could have $M^X_\mu(y) > M^X_\mu(x)$.

5   Conclusions

Some properties of different kinds of definable sets in an MAS μ over a domain U are studied. Two membership functions M^X_μ, G^X_μ for μ are defined, following two different approaches. The intersection ind_μ of all constituent indiscernibilities of μ is considered, to reflect the combined information from all the approximation spaces of μ. (U, ind_μ) is finer than the individual approximation spaces of μ, and points to other possibilities for defining composite indiscernibilities. Following this line of thought, a fuzzy relation is presented which represents the 'degree of indiscernibility' of two objects in μ.

   To define M^X_μ, we have used the rough membership functions of the constituent approximation spaces of μ. But there are limitations in this approach, and one may adopt alternatives. For example, the (probabilistic) membership function [3] P^X of the constituent approximation spaces $P^X(x) = \frac{Pr(X \cap [x])}{Pr([x])}$ may be used, where Pr is a probability distribution on the domain U. In some applications, a 'preference' may need to be accorded to information provided by one source over that by another. In such cases, one may define membership functions by considering weighted mean.

   In this paper, we have worked with equivalence relations. However, the general notion of dynamic spaces [4] would be more appropriate for many applications. For example, in [2], a conditional probabilistic relation is regarded to be more suitable in representing the relationship between two elements of data.

For generalized approximation spaces, there has been much work on exploring membership functions [12,6,15,14,2,11]. Depending on the collection of relations and the desired application, a study of appropriate membership functions and composite indiscernibilities for different dynamic spaces (generalized MASs) may be a worthwhile pursuit.

References

1. Banerjee, M., Khan, M.A.: Propositional Logics from Rough Set Theory. Transactions on Rough Sets VI, 1–25 (2007)
2. Intan, R., Mukaidono, M.: Generalization of Rough Membership Function Based on $\alpha-$Coverings of the Universe. In: Pal, N.R., Sugeno, M. (eds.) AFSS 2002. LNCS (LNAI), vol. 2275, pp. 129–135. Springer, Heidelberg (2002)
3. Liau, C.J.: An Overview of Rough Set Semantics for Modal and Quantifier Logics. Int. J. Uncertainty, Fuzziness and Knowledge Based Systems 8(1), 93–118 (2000)
4. Pagliani, P.: Pretopologies and Dynamic Spaces. Fundamenta Informaticae 59(2-3), 221–239 (2004)
5. Pawlak, Z.: Rough Sets. Int. J. Comp. Inf. Sci. 11(5), 341–356 (1982)
6. Pawlak, Z., Wong, S.K.M., Ziarko, W.: Rough sets: Probabilistic Versus Deterministic Approach. Int. J. Man-Machine Studies 29, 81–95 (1988)
7. Pawlak, Z., Skowron, A.: Rough Membership Functions. In: Yager, R., Fedrizzi, M., Kacprzyk, J. (eds.) Advances in the Dempster-Shafer Theory of Evidence, pp. 251–271. Wiley, New York (1994)
8. Pawlak, Z.: Rough Sets. Theoretical Aspects of Reasoning about Data. Kluwer Academic Publishers, Dordrecht (1991)
9. Rasiowa, R.: Mechanical Proof Systems for Logic of Reaching Consensus by Groups of Intelligent Agents. Int. J. Approximate Reasoning. 5(4), 415–432 (1991)
10. Rauszer, C.: Knowledge Representation Systems for Groups of Agents. In: Woleński, J. (ed.) Philosophical Logic in Poland, pp. 217–238. Kluwer, Dordrecht (1994)
11. Skowron, A., Stepaniuk, J.: Tolerance Approximation Spaces. Fundamenta Informaticae 27, 245–253 (1996)
12. Wong, S.K.M., Ziarko, W.: Comparison of the Probabilistic Approximate Classification and Fuzzy Set Model. Fuzzy Sets and Systems 21, 357–362 (1986)
13. Wong, S.K.M.: A Rough Set Model for Reasoning about Knowledge. In: Polkowski, L., Skowron, A. (eds.) Rough Sets in Knowledge Discovery 1: Methodology and Applications, pp. 276–285. Physica-Verlag, New York (1998)
14. Yao, Y.Y., Wong, S.K.M., Lin, T.Y.: A Review of Rough Set Models. In: Lin, T.Y., Cercone, N. (eds.) Rough Sets and Data Mining, pp. 47–75. Kluwer Academic Publishers, Boston (1997)
15. Ziarko, W.: Variable Precision Rough Set Model. J. Computer and System Sciences 46, 39–59 (1993)

Stability Analysis on Rough Set Based Feature Evaluation

Qinghua Hu, Jinfu Liu, and Daren Yu

Harbin Institute of Technology, Harbin 150001, China
huqinghua@hcms.hit.edu.cn
http://huqinghua.tooboo.com.cn

Abstract. Rough sets are widely used in feature evaluation and attribute reduction and a number of rough set based evaluation functions and search algorithms were reported. However, little attention has been paid to compute and compare stability of feature evaluation functions. In this work, we introduce three coefficients to calculate the stabilities of feature significance via perturbing samples. Experimental results show that entropy and fuzzy entropy based evaluation functions are more stable than the others and fuzzy rough set based functions are stable compared with the crisp functions. These results give a guideline to select feature evaluation for different applications.

1 Introduction

As the capability of gathering data increases, we are usually confronted data with hundreds, even ten thousands of features and millions of samples nowadays. It is believed that learning algorithms will be confused by the superfluous irrelevant features. Feature selection as a common preprocessing step for pattern recognition and machine learning is attracting much attention from theoretical and application domains [1,2,5,6,7].

Rough set theory, first proposed by Pawlak, is proven to be a powerful tool for dealing with imperfect information [3]. Feature evaluation and attribute reduction are considered as the most important application of rough sets. A number of algorithms were developed to evaluate features and search reducts based on rough sets. An early review about rough set based feature selection and attribute reduction was given in [9]. The algorithms discussed in this literature were developed based on Pawlak's rough sets, which take equivalence relations as the foundation of the model. However, the databases in real-world applications usually come with heterogeneous features. Thus fuzzy rough sets, covering based rough sets, neighborhood rough set and Tolerance-based approaches were introduced and some new algorithms based on these models were proposed [6,10,13,14]. A number of feature evaluating functions were developed based on rough sets. [4,6,8,11,12,15,16].

Although a number of coefficients for feature evaluation were proposed, little work has been devoted to compare them in practical situations. In 1996, Choubey

G. Wang et al. (Eds.): RSKT 2008, LNAI 5009, pp. 88–96, 2008.

and Deogun et al compared the classification accuracies and feature subsets between different evaluation coefficient. Their work was limited in Pawlak's rough set model; yet, only accuracy is used in comparing. Obviously, there are different viewpoints to compare them, such as numbers of selected features, complexity of the model learned from the reduced data, computational complexity and stability [19], where stability is an important measure to analyze the properties of feature evaluation functions. On one side, one requires a stable function for getting consistent results. On the other side, one needs an instable function to get diverse subsets of features for ensemble learning [18,19,20].

The stability of classification algorithms attracted much attention; However, little attention has been paid to the stability of feature evaluation and selection. Kalousis et al designed some measures to estimate the stability of feature evaluation functions in [17]. They considered the correlation or similarity degrees between different experiments via bootstrapping and obtained a relation matrix, where the element is the correlation coefficient or similarity of the ith and jth estimates. Then the grand average of the matrix is taken as the final measure of stability. Intuitively Correlation should be high if the evaluation function is stable because it is not sensitive to the variation of samples. We name this as sample-stability. In this work, we will introduce these measures to discuss the property of rough set based feature evaluation. Moreover, we introduce an information function to compute the stability of relation matrix [21].

2   Stability Coefficients of Feature Evaluation

Generally speaking, classification learning can be understood as a task of building a predictive model $f : C \rightarrow d$ from a set of samples which can be formulated as $< U, C, d >$, where U is the set of samples $\{x_1, x_2, \cdots, x_n\}$; C is the set of condition attributes $\{a_1, a_2, \cdots, a_m\}$ and d is the decision of samples. Usually, the elements in C are not equivalently important for constructing the predictive model f. Some of features are even irrelevant and should be eliminated from C. Therefore, an evaluation function is introduced to estimate the relevance between conditions and decision. With the evaluation function, a weight vector $\{w_1, w_2, \cdots, w_m\}$ can be associated with the attributes.

As to the same learning task, we examine the variation of weight vectors if the samples or the parameters in learning models are perturbed. We here require a measure to calculate the variation. Pearson's correlation coefficient can be introduced. Given two estimates of weight vectors $W = \{w_1, w_2, \cdots, w_m\}$ and $W' = \{w'_1, w'_2, \cdots, w'_m\}$, correlation coefficient is computed as

$$S_w(W, W') = \frac{\sum\limits_{i=1}^{m} (w_i - \overline{w_i})(w'_i - \overline{w'_i})}{\sqrt{\sum\limits_{i=1}^{m} (w_i - \overline{w_i})^2 \sum\limits_{i=1}^{m} (w'_i - \overline{w'_i})^2}}.$$

If there are k estimates of feature vector, we get a k×k matrix $M = (m_{ij})_{k \times k}$, where element m_{ij} is the correlation coefficient of the ith and jth estimates.

In some cases, one may care the ranking of the feature weights, rather than the weights themselves. Given two vectors of weights $W = \{w_1, w_2, \cdots, w_m\}$ and $W' = \{w'_1, w'_2, \cdots, w'_m\}$, two rankings $R = (r_1, r_2, \cdots, r_m)$ and $R' = (r'_1, r'_2, \cdots, r'_m)$ can be obtained via ranking the features with the weights in ascending order or descending order, where r_i and r'_i are the orders of feature a_i. To measure correlation between two rankings, we use Spearman's rank correlation coefficient:

$$S_R(R, R') = 1 - 6 \sum_{i=1}^{m} \frac{(r_i - r'_i)^2}{m(m^2 - 1)}.$$

Value domain of $S_R(R, R')$ is [-1, 1]. $S_R(R, R') = 1$ means that the two rankings are identical; $S_R(R, R') = 0$ means that there is no correlation between these rankings, while $S_R(R, R') = -1$ means the rankings have exactly inverse orders.

Generally speaking, the output of a feature selection algorithm is a subset of features. Using an unstable algorithm the selected subset may be different if we perturb the samples or model parameters. Assumed $B = \{a_1, a_2, \cdots\}$ and $B' = \{a'_1, a'_2, \cdots\}$ are two subsets of features, the similarity between these subsets can be calculated with a straightforward adaptation of Tanimoto distance:

$$S_S(B, B') = 1 - \frac{|B| + |B'| - 2|B \cap B'|}{|B| + |B'| - |B \cap B'|} = \frac{|B \cap B'|}{|B| + |B'| - |B \cap B'|},$$

where $|\bullet|$ means the cardinality of a set. S_S reflects the overlap of two sets. $S_S \in [0, 1]$. $S_S(B, B') = 1$ if $B = B'$, while $S_S(B, B') = 0$ if $B \cap B' = \emptyset$. As to a stable feature selection algorithm, the selected features may be similar even if the learning samples or model parameters are different. In this situation, a great value of S_S is obtained.

A resampling technique, such as bootstrapping, can be introduced for estimating multiple weight vectors. Bootstrapping is an effective method to produce diversity in constructing multiple classifier systems. This approach randomly selects n patterns with replacement from the original set of n patterns. Another well-known resampling method is k-fold cross-validation. By dividing data set randomly into k disjoint subsets, new overlapping training sets can be created for each estimate, by leaving out one of these k subsets and training on the remainder. K-fold cross validation (CV) is widely applied in estimating classification performance of a learning system. Usually k is specified with 3, 5 or 10. In this work, we are going to select 10-fold CV to estimate the stability of a feature evaluation algorithm. And we can obtain 10 estimates of significance of feature vector.

A 10 × 10 matrix is produced by computing the correlation coefficient or similarity of each pair of estimates. We denote it by $(S_{ij})_{10 \times 10}$. Obviously, we have $S_{ii} = 1$ and $S_{ij} = S_{ji}$. Moreover, $S_{ij} = 1$ if the estimated weights or the selected features are all the same. We consider that the evaluation function is stable in this case. $S_{ij} = 0$ $(i \neq j)$ if the estimated weights are irrelevant or the selected features are disjoint. Correspondingly, the evaluation function is unstable. We need to estimate the stability of feature evaluation function with the matrix $(S_{ij})_{10\times 10}$. It is notable that S_{ij} usually takes values in unit interval $[0, 1]$ in practice. Therefore, we can consider the matrix as a correlation coefficient based relation matrix. Yu, Hu and Wu discussed the uncertainty measure of this kind of matrix, respectively [21]. Here we introduce the fuzzy entropy to compute the overall similarity represented in the matrix as the measure of stability.

Given a fuzzy relation matrix $(S_{ij})_{k\times k}$, the uncertainty is computed as

$$H(S) = -\sum_{i=1}^{k} \frac{1}{k} \log |S_i|, \text{where } |S_i| = \sum_{l=1}^{k} S_{ik}.$$

$H(S) \in [0, 1]$, $H(S) \le H(S')$ if $S \supseteq S'$, where $S \supseteq S'$ means that $\forall ij$, $S_{ij} \ge S'_{ij}$. Furthermore we have $H(S) = \log k$ if $S_{ij} = 0, i \ne j$ and $H(S) = 0$ if $\forall ij, S_{ij} = 1$. In fact, $H(S)$ can be understood as a measure of instability in this context because a great value of $H(S)$ means an instable evaluation function.

3   Rough Set Based Feature Evaluation Functions

Given a decision table $< U, C, d >$, an optimal subset $B \subseteq C$, called a reduct, should satisfy that 1) $POS_B(D) = POS_C(D)$; 2) $\forall a \in B$, $POS_B(D) \supset POS_{B-a}(D)$. The first term guarantees that the selected features B produce the same decision positive region as the original features. The second condition shows features in the reduct is indispensable. In this work, we discuss seven evaluation functions: dependency, information entropy, consistency, neighborhood dependency (ND), fuzzy entropy (FE), Gaussian kernel approximation based fuzzy dependency (GKA) and neighborhood consistency (NC).

Given a decision table $< U, C, d >$, $C = \{a_1, a_2, \cdots, a_m\}$, the dependency of decision d to a_i is defined as $r_{a_i}(d) = \frac{|POS_{a_i}(d)|}{|U|}$, $i = 1, \cdots, m$, where $POS_{a_i}(d) = \{x | x \in U, [x]_{a_i} \subseteq [x]_d\}$.

Assumed that B is a subset of features, and $U/B = \{X_1, X_2, \cdots, X_l\}$ and $U/d = \{Y_1, Y_2, \cdots, Y_l\}$ are two families of equivalent classes induced by B and d, if we view $P(X_i) = |X_i|/|U|$ as a probability measure, then the entropy of the system can be computed by $H(B) = -\sum_i p(X_i) \log p(X_i)$. The conditional entropy, is calculated with $H(B|d) = H(B \cup d) - H(B)$. In terms of information entropy, the significance of attribute a is defined as $MI(B, d)$. It is easy to show $MI(B, d) = H(d) + H(B) - H(B \cup d)$. Given a set of features $C = \{a_1, a_2, \cdots, a_m\}$, the significance of features can be written as $MI(a_i, d) = H(d) + H(a_i) - H(a_i \cup d)$.

Hu et al. [8] introduced consistency to measure the significance of features. Consistency is the percentage of samples which can be correctly classified according to majority decision rule. As we know, the equivalent classes with samples from the same label belong to the decision positive region, while equivalent classes with samples from different decision labels constitute the decision boundary region. In computing dependency, the samples in the boundary region are not taken into consideration. However, not all the samples in the boundary region will be misclassified in practice. According to majority decision rule, the equivalent classes with samples from different decision labels will be assigned with the majority class. Correspondingly, the samples with the minority classes is misclassified in this case. Slezak first captured this idea and named the percentage of samples of majority classes as attribute quality measure [16]

Pawlak's rough sets, information entropy and consistency take equivalence relationes and equivalent classes as the foundation of the methods. These ideas work when the available features are categorical. Now three approaches were developed to deal with numerical features.

First, Neighborhood relationes and neighborhood rough sets were introduced to deal with numerical features, where we take the neighborhood information granules, to approximate the decision. Formally, the neighborhood granule around sample x is the subset: $N(x_i) = \{x|\|x - x_i\| \leq \delta, x \in U\}$, where $\| \bullet \|$ stands for some distance function. In neighborhood model, the decision positive region is defined as $POS_B(d) = \{x|x \in U, N(x) \subseteq [x]_d\}$, where $N(x)$ is the neighborhood granule in feature space B and $[x]_d$ is the set of samples with the same decision as x. The neighborhood dependency of d to numerical features B is defined as $|POS_B(d)|/|U|$. The neighborhood model is a natural generalization of Pawlak's model and it will degrade to Pawlak's model if $\delta = 0$.

Like Pawlak's rough sets, there is a similar problem with the function of neighborhood dependency. As mentioned above, not all the samples in the boundary region are misclassified; only the samples in the minority classes are mislabeled according to majority decision rule. Therefore, neighborhood consistency was introduced to evaluate the numerical features.

Finally, fuzzy set theory is integrated with rough sets in dealing with numerical information. Given a decision table $< U, C, d >$, R is a fuzzy relation over U and X is a fuzzy subset of U. The lower approximation and upper approximation is denoted by $\underline{R}X$ and $\overline{R}X$. $\forall x \in U$, are given by

$$\begin{cases} \mu_{\underline{R}X}(x) = \wedge\{\mu_X(y) \vee (1 - R(x, y)) : y \in U\}, x \in U \\ \mu_{\underline{R}X}(x) = \vee\{(\mu_X(y) \wedge R(x, y) : y \in U\}, x \in U \end{cases}$$

Jensen and Shen extended the dependency in Pawlak's model into the fuzzy case and proposed the definition of fuzzy dependency in [6].

How to extract the fuzzy relation from the data is one of the key problems in a fuzzy rough set based feature evaluation. Chen, Hu et al. compute the similarity degree between two samples with Gaussian kernel [22]. They found that the membership of sample $x \in d_t$ to the decision positive region is determined by the nearest sample u with a different class. The significance of B computed with Gaussian kernel approximation (**GKA**) is

$$r = \sum_{x \in U} \inf_{u \notin d_t} \sqrt{1 - R(x, u)^2}/|U|.$$

The similarity relation extracted with a fuzzy similarity function can be written as a fuzzy matrix $M(R) = (r_{ij})_{n \times n}$, where r_{ij} is the similarity degree between samples x_i and x_j. The uncertainty quantity of the fuzzy relation is defined in [12,21], and attribute reduction was constructed based on the proposed fuzzy entropy.

With the evaluation functions, we can construct various feature selection algorithms by using different search strategies. We here select the greedy search strategy in comparing the stability of feature selection for efficiency.

4   Experimental Analysis

To empirically estimate the stability of feature evaluation functions, 12 databases are gathered from UCI Machine Learning Repository. There are only categorical features in databases soybean and zoo, and only numerical features in databases iono, sonar, wdbc, wine and wpbc, while anneal, credit, heart hepatitis and horse have both categorical and numerical features. In computing the significance based on dependency, entropy and consistency, the numerical features should be discretized. Entropy-MDL discretization algorithm is used in preprocessing. Moreover, numerical features are standardized into [0, 1].In the below, the parameters used in computing neighborhood relation in neighborhood rough sets, the kernel parameter used in Gaussian kernel approximation is 0.1.

Table 1 shows the entropy value computed on different data with the seven evaluation functions, where ND for neighborhood dependency, FE for fuzzy entropy, GKA for Gaussian kernel fuzzy approximation, NC for neighborhood consistency, while D, E, C for dependency, entropy and consistency, respectively. As discussed above that the value of entropy reflects the instability of the evaluation functions. The greater the entropy is, the more the function is stable. Among these evaluation functions, entropy and fuzzy entropy produce the least value of entropy. This shows that entropy and fuzzy entropy is more robust to sample perturbation than other functions. By contraries, neighborhood consistency and consistency are instable. It is notable that dependency gets some NaN values in table 1 for data credit, heart, and sonar. It results from the zero weight for all

Table 1. Stability of feature evaluating indexes

Data	ND	FE	GKAD	NC	D	E	C
anneal	0.0001	0.0002	0.0001	0.0031	0.0001	0.0002	0.0019
credit	0.1285	0.0041	0.0208	0.5728	NaN	0.0046	0.6235
heart	0.1888	0.0310	0.0345	0.2927	NaN	0.0232	0.1694
hepatitis	0.1052	0.0165	0.0367	0.0330	0.0029	0.0171	0.0095
horse	0.1319	0.0060	0.0160	0.0415	0.0696	0.0027	0.0253
iono	0.1385	0.0389	0.1056	0.2918	0.0623	0.0210	0.1368
sonar	0.3022	0.0704	0.1910	1.0195	NaN	0.0208	0.4477
wdbc	0.0196	0.0018	0.0084	0.0707	0.1445	0.0017	0.1510
wine	0.0511	0.0079	0.0260	0.7752	0.0688	0.0066	0.4406
wpbc	0.0918	0.0920	0.0695	0.8214	0.5737	0.1260	0.7144
soybean	0.0290	0.0009	0.0290	0.1909	0.0290	0.0009	0.1909
zoo	0.0002	0.0054	0.0002	0.0637	0.0002	0.0054	0.0637
Average	**0.0989**	**0.0229**	**0.0448**	**0.3480**	**–**	**0.0192**	**0.2479**

Table 2. Stability of feature ranking index

Data	ND	FE	GKAD	NC	D	E	C
anneal	0.0222	0.0122	0.0190	0.0027	0.0222	0.0124	0.0009
credit	0.1013	0.0358	0.0599	0.6843	0.0141	0.0257	0.7221
heart	0.0918	0.0341	0.0326	0.4181	0.0549	0.0398	0.2992
hepatitis	0.0476	0.0460	0.0034	0.0126	0.0012	0.0438	0.0044
horse	0.0622	0.0099	0.0044	0.0536	0.0655	0.0082	0.0453
iono	0.1338	0.0819	0.1433	0.3574	0.0796	0.0377	0.1556
sonar	0.1958	0.1774	0.2230	1.0659	0.0351	0.0111	0.2739
wdbc	0.0842	0.0076	0.0165	0.1128	0.0693	0.0035	0.2909
wine	0.1533	0.0111	0.0347	0.7890	0.0549	0.0399	0.5293
wpbc	0.3127	0.1291	0.2254	0.5338	0.2352	0.2587	0.2831
soybean	0.0352	0.0052	0.0352	0.3378	0.0352	0.0052	0.3378
zoo	0	0.0077	0	0.0984	0	0.0077	0.0984
Average	**0.1033**	**0.0465**	**0.0664**	**0.3722**	**0.0556**	**0.0411**	**0.2534**

the single features. In this case, the correlation coefficient can not be computed because the denominator is zero. Therefore the corresponding value of entropy is also meaningless. As to other data, such as horse, iono, wpbc, etc, dependency is comparable with consistency. Gaussian kernel based fuzzy approximation is more stable than neighborhood dependency and neighborhood consistency with respect to data credit, heart, horse, iono, sonar, wdbc and wpbc. We rank the evaluation functions in terms of stability of weights: entropy > fuzzy entropy > GKA > neighborhood dependency > consistency > neighborhood consistency.

Sometimes users require the ranking of features for comparing the relatively importance. Raking is very important in feature ranking based selection. The variation of weights does not necessarily cause the variation of ranking. So, the instable evaluation functions of feature weights may produce stable ranking. Table 2 presents the stability of feature ranking. Among ND, FE, GKA and NC, FE and GKA are stable compared with ND and NC. E is the most stable among D, E and C. Comparing table 1 and table 2, we can get that the stability of feature ranking has the same rule as the stability of feature weight. The evaluation function with the great stability in weight estimation also gets the great stability in feature ranking.

Now let's discuss the stability of the selected features. Here the features are selected with the greedy search strategy. By perturbing the samples, we get 10 subsets of features. The similarity between two subsets is computed with S_S, and the overall similarity is the entropy of the similarity matrix. The value of zero in table 3 means that the selected features in 10 reducts are identical. If there are two empty sets of features, the similarity degree in this situation can not be computed. Then we get the entropy of stability is NaN.

ND, GKA, NC, D and C select the same features in 10 reductions as to data anneal, while D gets empty sets of features in some reducts for credit, heart and horse. Moreover, we can also get the conclusions that: as to stability of

Table 3. Stability of feature subsets

Data	ND	FE	GKAD	NC	D	E	C
anneal	0	0.3103	0	0	0	0.3103	0
credit	0.0984	0.0410	0.0790	0.0762	NaN	0.0644	0.3924
heart	0.2292	0.2876	0.0528	0.3413	NaN	0	0.0745
hepatitis	0.4898	0.1529	0.1327	1.1623	0.6473	0.8489	0.9242
horse	0.6169	0.4089	0.7945	0.9174	0.3981	0.2842	0.2842
iono	1.1287	1.0856	1.0665	1.6040	0.5572	1.0486	1.0699
sonar	2.2376	0.8374	1.6681	1.9947	NaN	0.5896	0.7410
wdbc	0.9794	0.3667	0.4874	1.2733	0.5044	0.7054	1.4939
wine	0.5973	0.9701	0.7139	0.8800	0.1174	0.3557	0.2336
wpbc	1.1481	1.1728	0.9656	1.2753	1.4681	1.2950	1.2790
soybean	0.3567	0.1696	0.3567	0.1539	0.3567	0.1696	0.3567
zoo	0.2826	0.5135	0.2826	0.0549	0.2826	0.5135	0.2826
Average	**0.6804**	**0.5264**	**0.5500**	**0.8111**	–	**0.5154**	**0.5943**

selected features, entropy > fuzzy entropy > GKA > neighborhood dependency > consistency > neighborhood consistency.

5   Conclusion

Although the stability of classification learning algorithms is widely discussed and used in comparing and selecting classification algorithms, little work has been devoted to analyzing the stability of feature evaluation and selection. Rough set theory is a hot topic in feature selection and attribute reduction these years. A number of feature evaluation functions have been proposed based on rough set methodology. In this work, we introduce three measures to calculate and compare the stability of rough set based feature evaluation. We empirically study the influence of sample perturbation on the stability and find entropy and fuzzy entropy based evaluation functions are more stable than consistency, neighborhood dependency, and neighborhood consistency functions, while Gaussian kernel approximation based fuzzy rough sets is similar to entropy functions.

Acknowledgement

This work is supported by Natural Science Foundation of China under Grant 60703013, Development Program for Outstanding Young Teachers in Harbin Institute of Technology under Grant HITQNJS.2007.017.

References

1. Liu, H., Yu, L.: Toward integrating feature selection algorithms for classification and clustering. IEEE TKDE 17(4), 491–502 (2005)
2. Guyon, I., Elisseeff, A.: An introduction to variable and feature selection. Journal of machine learning research 3, 1157–1182 (2003)

3. Pawlak, Z.: Rough sets: theoretical aspects of reasoning about data. Kluwer Academic Publishers, Dordrecht (1991)
4. Hu, X., Cercone, N.: Learning in relational databases: a rough set approach. Computational Intelligence 11, 323–338 (1995)
5. Wang, J.: Reduction algorithms based on discernibility matrix: The ordered attributes method. Journal of computer science and technology 16, 489–504 (2001)
6. Jensen, R., Shen, Q.: Fuzzy-rough sets assisted attribute selection. IEEE Transactions on fuzzy systems 15, 73–89 (2007)
7. Zhao, M., Han, S.Q., Wang, J.: Tree expressions for information systems. Journal of computer science and technology 22, 297–307 (2007)
8. Hu, Q.H., Zhao, H., Xie, Z.X., Yu, D.R.: Consistency based attribute reduction. In: Zhou, Z.-H., Li, H., Yang, Q. (eds.) PAKDD 2007. LNCS (LNAI), vol. 4426, pp. 96–107. Springer, Heidelberg (2007)
9. Swiniarski, R.W., Skowron, A.: Rough set methods in feature selection and recognition. Pattern recognition letters 24, 833–849 (2003)
10. Yeung, D.S., Chen, D.G., Tsang, E.C.C., et al.: On the generalization of fuzzy rough sets. IEEE Trans. Fuzzy Systems. 13(3), 343–361 (2005)
11. Hu, Q.H., Yu, D.R., Xie, Z.X.: Neighborhood classifier. Expert Systems with Applications 34, 866–876 (2008)
12. Hu, Q.H., Yu, D.R., Xie, Z.X.: Information-preserving hybrid data reduction based on fuzzy-rough techniques. Pattern recognition letters 27, 414–423 (2006)
13. Bhatt, R.B., Gopal, M.: On the extension of functional dependency degree from crisp to fuzzy partitions. Pattern recognition letters 27, 487–491 (2006)
14. Hu, Q.H., Xie, Z.X., et al.: Hybrid attribute reduction based on a novel fuzzy-rough model and information granulation. Pattern Recognition 40, 3509–3521 (2007)
15. Slezak, D.: Approximate Entropy Reducts. Fundam. Inform. 53, 365–390 (2002)
16. Slezak, D.: Normalized Decision Functions and Measures for Inconsistent Decision Tables Analysis. Fundam. Inform. 44(3), 291–319 (2000)
17. Kalousis, A., et al.: Stability of feature selection algorithms: a study on high-dimensional spaces. Knowledge and information systems 12, 95–116 (2007)
18. Tang, E.K., Suganthan, P.N., Yao, X.: An analysis of diversity measures. Machine Learning 65, 247–271 (2006)
19. Brown, G., Wyatt, J., Harris, R., Yao, X.: Diversity creation methods: a survey and categorization. Information Fusion 6, 5–20 (2005)
20. Hu, Q., Yu, D., Xie, Z., Li, X.: EROS: ensemble rough subspaces. Pattern recognition 40, 3728–3739 (2007)
21. Yu, D.R., Hu, Q.H., Wu, C.X.: Uncertainty measures on fuzzy relations and their applications. Applied soft computing 7, 1135–1143 (2007)
22. Chen, D.G., Hu, Q.H., Wang, X.Z.: A novel feature selection method based on fuzzy rough sets for Gaussian kernel SVM. Neurocomputing (submitted, 2007)

Apply a Rough Set-Based Classifier to Dependency Parsing

Yangsheng Ji, Lin Shang, Xinyu Dai, and Ruoce Ma

State Key Laboratory for Novel Software Technology, Nanjing University, China
jiyangsheng@ai.nju.edu.cn, shanglin@nju.edu.cn, {dxy,marc}@nlp.nju.edu.cn

Abstract. A rough set-based semi-naive Bayesian classification method is applied to dependency parsing, which is an important task in syntactic structure analysis of natural language processing. Many parsing algorithms have emerged combined with statistical machine learning techniques. The rough set-based classifier is embedded with Nivre's deterministic parsing algorithm to conduct dependency parsing task on a Chinese corpus. Experimental results show that the method has a good performance on dependency parsing task. Moreover, the experiments have justified the effectiveness of the classification influence.

Keywords: Rough set, Attribute dependency, Semi-naive Bayesian classifier, Dependency parsing.

1 Introduction

In natural language processing, dependency parsing is an important task as classification in data mining, and interest in dependency parsing has been increasing fast in recent years. Many syntactic parsing methods have been proposed, among which syntactic representation, parsing algorithms and machine learning are three mainstream dependency parsing techniques. Machine learning techniques and parsing algorithms are often combined together to carry out the parsing task. Nivre's memory based parsing algorithm [1] is such a kind of method. As Yuret [2] says, dependency parsing can be regarded as a classification task. So in this paper, the approach is designed to perform Nivres dependency parsing algorithm [1] utilizing a rough set-based semi-naive Bayesian classifier [3]. Experimental results show that the proposed method achieves a comparable performance on the Chinese corpus.

In data mining, classification is an important task, and many classification techniques have been developed, among which Naive-Bayesian classifier and decision tree have come into wide use because of their good performance. Naive-Bayesian classification is a simple, computationally efficient and accurate classifier. Its characteristics have been shown in many contributions [4]. Naive-Bayesian classification is based on Bayes theorem and an assumption that all attributes are mutually independent within each class. This classifier predicts that a test example A with a set of attribute values $A = \langle a_1, a_2, \ldots, a_n \rangle$ belongs to the class C_i, which maximizes the posterior probability $P(C_i|A)$

G. Wang et al. (Eds.): RSKT 2008, LNAI 5009, pp. 97–105, 2008.

$= \frac{P(C_i) \prod_j P(a_j|C_i)}{P(A)}$, where $P(C_i)$ is the prior probability of class C_i, and $P(a_j|C_i)$ is the conditional probability of examples with attribute value a_j occurs in class C_i. Although Naive Bayesian classifier has been proved to be the optimal classifier [5] in many domains, we cannot make sure that Naive Bayesian classifier performs better than others when the attribute dependence violates the mutual independent assumption. Thus, many kinds of semi-Naive Bayesian classifiers were developed to alleviate the strict assumption, e.g., TAN [6], LBR [7], AODE [8], NBTree [9] are among the well-known semi-Naive Bayesian classifiers. Decision tree learning is another classical classification technique. It is also very fast and comprehensible. Conducting the attribute selection is the main part of the inducing process at the current tree node. The process of inducing a decision tree classifier often uses the top down algorithms which follow the divide and conquer strategy [10,11]. Along with the top down process, replication and fragmentation [12] may decrease decision tree learner's performance.

Thus, we design an approach to perform Nivres dependency parsing algorithm [1] utilizing a rough set-based semi-naive Bayesian classifier [3]. Experimental results illustrate that the proposed method achieves a comparable performance on the Chinese corpus.

This paper is organized as follows: section 2 describes the related works on dependency parsing in natural language processing and rough sets dependence measure; section 3 describes the rough set-based classifier algorithm; section 4 illustrates the performance of Nivre's parsing algorithm embedded with the rough set-based classifier on a Chinese corpus; section 5 makes a conclusion and discusses the future work.

2   Preliminaries

2.1   Dependency Parsing

Dependency parsing, which is a hot and important topic in natural language processing, is a kind of syntactic structure, consisting of lexical terms, linked by binary asymmetric relations. The Fig. 1 below shows an example of dependency between the words for a Swedish sentence. "Dependency" means a kind of relationships between two words according to "head rules". If there exists a rule that nouns are dependent on verbs, so "John hit it" implies two dependencies: "John" is dependent on "hit" and "it" is dependent on "hit".

In recent years, many parsing algorithms and methods have emerged. Charniak [13] presents a parser which parses Wall Street Journal tree-bank down to Penn tree-bank style with 90% accuracy. McDonald [14] regards dependency parsing as a task to find a maximum spanning tree in a directed graph, and this parser can be naturally extended to non- projective parsing. Yamada [15] proposes a parsing algorithm with three parsing actions: shift, right and left. Similarly, Nivre [16] proposes another parsing algorithm with four parsing actions: shift, reduce, right and left. Also, Nivre [1] regards dependency parsing as a classification task and utilizes memory based method to guide parsing

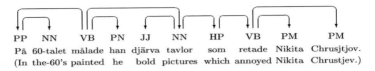

Fig. 1. Swedish: an example for dependency graph

process. Among them, Yamada and Nivre's algorithms are representative methods embedded with classification techniques.

To understand Yamada and Nivre's algorithms, we need to give a definition of a well-formed dependency graph [16]:

(1) A string of words W is represented as a list of tokens, where each token $n = (i, w)$ is a pair consisting of a position i and a word form W; the functional expressions $POS(n) = i$ and $LEX(n) = w$ can be used to extract the position and word form of a token. We let $<$ denote the complete and strict ordering of tokens in W, i.e. $n < n'$ iff $POS(n) < POS(n')$.

(2) A dependency graph for W is a directed graph $D=(N_W, A)$, where the set of nodes N_W is the set of tokens in W, and the arc relation A is binary, irreflexive relation on N_W. We write $n \to n'$ to say that there is an arc from n to n', i.e. $(n, n') \in A$; we use $\to^*$ to denote the reflexive and transitive closure of the arc relation A; and we use $\leftrightarrow$ and $\leftrightarrow^*$ for the corresponding undirected relations, i.e. $n \leftrightarrow n'$ iff $n \to n'$ or $n' \to n$.

(3) A dependency graph $D=(N_W, A)$ is well-formed iff the following conditions are satisfied:

Single head: $(\forall n n' n'')(n \to n' \land n'' \to n') \Rightarrow n = n''$

Acyclic: $(\forall n n')\neg(n \to n' \land n' \to^* n)$

Connected: $(\forall n n')n \leftrightarrow^* n'$

Projective: $(\forall n n' n'')(n \leftrightarrow n' \land n < n'' < n') \Rightarrow (n \to^* n'' \lor n' \to^* n'')$

Yamada and Nivre's algorithms are all based on the formal definition of dependency graph, and carry out the dependency parsing guided by several actions. There is a difference between Yamada and Nivre's parsing algorithms. At first, Yamada's algorithm [17] was designed for Japanese parsing with only two actions and later Yamada [15] added one more action to the algorithm so that when parsing English his method can achieve better results. Nivre's algorithm [16] has four actions : left, right, reduce, shift, with one more action "reduce" than Yamada's.

Nivre's deterministic parsing algorithm [16] is shown in Table 1. In the table, $\langle S, W, A \rangle$ is a triple which represents a state during the parsing process. W is the input string of one single sentence; S is a stack store the tokens removed from the stack of W; A records the set of arcs between the words in W. Every state in the parsing process has a triple $\langle S, W, A \rangle$ representation, and the parsing algorithm predicts an action based on the current state, and carries out the action which changes the state. This process repeats until W is empty.

Table 1. Formal Description of Actions

Initialization:	$\langle nil, W, \varnothing \rangle$
Termination:	$\langle S, nil, A \rangle$
Left Arc:	$\langle n\|S, n'\| W, A \rangle \rightarrow \langle S, n'\| W, A \cup \{(n', n)\} \rangle$
	if $LEX(n) \leftarrow LEX(n') \in R \ \neg \exists n''(n'', n) \in A$
Right Arc:	$\langle n\|S, n'\| W, A \rangle \rightarrow \langle n'\|n\|S, W, A \cup \{(n, n')\} \rangle$
	if $LEX(n) \rightarrow LEX(n') \in R \ \neg \exists n''(n'', n') \in A$
Reduce:	$\langle n\|S, W, A \rangle \rightarrow \langle S, W, A \rangle$
Shift:	$\langle S, n\| W, A \rangle \rightarrow \langle n\|S, W, A \rangle$

2.2 Rough Sets Dependence Measure

In rough set theory [18], an information table is denoted by $S = \langle U, A, V, f \rangle$, where U denotes a finite and nonempty set of instances, which is called universe; we have $A = C \cup D$, and $C \cap D = \emptyset$, where A denotes the attribute set, and C is the nonempty condition attribute set and D is the nonempty decision attribute set, and often $card(D) = 1$; $V = \bigcup_{a \in A} V_a$ is the domain of attribute values; f is an information function such that $f(u_i, a) \in V_a$, for every $a \in A$ and every $u_i \in U$.

Based on the above definitions and lower/upper approximations [18] in Rough Sets , we introduce Rough Sets dependence measure, which is utilized in Pawlak [19] to measure the dependencies among attributes.

Dependency measure

$$K(P, Q) = \frac{\sum_{i=1}^{m} card(\underline{P}(X_i))}{card(U)}, \quad where \ P \cap Q = \varnothing, X_i \in E(Q). \tag{1}$$

3 Rough Set-Based Classifier

Similar with Naive Bayesian classifiers and decision tree learners, the rough set-based classifier [3] utilizes the rough set methodology 1 to measure the attribute dependencies, and splits the training data into subsets in order to alleviate the interdependence using a tree model. The growth of the tree structure of the classifier continues until there are no dependencies in the subset of training data or the instances in the subset are less than a threshold. And finally, the classifier generates a local Naive Bayesian classifier in each leaf node. The procedure is similar with the decomposition methods in RSES, but different in the decomposition measure which is the rough set dependence measure in our method.

Let $A_1 = \{a_{11}, \ldots, a_{1g}\}$ denote the set of the test attributes on the path to the leaf, and let $A_2 = \{a_{21}, \ldots, a_{2(n-g)}\}$ denote the set of the remaining attributes.

The classifier in [3] classifies the test example by choosing the class with the maximum posterior probability:

$$\arg\max_{c_i} \left(P(c_i|a_{1t}) \prod_{j=1}^{n-g} P(a_{2j}|c_i, a_{1t}) \right) \tag{2}$$

where $a_{1t} \in A_1$, and it is a constant for a test example. Detailed algorithm is presented in Table 2.

Table 2. Rough Set-based Classifier Algorithm

Rough Set-based Classifier Algorithm
Input: Att, Data, TreeLevel, θ, P, $Q = Att - P$.
Output: A decision tree-like model.
Algorithm begin:
Initialization: $P = \emptyset, Q = Att - P$.
For$(a_i \in Q)$
{
$P_i = P \cup \{a_i\}$; $Q_i = Q - \{a_i\}$; compute $K(P_i, Q_i)$;
}
$index = \arg\max_i K(P_i, Q_i)$;
If(**stopping criterion==false**)
{
data[]=split(Data,$index$);
$P_{index} = P \cup \{a_{index}\}$;$Q_{index} = Q - \{a_{index}\}$;
For(data[i])
RSClassifier(Att,data[i],TreeLevel+1,θ,P_{index},Q_{index});
}
Else
NaiveBayes classifier=new NaiveBayes(Data);
Algorithm End.
Note: Stopping criterion–(Data.instanceNo$\leq$50) or (TreeLevel$\geq$4) or $(K(P_{index}, Q_{index})\leq\theta)$ or $(\frac{Data.instanceNO}{Data.branchNO} \leq 50)$

In our algorithm implementation, **Att** denotes the set of all attributes in the data set, while **P** and **Q** represent two subsets of **Att** whose union is equivalent to **Att**. **Data** represents the training data and **TreeLevel** means the depth of the current node in the tree structure of the classifier [3]. **Data[ ]** is equal to the partition on Data according to the attribute $\mathbf{a_{index}}$. Similar with Decision Tree, the key of the rough set-based classifier relies on the attribute selection by Equation (1) to partition **Data**. With checking stopping criterion, if not stopping, we recursively repeat the construction process. Otherwise, we stop splitting **Data** and train a NaiveBayes classifier on **Data**.

4   Experiments

Two experiments are designed in this section: (1) Both of the rough set-based classifier and Naive Bayesian classifier are embedded in Nivre's deterministic dependency parsing process. (2) Conduct dependency parsing task by combining Nivre's deterministic method with the rough set-based classifier.

4.1   Data Set and Task Description

Data set–The Chinese Corpus is from the CoNLL-X Shared Task on Multilingual Dependency Parsing of the Tenth Conference on Computational Natural Language Learning, and the corpus is especially in Chinese. Table 3 represents the information of the corpus in use.

Table 3. Data Set description

	Training data		Testing data	
	#Instance	#Attribute	#Instance	#Attribute
corpus	354719	5	39405	5
data set	451994	13	50806	13

Task description–The original corpus represents the information of every single word, while the single line in the data set contains the information of a word pair. So the original corpus needs transforming into the data set which is well formatted for the rough set-based classifier. While different parsers utilize different transformations, in this paper, the proposed parser utilizes a way of transformation in [16]. And the parsing process regarded as a classification task is illustrated in Fig.2.

4.2   Experimental Analysis

In dependency parsing, Dependency Accuracy(DA) and Root Accuracy(RA) are two important measures on evaluation of parsing performance. And they are defined as:

$$Dependency\ Accuracy = \frac{\#correct\ parsed\ word}{\#word} \qquad (3)$$

$$Root\ Accuracy = \frac{\#correct\ parsed\ root}{\#root} \qquad (4)$$

(1)Rough Set-based Classifier vs NaiveBayes:

In Table 4, values in bold face get a higher accuracy than the other. It shows that Rough Set-based Classifier can often get higher DA and RA values than Naive-Bayes, on almost all situations. However, Rough Set-based Classifier has only a slight improvement on Naive Bayesian Classifier. Maybe, this is because attribute dependency assumption is not violated in the data set. But in general, Rough Set-based Classifier performs better than NaiveBayes on this classification task.

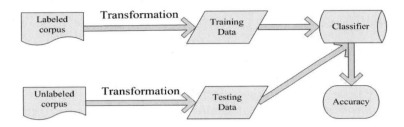

Fig. 2. Dependency parsing process

Table 4. Rough Set-based Classifier vs NaiveBayes

NO of Sentences		1000	2000	3000	4000	5000
Naive	RA	**0.854**	0.8605	0.8533	0.8555	0.8562
Bayesian	DA	0.5941	0.6063	0.6039	0.6017	0.6026
RS-based	RA	0.853	**0.8605**	**0.855**	**0.8567**	**0.8582**
Classifier	DA	**0.5960**	**0.6073**	**0.6056**	**0.6033**	**0.6040**

Table 5. Rough Set-based Parser vs Other Parsers

English corpora	Charniak	Collins	Yamada	Action Model	RS-based Parser
DA	0.921	0.91	0.903		
RA	0.952	0.952	0.916		
Chinese corpora	Charniak	Collins	Yamada	Action Model	RS-based Parser
DA		0.8013	0.8282	**0.8347**	0.6040
RA		0.7009	0.7013	0.6823	**0.8582**

(2)Rough Set-based Parser vs Other Parsers:
Action model [20] is a new deterministic parser, which achieves the best dependency accuracy on Chinese corpora. However, our rough set-based classifier combined with Niver's deterministic parsing algorithm performs much better on root accuracy than Action model and all other parsers. The reason is that, in the parsing process, our parser tends to classify a word to be dependent on the root of the sentence. So our parser usually performs a bias towards root accuracy, and as a result of this, our dependency accuracy is decreased. One of our future works will focus on how to improve our dependency accuracy without breaking the balance between these two measures.

5   Conclusion and Future Work

The experimental results prove that the rough set-based classifier gets better performance than Naive-Bayes classifier on dependency parsing task, and combining the rough set-based classifier classifier with Nivre's deterministic parser achieves comparable performance with other parsers. However, there are some issues we will work on.

(1) Constructing the rough set-based classifier model is a time consuming process. In training phase, the computation of attribute dependence measure is costly. So our future work will focus on how to decrease the time complexity of the rough set-based classifier.

(2) Improvement of the attribute dependence measure. The quality of dependence measure has direct influences on the classification performance. It is important to study how to improve the rough sets dependence measure to get better performance.

(3) Modification on Nivre's parsing algorithm to parsing Chinese sentences. Because Nivre's algorithm is not designed for Chinese dependency structure and the process performs a bias towards root accuracy, we need not only to design a classifier with better performance, but also to improve Nivre's parsing algorithm.

Acknowledgement

This work is supported by the National 863 High-Tech Projects of China under Grant No. 2006AA01Z143 and the National Natural Science Foundation of China under Grant No. 60503022 and No. 60673043.

References

1. Nivre, J., Hall, J., Nilsson, J.: Memory-Based Dependency Parsing. In: Ng, H.T., Riloff, E. (eds.) Proceedings of the Eighth Conference on Computational Natural Language Learning (CoNLL), Boston, Massachusetts, 6-7 May 2004, pp. 49–56 (2004)
2. Yuret, D.: Dependency Parsing as a Classification Problem. In: Proceedings of the Tenth Conference on Computational Natural Language Learning (CoNLL) (2006)
3. Ji, Y.S., Shang, L.: RoughTree A Classifier with Naive-Bayes and Rough Sets Hybrid in Decision Tree Representation. In: 2007 IEEE International Conference on Granular Computing, Silicon, Valley, USA, pp. 221–227 (2007)
4. Kononenko, I.: Comparison of inductive and Naive Bayesian learning approaches to automatic knowledge acquisition. In: Wielinga, B., et al. (eds.) Current Trends in Knowledge Acquisition, IOS Press, Amsterdam (1990)
5. Domingos, P., Pazzani, M.: Beyond independence: Conditions for the optimality of the simple Bayesian classifier. In: Proc. 13th Intl. Conf. on Machine Learning, pp. 105–112. Morgan Kaufmann, San Francisco (1996)
6. Friedman, N., Geiger, D., Goldszmidt, M.: Bayesian network classifiers. Machine Learning 29, 131–163 (1997)
7. Zheng, Z., Webb, G.I.: Lazy Learning of Bayesian Rules. Machine Learning 41(1), 53–84 (2000)
8. Webb, G.I., Boughton, J., Wang, Z.: Not So Naive Bayes: Aggregating One-Dependence Estimators. Machine Learning 58(1), 5–24 (2005)

9. Kohavi, R.: Scaling up the accuracy of naive-Bayes classifiers: A decision-tree hybrid. In: Proceedings of the Second International Conference on Knowledge Discovery and Data Mining, pp. 202–207. The AAAI Press, Menlo Park (1996)
10. Quinlan, J.R.: C4.5: Programs for Machine Learning. Morgan Kaufmann Publishers, Los Altos, California (1993)
11. Breiman, L., Friedman, J.H., Olshen, R.A., Stone, C.J.: Classification and Regression Trees. Wadsworth International Group (1984)
12. Pagallo, G., Haussler, D.: Boolean feature discovery in empirical learning. Machine Learning 5, 71–79 (1990)
13. Charniak, E.: A Maximum-Entropy-Inspired Parser. In: Proceedings of the Second Meeting of North American Chapter of Association for Computational Linguistics (NAACL 2000), pp. 132–139 (2000)
14. Donald, M.: Spanning Tree Methods for Discriminative Training of Dependency Parsers. In: Proceedings of COLING 2004, Geneva, Switzerland, August 23-27, pp. 105–110 (2004)
15. Yamada, H., Matsumoto, Y.: Statistical dependency analysis with support vector machines. In: Proceedings of IWPT, pp. 195–206 (2003)
16. Nivre, J.: An Efficient Algorithm for Projective Dependency Parsing. In: Proceedings of the 8th International Workshop on Parsing Technologies (IWPT 2003), Nancy, France, April 23-25, pp. 149–160 (2003)
17. Nakagawa, T., Kudoh, T., Matsumoto, Y.: Revision learning and its application to part-of-speech tagging. In: Proceedings of Association for Computational Linguistics, pp. 497–504 (2002)
18. Pawlak, Z.: Rough Sets. International Journal of Information and Computer Sciences 11(5), 341–356 (1982)
19. Pawlak, Z.: Rough Sets: Theoretical aspects of reasoning about data. Kluwer Academic Publishers, Dordrecht, Netherlands (1991)
20. Duan, X.Y., Zhao, J., Xu, B.: Chinese Dependency Parsing Based on Action Modeling. In: 9th Chinese National Conference on Computational Linguistics, Dalian, China, August 6-8, pp. 133–139 (2007)

Four-Valued Extension of Rough Sets*

Aida Vitória[1], Andrzej Szałas[2,3], and Jan Małuszyński[3]

[1] Department of Science and Technology, Linköping University
S 601 74 Norrköping, Sweden
aidvi@itn.liu.se
[2] Institute of Informatics, Warsaw University
02-097 Warsaw, Poland
andsz@mimuw.eu.pl
[3] Department of Computer and Information Science, Linköping University
581 83 Linköping, Sweden
janma@ida.liu.se

Abstract. Rough set approximations of Pawlak [15] are sometimes generalized by using similarities between objects rather than elementary sets. In practical applications, both knowledge about properties of objects and knowledge of similarity between objects can be incomplete and inconsistent. The aim of this paper is to define set approximations when all sets, and their approximations, as well as similarity relations are four-valued. A set is four-valued in the sense that its membership function can have one of the four logical values: unknown (**u**), false (**f**), inconsistent (**i**), or true (**t**). To this end, a new implication operator and set-theoretical operations on four-valued sets, such as set containment, are introduced. Several properties of lower and upper approximations of four-valued sets are also presented.

1 Introduction

Rough sets [15] are constructed by means of approximations obtained by using elementary sets which partition a universe. The assumption as to partitioning of the universe has been relaxed by many authors (see, e.g., [1,6,9,17,18,19]), however the Pawlak's idea of approximations has remained the same. Namely, an object o belongs to the lower approximation of a given set A whenever all objects indiscernible from o belong to A and o belongs to the upper approximation, when there are objects indiscernible from o belonging to A. Indiscernibility is modeled by similarity relations reflecting limited perceptual capabilities as well as incomplete and imprecise knowledge. Such approximations naturally lead to three- and four-valued logics (see, e.g., [3,8,10,14]).

The goal of the current paper is to ground Pawlak's ideas in a four-valued framework. To this end, we define a four-valued set theory in the sense that the membership function, set containment as well as union, intersection and complement of sets are four-valued. We may then have that either an element belongs to a given set, or it does not belong to the set, or its membership in the set may be unknown or inconsistent, perhaps, due to contradictory evidence. Notice that we assume that *all* sets and relations are four-valued, in particular the underlying similarity relation. To our knowledge, this is

---

* Supported in part by the MNiSW grant N N206 399334 and by the 6th Framework Programme project REWERSE number 506779 (cf. http://rewerse.net).

G. Wang et al. (Eds.): RSKT 2008, LNAI 5009, pp. 106–114, 2008.

a fundamental difference between the work we present and the work discussed by other authors in the rough-set field. Therefore, we take into account the fact that in practical applications our knowledge about similarity between objects of the universe can also be incomplete and inconsistent. This novel aspect of our work has led us to re-think and extend the usual notions of upper and lower approximations. Since we consider similarity relations as four-valued sets, there are cases, when we cannot establish with certainty whether the neighborhood of an element is a subset of (or whether is disjoint with) a given set. To tackle this problem, we propose upper and lower approximations that are also four-valued sets. It is important to note that in cases, when only the standard truth values **t** and **f** are used, all notions we define reduce to the standard operations on rough sets.

The work presented in [11] also captures the same type of knowledge uncertainty, as described above. However, uncertainty in the properties of objects is captured by fuzzy sets and (two-valued) equivalence relations are used to capture similarity between objects of the universe. Upper and lower approximation of a fuzzy set are defined in [11]. Unlike [11], in our framework, similarity between objects is captured by four-valued relations and all properties correspond to four-valued sets.

When considering similarities as well as approximations to be four-valued, as we do in the current paper, it appears that the truth ordering proposed in [14] makes definitions of set containment and approximations problematic. Therefore, we propose to slightly modify the truth ordering (by changing the relationship between **f** and **u** only).

The paper is structured as follows. First, in Section 2, we formalize the notion of four-valued set. Section 3 is devoted to four-valued approximations. In Section 4, we provide an example illustrating the approach. Finally, Section 5 concludes the paper.

2   The Framework

In this section, we first define four-valued sets together with the notions of four-valued set containment and four-valued set intersection and union. To formalize four-valued set containment, we introduce a new implication connective.

2.1   Four-Valued Sets

Let $\mathcal{B} = \{\mathbf{t}, \mathbf{f}, \mathbf{i}, \mathbf{u}\}$ be the set of truth values, where **t** stands for *true*, **f** stands for *false*, **i** stands for *inconsistent* and **u** stands for *unknown*. Any of these logical values can be negated: $\neg\mathbf{t} = \mathbf{f}$, $\neg\mathbf{i} = \mathbf{i}$, $\neg\mathbf{u} = \mathbf{u}$, and $\neg\mathbf{f} = \mathbf{t}$.

Let us now formalize the notion of four-valued sets. Given a universe U, we introduce a new set, disjoint with U, denoted by $\neg U$ and defined by $\neg U \stackrel{\text{def}}{=} \{\neg x \mid x \in U\}$, where $\neg x$ denotes elements in $\neg U$. Intuitively, $x \in A$ represents the fact that there is an evidence that x is in A and $(\neg x) \in A$ represents the fact that there is an evidence that x is not in A. A *four-valued set* A on U is any subset of $U \cup \neg U$.

In our framework, set membership is four-valued and it extends the usual two-valued membership. We assume that $\neg(\neg x)$ is equivalent to x.

Definition 1. *Set membership*, denoted as $\epsilon : U \times 2^{U \cup \neg U} \to \mathcal{B}$, is defined by:

$$x \,\epsilon\, A = \begin{cases} \mathbf{t} & \text{if } x \in A \text{ and } (\neg x) \notin A \\ \mathbf{i} & \text{if } x \in A \text{ and } (\neg x) \in A \\ \mathbf{u} & \text{if } x \notin A \text{ and } (\neg x) \notin A \\ \mathbf{f} & \text{if } x \notin A \text{ and } (\neg x) \in A. \end{cases}$$

The *complement* $\neg A$ *of a four-valued set* A, is defined by $\neg A \stackrel{\text{def}}{=} \{\neg x \mid x \,\epsilon\, A\}$. ◁

Let P be a four-valued set on a universe U. To simplify the notation, $P(x)$ stands for $x \,\epsilon\, P$. For instance, let $U = \{a, b, c\}$ and $\text{Red} = \{a, \neg a, b\}$, In this case, $\text{Red}(b) = \mathbf{t}$ and $\text{Red}(a) = \mathbf{i}$, since both $a \in \text{Red}$ and $\neg a \in \text{Red}$. Moreover, $\neg \text{Red} = \{a, \neg a, \neg b\}$.

2.2 Four-Valued Calculus

In our work, we use two orderings of the logical values: truth ordering and knowledge ordering. Truth ordering is used for calculations within a single information source while knowledge ordering is used for gathering knowledge from different sources. This approach has been considered by in [4] and in the framework of bilattices, in [13,12].

The *truth ordering* $\leq_t$ and the *knowledge ordering* $\leq_k$ on $\mathcal{B}$ are defined as the smallest reflexive and transitive relations satisfying $\mathbf{f} \leq_t \mathbf{u} \leq_t \mathbf{i} \leq_t \mathbf{t}$, $\mathbf{u} \leq_k \mathbf{f} \leq_k \mathbf{i}$, and $\mathbf{u} \leq_k \mathbf{t} \leq_k \mathbf{i}$. The disjunctions $\vee_t$ and $\vee_k$ (conjunctions $\wedge_t$ and $\wedge_k$) are defined to be the least upper bounds (greatest lower bounds) of their arguments w.r.t. $\leq_t$ and $\leq_k$, respectively.

Let us relate the orderings above with Belnap's truth and knowledge ordering [4]. The knowledge ordering we defined above coincides with Belnap's knowledge ordering. However, our truth ordering is different from the Belnap's truth ordering. This change is motivated by the fact that Belnap's truth ordering can give counterintuitive results when used for reasoning, as shown in [14].

We also need to extend the notion of set containment and set intersection to four-valued sets. We discuss next these two ideas.

Given two four-valued sets A_1 and A_2 over a universe U, $A_1 \Subset A_2$ stands for A_1 being contained in A_2. Notice that our notion of set containment is also four-valued. Thus, $A_1 \Subset A_2$ can be evaluated to one of the four logical values in $\mathcal{B}$. If A and B are two-valued sets, then set containment is defined as $A \subseteq B \stackrel{\text{def}}{=} \forall x \in U(x \in A \to x \in B)$. This definition relies on the notion of universal quantification and logical implication $\to$. We follow a similar idea. Therefore, we first extend the notion of universal and existential quantification. We then propose a new implication connective, $\hookrightarrow$, for our four-valued logic.

Since $\forall$ ($\exists$) is a generalized conjunction (a generalized disjunction), we define

$$\forall x[P(x)] \stackrel{\text{def}}{=} \underset{x \in U}{\text{GLB}^t}\{P(x)\} \quad \text{and} \quad \exists x[P(x)] \stackrel{\text{def}}{=} \underset{x \in U}{\text{LUB}^t}\{P(x)\} \,,$$

where the superscript ^{t} indicates that the greatest lower bound (GLB) and least upper bound (LUB) are computed w.r.t. truth ordering.

We define now the semantics of a new implication operator $\hookrightarrow$. To define this operator, we must have in mind that it should be suitable for determining set containment, in the case of four-valued sets. Obviously, $\hookrightarrow$ should also be an extension of the usual logical implication $\rightarrow$. We present below a table defining $\hookrightarrow$ and then, we provide the motivation.

$\hookrightarrow$	**f**	**u**	**i**	**t**
f	**t**	**t**	**t**	**t**
u	**u**	**u**	**i**	**t**
i	**i**	**i**	**i**	**t**
t	**f**	**u**	**i**	**t**

Let X and Y be two four-valued sets and assume we want to verify whether $X \in Y$. If we know that an element x does not belong to X (i.e., $x \in X = \mathbf{f}$) then the membership of x in Y should have no influence on the result of $X \in Y$. This motivates the first line of the table. If x belongs to Y (i.e., $x \in Y = \mathbf{t}$) then the truth valued of $x \in X$ is irrelevant. This explains the last column of the table above. If x belongs to X (i.e., $x \in X = \mathbf{t}$) then our conclusions w.r.t. element x should depend only on our knowledge about $x \in Y$. This is the intuition behind the last line of the table. It seems obvious that if we know nothing about membership of x in X and nothing about membership of x in Y (i.e., both $x \in X$ and $x \in Y$ are evaluated to $\mathbf{u}$) then, we cannot conclude anything about the contribution element x gives to determine whether X is a subset of Y. Thus, $\mathbf{u} \hookrightarrow \mathbf{u}$ is evaluated to $\mathbf{u}$. Similarly, $\mathbf{i} \hookrightarrow \mathbf{i}$ is evaluated to $\mathbf{i}$.

Let us now discuss the remaining cases. If $\hookrightarrow$ is going to be used to determine set containment, then it is desirable that it has the following property: $(b_1 \hookrightarrow b_2) = (\neg b_2 \hookrightarrow \neg b_1)$. Since $\mathbf{t} \hookrightarrow \mathbf{u}$ is evaluated to $\mathbf{u}$ and $\mathbf{t} \hookrightarrow \mathbf{i}$ is evaluated to $\mathbf{i}$, we should then have that $\mathbf{u} \hookrightarrow \mathbf{f}$ and $\mathbf{i} \hookrightarrow \mathbf{f}$ should be evaluated to $\mathbf{u}$ and $\mathbf{i}$, respectively.

If we have no information about whether x belongs to X (i.e., $x \in X = \mathbf{u}$) and we have contradictory evidence about the membership of x in Y (i.e., $x \in Y = \mathbf{i}$), then future information about membership of x in X can lead us to the conclusion that

– element x contributes with $\mathbf{i}$, if $x \in X$ is evaluated to $\mathbf{t}$ (note that $\mathbf{t} \hookrightarrow \mathbf{i} = \mathbf{i}$), or
– element x contributes with $\mathbf{t}$, if $x \in X$ is evaluated to $\mathbf{f}$ (note that $\mathbf{f} \hookrightarrow \mathbf{i} = \mathbf{t}$), or
– element x contributes with $\mathbf{i}$, if $x \in X$ is evaluated to $\mathbf{i}$ (note that $\mathbf{i} \hookrightarrow \mathbf{i} = \mathbf{i}$).

Thus, we define $\mathbf{u} \hookrightarrow \mathbf{i}$ to be $\mathbf{i}$ since that is a possible value for the contribution of element x in the future, when more knowledge is gathered, but it still conveys a degree of uncertainty. Moreover, we have then that $\mathbf{i} \hookrightarrow \mathbf{u}$ should also be defined as $\mathbf{i}$.

Other implication connectives have been proposed by other authors (see, e.g., [2]). Let us make a brief comparison. Obviously, the proposed implication $\hookrightarrow$ extends the usual two-valued logical implication, i.e. when we only consider the logical values $\mathbf{t}$ and $\mathbf{f}$. It also extends the implication of the Kleene three-valued logic [5,16] in the sense that when we restrict truth values to $\{\mathbf{t}, \mathbf{f}, \mathbf{u}\}$ (or $\{\mathbf{t}, \mathbf{f}, \mathbf{i}\}$), we obtain Kleene's implication.

On the other hand, our implication differs from the material implication $\mapsto$ proposed in [2], on the following two cases: $\mathbf{u} \mapsto \mathbf{i} = \mathbf{t}$ while $\mathbf{u} \hookrightarrow \mathbf{i} = \mathbf{i}$ and $\mathbf{i} \mapsto \mathbf{u} = \mathbf{t}$ while $\mathbf{i} \hookrightarrow \mathbf{u} = \mathbf{i}$. The material implication $\mapsto$ can be defined by means of negation and disjunction. The same property turns out to be true for our implication connective, i.e. $(b_1 \hookrightarrow b_2) = (\neg b_1 \vee_t b_2)$. In contrast to the internal implication $\supset$ proposed in [2], our implication does not satisfy the Modus Ponens, if we assume that $\{\mathbf{t}, \mathbf{i}\}$ is the set of designated values. For instance, $\mathbf{i} \hookrightarrow \mathbf{f} = \mathbf{i}$.

Observe that given two four-valued sets X and Y over a universe U, we can define a new four-valued set P, such that $(x \in P)$ has the same truth value as $(x \in X \hookrightarrow x \in Y)$, for every object $x \in U$. We can now formally define set containment, for four-valued sets.

Definition 2. Assume that X and Y are four-valued sets on a universe U. Then,

$$X \Subset Y \stackrel{\text{def}}{=} \forall x \in U (x \in X \hookrightarrow x \in Y)$$

We say X is a *subset of* Y iff $X \Subset Y = \mathbf{t}$ and X is not a subset of Y, denoted by $X \not\Subset Y$, iff $X \Subset Y = \mathbf{f}$.   ◁

The following proposition can be proved.

Proposition 1. Let X and Y be four-valued sets on a universe U. Then, $(X \Subset Y) = (\neg Y \Subset \neg X)$.   ◁

Let us now define the notion of four-valued set intersection ⋒ and union ⋓.

Definition 3. Let X and Y be four-valued sets on a universe U. Then,

$$x \in (X \Cap Y) \stackrel{\text{def}}{=} (x \in X) \wedge_t (x \in Y) \text{ and } x \in (X \Cup Y) \stackrel{\text{def}}{=} (x \in X) \vee_t (x \in Y).$$ ◁

Note that ⋒ and ⋓ reduce to the standard two-valued set intersection and union when only values $\mathbf{t}$ and $\mathbf{f}$ are present.

3   Four-Valued Set Approximations

In the usual rough-set framework, given a set and a similarity relation, lower and upper approximations of the set can be obtained. We extend these ideas to four-valued sets. In contrast to previous work, we deal with four-valued sets on a given universe and four-valued similarity relations. Moreover, upper and lower approximations of four-valued sets are themselves four-valued sets.

Definition 4. By a *four-valued similarity relation* σ we mean any four-valued binary relation on a universe U, satisfying the reflexivity condition, i.e., for any element x of the universe $(x, x) \in \sigma = \mathbf{t}$. By the *neighborhood of element* $x \in U$ w.r.t. σ, we understand the four-valued set $\sigma(x)$ such that $y \in \sigma(x) \stackrel{\text{def}}{=} (x, y) \in \sigma$.   ◁

We proceed now with the definitions of four-valued approximations.

Definition 5. Let A be a four-valued set. Then, the *lower and upper approximations of* A w.r.t. σ, denoted by A_σ^+ and $A_\sigma^\oplus$, respectively, are defined by $x \in A_\sigma^+ \stackrel{\text{def}}{=} \sigma(x) \Subset A$ and $x \in A_\sigma^\oplus \stackrel{\text{def}}{=} \exists y \in U [y \in (\sigma(x) \Cap A)]$.   ◁

The definitions above naturally extend the usual definitions of lower and upper approximations presented in the rough set literature. We determine the membership of an element x in the lower approximation by verifying (set) containment of its neighborhood, $\sigma(x)$, in set A. We determine the membership of an element x in the upper approximation by computing the largest membership value that an element of the universe can have in the intersection of $\sigma(x)$ and A, w.r.t. to $\leq_t$.

The next theorems summarize some important properties of lower and upper approximations of four-valued sets.

Theorem 1. Let A be a four-valued set on a universe U, σ be a four-valued similarity relation. Then, $(x \; \epsilon \; A_\sigma^+) \leq_t (x \; \epsilon \; A_\sigma^\oplus)$, for all $x \in U$.   ◁

The above theorem lead us to the conclusion that it is never the case that $A_\sigma^+ \not\subseteq A_\sigma^\oplus$ and it is formalized below. From an intuitive point of view, this conclusion is the counterpart of a known property of rough-set approximations: the lower approximation is always contained in the upper approximation, when usual (two-valued) sets are used.

Theorem 2. Let A be a four-valued set on a universe U, σ be a four-valued similarity relation. Then, $(A_\sigma^+ \subseteq A_\sigma^\oplus) >_t \mathbf{f}$.   ◁

4   An Example

In this section, we present an example that illustrates the discussed ideas.

Perception can be modelled by similarity relations in the sense that objects indiscernible due to the perceptive limitations are considered similar to each other (see, e.g., [7], where incomplete knowledge about similarities has been taken into account in the context of perception). However, in our framework, in addition to incompleteness, the knowledge of similarity between objects can also be inconsistent. For instance, two different sensors, may give contradictory evidence about the similarity of two objects.

The universe $U = \{a, b, c, d\}$ consists of objects classified as being dangerous. The four-valued set Danger represents this property. Note that this classification may be incomplete in some cases and uncertain in others. For instance, for object d there is no information about its danger, while for for object c there is contradictory evidence about whether it is dangerous (i.e. $c \; \epsilon$ Danger $= \mathbf{i}$).

Suppose that we have four information sources, denoted by I_i ($i \in \{1, 2, 3, 4\}$), about the objects similarity, modelled by the similarity relations σ_i, respectively.

The set Danger and the similarities of a to other elements of the universe, $\sigma_i(a)$, are given below.

	a	b	c	d
$\sigma_1(a)$	t	i	u	f
$\sigma_2(a)$	t	f	f	f
$\sigma_3(a)$	t	t	u	t
$\sigma_4(a)$	t	u	f	t
Danger	t	f	i	u

Membership of a in the lower approximation of $\texttt{Danger}$ for information source I_1, is obtained by computing $\forall y \in U[y \ \epsilon \ \sigma_1(a) \hookrightarrow y \ \epsilon \ \texttt{Danger}]$.

$$a \ \epsilon \ \texttt{Danger}^+_{\sigma_1} = \text{GLB}^t_{y \ \epsilon \ U}\{y \in \sigma_1(a) \hookrightarrow y \in \texttt{Danger}\} =$$
$$\text{GLB}^t\{\mathbf{t} \hookrightarrow \mathbf{t}, \mathbf{i} \hookrightarrow \mathbf{f}, \mathbf{u} \hookrightarrow \mathbf{i}, \mathbf{f} \hookrightarrow \mathbf{u}\} = \text{GLB}^t\{\mathbf{t}, \mathbf{i}, \mathbf{i}, \mathbf{t}\} = \mathbf{i}$$

Membership of a in the upper approximation of $\texttt{Danger}$ for information source I_1, is obtained by computing $\exists y \in U[y \ \epsilon \ \sigma_1(a) \wedge_t y \ \epsilon \ \texttt{Danger}]$.

$$a \ \epsilon \ \texttt{Danger}^{\oplus}_{\sigma_1} = \text{LUB}^t_{y \ \epsilon \ U}\{y \in \sigma_1(a) \wedge_t y \ \epsilon \ \texttt{Danger}\} =$$
$$\text{LUB}^t\{\mathbf{t} \wedge_t \mathbf{t}, \mathbf{i} \wedge_t \mathbf{f}, \mathbf{u} \wedge_t \mathbf{i}, \mathbf{f} \wedge_t \mathbf{u}\} = \text{LUB}^t\{\mathbf{t}, \mathbf{f}, \mathbf{u}, \mathbf{f}\} = \mathbf{t}$$

Membership of a in the lower approximation of $\texttt{Danger}$ for the other information sources, I_2, I_3, and, I_4, is shown below.

$$a \ \epsilon \ \texttt{Danger}^+_{\sigma_2} = \text{GLB}^t_{y \in U}\{y \ \epsilon \ \sigma_2(a) \hookrightarrow y \ \epsilon \ \texttt{Danger}\} = \mathbf{t}$$
$$a \ \epsilon \ \texttt{Danger}^+_{\sigma_3} = \text{GLB}^t_{y \in U}\{y \ \epsilon \ \sigma_3(a) \hookrightarrow y \ \epsilon \ \texttt{Danger}\} = \mathbf{f}$$
$$a \ \epsilon \ \texttt{Danger}^+_{\sigma_4} = \text{GLB}^t_{y \ \epsilon \ U}\{y \ \epsilon \ \sigma_4(a) \hookrightarrow y \ \epsilon \ \texttt{Danger}\} = \mathbf{u}.$$

For any of the information sources above, $a \ \epsilon \ \texttt{Danger}^{\oplus}_{\sigma_i} = \mathbf{t}$, with $i \in \{2, 3, 4\}$.

We may compare the conclusions obtained with different information sources about the level of danger of object a. For instance, with the information from I_2, we are more certain that object a is dangerous than with the information provided by I_1 or I_4, since $a \ \epsilon \ \texttt{Danger}^+_{\sigma_1} = \mathbf{i}$ and $a \ \epsilon \ \texttt{Danger}^+_{\sigma_4} = \mathbf{u}$, while $a \ \epsilon \ \texttt{Danger}^+_{\sigma_2} = \mathbf{t}$. However, all information sources indicate that object a may be dangerous, since $a \ \epsilon \ \texttt{Danger}^{\oplus}_{\sigma_i} = \mathbf{t}$, with $i \in \{1, 2, 3, 4\}$. Note that the fact lower and upper approximations are four-valued sets, allow us a finer comparison of the degree of evidence each information source has to support its conclusions.

The discussion in the previous paragraph brings us to the problem of collecting and combining knowledge from different sources (agents), which is a relevant point from the practical point of view. Note that this aspect is work in progress and it is not formalized in the paper. However, we use this example to illustrate our point and the relevance of knowledge ordering, $\leq_k$, in this context. For example, according to sources I_2 and I_4, we can conclude that a is surely dangerous, since

$$\text{LUB}^k\{a \ \epsilon \ \texttt{Danger}^+_{\sigma_2}, a \ \epsilon \ \texttt{Danger}^+_{\sigma_4}\} = \text{LUB}^k\{\mathbf{t}, \mathbf{u}\} = \mathbf{t}.$$

On the other hand, sources I_2 and I_3 provide a contradictory information about membership of a in the lower approximation of $\texttt{Danger}$ and accordingly,

$$\text{LUB}^k\{a \ \epsilon \ \texttt{Danger}^+_{\sigma_2}, a \ \epsilon \ \texttt{Danger}^+_{\sigma_3}\} = \text{LUB}^k\{\mathbf{t}, \mathbf{f}\} = \mathbf{i}.$$

Similarly, fusing information from all sources or from sources including I_1 result in common knowledge being $\mathbf{i}$.

5 Conclusions

We introduced a notion of four-valued set to model situations where both knowledge about properties of objects and knowledge of similarity between objects can be incomplete or inconsistent. For modelling inclusion of four-valued sets, we proposed a new implication operator. We have shown how the similarity-based notions of lower- and upper approximation used in the usual rough-set framework can be extended in a natural way to our four-valued setting.

As future work, we plan to develop and to implement a language that allows users to define and reason about vague relations. Vague relations will be represented as four-valued sets and underlying similarity relations will be used to build relation (set) approximations, as we discuss in the paper. A knowledge base of such relations can then be queried by applications.

References

1. Andersson, R., Vitória, A., Małuszyński, J., Komorowski, J.: Rosy: A rough knowledge base system. In: Ślkezak, D., Yao, J., Peters, J.F., Ziarko, W., Hu, X. (eds.) RSFDGrC 2005. LNCS (LNAI), vol. 3642, pp. 48–58. Springer, Heidelberg (2005)
2. Arieli, O., Avron, A.: The value of the four values. Artificial Intelleligence 102(1), 97–141 (1998)
3. Banerjee, M.: Rough sets and three-valued łukasiewicz logic. Fundamenta Informaticae 32, 213–220 (1997)
4. Belnap, N.: A useful four-valued logic. In: Eptein, G., Dunn, J. (eds.) Modern Uses of Many Valued Logic, Reidel, pp. 8–37 (1977)
5. Bolc, L., Borowik, P.: Many-Valued Logics, 1. Theoretical Foundations. Springer, Berlin (1992)
6. Doherty, P., Łukaszewicz, W., Szałas, A.: Tolerance spaces and approximative representational structures. In: Günter, A., Kruse, R., Neumann, B. (eds.) KI 2003. LNCS (LNAI), vol. 2821, pp. 475–489. Springer, Heidelberg (2003)
7. Doherty, P., Łukaszewicz, W., Szałas, A.: Communication between agents with heterogeneous perceptual capabilities. Journal of Information Fusion 8(1), 56–69 (2007)
8. Doherty, P., Łukaszewicz, W., Skowron, A., Szałas, A.: Knowledge Representation Techniques. In: A Rough Set Approach. Studies in Fuziness and Soft Computing, vol. 202, Springer, Heidelberg (2006)
9. Doherty, P., Szałas, A.: On the correspondence between approximations and similarity. In: Tsumoto, S., Słowiński, R., Komorowski, J., Grzymała-Busse, J.W. (eds.) RSCTC 2004. LNCS (LNAI), vol. 3066, pp. 143–152. Springer, Heidelberg (2004)
10. Doherty, P., Szałas, A.: A correspondence framework between three-valued logics and similarity-based approximate reasoning. Fundamenta Informaticae 75(1-4), 179–193 (2007)
11. Dubois, D., Prade, H.: Putting rough sets and fuzzy sets together. Intelligent Decision Support: Handbook of Applications and Advances of the Rough Sets Theory, pp. 203–232. Kluwer, Dordrecht (1992)
12. Fitting, M.C.: Bilattices in logic programming. In: Epstein, G. (ed.) 20th International Symposium on Multiple-Valued Logic, pp. 238–247. IEEE CS Press, Los Alamitos (1990)
13. Ginsberg, M.: Multivalued logics: a uniform approach to reasoning in AI. Computational Intelligence 4, 256–316 (1988)

14. Małuszyński, J., Szałas, A., Vitória, A.: A four-valued logic for rough set-like approximate reasoning. In: Peters, J.F., Skowron, A., Düntsch, I., Grzymała-Busse, J.W., Orłowska, E., Polkowski, L. (eds.) Transactions on Rough Sets VI. LNCS, vol. 4374, pp. 176–190. Springer, Heidelberg (2007)

15. Pawlak, Z.: Rough Sets. Theoretical Aspects of Reasoning about Data. Kluwer Academic Publishers, Dordrecht (1991)

16. Rescher, N.: Many-Valued Logic. McGraw Hill, New York (1969)

17. Skowron, A., Stepaniuk, J.: Tolerance approximation spaces. Fundamenta Informaticae 27, 245–253 (1996)

18. Słowiński, R., Vanderpooten, D.: A generalized definition of rough approximations based on similarity. IEEE Trans. on Data and Knowledge Engineering. 12(2), 331–336 (2000)

19. Vitória, A.: A framework for reasoning with rough sets. In: Peters, J.F., Skowron, A. (eds.) Transactions on Rough Sets IV. LNCS, vol. 3700, pp. 178–276. Springer, Heidelberg (2005)

A Note on Characteristic Combination Patterns about How to Combine Objects in Object-Oriented Rough Set Models

Yasuo Kudo[1] and Tetsuya Murai[2]

[1] Dept. of Computer Science and Systems Eng., Muroran Institute of Technology
Mizumoto 27-1, Muroran 050-8585, Japan
kudo@csse.muroran-it.ac.jp
[2] Graduate School of Information Science and Technology, Hokkaido University
Kita 14, Nishi 9, Kita-ku, Sapporo 060-0814, Japan
murahiko@main.ist.hokudai.ac.jp

Abstract. We propose a concept of characteristic combination patterns to treat characteristics about how to combine objects in the object-oriented rough set model proposed by the authors. The object-oriented rough set model treats semi-structured data in the framework of rough sets by using structural hierarchies among objects, and semi-structured decision rules represent structural characteristics among objects, which enable us to capture what to combine objects. However, it is generally difficult to capture characteristics about how to combine objects by semi-structured decision rules. Thus, in this paper, we consider to capture how to combine objects by characteristic combination patterns.

Keywords: Semi-structured decision rules, Object-oriented rough sets, Characteristic combination patterns.

1 Introduction

Rough set theory proposed by Prof. Z. Pawlak [4] provides an interesting theoretical framework and useful tools for data mining based on approximation of concepts and reasoning about data. In applications of rough sets, generating decision rules from various kinds of data is one of the most hot topics. Usually, target data of Pawlak's rough set theory is illustrated by decision tables with fixed attributes and no hierarchy among data. On the other hand, the authors have proposed the object-oriented rough set model (for short, OORS) [1], and semi-structured decision rules in OORS [3]. The object-oriented rough set model illustrates hierarchical structures among data by using the concepts of classes, names, objects, and is-a and has-a relationships. Moreover, semi-structured decision rules in OORS illustrate characteristic combination of objects as parts of some objects, which enable us to capture characteristics about *what to combine objects*. However, it is generally difficult to capture characteristics about *how to combine objects* by semi-structured decision rules. In this paper, we propose a concept of characteristic combination patterns to treat characteristics about how

G. Wang et al. (Eds.): RSKT 2008, LNAI 5009, pp. 115–123, 2008.

to combine objects in OORS, and illustrate a method to calculate characteristic combination patterns from semi-structured decision rules.

2   The Object-Oriented Rough Set Model

We briefly review the object-oriented rough set model. Note that the contents of this section are entirely based on the authors' previous papers [1,2,3].

OORS consists of the following three triples: a *class structure* $\mathcal{C}$, a *name structure* $\mathcal{N}$ and an *object structure* $\mathcal{O}$, respectively:

$$\mathcal{C} = (C, \ni_C, \sqsupseteq_C), \quad \mathcal{N} = (N, \ni_N, \sqsupseteq_N), \quad \mathcal{O} = (O, \ni_O, \sqsupseteq_O),$$

where C, N and O are finite and disjoint non-empty sets such that $|C| \leq |N|$ ($|X|$ is the cardinality of X). Each element $c \in C$ is called a *class*. Similarly, each $n \in N$ is called a *name*, and each $o \in O$ is called an *object*. The relation $\ni_X$ ($X \in \{C, N, O\}$) is an acyclic binary relation on X, and the relation $\sqsupseteq_X$ is a reflexive, transitive, and asymmetric binary relation on X. Moreover, the relations $\ni_C$ and $\sqsupseteq_C$ of the class structure $\mathcal{C}$ satisfy the property:

- $\forall c_i, c_j, c_k \in C, \ c_i \sqsupseteq_C c_j, \ c_j \ni_C c_k \ \Rightarrow \ c_i \ni_C c_k.$

These three structures have the following characteristics, respectively:

- The class structure illustrates abstract data forms and those hierarchical structures based on part / whole relationship (has-a relation) and specialized / generalized relationship (is-a relation).
- The name structure introduces numerical constraint of objects and those identification, which provide concrete design of objects.
- The object structure illustrates actual combination of objects.

Two relations $\ni_X$ and $\sqsupseteq_X$ on $X \in \{C, N, O\}$ illustrate hierarchical structures among elements in X. The relation $\ni_X$ is called a *has-a relation*, and $x_i \ni_X x_j$ means "x_i has-a x_j", or "x_j is a part of x_i". On the other hand, the relation $\sqsupseteq_X$ is called an *is-a relation*, and $x_i \sqsupseteq_X x_j$ means that "x_i is-a x_j".

Each object $o \in O$ is defined as an instance of some class $c \in C$, and the class of o is identified by the *class identifier* function. The class identifier id_C is a *p-morphism* between $\mathcal{O}$ and $\mathcal{C}$ (cf. [5], p.142), that is, the function $id_C : O \longrightarrow C$ satisfies the following conditions:

1. $\forall o_i, o_j \in O, o_i \ni_O o_j \Rightarrow id_C(o_i) \ni_C id_C(o_j).$
2. $\forall o_i \in O, \ \forall c_j \in C, \ id_C(o_i) \ni_C c_j \Rightarrow \exists o_j \in O$ s.t. $o_i \ni_O o_j$ and $id_C(o_j) = c_j,$

and the same conditions are also satisfied for $\sqsupseteq_O$ and $\sqsupseteq_C$. $id_C(o) = c$ means that the object o is an instance of the class c.

The object structure $\mathcal{O}$ and the class structure $\mathcal{C}$ are also connected through the name structure $\mathcal{N}$ by the *naming function* $nf : N \longrightarrow C$ and the *name assignment* $na : O \longrightarrow N$. The naming function provides names to each class, which enable us to use plural instances of the same class simultaneously. On the other hand, the name assignment provides names to every objects, thus we can treat objects by using their names. Formally, the naming function $nf : N \longrightarrow C$ is a surjective p-morphism between $\mathcal{N}$ and $\mathcal{C}$, and satisfies the following *name preservation constraint*:

- If $nf(n_i) = nf(n_j)$, $\forall n_i, n_j \in N$, then $H_N(c|n_i) = H_N(c|n_j)$, $\forall c \in C$,

where $H_N(c|n) = \{n_j \in N \mid n \ni_N n_j, f(n_j) = c\}$ is the set of names of c that n has. These characteristics of the naming function nf imply that (1) there is at least one name for each class, (2) the name structure reflects all structural characteristics of the class structure, and (3) all names of the parts of any class are uniquely determined.

On the other hand, the name assignment $na : O \longrightarrow N$ is a p-morphism between $\mathcal{O}$ and $\mathcal{N}$, and satisfies the following *uniqueness condition*:

- For any $x \in O$, if $H_O(x) \neq \emptyset$, the restriction of na into $H_O(x)$:
  $na|_{H_O(x)} : H_O(x) \longrightarrow N$ is injective,

where $H_O(x) = \{y \in O \mid x \ni_O y\}$ is the set of objects that x has. $na(x) = n$ means that the name of the object x is n. The uniqueness condition requires that all distinct parts $y \in H_O(x)$ have different names.

We say that $\mathcal{C}$, $\mathcal{N}$ and $\mathcal{O}$ are *well-defined* if and only if there exist a naming function $nf : N \longrightarrow C$ and a name assignment $na : O \longrightarrow N$ such that $id_C = nf \circ na$, that is, $id_C(x) = nf(na(x))$ for all $x \in O$.

In well-defined structures, if a class c_i has m objects of a class c_j, then any instance o_i of the class c_i has exactly m instances $o_{j1}, \cdots, o_{jm}$ of the class c_j [1]. This good property enables us the following description for clear representation of objects. Suppose we have $o_1, o_2 \in O$, $n_1, n_2 \in N$, and $c_1, c_2 \in C$ such that $o_1 \ni_O o_2$, and $na(o_i) = n_i$, $nf(n_i) = c_i$ for $i \in \{1, 2\}$. We denote $o_1.n_2$ instead of o_2 by means of "the instance of c_2 named n_2 as a part of o_1".

In OORS, an indiscernibility relations $\sim$ on O is introduced by using the concept of *equivalence as instances* as follows [1]:

$$x \sim y \iff \begin{array}{l} x \text{ and } y \text{ satisfy the following two conditions:} \\ 1.\ id_C(x) = id_C(y), \text{ and,} \\ 2.\ \begin{cases} x.n \sim y.n, \ \forall n \in H_N(na(x)) & \text{if } H_N(na(x)) \neq \emptyset, \\ Val(x) = Val(y) & \text{otherwise,} \end{cases} \end{array} \quad (1)$$

where $H_N(na(x))$ is the set of names that $na(x)$ has. $Val(x)$ is the "value" of the "value object" x. Because C is a finite non-empty set and $\ni_C$ is acyclic, there is at least one class a such that a has no other class c. We call such class a an *attribute*, and for any instance x of the attribute a, we call x a *value object* of a. The value object x of a represents a "value" of the attribute a. Moreover, we assume that we can compare "values" of value objects of the same attribute.

Moreover, to capture hierarchical characteristics among objects by indiscernibility relations in OORS, *consistent sequences of names* are introduced [3]. Formally, a sequence of names $n_1. \cdots .n_k$ with length k ($k \geq 1$) such that

$n_i \in N$ $(1 \leq i \leq k)$ is called a consistent sequence of names if and only if either (1) $k = 1$, or (2) $k \geq 2$ and $n_{j+1} \in H_N(n_j)$ for each name n_j $(1 \leq j \leq k-1)$. We denote the set of all consistent sequences of names in $\mathcal{N}$ by N^+. Hereafter, we concentrate consistent sequences of names. Note that all names $n \in N$ are consistent sequences with length 1, thus we have $N \subset N^+$.

For any non-empty set of sequences of names $S \subseteq N^+$, indiscernibility relations $\approx_S$ on O which directly treat "nested" parts of objects are defined [3]:

$$x \approx_S y \iff \begin{array}{l} x \text{ and } y \text{ satisfy the following two conditions:} \\[4pt] 1.\ \forall n_1.\cdots.n_k \left[\begin{array}{l} n_1.\cdots.n_k \in S \Rightarrow \\ \{n_1 \in H_N(na(x)) \Leftrightarrow n_1 \in H_N(na(y))\} \end{array} \right], \\[12pt] 2.\ \forall n_1.\cdots.n_k \left[\begin{array}{l} n_1.\cdots.n_k \in S \text{ and } n_1 \in H_N(na(x)) \\ \Rightarrow x.n_1.\cdots.n_k \sim y.n_1.\cdots.n_k \end{array} \right]. \end{array} \quad (2)$$

The condition 1 in (2) requires that the object x and y concern the same sequences in S. Thus, if the objects x and y satisfy the condition 1, we say that x and y are *comparable* by the relation $\approx_S$. The condition 2 requires that, for all sequences $n_1.\cdots.n_k \in S$ that connect both x and y, the object $x.n_1.\cdots.n_k$ as a nested part of x is equivalent to the object $y.n_1.\cdots.n_k$ as a nested part of y.

Let N_{CON}^+ and N_{DEC}^+ be the sets of consistent sequences which may appear in antecedents and conclusions, respectively. *Semi-structured decision rules* in OORS are introduced with the following form [3]:

$$c \wedge c.n_{11}.\cdots.n_{1k_1} \sim x.n_{11}.\cdots.n_{1k_1} \wedge \cdots \wedge c.n_{s1}.\cdots.n_{sk_s} \sim x.n_{s1}.\cdots.n_{sk_s}$$
$$\Rightarrow c.m_{11}.\cdots.m_{1l_1} \sim x.m_{11}.\cdots.m_{1l_1} \wedge \cdots \wedge c.m_{t1}.\cdots.m_{tl_t} \sim x.m_{t1}.\cdots.m_{tl_t}$$
$$(3)$$

where $c \in C$, $x \in O$ such that $id_C(x) = c$, $n_{i1}.\cdots.n_{ik_i} \in N_{CON}^+$ $(1 \leq i \leq s, k_i \geq 1)$, $m_{j1}.\cdots.m_{jl_j} \in N_{DEC}^+$ $(1 \leq j \leq t, l_j \geq 1)$, and all sequences which appear in (3) have to connect to the object x. We call this rule a *semi-structured decision rule of the class c by the object x*. As a special case, we allow semi-structured decision rules with no sequences of names at the antecedent part as follows:

$$c \Rightarrow c.m_{11}.\cdots.m_{1l_1} \sim x.m_{11}.\cdots.m_{1l_1} \wedge \cdots \wedge c.m_{t1}.\cdots.m_{tl_t} \sim x.m_{t1}.\cdots.m_{tl_t}.$$
$$(4)$$

We say that a semi-structured decision rule is *consistent* if all the other semi-structured decision rules with the same antecedents to the semi-structured decision rule are also have the same conclusions. Hereafter, all semi-structured decision rules we treat in this paper are consistent.

Example 1. Suppose that we need to check a client person Mr. Foo's breakfast menus in eight days, and extract Mr. Foo's favorite patterns in breakfast menus. Here, we consider the following class structure $\mathcal{C} = (C, \ni_C, \sqsupseteq_C)$ with $C = \{\mathsf{Breakfast}i \ (1 \leq i \leq 6), \cdots, \mathsf{Favorite}\}$, where Kind, Taste and $\mathsf{Favorite}$ are attributes. Next, the name structure is $\mathcal{N} = (N, \ni_N, \sqsupseteq_N)$ with $N = \{\mathsf{breakfast}i, \cdots, \mathsf{favorite}\}$, Finally, the object structure that actually illustrates breakfast menus in eight days is $\mathcal{O} = (O, \ni_O, \sqsupseteq_O)$ with $O = \{\mathsf{bf}i, \mathsf{b}i, \cdots, \mathsf{yes}, \mathsf{no}\}$. Suppose also that these structures are well-defined.

Figure 1 illustrates has-a relationship among objects, and assignment of names to objects. For example, the object bf1 is an instance of the class Breakfast1, and bf1 consists of an instance b1(=bf1.bread) of the class Bread, an instance e1(=bf1.egg) of the class Egg, and so on, respectively. Note that the value of the attribute Favorite means whether Mr. Foo likes the menu, thus Mr. Foo likes the breakfast menus bf1, bf3 and bf4.

Now, to extract Mr. Foo's favorite patterns in breakfast menus, we consider to generate semi-structured decision rules from these well-defined structures. Using the set $N_{DEC}^+ = \{\text{favorite}\}$, and the following set:

$$N_{CON}^+ = (N \setminus \{\text{favorite}\}) \cup \left\{ \begin{array}{l} \text{breakfast}i.\text{bread, breakfast}i.\text{egg, breakfast}i.\text{salad,} \\ \text{breakfast}i.\text{drink, breakfast}i.\text{bread.kind,} \\ \text{breakfast}i.\text{bread.taste, breakfast}i.\text{egg.kind,} \cdots \end{array} \right\},$$

Therefore, we can consider the following semi-structured decision rules about Mr. Foo's favorites in breakfast menus:

- Breakfast1∧ Breakfast1.bread.kind ∼ toast ⇒ Breakfast1.favorite ∼ yes,
- Breakfast1∧ Breakfast1.salad.kind ∼ ham ⇒ Breakfast1.favorite ∼ yes,
- Breakfast2 ⇒ Breakfast2.favorite ∼ yes,
- etc. ⋯.
- Breakfast1∧ Breakfast1.bread.kind ∼ raisin ⇒ Breakfast1.favorite ∼ no,
- Breakfast1∧ Breakfast1.salad.kind ∼ vegetable ⇒ Breakfast1.favorite ∼ no,
- Breakfast$i$ ⇒ Breakfast$i$.favorite ∼ no $(3 \leq i \leq 6)$,
- etc. ⋯.

By these semi-structured decision rules, we can consider characteristic combinations of foods in *each breakfast menu*, however, it is difficult to interpret Mr. Foo's favorite patterns about *how to* combine foods.

3 Characteristic Combination Patterns as Essential Part of How to Combine Objects

In this section, we propose a concept of *characteristic combination patterns* which illustrate essential part of how to combine objects based on semi-structured decision rules in OORS. As we have mentioned, generally, semi-structured decision rules for a decision class describe what objects should be checked for correctly discerning objects in the decision class from other objects in different decision classes. Thus, we can consider characteristics about *what to combine objects* by such semi-structured decision rules. On the other hand, in some cases, we can discern objects in a decision class from other objects in different decision classes by just checking classes of objects. We think that this discernibility based on checking classes is due to some parts of the name structure related to the class. We regard such essential part as *characteristic combination patterns* which represent uniqueness of *how to combine objects* in the decision class.

By the above discussion, we introduce the concept of *characteristic combination patterns* for decision classes to capture the concept of characteristics about how to combine objects.

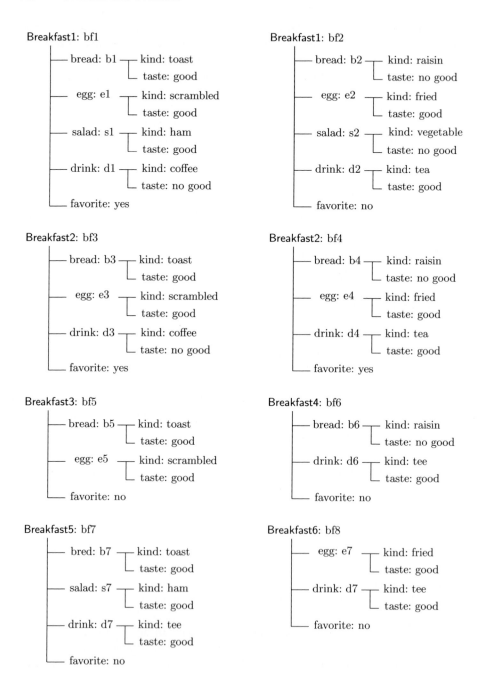

Fig. 1. Breakfast menus in Example 1

Definition 1. *Let N_{DEC}^+ be the set of consistent sequences which may appear in conclusions, $D \subseteq O$ be a decision class based on the equivalence relation $\approx_{N_{DEC}^+}$, $x \in D$ be an instance of a class c, that is, $id_C(x) = c$. Moreover, suppose there is a consistent semi-structured decision rule of the class c by the object x with no sequences of names in the antecedent part of the rule like (4). A characteristic combination pattern $CCP(c; D)$ of the decision class D by the class c is a non-empty subset of names defined as follows:*

$$CCP(c; D) \stackrel{\text{def}}{=} \left(\bigcup_{y \in T(c)} \{H_N(c) \setminus H_N(id_C(y))\} \right) \setminus N_{DEC}^+, \qquad (5)$$

where $A \setminus B \stackrel{\text{def}}{=} \{x \mid x \in A \text{ and } x \notin B\}$. The set $T(c)$ is a set of "target" objects discerned from instances of the class c, and defined as follows:

$$T(c) \stackrel{\text{def}}{=} \left\{ y \; \middle| \; \begin{array}{l} y \in O, id_C(y) \neq c \text{ and } \exists n_1. \cdots .n_k \in N_{DEC}^+ \\ \text{such that } n_1 \in H_N(c) \cap H_N(na(y)) \end{array} \right\}. \qquad (6)$$

For example, in the case of the class Breakfast2 and a non-favorite menu bf8 of the class Breakfast6 in Example 1, we have the following sets, respectively:

- $H_N(\text{Breakfast2}) = \{\text{bread}, \text{egg}, \text{drink}, \text{favorite}\}$,
- $H_N(id_C(\text{bf8})) = \{\text{egg}, \text{drink}, \text{favorite}\}$,
- $H_N(\text{Breakfast2}) \setminus H_N(id_C(\text{bf8})) = \{\text{bread}\}$.

Then, the existence of the name "bread" in $H_N(\text{Breakfast2})$ is essential to discern Breakfast2 from the class $id_C(\text{bf8})$ (=Breakfast6).

More generally, in Definition 1, the set $H_N(c) \setminus H_N(id_C(y))$ for any object $y \in T(c)$ illustrates essential parts to discern the class c from the class $id_C(y)$, therefore the set $H_N(c) \setminus H_N(id_C(y))$ captures the essential part of discernibility by checking the class c as follows:

- If there is some name n in $H_N(c) \setminus H_N(id_C(y))$: existence of the name n in $H_N(c)$ is essential to discern the class c from the class $id_C(x)$.
- If $H_N(c) \setminus H_N(id_C(y)) = \emptyset$ and there is some name n in $H_N(id_C(y))$: absence of the name n in $H_N(c)$ is essential to discern c from $id_C(y)$.
- Otherwise: we have no need to discern c from $id_C(y)$.

Thus, combining all the essential parts $H_N(c) \setminus H_N(id_C(y))$ of all $y \in T(c)$, and removing all names related to decision classes, we can get the characteristic combination pattern $CCP(c; D)$ of the decision class D by the class c.

Example 2. This example is continuation of Example 1. Using the following consistent semi-structured decision rule, we construct a characteristic combination pattern $CCP(\text{Breakfast2}; D)$ of the decision class of favorite menus $D \stackrel{\text{def}}{=} \{\text{bf1}, \text{bf3}, \text{bf4}\}$ by the class Breakfast2 based on the equivalence relation $\approx_{\text{favorite}}$:

- Breakfast2 $\Rightarrow$ Breakfast2.favorite $\sim$ yes.

First, we need to construct the set $T(\mathsf{Breakfast2})$ of objects. Each element in $T(\mathsf{Breakfast2})$ is comparable to any instance of $\mathsf{Breakfast2}$ by the equivalence relation $\approx_{\mathrm{favorite}}$, but the element is not an instance of $\mathsf{Breakfast2}$. Because objects to which the name "favorite" connects are instances of the classes $\mathsf{Breakfast}i$ $(1 \leq i \leq 6)$, thus, the set $T(\mathsf{Breakfast2})$ is the set of non-favorite menus:

- $T(\mathsf{Breakfast2}) \stackrel{\mathrm{def}}{=} \{\mathrm{bf2}, \mathrm{bf5}, \mathrm{bf6}, \mathrm{bf7}, \mathrm{bf8}\}$.

Next, we calculate essential parts to discern $\mathsf{Breakfast2}$ from classes of non-favorite menus as follows:

$$H_N(\mathsf{Breakfast2}) \setminus H_N(id_C(\mathrm{bf2})) = H_N(\mathsf{Breakfast2}) \setminus H_N(id_C(\mathrm{bf8})) = \{\mathrm{bread}\},$$
$$H_N(\mathsf{Breakfast2}) \setminus H_N(id_C(\mathrm{bf5})) = \{\mathrm{drink}\}, \ H_N(\mathsf{Breakfast2}) \setminus H_N(id_C(\mathrm{bf7})) = \emptyset,$$
$$H_N(\mathsf{Breakfast2}) \setminus H_N(id_C(\mathrm{bf6})) = \{\mathrm{egg}\}.$$

Combining these essential parts, we get a characteristic combination pattern:

- $CCP(\mathsf{Breakfast2}, D) = \{\mathrm{bread}, \mathrm{egg}, \mathrm{drink}\}$.

The constructed characteristic combination pattern indicates that the names "bread", "egg" and "drink" are essential to discern the class $\mathsf{Breakfast2}$ from the other classes. Therefore, by this result, we can guess that Mr. Foo likes the combination of bread, egg and drink.

4   Conclusion

In this paper, we have proposed characteristic combination patterns to treat characteristics about how to combine objects in OORS. Characteristic combination patterns of a fixed decision class illustrate essential parts to discern the decision class from other decision classes, thus we consider that the characteristic combination patterns capture the uniqueness of how to combine objects in the decision class. As future issues, we need to refine the proposed method to calculate characteristic combination patterns in more general cases, because the calculation method in this paper requires existence of semi-structured decision rules with special forms. Moreover, implementation of the proposed method and evaluation experiments using real-life data are very important future issues.

Acknowledgments

We would like to thank to anonymous reviewers for their helpful comments.

References

1. Kudo, Y., Murai, T.: A Theoretical Formulation of Object-Oriented Rough Set Models. Journal of Advanced Computational Intelligence and Intelligent Informatics 10(5), 612–620 (2006)

2. Kudo, Y., Murai, T.: A Method of Generating Decision Rules in Object-Oriented Rough Set Models. In: Greco, S., Hata, Y., Hirano, S., Inuiguchi, M., Miyamoto, S., Nguyen, H.S., Słowiński, R. (eds.) RSCTC 2006. LNCS (LNAI), vol. 4259, pp. 338–347. Springer, Heidelberg (2006)
3. Kudo, Y., Murai, T.: Semi-Structured Decision Rules in Object-Oriented Rough Set Models for Kansei Engineering. In: Yao, J., Lingras, P., Wu, W.-Z., Szczuka, M., Cercone, N.J., Ślęzak, D. (eds.) RSKT 2007. LNCS (LNAI), vol. 4481, pp. 219–227. Springer, Heidelberg (2007)
4. Pawlak, Z.: Rough Sets. International Journal of Computer and Information Science 11, 341–356 (1982)
5. Popkorn, S.: First Steps in Modal Logic. Cambridge University Press, Cambridge (1994)

Induced Intuitionistic Fuzzy Ordered Weighted Averaging Operator and Its Application to Multiple Attribute Group Decision Making

Guiwu Wei

Department of Economics and Management, Chongqing University of Arts and Sciences,
Yongchuan, Chongqing, 402160, P.R. China
weiguiwu@163.com

Abstract. With respect to multiple attribute group decision making (MAGDM) problems in which both the attribute weights and the expert weights take the form of real numbers, attribute values take the form of intuitionistic fuzzy numbers, a new group decision making analysis method is developed. Firstly, some operational laws of intuitionistic fuzzy numbers, score function and accuracy function of intuitionistic fuzzy numbers are introduced. Then a new aggregation operator called induced intuitionistic fuzzy ordered weighted averaging (I-IFOWA) operator is proposed, and some desirable properties of the I-IFOWA operators are studied, such as commutativity, idempotency and monotonicity. An I-IFOWA and IFWA (intuitionistic fuzzy weighted averaging) operators-based approach is developed to solve the MAGDM under the intuitionistic fuzzy environment. Finally, an illustrative example is given to verify the developed approach and to demonstrate its practicality and effectiveness.

Keywords: Group decision making, Intuitionistic fuzzy numbers, Induced intuitionistic fuzzy ordered weighted averaging (I-IFOWA) operator.

1 Introduction

Atanassov [1,2] introduced the concept of intuitionistic fuzzy set(IFS), which is a generalization of the concept of fuzzy set [3]. The intuitionistic fuzzy set has received more and more attention since its appearance. Gau and Buehrer [4] introduced the concept of vague set. But Bustince and Burillo [5] showed that vague sets are intuitionistic fuzzy sets. In [6], Xu developed some geometric aggregation operators. In [7], Xu developed some arithmetic aggregation operators. In this paper, we shall develop a new operator called induced intuitionistic fuzzy ordered weighted averaging (I-IFOWA) operator which is an extension of IOWA operator proposed by Yager and Filev[8]. Based on the I-IFOWA and IFWA operators, we shall develop an approach to MAGDM under intuitionistic fuzzy environment. The remainder of this paper is set out as follows. In the next section, we introduce some basic concepts related to intuitionistic fuzzy sets. In Section 3 a new aggregation operator called I-IFOWA operator is proposed, and some desirable properties of the I-IFOWA operators are studied, such as commutativity, idempotency and monotonicity. In Section 4, An

G. Wang et al. (Eds.): RSKT 2008, LNAI 5009, pp. 124–131, 2008.

I-IFOWA and IFWA operators-based approach is developed to solve the MAGDM under the intuitionistic fuzzy environment. In Section 5, an illustrative example is given. In Section 6 we conclude the paper and give some remarks.

2 Preliminaries

In the following, we introduce some basic concepts related to IFS.

Definition 1. An IFS A in X is given by

$$A = \left\{ \left\langle x, \mu_A(x), \nu_A(x) \right\rangle \middle| x \in X \right\} \tag{1}$$

where $\mu_A : X \to [0,1]$ and $\nu_A : X \to [0,1]$, with the condition

$$0 \leq \mu_A(x) + \nu_A(x) \leq 1, \ \forall \, x \in X$$

The numbers $\mu_A(x)$ and $\nu_A(x)$ represent, respectively, the membership degree and non- membership degree of the element x to the set A [1,2].

Definition 2. For each IFS A in X, if

$$\pi_A(x) = 1 - \mu_A(x) - \nu_A(x), \ \forall \, x \in X . \tag{2}$$

Then $\pi_A(x)$ is called the degree of indeterminacy of x to A [1,2].

Definition 3. Let $\tilde{a} = (\mu, \nu)$ be an intuitionistic fuzzy number, a score function S of an intuitionistic fuzzy value can be represented as follows [9]:

$$S(\tilde{a}) = \mu - \nu, \ \ S(\tilde{a}) \in [-1,1] . \tag{3}$$

Definition 4. Let $\tilde{a} = (\mu, \nu)$ be an intuitionistic fuzzy number, a accuracy function H of an intuitionistic fuzzy value can be represented as follows [10]:

$$H(\tilde{a}) = \mu + \nu, \ \ H(\tilde{a}) \in [0,1] . \tag{4}$$

to evaluate the degree of accuracy of the intuitionistic fuzzy value $\tilde{a} = (\mu, \nu)$, where $H(\tilde{a}) \in [0,1]$. The larger the value of $H(\tilde{a})$, the more the degree of accuracy of the intuitionistic fuzzy value $\tilde{a}$.

Based on the score function S and the accuracy function H, Xu[6] give an order relation between two intuitionistic fuzzy values, which is defined as follows:

Definition 5. Let $\tilde{a}_1 = (\mu_1, \nu_1)$ and $\tilde{a}_2 = (\mu_2, \nu_2)$ be two intuitionistic fuzzy values, $s(\tilde{a}_1) = \mu_1 - \nu_1$ and $s(\tilde{a}_2) = \mu_2 - \nu_2$ be the scores of $\tilde{a}$ and $\tilde{b}$, respectively, and let $H(\tilde{a}_1) = \mu_1 + \nu_1$ and $H(\tilde{a}_2) = \mu_2 + \nu_2$ be the accuracy degrees of $\tilde{a}$ and $\tilde{b}$, respectively, then if $S(\tilde{a}) < S(\tilde{b})$, then $\tilde{a}$ is smaller than $\tilde{b}$, denoted by $\tilde{a} < \tilde{b}$; if

$S(\tilde{a}) = S(\tilde{b})$, then, (1) if $H(\tilde{a}) = H(\tilde{b})$, then $\tilde{a}$ and $\tilde{b}$ represent the same information, denoted by $\tilde{a} = \tilde{b}$; (2) if $H(\tilde{a}) < H(\tilde{b})$, $\tilde{a}$ is smaller than $\tilde{b}$, denoted by $\tilde{a} < \tilde{b}$ [6].

3 I-IFOWA Operator

Definition 6. Let $\tilde{a}_j = (\mu_j, v_j)(j = 1, 2, \cdots, n)$ be a collection of intuitionistic fuzzy values, and let IFWA: $Q^n \to Q$, if

$$\text{IFWA}_\omega(\tilde{a}_1, \tilde{a}_2, \cdots, \tilde{a}_n) = \sum_{j=1}^{n} \omega_j \tilde{a}_j = \left(1 - \prod_{j=1}^{n}(1 - \mu_j)^{\omega_j}, \prod_{j=1}^{n} v_j^{\omega_j}\right) \quad (5)$$

where $\omega = (\omega_1, \omega_2, \cdots, \omega_n)^T$ be the weight vector of $\tilde{a}_j (j = 1, 2, \cdots, n)$, and $\omega_j > 0$, $\sum_{j=1}^{n} \omega_j = 1$, then IFWA is called the intuitionistic fuzzy weighted averaging (IFWA) operator [7].

Definition 7. Let $\tilde{a}_j = (\mu_j, v_j)(j = 1, 2, \cdots, n)$ be a collection of intuitionistic fuzzy values, An intuitionistic fuzzy ordered weighted averaging (IFOWA) operator of dimension n is a mapping IFOWA: $Q^n \to Q$, that has an associated weight vector $w = (w_1, w_2, \cdots, w_n)^T$ such that $w_j > 0$ and $\sum_{j=1}^{n} w_j = 1$. Furthermore,

$$\text{IFOWA}_w(\tilde{a}_1, \tilde{a}_2, \cdots, \tilde{a}_n) = \sum_{j=1}^{n} w_j \tilde{a}_{\sigma(j)} = \left(1 - \prod_{j=1}^{n}(1 - \mu_{\sigma(j)})^{w_j}, \prod_{j=1}^{n} v_{\sigma(j)}^{w_j}\right) \quad (6)$$

where $(\pi(1), \pi(2), \cdots, \pi(n))$ is a permutation of $(1, 2, \cdots, n)$, such that $\tilde{\alpha}_{\pi(j-1)} \geq \tilde{\alpha}_{\pi(j)}$ for all $j = 2, \cdots, n$. [7].

In the following, we shall develop an induced intuitionistic fuzzy ordered weighted averaging (I-IFOWA) operator.

Definition 8. An induced intuitionistic fuzzy ordered weighted averaging (I-IFOWA) operator is defined as follows:

$$\text{I-IFOWA}_w(\langle u_1, \tilde{a}_1 \rangle, \langle u_2, \tilde{a}_2 \rangle, \cdots, \langle u_n, \tilde{a}_n \rangle)$$

$$= \sum_{j=1}^{n} w_j \tilde{g}_j = \left(1 - \prod_{j=1}^{n}(1 - \bar{\mu}_j)^{w_j}, \prod_{j=1}^{n} \bar{v}_j^{w_j}\right) \quad (7)$$

where $w = (w_1, w_2, \cdots, w_n)^T$ is a weighting vector, such that $w_j \in [0,1]$,

$\sum_{j=1}^{n} w_j = 1$, $j = 1, 2, \cdots, n$, $\tilde{g}_j = (\bar{\mu}_j, \bar{V}_j)$ is the $\tilde{a}_i$ value of the IFOWA pair

$\langle u_i, \tilde{a}_i \rangle$ having the jth largest u_i $(u_i \in [0,1])$,and u_i in $\langle u_i, \tilde{a}_i \rangle$ is referred to as the

order inducing variable and $\tilde{a}_i$ $(\tilde{a}_i = (\mu_i, V_i))$ as the intuitionistic fuzzy values.

The I-IFOWA operator has the following properties similar to those of the IOWA operator[8].

Theorem 1 (Commutativity)
$$\text{I-IFOWA}_w \left(\langle u_1, \tilde{a}_1 \rangle, \langle u_2, \tilde{a}_2 \rangle, \cdots, \langle u_n, \tilde{a}_n \rangle \right)$$
$$= \text{I-IFOWA}_w \left(\langle u_1, \tilde{a}_1' \rangle, \langle u_2, \tilde{a}_2' \rangle, \cdots, \langle u_n, \tilde{a}_n' \rangle \right)$$
where $\left(\langle u_1, \tilde{a}_1' \rangle, \langle u_2, \tilde{a}_2' \rangle, \cdots, \langle u_n, \tilde{a}_n' \rangle \right)$ is any permutation of $\left(\langle u_1, \tilde{a}_1 \rangle, \langle u_2, \tilde{a}_2 \rangle, \cdots, \langle u_n, \tilde{a}_n \rangle \right)$.

Theorem 2. (Idempotency) If $\tilde{a}_j \left(\tilde{a}_j = (\mu_j, V_j) \right) = \tilde{a} \left(\tilde{a} = (\mu, V) \right)$ for all j , then
$$\text{I-IFOWA}_w \left(\langle u_1, \tilde{a}_1 \rangle, \langle u_2, \tilde{a}_2 \rangle, \cdots, \langle u_n, \tilde{a}_n \rangle \right) = \tilde{a}$$

Theorem 3. (Monotonicity) If $\tilde{a}_j \leq \tilde{a}_j'$ for all j, then
$$\text{I-IFOWA}_w \left(\langle u_1, \tilde{a}_1 \rangle, \langle u_2, \tilde{a}_2 \rangle, \cdots, \langle u_n, \tilde{a}_n \rangle \right)$$
$$\leq \text{I-IFOWA}_w \left(\langle u_1, \tilde{a}_1' \rangle, \langle u_2, \tilde{a}_2' \rangle, \cdots, \langle u_n, \tilde{a}_n' \rangle \right)$$

4 An Approach to Group Decision Making with Intuitionistic Fuzzy Information

Let $A = \{A_1, A_2, \cdots, A_m\}$ be a discrete set of alternatives, and $G = \{G_1, G_2, \cdots, G_n\}$ be the set of attributes, $\omega = (\omega_1, \omega_2, \cdots, \omega_n)$ is the weighting vector of the attribute G_j $(j = 1, 2, \cdots, n)$, where $\omega_j \in [0,1]$, $\sum_{j=1}^{n} \omega_j = 1$. Let $D = \{D_1, D_2, \cdots, D_t\}$ be the set of decision makers, $V = (V_1, V_2, \cdots, V_n)$ be the weighting vector of decision makers, with $V_k \in [0,1]$, $\sum_{k=1}^{t} V_k = 1$. Suppose that

$\tilde{R}_k = \left(\tilde{r}_{ij}^{(k)}\right)_{m \times n} = \left(\mu_{ij}^{(k)}, v_{ij}^{(k)}\right)_{m \times n}$ is the intuitionistic fuzzy decision matrix, where $\mu_{ij}^{(k)}$ indicates the degree that the alternative A_i satisfies the attribute G_j given by the decision maker D_k, $v_{ij}^{(k)}$ indicates the degree that the alternative A_i doesn't satisfy the attribute G_j given by the decision maker D_k, $\mu_{ij}^{(k)} \subset [0,1]$, $v_{ij}^{(k)} \subset [0,1]$, $\mu_{ij}^{(k)} + v_{ij}^{(k)} \leq 1$, $i = 1, 2, \cdots, m$, $j = 1, 2, \cdots, n$, $k = 1, 2, \cdots, t$.

In the following, we apply the I-IFOWA and IFWA operator to MAGDM based on intuitionistic fuzzy information.

Step 1. Utilize the decision information given in matrix $\tilde{R}_k$, and the I-IFOWA operator which has associated weighting vector $w = \left(w_1, w_2, \cdots, w_n\right)^T$

$$\tilde{r}_{ij} = \left(\mu_{ij}, v_{ij}\right) = \text{I-IFOWA}_w\left(\left\langle v_1, \tilde{r}_{ij}^{(1)}\right\rangle, \left\langle v_2, \tilde{r}_{ij}^{(2)}\right\rangle, \cdots, \left\langle v_t, \tilde{r}_{ij}^{(t)}\right\rangle\right)$$
$$i = 1, 2, \cdots, m, \, j = 1, 2, \cdots, n. \tag{8}$$

to aggregate all the decision matrices $\tilde{R}_k \, (k = 1, 2, \cdots, t)$ into a collective decision matrix $\tilde{R} = \left(\tilde{r}_{ij}\right)_{m \times n}$, where $V = \{v_1, v_2, \cdots, v_t\}$ be the weighting vector of decision makers.

Step 2. Utilize the decision information given in matrix $\tilde{R}$, and the IFWA operator

$$\tilde{r}_i = \left(\mu_i, v_i\right) = \text{IFWA}_\omega\left(\tilde{r}_{i1}, \tilde{r}_{i2}, \cdots, \tilde{r}_{in}\right), \, i = 1, 2, \cdots, m. \tag{9}$$

to derive the collective overall preference values $\tilde{r}_i \, (i = 1, 2, \cdots, m)$ of the alternative A_i, where $\omega = \left(\omega_1, \omega_2, \cdots, \omega_n\right)^T$ is the weighting vector of the attributes.

Step 3. calculate the scores $S(\tilde{r}_i) \, (i = 1, 2, \cdots, m)$ of the collective overall intuitionistic fuzzy preference values $\tilde{r}_i \, (i = 1, 2, \cdots, m)$ to rank all the alternatives $A_i \, (i = 1, 2, \cdots, m)$ and then to select the best one(s) (if there is no difference between two scores $S(\tilde{r}_i)$ and $S(\tilde{r}_j)$, then we need to calculate the accuracy degrees $H(\tilde{r}_i)$ and $H(\tilde{r}_j)$ of the collective overall intuitionistic fuzzy preference values $\tilde{r}_i$ and $\tilde{r}_j$, respectively, and then rank the alternatives A_i and A_j in accordance with the accuracy degrees $H(\tilde{r}_i)$ and $H(\tilde{r}_j)$.

Step 4. Rank all the alternatives $A_i \, (i = 1, 2, \cdots, m)$ and select the best one(s) in accordance with $S(\tilde{r}_i)$ and $H(\tilde{r}_i) \, (i = 1, 2, \cdots, m)$.

5 Illustrative Example

Let us suppose there is an investment company, which wants to invest a sum of money in the best option [11]. There is a panel with five possible alternatives to invest the money:①A_1 is a car company;②A_2 is a food company;③A_3 is a computer company;④A_4 is an arms company;⑤A_5 is a TV company. The investment company must take a decision according to the following four attributes:①G_1 is the risk analysis;②G_2 is the growth analysis;③G_3 is the social-political impact analysis;④G_4 is the environmental impact analysis.

The five possible alternatives $A_i(i=1,2,\cdots,5)$ are to be evaluated using the intuitionistic fuzzy numbers by the three decision makers (whose weighting vector $v=(0.35,0.40,0.25)^T$) under the above four attributes (whose weighting vector $\omega=(0.2,0.1,0.3,0.4)^T$), and construct, respectively, the decision matrices as listed in the following matrices $\tilde{R}_k=\left(\tilde{r}_{ij}^{(k)}\right)_{5\times4}$ $(k=1,2,3)$ as follows:

$$\tilde{R}_1=\begin{bmatrix}(0.4,0.3)&(0.5,0.2)&(0.2,0.5)&(0.1,0.6)\\(0.6,0.2)&(0.6,0.1)&(0.6,0.1)&(0.3,0.4)\\(0.5,0.3)&(0.4,0.3)&(0.4,0.2)&(0.5,0.2)\\(0.7,0.1)&(0.5,0.2)&(0.2,0.3)&(0.1,0.5)\\(0.5,0.1)&(0.3,0.2)&(0.6,0.2)&(0.4,0.2)\end{bmatrix}$$

$$\tilde{R}_2=\begin{bmatrix}(0.5,0.4)&(0.6,0.3)&(0.3,0.6)&(0.2,0.7)\\(0.7,0.3)&(0.7,0.2)&(0.7,0.2)&(0.4,0.5)\\(0.6,0.4)&(0.5,0.4)&(0.5,0.3)&(0.6,0.3)\\(0.8,0.1)&(0.6,0.3)&(0.3,0.4)&(0.2,0.6)\\(0.6,0.2)&(0.4,0.3)&(0.7,0.1)&(0.5,0.3)\end{bmatrix}$$

$$\tilde{R}_3=\begin{bmatrix}(0.4,0.5)&(0.5,0.4)&(0.2,0.7)&(0.1,0.8)\\(0.6,0.4)&(0.6,0.3)&(0.6,0.3)&(0.3,0.6)\\(0.5,0.5)&(0.4,0.5)&(0.4,0.4)&(0.5,0.4)\\(0.7,0.2)&(0.5,0.4)&(0.2,0.5)&(0.1,0.7)\\(0.5,0.3)&(0.3,0.4)&(0.6,0.2)&(0.4,0.4)\end{bmatrix}$$

Then, we utilize the approach developed to get the most desirable alternative(s).

Step 1. Utilize the decision information given in matrix $\tilde{R}_k$, and the I-IFOWA operator which has associated weighting vector $w = (0.2, 0.35, 0.45)^T$, we get a collective decision matrix $\tilde{R} = \left(\tilde{r}_{ij}\right)_{m \times n}$ as follows:

$$\tilde{R} = \begin{bmatrix} (0.421, 0.400) & (0.522, 0.296) & (0.221, 0.603) & (0.121, 0.704) \\ (0.622, 0.296) & (0.622, 0.188) & (0.622, 0.188) & (0.321, 0.502) \\ (0.522, 0.400) & (0.421, 0.400) & (0.421, 0.296) & (0.522, 0.296) \\ (0.723, 0.137) & (0.522, 0.296) & (0.221, 0.400) & (0.121, 0.603) \\ (0.522, 0.188) & (0.321, 0.296) & (0.622, 0.174) & (0.421, 0.296) \end{bmatrix}$$

Step 2. Utilize the IFWA operator, we obtain the collective overall preference values $\tilde{r}_i$ of the alternatives $A_i \, (i = 1, 2, \cdots, 5)$.

$$\tilde{r}_1 = (0.266, 0.551), \tilde{r}_2 = (0.523, 0.305), \tilde{r}_3 = (0.484, 0.324)$$
$$\tilde{r}_4 = (0.367, 0.369), \tilde{r}_5 = (0.502, 0.231)$$

Step 3. calculate the scores $S(\tilde{r}_i)(i=1,2,\cdots,5)$ of the collective overall intuitionistic fuzzy preference values $\tilde{r}_i \, (i = 1, 2, \cdots, 5)$

$$S(\tilde{r}_1) = -0.284, S(\tilde{r}_2) = 0.217, S(\tilde{r}_3) = 0.160, S(\tilde{r}_4) = -0.002, S(\tilde{r}_5) = 0.271$$

Step 4. Rank all the alternatives $A_i \, (i = 1, 2, 3, 4, 5)$ in accordance with the scores $S(\tilde{r}_i) \, (i=1,2,\cdots,5)$ of the collective overall intuitionistic fuzzy preference values $\tilde{r}_i (i=1,2,\cdots,5)$: $A_5 \succ A_2 \succ A_3 \succ A_4 \succ A_1$, and thus the most desirable alternative is A_5.

6 Conclusion

In this paper, we have developed an induced intuitionistic fuzzy ordered weighted averaging (I-IFOWA) operator, which take as their argument pairs, called IFOWA pairs, in which one component is used to induce an ordering over the second components which are intuitionistic fuzzy values and then aggregated. We have studied some desirable properties of the I-IFOWA operators, such as commutativity, idempotency and monotonicity, and applied the I-IFOWA operators to group decision making with intuitionistic fuzzy information. Finally an illustrative example has been given to show the developed method. In the future, we shall continue working in the extension and application of the I-IFOWA operators to other domains.

References

1. Atanassov, K.: Intuitionistic Fuzzy Sets. Fuzzy Sets and Systems 20, 87–96 (1986)
2. Atanassov, K.: More on Intuitionistic Fuzzy Sets. Fuzzy Sets and Systems 33, 37–46 (1989)
3. Zadeh, L.A.: Fuzzy Sets. Information and Control 8, 338–356 (1965)
4. Gau, W.L., Buehrer, D.J.: Vague Sets. IEEE Transactions on Systems, Man and Cybernetics 23(2), 610–614 (1993)
5. Bustine, H., Burillo, P.: Vague Sets are Intuitionistic Fuzzy Sets. Fuzzy Sets and Systems 79, 403–405 (1996)
6. Xu, Z.S., Yager, R.R.: Some Geometric Aggregation Operators Based on Intuitionistic Fuzzy Sets. International Journal of General System 35(4), 417–433 (2006)
7. Xu, Z.S.: Intuitionistic Fuzzy Aggregation Operators. IEEE Transations on Fuzzy Systems 15(6), 1179–1187 (2007)
8. Yager, R.R., Filev, D.P.: Induced Ordered Weighted Averaging Operators. IEEE Transactions on Systems, Man, and Cybernetics- Part B 29, 141–150 (1999)
9. Chen, S.M., Tan, J.M.: Handling Multicriteria Fuzzy Decision-making Problems Based on Vague Set Theory. Fuzzy Sets and Systems 67, 163–172 (1994)
10. Hong, D.H., Choi, C.H.: Multicriteria Fuzzy Problems Based on Vague Set Theory. Fuzzy Sets and Systems 114, 103–113 (2000)
11. Herrera, F., Herrera-Viedma, E.: Linguistic Decision Analysis: Steps for Solving Decision Problems under Linguistic Information. Fuzzy Sets and Systems 115, 67–82 (2000)
12. Yager, R.R., Kacprzyk, J.: The Ordered Weighted Averaging Operators: Theory and Applications. Kluwer, Boston (1997)

Game-Theoretic Risk Analysis in Decision-Theoretic Rough Sets

Joseph P. Herbert and JingTao Yao

Department of Computer Science, University of Regina
Regina, Saskatchewan, Canada S4S 0A2
[herbertj,jtyao]@cs.uregina.ca

Abstract. Determining the correct threshold values for probabilistic rough set models has been a heated issue among the community. This article will formulate a game-theoretic approach to calculating these thresholds to ensure correct approximation region size. By finding equilibrium within payoff tables created from approximation measures and modified conditional risk strategies, we provide the user with tolerance levels for their loss functions. Using the tolerance values, new thresholds are calculated to provide correct classification regions. Better informed decisions can be made when utilizing these tolerance values.

1 Introduction

In rough sets [10], a set within the universe of discourse is approximated. Rough set regions are defined with these approximations. One of the goals of improving the classification ability of rough sets is to reduce the boundary region, thus, reducing the impact that this uncertainty has on decision making. The decision-theoretic rough set [16] and variable-precision rough set [17] models were proposed solutions to this problem of decreasing the boundary region.

The decision-theoretic rough set model (DTRS) [14] utilizes the Bayesian decision procedure to calculate rough set classification regions. Loss functions correspond to the risks involved in classifying an object into a particular classification region. This gives the user a scientific means for linking their risk tolerances with the probabilistic classification ability of rough sets [12].

The decision-theoretic model observes a lower and upper-bound threshold for region classification [13]. The thresholds α and β provide the probabilities for inclusion into the positive, negative, and boundary regions. The α and β thresholds are calculated through the analysis of loss function relationships, thus, a method of reducing the boundary region materializes from the modification of the loss functions. Utilizing game theory to analyze the relationships between classification ability and the modification of loss functions, we can provide the user with a means for changing their risk tolerances.

Classification ability of a rough set analysis system is a measurable characteristic [4]. In this article, we introduce a method for calculating loss tolerance using game theory to analyze the effects of modifying the classification risk. This also provides an effective means of determining how much a loss function can fluctuate in order to maintain effective classification ability.

G. Wang et al. (Eds.): RSKT 2008, LNAI 5009, pp. 132–139, 2008.

2   Decision-Theoretic Rough Sets

The decision-theoretic approach is a robust extension of rough sets for two reasons. First, it calculates approximation parameters by obtaining easily understandable notions of risk or loss from the user [14,15].

2.1   Loss Functions

Let $P(w_j|\mathbf{x})$ be the conditional probability of an object x being in state w_j given the object description $\mathbf{x}$. The set of actions is given by $\mathcal{A} = \{a_P, a_N, a_B\}$, where a_P, a_N, and a_B represent the three actions to classify an object into $POS(A)$, $NEG(A)$, and $BND(A)$ respectively. Let $\lambda(a_\diamond|A)$ denote the loss incurred for taking action $a_\diamond$ when an object is in A, and let $\lambda(a_\diamond|A^c)$ denote the loss incurred by taking the same action when the object belongs to A^c. This can be given as loss functions $\lambda_{\diamond P} = \lambda(a_\diamond|A)$, $\lambda_{\diamond N} = \lambda(a_\diamond|A^c)$, and $\diamond = P$, N, or B. Through the combination of the set of loss functions, α, β, and γ parameters can be calculated to define the regions.

A crucial assumption when using this model is that the set of loss functions is provided by the user. This is a drawback, as it is still dependant upon user-provided information for calculating rough set region boundaries. In order to pass this obstacle, a method of calculating loss functions from the relationships found within the actual data must be found. Although this is beyond the scope of this article, we can provide a method for determining how much these loss functions can change, an equally important problem.

2.2   Conditional Risk

The expected loss $R(a_\diamond|[x])$ associated with taking the individual actions can be expressed as:

$$
\begin{aligned}
R_P = R(a_P|[x]) &= \lambda_{PP}P(A|[x]) + \lambda_{PN}P(A^c|[x]), \\
R_N = R(a_N|[x]) &= \lambda_{NP}P(A|[x]) + \lambda_{NN}P(A^c|[x]), \\
R_B = R(a_B|[x]) &= \lambda_{BP}P(A|[x]) + \lambda_{BN}P(A^c|[x]),
\end{aligned}
\tag{1}
$$

where $\lambda_{\diamond P} = \lambda(a_\diamond|A)$, $\lambda_{\diamond N} = \lambda(a_\diamond|A^c)$, and $\diamond = P$, N, or B. R_P, R_N, and R_B are the expected losses of classifying an object into the positive region, negative region, and boundary region respectively. The Bayesian decision procedure leads to the following minimum-risk decision rules (PN-BN):

(PN)   If $R_P \leq R_N$ and $R_P \leq R_B$,   decide $POS(A)$;
(NN)   If $R_N \leq R_P$ and $R_N \leq R_B$,   decide $NEG(A)$;
(BN)   If $R_B \leq R_P$ and $R_B \leq R_N$,   decide $BND(A)$;

These minimum-risk decision rules offer us a foundation in which to classify objects into approximation regions. They give us the ability to not only collect decision rules from data frequent in many rough set applications [6], but also the calculated risk that is involved when discovering (or acting upon) those rules.

3   A Game-Theoretic Calculation for Conditional Risk

We stated previously that the user could make use of a method of linking their notions of cost (risk) in taking a certain action and classification ability of the classification system. Game theory can be a powerful mathematical paradigm for analyzing these relationships and also provides methods for achieving optimal configurations for classification strategies. It could also provide a means for the user to change their beliefs regarding the types of decisions they can make [7]. They would not have to change the probabilities themselves, only their risk beliefs. This is beneficial as many users cannot intuitively describe their decision needs in terms of probabilities.

3.1   The Boundary Region and Conditional Risk

We wish to emphasize the relationship between the conditional risk, loss functions, and boundary region. Classification can be performed by following the minimum risk decision rules PN, NN, and BN or by using the α and β parameters to define region separation. We wish to make the boundary region smaller by modifying either method so that the positive region can be increased. To measure the changes made to the regions, we use two measures: approximation accuracy (ϕ) and approximation precision (ψ).

When increasing the size of the positive region, the size of the lower approximation is made larger. By recording the accuracy and precision measures, we can directly see the impact this has on classification ability. To increase the size of the lower approximation, measured by ϕ and ψ, we can observe the changes in the conditional risk found in Equation 1. That is, to increase the size of the lower approximation, we can reduce the risk associated with classifying an object into the positive region. This can be done by modifying the loss functions.

Furthermore, while doing this, we need to maintain the size of $|\overline{apr}(A)|$. Recalling rules (PN, NN, BN), we see that in order to increase the size of the lower approximation, we need decrease the expected loss R_P. This results in more objects being classified into the positive region since it is less "risky" to do so. An increase R_N and R_B may also have the desired effect. This is intuitive when considering that in order for more objects to be classified into $POS(A)$, we need to lower the risk involved in classifying an object into this region.

We see that in order to decrease the value of R_P, we need to decrease one or both of the loss functions λ_{PP} and λ_{PN} (Equation 1: R_P). Likewise, to increase

Table 1. The strategy scenario of increasing approximation accuracy

Action (Strategy)	Goal	Method	Result
a_1 ($-R_P$)	Decrease R_P	Decrease λ_{PP} or λ_{PN}	Larger POS region
a_2 ($+R_N$)	Increase R_N	Increase λ_{NP} or λ_{NN}	Smaller NEG region
a_3 ($+R_B$)	Increase R_B	Increase λ_{BP} or λ_{BN}	Smaller BND region

R_N, we need to increase either λ_{NP} or λ_{NN}. Finally, to increase R_B, we need to increase λ_{BP} or λ_{BN}. This is summarized in Table 1.

We want to increase approximation precision when considering the second measure, ψ. For the deterministic case, in order to increase precision, we need to make $|\underline{apr}(A)|$ as large as possible. Again, recalling rules (PN, NN, BN), we see that in order to increase the size of the lower approximation, we need to decrease the expected loss R_P and to increase R_N and R_B. It has the same strategy set as the first player because we wish to increase the size of the lower approximation.

Of course, there may be some tradeoff between the measures ϕ and ψ. An increase in one will not have a similar increase in the other. This implies some form of conflict between these measures. We can now use game theory to dictate the increases/decreases in conditional risk for region classification and as a method for governing the changes needed for the loss functions.

3.2   Game-Theoretic Specification

Game theory [9] has been one of the core subjects of the decision sciences, specializing in the analysis of decision-making in an interactive environment. The disciplines utilizing game theory include economics [8,11], networking [1], and machine learning [5].

When using game theory to help determine suitable loss functions, we need to correctly formulate the following: a set of players, a set of strategies for each player, and a set of payoff functions. Game theory uses these formulations to find an optimal strategy for a single player or the entire group of players if cooperation (coordination) is wanted. A single game is defined as,

$$G = \{O, S, F\}, \tag{2}$$

where G is a game consisting of a set of players O using strategies in S. These strategies are measured using individual payoff functions in F.

To begin, the set of players should reflect the overall purpose of the competition. In a typical example, a player can be a person who wants to achieve certain goals. For simplicity, we will be using competition between two players. With improved classification ability as the competition goal, each player can represent a certain measure such as accuracy (ϕ) and precision (ψ). With this in mind, a set of players is formulated as $O = \{\phi, \psi\}$. Through competition, optimal values are attempting to appear for each measure. Although we are measuring accuracy and precision, the choice of measures is ultimately up to the user to decide. We wish to analyze the amount of movement or compromise loss functions can have when attempting to achieve optimal values for these two measures.

Each measure is effectively competing with the other to win the "game". Here, the game is to improve classification ability. To compete, each measure in O has a set of strategies it can employ to achieve payoff. Payoff is the measurable result of actions performed using the strategies. These strategies are executed by the player in order to better their position in the future, e.g., maximize payoff. Individual strategies, when performed, are called *actions*. The strategy set $S_i = \{a_1, \ldots, a_m\}$ for any measure i in O contains these actions. A total of m actions can be performed for this player. The strategic goal for ϕ would be along the lines of "acquire a maximal value for approximation accuracy as possible". Likewise, the strategy for ψ would be to "acquire a maximal value for approximation precision as possible".

Approximation accuracy (ϕ), is defined as the ratio measured between the size of the lower approximation of a set A to the upper approximation of a set A. A large value of ϕ indicates that we have a small boundary region. To illustrate the change in approximation accuracy, suppose we have player ϕ taking two turns in the competition. For the first turn, player ϕ executes action a_1 from it's strategy set. When it is time to perform another turn, the player executes action a_2. Ultimately, since the player's goal is to increase approximation accuracy, we should measure that $\phi_{a_1} \leq \phi_{a_2}$. If this is not the case ($\phi_{a_1} > \phi_{a_2}$), the player has chosen a poor second action from it's strategy set.

The second player, approximation precision (ψ), observes the relationship between the upper approximation and a set. In order to increase precision, we need to make $|\underline{apr}(A)|$ as large as possible. For non-deterministic approximations, Yao [13] suggested an alternative precision measure.

In general, the two measures ϕ and ψ show the impacts that the loss functions have on the classification ability of the DTRS model. Modifying the loss functions contribute to a change in risk (expected cost). Determining how to modify the loss functions to achieve different classification abilities requires a set of risk modification strategies.

3.3   Measuring Action Payoff

Payoff, or utility, results from a player performing an action. For a particular payoff for player i performing action a_j, the utility is defined as $\mu_{i,j} = \mu(a_j)$. A set of payoff functions F contains all μ functions acting within the game G. In this competition between accuracy and precision, $F = \{\mu_\phi, \mu_\psi\}$, showing payoff functions that measure the increase in accuracy and precision respectively.

A formulated game typically has a set of payoffs for each player. In our approach, given two strategy sets S_1 and S_2, each containing three strategies, the two payoff functions $\mu_\phi : S_1 \mapsto P_1$ and $\mu_\psi : S_2 \mapsto P_2$ are used to derive the payoffs for ϕ and ψ containing:

$$P_1 = \{\phi_{1,1}, \phi_{1,2}, \phi_{1,3}\}, \tag{3}$$
$$P_2 = \{\psi_{2,1}, \psi_{2,2}, \psi_{2,3}\}, \tag{4}$$

reflecting payoffs from the results of the three actions, i.e., $\mu_\phi(a_j) = \phi_{1,j}$. This is a simple approach that can be expanded to reflect true causal utility based on the opposing player's actions. This means that not only is an action's payoff dependant on the player's action, but also the opposing player's strategy.

After modifying the respective loss functions, the function μ_ϕ calculates the payoff via approximation accuracy. Likewise, the payoff function μ_ψ calculates the payoff with approximation precision for deterministic approximations. More elaborate payoff functions could be used to measure the state of a game G, including entropy or other measures according to the player's overall goals [2].

Table 2. Payoff table for ϕ, ψ payoff calculation (deterministic)

		ψ		
		$-R_P$	$+R_N$	$+R_B$
	$-R_P$	$< \phi_{1,1}, \psi_{1,1} >$	$< \phi_{1,2}, \psi_{1,2} >$	$< \phi_{1,3}, \psi_{1,3} >$
ϕ	$+R_N$	$< \phi_{2,1}, \psi_{2,1} >$	$< \phi_{2,2}, \psi_{2,2} >$	$< \phi_{2,3}, \psi_{2,3} >$
	$+R_B$	$< \phi_{3,1}, \psi_{3,1} >$	$< \phi_{3,2}, \psi_{3,2} >$	$< \phi_{3,3}, \psi_{3,3} >$

The payoff functions imply that there are relationships between the measures selected as players, the actions they perform, and the probabilities used for region classification. These properties can be used to formulate guidelines regarding the amount of flexibility the user's loss function can have to maintain a certain level of consistency in the data analysis. As we see in the next section, the payoffs are organized into a payoff table in order to perform analysis.

3.4   Payoff Tables and Equilibrium

To find optimal solutions for ϕ and ψ, we organize payoffs with the corresponding actions that are performed. A payoff table is shown in Table 2, and will be the focus of our attention.

The actions belonging to ϕ are shown row-wise whereas the strategy set belonging to ψ are column-wise. In Table 2, the strategy set S_1 for ϕ contains three strategies, $S_1 = \{-R_P, +R_N, +R_B\}$, pertaining to actions resulting in a decrease in expected cost for classifying an object into the positive region and an increase in expected cost for classifying objects into the negative and boundary regions. The strategy set for ψ contains the same actions for the second player.

Each cell in the table has a payoff pair $< \phi_{1,i}, \psi_{2,j} >$. A total of 9 payoff pairs are calculated. For example, the payoff pair $< \phi_{3,1}, \psi_{3,1} >$ containing payoffs $\phi_{3,1}$ and $\psi_{3,1}$ correspond to modifying loss functions to increase the risk associated with classifying an object into the boundary region and to decrease the expected cost associated with classifying an object into the positive region. Measures pertaining to accuracy and precision after the resulting actions are performed for all 9 cases. These payoff calculations populate the table with payoffs so that equilibrium analysis can be performed.

In order to find optimal solutions for accuracy and precision, we determine whether there is equilibrium within the payoff table [3]. This intuitively means that both players attempt to maximize their payoffs given the other player's chosen action, and once found, cannot rationally increase this payoff.

A pair $< \phi_{1,i}^*, \psi_{2,j}^* >$ is an equilibrium if for any action a_k, where $k \neq i, j$, $\phi_{1,i}^* \geq \phi_{1,k}$ and $\psi_{2,j}^* \geq \psi_{2,k}$. The $< \phi_{1,i}^*, \psi_{2,j}^* >$ pair is an optimal solution for determining loss functions since no actions can be performed to increase payoff.

Thus, once an optimal payoff pair is found, the user is provided with the following information: a suggested tolerance level for the loss functions and the amount of change in accuracy and precision resulting from the changed loss functions. Equilibrium is a solution to the amount of change loss functions can undergo to achieve levels of accuracy and precision noted by the payoffs.

3.5   Loss Tolerance Calculation

Observed from decision rules (PN, NN, BN), we can calculate how much the loss functions need to be modified to acquire a certain level of accuracy or precision. There is a limit to the amount of change allowable for loss functions. For example, the action of reducing the expected cost R_P. We can reduce this cost any amount and rule (PN) will be satisfied. However, the rules (NN) and (BN) are also sensitive to the modification of R_P, denoted R_P^*. R_P^* must satisfy $R_P^* \geq (R_P - R_N)$ and $R_P^* \geq (R_P - R_B)$. This results in upper limit of t_{PP}^{max} for λ_{PP} and lower limit of t_{PN}^{min} for λ_{PN}. Assuming that $\lambda_{PP} \leq \lambda_{BP} < \lambda_{NP}$ and $\lambda_{NN} \leq \lambda_{BN} < \lambda_{PN}$, we calculate the following,

$$t_{PP}^{max} \leq \frac{\lambda_{BP} - \lambda_{PP}}{\lambda_{PP}} \quad , \quad t_{PN}^{min} < \frac{\lambda_{PN} - \lambda_{BN}}{\lambda_{PN}}. \tag{5}$$

That is, t_{PP} is the *tolerance* that loss function λ_{PP} can have (t_{PP} for λ_{PN}). Tolerance values indicate how much change a user can have to their risk beliefs (loss functions) in order to maintain accuracy and precision measures of $< \phi_{1,i}^*, \psi_{2,j}^* >$. In brief, when selecting a strategy, i.e., $(+R_P)$, the game calculates payoffs by measuring the approximation accuracy and prediction that result from modifying the loss functions λ_{PP} and λ_{PN}. The new loss functions, λ_{PP}^* and λ_{PN}^* are used to calculate a new expected loss R_P^*. In order to maintain the levels of accuracy and precision stated in the payoffs, the user must have new loss functions within the levels of t_{PP} for λ_{PP} and t_{PN} for λ_{PN}.

For example, let $\lambda_{PP} = \lambda_{NN} = 4$, $\lambda_{BP} = \lambda_{BN} = 6$, and $\lambda_{PN} = \lambda_{NP} = 8$. The inequality restrictions for the loss functions hold. We calculate that $t_{PP} = 0.5$ and $t_{PN} = -0.125$. This means that we can increase the loss function λ_{PP} by 50% and decrease the loss function λ_{PN} by 12.5% and maintain the same classification ability. This new information was derived from the analysis of the conditional risk modifications made possible through the use of game theory.

4   Conclusions

We provide a preliminary study on using game theory for determining the relationships between loss function tolerance and conditional risk. By choosing measures of approximation accuracy and approximation precision as players in a game, with goals of maximizing their values, we set up a set of strategies that each can perform. We investigate the use of three strategies for the deterministic approximation case. The strategies involve decreasing or increasing the expected losses for classifying objects into rough set regions.

Ultimately, taking an action within the strategy set involves modifying user-provided loss functions. We provide a method for indicating how much a loss function can be modified in order to provide optimal approximation accuracy and precision. This is very useful for the users as determining the amount of tolerance they should have when modifying loss functions is difficult.

By finding an equilibrium in the payoff tables, we may find the correct values for the loss functions, and thus, the optimal values of α and β parameters for determining the region boundaries. Based on this, we express the consequences of an increased or decreased expected loss of classification with the approximation accuracy and precision measures.

References

1. Bell, M.G.F.: The use of game theory to measure the vulnerability of stochastic networks. IEEE Transactions on Reliability 52, 63–68 (2003)
2. Duntsch, I., Gediga, G.: Uncertainty measures of rough set prediction. Artificial Intelligence 106, 109–137 (1998)
3. Fudenberg, D., Tirole, J.: Game Theory. MIT Press, Cambridge (1991)
4. Gediga, G., Duntsch, I.: Rough approximation quality revisited. Artificial Intelligence 132, 219–234 (2001)
5. Herbert, J., Yao, J.T.: A game-theoretic approach to competitive learning in self-organizing maps. In: Wang, L., Chen, K., Ong S., Y. (eds.) ICNC 2005. LNCS, vol. 3610, pp. 129–138. Springer, Heidelberg (2005)
6. Herbert, J., Yao, J.T.: Time-series data analysis with rough sets. In: Proceedings of Computational Intelligence in Economics and Finance (CIEF 2005), pp. 908–911 (2005)
7. Herbert, J.P., Yao, J.T.: Rough set model selection for practical decision making. In: Proceeding of Fuzzy Systems and Knowledge Discovery (FSKD 2007). III, pp. 203–207 (2007)
8. Nash, J.: The bargaining problem. Econometrica 18, 155–162 (1950)
9. Von Neumann, J., Morgenstern, O.: Theory of Games and Economic Behavior. Princeton University Press, Princeton (1944)
10. Pawlak, Z.: Rough sets. International Journal of Computer and Information Sciences 11, 341–356 (1982)
11. Roth, A.: The evolution of the labor market for medical interns and residents: a case study in game theory. Political Economy 92, 991–1016 (1984)
12. Yao, J.T., Herbert, J.P.: Web-based support systems with rough set analysis. In: Kryszkiewicz, M., Peters, J.F., Rybinski, H., Skowron, A. (eds.) RSEISP 2007. LNCS (LNAI), vol. 4585, pp. 360–370. Springer, Heidelberg (2007)
13. Yao, Y.Y.: Probabilistic approaches to rough sets. Expert Systems 20, 287–297 (2003)
14. Yao, Y.Y.: Decision-theoretic rough set models. In: Yao, J., Lingras, P., Wu, W.-Z., Szczuka, M., Cercone, N.J., Ślęzak, D. (eds.) RSKT 2007. LNCS (LNAI), vol. 4481, pp. 1–12. Springer, Heidelberg (2007)
15. Yao, Y.Y., Wong, S.K.M.: A decision theoretic framework for approximating concepts. International Journal of Man-machine Studies 37, 793–809 (1992)
16. Yao, Y.Y., Wong, S.K.M., Lingras, P.: A decision-theoretic rough set model. In: Ras, Z.W., Zemankova, M., Emrich, M.L. (eds.) Methodologies for Intelligent Systems, vol. 5, pp. 17–24. North-Holland, New York (1990)
17. Ziarko, W.: Variable precision rough set model. Journal of Computer and System Sciences 46, 39–59 (1993)

Multi-agent Based Multi-knowledge Acquisition Method for Rough Set

Yang Liu[1,2], Guohua Bai[2], and Boqin Feng[1]

[1] Department of Computer Science and Technology,
Xi'an Jiaotong University, Xi'an 710049, P.R. China
liuyang2006@gmail.com, bqfeng@mail.xjtu.edu.cn
[2] School of Engineering, Blekinge Institute of Technology,
Ronneby, 372 25, Sweden
gba@bth.se

Abstract. The key problem in knowledge acquisition algorithm is how to deal with large-scale datasets and extract small number of compact rules. In recent years, several approaches to distributed data mining have been developed, but only a few of them benefit rough set based knowledge acquisition methods. This paper is intended to combine multi-agent technology into rough set based knowledge acquisition method. We briefly review the multi-knowledge acquisition algorithm, and propose a novel approach of distributed multi-knowledge acquisition method. Information system is decomposed into sub-systems by independent partition attribute set. Agent based knowledge acquisition tasks depend on universes of sub-systems, and the agent-oriented implementation is discussed. The main advantage of the method is that it is efficient on large-scale datasets and avoids generating excessive rules. Finally, the capabilities of our method are demonstrated on several datasets and results show that rules acquired are compact, having classification accuracy comparable to state-of-the-art methods.

Keywords: Attribute reduction, Multi-agent technology, Knowledge acquisition, Classification accuracy.

1 Introduction

Rough set theory provides a framework for knowledge acquisition [1,2]. There has been a great interest in designing efficient knowledge acquisition algorithm based on rough set theory [3,4]. Most methods treat knowledge acquisition task as a single reduction to a decision system. Because knowledge extracted is based on a single reduct, results of these methods always include large amounts of rules that are relatively hard for people to understand. In addition, if some attribute values are changed by environment, knowledge acquired by these methods becomes obsolete, contributing little to decision-making process. The idea of extracting multi-knowledge, which is based on multiple reducts of a decision system, is proposed to address this issue [5,6]. However, these centralized methods cannot avoid highly computing complexity in most cases. Modern information system

G. Wang et al. (Eds.): RSKT 2008, LNAI 5009, pp. 140–147, 2008.

usually maintains a large quantity of cases [7]. In decentralized data sources, centralized method needs to be modified for handling distributions. In such circumstances, centralized multi-knowledge acquisition methods are inefficient to scale up to massive datasets.

Decentralization research shows that multi-agent technology can improve the performance of particular data mining techniques [8]. Agent technology also has many intelligent properties such as autonomy, activity, reactivity, mobility and sociality, which make it adapt to data mining algorithms [9]. Agent collaboration [10] is used to address the scalability problem and enhance performance of a system. Multi-agents may operate independently on decentralized datasets, and then combine their respective obtained models. On the other hand, they may share potential knowledge that has been discovered, in order to benefit from the additional opinion of other agents. The core tasks concern two aspects: (1) how agents mine from individual datasets, (2) how agents refine mined results through collaborative work. These aspects are in the focus of this paper.

The remainder of this paper is organized as follows. First, a multi-reduction algorithm is proposed in Section 2. It is the primary task for multi-agent based knowledge acquisition. We then introduce the schema of agent cooperation to optimize extracted results and a multi-agent based knowledge acquisition method is presented in Section 3. In Section 4, we present results of the method on several public datasets. Finally, Section 5 concludes this paper and potential work is given.

2 Multi-reduction Algorithm for Decision Space

A decision system can be denoted by $S = (U, C \bigcup D, V, f)$ [2], where U is a non-empty, finite set of cases called universe, C is a set of conditional attributes, D is a set of decision attributes, where $C \bigcap D = \emptyset$, $C \bigcup D = A$, V is the domain of attributes, and f is a information function. For an attribute set $B \subseteq A$, its indiscernibility relation is denoted by $INDS(B) = \{(x, y) \mid \forall x \forall y \forall a (x \in U \wedge y \in U \wedge a \in B \wedge f(x, a) = f(y, a))\}$. Obviously, the indiscernibility relation is an equivalence relation on U. The equivalence class containing $x \in U$ related to $INDS(B)$ is denoted by $[x]_B = \{y \mid \forall y((y \in U) \wedge ((x, y) \in INDS(B)))\}$.

The B-lower approximation of set $X \subseteq U$ is $B_- = \{x \mid \forall x((x \in U) \wedge ([x]_B \subseteq X))\}$, and the B-upper approximation of X is $B^-(X) = \{x \mid \forall x((x \in U) \wedge ([x]_B \bigcap X \neq \emptyset))\}$. For $P, Q \subseteq C$, P-positive region of Q can be denoted by $POS_P(Q) = \bigcup_{X \in U/Q} P_-(X)$. If $POS_{RED}(D) = POS_P(D)$, where $RED \subseteq P$, and $\forall a \in RED$, $POS_{RED \setminus \{a\}}(D) \neq POS_P(D)$, we call RED is a P's reduct of S. If $P = C$, RED is a reduct of S. A set of reducts of S is denoted by $\Re(S)$.

An algorithm for construction of multiple reducts (MRA) is presented in Algorithm 1. $\gamma(B, D) = |POS_B(D)|/|U|$ is the degree of the dependency [11]. It is important to note that the addition strategy is adopted in MRA. The algorithm starts from a set of partial reducts that contain only one attribute, and adds attributes with maximal σ value in LINE 11 one by one into partial reducts until reducts are obtained. After the reducts construction process, MRA can obtain up to $|C|$ reducts.

Data: A decision system $S = (U, C \bigcup D, V, f)$.
Result: A set of reducts $\Re$.

1 *Queue* q;
2 **for** $\forall a \in C$ **do**
3 $\quad$ **if** $\gamma(\{a\}, D) \neq 0$ **then**
4 $\quad\quad$ | $Enqueue(q, \{a\})$;
5 $\quad$ **end**
6 **end**
7 **while** *not IsEmpty(q)* **do**
8 $\quad$ $B \Leftarrow Dequeue(q)$;
9 $\quad$ $max \Leftarrow 0$;
10 $\quad$ **for** $\forall a \in C \setminus B$ **do**
11 $\quad\quad$ $\sigma \Leftarrow \gamma(B \bigcup \{a\}, D) - \gamma(B, D)$;
12 $\quad\quad$ **if** $\sigma > max$ **then**
13 $\quad\quad\quad$ | $max = \sigma$;
14 $\quad\quad\quad$ | $CandidateAttr = a$;
15 $\quad\quad$ **end**
16 $\quad$ **end**
17 $\quad$ **if** $max = 0$ **then**
18 $\quad\quad$ | $\Re \Leftarrow \Re \cup \{B\}$;
19 $\quad$ **else**
20 $\quad\quad$ | $Enqueue(q, B \cup \{CandidateAttr\})$;
21 $\quad$ **end**
22 **end**
23 **return** $\Re$;

Algorithm 1. Multi-Reduction Alg. (MRA)

3 Multi-agent Based Knowledge Acquisition Method

Let S be a decision system, and $\pi = \{\phi_1, \phi_1, \ldots, \phi_n\}$ be a partition of U. An attribute set $\hat{P}$ is called as *independent partition attribute set* if it satisfies that $\hat{P} = \arg\max_{P \subseteq C} \sum_{i=1}^n \frac{|P_-(\phi_i)|}{|P^-(\phi_i)|}$. The boundary regions of all sub-universes related to independent partition attribute set contain a small set of cases. Therefore, the sum of approximation accuracy of sub-universes can maintain the highest quality of approximation of decision classes on the whole universe U. Decision system S is partitioned into sub-systems by $\hat{P}$, having sub-universe $U_i = \hat{P}^-(\phi_i)$, $i = 1, \ldots, n$. The set of sub-systems partitioned by $\hat{P}$ is denoted by $S_{\hat{P}}$.

The multi-agent system for knowledge acquisition can be constructed as $MAS =< S_{\hat{P}}, AG_{local}, AG_{global} >$. $S_{\hat{P}}$ is the set of sub-systems. There are two sets of agents. AG_{local} is a set of local agents, and AG_{global} is a set of global agents.

Figure 1 illustrates multi-agent architecture for knowledge acquisition. Local agents firstly extract multiple reducts from sub-systems as candidate knowledge concept. Knowledge acquisition tasks are based on cooperation between agents. In our system, local agents score and exchange interactive rules using

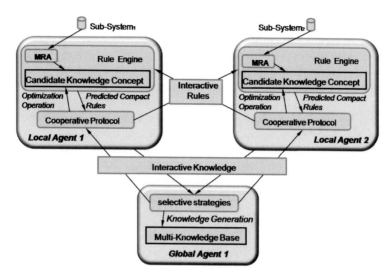

Fig. 1. Multi-Agent based Knowledge Acquisition Architecture

communication policies in [9]. For example, local agent 1 could decide to send predicted compact rules to local agent 2. The cooperative module in local agent 2 may report score of interactive rules on its sub-system to local agent 1. Rules in local agent 1 are optimized by means of agent cooperation. Global agents are responsible for constructing final multi-knowledge base through synthesizing scores of rules computed by local agents.

The multi-agent system firstly partitions the training data into n disjoint subsets, assigns a local agent to each subset, and provides the infrastructure for agent-communication when a local agent detects an acceptable rule. Local agents construct rules from multiple reducts of individual sub-systems. The candidate knowledge concept, which can be seen as a local rule base, is denoted by $MKB_i = \{B \longrightarrow_x D \mid x \in U_i \wedge \exists RED(RED \in \Re_i \wedge B \subseteq RED)\}$, where $\Re_i$ is a set of reducts of S_i, for $i = 1, ..., n$.

A local agent takes the role of mining local dataset and it engages the negotiation to agree with interactive rules by matching score of the rule. Local agents select rules with high confidence as the interactive rules and send them to other local agents for feedback information. When a rule meets the evaluation criterion for a sub-system, it becomes a candidate knowledge concept for testing the globally evaluation criterion. A possible implementation of interaction between local agents according communicative acts of ACL FIPA is based on the primitives such as *call-for-proposal, accept, inform*, etc. These primitives give a first level semantics to message exchange between local agents. Negotiation is initiated by local agents, and it contains handshaking primitives to ensure that all sub-systems participate and the negotiation ends only if an agreement has been reached. Each local agent computes quality of an interactive rule and sends results to initiative agent. After the negotiation procedure, local agents can obtain a set of rules with high confidence in sub-systems. These rules are then sent to global agents for global evaluation.

Local agents report the matching score of a case to a global agent using Definition 1. The certification and coverage factors of a rule are introduced by Pawlak [2]. In our system, *decentralized selective protocol* is used to match a test case with a rule. For a given rule, local agents attempt to match a case using complete matching score. If complete matching in local knowledge base is impossible, agents try to match the case with all conditions except one on individual sub-system. If all conditions except one in the rule can match the case, knowledge extraction operation is completed. If not, all conditions except two are searched for, etc. This strategy evaluates cases from decentralized knowledge bases. Therefore, complete knowledge can be discovered by our method.

Definition 1. Matching score of a rule

Let r be a decision rule and x be a case. The matching score of x to r is denoted by $MR(r, x) = \rho \times Cer(r) \times Cov(r)$, where Cer is rule certification factor, Cov is rule coverage factor, and ρ is the ratio of matching conditions to all conditions in rule r. If $\rho = 1$, all condition attributes of x is matched by r, and we called the matching score is *complete matching score*. Otherwise, $0 < \rho < 1$, we called the matching score is *partially matching score*.

Global agents aim at integrating the knowledge that is discovered out of sub-systems with a minimum amount of network communication, and maximum of local computation. They guarantee that each extracted rule is satisfactory over the entire dataset. As a global agent discovers an acceptable rule, it broadcasts the rule to other local agents to review its statistics over the rest of the dataset for combining and exchanging the locally mined candidate knowledge. If the rule meets the evaluation criterion globally, it is posted as a compact rule. Otherwise, its local statistics are replaced with the global statistics in classification module. Rules evaluated by global agents are ordered by rule performance on whole dataset. The global agents gradually select rules until the total performance of rule set is satisfactory.

In classification module, global agents decide how test cases match the concept of whole decision system. They estimate rule quality using the matching score of each concept in Definition 2. The concept with maximal matching score is selected as the final classification result.

Definition 2. Matching score of a concept

Let x be a case. The matching score of x to the concept is denoted by $MC(x) = \sum_{i=1}^{n} \sum_{r \in R_i} MR(r, x)$, where R_i is a set of rules belong to the concept extracted by local agent i, for $i = 1, ..., n$.

According to this approach, a multi-agent based multi-knowledge acquisition algorithm is presented in Algorithm 2. Given a decision system S, independent partition attribute set $\hat{P}$ is employed to decomposes S into sub-systems. Local agents are dispatched to sub-systems to perform multi-reduction algorithm and evaluate rules through agent cooperation. Extracted rule set is optimized through collaboratively exchanging interactive rules in the multi-agent system.

Data: A decision system $S = (U, C \bigcup D, V, f)$.
Result: A set of rules MKB.
1 Randomly partition U into π;
2 Compute independent partition attribute set $\hat{P}$, $\hat{P} \subseteq C$;
3 Partition decision system S into n sub-systems $S_{\hat{P}}$;
4 Construct a multi-agent system $MAS = < S_{\hat{P}}, AG_{local}, AG_{global} >$;
5 $MKB \Leftarrow$ Select top k rules $r_1 - r_k$ according to matching score, such that
$\sum_{i=1}^{k} \sum_{x \in U} MR(r_i, x) > \frac{1}{2}(\sum_{r \in MKB_1 \bigcup ... MKB_n} \sum_{x \in U} MR(r, x))$, where
MKB_i is a set of rule extracted by ith agent in AG_{local}, $i = 1, ..., n$;
6 **return** MKB;

Algorithm 2. Multi-Agent based Multi-Knowledge Acquisition Alg. (MAMKA)

4 Experimental Results and Discussion

LERS [12], MKE [5], and GR [6] are primarily designed to induce rules from training cases and classify new, unseen cases. Since they are extensively compared by most other rough set based knowledge acquisition algorithms, MAMKA algorithm is compared with these three algorithms. In LERS system, the LEM2 option is chosen. Four algorithms are tested on the same platform. Benchmark datasets are chosen from UCI ML repository [13]. Numerical attributes are firstly discretized by a modified Chi2 algorithm [14]. Conceptual reconstruction algorithm [15] is used to deal with missing values in datasets. We adopt ten-fold cross validation to validate experimental results.

Table 1. Comparison of four algorithms on eight datasets

Dataset[1]		1	2	3	4	5	6	7	8	Avg. Rank	Rank
MAMKA	Accuracy	82.6	72.4	70.7	90.8	92.3	78.5	84.6	91.5	2.1	
	Time	4	5	7	5	36	39	53	72	**1**	**1.6**
	Rules	10	8	14	8	14	16	20	34	1.6	
MKE	Accuracy	84.1	71.7	72.2	94.5	89.7	79.2	80.4	89.6	2.4	
	Time	10	18	29	12	132	165	180	245	3	2.8
	Rules	13	15	17	12	21	40	47	58	3	
LERS	Accuracy	85.6	72.0	68.7	96.3	91.6	80.4	85.3	94.1	**1.5**	
	Time	4	12	22	7	43	52	78	113	1.9	2.5
	Rules	16	24	29	23	34	42	53	62	4	
GR	Accuracy	77.0	70.1	54.3	81.4	68.2	72.1	78.2	82.3	4	
	Time	-	-	-	-	-	-	-	-	4	3.1
	Rules	6	10	12	6	12	8	24	16	**1.4**	

---

[1] Eight datasets are as follows: 1 - StatLog heart disease, 2 - BUPA liver disorder, 3 - TA evaluation , 4 - Congressional voting records, 5 - Wisconsin prognostic breast cancer, 6 - Pima indians diabetes, 7 - Waveform, 8 - Thyroid disease.

Table 1 shows percentage of accuracy, execution time in seconds and size of rule set of four algorithms. The Avg. Rank column averages each estimator's rank out of four algorithms on eight datasets, and the Rank column summarizes three Avg. Rank estimators of each algorithm. We use values in the Rank column to evaluate the overall performance of algorithms. Because GR algorithm has highly computing complexity, its execution time is not compared in this table, which means it gets the lowest rank of execution time. On average, the MAMKA algorithm archives the highest overall performance on eight benchmark datasets. The second best algorithm is LEM2 in LERS system. It achieves the highest accuracy on the majority of datasets, but the Avg. Rank for the size of rule set is much lower than MAMKA. Because MKE is a centralized multi-knowledge acquisition method, the running time and size of rules extracted are not competent on large datasets (e.g. Dataset 5, 6, 7 and 8).

Scalability is tested on PIMA database. The dataset is firstly separated into a half training set and a half testing set. We incrementally double the size of training dataset from 50 to 400. Each algorithm runs ten times and come up with the average results.

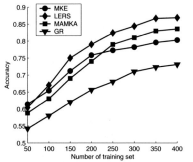

(a) Accuracy comparison between four algorithms with increasing training datasets

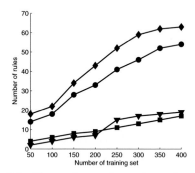

(b) Comparison of number of extracted rules between four algorithms with increasing training datasets

Fig. 2. Scalability of four algorithms on PIMA database

As we can see from Figure 2, MAMKA gradually improves classification accuracy but marginally increases the number of extracted rules. It is obvious that the size of rule set obtained by MAMKA is much smaller than MKE and LERS. That means in most cases MAMKA acquires complete knowledge containing the least redundant rules. Thus, MAMKA is rational on both the number of extracted rules and classification accuracy on new test cases.

5   Conclusion

A decentralized knowledge acquisition algorithm is presented to extract compact knowledge on large-scale datasets. It combines multi-agent technology and rough set based knowledge acquisition method. Information system is partitioned into sub-systems by independent partition attribute set. Local agents pursue multi-reduction tasks and optimize extracted results through communication. Two main advantages of this method are its scalability on huge datasets, and its robustness to avoiding generating large amounts of rules. The algorithm is extensively tested and compared with conventional algorithms. Experimental results show that our method is faster, and generates compact rule set without sacrificing classification accuracy on testing data. Our proposed method is thus well suited to distributed knowledge acquisition system, which commonly involves huge datasets.

One of the open issues is the capability of the proposed method to deal with noise in a decision system. Future work will aim to address this issue and perform experiments on noisy datasets.

References

1. Pawlak, Z.: Rough sets. International J. Comp. Inform. Science 11, 341–356 (1982)
2. Pawlak, Z.: Rough Sets. Theoretical Aspects of Reasoning About Data. Kluwer Academic Publishers, Dordrecht, Boston, London (1991)
3. Wang, G.Y., Liu, F.: The inconsistency in rough set based rule generation. In: Rough Sets and Current Trends in Computing, pp. 370–377 (2000)
4. Yasdi, R.: Combining rough sets learning-method and neural learning-method to deal with uncertain and imprecise information. Neuro-Computing 7(1), 61–84 (1995)
5. Wu, Q.: Bell, David: Multi-knowledge extraction and application. In: Wang, G., Liu, Q., Yao, Y., Skowron, A. (eds.) RSFDGrC 2003. LNCS (LNAI), vol. 2639, pp. 274–278. Springer, Heidelberg (2003)
6. Hang, X.S., Dai, H.: An optimal strategy for extracting probabilistic rules by combining rough sets and genetic algorithm. In: Grieser, G., Tanaka, Y., Yamamoto, A. (eds.) DS 2003. LNCS (LNAI), vol. 2843, pp. 153–165. Springer, Heidelberg (2003)
7. Klusch, M., Lodi, S., Moro, G.: Issues of agent-based distributed data mining. In: International Conference on Autonomous Agents, pp. 1034–1035 (2003)
8. Wooldridge, M.J., Jennings, N.R.: Intelligent agent: theory and practice. Knowledge Engineering Review 10(2), 115–152 (1995)
9. Pynadath, D.V., Tambe, M.: The communicative multiagent team decision problem: Analyzing teamwork theories and models. Journal of Artificial Intelligence Research, 389–423 (2002)
10. Yaskawa, S., Sakata, A.: The application of intelligent agent technology to simulation. Mathematical and Computer Modelling 37(9), 1083–1092 (2003)
11. Pawlak, Z., Skowron, A.: Rudiments of rough sets. Information Sciences, 3–27 (2007)
12. Grzymala-Busse, J.W.: MLEM2 - Discretization during rule induction. Intelligent Information Systems, 499–508 (2003)
13. UCI machine learning repository, http://archive.ics.uci.edu/ml/
14. Tay, F.E., Shen, L.: A modified Chi2 algorithm for discretization. IEEE Transactions on Knowledge and Data Engineering 14(3), 666–670 (2002)
15. Aggarwal, C.C., Parthasarathy, S.: Mining massively incomplete data sets by conceptual reconstruction. In: Proceedings of the seventh ACM SIGKDD international conference on Knowledge discovery and data mining, pp. 227–232 (2001)

Knowledge-Based Genetic Algorithms*

Gaowei Yan, Gang Xie, Zehua Chen, and Keming Xie

College of Information Engineering, Taiyuan University of Technology
Taiyuan, Shanxi, P.R. China, 030024
{yangaowei,xiegang,chenzehua,kmxie}@tyut.edu.cn

Abstract. In this paper, Rough Set Theory (RST) was introduced to discover knowledge hidden in the evolution process of Genetic Algorithm. Firstly it was used to analyze correlation between individual variables and their fitness function. Secondly, eigenvector was defined to judge the characteristic of the problem. And then the knowledge discovered was used to select evolution subspace and to realize knowledge-based evolution. Experiment results have shown that the proposed method has higher searching efficiency, faster convergent speed, and good performance for deceptive problem and multi-modal problems.

Keywords: Rough set theory (RST), Genetic Algorithms (GAs), Knowledge discovery, Knowledge evolution, Eigenvector.

1   Introduction

John Holland proposed binary Genetic Algorithm(GA) in 1960s. As an optimal searching method, it is preferable for complex systems[1,2]. However, its intelligent feature is due to natural selection strategy. Vose and Liepins[3] considered that GA operators do not have enough ability to guide individuals to approximate to best schema, and this is why exist the low convergent speed, the prematurity, and the local optimization problem.

Rough Sets[4] Theory (RST) proposed by Z.Pawlak in 1982 is one of data mining tools, which has now been widely used in inductive reasoning[5], automatic classification[6], pattern recognition[7], and learning algorithm[5]. There are also some fruits in the combination of rough set and fuzzy logic, Petri Net and granular computing[8,9]. Although there are some distinguished differences, Lingras Pawan provided another sort of rough genetic algorithms[10,11].

RST needs no transcendental knowledge for data analysis except data itself. In this paper, RST combined with binary granular computing is used to find the relation between individual variables and fitness function. It can discover the knowledge behind the data generated in evolution process, which will direct evolution orientation and decrease searching space, those knowledge can provide an ability to overcomes deceptive problem to a great extent.

---

* This paper is supported by the Youth Science Foundations Project of Shanxi Province (No.2006021016 and No.2007021018).

G. Wang et al. (Eds.): RSKT 2008, LNAI 5009, pp. 148–155, 2008.

The paper is organized as follows. In section 2, basic concepts of RST are described and relation matrix between individuals and their fitness function is defined. In section 3, the knowledge-based Genetic Algorithm is introduced. Testing function and example application are described in section 4. Conclusions and discussions for further study are given in section 5.

2 Some Basic Concepts

2.1 Indiscernibility Relation

Let $K=(U,R)$ be an information system, then in any $B \subseteq R$ there is an associated equivalence relation $IND_R(B)$:

$$IND_R(B) = \left\{ (x, x') \in U \times U : \forall a \in B \; a(x) = a(x') \right\}. \tag{1}$$

$IND_R(B)$ ($IND(B)$, for short) is called the B-indiscernibility relation and its equivalence classes are denoted by $[x]_B$.

$$[x]_B = \left\{ y \in U \,|\, yBx \right\}. \tag{2}$$

Each equivalence class can be viewed as a granule consisting of indistinguishable elements.

2.2 Binary Granule Definition by Equivalence [12,13]

Let $K = (U, R)$ be an information system, where $U=\{u_1, u_2, \cdots, u_l, \cdots, u_L\}$. For the subset $[Y]_i$ in $U_{\text{ind}(P)}$, the coding space is defined as a mapping function from integer domain to binary space $f : Z^+ \to \{0,1\}^L$, and the binary string of granule $[Y]_i$ can be respectively expressed as

$$Y_i = \{a_1, a_2, \cdots a_l \cdots, a_L\}. \tag{3}$$

$$a_l = \begin{cases} 1, & if \; u_l \in Y_i \\ 0, & if \; u_l \notin Y_i \end{cases} \quad 1 \leq l \leq L. \tag{4}$$

After the granular extension of RST, logic operation can be introduced into RST to extract knowledge quickly.

2.3 Space Partition of the GA Based on Indiscernibility Relation

The data generated in the genetic process can be regarded as a knowledge system. Based on the indiscernibility relation in RST, the searching space of GA can be divided into different regions by partitioning fitness function or object function.

Suppose that a multi-optimal problem is given a non-empty set S as the solution space of GA. Let S be a bounded set in the n-dimension Euclidean space R^n, as $S = \Pi_{i=1}^n [a_i, b_i]$. Marking any individual in GA operation as X, $X = \{x_1, x_2, \cdots x_n\}$, $x_i \in [a_i, b_i]$, x_i is called individual variable. X and its corresponding fitness F constitute an ordered pair $\{X, F\}$, hence consisting the universe U. If the interval $[a_i, b_i]$ of each variable x_i is divided probably, for instance, by m-division method, that is, the solution space S of GA can be divided into a series of solution subspace S', where $S' \in S$.

$$S' = \prod_{i=1}^{n} \prod_{k=1}^{m} \left[a_i + \frac{b_i - a_i}{m} * (k - 1), a_i + \frac{b_i - a_i}{m} * k \right]. \tag{5}$$

S' or the union of S' is called optimum-searching subspace.

If the fitness function being properly defined, all of the optimization problems can be converted into the maximum value finding problem. Therefore, adopting a set of thresholds α_j, $\beta_j (0 < \alpha_j < \beta_j)$, $j = 1, 2, \cdots, h$, where h is the number of the object function, we define the following equivalence relation.

Relation 1: The solutions in some searching spaces are excellent, that is, the index of each object function is greater than the corresponding threshold β_j:

$$R_1 = \bigwedge_{j=1}^{h} (f_{v_j} \geq \beta_j). \tag{6}$$

Relation 2: The object functions in some solution spaces are very bad, that is, the index of all the object functions are less than the threshold α_j:

$$R_2 = \bigwedge_{j=1}^{h} (f_{v_j} < \alpha_j). \tag{7}$$

Therefore, the whole solution space can be divided into three parts as follows:

$$S_{POS} = \bigcup \left\{ S' \left| \bigwedge_{j=1}^{h} (f_{v_j} \geq \beta_j), S' \in S \right. \right\}. \tag{8}$$

$$S_{NEG} = \bigcup \left\{ S' \left| \bigwedge_{j=1}^{h} (f_{v_j} < \alpha_j), S' \in S \right. \right\}. \tag{9}$$

$$S_{BON} = S - S_{POS} - S_{NEG}. \tag{10}$$

where S_{POS} is the subset of solution space where exist the excellent solutions, S_{NEG} is the subset of the bad solution space, S_{BON} is the subset of the solution space which has the optimum solution potentially.

2.4 Binary Relation Matrix of Individual Variables and Their Fitness Function

For granule $[x_i]_j$ $(j = 1, 2, \cdots, N)$ and fitness function granule $[f]_k$ $(k = 1, 2, \cdots, M)$, matrix-based granular computing algorithm[13] is introduced to compute the relation of each individual variable to its fitness function. Let

$$A_{x_i} = \left([x_i]_1 \cdots [x_i]_j \cdots [x_1]_N \right)' = \begin{pmatrix} a_{11} & \cdots & a_{1L} \\ \vdots & \ddots & \vdots \\ a_{N1} & \cdots & a_{NL} \end{pmatrix}. \tag{11}$$

$$A_f = ([f]_1 \cdots [f]_k \cdots [f]_M)' = \begin{pmatrix} b_{11} & \cdots & b_{1L} \\ \vdots & \ddots & \vdots \\ b_{M1} & \cdots & b_{ML} \end{pmatrix}. \tag{12}$$

$$T_{x_{i \to f}} = A_{x_i} \times A'_f = (t_{jk})_{N \times M}. \tag{13}$$

$$t_{jk} = \begin{cases} 1, if \ \sum\limits_{l=1}^{L} (a_{jl} \wedge b_{kl}) > 0 \\ 0, if \ \sum\limits_{l=1}^{L} (a_{jl} \wedge b_{kl}) = 0 \end{cases}. \tag{14}$$

where i is the series number of the variable, $j = 1, 2, \cdots, N$, $k = 1, 2, \cdots, M$, $l = 1, 2, \cdots, L$. And, "$\wedge$" is the "and" operation in logic, "$\times$" is the traditional Cartesian product, t_{jk} shows the subordinative relation between $[x_i]_j$ and $[f]_k$.

3   Knowledge-Based Genetic Algorithm(KGA)

3.1   Classification for the Problem Being Solved

We defined eigenvector $G(x_i)$ to judge the property of the problem and use it to measure the contribution of each subspace to fitness function as follows:

$$G(x_i) = \{G_1(x_i), \cdots, G_j(x_i), \cdots, G_N(x_i)\}, j = 1, 2, \cdots, N. \tag{15}$$

$$G_j(x_i) = MAX\{g_k | g_k = t_{jk} \times (k-1), k = 1, 2, \cdots, M\}. \tag{16}$$

where, $j = 1, 2, \cdots, N$ represents the different subspaces, x_i is the ith variable.

After some definite generations, we can conclude evolution process through data analysis no matter whether the global optimum is found. If the eigenvector is monotone, the problem is single modal, and if $G(x_i)$ is not monotone, however, it has only one maximum, it can also be considered to be a single modal problem. Otherwise, it will belong to the multi-modal problem which includes deception.

3.2   Algorithm Description

In most optimal searching problems, the distribution of the optimal solutions is unknown, especially for some complex problems such as multi-modal or deceptive problems. The optimal solutions can be distributed in different subspaces. In this proposed algorithm, the traditional GAs with large population, large mutation probability and small selection probability are taken in the initial stage of the evolution so as to make individual expansion in the whole space rapidly and to collect as much data as possible.

Then RST is used to analyze the data and to form S_{POS} region, S_{NEG} region and S_{BON} region by the fitness function thresholds $\alpha, \beta (0 < \alpha < \beta < M)$. Determining the subspace for further searching depends on computing correlation between individual variables and their corresponding fitness function. Eigenvec-tor is used to judge the property of the problem, and then different evolution strategies are applied.

Through data analysis by RST, the whole solution space can be divided into multiple subspaces. Multiple population evolution strategy keeps implicit paral-lelism of the GA, and at the same time, the natural population is also included to supplement the new knowledge. This process embodies the knowledge learning ability of human beings, from rough to fine and from false to true. The algorithm flow chart is shown as Fig. 1 .

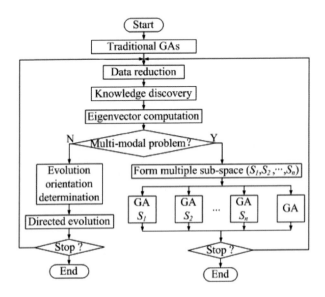

Fig. 1. Schema of knowledge-based Genetic Algorithm

4   Simulation Experiment

To evaluate the efficiency of the algorithm, the functions in the Table 1 are used for simulation. The parameters in the simulation process are set as follows: the initial population is 200, the cross probability is 0.1. In the beginning of the evolution, the mutation probability is 0.2 to keep the diversity of the population. And then the cross probability is taken 0.05 after the evolution space is identified.

In Table 1 function $\min f_5(x_1, x_2)$ has three global minimum values, which belong to the multi-modal function, and herein $\min f_5(x_1, x_2)$ is taken as an example to illustrate the proposed algorithm. The traditional GA can find only one of the global optimal values, which is 0.3979. In this paper $\max f_5'(x_1, x_2) = 3.98/\min f_5(x_1, x_2)$ is taken as the fitness value of the function which is converted into the maximum function optimization problem. After ten generations, data pairs $\{X_t, F_t\}$ are obtained. Variables $\{x_1, x_2\}$ and their fitness function F are both discretized into 7 grades ($M=N=7$) after eliminating redundant data, and

Table 1. Test function

Function	Optimal value
$\max f_1(x_1, x_2) = 10.0 + \frac{\sin(1/x_1)}{(x_1-0.16)^2+0.1} + 10.0 + \frac{\sin(1/x_2)}{(x_2-0.16)^2+0.1}$ $x_1, x_2 \in (0,1)$	39.7898
$\min f_2(x_1, x_2) = (x_1^2 + x_2^2)^{0.25} (\sin^2(50(x_1^2+x_2^2)^{0.1}) + 1)$ $x_1, x_2 \in [-10,10]$	0.00
$\max f_3(x_1, x_2) = \left(\frac{a}{b+(x_1^2+x_2^2)}\right)^2 + (x_1^2 + x_2^2)^2$ $x_1, x_2 \in [-5.12, 5.12],\ a{=}3.0,\ b{=}0.05$	3600
$\max f_4(x_1, x_2) = 0.5 - \frac{\sin^2\sqrt{x_1^2+x_2^2}-0.5}{(1+0.001(x_1^2+x_2^2))^2}$ $x_1, x_2 \in [-10,10]$	1.00
$\min f_5(x_1, x_2) = (x_2 - \frac{5.1}{4\pi^2}x_1^2 + \frac{5}{\pi}x_1 - 6)^2 + 10(1 - \frac{1}{8\pi})\cos x_1 + 10$ $x_1 \in [-5,10],\ x_2 \in [0.15]$	0.3979

Table 2. Binary granule table for $\max f_5'(x_1, x_2)$

U	x_1							x_2							F						
	0	1	2	3	4	5	6	0	1	2	3	4	5	6	0	1	2	3	4	5	6
u_1	1	0	0	0	0	0	0	0	0	0	0	0	0	1	0	1	0	0	0	0	0
u_2	0	1	0	0	0	0	0	0	0	0	0	1	0	0	0	0	0	1	0	0	0
u_3	0	0	0	1	0	0	0	1	0	0	0	0	0	0	0	1	0	0	0	0	0
u_4	0	1	0	0	0	0	0	0	0	0	0	0	1	0	0	0	0	0	0	0	1
$\vdots$	$\vdots$	$\vdots$	$\vdots$	$\vdots$	$\vdots$	$\vdots$	$\vdots$	$\vdots$	$\vdots$	$\vdots$	$\vdots$	$\vdots$	$\vdots$	$\vdots$	$\vdots$	$\vdots$	$\vdots$	$\vdots$	$\vdots$	$\vdots$	$\vdots$
u_L	0	0	0	1	0	0	0	0	1	0	0	0	0	0	0	0	0	0	0	0	1

expanded granularly, then the Table 2 is obtained. Each classification can be expressed in form of binary granule.

The subspace and individual variables x_1 and x_2 relation matrix with their fitness are obtained as follows.

$$T_{x_1 \to f} = \begin{pmatrix} 1 & 1 & 0 & 0 & 0 & 0 & 0 \\ 1 & 1 & 1 & 1 & 1 & 1 & 1 \\ 1 & 0 & 0 & 0 & 0 & 0 & 0 \\ 1 & 1 & 1 & 1 & 1 & 1 & 1 \\ 1 & 1 & 0 & 0 & 0 & 0 & 0 \\ 1 & 1 & 0 & 0 & 0 & 0 & 0 \\ 1 & 1 & 1 & 1 & 1 & 1 & 1 \end{pmatrix}_{7\times7} \qquad T_{x_2 \to f} = \begin{pmatrix} 1 & 1 & 0 & 0 & 0 & 0 & 0 \\ 1 & 1 & 1 & 1 & 1 & 1 & 0 \\ 1 & 1 & 0 & 0 & 0 & 0 & 0 \\ 1 & 0 & 0 & 0 & 0 & 0 & 0 \\ 1 & 1 & 1 & 1 & 0 & 0 & 0 \\ 1 & 1 & 1 & 1 & 1 & 1 & 1 \\ 1 & 1 & 1 & 0 & 0 & 0 & 0 \end{pmatrix}_{7\times7}$$

The eigenvectors $G(x_1) = \{1, 6, 0, 6, 1, 1, 6\}$ and $G(x_2) = \{1, 5, 1, 0, 3, 6, 2\}$ reflect the nature characteristic of the function of $\max f_5'(x_1, x_2)$. Let $\beta = 0.618M$, the corresponding individual subspace is taken into consideration in S_{POS} region. By matrix-based binary granular computing to those classifications in $F \geq \beta$, the solution space in the S_{POS} region can be found. Confirmed optimal searching

Table 3. Subspace distribution and optimization result

Index	Subspace		Independent variables		$\max f_5'(x_1, x_2)$	$\min f_5\ (x_1, x_2)$
	x_1	x_2	x_1	x_2		
1	{3}	{1}	3.14167	2.27488	10.00250	0.39790
2	{6}	{1}	9.42479	2.47687	10.00251	0.39790
3	{1}	{5}	-3.14155	12.2749	10.00251	0.39790

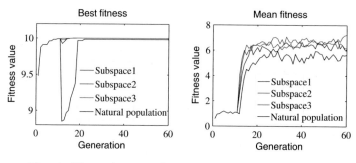

Fig. 2. The evolution process of the function $\max f_7'(x_1, x_2)$

subspaces are shown in Table 3. The optimal searching process of the function $\max f_5'(x_1, x_2)$ is shown in Fig. 2. There are three subspaces and natural population in Fig. 2. The three global optimums of the function $\max f_5'(x_1, x_2)$ are found in subspaces 1, 2 and 3. Compared with the optimal searching of the natural population, the convergent speed and the optimization precision are much higher in the limited iteration times.

The comparison of the optimization results is shown in Table 4 between the proposed algorithm and the traditional GA. The hardware configuration is Intel Pentium 725/1.6 GHz, 512M memory and the software requires Matlab 7.0 platform. The data in Table 4 are the mean values after 20 times of optimization, where the calculation times and average time are the calculation times of the fitness function and consuming time when obtaining the maximum. In the process of KGA algorithm, the mean time is 1.0127s when doing rough set calculation. From Table 4, the proposed KGA algorithm has more optimization precision and calculation efficiency than the traditional GA.

Table 4. Test results

Function	GA			KGA		
	Calculation times	Average time (s)	Optimum	Calculation times	Average time (s)	Optimum
F1	63666	13.3272	39.7800628	25103.2	5.34736	39.7897702
F2	61853.44	10.8576	0.0151739	29188	5.3878	0.0111307
F3	54627.52	11.8903	3599.997315	20359	4.1255	3599.999982
F4	50759.08	10.3656	0.9999958	19137	4.7252	0.99999995

5   Conclusion

A knowledge-based genetic algorithm is proposed in this paper. Rough set and granular computing method is introduced to explore knowledge hidden in the data generated in GA evolution process. The knowledge then can be used to guide the evolution orientation. The test function show that the proposed algorithm in this paper is good to increase the convergent speed of the genetic algorithm. It overcomes the deceptive problem to some extent in the genetic algorithm. The proposed algorithm shows the good performance for the multimodal function. Therefore the knowledge-based genetic algorithm is effective.

At the same time there are still some problems to be further studied. For example, the degrees of the data discretization and the size of the granule generation based on the discrete data have the great influence on the calculation efficiency of the rough set. The division thresholds α, β to the rough set has great influence on the performance of the proposed algorithm and directly on the number of the optimum searching subspaces. All the above is the research direction in the future.

References

1. Holland, J.H.: Genetic algorithms. Sci. Amer., 66–72 (1992)
2. Goldberg, D.E.: Genetic Algorithm in Search, Optimizaiton and Machine Learning. Addison Wesley, Reading (1989)
3. Vose, M.D., Liepins, G.E.: Schema Disruption. In: Proceedings of the Fourth International Conference on Genetic Algorithms, San Francisco, pp. 196–203 (1991)
4. Pawlak, Z.: Rough sets. International J. Comp. Inform. Science 11, 341–356 (1982)
5. Pawlak, Z., Wong, S.K.M., Ziarko, W.: Rough Sets: Probabilistic Versus Deterministic Approach. Int. J. Man-Machine Studies 29, 81–95 (1988)
6. Peters, J.F., Suraj, Z., et al.: Classification of meteorological volumetric radar data using rough set methods. Pattern Recognition Letters 24(6), 911–920 (2003)
7. Mrozek, A.: Rough Sets and Dependency Analysis Among Attributes in Computer Implementation of Expert's Inference Models. Int. J. Man-Machine Studies 30, 457–473 (1989)
8. James, F.P., Skowron, A., Ramanna, S., Synak, P.: Rough sets and information granulation. In: De Baets, B., Kaynak, O., Bilgiç, T. (eds.) IFSA 2003. LNCS, vol. 2715, pp. 370–377. Springer, Heidelberg (2003)
9. Yao, Y.Y.: Modeling data mining with granular computing. In: Proceedings of the 25th Annual International Computer Software and Applications Conference (COMPSAC 2001), Chicago, USA, pp. 638–643 (2001)
10. Lingras, P.W., Davies, C.: Applications of rough genetic algorithms. Computational Intelligence 17(3), 435–445 (2001)
11. Lingras, P., Davies, C.: Rough Genetic Algorithms. In: Franz, A. (ed.) Automatic Ambiguity Resolution in Natural Language Processing. LNCS, vol. 1171, pp. 38–46. Springer, Heidelberg (1996)
12. Liu, Q.: Rough sets and rough reasoning, pp. 105–112. Science press (2001)
13. Chen, Z.H., Xie, G., Yan, G.W., Xie, K.M.: Application of a Matrix-Based Binary Granular Computing Algorithm in RST. In: IEEE International Conference On Granular Computing, Beijing, pp. 409–412 (2005)

Dependent Uncertain Linguistic OWA Operator

Guiwu Wei

Department of Economics and Management, Chongqing University of Arts and Sciences,
Yongchuan, Chongqing, 402160, P.R. China
weiguiwu@163.com

Abstract. Yager [1] introduced several families of ordered weighted averaging (OWA) operators, in which the associated weights depend on the aggregated arguments. In this paper, we develop a new dependent uncertain linguistic OWA operator based on dependent OWA operator and dependent uncertain OWA operator in uncertain linguistic setting, and study some of its desirable properties. The prominent characteristic of this dependent uncertain linguistic OWA operator is that it can relieve the influence of unfair arguments on the aggregated results. Finally, an illustrative example is given.

Keywords: Group decision making (GDM), Uncertain linguistic variable, Dependent uncertain linguistic OWA operator.

1 Introduction

The ordered weighted aggregation operator as an aggregation technique has received more and more attention since its appearance [2]. One important step of the OWA operator is to determine its associated weights. Many authors have developed some useful approaches to obtaining the OWA weights [1-7]. Especially, Xu and Da [4] established a linear objective-programming model for obtaining the weights of the OWA operator. Xu [5] developed a normal distribution based method. Xu [6] developed a new dependent OWA (DOWA) operator. Xu [7] developed some dependent uncertain ordered weighted aggregation operators. However, in many situations, the input arguments take the form of uncertain linguistic variables because of time pressure, lack of knowledge, and people's limited expertise related with problem domain. In this paper, we will develop a new argument-dependent approach to determining the uncertain linguistic OWA weights. The remainder of this paper is set out as in follows. In the next section, we introduce some basic concepts of uncertain linguistic variables and develop a dependent uncertain linguistic ordered weighted averaging (DULOWA) operator in which the associated weights only depend on the aggregated uncertain linguistic arguments and can relieve the influence of unfair uncertain linguistic arguments on the aggregated results by assigning low weights to those "false" and "biased" ones. In Section 3 we develop a practical method based on the ULWA and the DULOWA operators for GDM problem. In Section 4, we give an illustrative example. In Section 5 we conclude the paper and give some remarks.

G. Wang et al. (Eds.): RSKT 2008, LNAI 5009, pp. 156–163, 2008.
© Springer-Verlag Berlin Heidelberg 2008

2 DULOWA Operator

We consider a finite and totally ordered discrete linguistic label set $S = \{s_i | i = -t, \cdots, -1, 0, 1, \cdots, t\}$, where t is a positive integer, s_i represents a linguistic variable and satisfy $s_i > s_j$, if $i > j$. For example, S can be defined as [8].

$$S = \{s_{-4} = extremely\ poor, s_{-3} = very\ poor, s_{-2} = poor, s_{-1} = slightly\ poor,$$
$$s_0 = fair, s_1 = slightly\ good, s_2 = good, s_3 = very\ good, s_4 = extremely\ good\}$$

To preserve all the given information, Xu[8] extend the discrete term set S to a continuous term set $\overline{S} = \{s_a | a \in [-q, q]\}$, where $q(q > t)$ is a sufficiently large positive integer. If $s_a \in S$, then we call s_a the original linguistic term, otherwise, we call s_a the virtual linguistic term. In general, the decision maker uses the original linguistic term to evaluate attributes and alternatives, and the virtual linguistic terms can only appear in calculation [8].

Definition 1. Let $\tilde{s} = [s_\alpha, s_\beta]$, $s_\alpha, s_\beta \in \overline{S}$, s_α and s_β are the lower and the upper limits, respectively, we call $\tilde{s}$ the uncertain linguistic variable [8].

Let $\tilde{S}$ be the set of all the uncertain linguistic variables. Consider any three uncertain linguistic variables $\tilde{s} = \left[s_\alpha, s_\beta\right]$, $\tilde{s}_1 = \left[s_{\alpha_1}, s_{\beta_1}\right]$ and $\tilde{s}_2 = \left[s_{\alpha_2}, s_{\beta_2}\right]$, $\tilde{s}, \tilde{s}_1, \tilde{s}_2 \in \tilde{S}, \lambda \in [0,1]$, then their operational laws are defined as [8]:

(1) $\tilde{s}_1 \oplus \tilde{s}_2 = \left[s_{\alpha_1}, s_{\beta_1}\right] \oplus \left[s_{\alpha_2}, s_{\beta_2}\right] = \left[s_{\alpha_1} \oplus s_{\alpha_2}, s_{\beta_1} \oplus s_{\beta_2}\right] = \left[s_{\alpha_1 + \alpha_2}, s_{\beta_1 + \beta_2}\right]$;

(2) $\lambda\tilde{s} = \lambda\left[s_\alpha, s_\beta\right] = \left[\lambda s_\alpha, \lambda s_\beta\right] = \left[s_{\lambda\alpha}, s_{\lambda\beta}\right]$.

Definition 2. Let $\tilde{s}_1 = \left[s_{\alpha_1}, s_{\beta_1}\right]$ and $\tilde{s}_2 = \left[s_{\alpha_2}, s_{\beta_2}\right]$ be two uncertain linguistic variables, and let $len(\tilde{s}_1) = \beta_1 - \alpha_1, len(\tilde{s}_2) = \beta_2 - \alpha_2$, then the degree of possibility of $\tilde{s}_1 \geq \tilde{s}_2$ is defined as[8]

$$p(\tilde{s}_1 \geq \tilde{s}_2) = \frac{\max\left(0, len(\tilde{s}_1) + len(\tilde{s}_2) - \max\left(\beta_2 - \alpha_1, 0\right)\right)}{len(\tilde{s}_1) + len(\tilde{s}_2)} \quad (1)$$

From Definition 3, we can easily get the following results easily:

(1) $0 \leq p(\tilde{s}_1 \geq \tilde{s}_2) \leq 1, 0 \leq p(\tilde{s}_2 \geq \tilde{s}_1) \leq 1$;

(2) $p(\tilde{s}_1 \geq \tilde{s}_2) + p(\tilde{s}_2 \geq \tilde{s}_1) = 1$. Especially, $p(\tilde{s}_1 \geq \tilde{s}_1) = p(\tilde{s}_2 \geq \tilde{s}_2) = 0.5$.

Definition 3. Let $\text{ULWA}:\left(\tilde{S}\right)^{n}\rightarrow\tilde{S}$, if

$$\text{ULWA}_{\omega}\left(\tilde{s}_{1},\tilde{s}_{2},\cdots,\tilde{s}_{n}\right)=\omega_{1}\tilde{s}_{1}\oplus\omega_{2}\tilde{s}_{2}\oplus\cdots\otimes\omega_{n}\tilde{s}_{n} \qquad (2)$$

where $\omega=\left(\omega_{1},\omega_{2},\cdots,\omega_{n}\right)^{T}$ is the weighting vector of uncertain linguistic variables $\tilde{s}_{j}\left(j=1,2,\cdots,n\right)$, and $\omega_{j}\in\left[0,1\right]$, $j=1,2,\cdots,n$, $\sum_{j=1}^{n}\omega_{j}=1$, then ULWA is called the uncertain linguistic weighted averaging (ULWA) operator [8].

Definition 4. An uncertain linguistic ordered weighted averaging (ULOWA) operator of dimension n is a mapping $\text{ULOWA}:\left(\tilde{S}\right)^{n}\rightarrow\tilde{S}$, which has associated weighting vector $w=\left(w_{1},w_{2},\cdots,w_{n}\right)^{T}$ such that $w_{j}\in\left[0,1\right]$, $\sum_{j=1}^{n}w_{j}=1$, $j=1,2,\cdots,n$. Furthermore:

$$\text{ULOWA}_{w}\left(\tilde{s}_{1},\tilde{s}_{2},\cdots,\tilde{s}_{n}\right)=w_{1}\tilde{s}_{\pi(1)}\oplus w_{2}\tilde{s}_{\pi(2)}\oplus\cdots\otimes w_{n}\tilde{s}_{\pi(n)} \qquad (3)$$

where $\left(\pi\left(1\right),\pi\left(2\right),\cdots,\pi\left(n\right)\right)$ is a permutation of $\left(1,2,\cdots,n\right)$, such that $\tilde{s}_{\pi(j-1)}\geq\tilde{s}_{\pi(j)}$ for all $j=2,\cdots,n$ [8].

Definition 5. Let $\tilde{s}_{1},\tilde{s}_{2},\cdots,\tilde{s}_{n}$ be a collection of uncertain linguistic variables, where $\tilde{s}_{j}\in\tilde{S}$, $j=1,2,\cdots,n$, then we define the mean of these uncertain linguistic variables as

$$\tilde{s}^{*}=\left(\tilde{s}_{1}\oplus\tilde{s}_{2}\oplus\cdots\oplus\tilde{s}_{n}\right)/n \qquad (4)$$

Definition 6. Let $\tilde{s}_{1}=\left[s_{\alpha_{1}},s_{\beta_{1}}\right]$ and $\tilde{s}_{2}=\left[s_{\alpha_{2}},s_{\beta_{2}}\right]$ be two uncertain linguistic variables, then we call [9]

$$d\left(\tilde{s}_{1},\tilde{s}_{2}\right)=\left(\left|\alpha_{1}-\alpha_{2}\right|+\left|\beta_{1}-\beta_{2}\right|\right)/2 \qquad (5)$$

the distance between $\tilde{s}_{1}$ and $\tilde{s}_{2}$.

Definition 7. Let $\tilde{s}_{1},\tilde{s}_{2},\cdots,\tilde{s}_{n}$ be a collection of uncertain linguistic variables, and let $\tilde{s}^{*}$ the mean of these uncertain linguistic variables, then we call

$$sd\left(\tilde{s}_{\pi(j)},\tilde{s}^{*}\right)=1-d\left(\tilde{s}_{\pi(j)},\tilde{s}^{*}\right)\Big/\sum_{j=1}^{n}d\left(\tilde{s}_{\pi(j)},\tilde{s}^{*}\right), \quad j=1,2,\cdots,n. \qquad (6)$$

the degree of similarity between the jth largest uncertain linguistic variables $\tilde{s}_{\pi(j)}$ and the mean $\tilde{s}^{*}$, where $\left(\pi\left(1\right),\pi\left(2\right),\cdots,\pi\left(n\right)\right)$ is a permutation of $\left(1,2,\cdots,n\right)$, such that $\tilde{s}_{\pi(j-1)}\geq\tilde{s}_{\pi(j)}$ for all $j=2,\cdots,n$.

In real-life situations, the uncertain linguistic variables $\tilde{s}_1, \tilde{s}_2, \cdots, \tilde{s}_n$ usually take the form of a collection of n preference values provided by n different individuals. Some individuals may assign unduly high or unduly low preference values to their preferred or repugnant objects. In such a case, we shall assign very low weights to these "false" or "biased" opinions, that is to say, the closer a preference value is to the mid one(s), the more the weight [5]. As a result, based on (3), we define the ULOWA weights as

$$w_j = sd\left(\tilde{s}_{\pi(j)}, \tilde{s}^*\right) \Big/ \sum\nolimits_{j=1}^{n} sd\left(\tilde{s}_{\pi(j)}, \tilde{s}^*\right), \quad j = 1, 2, \cdots, n \tag{7}$$

Obviously, $w_j \geq 0$, $j = 1, 2, \cdots, n$ and $\sum\nolimits_{j=1}^{n} w_j = 1$. Especially, if $\tilde{s}_i = \tilde{s}_j$, for all $i, j = 1, 2, \cdots, n$, then by (7), we have $w_j = 1/n$, for all $j = 1, 2, \cdots, n$.

By (3), we have

$$\text{ULOWA}_w\left(\tilde{s}_1, \tilde{s}_2, \cdots, \tilde{s}_n\right) = \sum_{j=1}^{n} \frac{sd\left(\tilde{s}_{\pi(j)}, \tilde{s}^*\right)}{\sum_{j=1}^{n} sd\left(\tilde{s}_{\pi(j)}, \tilde{s}^*\right)} \tilde{s}_{\pi(j)} = \frac{\sum_{j=1}^{n} sd\left(\tilde{s}_{\pi(j)}, \tilde{s}^*\right)\tilde{s}_{\pi(j)}}{\sum_{j=1}^{n} sd\left(\tilde{s}_{\pi(j)}, \tilde{s}^*\right)} \tag{8}$$

Since $\sum\nolimits_{j=1}^{n} sd\left(\tilde{s}_{\pi(j)}, \tilde{s}^*\right)\tilde{s}_{\pi(j)} = \sum\nolimits_{j=1}^{n} sd\left(\tilde{s}_j, \tilde{s}^*\right)\tilde{s}_j$

and $\sum\nolimits_{j=1}^{n} sd\left(\tilde{s}_{\pi(j)}, \tilde{s}^*\right) = \sum\nolimits_{j=1}^{n} sd\left(\tilde{s}_j, \tilde{s}^*\right)$

then we replace (8) by

$$\text{ULOWA}_w\left(\tilde{s}_1, \tilde{s}_2, \cdots, \tilde{s}_n\right) = \sum\nolimits_{j=1}^{n} sd\left(\tilde{s}_j, \tilde{s}^*\right)\tilde{s}_j \Big/ \sum\nolimits_{j=1}^{n} sd\left(\tilde{s}_j, \tilde{s}^*\right) \tag{9}$$

We call (9) a dependent uncertain linguistic ordered weighted averaging (DU-LOWA) operator, which is a generalization of the dependent uncertain ordered weighted averaging (DUOWA) operator [7]. Similar to DOWA operator[6] and DUOWA[7], consider that the aggregated value of the DULOWA operator is independent of the ordering, thus it is also a neat operator.

Similar to [7], we have the following result:

Theorem 1. Let $\tilde{s}_1, \tilde{s}_2, \cdots, \tilde{s}_n$ be a collection of uncertain linguistic variables, and let $\tilde{s}^*$ the mean of these linguistic arguments, $\left(\pi(1), \pi(2), \cdots, \pi(n)\right)$ is a permutation of $\left(1, 2, \cdots, n\right)$, such that $\tilde{s}_{\pi(j-1)} \geq \tilde{s}_{\pi(j)}$ for all $j = 2, \cdots, n$. If $sd\left(\tilde{s}_{\pi(i)}, \tilde{s}^*\right) \geq sd\left(\tilde{s}_{\pi(j)}, \tilde{s}^*\right)$, then $w_i \geq w_j$.

3   An Approach to GDM under Uncertain Linguistic Environment

For the GDM making problems, in which both the attribute weights and the expert weights take the form of real numbers, and the attribute preference values take the form of uncertain linguistic variables, we shall develop an approach based on the ULWA and DULOWA operators to GDM under uncertain linguistic environment.

Let $A = \{A_1, A_2, \cdots, A_m\}$ be a discrete set of alternatives, $G = \{G_1, G_2, \cdots, G_n\}$ be the set of attributes, $\omega = (\omega_1, \omega_2, \cdots, \omega_n)$ is the weighting vector of the attributes $G_j (j = 1, 2, \cdots, n)$, where $\omega_j \in [0,1], \sum_{j=1}^{n} \omega_j = 1$, $D = \{D_1, D_2, \cdots, D_t\}$ be the set of decision makers. Suppose that $\tilde{R}_k = \left(\tilde{r}_{ij}^{(k)} \right)_{m \times n}$ is the decision matrix, where $\tilde{r}_{ij}^{(k)} \in \tilde{S}$ is a preference values, which take the form of uncertain linguistic variable, given by the decision maker $D_k \in D$, for the alternative $A_i \in A$ with respect to the attribute $G_j \in G$.

Step 1. Utilize the decision information given in matrix $\tilde{R}_k$, and the ULWA operator

$$\tilde{r}_i^{(k)} = \text{ULWA}_\omega \left(\tilde{r}_{i1}^{(k)}, \tilde{r}_{i2}^{(k)}, \cdots, \tilde{r}_{in}^{(k)} \right), i = 1, 2, \cdots, m, k = 1, 2, \cdots, t.$$

to derive the individual overall preference value $\tilde{r}_i^{(k)}$ of the alternative A_i.

Step 2. Utilize the DULOWA operator:

$$\tilde{r}_i = \text{DULOWA}_w \left(\tilde{r}_i^{(1)}, \tilde{r}_i^{(2)}, \cdots, \tilde{r}_i^{(t)} \right), i = 1, 2, \cdots, m.$$

to derive the collective overall preference values $\tilde{r}_i (i = 1, 2, \cdots, m)$ of the alternative A_i, $w = (w_1, w_2, \cdots, w_n)$ is the associated weighting vector of the DULOWA operator, with $w_j \in [0,1]$, $\sum_{j=1}^{n} w_j = 1$.

Step 3. To rank these collective overall preference values $\tilde{r}_i (i = 1, 2, \cdots, m)$, we first compare each $\tilde{r}_i$ with all the $\tilde{r}_j (j = 1, 2, \cdots, m)$ by using Eq.(1). For simplicity, we let $p_{ij} = p(\tilde{r}_i \geq \tilde{r}_j)$, then we develop a complementary matrix as $P = \left(p_{ij} \right)_{m \times m}$, where $p_{ij} \geq 0$, $p_{ij} + p_{ji} = 1$, $p_{ii} = 0.5$, $i, j = 1, 2, \cdots, n$.

Summing all the elements in each line of matrix P, we have

$$p_i = \sum_{j=1}^{m} p_{ij}, i = 1, 2, \cdots, m.$$

Then we rank the collective overall preference values $\tilde{r}_i (i = 1, 2, \cdots, m)$ in descending order in accordance with the values of $p_i (i = 1, 2, \cdots, m)$.

Step 4. Rank all the alternatives $A_i (i = 1, 2, \cdots, m)$ and select the best one(s) in accordance with the collective overall preference values $\tilde{r}_i (i = 1, 2, \cdots, m)$.

4 Illustrative Example

Let us suppose there is an investment company, which wants to invest a sum of money in the best option (adapted from [10]). There is a panel with five possible alternatives to invest the money:①A1 is a car company;②A2 is a food company;③ A3 is a computer company;④A4 is a arms company;⑤A5 is a TV company. The investment company must take a decision according to the following four attributes: ①G1 is the risk analysis;②G2 is the growth analysis;③G3 is the social-political impact analysis;④G4 is the environmental impact analysis. The five possible alterna-

tives are to be evaluated using the linguistic term set S by the three decision makers under the above four attributes, and construct, respectively, the decision matrices as

follows $\tilde{R}_k = \left(\tilde{r}_{ij}^{(k)} \right)_{5 \times 4} (k = 1, 2, 3)$:

$$\tilde{R}_1 = \begin{pmatrix} [s_2, s_3] & [s_3, s_4] & [s_1, s_2] & [s_1, s_2] \\ [s_1, s_2] & [s_{-2}, s_3] & [s_3, s_4] & [s_2, s_3] \\ [s_2, s_4] & [s_1, s_3] & [s_0, s_3] & [s_{-3}, s_1] \\ [s_0, s_2] & [s_1, s_2] & [s_2, s_3] & [s_2, s_3] \\ [s_3, s_4] & [s_{-1}, s_3] & [s_1, s_4] & [s_3, s_4] \end{pmatrix}$$

$$\tilde{R}_2 = \begin{pmatrix} [s_1, s_2] & [s_1, s_4] & [s_1, s_2] & [s_{-1}, s_3] \\ [s_1, s_3] & [s_2, s_4] & [s_2, s_3] & [s_2, s_3] \\ [s_2, s_4] & [s_1, s_2] & [s_1, s_2] & [s_{-4}, s_2] \\ [s_2, s_3] & [s_3, s_4] & [s_{-2}, s_1] & [s_1, s_3] \\ [s_{-3}, s_2] & [s_0, s_2] & [s_1, s_4] & [s_1, s_3] \end{pmatrix}$$

$$\tilde{R}_3 = \begin{pmatrix} [s_2, s_4] & [s_1, s_3] & [s_1, s_4] & [s_1, s_2] \\ [s_1, s_2] & [s_2, s_4] & [s_1, s_2] & [s_2, s_4] \\ [s_{-3}, s_1] & [s_{-3}, s_4] & [s_1, s_4] & [s_1, s_3] \\ [s_2, s_4] & [s_1, s_3] & [s_{-1}, s_3] & [s_2, s_4] \\ [s_0, s_3] & [s_2, s_4] & [s_2, s_4] & [s_{-2}, s_2] \end{pmatrix}$$

In the following, we shall utilize the proposed approach in this paper getting the most desirable alternative(s):

Step 1. Utilize the ULWA operator (Let $\omega = (0.3, 0.1, 0.2, 0.4)$) to derive the individual overall preference value $\tilde{r}_i^{(k)}$ of the alternative A_i.

$$\tilde{r}_1^{(1)} = [s_{1.5}, s_{2.5}], \tilde{r}_2^{(1)} = [s_{1.5}, s_{2.9}], \tilde{r}_3^{(1)} = [s_{-0.5}, s_{2.5}], \tilde{r}_4^{(1)} = [s_{1.3}, s_{2.6}]$$

$$\tilde{r}_5^{(1)} = [s_{2.2}, s_{3.9}], \tilde{r}_1^{(2)} = [s_{0.2}, s_{2.6}], \tilde{r}_2^{(2)} = [s_{1.7}, s_{3.1}], \tilde{r}_3^{(2)} = [s_{-0.7}, s_{2.6}]$$

$$\tilde{r}_2^{(2)} = [s_{1.7}, s_{3.1}], \tilde{r}_3^{(2)} = [s_{-0.7}, s_{2.6}], \tilde{r}_4^{(2)} = [s_{0.9}, s_{2.7}], \tilde{r}_5^{(2)} = [s_{-0.3}, s_{2.8}]$$

$$\tilde{r}_1^{(3)} = [s_{1.3}, s_{3.1}], \tilde{r}_2^{(3)} = [s_{1.5}, s_{3}], \tilde{r}_3^{(3)} = [s_{-0.6}, s_{2.7}], \tilde{r}_4^{(3)} = [s_{1.3}, s_{3.7}]$$

$$\tilde{r}_5^{(3)} = [s_{-0.2}, s_{2.9}]$$

Step 2. Utilize the DULOWA operator to derive the collective overall preference values $\tilde{r}_i \ (i = 1, 2, \cdots, 5)$ of the alternative A_i.

$$\tilde{r}_1 = 0.343 \times [s_{1.5}, s_{2.5}] \oplus 0.300 \times [s_{0.2}, s_{2.6}] \oplus 0.357 \times [s_{1.3}, s_{3.1}] = [s_{1.039}, s_{2.744}]$$

$$\tilde{r}_2 = 0.321 \times [s_{1.5}, s_{2.9}] \oplus 0.250 \times [s_{1.7}, s_{3.1}] \oplus 0.429 \times [s_{1.5}, s_3] = [s_{1.550}, s_{2.993}]$$

$$\tilde{r}_3 = 0.250 \times [s_{-0.5}, s_{2.5}] \oplus 0.375 \times [s_{-0.7}, s_{2.6}] \oplus 0.375 \times [s_{-0.6}, s_{2.7}] = [s_{-0.613}, s_{2.613}]$$

$$\tilde{r}_4 = 0.362 \times [s_{1.3}, s_{2.6}] \oplus 0.353 \times [s_{0.9}, s_{2.7}] \oplus 0.284 \times [s_{1.3}, s_{3.7}] = [s_{1.159}, s_{2.948}]$$

$$\tilde{r}_5 = 0.250 \times [s_{2.2}, s_{3.9}] \oplus 0.364 \times [s_{-0.3}, s_{2.8}] \oplus 0.386 \times [s_{-0.2}, s_{2.9}] = [s_{0.364}, s_{3.114}]$$

Step 3. By using Eq.(1), and then we develop a complementary matrix:

$$P = \begin{bmatrix} 0.500 & 0.379 & 0.681 & 0.454 & 0.534 \\ 0.621 & 0.500 & 0.772 & 0.567 & 0.627 \\ 0.319 & 0.228 & 0.500 & 0.290 & 0.376 \\ 0.546 & 0.433 & 0.710 & 0.500 & 0.569 \\ 0.466 & 0.373 & 0.624 & 0.431 & 0.500 \end{bmatrix}$$

Summing all the elements in each line of matrix P, we have

$$p_1 = 2.548, p_2 = 3.008, p_3 = 1.713, p_4 = 2.758, p_5 = 2.393.$$

Step 4. Rank all the alternatives $A_i \ (i = 1, 2, \cdots, 5)$ in accordance with the values $p_i \ (i = 1, 2, \cdots, 5)$: $A_2 \succ A_4 \succ A_1 \succ A_5 \succ A_3$, and thus the most desirable alternative is A_2.

5   Conclusion

In this paper, we have investigated the dependent uncertain linguistic OWA operators in uncertain linguistic setting, and developed a new argument-dependent approach to determining the uncertain linguistic OWA weights, which can relieve the influence of unfair arguments on the aggregated results. We have verified the practicality and effectiveness of the approach with a numerical example.

References

1. Yager, R.R.: Families of OWA Operators. Fuzzy Sets and Systems 59, 125–148 (1993)
2. Yager, R.R.: On Ordered Weighted Averaging Aggregation Operators in Multicriteria Decision Making. IEEE Transactions on Systems, Man, and Cybernetics 18, 183–190 (1988)
3. Filev, D.P., Yager, R.R.: On the Issue of Obtaining OWA Operator Weights. Fuzzy Sets and Systems 94, 157–169 (1998)
4. Xu, Z.S., Da, Q.L.: The Uncertain OWA Operator. International Journal of Intelligent Systems 17, 569–575 (2002)
5. Xu, Z.S.: An Overview of Methods for Determining OWA Weights. International Journal of Intelligent Systems 20(8), 843–865 (2005)
6. Xu, Z.S.: Dependent OWA operators. In: Torra, V., Narukawa, Y., Valls, A., Domingo-Ferrer, J. (eds.) MDAI 2006. LNCS (LNAI), vol. 3885, pp. 172–178. Springer, Heidelberg (2006)
7. Xu, Z.S.: Dependent Uncertain Ordered Weighted Aggregation Operators. Information Fusion (in press, 2006)
8. Xu, Z.S.: Uncertain Linguistic Aggregation Operators Based Approach to Multiple Attribute Group Decision Making under Uncertain Linguistic Environment. Information Science 168, 171–184 (2004)
9. Xu, Z.S.: An Approach to Pure Multiple Attribute Decision Making under Uncertainty. International Journal of Information Technology & Decision Making 4(2), 197–206 (2005)
10. Delgado, M., Herrera, F., Herrera-Viedma, E., Martinez, L.: Combining Numerical and Linguistic Information in Group Decision Making. Information Sciences 107, 177–194 (1998)

Efficient Gene Selection with Rough Sets from Gene Expression Data

Lijun Sun[1], Duoqian Miao[2], and Hongyun Zhang[3]

[1] Department of Computer Science and Technology,
Tongji University, Shanghai, 201804, P.R. China
Sunlj1028@yahoo.com.cn
[2] Department of Computer Science and Technology,
Tongji University, Shanghai, 201804, P.R. China
Miaoduoqian@163.com
[3] Department of Computer Science and Technology,
Tongji University, Shanghai, 201804, P.R. China
Zhanghongyun583@sina.com

Abstract. The main challenge of gene selection from gene expression dataset is to reduce the redundant genes without affecting discernibility between objects. A pipelined approach combining feature ranking together with rough sets attribute reduction for gene selection is proposed. Feature ranking is used to narrow down the gene space as the first step, top ranked genes are selected; the minimal reduct is induced by rough sets to eliminate the redundant attributes. An exploration of this approach on Leukemia gene expression data is conducted and good results are obtained with no preprocessing to the data. The experiment results show that this approach is successful for selecting high discriminative genes for cancer classification task.

Keywords: Gene selection, Feature ranking, Rough sets, Attributes reduction.

1   Introduction

The emergence of cDNA microarray technologies makes it possible to record the expression levels of thousands of genes simultaneously. Generally, different cells or a cell under different conditions yield different microarray results, thus comparisons of gene expression data derived from microarray results between normal and tumor cells can provide the important information for tumor classification [1]. A reliable and precise classification of tumors based on gene expression data may lead to a more complete understanding of molecular variations among tumors, and hence, to better diagnosis and treatment strategies.

Gene expression data set has very unique characteristics that are very different from all the previous data used for classification. Most publicly available gene expression data usually has the following properties:

- high dimensionality: Up to tens of thousands of genes,

G. Wang et al. (Eds.): RSKT 2008, LNAI 5009, pp. 164–171, 2008.
© Springer-Verlag Berlin Heidelberg 2008

- very small data set size: Not more than a few dozens of samples,
- most genes are not related to tumor classification.

With such a huge attribute space, it is almost certain that very accurate classification of tissue samples is difficult and among a large amount of genes, only a very small fraction of them are informative for classification task [1] [2] [3] [4] [5] [12] [14] [15], thus performing gene selection prior to classification makes help to narrowing down the attribute number and improving classification accuracy and time-complexity of classification algorithms. More importantly, with the "noise" from the large number of irrelevant genes removed, the biological information hidden within will be less obstructed; this would assist in drug discovery and early tumor discovery. How to select the most useful genes for cancer classification is becoming a very challenging task.

A good number of algorithms have been developed for this purpose [1] [2] [3] [5] [11] [12] [14] [15]; feature-ranking approach is most widely used. In this approach, each feature/attribute is measured for correlation with the class according to some measuring criteria. The features/attributes are ranked and the top ones or those that satisfy a certain criterion are selected. The main characteristic of feature ranking is that it is based on individual feature correlation with respect to class separately. Simple method such as statistical tests (t-test, F-test) has been shown to be effective [1] [6]. This kind of approach also has the virtue of being easily and very efficiently computed.

Feature sets so obtained have certain redundancy because genes in similar pathways probably all have very similar scores and therefore no additional information gain, rough sets attribute reduction can be used to eliminate such redundancy and minimize the feature sets. The theory of rough sets [7] , as a major mathematical tool for managing uncertainty that arises from granularity in the domain of discourse-that is, from the indiscernibility between objects in a set, has been applied mainly in data mining tasks like classification, clustering and feature selection. Recent years, rough sets theory has been used in gene selection task by some researchers. Evolutionary rough feature selection is employed on three gene expression datasets in [19], not more than 10 genes are selected on each data set while high classification accuracies are obtained; In [20], with the positive region based reduct algorithm, more than 90% of redundant genes are eliminated.

In this paper, we introduce a pipelined method using feature ranking and rough sets attribute reduction for gene selection. This paper is organized as follows. The next section gives the background of rough sets. Then, our method is detailed in Section 3. And in Section 4, experimental results are listed. The discussions of these results are given. Finally, the conclusions are drawn in Section 5.

2   Rough Sets Based Feature Selection

In rough sets theory, a decision table is denoted by $T = \{U, A)$, where $A = C \cup D$,C is called condition attribute sets,$D = \{d\}$ is decision feature, and U

is universe of discourse. Rows of the decision table correspond to objects, and columns correspond to attributes [7].

Definition 1. Indiscernibility Relation. Let $a \in A, P \subseteq A$, a binary relation $IND(P)$, called the indiscernibility relation, is defined as the following:

$$IND(P) = \{(x, y) \in U \times U | \forall a \in P, a(x) = a(y)\}$$

Let $U/IND(P)$ denotes the family of all equivalence classes of the relation $IND(P), U/IND(P)$ is also a definable partition of the universe induced by P.

Definition 2. Indispensable and Dispensable Attribute. An attribute $c \in C$ is an indispensable attribute if

$$Card(U/IND(C - \{c\})) \neq Card(U/IND(C - \{c\} \cup D))$$

An attribute $c \in C$ is a dispensable attribute if

$$Card(U/IND(C - \{c\})) = Card(U/IND(C - \{c\} \cup D))$$

Definition 3. Reduct. The subset of attributes $R \subseteq C$ is a reduct of attribute set C if

$$Card(U/IND(R \cup D)) = Card(U/IND(C \cup D))$$

And $\forall Q \subset R$

$$Card(U/IND(Q \cup D)) \neq Card(U/IND(C \cup D))$$

Definition 4. Core. The set of all indispensable features in C is

$$CORE(C) = \cap RED(C)$$

where $RED(C)$ is the set of all reducts of C with respect to D

The reduct represent the minimal set of non-redundant features that are capable of discerning objects in a decision table. An optimal feature subset selection based on the rough set theory can be viewed as finding such a reduct $R, R \subseteq C$ with the best classifying properties. R, instead of C, will be used in a rule discovery algorithm. It is obvious that all of indispensable features in core cannot be deleted from C without losing the accuracy of a decision table; the feature(s) in core must be the member of feature subsets. Therefore, the problem of feature subset selection will become how to select the features from dispensable features for forming the best reduct with core. Obtaining all reducts of a decision table is a NP-hard problem, thus heuristic knowledge deriving from the dependency relationship between condition attributes and decision attributes in a decision table is mainly utilized to assist the attribute reduction. Many methods have been proposed to search for the attribute reducts, which are classified into several categories: 1) positive region [7]; 2) frequency function [8]; 3) information entropy [9]; etc. .

3   Rough Sets Based Gene Selection Method

Our learning problem is to select high discriminate genes for cancer classification from gene expression data. We may formalize this problem as a decision system, where universe $U = \{x_1, x_2, ..., x_m\}$ is a set of tumors, the conditional attributes set $C = \{g_1, g_2, ..., g_n\}$ contains each gene; the decision attribute d corresponds to class label of each sample. Each attribute $g_i \in C$ is represented by a vector $g_i = \{x_{1,i}, x_{2,i}, ..., x_{m,i}\}$, $i = 1, 2, ..., n$, where $x_{k,i}$ is the expression level of gene i at sample k, $k = 1, 2, ..., m$.

To select genes, t-test is widely used in the literature [1] [6]. Assuming that there are two classes of samples in a gene expression data set, the t-value for gene g is given by:

$$t = \frac{\mu_1 - \mu_2}{\sqrt{\sigma^2/n_1 + \sigma^2/n_2}} \tag{1}$$

Where μ_i and σ_i are the mean and the standard deviation of the expression levels of a gene g for class i respectively, and n_i is the number of samples in class i for $i = 1, 2$. When there are multiple classes of samples, the t-value is typically computed for one class versus all the other classes. The top genes ranked by t-value can then be selected for data mining. Feature sets so obtained have certain redundancy because genes in similar pathways probably all have very similar score and therefore no additional information gain. If several pathways involved in perturbation but one has main influence it is possible to describe this pathway with fewer genes, therefore Rough Sets attribute reduction is used to minimize the feature sets.

Reduct is constructed from core because it represents the set of indispensable features, thus all attributes in core must be in the reduct, then we adding attributes using information entropy as the heuristic information until a reduct is find. The attribute with lowest information entropy will be selected first because the higher attribute entropy means the more expected information is needed using the attribute to classify the samples. Given the partition by D, $U/IND(D)$, of U, the entropy based on the partition by $c \in C$, $U/IND(c)$, of U, is given by

$$E(c) = -\frac{1}{|U|} \sum_{X \in U/IND(D)} \sum_{Y \in U/IND(c)} |X \cap Y| \log_2 \frac{|X \cap Y|}{|Y|} \tag{2}$$

We can formulate our method as the following steps:

1. Calculate t-value of each gene, select the top ranked n genes to form the attribute set C.
2. Calculate core attribute sets of C using Discernibly Matrix [8], denoted by $CORE(C)$.
3. Calculate the reduct of C using information entropy as the heuristic information. Let $RED(C) \leftarrow CORE(C)$, while $Card(U/IND(C \cup D)) \neq Card(U/IND(RED(C) \cup D))$, we calculate information entropy of each gene $g \in C - RED(C)$, denoted by $E(g)$, if $E(g_1) = \min_{g \in C - RED(C)} E(g)$ then we assign g_1 to $RED(C)$. Repeat the above operation until we find a reduct of C.

4   Experimental Results

A well known gene expression data sets, leukemia data set of Golub et al. (1999), which is the same data sets used in many publications for gene selection and cancer classification [1] [3] [5] [11] [12] [19] [20], is used to evaluate the performance of our method. The acute leukemia dataset (http://www.genome.wi.mit.edu/MPR) consists of 38 samples including 27 cases of acute lymphoblastic leukemia (ALL) and 11 cases of acute myeloid leukemia (AML). The gene expression measurements were taken from high-density oligonucleotide microarrays containing 7129 genes. An independent test set of 20 ALL and 14 AML samples also exists.

First t-test is employed as a filter on the training set; the top ranked 50 genes are selected. Then entropy based discretization introduced in [16] is used to discretize the domain of each attribute because rough sets methods require discrete input. Entropy based attribute reduction algorithm is employed on the data set to find a minimal reduction. As the result, X95735 is the only gene to be selected in the reduction. A box plot of X95735 expression levels in the training set is presented in Fig. 1. This figure clearly indicates that the expression levels of X95735 can be used to distinguish ALL from AML in the training set.

Two rules are induced by Rough sets: if the expression level of X95735 $\geqslant$ 938 then the sample is classified as AML; If the expression level of X95735 <938 then the sample is classified as ALL. With the simples rules induced by Rough sets, 31 test samples are correctly classified; there are only three mistakes, one for AML, and two for ALL.

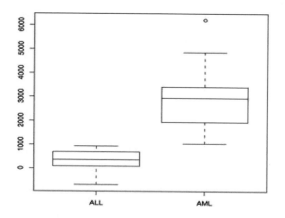

Fig. 1. Expression Levels of X95735 in Training Set

It is interesting that X95735 is also selected by many other methods. It is reported in [5] that X95735 is the only gene identified by J48 pruned tree and the emerging patterns algorithm, and X95735 is also selected by voting machine [1], SVM [10], Deb's NSGA-algorithm [21] and Cho's work [22]. An approach using clustering in combination with Rough Sets and neural networks has been investigated in [11], X95735 is repeated selected, and the classification accuracy is 91.2% on test data set. For comparison, the feature selection and classification results obtained by our method and some results in previous publishers are shown in table 1.

Table 1. The Comparison of Feature Selection and Classification Results

Method	Number of features	Classification Results
Rough sets	1	31
J48	1	31
Emerging Patterns	1	31
SVM	7	34
NSGA-II	3	34

The results obtained by us suggest that the expression level of X95735 plays an important role in distinguishing two types of acute leukemia. Role of X95735 in discerning between two types of acute leukemia samples is also verified by biological researchers [17] [18].

5   Conclusions

Gene expression data set usually has thousands of genes while a few dozens of samples, among a large amount of genes, only a very small fraction of them are informative for classification task. In order to achieve good classification performance, and obtain more useful insight about the biological related issues in cancer classification, gene selection should be well explored to reduce the noise and avoid overfitting of classification algorithm.

In this paper, a successful gene selection method based on rough sets theory is presented. Filter kind of method is done first as a preprocessing to select top ranked genes; the minimal reduct of the filtered attribute sets is induced by rough sets. Acute leukemia gene expression dataset is used to test the performance of this novel method; only one gene X95735 is selected, and high prediction accuracies have been achieved on the test data set. Gene X95735 is also selected by many other methods, and has been verified by biological researchers to play an important role in distinguish two different types of acute leukemia, AML and ALL.

Acknowledgements

This research support by The National Natural Science Foundation of China (Granted No. 60475019, No. 60775036) and The Research Fund for the Doctoral Program of Higher Education (Granted No. 20060247039).

References

1. Golub, T.R., Slonim, D.K., Tamayo, P., Huard, C., Gaasenbeek, M., Mesirov, J.P., Coller, H., Loh, M.L., Downing, J.R., Caligiuri, M.A., Bloomfield, C.D., Lander, E.S.: Molecular Classification of Cancer: Class Discovery and Class Prediction by Gene Expression Monitoring. Science 286, 531–537 (1999)
2. Wang, L.P., Feng, C., Xie, X.: Accurate Cancer Classification Using Expressions of Very Few Genes. IEEE/ACM Transactions on Computational Biology and Bioinformatics 4, 40–53 (2007)
3. Au, A., Chan, K.C.C., Wong, A.K.C., Wang, Y.: Attribute Clustering for Grouping, Selection, and Classification of Gene Expression Data. IEEE/ACM Transactions on Computational Biology and Bioinformatics 2, 83–101 (2005)
4. Smet, F.D., Pochet, N.L.M.M., Engelen, K., Gorp, T.V., Hummelen, P.V., Marchal, K., Amant, F., Timmerman, D., Moor, B.D., Vergote, I.: Predicting the Clinical Behavior of Ovarian Cancer from Gene Expression Profiles. International Journal of Gynecological Cancer 16, 147–151 (2006)
5. Wang, Y., Tetko, I.V., Hall, M.A., Frank, E., Facius, A., Mayer, K.F.X., Mewes, H.W.: Gene Selection from Microarray Data for Cancer Classification-A Machine Learning Approach. Computational Biology and Chemistry 29, 37–46 (2005)
6. Ding, C.: Analysis of Gene Expression Profiles: Class Discovery and Leaf Ordering. In: 6th Annual Conference on Research in Computational Molecular Biology, pp. 127–136. ACM Press, New York (2002)
7. Pawlak, Z.: Rough Set- Theoretical Aspects of Reasoning about Data. Kluwer Academic Publishers, Dorderecht (1991)
8. Wang, J., Waog, J.: Reduction Algorithms Based on Discernibly Matrix: The Ordered Attributes Method. Journal of Computer Science And Technology 16, 489–504 (2002)
9. Miao, D.Q., Hu, G.R.: A Heuristic Algorithm for Reduction of Knowledge. Journal of Computer Research and Development 36, 681–684 (1999)
10. Furey, T.S., Cristianini, N., Duffy, N., Bednarski, D.W., Schummer, M., Haussler, D.: Support Vector Machine Classification and Validation of Cancer Tissue Samples Using Microarray Expression Data. Bioinformatics 16, 906–914 (2000)
11. Valdes, J.J., Barton, A.J.: Gene Discovery in Leukemia Revisited: A Computational Intelligence Perspective. In: Orchard, B., Yang, C., Ali, M. (eds.) IEA/AIE 2004. LNCS (LNAI), vol. 3029, pp. 118–127. Springer, Heidelberg (2004)
12. Ding, C., Peng, H.C.: Minimum Redundancy Feature Selection from Microarray Gene Expression Data. Journal of Bioinformatics and Computational Biology 3, 185–205 (2003)
13. Ben-Dor, A., Bruhn, L., Friedman, N., Nachman, I., Schummer, M., Yakhini, Z.: Tissue Classification with Gene Expression Profiles. In: 4th Annual International Conference on Computational Molecular Biology (RECOMB), pp. 54–64. Universal Academy Press, Tokyo (2000)
14. Tseng, V.S., Kao, C.P.: Efficiently Mining Gene Expression Data via a Novel Parameterless Clustering Method. IEEE/ACM Transactions on Computational Biology and Bioinformatics 2, 355–365 (2005)
15. Mitra, S., Hayashi, Y.: Bioinformatics with Soft Computing. IEEE Transactions on Systems, Man and Cybernetics-Part C: Applications and Reviews 36, 616–635 (2006)

16. Fayyad, U.M., Irani, K.B.: Multi-Interval Discretization of Continuous-Valued Attributes for Classification Learning. In: Proceedings of the 13th International Joint Conference of Artificial Intelligence, pp. 1022–1027. Morgan Kaufmann, Chambery, France (1993)

17. Van, D.G.E., Leccia, M., Dekker, S., Jalbert, N., Amodeo, D., Byers, H.: Role of Zyxin in Differential Cell Spreading and Proliferation of Melanoma Cells and Melanocytes. J. Invest. Dermatol. 118, 246–254 (2002)

18. Yagi, T., Morimoto, A., Eguchi, M., Hibi, S., Sako, M., Ishii, E., Mizutani, S., Imashuku, S., Ohki, M., Ichikawa, H.: Identification of a Gene Expression Signature Associated with Pediatric AML Prognosis. Blood 102, 1849–1856 (2003)

19. Banerjee, M., Mitra, S., Banka, H.: Evolutinary-Rough Feature Selection in Gene Expression Data. IEEE Transaction on Systems, Man, and Cybernetics, Part C: Application and Reviews 37, 622–632 (2007)

20. Momin, B.F., Mitra, S., Datta Gupta, R.: Reduct Generation and Classification of Gene Expression Data. In: Proceeding of First International Conference on Hybrid Information Technology (ICHICT 2006), pp. 699–708. IEEE Press, New York (2006)

21. Deb, K., Reddy, A.R.: Reliable Classification of Two Class Cancer Data Using Evolutionary Algorithms. BioSystems 72, 111–129 (2003)

22. Cho, S.B., Ryu, J.: Classification Gene Expression Data of Cancer Using Classifier Ensemble with Mutually Exclusive Features. In: Proceedings of the IEEE, Special Issue on Bioinformatics Part-I: Advances and Challenges, pp. 1744–1753. IEEE Press, New York (2002)

Rough Cluster Algorithm Based on Kernel Function

Tao Zhou[1,2], Yanning Zhang[1], Huiling Lu[3], Fang'an Deng[2],
and Fengxiao Wang[2]

[1] School of Computer Science, Northwestern Polytechnical Univ.,710072 Xi'an, China
[2] Department of Maths, Shaanxi Univ. of Tech., 723000 Hanzhong, Shaanxi, China
[3] Department of Comp., Shaanxi Univ. of Tech. 723000 Hanzhong, Shaanxi, China

Abstract. By means of analyzing kernel clustering algorithm and rough set theory, a novel clustering algorithm, rough kernel k-means clustering algorithm, was proposed for clustering analysis. Through using Mercer kernel functions, samples in the original space were mapped into a high-dimensional feature space, which the difference among these samples in sample space was strengthened through kernel mapping, combining rough set with k-means to cluster in feature space. These samples were assigned into up-approximation or low-approximation of corresponding clustering centers, and then these data that were in up-approximation and low-approximation were combined and to update cluster center. Through this method, clustering precision was improved, clustering convergence speed was fast compared with classical clustering algorithms The results of simulation experiments show the feasibility and effectiveness of the kernel clustering algorithm.

Keywords: Kernel methods, Kernel clustering algorithm, K-means, Rough set, Rough clustering.

1 Introduction

Clustering has been applied in a wide variety of fields, ranging from engineering(machine learning, artificial intelligence, pattern recognition), computer sciences and medical sciences to social science. Accordingly, clustering is also known as numerical taxonomy, learning without a teacher (or unsupervised learning), typological analysis and partition. The diversity reflects the important position of clustering in scientific research. Clustering algorithms partition data into a certain number of clusters. There is no universally agreed upon definition. Most researchers describe a cluster by considering the internal homogeneity and the external separation. i.e. patterns in the same cluster should be similar and the dissimilarity should be examinable in a clear and meaningful way. Here we give some simple mathematical description of several types of clustering, based on the description in [1].

These clustering algorithms don't consider sample characters and cluster these samples, hence clustering performance depend on distribution of samples deeply.

G. Wang et al. (Eds.): RSKT 2008, LNAI 5009, pp. 172–179, 2008.

In order to solve this bottleneck, we put forward a novel clustering algorithm, kernel-based rough k-means clustering algorithm. By using kernel learning theory, firstly, we can transform these samples in the samples space to a high-dimensional feature space, and the data can be classified linearly (maybe usually nonlinear). Secondly, we can use Lingras' algorithm-rough k-means to perform clustering in feature space. Efficiently, Supposing the non-linear mapping is continuum and velvet, the geometrical structure and sequence of samples are preserved in feature space, and distribution of samples are changed, therefore assemble data in original space will be assemble data in feature space. Besides these, the character of data in sample space is enhanced through kernel mapping, clustering precision of the method is improved. In order to improve clustering preciseness rate furtherly, we adopt rough k-means algorithm [5]-[6] to cluster in feature space, comparing with k-means, clustering precision can be increased obviously.

2   Related Basic Theory

2.1   Kernel Method

Kernel methods [11]-[13] are appellation of a serious advanced nonlinear data dealing technology, and using kernel map are their common characteristics. From idiographic operating, firstly, samples are mapped into feature space using non-linear method; secondly, dealing with these samples in feature space in linear operating. The process can be illustrated in Fig 1. For using nonlinear mapping and these nonlinear mapping are often rather complex, kernel method possess strong ability of dealing with nonlinear data.

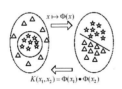

Fig. 1. Kernel mapping process

Kernel functions are nonlinear map function $\Phi : \Re^d \rightarrow F$ which for all pattern sets,

$$x_j \in \Re^d, j = 1, 2, \cdots, N, . \Leftrightarrow \{x_1, x_2, \cdots, x_N\} \subset \Re^d, \tag{1}$$

$$\Phi(x_1), \Phi(x_2), \cdots, \Phi(x_N) \subset F.$$

Here, $\Re^d$ is sample space .F represents a feature space with arbitrary high dimensionality, In feature space, kernels are often referred to as Mercer kernels or others .They provide an element way of dealing with nonlinear algorithms by reducing them to linear ones in some feature space F. Φ may not be explicitly known, but is defined implicitly in terms of the positive (semi-) definite kernel function satisfying the Mercer Condition:

$$K(x_1, x_2) = \Phi(x_1) \bullet \Phi(x_2)^T, \tag{2}$$

Kernel-based clustering have many advantages:

1) It is more possible to obtain a linearly separable hyperplane in the high-dimensional, or even infinite feature space.

2) They can form arbitrary clustering shapes other than hyperellipsoid and hypersphere.

3) Kernel-based clustering algorithms, like SVC, have the capability of dealing with noise and outliers.

4) There is no requirement for priori knowledge to determine the system topological means to estimate the number of clusters. In order to illustrate kernel method, we give an example to show the process.

2.2   Kernel Clustering Algorithm

Given an unlabeled data set $X = \{x_1, x_2, \cdots, x_n\}$ in the d-dimension input space $\Re^d$, let $\Phi : \Re^d \rightarrow F$ be a nonlinear map function from input space to a high dimensional feature space F. we can use k-means to perform clustering in feature space. efficientlyThe character of data in sample space is enhanced through kernel mapping [11]-[13].

2.3   Rough Clustering Algorithm

There are two different points between Lingras' algorithm[4]-[6] and k-means.

1) An sample is not only assigned to one class. According to the distance between sample and clustering center, and assigning it to up approximation and low approximation of different cluster center;

2) Adjustment of cluster centroids depend on samples in up-approximation and low-approximation.

3   Rough Kernel Clustering Algorithm

Main idea about rough kernel clustering is firstly, Given an unlabeled data set $X = \{x_1, x_2, \cdots, x_n\}$ in the d-dimension input space $\Re^d$, let $\Phi : \Re^d \rightarrow F$ be a nonlinear map function from input space to a high dimensional feature space F, differences of these samples are enhanced and these samples can be classified linearly (maybe close to linear).Secondly, we can use k-means clustering algorithm to perform clustering in feature space.

Object function is following:

$$J^\phi = \sum_{j=1}^{C} \sum_{i=1}^{N_K} ||\Phi(x_i) - m_j^\phi||^2$$

$$= \sum_{j=1}^{C}\sum_{i=1}^{N_K} K(x_i, X_j) - \frac{2}{N}\sum_{k=1}^{N_j} K(x_i, X_k) + \frac{1}{N_i^2}\sum_{k,p=1}^{N_j} K(x_k, x_p).$$

K-means algorithm in the high dimension feature space F iteratively searches for k clusters by minimizing the object function of the algorithm.

3.1   Basic Process of Rough Kernel Clustering Algorithm

1) Given an unlabeled data set in sample space $\Re^d$: $\{\boldsymbol{x_1}, \boldsymbol{x_2}, \cdots, \boldsymbol{x_N} \subset \Re^d$;

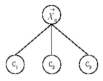

Fig. 2. An example of assigning

2) Given a kernel function $\varPhi$ and map these samples into Hilbert space F: $\{\varPhi(\boldsymbol{x_1}), \varPhi(\boldsymbol{x_2}), \cdots, \varPhi(\boldsymbol{x_N}) \subset F$.

3) Initializing clustering centroids $\boldsymbol{m}_k^\phi = \{\boldsymbol{m}_1^\phi, \boldsymbol{m}_2^\phi, \cdots, \boldsymbol{m}_K^\phi\}$, where K is number of class;

4) In kernel space, assigning each sample $\varPhi(\boldsymbol{x_i})$ into up-approximation and low-approximation according to near distance principle

a) $d(\varPhi(\boldsymbol{x_i}), \boldsymbol{m}_k)$,calculating the distance between $\varPhi(\boldsymbol{x_i})$ and clustering centroids $\boldsymbol{m}_k)$, $k = 1, 2, \cdots, K$

$$d_{i,h}^{min} = d_{i,h}(\varPhi(\boldsymbol{x_i}), \boldsymbol{m}_k) = min_{k=1,2,\cdots,K}\{\varPhi(\boldsymbol{x_i}), \boldsymbol{m}_k\}. \tag{4}$$

and assigning $\varPhi(\boldsymbol{x_i})$ to up- approximation of $\boldsymbol{m}_k)$, means $\varPhi(\boldsymbol{x_i}) \in \bar{C}_h$;

b) the distance $d(\varPhi(\boldsymbol{x_i}), \boldsymbol{m}_k)$ between $\varPhi(\boldsymbol{x_i})$ and $\boldsymbol{m}_k)$ is less than $d_{i,h}^{min}$ and $\exists \varepsilon$:

$$T = \{k : (d(\varPhi(\boldsymbol{x_i}), \boldsymbol{m}_k) - d(\varPhi(\boldsymbol{x_i}), \boldsymbol{m}_h) \leq \varepsilon) \wedge (h \neq k)\}. \tag{5}$$

if $T \neq \emptyset$,

then $\varPhi(\boldsymbol{x_i}) \in \bar{(C_i)}, \forall t \in T$;

else $\varPhi(\boldsymbol{x_i}) \in \underline{C_h}$;

5) Re-calculating clustering centroids $\boldsymbol{m}_k^\phi = \{\boldsymbol{m}_1^\phi, \boldsymbol{m}_2^\phi, \cdots, \boldsymbol{m}_K^\phi\}$ and object function J^ϕ;

Because clustering center cannot calculated explicitly in feature space, calculation of cluster centroids is a puzzle question in this method. In this paper, we adopt following method that every sample can be assigned into up-approximation and low-approximation of each cluster centroids and Lingras algorithm is adopted to update cluster centroids, we can use following formulation to do this work:

$$\boldsymbol{m}_k^\phi := \begin{cases} w_l \cdot \sum_{\Phi(\boldsymbol{x_n}) \in |C_k|} \frac{\Phi(\boldsymbol{x_n})}{|C_k|} + w_B \cdot \sum_{\Phi(\boldsymbol{x_n}) \in |C_k^B|} \frac{\Phi(\boldsymbol{x_n})}{|\overline{C_k}|} & \text{for } \underline{C_k} \neq \phi \\ w_l \cdot \sum_{\Phi(\boldsymbol{x_n}) \in |\overline{C_k}|} \frac{\Phi(\boldsymbol{x_n})}{|\overline{C_k}|} & otherwise \end{cases}$$

6) repeat step 4 and step 5until the value diversification of J^ϕ is very small.

3.2   An Example of Rough Assigning

About step 4, we show a example to illustrate the process (Fig. 2). Let a sample and three clustering centroids C_1, C_2, C_3.

Let $d(\boldsymbol{x_n}, \boldsymbol{C_1}) = 1.665, d(\boldsymbol{x_n}, \boldsymbol{C_2}) = 1.5, d(\boldsymbol{x_n}, \boldsymbol{C_3}) = 1.65, \varepsilon = 0.2$

$$d_{n,2}^{min} = d(\boldsymbol{x_n}, \boldsymbol{m_2}) = min_{k=1,\cdots,3} d(\boldsymbol{x_n}, \boldsymbol{m_k})$$

Hence $h = 2, k = 1, 2, 3, \boldsymbol{x_n} \in \overline{C_2}$
If $d_1 = d(\boldsymbol{x_n}, \boldsymbol{C_1}) - d(\boldsymbol{x_n}, \boldsymbol{C_2}) = 0.165 < 0.2 = \varepsilon$, then $\boldsymbol{x_n} \in \overline{C_1}$;
If $d_2 = d(\boldsymbol{x_n}, \boldsymbol{C_3}) - d(\boldsymbol{x_n}, \boldsymbol{C_2}) = 0.15 < 0.2 = \varepsilon$, then $\boldsymbol{x_n} \in \overline{C_3}$;
If $d_1 > 0.2$ and $d_2 > 0.2$, then $\boldsymbol{x_n} \in \underline{C_2}$.

For outlier, taking into account this assigning method about up-approximation and low-approximation, there are little opportunities that outlier are assigned into low-approximation in clustering later process. Otherwise, outlier maybe affect up-approximation of cluster center are more than low-approximation, cluster centroids can be calculated by combination of up-approximation and low-approximation, Hence this cluster method have a thick thin about outlier.

3.3   Rough Kernel Cluster Algorithm

Algorithm 1 Rough kernel cluster algorithm;
   Input1) $\{x_1, x_2, \cdots, x_N\} \subset \Re^d$; 2) Gauss RBF 3) $w_1, w_u, \varepsilon, \delta$;
   OutputClustering result (K cluster centroids and K classes)
   Begin
   flag=1; num=1; for $i = 1 : K, \boldsymbol{m}_i^\phi = \Phi(\boldsymbol{x}_i),$// initializing clustering centroids;
   initializing $J^\phi(1)$;
   for $i = 1 : N$ //calculating kernel metrices among samples for $j = 1 : N, k(\boldsymbol{x}_i, \boldsymbol{x}_j) = d(\Phi(\boldsymbol{x}_i) \bullet \Phi(\boldsymbol{x}_j))$; end for end for;
   While (flag==1) for i=1:N // calculating kernel metrices between samples and clustering centroids for $j = 1 : K, k(\boldsymbol{x}_i, \boldsymbol{m}_j) = d(\Phi(\boldsymbol{x}_i) \bullet \Phi(\boldsymbol{m}_j))$; end for end for;
   for $i = 1 : K$ // calculating kernel metrices among clustering centroids for $j = 1 : K, k(\boldsymbol{x}_i, \boldsymbol{x}_j) = d(\Phi(\boldsymbol{x}_i) \bullet \Phi(\boldsymbol{x}_j))$; end for end for;

$$d_{i,h}^{min} = d_{i,h}(\Phi(\boldsymbol{x}_i), m_h) = min_{k=1,\cdots,K}\{\Phi(\boldsymbol{x}_i), m_h\},$$

$$\Phi(\boldsymbol{x}_i) \in \overline{C_h};$$

$$T = \{k : (d(\Phi(\boldsymbol{x}_i), \boldsymbol{m}_k) - d(\Phi(\boldsymbol{x}_i), \boldsymbol{m}_h) \leq \varepsilon) \wedge (h \neq k)\};$$

if $T \neq \emptyset$,

then $\Phi(\boldsymbol{x_i}) \in \overline{C_i}, \forall t \in T$;

else $\Phi(\boldsymbol{x_i}) \in \underline{C_h}$; end if;

for $k = 1 : K$ //update cluster centroids

$$
\boldsymbol{m}_k^\phi := \begin{cases} w_l \cdot \sum_{\Phi(\boldsymbol{x_n}) \in |\underline{C_k}|} \frac{\Phi(\boldsymbol{x_n})}{|C_k|} + w_B \cdot \sum_{\Phi(\boldsymbol{x_n}) \in |C_k^B|} |\frac{\Phi(\boldsymbol{x_n})}{|C_k|}, & \text{for } \underline{C_k} \neq \phi \\ w_l \cdot \sum_{\Phi(\boldsymbol{x_n}) \in |\overline{C_k}|} \frac{\Phi(\boldsymbol{x_n})}{|C_k|}, & otherwise \end{cases}
$$

end if

num=num+1;

$$
J^\phi(num) = \sum_{j=1}^{C} \sum_{i=1}^{N_K} ||\Phi(x_i) - m_j^\phi||^2
$$

$$
= \sum_{j=1}^{C} \sum_{i=1}^{N_K} K(x_i, X_j) - \frac{2}{N} \sum_{k=1}^{N_j} K(x_i, X_k) + \frac{1}{N_i^2} \sum_{k,p=1}^{N_j} K(x_k, x_p).
$$

if $(J^\phi(num) - J^\phi(num - 1) < \delta$ flag=0;

4   Experiments

In order to validate the feasibility and validity about this algorithm, we realize the algorithm about one typical data sets, and kernel function is Gauss RBF. Our experiment environment is Pentium IV computer, 256M memory, MATLAB 7.0.

Sample is Set1(Fig3), and means value of samples is (0.0029,-0.2686), there are 200 samples in this data set and the distribution of these samples are two nested cycles. It is difficult to cluster about Set1. We adopt rough kernel cluster algorithm to do this work and get the result in Fig 4. Note there are 7 times better results among 10 experiments. Close curve denote equidistance thread.

From above 3 experiments, we can see kernel clustering algorithm is better than k-means in clustering performance. Note the reason that rough k-means algorithm performance is worse than k-means is due to these data sets. Adopting rough kernel k-means to do this work and get best performance than others

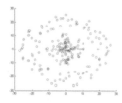

Fig. 3. Sample Set1

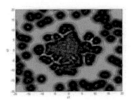

Fig. 4. Clustering result about Set1

Table 1. Comparing with 3 algorithm clustering result about 1 data set

Data set set	k-means (Error rate %)	Rough k-means (Error rate %)	Kernel k-means (Error rate %)	Rough kernel k-means (Error rate %)
Set1	Class 1:54	Class 1:47	Class 1:3	Class 1:0
	Class 2:49	Class 2:45	Class 2:4	Class 2:1

algorithm. Via experiment, we know that kernel clustering result strongly depend on initializing clustering centers. Rough kernel k-means algorithm has a better stability, clustering performance and clustering precision is improved.

5   Conclusions

Generally, the stability, precision and performance of clustering algorithm are depended on geometrical characteristics of the training data. If the difference is evidence and can be clustered easily. However in practical, this difference is not obvious among training data and even different samples in different are cross, traditional clustering algorithm based on distance can not resolve this problem. Kernel clustering algorithm can transform samples in original space into Hilbert space, the difference among samples can be preserved and even magnified, and therefore we can use k-means to do clustering in Hilbert space and get a better clustering performance. In this paper, we think different samples can have different affection about different clustering centroids, assigning each sample into up-approximation and low-approximation according to some principles. And updating clustering centroids according to the linear combination of up-approximation and low-approximation. Via experiment, we can know that this method has a better clustering performance and precision than kernel clustering algorithm, rough clustering algorithm. Another merit is that it can restrain outlier role about clustering results.

Acknowledgments

This work was supported by the National Natural Science Foundation of China (NSFC) under Grants 60472072, the Astronautics Science Foundation of China under Grant 06CASC0404 and the Shaanxi University of Technology Research Foundation Project (SLG0631).

References

1. Hansen, P., Jaumard, B.: Cluster analysis and mathematical programming. Mathematical programming, pp. 191–215. Springer, Berlin (1997)
2. Pawlak, Z.: Rough Sets theoretical Aspects of Reasoning About Data. Kluwer Academic Publishers, Dordrecht (1991)
3. Hartigan, J.A., Wong, M.A.: Algorithm AS136: A k-means clustering algorithm. Applied statistics 28(1), 100–108 (1979)
4. Gestel, T.V., Baesses, B., Suykens, J., et al.: Bayesian Kernel based classification for financial distress detection. European Journal of Operational research 172(3), 979–1003 (2006)
5. Xu, R.: Donald Wunsch II, Survey of clustering algorithm. IEEE transaction on neural networks 10(3), 645–678 (2005)
6. Rajen, B.B., Gopal, M.: On fuzzy-rough sets approach to feature selection. Pattern Recognition Letter 26(7), 865–975 (2005)
7. Lingras, P., West, J.: Interval set clustering of web users with rough k-means. Journal of Intelligent Information Systems 23(1), 5–16 (2004)
8. Peters, G.: Some refinements of rough k-means clustering. Pattern Recognition 39(8), 1481–1491 (2006)
9. Mitra, S.: An evolutionary rough partitive clustering. Pattern Recognition Letters 25(12), 1439–1449 (2004)
10. Davies, D.L., Bouldin, D.W.: A cluster separation measure. IEEE transaction on pattern analysis and Machine Intelligence 1(4), 224–227 (1979)
11. Su, C.-T., Chen, L., Yih, Y.: Knowledge acquisition through information granulation for imbalanced data. Expert Systems with Applications 31(3), 531–541 (2006)
12. Alexei, P., Samy, B.: Invariances in kernel methods: From samples to objects. Pattern Recognition Letters 27(10), 1087–1097 (2006)
13. Stphane, C., Alex, S.: Kernel methods and the exponential family. Neurocomputing 69(7), 714–720 (2006)
14. Zhang, L., Zhou, W., Jiao, L.: Kernel clustering Algorithm. Chinese Journal of Computers 25(6), 587–590 (2002)

New Reduction Algorithm Based on Decision Power of Decision Table

Jiucheng Xu and Lin Sun

College of Computer and Information Technology
Henan Normal University, Xinxiang Henan 453007, China
xjch3701@sina.com, slinok@126.com

Abstract. The current reduction algorithms based on rough sets still have some disadvantages. First, we indicated their limitations for reduct generation. We modified the mean decision power, and proposed to use the algebraic definition of decision power. To select optimal attribute reduction, the judgment criterion of decision with inequality was presented and some important conclusions were obtained. A complete algorithm for the attribute reduction was designed. Finally, through analyzing the given example, it was shown that the proposed heuristic information was better and more efficient than the others, and the presented in the paper method reduces time complexity and improves the performance. We report experimental results with several data sets from UCI repository and we compare the results with some other methods. The results prove that the proposed method is promising.

Keywords: Rough set, Decision table, Reduction, Decision power.

1 Introduction

Rough set theory is a valid mathematical tool that deals with imprecise, uncertain, vague or incomplete knowledge of a decision system (see [1]). Reduction of knowledge is always one of the most important topics. Pawlak (see [1]) first proposed attribute reduction from the algebraic point of view. Wang (see [2, 3]) proposed some reduction theories based on the information point of view, and introduced two novel heuristic algorithms of knowledge reduction with the time complexity $O(|C||U|^2) + O(|U|^3)$ and $O(|C|^2|U|) + O(|C||U|^3)$ respectively, where $|C|$ denotes the number of conditional attributes and $|U|$ is the number of objects in U, and the heuristic algorithm based on the mutual information (see [4]) with the time complexity $O(|C||U|^2) + O(|U|^3)$. These presented reduction algorithms have still their own limitations, such as sensitivity to noises, relatively high complexities, nonequivalence in the representation of knowledge reduction and some drawbacks in dealing with inconsistent decision tables.

It is known that reliability and coverage of a decision rule are all the most important standards for estimating the decision quality (see [5, 6]), but these algorithms (see [1, 2, 3, 7, 8, 9]) can't reflect the change of decision quality objectively. To compensate for their limitations, we construct a new method for

G. Wang et al. (Eds.): RSKT 2008, LNAI 5009, pp. 180–188, 2008.

separating consistent objects from inconsistent objects, and the corresponding judgment criterion with an inequality used in searching for the minimal or optimal reducts. Then we design a new heuristic reduction algorithm with relatively lower time complexity. For the large decision tables, since usually $|U| \gg |C|$, the reduction algorithm is more efficient than the algorithms discussed above. Finally, six data sets from UCI repository are used to illustrate the performance of the proposed algorithm and a comparison with the existing methods is reported.

2 Rough Set Theory Preliminaries

A decision table is defined as $S = (U, C, D, V, f)$. Let P, Q be equivalence relations on U. Then the partition U/Q is coarser than the partition U/P, in symbols $U/P \leq U/Q$ if and only if $\forall P_i \in U/P$ and $\exists Q_j \in U/Q$ such that $P_i \subseteq Q_j$. If $U/P \leq U/Q$ and $U/P \geq U/Q$, then $U/P = U/Q$. If $U/P \leq U/Q$ and $U/P \neq U/Q$, then U/Q is strictly coarser than U/P, i.e., $U/P < U/Q$.

Let $U/P=\{X_1, X_2, \ldots, X_r\}$, and $U/Q=\{X_1, X_2, \ldots, X_{i-1}, X_{i+1}, \ldots, X_{j-1}, X_{j+1}, \ldots, X_r, X_i \cup X_j\}$ be a new partition formed by unifying X_i and X_j in U/P to $X_i \cup X_j$, then $U/P \leq U/Q$. Moreover, let us assume $U/B = \{Y_1, Y_2, \ldots, Y_s\}$, then for the conditional entropy we have $H(B|P) \leq H(B|Q)$. The equation holds iff $\dfrac{|X_i \cap Y_k|}{|X_i|} = \dfrac{|X_j \cap Y_k|}{|X_j|}$, $\forall Y_k \in U/B$. If $P, Q \subseteq C$, then $H(D|P) \leq H(D|Q)$. The equation holds iff $\dfrac{|X_i \cap D_r|}{|X_i|} = \dfrac{|X_j \cap D_r|}{|X_j|}$, $\forall D_r \in U/D$ (see [10]).

Thus let $r \in P$, and if $H(D|P) = H(D|P - \{r\})$, then r in P is unnecessary for D, else r is necessary. If every element in P is necessary for D, then P is independent relative to D.

For $P \subseteq C$, $POS_P(D) = \cup\{PY | Y \in U/D\}$ is called the P-positive region of D, where $\underline{P}Y = \cup\{[x]_P | [x]_P \subseteq Y\}$ indicates the P-lower approximation of Y.

If $POS_C(D) = U$, then the decision table is called a consistent one, otherwise an inconsistent one. The set $POS_C(D)$ is called the (positive region) consistent object set, and $U - POS_C(D)$ is called the inconsistent object set.

3 The Proposed Approach

3.1 Limitations of Current Reduction Algorithms

Hence, one can analyze algorithms based on the positive region and the conditional entropy deeply. Firstly, if for any $P \subseteq C$, the P-quality of approximation relative to D is equal to the C-quality of approximation relative to D, i.e., $\gamma_P(D) = \gamma_C(D)$, and there is no $P^* \subset P$ such that $\gamma_{P*}(D) = \gamma_C(D)$, then P is called the reduct of C relative to D (see [1, 7, 8, 9]). In these algorithms, whether or not any conditional attributes is redundant depends on whether the lower approximation corresponding to decision set is changed or not after the attribute is deleted. Accordingly if new inconsistent objects are added to the decision table, it is not taken into account whether the conditional probability distribution of the primary inconsistent objects are changed in every corresponding decision class (see [10]). Hence, if the generated deterministic decision rules are the same, they will support the same important standards for estimating decision quality. Suppose the generated deterministic decision rules are the same, that is, the prediction of these rules is not changing. Thus it is seen that these presented algorithms only take into account whether or not the prediction of deterministic decision rules is changing after reduction.

Secondly, if for any $P \subseteq C$, $H(D|P) = H(D|C)$ and P is independent relative to D, then P is called the reduct of C relative to D (see [2, 3, 10, 11]). Hence, whether any conditional attributes is redundant or not depends on whether the conditional entropy of decision table is changed or not, after the attribute is deleted. It is known that the conditional entropy generated by $POS_C(D)$ is 0, thus $U - POS_C(D)$ can lead to a change of conditional entropy. Due to the new added and primary inconsistent objects in every corresponding decision class, if their conditional probability distribution changes, it will cause the change of conditional entropy of the whole decision table. Therefore, as it goes, the main criterions of these algorithms for estimating decision quality include two aspects, the invariability of the deterministic decision rules, the invariability of the reliability of nondeterministic decision rules.

So, some researchers above only think about the change of reliability for all decision rules after reduction. However, in decision application, besides the reliability of decision rules, the object coverage of decision rules is also one of the most important standards of estimating decision quality. So these current reduction algorithms above can't reflect the change of decision quality objectively. Meanwhile, the significance of attribute is regarded as the quantitative computation of radix for the positive region, which merely describes the subsets of certain classes in U, while from the information point of view, the significance of attribute only indicates the detaching objects of different decision classes in the equivalence relation of conditional attribute subset. However, for the inconsistent objects, these current measures for attribute reduction lack of dividing U into consistent object sets and inconsistent object sets for the inconsistent decision table. Therefore, these algorithms will not be equivalent in the representation of of knowledge reduction for inconsistent decision tables (see [12]). It is necessary to seek for a new kind of measure to search for the precise reducts effectively.

3.2   Representation of Decision Power on Decision Table

If $A \subseteq C$, then $POS_A(D) = POS_C(D)$ if and only if the A-lower approximation of D_i is equal to the C-lower approximation of D_i, i.e., $\underline{A}D_i = \underline{C}D_i$, for any $D_i \in U/D$ (see [12]). Thus suppose $D_0 = U - POS_C(D)$, we have $\underline{C}D_0 = D_0$. If all sets from $\{\underline{A}D_0, \underline{A}D_1, \underline{A}D_2, \ldots, \underline{A}D_m\}$ are nonempty, then the sets create a partition of U. If $\underline{A}D_i$ is empty, then $\underline{A}D_i$ is called a redundant set of the new partition. After all redundant sets are taken out, we obtain a partition of U.

Suppose $A \subset C$, then in the partition $\{\underline{A}D_0, \underline{A}D_1, \underline{A}D_2, \ldots, \underline{A}D_m\}$, all inconsistent objects are the set $\underline{A}D_0$. Meanwhile, the new partition of conditional attribute set C is $\{\underline{C}D_0, \underline{C}D_1, \underline{C}D_2, \ldots, \underline{C}D_m\}$, then we have a new equivalence relation generated by this new partition, denoted by R_D, i.e., $U/R_D=\{\underline{C}D_0,$ $\underline{C}D_1, \underline{C}D_2, \ldots, \underline{C}D_m\}$. Accordingly the decision partition U/R_D not only covers consistent objects from different decision classes in U, but also separates consistent objects from inconsistent objects, while U/D is gained through extracting objects from different decision classes relative to equivalence classes.

The most concise decision rules set that satisfies condition of the mean decision power discussed in the paper is regarded as the final reduction result in the new reduction model (see [13, 14]). Some experiments show that the mean decision power can acquire good standards. At all points of attribute reduction on decision table based on the mean decision power, we suppose that a new measure to knowledge reduction is presented without the number of original decision rules, compared with classical reduction algorithms. Thus, it not only occupies much smaller storage space and requires much lower computational costs and implementation complexity, but also has no effects on helping to get the minimal or optimal reducts. Thereby, based on U/R_D, the mean decision power is introduced and modified to discuss the roughness and attribute reduction based on rough sets. Thus we propose the algebraic definition of the decision power, which not only has effects on the subsets of the certain classes but also on the subsets of the uncertain (relative to the decision) classes in U.

Definition 1. Let $P \subseteq C$, $U/P = \{X_1, X_2, \ldots, X_t\}$, $D = \{d\}$, $U/D = \{Y_1, Y_2, \ldots, Y_m\}$, and $U/R_D=\{\underline{C}Y_0, \underline{C}Y_1, \underline{C}Y_2, \ldots, \underline{C}Y_m\}$, then the decision power of equivalence relation R_D relative to P is defined as

$$S\left(R_D; P\right) = \sum_{i=1}^{t}\sum_{j=0}^{m}\left(\frac{|X_i \cap \underline{C}Y_j|}{|X_i|} \times \frac{|X_i \cap \underline{C}Y_j|}{|U|}\right) = \sum_{i=1}^{t}\sum_{j=0}^{m}\left(\frac{|X_i \cap \underline{C}Y_j|^2}{|X_i||U|}\right). \quad (1)$$

From Definition 1, we know that any X_i is C-definable. Hence, X_i is a union of some C-equivalence classes. C-lower approximation of Y_j is also C-definable. Hence, the C-lower approximation of Y_j is also a union of some C-equivalence classes. We obtain that the intersection of X_i and the C-lower approximation of Y_j is equal to the union of all C-equivalence classes which are included in the C-lower approximation of Y_j and in X_i. Meanwhile, $|X_i \cap \underline{C}Y_j|/|X_i|$ and $|X_i \cap \underline{C}Y_j|/|U|$ represent the reliability of a decision rule and the object coverage corresponding to the rule respectively (see [6]), while it is only taken into account the change of reliability of all decision rules for the conditional entropy.

Theorem 1. Let $r \in P \subseteq C$, then we have $S(R_D; P) \geq S(R_D; P - \{r\})$.

Proof. To simplify notation in the proof, we consider only a special case. The proof in the general case goes in the analogous way. We assume that $U/P = \{X_1, \ldots, X_t\}$ and $U/P - \{r\}$ contains the same classes as in U/P with the only one exception that X_p and X_q are joined, i.e., $X_p \cup X_q$ is a class of $U/P - \{r\}$ and X_p, X_q are not. If many classes in U/P will be also joined after r in P is deleted, the coalition may be considered as automatically comprising more two partitions continually. So we have

$$S_\triangle = S(R_D; P) - S(R_D; P - \{r\})$$

$$= \sum_{j=0}^{m} \left(\frac{|X_p \cap \underline{C}Y_j|^2}{|X_p||U|} \right) + \sum_{j=0}^{m} \left(\frac{|X_q \cap \underline{C}Y_j|^2}{|X_q||U|} \right) - \sum_{j=0}^{m} \left(\frac{|(X_p \cap \underline{C}Y_j) \cup (X_q \cap \underline{C}Y_j)|^2}{|X_p \cup X_q||U|} \right).$$

Suppose $|X_p| = x$, $|X_q| = y$, $|X_p \cap \underline{C}Y_j| = ax$, $|X_q \cap \underline{C}Y_j| = by$, there must be $x > 0$, $y > 0$, $0 \le a \le 1$, and $0 \le b \le 1$, thus we have

$$S_\triangle = \sum_{j=0}^{m} \frac{(ax)^2}{x|U|} + \sum_{j=0}^{m} \frac{(by)^2}{y|U|} - \sum_{j=0}^{m} \frac{(ax+by)^2}{(x+y)|U|} = \frac{1}{|U|} \sum_{j=0}^{m} \frac{xy(a-b)^2}{x+y}.$$

Assume a function $f_j = \dfrac{xy(a-b)^2}{x+y}$, for any j $(j = 0, 1, \ldots, m)$. Hence, it is obviously true that when $a = b$, we have $\dfrac{|X_p \cap \underline{C}Y_j|}{|X_p|} = \dfrac{|X_q \cap \underline{C}Y_j|}{|X_q|}$, then $f_j = 0$.

So, when any attribute r is deleted from decision table, there must exist $S_\triangle \ge 0$. Hence, the proposition $S(R_D; P) \ge S(R_D; P - \{r\})$ is true.

Thus, we obtain the conclusion that the decision power of knowledge decreases non-monotonously as the information granularities become finer.

Theorem 2. If S is a consistent one, then $U/R_D = U/D$. Assume that $\dfrac{|X_p \cap \underline{C}Y_j|}{|X_p|} = \dfrac{|X_q \cap \underline{C}Y_j|}{|X_q|}$ such that $\dfrac{|X_p \cap Y_j|}{|X_p|} = \dfrac{|X_q \cap Y_j|}{|X_q|}$, then $S(R_D; P) = S(R_D; P - \{r\}) \Leftrightarrow H(D|P) = H(D|P - \{r\}) \Leftrightarrow \gamma_P(D) = \gamma_{p-\{r\}}(D)$. If S is inconsistent, then $\underline{C}Y_0 = Y_0$. Assume that $\dfrac{|X_p \cap \underline{C}Y_0|}{|X_p|} = \dfrac{|X_q \cap \underline{C}Y_0|}{|X_q|}$ such that $\dfrac{|X_p \cap Y_0|}{|X_p|} = \dfrac{|X_q \cap Y_0|}{|X_q|}$, then $S(R_D; P) = S(R_D; P-\{r\}) \Leftrightarrow \gamma_P(D) = \gamma_{p-\{r\}}(D)$.

Proof. One can prove Theorem 2 easily from Lemma 1 in [3].

Theorem 3. Let $P \subseteq C$, then any attribute r in P is said to be unnecessary relative to D if and only if $S(R_D; P) = S(R_D; P - \{r\})$.

Definition 2. If $P \subseteq C$, then the significance of any attribute $r \in C - P$ relative to D is defined by

$$SGF(r, P, D) = S(R_D; P \cup \{r\}) - S(R_D; P). \tag{2}$$

Notice that when $P = \emptyset$, $SGF(r, \emptyset, D) = S(R_D; \{r\})$.

From Theorem 2 and (2), we know that if $SGF(r, P, D) = 0$, then the significance of attribute based on the positive region is also 0, on the other hand, if the radix of positive region fills out after adding any attributes, then that significance of attribute isn't 0. Meanwhile, we also have $SGF(r, P, D) \ne 0$. Hence, $SGF(r, P, D)$ can not only include more information than that based on the positive region, but also compensate for some limitations of the significance of attribute based on the algebraic point of view and the information point of view.

Definition 3. Let $P \subseteq C$, then P is an attribute reduction of C relative to D, if $S(R_D; P) = S(R_D; C)$ and $S(R_D; P^*) < S(R_D; P)$, for any $P^* \subset P$.

3.3   Design of Reduction Algorithm Based on Decision Power

We know that the calculated $S(R_D; P)$ is the same every time, then calculating any attribute r with the maximum of $SGF(r, P, D)$, when used as heuristic information, is in fact to calculate that with the maximum of $S(R_D; P \cup \{r\})$. Therefore, we only need calculate $S(R_D; P \cup \{r\})$ except $S(R_D; P)$. However, calculating $S(R_D; P \cup \{r\})$ is in fact to calculate corresponding partitions and positive region principally. Then we make use of the effective computational methods of equivalence (indiscernibility) classes, positive region and attribute importance in [9], attribute core in [15]. Thus it will help to reduce the quantity of computation and the time and space of search.

The reduct consists of the following steps, first, detaching objects from equivalence classes step by step, then determining attribute core of objects, finally, obtaining the minimum relative reducts through adding attributes bottom-up.

Input: Decision table $S = (U, C, D, V, f)$.
Output: A relative attribute reduction P.
(1) Calculate $POS_C(D)$ and $U - POS_C(D)$ for the partition U/R_D.
(2) Calculate $S(R_D; C)$, $CORE_D(C)$, and let $P = CORE_D(C)$.
(3) If $P = \varnothing$, then turn to (4), and if $S(R_D; P) = S(R_D; C)$, then turn to (6).
(4) Calculate $S(R_D; P \cup \{r\})$, for any $r \in C - P$, select an attribute r with the maximum of $S(R_D; P \cup \{r\})$, and if this r is not only, then select that with maximum of $|U/(P \cup \{r\})|$.
(5) Let $P = P \cup \{r\}$, and if $S(R_D; P) \neq S(R_D; C)$, then turn to (4), else $\{P^* = P - CORE_D(C);\ t = |P^*|;$
for$(i=1;\ i \leq t;\ i++)$
$\{\ r_i \in P^*;\ P^* = P^* - \{r_i\};$
and if $S(R_D; P^* \cup CORE_D(C)) < S(R_D; P)$, then $P^* = P^* \cup \{r_i\};\}$
$P = P^* \cup CORE_D(C);\}$
(6) The output P is a minimum relative attribute reduction.

It is clear that this algorithm is complete. In other words, none of the attributes in P can be eliminated again without decreasing its discriminating quality, whereas many algorithms are incomplete, which can't ensure that the final reducts will be obtained (see [8]). Thus the algorithms in [8, 12, 14] are complete, but the algorithms in [3, 4] are not. Meanwhile, we can easily see that the time complexity of algorithm is $O(|C|^2|U|)$, which is less than that of [3, 4, 6, 8, 12].

4   Experimental Results

In Table 1 below, we give an inconsistent decision table $S = (U, C, D, V, f)$, where $U = \{x_1, x_2, \ldots, x_{10}\}$, $C = \{a_1, a_2, \ldots, a_5\}$, and $D = \{d\}$.

In Table 2 below, there is the significance of attribute relative to the core $\{a_2\}$ and the relative reducts, the Algorithm in [7],CEBARKCC in [3], Algorithm 2 in [12], and the proposed Algorithm are denoted by A1, A2, A3, and A4 respectively, and let m, n be the number of attributes and universe respectively.

Table 1. Inconsistent Decision Table S

U	a_1	a_2	a_3	a_4	a_5	d
x_1	0	1	0	0	1	0
x_2	0	0	0	1	0	0
x_3	0	1	0	0	1	1
x_4	0	0	1	0	0	0
x_5	1	0	1	1	1	1
x_6	0	1	1	0	0	0
x_7	0	0	0	0	1	1
x_8	0	0	1	0	0	1
x_9	1	0	0	1	1	1
x_{10}	0	0	0	0	1	0

Table 2. Comparison of Experimental Results

Algorithm	Relative $\{a_2\}$				Reduction Result	Time Complexity
	a_1	a_3	a_4	a_5		
A1	0.200	0.100	0	0.100	$\{a_1,a_2,a_3,a_5\}$	$O(m^3 n^2)$
A2	0.204	0.089	0.014	0.165	$\{a_1,a_2,a_3,a_5\}$	$O(mn^2)+O(n^3)$
A3	0.604	0.365	0.689	0.565	$\{a_2,a_4,a_5\}$	$O(m^2 n\log(n))$
A4	0.240	0.200	0.267	0.200	$\{a_2,a_4,a_5\}$	$O(m^2 n)$

Table 3. Comparison of Reducts for Data Sets

Database	T or F	Objects	Radix	m	A1		A2		A3		A4		A5	
					n	t	n	t	n	t	n	t	n	t
Balloon(1)	T	20		4	2	0.20	2	0.06	2	0.04	2	0.02	2	0.12
Zoo	F	101		17	10	0.36	11	0.29	10	0.14	10	0.09	10	5.83
Voting-records	T	435		16	10	0.98	9	0.51	9	0.27	9	0.26	9	6.75
Tic-tac-toe	T	958		9	8	0.95	8	1.38	8	0.56	8	0.52	8	9.65
Chess end-game	T	3196		36	29	275.27	29	23.15	29	5.56	29	5.25	29	32.28
Mushroom	T	8124		22	5	486.36	4	16.58	4	6.68	4	6.56	4	29.87

From Table 2, the significance of attribute in [3, 7] a_4 is relatively minimum, and their reducts are $\{a_1, a_2, a_3, a_5\}$, rather than the minimum relative reduct $\{a_2, a_4, a_5\}$. However, the $SGF(a_4, \{a_2\}, D)$ is relatively maximum. Thus we get the minimum relative reduction $\{a_2, a_4, a_5\}$ generated by A3 and A4. Compared with A1 and A2, the new proposed algorithm does not need much mathematical computation, logarithm computation in particular. Meanwhile, we know that the general schema of adding attributes is typical for old approaches to forward selection of attributes although they are using different evaluation measures, but it is clear that on the basis of U/R_D, the proposed decision power is feasible to discuss the roughness of rough sets. Hence, the new heuristic information will compensate for the proposed limitations of those current algorithms. Therefore, this algorithm's effects on reduction of knowledge are well remarkable.

Here we choose six discrete data sets from UCI repository and five algorithms to do more experiments on PC (P4 2.6G, 256M RAM, WINXP) under JDK1.4.2 in Table 3 below, where T or F indicates that the data sets are consistent or not, m, n are the number of primal attributes and after reduction respectively, t is the time of operation, and A5 denotes the algorithm in [6].

5   Conclusion

In this paper, to reflect the change of decision quality objectively, a measure for reduction of knowledge and its judgment theorem with an inequality are established by introducing the decision power from the algebraic point of view. To compensate for these current disadvantages of classical algorithms, we design an efficient complete algorithm for reduction of knowledge with the time complexity reduced to $O(|C|^2|U|)$ (In preprocessing, the complexity for computing U/C based on radix sorting is cut down to $O(|C||U|)$, and the complexity for measuring attribute importance based on the positive region is descended to $O(|C - P||U' - U'_P|)$ (see [9]).), and the result of this method is objective.

Acknowledgement. This work was supported by grants from Natural Science Fund of Henan Province (No. 0511011500) and New Century Excellence Genius Support Plan of Henan Province of China (No. 2006HANCET-19).

References

1. Pawlak, Z.: Rough Sets and Intelligent Data Analysis. Int. J. of Inf. Sci. 147, 1–12 (2002)
2. Wang, G.Y.: Rough Reduction in Algebra View and Information View. Int. J. of Int. Syst. 18, 679–688 (2003)
3. Wang, G.Y., Yu, H., Yang, D.C.: Decision Table Reduction Based on Conditional Information Entropy. J. of Comp. 25(7), 759–766 (2002)
4. Miao, D.Q., Hu, G.R.: A Heuristic Algorithm for Reduction of Knowledge. Journal of Computer Research and Development 36(6), 681–684 (1999)
5. Liang, J.Y., Shi, Z.Z., Li, D.Y.: Applications of Inclusion Degree in Rough Set Theory. Int. J. of Comp. Cogn. 1(2), 67–68 (2003)
6. Jiang, S.Y., Lu, Y.S.: Two New Reduction Definitions of Decision Table. Mini-Micro Systems 27(3), 512–515 (2006)
7. Guan, J.W., Bell, D.A.: Rough Computational Methods for Information Systems. Int. J. of AI. 105, 77–103 (1998)
8. Liu, S.H., Sheng, Q.J., Wu, B., et al.: Research on Efficient Algorithms for Rough Set Methods. J. of Comp. 26(5), 524–529 (2003)
9. Xu, Z.Y., Liu, Z.P., et al.: A Quick Attribute Reduction Algorithm with Complexity of Max($O(|C||U|)$,$O(|C|^2|U/C|)$). J. of Comp. 29(3), 391–399 (2006)
10. Ślęzak, D.: Approximate Entropy Reducts. Fundam. Inform. 53, 365–390 (2002)
11. Ślęzak, D., Wróblewski, J.: Order Based Genetic Algorithms for the Search of Approximate Entropy Reducts. In: Wang, G., Liu, Q., Yao, Y., Skowron, A. (eds.) RSFDGrC 2003. LNCS (LNAI), vol. 2639, p. 570. Springer, Heidelberg (2003)
12. Liu, Q.H., Li, F., et al.: An Efficient Knowledge Reduction Algorithm Based on New Conditional Information Entropy. Control and Decision 20(8), 878–882 (2005)

13. Jiang, S.Y.: An Incremental Algorithm for the New Reduction Model of Decision Table. Comp. Eng.g and Appl. 28, 21–25 (2005)
14. Ślęzak, D.: Various Approaches to Reasoning with Frequency-Based Decision Reducts: A Survey. In: Polkowski, L., Lin, T.Y., Tsumoto, S. (eds.) Rough Set Methods and Applications: New Developments in Knowledge Discovery in Information Systems, vol. 56, pp. 235–285. Springer, Heidelberg (2000)
15. Han, J.C., Hu, X.H., Lin, T.Y.: An Efficient Algorithm for Computing Core Attributes in Database Systems. In: Zhong, N., Raś, Z.W., Tsumoto, S., Suzuki, E. (eds.) ISMIS 2003. LNCS (LNAI), vol. 2871, pp. 663–667. Springer, Heidelberg (2003)

A Heuristic Algorithm Based on Attribute Importance for Feature Selection

Xingbo Sun[1,*], Xiuhua Tang[2], Huanglin Zeng[1], and Shunyong Zhou[1]

[1] Dept. of Electronic Engineering
Sichuan University of Science & Engineering
Zigong, Sichuan 643000, P.R. China
sxb741021@163.com
[2] Dept. of Material & Chemical Engineering
Sichuan University of Science & Engineering
Zigong, Sichuan 643000, P.R. China

Abstract. In this paper we devote to study some feature selection of an information system in which redundant or insignificant attributes in data sets can be eliminated. An approach of importance gain function is suggested to evaluate the global average information gain associated with a subset of features. A heuristic algorithm on iterative criterion of feature selection on the significance of attributes is proposed to get the least reduction of attribute set in knowledge discovery. The feasibility of feature selection proposed here is validated by some of examples.

Keywords: Rough set, Importance, Feature selection.

1 Introduction

Feature selection is a process of finding a subset of features from the original set of features, forming patterns in a given data set, optimally according to the given goal of processing and criterion. See, for example, [1]-[2].Reduction of pattern dimensionality via feature extraction and feature selection belongs to the most fundamental steps in data preprocessing. Feature selection is often isolated as a separate step in processing sequence. Features constituting the object's pattern may be irrelevant (having no effect on processing performance) or relevant (having an impact on processing performance). Features can be redundant (dependent), and may have a different discriminatory or predictive power. In the virtue of the minimum construction idea, one of the techniques for the best feature selection could be based on choosing a minimal feature subset that fully describes all concepts (for example classes in prediction-classification) in a given data set. Let us call this paradigm a minimum concept description.

The concept of a rough set, which was presented by Pawlak et al. [3] and Pawlak [4], is an important concept since it is applicable in many fields such as artificial intelligence, expert systems, data mining, pattern recognition and decision theory. See, for example, [8]-[9].

---

* The corresponding author.

G. Wang et al. (Eds.): RSKT 2008, LNAI 5009, pp. 189–196, 2008.

The application of rough sets in feature selection was proposed in several contributions. The simplest approach is based on calculation of a core for discrete attribute data set, containing strongly relevant features, and reduction, containing a core plus additional weakly relevant features, so that each reduction is satisfactory to determine concepts in the data set. Based on a set of reduction for a data set, some criteria for feature selection can be formed, for example selecting features from a minimal reduction, i.e., a reduction containing minimal set of attributes.

In this paper, we concentrate on the issue of the attribute reduction which is addressed in the theory of rough sets in terms of (approximate) decision reduction [10]-[13]. We give the importance gain function, to evaluate the global average information gain associated with a subset of features. We also formulate criteria for maintaining the level of the importance gain during the process of attribute reduction.

2   Preliminaries

In rough set, the decision table expression is $T = (U, A, C, D)$, where U is a set of objects, A is a nonempty finite set of attribute, $A = C \cup D$, $C \cap D$, and C and D are called as the condition attribute set and the decision attribute set, respectively.

Definition 1. The indiscernible binary relation is defined as: for $a \in A$, $P \subset A$,

$$IND(P) = \{(x, y) \in U \times U \,|\, a(x) = a(y), \forall a \in P\}.$$

All the classes of equivalence relations $IND(P)$ are expressed by $U/IND(P)$, and $U/IND(P)$ is denoted by U/P. $U/IND(C)$ and $U/IND(D)$ are called as the condition class and the decision class, respectively.

Definition 2. $R_-(x)$ is the lower approximation of R of X

$$R_-(x) = \bigcup \{Y \in U/R \,|\, Y \subseteq X\}, \quad \text{for} \quad R \subseteq C, X \subseteq U.$$

Definition 3. The positive region of C of D:

$$POS_C(D) = \bigcup_{X \in U/D} C_-(X),$$

where $C_-(X)$ expresses the lower approximation of C to X.

Definition 4. For $c \in C$, feature c is called omit-able in T when $POS_{(C-\{c\})}(D) = POS_C(D)$, otherwise cannot be omitted. T is independent if all $c \in C$ can't be omitted.

Definition 5. If $T' = (U, A, C, D)$ is independent and $POS_R(D) = POS_C(D)$, then R is the reduction of C. In other words, the least reduction of attribute-set satisfies the above conditions.

Definition 6. All of the independent attribute set of C is denoted by $CORE(C)$, namely, $CORE(C) = \bigcap RED(C)$, where $RED(C)$ is the reduction of C.

3   Importance

Let $I = (U, A)$ is an information system, $X \subseteq A$ is an attribute subset, $x \in A$ is an attribute, we consider the importance of x for X, namely the attribute x is added to X to enhance the degree of resolution, the bigger degree to be enhanced, the more important x is to X, so the definition is as the following:

Definition 7. Assume $X \subseteq A$ is an attribute subset, $x \in A$ is an attribute, the importance of x for X is denoted by $Sig_X(x)$, the definition is:

$$Sig_X(x) = 1 - |X \cup \{x\}|/|X|,$$

where $|X| = |IND(X)|$. Suppose $U/IND(X) = U/X = \{X_1, X_2, \cdots, X_n\}$, then $|X| = |IND(X)| = \sum_{i=1}^{n} |X_i|^2$.

$|X| - |X \cup \{x\}|$ represents the decrement of indiscernibility and also the increment of discernibility as attribute x is added to X. Namely, the number of selection methods is originally indiscernible in X, but it is discernible in $X \cup \{x\}$, and the increment of indiscernibility is expressed by

$$(|X| - |X \cup \{x\}|)/X = 1 - |X \cup \{x\}|/X.$$

Proposition 1. The following conclusions are equal:

1) x for X is unimportant, it means $Sig_X(x) = 0$.
2) x for X redundancy, namely:

$$X \cup \{x\} \leftrightarrow X.$$

3) $X \to x$.
4) $IND(X \cup \{x\}) = IND(X)$.
5) For $u, v \in U$, if u, v are indiscernible in X, then u, v are indiscernible in x. It means that if $uIND(X)v$, then $u\theta_x v$ (θ_x expresses the equal relations produced by x).

Proposition 2. Let $X, Y \subseteq A$. If $X \leftrightarrow Y$, then $Sig_X(x) = Sig_Y(x)$, for $x \in A$.

Proposition 3. Let $X \subseteq A$. Then

1) $Sig_{(RED(X) - \{x\})}(x) > 0$, for $x \in RED(X)$, where $RED(X)$ is the reduction of X. Namely, each element is important to other elements of the reduction.
2) $Sig_{RED(X)}(x) = 0$ for $x \in X - RED(X)$ when $RED(X) \neq \emptyset$.

Namely each outside element of the reduction is unimportant to the reduction.

Proposition 4. Let $X \subseteq A$. if $X - CORE(X) \neq \emptyset$, the necessary and sufficient condition of $CORE(X) = RED(X)$ is $Sig_{CORE(X)}(x) = 0$ for $x \in X - CORE(X)$.

4   Heuristic Search of the Selection of Feature Subset

Feature selection's final goal is to archive optimal feature subset according to some criterion without damaging sort power. The heuristic search is used to get the subset when the data processed is massive and the features are too many. The best feature will be selected gradually from the feature space, until a reduction is gained.

4.1   Heuristic

The importance may serve as the selection criteria for attribute, the least reduction of X is constituted by adding attributes in $CORE(X)$ one by one according to importance of attributes. The detail is as the following:

1) The selection of feature a. Add feature a in subset R, if its value of importance Sig is the biggest one.
2) If two features obtained the same importance value, then the one has less discrete value is selected.

Hence, the heuristic algorithm of approximate-optimization reduction R can be described as: Take $CORE$ as the initialization subset, the feature is selected from the space of omit-able feature and is added in the subset one by one, until the near-optimization reduction is gained.

4.2   Heuristic Method

According to the above discussion, the algorithm is described as the following:
    Assume $X \subseteq A$ is an attribute subset, $x \in X$ is an attribute.

1) Calculation of nucleus $CORE(X)$: Compute $Sig_{X-\{x\}}(x)$ for $x \in X$. Thus, $CORE(X)$ is constructed by all of the attributes which Sig values are bigger than zero, maybe $CORE(X) = \emptyset$.
2) $RED(X) \leftarrow CORE(X)$.
3) Go to 6) when $IND(RED(X)) = IND(X)$. Otherwise go to 4).
4) Calculate all $Sig_{RED(X)}(x)$ value for $x \in X - RED(X)$, take x_1 to satisfy:

$$Sig_{RED(X)}(x_1) = \max \max_{x \in RED(X)} \{Sig_{RED(X)}(x)\}.$$

If there are two features obtained the same value, then select the one that has the less feature value.
5) $RED(X) \leftarrow RED(X) \cup \{x_1\}$, go to 3).
6) Output least reduct $RED(X)$.

5   An Example

Let the condition attribute $A = \{a, b, c, d\}$ and the decision attribute $D = \{e\}$. Then, the data of an information system are shown in Table 1.

Table 1. Data of an information system

U	a	b	c	d	e
1	0	2	1	0	1
2	0	0	1	2	2
3	1	2	1	2	0
4	0	0	2	1	2
5	0	1	2	1	2
6	1	2	2	1	1

1) the calculation of nucleus

$$Sig_{A-\{a\}}(a) = 1 - \frac{|A|}{|A - \{a\}|} = 1 - \frac{1+1+1+1+1+1}{1+1+1+1+1+1} = 0,$$

$$Sig_{A-\{b\}}(b) = 1 - \frac{|A|}{|A - \{b\}|} = 1 - \frac{1+1+1+1+1+1}{1+2\times2+1+1+1} = 0.25,$$

$$Sig_{A-\{c\}}(c) = 1 - \frac{|A|}{|A - \{c\}|} = 1 - \frac{1+1+1+1+1+1}{1+1+1+1+1+1} = 0,$$

$$Sig_{A-\{d\}}(d) = 1 - \frac{|A|}{|A - \{d\}|} = 1 - \frac{1+1+1+1+1+1}{1+1+1+1+1+1} = 0.$$

Therefore, $CORE(A) = \{b\}$.

2) Calculation of the least reduction.

$$Sig_{CORE(A)}(a) = 1 - \frac{|\{a,b\}|}{|\{b\}|} = 1 - \frac{1+2\times2+2\times2+1}{1+2\times2+3\times3} = \frac{2}{7},$$

$$Sig_{CORE(A)}(c) = 1 - \frac{|\{b,c\}|}{|\{b\}|} = 1 - \frac{1+2\times2+1+1+1}{1+2\times2+3\times3} = \frac{3}{7},$$

$$Sig_{CORE(A)}(d) = 1 - \frac{|\{b,d\}|}{|\{b\}|} = 1 - \frac{1+1+1+1+1+1}{1+2\times2+3\times3} = \frac{4}{7}.$$

$Sig_{\{b\}}(d)$ is the biggestso d is added to subset.

Let $A_1 = CORE(A) \cup \{d\} = \{b,d\}$.

$$Sig_{A_1}(a) = 1 - \frac{|\{a,b,c\}|}{|\{b,d\}|} = 1 - \frac{1+1+1+1+1+1}{1+1+1+1+1+1} = 0,$$

$$Sig_{A_1}(c) = 1 - \frac{|\{b,c,d\}|}{|\{b,d\}|} = 1 - \frac{1+1+1+1+1+1}{1+1+1+1+1+1} = 0.$$

Therefore, this information system least reduction is $\{b,d\}$.

6   Application in Image Classification

The method has been used in our research project of provincial educational committee, it works well. In this experiment, we use the micrograph of Chinese wine. These images are shown in Figure 1.

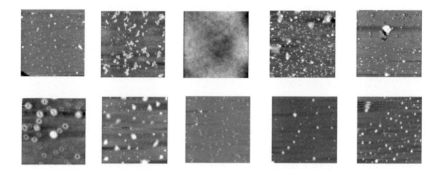

Fig. 1. Part of Chinese Wine micrograph

These images have been converted into gray-scale format, and resized to 256256.We sub-divide the image into a number of equal sized sub-images. For Feature Extraction methodology, we select most of 26 characteristic feature specified in [14]. The 26 features are used to compute the meaningful characteristics of each sub-image. We apply the Statistic Zoning method to transforming continuous numbers into discrete categories. Figure 2 shows part of the discrete categories.

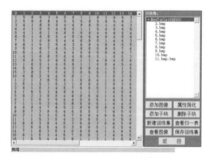

Fig. 2. Part of the discrete categories

We collect the feature extracted values for each micrograph and build the training DB. We use 10 images, and these images are taken from various Chinese Wine. We compute the reducts, cores, and rule generation by using the method introduced in the paper. Figure 3 shows part of the result.

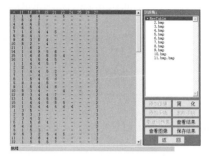

Fig. 3. Part of the feature selection result

7   Feature Selection Results

The method of heuristic attribute reduction is proposed based on the attribute importance in this paper. Take $CORE(A)$ as initial condition, we select the feature and add it in subset according to the Sig value of attribute importance, until the least reduction is obtained. In the example we use sub-image size 64×64, and the system generates 16 sub-images. We extract 26 features in each sub-image, so there are totally 4160 features for ten Chinese Wine micrographs. We compute the least reduct of training DB through using the method in the paper and traditional method specified in [15]. Table 2 shows the results.

Table 2. Result of feature selection

Items	Traditional	New
Feature number	480(4160)	384(4160)
Selection ratio	11.54%	9.23%
Run time	25 seconds	12 seconds

With the results from the computed reducts, core, and rule generations, we create the appropriate knowledge and techniques to perform image recognition. Using the classification rules, the automated process sequentially examines each image representative object in testing set, applying the optimal classified rule and declaring the result. The testing set collecting is similar to the training database, but we use other micrographs of these ten kinds of Chinese Wine. We compare the two feature selection methods by the final classification result and show as Table 3.

From Tables 2 and 3, it follows that we have obtained about 9.23% selection ratio and 84.37% classification results in this experiment by using our feature selection method, and the method is more efficient than traditional rough set method specified in [15].

Table 3. Result of image classification

Items	Traditional	New
Error sub-image number	27(160)	25(160)
Error ratio	16.88%	15.63%

Acknowledgments

This work was supported by the Scientific Research Fund of Sichuan Provincial Education Department (07ZZ017).

References

1. Banerjee, M., Mitra, S., Banka, H.: Evolutionary-Rough Feature Selection in Gene Expression Data. IEEE Transactions on Systems, Man, and Cybernetics, Part C: Applications and Reviews (37), 622–632 (2007)
2. Momin, B.F., Mitra, S., Gupta, R.D.: Reduct Generation and Classification of Gene Expression Data. In: Proceedings of First International Conference on Hybrid Information Technology (ICHIT 2006), Cheju Island, Korea, pp. 699–707 (2006)
3. Pawlak, Z., Slowinski, K., Slowinski, R.: Rough Classification of Patients after Highly Selective Vagotomy for Duodenal Ulcer. International Journal of Man-Machine Studies 24, 413–433 (1986)
4. Pawlak, Z.: Rough Sets. Kluwer Academic Publishers, Dordrecht (1991)
5. Datcu, M.: Information Mining in Remote Sensing Image Archives Part A: System Concepts. IEEE Trans. on Geosciences and Remote Sensing 41(12), 2923–2936 (2003)
6. Lashin, E.F., Medhat, T.: Topological Reduction of Information Systems. Chaos, Solitons and Fractals 25, 277–286 (2005)
7. Shao, X.Y., Wang, Z.H., Li, P.G., et al.: Integrating Rough Set and Data Mining for Group Based Discovery of Configuration rules. International Journal of Production Research 44(16), 2789–2811 (2006)
8. Swiniarski, R., Skowron, A.: Rough Set Methods in Feature Selection and Recognition. Pattern Recog. Lett. 24(6), 833–849 (2003)
9. Wei, J.M.: Rough Set Based Approach to Selection of Node. International Journal Computational Cognition 1(2), 25–40 (2003)
10. Ślęzak, D.: Approximate Entropy Reducts. Fundamenta Informaticae 53(3-4), 365–387 (2002)
11. Ślęzak, D.: Normalized Decision Functions and Measures for Inconsistent Decision Tables Analysis. 44(3), 291–319 (2000)
12. Tsumoto, S.: Statistical Extension of Rough Set Rule Induction. In: Proc. Of SPIE: Data Mining and Knowledge Discovery III (2001)
13. Li, R., Wang, Z.: Mining Classification Rules Using Rough Sets and Neural Networks. European Journal of Operational Research 157, 439–448 (2004)
14. Liu, S.S., Jernigan, M.E.: Texture Analysis and Discrimination in Additive Noise. Computer Vision, Graphics, and Image Processing, pp. 52–57 (1990)
15. Zeng, H.L.: Intelligent Computing (About Rough Sets, Fuzzy logic, Neutral Network and Applications) (in Chinese), pp. 90–140. Chongqing University Press (2004)

Rough Mereology in Analysis of Vagueness

Lech Polkowski

Polish–Japanese Institute of Information Technology
Koszykowa 86, Warsaw, Poland
polkow@pjwstk.edu.pl

Abstract. This work aims at presenting to a wider audience fundamental notions and ideas of rough mereology. We discuss various methods for constructing rough inclusions in data sets, then we show how to apply them to the task of knowledge granulation, and finally, we introduce granular reflections of data sets with examples of classifiers built on them.

Keywords: rough sets, knowledge granulation, rough mereology, rough inclusions.

1 Motivations: Rough Set Analysis of Vagueness

Rough set analysis of vague concepts [4], begins with the idea of saturation by classes of indiscernibility: given an information function $Inf : U \to V$ defined on objects in a set U with values in a set V which induces an indiscernibility relation Ind on the set $U \times U$ with $Ind(u,v)$ iff $Inf(u) = Inf(v)$, concepts $X \subseteq U$ are divided into two categories: the category of Inf–definable concepts which are representable as unions of classes $[u]_{Ind} = \{v \in U : Ind(u,v)\}$ of the relation Ind, and the category of Inf–non–definable (or, Inf–rough) concepts which do not possess the definability property.

Definable concepts are the concepts which can be described with certainty: for objects $u, v \in U$ with $Ind(u,v)$, and a definable concept X, either u, v belong in X or u, v do not belong in X; whereas for a non–definable concept Y, there exist objects u, v such that $Ind(u,v)$ and u belongs in Y but v belongs in $U \setminus Y$.

Rough set theory solves the problem of non–definable concepts with the idea of an approximation: given a concept Y, there exist by completeness of the containment relation $\subseteq$, two definable concepts $\underline{Y}$ and $\overline{Y}$ such that $\underline{Y} \subseteq Y \subseteq \overline{Y}$, $\underline{Y}$ is the largest definable subset of Y and $\overline{Y}$ is the smallest definable superset of Y.

The following points deserve attention in the above presented scheme:
1. Definable concepts are unions of atomic concepts: indiscernibility classes.
2. Non–definable concepts are approached with definable ones by means of containment.

Both operations involved in 1, 2, above, are particular cases of general constructs of mereology [3]: the union of sets is a particular class operator and containment is a particular ingredient relation. It follows that setting the rough set context in the realm of mereology, one obtains a more general and formally

G. Wang et al. (Eds.): RSKT 2008, LNAI 5009, pp. 197–204, 2008.

adequate means of analysis of vagueness on the lines of rough set theory. This is what we are going to present in the sequel.

2   A Mereological Content of Rough Set Analysis of Vagueness

The relation π of being a part is [3] a non–reflexive and transitive relation on objects, i.e.,

P1. $\pi(u, u)$ for no u.
P2. $\pi(u, v)$ and $\pi(v, w)$ imply $\pi(u, w)$.

An example is the proper containment relation $\subset$ on sets.

It is easy to make π into a partial order relation ing of an *ingredient*: v ing u if and only if either $\pi(v, u)$ or $v = u$. Clearly, ing is reflexive, weakly–antisymmetric and transitive. An example is the containment relation $\subseteq$ on sets.

The union of sets operator used in constructions of approximations, has its counterpart in the mereological class operator Cls [3]; it is applied to any non–empty collection F of objects to produce the object $ClsF$; the formal definition is given in terms of the ingredient relation: an entity X is the class $ClsF$ if and only if the two conditions are satisfied,

C1. u ing X for each $u \in F$.
C2. u ing X implies the existence of entities v, w with the properties:

i. v ing u;
ii. v ing w;
iii. $w \in F$.

It is easy to verify that in the case when π is $\subset$, the relation ing is $\subseteq$, and for F, a non–empty collection of sets, $ClsF$ is $\bigcup F$, the union of F.

3   Rough Mereology: Motivation

In the process of development of rough set theory, it has turned out that indiscernibility could rather be relaxed to similarity: in [13] attention was focused on tolerance relations, i.e., relations which are reflexive and symmetric but need not be transitive. An example of such relation was given in [5]: given a metric ρ and a fixed small positive δ, one declares points x, y in the relation $sim_{\rho,\delta}$ if and only if $\rho(x, y) < \delta$. The relation $sim_{\rho,\delta}$ is a tolerance relation but it is not any equivalence, save, e.g., for non–archimedean ρ's.

We continue this example by introducing a graded version of $sim_{\rho,\delta}$, viz., for a real number $r \in [0, 1]$, we define the relation $sim_{\rho,\delta,r}$ by letting,

$$sim_{\rho,\delta,r}(x, y) \text{ iff } \rho(x, y) \leq 1 - r. \tag{1}$$

The collection $sim_{\rho,\delta,r}$ of relations have the following properties evident by the properties of the metric ρ:

SIM1. $sim_{\rho,\delta,1}(x,y)$ iff $x = y$.
SIM2. $sim_{\rho,\delta,1}(x,y)$ and $sim_{\rho,\delta,r}(z,x)$ imply $sim_{\rho,\delta,r}(z,y)$.
SIM3. $sim_{\rho,\delta,r}(x,y)$ and $s < r$ imply $sim_{\rho,\delta,s}(x,y)$.

Properties SIM1–SIM3 induced by the metric ρ refer to the ingredient relation $=$ whose corresponding relation of part is empty; a generalization can thus be obtained by replacing the identity with an ingredient relation ing in a mereological universe (U, π).

In consequence a relation $\mu(u, v, r)$ is defined that satisfies the following conditions:

RI1. $\mu(u, v, 1)$ iff $u\ ing\ v$.
RI2. $\mu(u, v, 1)$ and $\mu(w, u, r)$ imply $\mu(w, v, r)$.
RI3. $\mu(u, v, r)$ and $s < r$ imply $\mu(u, v, s)$.

Any relation μ which satisfies the conditions RI1–RI3 is called a *rough inclusion*, see [7], [12]. This relation is a similarity relation which is not necessarily symmetric, but it is reflexive. It is read as "the relation of a part to a degree".

4   Rough Inclusions: Case of Information Systems

The problem of methods by which rough inclusions could be introduced in information/decision systems has been studied in [7], [8], [9] among others. Here we recapitulate the results and add new ones. We recall that an *information system* is a method of representing knowledge about a certain phenomenon in the form of a table of data; formally, it is a pair (U, A) where U is a set of *objects* and A is a set of *conditional attributes*; any object $u \in U$ is described by means of its *information set* $Inf(u) = \{(a, a(u)) : a \in A\}$. The indiscernibility relation Ind, definable sets and non–definable sets are defined from Inf as indicated in Sect. 1.

We discuss some methods for inducing rough inclusions:

Case 1. Rough inclusions from metrics
Case 2. Rough inclusions from t–norms:

Subcase a. Archimedean t–norms
Subcase b. Continuous t–norms

Case 3. Weak variants of rough inclusions.

Case 1

As Sect. 3 shows, any metric ρ defines a rough inclusion μ_ρ by means of the equivalence $\mu_\rho(u, v, r) \Leftrightarrow \rho(u, v) \leq 1 - r$. A very important example of a rough inclusion obtained on these lines is the rough inclusion μ_h with $h(u, v)$

being the reduced Hamming distance on information vectors of u and v, i.e., $h(u,v) = \frac{|\{a \in A : (a, a(u)) \neq (a, a(v))\}|}{|A|}$, $|X|$ denoting the cardinality of the set X.

Thus, $\mu_h(u, v, r)$ iff $h(u, v) \leq 1 - r$; introducing sets $DIS(u, v) = \{a \in A : (a, a(u)) \neq (a, a(v))\}$ and $IND(u, v) = A \setminus DIS(u, v) = \{a \in A : a(u) = a(v)\}$, one can write down the formula for μ_h either as,

$$\mu_h(u, v, r) \Leftrightarrow \frac{|DIS(u, v)|}{|A|} \leq 1 - r, \tag{2}$$

or,

$$\mu_h(u, v, r) \Leftrightarrow \frac{|IND(u, v)|}{|A|} \geq r. \tag{3}$$

The formula (3) witnesses that the rough inclusion μ_h is an extension of the indiscernibility relation Ind to a graded indiscernibility.

In a similar manner one should be able to compute rough inclusions induced by other metrics standardly used on information sets like Euclidean, Manhattan, or l^p.

Rough inclusions induced by metrics possess an important property of *functional transitivity* expressed in general form by the rule,

$$\frac{\mu_\rho(u, v, r), \mu_\rho(v, w, s)}{\mu_\rho(u, w, L(r, s))}, \tag{4}$$

where $L(r, s) = max\{0, r + s - 1\}$ is the Lukasiewicz t–norm, see, e.g. [6]. We offer a short proof of this fact: assuming that $\mu_\rho(u, v, r), \mu_\rho(v, w, s)$ which means in terms of the metric ρ that $\rho(u, v) \leq 1 - r, \rho(v, w) \leq 1 - s$; by the triangle inequality, $\rho(u, w) \leq (1 - r) + (1 - s)$, i.e., $\mu_\rho(u, w, r + s - 1)$.

Case 2a

A functor (t–norm) $t : [0, 1] \times [0, 1]$ is Archimedean in case the equality $t(x, x) = x$ holds for $x = 0, 1$ only; it is known, see e.g. [6] that such t–norms are the Lukasiewicz L and the product t–norm $P(x, y) = x \cdot y$.

Each of these t–norms admits a functional representation: $t(x, y) = g(f(x) + f(y))$, see, e.g., [6].

One defines a rough inclusion μ_t by letting, see [7],

$$\mu_t(u, v, r) \Leftrightarrow g\left(\frac{|DIS(u, v)|}{|A|}\right) \geq r. \tag{5}$$

In particular, in case of the t–norm L, one has $g(x) = 1 - x$, see, e.g., [6], and thus the rough inclusion μ_L is expressed by means of the formula (3).

Case 2b

Other systematic method for defining rough inclusions is by means of residual implications of continuous t–norms, see [8]. For a continuous t–norm t, the residual implication $x \Rightarrow_t y$ is a mapping from the square $[0, 1]^2$ into $[0, 1]$ defined as follows, see, e.g., [6],

$$x \Rightarrow_t y \geq z \text{ iff } t(x, z) \leq y; \qquad (6)$$

thus, $x \Rightarrow_t y = max\{z : t(x, z) \leq y\}$.

Proposition 1. *The residual implication $x \Rightarrow_t y$ does induce a rough inclusion $\mu_t^{\rightarrow}$ by means of the formula: $\mu_t^{\rightarrow}(x, y, r)$ if and only if $x \Rightarrow_t y \geq r$ for every continuous t–norm t.*

Proof. We include a short argument for the sake of completeness; clearly, $\mu_t^{\rightarrow}(x, x, 1)$ holds as $x \Rightarrow_t y \geq 1$ is equivalent to $x \leq y$. Assuming $\mu_t^{\rightarrow}(x, y, 1)$, i.e., $x \leq y$, and $\mu_t^{\rightarrow}(z, x, r)$, i.e., $z \Rightarrow x \geq r$ hence $t(z, r) \leq x$, we have $t(z, r) \leq y$, i.e., $z \Rightarrow y \geq r$ so finally $\mu_t^{\rightarrow}(x, y, r)$. Clearly, by definition, from $\mu_t^{\rightarrow}(x, y, r)$ and $s < r$ it does follow that $\mu_t^{\rightarrow}(x, y, s)$.

We list here some rough inclusions obtained from most frequently applied t–norms. In all cases, $\mu_t^{\rightarrow}(x, y, 1)$ iff $x \leq y$ so the associated *ing* relation is $\leq$ and the underlying part relation is $<$. For $r < 1$, i.e., $x > y$, one has

For $t = L$: $x \Rightarrow_L y = min\{1, 1 - x + y\}$, hence $\mu_L^{\rightarrow}(x, y, r)$ if and only if $1 - x + y \geq r$.

For $t = P$: $x \Rightarrow_P y = \frac{y}{x}$ when $x \neq 0$ and 1 when $x = 0$ hence $\mu_P^{\rightarrow}(x, y, r)$ if and only if $y \geq x \cdot r$.

For $t = min(x, y)$: $x \Rightarrow_{min} y$ is y hence $\mu(x, y, r)$ if and only if $y \geq r$.

Case 3

In applications to be presented in works in this special session, some modified rough inclusions or weaker similarity measures will be instrumental, and we include a discussion of them here.

For the rough inclusion μ_L, the formula $\mu_L(v, u, r)$ means that $\frac{|IND(v,u)|}{|A|} \geq r$, i.e., at least $r \cdot 100$ percent of attributes agree on u and v; an extension of this rough inclusion depends on a chosen metric ρ bounded by 1 in the attribute value space V (we assume a simple case that ρ works for each attribute).

Then, given an $\varepsilon \in [0, 1]$, we let $\mu^{\varepsilon}(v, u, r)$ iff $|\{a \in A : \rho(a(v), a(u)) < \varepsilon\}| \geq r \cdot |A|$; it is manifest that μ^{ε} is a rough inclusion if ρ is a non–archimedean metric, i.e., $\rho(u, w) \leq max\{\rho(u, v), \rho(v, w)\}$; otherwise the monotonicity condition RI2 of Sect. 3 need not be satisfied and this takes place with most popular metrics like Euclidean, Manhattan, or l^p.

In this case, a remedy is to define a rough inclusion μ^* as follows: $\mu^*(v, u, r)$ if and only if there exists an ε such that $\mu^{\varepsilon}(v, u, r)$. Then it is easy to check that μ^* is a rough inclusion.

Assume that a residual implication $\Rightarrow_t$ is chosen, and for an information system (U, A), with an ingredient relation *ing* on U, a mapping $\phi : U \rightarrow [0, 1]$ is given such that $\phi(u) \leq \phi(v)$ iff u *ing* v. Then, the relation,

$$\mu_{\phi}(v, u, r) \text{ iff } \phi(u) \Rightarrow_t \phi(v) \geq r, \qquad (7)$$

is a rough inclusion on U. We include a short proof of this fact: $\mu_{\phi}(u, v, 1)$ is equivalent to $\phi(u) \leq \phi(v)$ hence to u *ing* v.

Assuming that $\mu_\phi(u, v, 1)$ and $\mu_\phi(w, u, r)$, the proof that RI2 holds, i.e., $\mu_\phi(w, v, r)$ goes like proof of RI2 in the proof of Prop. 1. Finally, the property RI3 is evident.

As candidates for ϕ, relative to a *standard object* $s \in U$, we consider,

1. $\phi_1 = dis(u) = \frac{|\{a \in A : a(s) \neq a(u)\}|}{|A|}$.

2. $\phi_2 = ind(u) = \frac{|\{a \in A : a(s) = a(u)\}|}{|A|}$.

3. $\phi_3 = dis_\varepsilon(u) = \frac{|\{a \in A : \rho(a(s), a(u)) \geq \varepsilon\}|}{|A|}$.

4. $\phi_4 = ind_\varepsilon(u) = \frac{|\{a \in A : \rho(a(s), a(u)) \leq \varepsilon\}|}{|A|}$,

where ρ is a chosen metric on the set of attribute values V, and ε is a chosen threshold in $[0, 1]$.

Proposition 2. *In all cases $i = 1, 2, 3, 4$, the relation $\mu_{\phi_i}(v, u, r)$ defined with ϕ_i as above is a rough inclusion.*

The reference object s can be chosen as a "standard object", e.g., such that its conditional class is contained in its decision class.

Comparison od objects u, v on lines of Case 3, need not lead to rough inclusions due to a possible violation of the property RI2; yet, such variants are of importance as they allow for a direct comparison among objects, rules and granules.

We introduce for given objects u, v, and $\varepsilon \in [0, 1]$, factors: $dis_\varepsilon(u, v) = \frac{|\{a \in A : \rho(a(u), a(v)) \geq \varepsilon\}|}{|A|}$, and $ind_\varepsilon(u, v) = \frac{|\{a \in A : \rho(a(u), a(v)) < \varepsilon\}|}{|A|}$, where ρ is a metric on attribute value sets.

Then, we modify the formula (7) to the form,

$$\nu(u, v, r) \text{ iff } dis_\varepsilon(u, v) \Rightarrow_t ind_\varepsilon(u, v) \geq r. \tag{8}$$

Clearly, ν has properties: 1. $\nu(u, u, 1)$; 2. $\nu(u, v, r)$ and $s < r$ imply $\nu(u, v, s)$ but monotonicity property RI2 need not hold. Yet, ν has an advantage of accounting for oscillations in attribute values on objects. Rough inclusions and their weak variants will be essentially exploited in data mining tasks presented in other contributions in this session, e.g., [1],[2],[11].

5   Applications: Granulation of Knowledge

Formal theory of rough inclusions allows for a formal mechanism of granulation of knowledge; we assume an information system (U, A) is given. Granulation of knowledge, proposed as a paradigm by L. A. Zadeh, means grouping objects into collections called granules, objects within a granule being similar with respect to a chosen measure; granular computing means computing with granules in place of objects.

The mechanism of granule formation based on rough inclusions has been presented by the author in a few works, see, e.g. [8], [9], and we recall it here. The basic tool in establishing properties of granules is the class operator of mereology, see Sect. 2 along with the Lesniewski Inference Rule (IR), see [3]:

(IR) For objects x, y, if for each z, from z *ing* x it follows that there exists w such that w *ing* z, w *ing* y, then x *ing* y.

Given a rough inclusion μ on the universe U, for each object u and each $r \in [0, 1]$, the *granule* $g_\mu(u, r)$ *of the radius* r *about* u *relative to* μ is defined as the class of the property $\Phi(u, r, \mu) = \{v : \mu(v, u, r)\}$:

$$g_\mu(u, r) \text{ is } Cls\Phi(u, r, \mu). \tag{9}$$

In case of symmetric and transitive rough inclusions, the following holds, see [7],

$$u \text{ } ing \text{ } g_\mu(v, r) \text{ iff } \mu(u, v, r). \tag{10}$$

This fact allows for representing the granule $g_\mu(v, r)$ as the list of those u for which $\mu(u, v, r)$ holds.

5.1   Applications: Granular Data Sets

Given a decision system (U, A, d), a rough inclusion μ on the universe U, and a radius r, one can find granules $g_\mu(u, r)$ for all $u \in U$ and make them into the set $Gran(U, r, \mu)$. From this set, a covering $Cov(U, r, \mu)$ of the universe U can be selected by means of a strategy $\mathcal{G}$, i.e., $Cov(U, r, \mu) = \mathcal{G}(Gran(U, r, \mu))$. Each granule g in $Cov(U, r, \mu)$ is a collection of objects; attributes in the set $A \cup \{d\}$ can be factored through the granule g by means of a chosen strategy $\mathcal{S}$, i.e., for each attribute $f \in A \cup \{d\}$, the new factored attribute $\overline{f}$ is defined by means of the formula,

$$\overline{f}(g) = \mathcal{S}(\{a(u) : u \text{ } ing \text{ } g_\mu(u, r)\}). \tag{11}$$

In effect, a new decision system $\mathcal{F}(U) = (Cov(U, r, \mu), \{\overline{a} : a \in A\}, \overline{d})$ is defined which is called the *granular reflection of the original system*. The object v with $Inf(v) = \{(\overline{a}, \overline{a}(g)) : a \in A\}$ is called the *granular reflection of* g. Granular reflections of granules need not be objects found in data set; yet, the results show, see, e.g., [1], [2], [10], that they mediate very well between the training and test objects. Granular data sets were proposed in [8], [9] and their usefulness in data classification was pointed there to.

6   Conclusions

We have surveyed basic means for inducing rough inclusions in data sets. Applications of them to classification of data have been tested with real data with very good results, witness [10]. This is confirmed also by papers [1], [2] in these Proceedings.

References

1. Artiemjew, P.: On classification of data by means of rough mereological granules of objects and rules. In: RSKT 2008. LNCS (LNAI), vol. 5009, Springer, Heidelberg (in print, 2008)
2. Artiemjew, P.: Rough mereological classifiers obtained from weak rough inclusions. In: RSKT 2008. LNCS (LNAI), vol. 5009, Springer, Heidelberg (in print, 2008)
3. Leśniewski, S.: On the foundations of set theory. Topoi 2, 7–52 (1982)
4. Pawlak, Z.: Rough Sets: Theoretical Aspects of Reasoning about Data. Kluwer, Dordrecht (1991)
5. Poincare, H.: Science et Hypothèse, Paris (1905)
6. Polkowski, L.: Rough Sets. In: Mathematical Foundations, Physica Verlag, Heidelberg (2002)
7. Polkowski, L.: Toward rough set foundations. Mereological approach (a plenary lecture). In: Tsumoto, S., Słowiński, R., Komorowski, J., Grzymała-Busse, J.W. (eds.) RSCTC 2004. LNCS (LNAI), vol. 3066, pp. 8–25. Springer, Heidelberg (2004)
8. Polkowski, L.: Formal granular calculi based on rough inclusions (a feature talk). In: Hu, X., Liu, Q., Skowron, A., Lin, T.Y., Yager, R.R., Zhang, B. (eds.) IEEE GrC 2005, pp. 57–62. IEEE Press, Piscataway (2005)
9. Polkowski, L.: Formal granular calculi based on rough inclusions (a feature talk). In: Zhang, Y.-Q., Lin, T.Y. (eds.) IEEE GrC 2006, pp. 9–18. IEEE Press, Piscataway (2006)
10. Polkowski, L.: The paradigm of granular rough computing. In: Zhang, D., Wang, Y., Kinsner, W. (eds.) ICCI 2007, pp. 145–163. IEEE Computer Society, Los Alamitos (2007)
11. Polkowski, L.: On the idea of using granular rough mereological structures in classification of data. In: RSKT 2008. LNCS (LNAI), vol. 5009, Springer, Heidelberg (in print, 2008)
12. Polkowski, L., Skowron, A.: Rough mereology: a new paradigm for approximate reasoning. International Journal of Approximate Reasoning 15(4), 333–365 (1997)
13. Polkowski, L., Skowron, A., Żytkow, J.: Tolerance based rough sets. In: Lin, T.Y., Wildberger, M.A. (eds.) Soft Computing:Rough Sets, Fuzzy Logic, Neural Networks, Uncertainty Management, pp. 55–58. Simulation Councils, Inc., San Diego (1995)

Reasoning about Concepts by Rough Mereological Logics

Lech Polkowski[1] and Maria Semeniuk–Polkowska[2]

[1] Polish–Japanese Institute of Information Technology
Koszykowa 86, 02008 Warsaw, Poland
[2] Warsaw University, Browarna 8/12, 00650 Warsaw,Poland
polkow@pjwstk.edu.pl, m.polkowska@uw.edu.pl

Abstract. Rough mereology allows for similarity measures (called rough inclusions) which in turn form a basis for the mechanism of granulation of knowledge. Granules of knowledge, defined as classes of satisfactorily similar objects, can be regarded as worlds in which properties of entities are evaluated. Obtained in this way granular rough mereological intensional logics reveal essential properties of rough set based reasoning. We present in this work the essential facts about these logics.

Keywords: rough sets, granulation of knowledge, rough mereology, logics for reasoning about knowledge.

1 Introductory Notions

We assume that the reader is familiar with basics of rough sets, see, e.g., [4]; the context in which our considerations are set is an information system (U, A) or a decision system (U, A, d).

We recall that a rough inclusion μ is a relation $\mu \subseteq U \times U \times [0,1]$ which satisfies the conditions,

RI1. $\mu(u, v, 1)$ iff $ing(u, v)$.
RI2. $\mu(u, v, 1), \mu(w, u, r)$ imply $\mu(w, v, r)$.
RI3. $\mu(u, v, r), s < r$ imply $\mu(u, v, s)$.

The relation ing is an ingredient (element) relation of mereology [2], see also [8].

Rough inclusions can be regarded as similarity measures on objects, $\mu(u, v, r)$ meaning that the object u is similar (is part of) to the object v to a degree of r. Formal definition of rough inclusions along with technical features of mereology, see [2] allows for a formalization of rough set based reasoning. The first step in this process is granulation of knowledge.

2 Granulation of Knowledge

The paradigm of granulation was proposed by Zadeh [11]; within rough set framework, the attention to it was brought for in Lin [3].

G. Wang et al. (Eds.): RSKT 2008, LNAI 5009, pp. 205–212, 2008.
© Springer-Verlag Berlin Heidelberg 2008

Our approach to granulation differs from those by other authors because we exploit rough inclusions as a basic tool, not using directly indiscernibility relations or inverse images of binary relations. We begin with an inventory of technical means for forming granules and establishing their basic properties. This discussion is divided into parts: mereology, rough mereology, granule formation.

2.1   Mereological Tools

In mereology theory of concepts proposed by Lesniewski [2], the basic notion of a part is a relation p which is transitive and non–reflexive, i.e.,

P1. $p(x, y), p(y, z)$ imply $p(x, z)$.
P2. $p(x, x)$ holds for no x.
   The associated with p ingredient relation ing_p is defined as

$$ing_p(x, y) \text{ iff } p(x, y) \text{ or } x = y. \tag{1}$$

The important property of the ingredient relation is the rule established by Lesniewski [2],

If for each z $[ing(z, x)$ implies w s.t. $ing(w, z), ing(w, y)]$ then $ing(x, y)$.   (2)

This means that if for each z such that z is an ingredient of x there exists an object w such that w is an ingredient of both z and y then x is an ingredient of y.
   In turn, the ingredient relation is essential in definition of the class operator Cls whose role is to make collections (properties) of objects into single objects representing those collections. The definition of the class operator is as follows,

$$\begin{aligned} C1. &\text{ If } x \in F \text{ then } ing(x, ClsF); \\ C2. &\text{ If } ing(x, ClsF) \text{ then } \exists.w, z \text{ such that} \\ &ing(w, x), ing(w, z), z \in F. \end{aligned} \tag{3}$$

This defines the class $ClsF$ for each non–empty collection F.
   To visualize the working of the operator Cls, we remind that the strict containment relation $\subset$ is a part relation, the corresponding ingredient relation is the containment relation $\subseteq$, and the class $ClsF$ for a non–empty family of sets is then according to (3) the union $\bigcup F$ of F.

2.2   Rough Mereological Tools

The part relation of mereology was given an extension in the form of a relation μ of part to a degree, see [8]. The relation μ was required to satisfy conditions RI1–RI3 of Sect. 1.
   Rough inclusions in information systems are described in, e.g., [8]. In applications, presented in this session, the most important rough inclusion is defined as follows,

$$\mu_L(u, v, r) \text{ iff } \frac{|\{a \in A : a(u) = a(v)\}|}{|A|} \geq r, \tag{4}$$

and it is a graded extension of indiscernibility relation.

An important property of this rough inclusion is its transitivity, which means,

$$\mu_L(u, v, r), \mu_L(v, w, s) \text{ imply } \mu_L(u, w, L(r, s)), \tag{5}$$

where $L(r, s) = max\{0, r + s - 1\}$ is the Lukasiewicz tensor product (t–norm), see, e.g., [5].

We also recall, see [8], that μ_L can be obtained from the t–norm L by means of the formula: $\mu_L(v, u, r)$ iff $g(dis_A(u, v)) \geq r$ where g is a mapping in the representation $L(r, s) = g(f(r) + f(s))$, see [5], and $dis_A(u, v) = \frac{|\{a \in A: a(u) \neq a(v)\}|}{|A|}$.

Rough Inclusions on Sets. For our purpose it is essential to extend rough inclusions to sets; we follow the method described already in Sect. 2.2, and we use the t–norm L along with the representation $L(r, s) = g(f(r) + f(s))$ already mentioned. We denote these kind of inclusions with the generic symbol ν.

For finite sets X, Y, we let,

$$\nu_L(X, Y, r) \text{ iff } g(\frac{|X \setminus Y|}{|X|}) \geq r; \tag{6}$$

as $g(x) = 1 - x$, see [5], we have that $\nu_L(X, Y, r)$ holds if and only if $\frac{|X \cap Y|}{|X|} \geq r$. Let us observe that ν_L is *regular*, i.e., $\nu_L(X, Y, 1)$ if and only if $X \subseteq Y$ and $\nu_L(X, Y, r)$ only with $r = 0$ if and only if $X \cap Y = \emptyset$.

Thus, the ingredient relation associated with a regular rough inclusion is the improper containment $\subseteq$ whereas the underlying part relation is the strict containment $\subset$.

Other rough inclusion on sets we we exploit is the 3–valued rough inclusion ν_3 defined via the formula, see [6],

$$\nu_3(X, Y, r) \text{iff} \begin{cases} X \subseteq Y \text{ and } r = 1 \\ X \cap Y = \emptyset \text{ and } r = 0 \\ r = \frac{1}{2} \text{ otherwise,} \end{cases} \tag{7}$$

The rough inclusion ν_3 is regular.

2.3   Granule Formation

Granules of knowledge are defined by means of a rough inclusion μ in the universe U of an information/decision system (U, A, d) as classes of appropriate similarity property, see [7],

$$g(u, r, \mu) \text{ is } Cls(P_\mu(u, r)), \tag{8}$$

where $g(u, r, \mu)$ is the granule about an object u of the radius r, and $P_\mu(u, r)$ is the property (collection) $\{v \in U : \mu(v, u, r)\}$.

Of particular interest in this work are granules induced by the rough inclusion μ_L. Their important property relies in the lack of synergic effects which could be caused by general definition of the class, see [7],

$$ing(v, g(u, r, \mu_L)) \text{ iff } \mu_L(v, u, r) \text{ iff } P_\mu(u, r)(v). \tag{9}$$

On the strength of this property, one can express the granule $g(u, r, \mu_L)$ as a list of those objects v which satisfy $\mu_L(v, u, r)$. In order to show the reader the flavor of reasoning by mereological notions, we offer a short proof of (9).

Assume that an object v is an ingredient of the granule $g(u, r, \mu_L)$; by definition (8) of a granule, and by definition (3) of the class operator, there exist objects w, z such that: (a) $ing(w, v)$; (b) $ing(w, z)$; (c) $z \in P_{\mu_L}(u, r)$, i.e., $\mu_L(z, u, r)$, where ing is the ingredient relation induced by μ_L. By condition RI1 in Sect. 1, it follows from (a) and (b) respectively that (a') $\mu_L(w, v, 1)$; (b') $\mu_L(w, z, 1)$. By symmetry of μ_L, one obtains from (a') that $\mu_L(v, w, 1)$ which together with (b') yields by the transitivity property (5) that $\mu_L(v, z, L(1, 1))$, i.e., $\mu_L(v, z, 1)$. The last fact combined with (c) by means of the transitivity property, gives that $\mu_L(v, u, L(r, 1))$, i.e., $\mu_L(v, u, r)$, as required. The converse implication holds obviously by condition C1 in (3).

This pattern of reasoning can be applied as well in establishing a number of properties of granules, see in this respect [7].

3   Granular Rough Mereological Logics

The idea of a granular rough mereological logic, see [9], [7], consists in measuring the meaning of a predicate (unary) in the model which is a universe of an information system against a granule defined to a certain degree by means of a rough inclusion. The result can be regarded as the degree of truth (the logical value) of the predicate with respect to the given granule. The obtained logics are intensional as they can be regarded as mappings from the set of granules (possible worlds) to the set of logical values in the interval $[0, 1]$, the value at a given granule regarded as the extension at that granule of the generally defined intension, see [1] for a general introduction to intensional logics.

We assume that an information/decision system (U, A, d) is given, along with a rough inclusion ν on the subsets of the universe U; for a collection of predicates (unary) Pr, interpreted in the universe U (meaning that for each predicate $\phi \in Pr$ the meaning $[\phi]$ is a subset of U), we define the intensional logic grm_ν on Pr by assigning to each predicate ϕ in Pr its intension $I_\nu(\phi)$ defined by the family of extensions $I_\nu^\vee(g)$ at particular granules g, as,

$$I_\nu^\vee(g)(\phi) \geq r \text{ iff } \nu(g, [\phi], r). \tag{10}$$

With respect to the rough inclusion ν_L, the formula (10) becomes,

$$I_{\nu_L}^\vee(g)(\phi) \geq r \text{ iff } \frac{|g \cap [\phi]|}{|g|} \geq r. \tag{11}$$

The counterpart for ν_3 is specified by definition (7).

Descriptor Logic. An important logic for information systems is the descriptor logic, see, e.g., [4]. A descriptor is a formula of the form (a, v) with the meaning $[a, v] = \{u \in U : a(u) = v\}$; from descriptors, formulas are formed by means of connectives $\vee, \wedge, \Rightarrow, \neg$ with meanings defined by recursion,

- $[\alpha \vee \beta] = [\alpha] \cup [\beta]$.
- $[\alpha \wedge \beta] = [\alpha] \cap [\beta]$.
- $[\neg\alpha] = U \setminus [\alpha]$.
- $[\alpha \Rightarrow \beta] = (U \setminus [\alpha]) \cup [\beta]$.

In the language of descriptor logic, *decision rules* are written, in the form,

$$\bigwedge\{(a, a(u)) : a \in B\} \Rightarrow (d, d(u)). \tag{12}$$

The Notion of Truth. We say that a formula ϕ interpreted in the universe U of an information system (U, A) is *true* at a granule g with respect to a rough inclusion ν if and only if $I_\nu^\vee(g)(\phi) = 1$.

Thus, for every regular rough inclusion ν, a formula ϕ interpreted in the universe U, with meaning $[\phi]$, is true at a granule g with respect to ν if and only if $g \subseteq [\phi]$. In particular, for a decision rule $r : p \Rightarrow q$ in the descriptor logic, the rule r is true at a granule g with respect to a regular rough inclusion ν if and only if $g \cap [p] \subseteq [q]$.

The formula $\nu(g, [\phi], r) = 1$ stating the truth of ϕ at g, ν with ν regular can be regarded as a condition of orthogonality type, with the usual consequences.

1. If ϕ is true at granules g, h then it is true at $g \cup h$.
2. If ϕ is true at granules g, h then it is true at $g \cap h$.
3. If ϕ, ψ are true at a granule g then $\phi \vee \psi$ is true at g.
4. If ϕ, ψ are true at a granule g then $\phi \wedge \psi$ is true at g.
5. If ψ is true at a granule g then $\phi \Rightarrow \psi$ is true at g for every formula ϕ.
6. If ϕ is true at a granule g then $\phi \Rightarrow \psi$ is true at g if and only if ψ is true at g.

The graded relaxation of truth is given obviously by the condition, a formula ϕ is *true to a degree at least r at g, ν* if and only if $I_\nu^\vee(g)(\phi) \geq r$, i.e., $\nu(g, [\phi], r)$ holds. In particular, ϕ is *false* at g, ν if and only if $I_\nu^\vee(g)(\phi) \geq r$ implies $r = 0$, i.e. $\nu(g, [\phi], r)$ implies $r = 0$.

The Standard Semantics of Connectives. The properties 1–6 above do suggest that the notion of truth in rough mereological logics has with respect to connectives of logical calculi similar properties to those of classical sentential calculus. Therefore, we introduce the following semantics of sentential connectives $\neg, \vee, \wedge, \Rightarrow$.

1. $[\neg\alpha] = U \setminus [\alpha]$.
2. $[\alpha \vee \beta] = [\alpha] \cup [\beta]$.
3. $[\alpha \wedge \beta] = [\alpha] \cap [\beta]$.
4. $[\alpha \Rightarrow \beta] = (U \setminus [\alpha]) \cup [\beta]$.

With respect to this semantics, the following properties hold.

1. For each regular ν, a formula α is true at g, ν if and only if $\neg\alpha$ is false at g, ν.
2. For $\nu = \nu_L, \nu_3$, $I_\nu^\vee(g)(\neg\alpha) \geq r$ if and only if $I_\nu^\vee(g)(\alpha) \geq s$ implies $s \leq 1 - r$.

3. For $\nu = \nu_L, \nu_3$, the implication $\alpha \Rightarrow \beta$ is true at g if and only if $g \cap [\alpha] \subseteq [\beta]$ and $\alpha \Rightarrow \beta$ is false at g if and only if $g \subseteq [\alpha] \setminus [\beta]$.

4. For $\nu = \nu_L$, if $I_\nu^\vee(g)(\alpha \Rightarrow \beta) \geq r$ then $\Rightarrow_L (t, s) \geq r$ where $I_\nu^\vee(g)(\alpha) \geq t$ and $I_\nu^\vee(g)(\beta) \geq s$.

The functor $\Rightarrow_L$ in 4. is the Lukasiewicz implication of many–valued logic: $\Rightarrow_L (t, s) = min\{1, 1 - t + s\}$.

Further analysis should be split into the case of ν_L and the case of ν_3 as the two differ essentially with respect to the form of reasoning they imply.

4   Reasoning with ν_L

The last property 4. shows in principle that the value of $I_\nu^\vee(g)(\alpha \Rightarrow \beta)$ is bounded from above by the value of $\Rightarrow_L (I_\nu^\vee(g)(\alpha), I_\nu^\vee(g)(\beta))$.

This suggests that the idea of collapse attributed to S. Lesniewski can be applied to formulas of rough mereological logic in the following form: for a formula $q(x)$ we denote by the symbol q^* the formula q regarded as a sentential formula (i.e., with variable symbols removed) subject to relations $\neg q(x)^*$ is $\neg(q(x)^*)$ and $p(x) \Rightarrow q(x)^*$ is $p(x)^* \Rightarrow q(x)^*$. As the value $[q^*]_g$ of the formula $q(x)^*$ we admit the value of $\frac{|g \cap [q(x)]|}{|g|}$. Thus, the item 4 can be rewritten in the form.

$$I_\nu^\vee(g)(\alpha \Rightarrow \beta) \leq \Rightarrow_L ([\alpha^*]_g, [\beta^*]_g). \tag{13}$$

The following statement is then obvious: if $\alpha \Rightarrow \beta$ is true at g then the collapsed formula has the value 1 of truth at the granule g. This gives a necessity condition for verification of implications of rough mereological logics: if $\Rightarrow_L ([\alpha^*]_g, [\beta^*]_g) < 1$ then the implication $\alpha \Rightarrow \beta$ is not true at g. This concerns in particular decision rules: for a decision rule $p(v) \Rightarrow q(v)$, it follows that the decision is true on a granule g if and only if $[p^*]_g \leq [q^*]_g$.

4.1   Rough Set Reasoning: Possibility and Necessity

Possibility and necessity are introduced in rough set theory by means of approximations: the upper and the lower, respectively. A logical rendering of these modalities in rough mereological logics exploits the approximations. Denoting the lower approximation as $\underline{X} = \{u \in U : [u]_A \subseteq X\}$ and the upper approximation as $\overline{X} = \{u \in U : [u]_A \cap X \neq \emptyset\}$, we define two modal operators: M (possibility) and L (necessity) by means of their semantics.

To this end, we let

$$I_\nu^\vee(g)(M\alpha) \geq r \text{ iff } \nu_L(g, \overline{[\alpha]}, r) \\ I_\nu^\vee(g)(L\alpha) \geq r \text{ iff } \nu_L(g, \underline{[\alpha]}, r). \tag{14}$$

Then we have the following criteria for necessarily or possibly true formulas.

A formula α is *necessarily true at a granule* g if and only if $g \subseteq \underline{[\alpha]}$; α is *possibly true at* g if and only if $g \subseteq \overline{[\alpha]}$.

This semantics of modal operators M, L can be applied to show that rough set structures carry the semantics of S5 modal logics, i.e., the following relations hold at each granule g.

1. $L(\alpha \Rightarrow \beta) \Rightarrow [(L\alpha) \Rightarrow L(\beta)]$.
2. $L\alpha \Rightarrow \alpha$.
3. $L\alpha \Rightarrow LL\alpha$.
4. $M\alpha \Rightarrow LM\alpha$.

Proofs can be found in [9].

5   Reasoning with ν_3

In case of $\nu = \nu_3$, one can check on the basis of definitions that $I_\nu^\vee(g)(\neg\alpha) \geq r$ if and only if $I_\nu^\vee(g)(\alpha) \leq 1-r$; thus the negation functor in rough mereological logic based on ν_3 is the same as the negation functor in the 3–valued Łukasiewicz logic. For implication, the relations between rough mereological and 3–valued logics L_3 follow from tables of values of truth.

Table 1 shows truth values for the 3–valued logic L_3 and Table 2 does the same for rough mereological logic based on ν_3.

Table 1. Truth values for implication in L_3

$\Rightarrow$	0	1	$\frac{1}{2}$
0	1	1	1
1	0	1	$\frac{1}{2}$
$\frac{1}{2}$	$\frac{1}{2}$	1	1

Table 2. Truth values for implication $\alpha \Rightarrow \beta$ in logic based on ν_3

$\Rightarrow$	$I_\nu^\vee(g)(\beta) = 0$	$I_\nu^\vee(g)(\beta) = 1$	$I_\nu^\vee(g)(\beta) = \frac{1}{2}$
$I_\nu^\vee(g)(\alpha) = 0$	1	1	1
$I_\nu^\vee(g)(\alpha) = 1$	0	1	$\frac{1}{2}$
$I_\nu^\vee(g)(\alpha) = \frac{1}{2}$	$\frac{1}{2}$	1	1 when $g \cap [\alpha] \subseteq [\beta]$; $\frac{1}{2}$ otherwise

To express the relation between the two implications, we introduce a new notion: we say that a formula ϕ is *acceptable* in either logic (L_3 or grm_{ν_3}) at a granule g if and only if $[\phi^*]_g \geq \frac{1}{2}$, respectively, $I_\nu^\vee(g)(\phi) \geq \frac{1}{2}$.

From Tables 1,2, one infers that $I_\nu^\vee(g)(\phi) \geq [\phi^*]_g$. This crucial relationship implies that: if ϕ^* is acceptable at g then ϕ is acceptable at g; if ϕ^* is true at g then ϕ is true at g. Also, if ϕ at false at g then ϕ^* is false at g.

An another application of logics presented here can be found in [10] where a formalization of perception calculus was given.

References

1. Van Bentham, J.: A Manual of Intensional Logic. CSLI Stanford University (1988)
2. Leśniewski, S.: On the foundations of set theory. Topoi 2, 7–52 (1982)
3. Lin, T.Y.: From rough sets and neighborhood systems to information granulation and computing with words. In: Proceedings of the European Congress on Intelligent Techniques and Soft Computing, Verlag Mainz, Aachen, pp. 1602–1606 (1997)
4. Pawlak, Z.: Rough Sets: Theoretical Aspects of Reasoning about Data. Kluwer, Dordrecht (1991)
5. Polkowski, L.: Rough Sets. Mathematical Foundation. Physica Verlag, Heidelberg (2002)
6. Polkowski, L.: A note on 3–valued rough logic accepting decision rules. Fundamenta Informaticae 61, 37–45 (2004)
7. Polkowski, L.: Formal granular calculi based on rough inclusions (a feature talk). In: Hu, X., Liu, Q., Skowron, A., Lin, T.Y., Yager, R.R., Zhang, B. (eds.) IEEE GrC 2005, pp. 57–62. IEEE Press, Piscataway (2005)
8. Polkowski, L.: Rough mereology in analysis of vagueness. In: RSKT 2008. LNCS (LNAI), vol. 5009, Springer, Heidelberg (in print, 2008)
9. Polkowski, L., Semeniuk–Polkowska, M.: On rough set logics based on similarity relations. Fundamenta Informaticae 64, 379–390 (2005)
10. Polkowski, L., Semeniuk–Polkowska, M.: A formal approach to perception calculus of Zadeh by means of rough mereological logic. In: Actes 11th International Conference on Information Processing and Management in Knowledge–Based Systems IPMU 2006, Paris, Univ. Marie Curie, pp. 1468–1473 (2006)
11. Zadeh, L.A.: Fuzzy sets and information granularity. In: Gupta, M., Ragade, R., Yager, R.R. (eds.) Advances in Fuzzy Set Theory and Applications, pp. 3–18. North–Holland, Amsterdam (1979)

On the Idea of Using Granular Rough Mereological Structures in Classification of Data

Lech Polkowski

Polish-Japanese Institute of Information Technology
Koszykowa 86, 02008 Warsaw, Poland
polkow@pjwstk.edu.pl

Abstract. This paper is devoted to an exposition of the idea of using granular structures obtained from data in the classification tasks of these data into decision classes. Classifiers are induced from granular reflections of data sets.

Keywords: rough sets, granulation of knowledge, rough inclusions, granular classifiers.

1 Introduction: Rough Inclusions, Granulation of Knowledge

We begin with an information system $I = (U, A)$ where U is the set of objects and A is the set of attributes. The value of an attribute a on an object u is denoted $a(u)$. A rough inclusion is a relation $\mu \subseteq U \times U \times [0,1]$ with $\mu(u, v, r)$ interpreted as object u being a part of object v to degree r at least. Specific recipes for μ are given in [Po3].

Granulation of knowledge by means of μ consists in forming for each $r \in [0,1]$ and each $u \in U$, of a granule $g_\mu(u, r) = \{v \in U : \mu(v, u, r)\}$.

Granular data sets were proposed by L.Polkowski in [Po1], [Po2]. Given $r \in [0,1]$, the set of all granules $G_r^\mu = \{g_\mu(u, r) : u \in U\}$ is defined. From this set, a covering $Cov_r^\mu(\mathcal{G})$ is chosen according to a strategy $\mathcal{G}$. Granules in $Cov_r^\mu(\mathcal{G})$ form a new universe of objects. For each $g \in Cov_r^\mu(\mathcal{G})$, and each attribute $a \in A$, a factored attribute $\bar{a}$ is defined as $\bar{a} = \mathcal{S}(\{a(u) : u \in g\})$ for a chosen strategy $\mathcal{S}$.

The new information system $I_r^\mu = (Cov_r^\mu(\mathcal{G}), \{\bar{a} : a \in A\})$ is a *granular reflection* of the original information system I. The object $o(g)$ defined for a granule g by means of $Inf_{\bar{A}}(o(g)) = \{(\bar{a}, \bar{a}(g)) : a \in A\}$ according to a strategy $\mathcal{S}$ is called an $\mathcal{S}$–*reflection of the granule* g; clearly, $o(g)$ need not be a real object in the training or test sets. Its role is to be a link between training and test sets.

2 The Case of Rough Inclusions from Hamming Metrics on Information Sets

This rough inclusion, see [Po3], is of the form,

$$\mu_h(u, v, r) \text{iff } \frac{|IND(u, v)|}{|A|} \geq r, \tag{1}$$

G. Wang et al. (Eds.): RSKT 2008, LNAI 5009, pp. 213–220, 2008.

where $IND(u, v) = \{a \in A : a(u) = a(v)\}$ and $h(u, v) = \frac{|\{a \in A: a(u) \neq a(v)\}|}{|A|}$. It is manifest that μ_h is a direct extension of indiscernibility.

We show results of tests with Australian credit data set, see [UCI] and we include in Table 1, for comparison, some best results obtained by other authors. Classification quality is expressed by means of two factors: accuracy which is the ratio of the number of correctly classified objects to the number of recognized test objects and coverage, $\frac{rec}{test}$, where rec is the number of recognized test cases and $test$ is the number of test cases.

Table 1. Best results for Australian credit by some rough set based algorithms

source	method	accuracy	coverage
[Ba]	$SNAPM(0.9)$	$error = 0.130$	–
[H]	simple.templates	0.929	0.623
[H]	general.templates	0.886	0.905
[H]	tolerance.gen.templ.	0.875	1.0
[Wr]	adaptive.classifier	0.863	–

Tests on this data with granular approach indicated above were carried by splitting the Australian credit data set into the training and test sets in ratio 1:1; the training sample was granulated and a granular reflection was formed from which by means of RSES exhaustive algorithm a classifier was produced which was applied to the test part of data to find quality of classification.

Granules were calculated in a twofold way: first as indicated above and second, by a modified procedure of *concept dependent granulation*, see [A1]: in the latter procedure, the granule $g_h^c(u, r) = g_h(u, r) \cap [u]_d$ was computed relative to the *concept*, i.e., decision class, to which u belonged. The results of tests are given in Table 2 in which the best results obtained with various granulation radii are shown.

Table 2. Best results for Australian credit by granular approach

source	method	accuracy	coverage
[PoA]	$granular^*.r = 0.642857$	0.867	1.0
[PoA]	$granular^{**}.r = 0.714826$	0.875	1.0
[A1]	$granular^{***}.concept.r = 0.785$	0.9970	0.9995

Results in Table 2 do witness that granular approach gives results fully comparable with other results for satisfactorily large radii of granulation whereas the concept dependent granulation gives results better than any other existing approach.

3   Parameterized Variants of Rough Inclusions μ_h in Classification of Data

As discussed in [Po3], for the formula $\mu_h(v, u, r)$ an extension is proposed which depends on a chosen metric ρ bounded by 1 in the attribute value space V (we assume for simplicity that ρ is suitable for all attributes).

Then, given an $\varepsilon \in [0, 1]$, we let $\mu_h^\varepsilon(v, u, r)$ iff $|\{a \in A : \rho(a(v), a(u)) < \varepsilon\}| \geq r \cdot |A|$. The parameter r is called the *catch radius*.

Granules induced by μ_h^ε with $r = 1$ have a simple structure: a granule $g_h^\varepsilon(u, 1)$ consists of all $v \in U$ such that $\rho(a(u), a(v)) \leq \varepsilon$.

We use these granules to assign a decision class to an object u in the test set.

First, rules are induced from the training set by an exhaustive algorithm. Then, given a set Rul of these rules, and an object u in the test set, a granule $g_h^\varepsilon(u, 1)$ is formed in the set Rul: $g_h^\varepsilon(u, 1) = \{r \in Rul : \rho(a(u), a(r)) \leq \varepsilon$ for each attribute $a \in A$ where $a(r)$ is the value of the attribute a in the premise of the rule.

Rules in the granule $g_h^\varepsilon(u, 1)$ are taking part in voting: for each value c of a decision class, the following factor is computed,

$$\text{param}(c) = \frac{\text{sum of supports of rules pointing to c}}{\text{cardinality of c in the training set}}, \tag{2}$$

cf., [Ba] for a discussion of various strategies of voting for decision values.

The class c_u assigned to u is decided by

$$param(c_u) = max_c param(c), \tag{3}$$

with random resolution of ties.

In computing granules, the parameter ε is normalized to the interval $[0, 1]$ as follows: first, for each attribute $a \in A$, the value $train(a) = max_{training\ set}a - min_{training\ set}a$ is computed and the real line $(-\infty, +\infty)$ is contracted to the interval $[min_{training\ set}a, max_{training\ set}a]$ by the mapping f_a,

$$f_a(x) = \begin{cases} min_{training\ set}a & \text{in case } x \leq min_{training\ set}a \\ x & \text{in case } x \in [min_{training\ set}a, max_{training\ set}a] \\ max_{training\ set}a & \text{in case } x \geq max_{training\ set}a. \end{cases} \tag{4}$$

When the value $a(u)$ for a test object u is off the range $[min_{training\ set}a, max_{training\ set}a]$, it is replaced with the value $f_a(a(u))$ in the range. For an object v, or a rule r with the value $a(v)$, resp., $a(r)$ of a denoted $a(v, r)$, the parameter ε is computed as $\frac{|a(v,r) - f_a(a(u))|}{train(a)}$. The metric ρ was chosen as the metric $|x - y|$ in the real line. We show results of experiments with rough inclusions discussed in this work. Our data set was a subset of Australian credit data in which training set had 100 objects from class 1 and 150 objects from class 0 (which approximately yields the distribution of classes in the whole data set). T he test set had 100 objects, 50 from each class. The RSES exhaustive classifier [RSES] applied to this data set gave accuracy of 0.79 and coverage of 1.0.

3.1   Results of Tests with Granules of Training Objects According to $\mu_h^\varepsilon(v, u, 1)$ Voting for Decision

In Fig. 1 results of classification are given in function of ε for accuracy as well as for coverage.

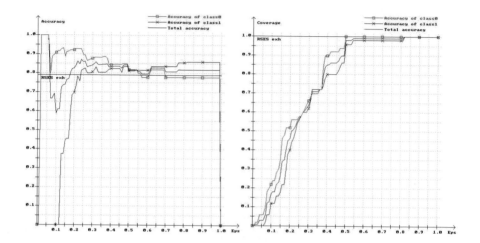

Fig. 1. Results for algorithm 1_v1, Best result for $\varepsilon = 0.62$: accuracy $= 0.828283$, coverage $= 0.99$

Table 3. (40%-60%)(1-0); Australian credit; Algorithm 1_v2. r_catch=catch radius, optimal_eps=Best ε, acc= accuracy, cov= coverage

r_catch	optimal eps	acc	cov
nil	nil	0.79	1.0
0.071428	0	0.06	1.0
0.142857	0	0.66	1.0
0.214286	0.01	0.74	1.0
0.285714	0.02	0.83	1.0
0.357143	0.07	0.82	1.0
0.428571	0.05	0.82	1.0
0.500000	0	0.82	1.0
0.571429	0.08	0.84	1.0
0.642857	0.09	0.84	1.0
0.714286	0.16	0.85	1.0
0.785714	0.22	0.86	1.0
0.857143	0.39	0.84	1.0
0.928571	0.41	0.828283	0.99
1.000000	0.62	0.828283	0.99

3.2 Results of Tests with Granules of Training Objects According to $\mu_h^\varepsilon(v, u, r)$ Voting for Decision

The procedure applied in case of $\mu_h^\varepsilon(v, u, 1)$ can be repeated with r variable. The resulting classifier is a function of two parameters ε, r. In Table 3, results are included where against values of the catch radius r the best value for ε's marked by the optimal value *optimal eps* is given for accuracy and coverage.

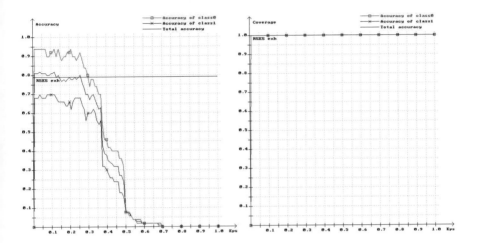

Fig. 2. Results for algorithm 5_v1, Best result for $\varepsilon = 0.04$, accuracy $= 0.82$, coverage $= 1$

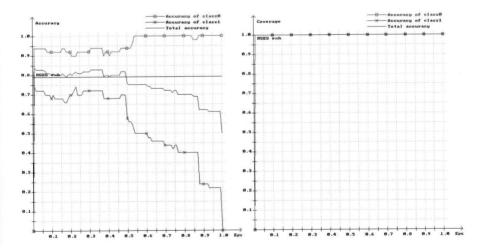

Fig. 3. Results for algorithm 5_v2, Best result for $\varepsilon = 0.01$, accuracy $= 0.84$, coverage $= 1$

4   Rough Inclusions and Their Weaker Variants from Residual Implications in Classification of Data

As shown in [Po3], residual implications of continuous t–norms can supply rough inclusions according to a general formula,

$$\mu_\phi(v,u,r) \text{ iff } \phi(u) \Rightarrow_t \phi(v) \geq r, \tag{5}$$

where ϕ maps the set U of objects into $[0,1]$ and $\phi(u) \leq \phi(v)$ if and only if $u \; ing \; v$ (ing is an ingredient relation of the underlying mereology, see e.g., [Po3]); $\Rightarrow_t$ is the residual implication induced by the t–norm , see [Po3].

Candidates for ϕ are proposed in [Po3], and a weak interesting variant of this class of rough inclusions is indicated. This variant uses sets $dis_\varepsilon(u,v) = \frac{|\{a \in A: \rho(a(u),a(v)) \geq \varepsilon\}|}{|A|}$, and $ind_\varepsilon(u,v) = \frac{|\{a \in A: \rho(a(u),a(v)) < \varepsilon\}|}{|A|}$, for $u,v \in U$, $\varepsilon \in [0,1]$, where ρ is a metric $|x-y|$ on attribute value sets.

The resulting weak variant of the rough inclusion μ_ϕ is,

$$\mu_t(u,v,r) \text{ iff } dis_\varepsilon(u,v) \Rightarrow_t ind_\varepsilon(u,v) \geq r. \tag{6}$$

Objects in the class c in the training set vote for decision at the test object u according to the formula: $p(c) = \frac{\sum_{v \in c} w(v,t)}{|c| \text{ in the training set}}$ where weight $w(v,t)$ is $dis_\varepsilon(u,v) \rightarrow_t ind_\varepsilon(u,v)$; rules induced from the training set pointing to the class c vote according to the formula $p(c) = \frac{\sum_r w(r,t) \cdot support(r)}{|c| \text{ in the training set}}$. In either case, the class c* with $p(c^*) = \max p(c)$ is chosen. We include here results of tests with training objects and t=min (Fig.2)and rules and t=min (Fig.3); a detailed presentation of results is given in [A2].

Similarly, we include in Figs. 3,4 results of tests with granules of training objects and rules for t=P, the product t–norm.

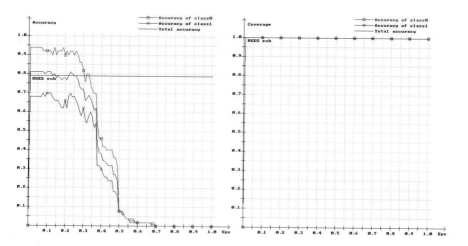

Fig. 4. Results for algorithm 6_v1, Best result for $\varepsilon = 0.01$, accuracy=0.81, coverage=1

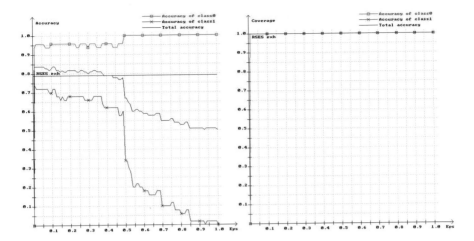

Fig. 5. Results for algorithm 6_v2, Best result for $varepsilon = 0.01$, accuracy $= 0.84$, coverage $= 1$

The results of tests in best cases for optimal values of ε exceed results obtained with the standard exhaustive algorithm. More detailed tests and results are given in [A2], [A3].

5 Conclusion

The results shown here do witness that the proposed approach is a valid method for building effective classifiers and validates the hypothesis put forth by the author in [Po1].

References

[A1]   Artiemjew, P.: Classifiers from granulated data sets: Concept dependent and layered granulation. In: Rough Sets in Knowledge Discovery RSKD 2007. Workshop at European Conference on Machine Learning/International Conference on Principles of Knowledge Discovery in Data ECML/PKDD 2007, pp. 1–9. Warsaw Univ. Press, Warsaw (2007)

[A2]   Artiemjew, P.: On classification of data by means of rough mereological granules of objects and rules. In: RSKT 2008. LNCS (LNAI) (in print, 2008)

[A3]   Artiemjew, P.: Rough mereological classifiers obtained from weak rough inclusions. In: RSKT 2008. LNCS (LNAI), vol. 5009, Springer, Heidelberg (in print, 2008)

[Ba]   Bazan, J.G.: A comparison of dynamic and non–dynamic rough set methods for extracting laws from decision tables. In: Polkowski, L., Skowron, A. (eds.) Rough Sets in Knowledge Discovery, vol. 1, pp. 321–365. Physica Verlag, Heidelberg (1998)

220   L. Polkowski

[H]   Nguyen, S.H.: Regularity analysis and its applications in Data Mining. In:
           Polkowski, L., Tsumoto, S., Lin, T.Y. (eds.) Rough Set Methods and Appli-
           cations, pp. 289–378. Physica Verlag, Heidelberg (2000)
[Po1]   Polkowski, L.: Formal granular calculi based on rough inclusions (a feature
           talk). In: Hu, X., Liu, Q., Skowron, A., Lin, T.Y., Yager, R.R., Zhang, B.
           (eds.) IEEE GrC 2005, pp. 57–62. IEEE Press, Piscataway (2005)
[Po2]   Polkowski, L.: Formal granular calculi based on rough inclusions (a feature
           talk). In: Zhang, Y.-Q., Lin, T.Y. (eds.) IEEE GrC 2006, pp. 9–18. IEEE
           Press, Piscataway (2006)
[Po3]   Polkowski, L.: On the idea of using granular rough mereological structures
           in classification of data. In: RSKT 2008. LNCS (LNAI), vol. 5009, Springer,
           Heidelberg (in print, 2008)
[PoA]   Polkowski, L., Artiemjew, P.: On granular rough computing: Factoring clasi-
           fiers through granular structures. In: Kryszkiewicz, M., Peters, J.F., Ry-
           binski, H., Skowron, A. (eds.) RSEISP 2007. LNCS (LNAI), vol. 4585, pp.
           280–290. Springer, Heidelberg (2007)
[RSES]   RSES, http://logic.mimuw.edu.pl/rses
[UCI]   UCI Repository, http://www.ics.uci.edu/mlearn/~databases/
[Wr]   Wróblewski, J.: Adaptive aspects of combining approximation spaces. In:
           Pal, S.K., Polkowski, L., Skowron, A. (eds.) Rough Neural Computing, pp.
           139–156. Springer, Berlin (2004)

On Classification of Data by Means of Rough Mereological Granules of Objects and Rules

Piotr Artiemjew

University of Warmia and Mazury
Zolnierska 14, 10-560 Olsztyn, Poland
artem@matman.uwm.edu.pl

Abstract. Granulation of knowledge has turned an effective tool in data classification. We propose the approach to classification of data which extends our earlier methods by considering granules of either objects or decision rules obtained either from the original training set or from its granular reflection. Members of a granule vote for the decision class of that object. We present results of tests which show that this method usually gives results at least as good as the exhaustive classifier built on rough set principles.

Keywords: rough sets, granulation of knowledge, rough inclusions, data classification.

1 Introduction

The idea of granulation of knowledge encoded in information systems by means of rough inclusions is presented in [Po1] in detail. We recall here its main lines. We assume a given information system $I = (U, A)$ with the object set U and the attribute set A.

A rough inclusion is a ternary relation among pairs of objects and real numbers in the interval $[0, 1]$, written down in symbolic form as $\mu(u, v, r)$ and read as "u is a part of v to a degree not less than r", see [Po1]. Rough inclusions which are used in this paper stem from a class of rough inclusions defined by metrics see [Po1].

In particular, see [Po1], the Hamming metric δ on information sets of objects in an information system, relative to size $|A|$ of the attribute set, does induce the rough inclusion μ_δ,

$$\mu_\delta(u, v, r) \text{ if and only if } \frac{|IND(u, v)|}{|A|} \geq r, \tag{1}$$

where $IND(u, v) = \{a \in A : a(u) = a(v)\}$, meaning that u is a part of v (similar to v) to degree not less than r if and only if at least $r \cdot 100$ percent of attributes take the same value on u and v.

In [Po4], the idea of a granular reflection of a data set was first put forth with the hypothesis that granulated data sets should preserve to a satisfactory degree knowledge contained in the original data set.

G. Wang et al. (Eds.): RSKT 2008, LNAI 5009, pp. 221–228, 2008.

In this contribution, we report results of data classification tests in which some modifications of this rough inclusion have been applied. In order to present the results, some necessary notions should be introduced first. The reader will find a detailed introduction in [Po1].

2 Granulation of Knowledge and Granular Reflections of Data Sets

The idea of granulation of knowledge used here was proposed and studied in [Po3], [Po4]. Granules of knowledge are formed essentially in this approach as sets of objects close to a specified degree to the object designated as the granule center; formally, for a radius of granulation r, an object u and a given rough inclusion μ, the granule of radius r about u is given as

$$g(u, r, \mu) = \{v \in U : \mu(v, u, r)\}. \tag{2}$$

The idea of a granular reflection of a data set was proposed in [Po4]: given a granulation radius r, the set $G(r, \mu)$ of all granules of the radius r is formed. From this set, a covering $C(r, \mu, \mathcal{G})$ of the set of objects U is chosen by means of a strategy $\mathcal{G}$, which is usually a random choice of granules with irreducibility checking.

Given the covering $C(r, \mu, \mathcal{G})$, attributes in the set A are factored through granules to make a new attribute set. For an attribute $a \in A$, and a granule g, the value of the new attribute a_G on g is defined as,

$$a_G(g) = \mathcal{S}(\{a(v) : v \in g\}), \tag{3}$$

and a new information system is formed: $(U^G = C(r, \mu, \mathcal{G}), A_G = \{a_G : a \in A\})$ called a *granular reflection* of the given information system, see [Po1], [Po4].

3 A Modified Rough Inclusion and Applications to Data Classification

As introduced in [Po1], the graded variant of μ_δ assumes that in deciding the decision value at a test object u, collections of rules or objects take part; these collections are built as granules with respect to a modified rough inclusion μ_δ^ε see [Po1].

Given $\varepsilon \in [0, 1]$, we let $\mu_\delta^\varepsilon(v, u, r)$ iff

$$|\{a \in A : \delta(a(v), a(u)) \leq \varepsilon\}| \geq r \cdot |A|. \tag{4}$$

The parameter r is called the *catch radius*.

With $\delta = l^1$, the granules are defined as,

$$g(u, r, \mu_\delta^\varepsilon) = \{v \in U : |\{a \in A : |a(v) - a(u)| \leq \varepsilon\}| \geq r \cdot |A|. \tag{5}$$

In particular, in case $r = 1$, the granule $g(u, 1, \mu_\delta^\varepsilon)$ consists of v such that $|a(v) - a(u)| \leq \varepsilon$ for each attribute a.

In computing granules, the parameter ε is normalized to the interval $[0, 1]$ as follows: first, for each attribute $a \in A$, the value $train(a) = max_{training\ set}a - min_{training\ set}a$ is computed and the real line $(-\infty, +\infty)$ is contracted to the interval $[min_{training\ set}a, max_{training\ set}a]$ by the mapping f_a,

$$f_a(x) = \begin{cases} min_{training\ set}a \text{ in case } x \leq min_{training\ set}a \\ x \text{ in case } x \in [min_{training\ set}a, max_{training\ set}a] \\ max_{training\ set}a \text{ in case } x \geq max_{training\ set}a. \end{cases} \tag{6}$$

When the value $a(u)$ for a test object u is off the range $[min_{training\ set}a, max_{training\ set}a]$, it is replaced with the value $f_a(a(u))$ in the range. The value of ε in this case is defined as $\frac{f_a(a(u)) - min_{training\ set}a}{train(a)}$.

3.1 Voting by Granules on Decision Values

Given a granule g of either decision rules or objects in a decision system (U, A, d), for each test object u, the value of decision assigned to u by the granule g is defined as,

$$d(u) = c^* \text{ iff } \frac{\text{sum of supports of rules pointing to } c^*}{\text{cardinality of } c^* \text{ in the training set}} = max_c \frac{\text{sum of supports of rules pointing to } c}{\text{cardinality of } c \text{ in the training set}}. \tag{7}$$

where c denotes a decision value and $c*$ is the decision value assigned to u.

4 Results of Tests with Classification Based on Granulation by Means of $\mu_\delta^\varepsilon(v, u, 1)$

We present here results of tests in which the granulation with $\mu_\delta^\varepsilon(v, u, 1)$ has been applied in four cases:
1. Granules of training objects have been used.
2. Granules of granular objects have been used.
3. Granules of rules from the training set have been used.
4. Granules of rules from the granulated training set have been used.

All tests have been done with Australian credit data [UCI], split into the training and test sets. The 5–fold cross validation has been applied. results are expressed in terms of accuracy and coverage, cf., [RS].

4.1 Results of Tests

In Fig. 1, below the results of the test with granules of training objects (test 1-v.1) are shown for accuracy and coverage respectively in function of ε applied. For comparison, the result obtained with the standard RSES exhaustive classifier

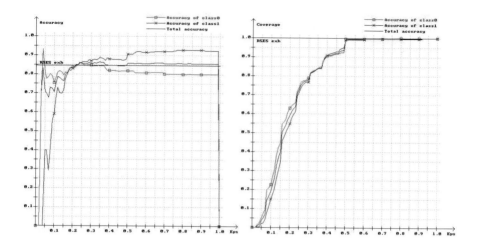

Fig. 1. Results for test with granules of training objects. Best result for $\varepsilon_{opt} = 0.83$: accuracy = 0.859219, coverage = 0.998551.

[RS] (marked with the horizontal line in Fig.1) is 0.845 for accuracy and 1.0 for coverage.

In Table 1 results are shown of experiments with granules of granular objects where for each value of the granulation radius r, the optimal (best) result for ε's is given as ε_{opt}.

Fig. 2 shows the results of experiments with granules of rules from the training set.

In Table 2, we give results of tests with granules of rules from a granular reflection of data set.

Table 1. CV-5; Australian credit; Algorithm 2_v1. r_gran=granulation radius, ε_{opt}= optimal epsilon, acc= accuracy, cov= coverage, m_trn=mean training sample. Best result for $r_{gran} = 0.7857$: $\varepsilon_{opt} = 0.54$, accuracy=0.8616.

r_gran	ε_{opt}	acc	cov	m_trn
nil	nil	0.845	1.0	552
0.428571	0.96	0.858144	0.742029	21.6
0.500000	0.95	0.838491	0.95942	52.8
0.571429	0.93	0.82871	0.998551	134.8
0.642857	0.95	0.831884	1.0	295.8
0.714286	0.71	0.858987	0.997102	456.4
0.785714	0.54	0.861673	0.995652	533.2
0.857143	0.83	0.859219	0.998551	546.2
0.928571	0.83	0.859219	0.998551	548
1.000000	0.83	0.859219	0.998551	552

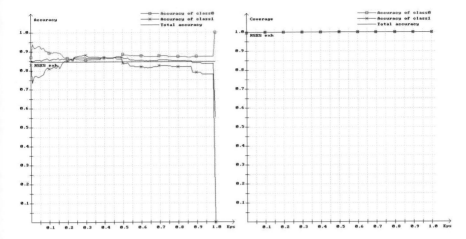

Fig. 2. Results for algorithm 3_v1. Best result for $\varepsilon_{opt} = 0.46$: accuracy $= 0.871015$, coverage $= 1$.

Table 2. CV-5; Australian credit; Algorithm 4_v1. r_gran=granulation radius, ε_{opt}=optimal_eps= optimal epsilon, acc= accuracy, cov= coverage. Best result for $r_{gran} = 0.7857$: $\varepsilon_{opt} = 0.01$, accuracy=0.85072.

r_gran	ε_{opt}	acc	cov	m_trn
nil	nil	0.845	1.0	552
0.428571	0.01	0.765217	1.0	23.8
0.500000	0	0.771014	1.0	53.2
0.571429	0	0.824638	1.0	130.6
0.642857	0	0.834783	1.0	294.8
0.714286	0.02	0.83913	1.0	454.8
0.785714	0.01	0.850724	1.0	533
0.857143	0.01	0.850724	1.0	546.2
0.928571	0.01	0.849275	1.0	548
1.000000	0.1	0.850725	1.0	552

5 Results of Tests with Classification Based on Granulation by Means of $\mu_\delta^\varepsilon(v, u, r)$

We present results of tests with granules based on $\mu_\delta^\varepsilon(v, u, r)$ with all values of r. Thus, a granule $g(u, r, \mu_\delta^\varepsilon)$ consists of objects v such that at least $r \cdot 100$ percent of attributes a satisfy $|a(u) - a(v)| \leq \varepsilon$. We present results of experiments with granules of objects in training sets as well as rules from the training set.

In Table 3, results are given of test with granules of training objects. Best classification results for distinct values of ε are shown against values of r.

Table 3. CV-5; Australian credit; Algorithm 1_v2. r_catch=catch radius, $\varepsilon_{opt}=$ optimal epsilon, acc= accuracy, cov= coverage. Best result for $r_{catch} = 0.7857$, $\varepsilon_{opt} = 0.18$, accuracy=0.872.

r_catch	ε_{opt}	acc	cov
nil	*nil*	0.845	1.0
0.071428	0	0.155073	1.0
0.142857	0	0.750725	1.0
0.214286	0.01	0.823188	1.0
0.285714	0.01	0.853623	1.0
0.357143	0.02	0.844927	1.0
0.428571	0.04	0.842029	1.0
0.500000	0.05	0.852174	1.0
0.571429	0.03	0.860869	1.0
0.642857	0.05	0.870866	0.998551
0.714286	0.24	0.868116	1.0
0.785714	0.18	0.872136	0.997102
0.857143	0.5	0.869565	1.0
0.928571	0.57	0.868116	1.0
1.000000	0.84	0.859219	0.998551

Table 4. CV-5; Australian credit; Algorithm 2_v2. r_gran=granulation radius, optimal_r_catch=optimal catch radius, $\varepsilon_{opt}=$ optimal epsilon, acc= accuracy, cov= coverage. Best result for $r_{gran} = 0.7142$, optimal $r_{catch} = 0.7142$, $\varepsilon_{opt} = 0.08$, accuracy=0.8747.

r_gran	$optimal\_r\_catch$	$optimal\ eps$	acc	cov
0	0.357143	0.11	0.556755	0.997102
0.0714286	0.357143	0.11	0.556755	0.997102
0.142857	0.428571	0.15	0.556755	0.997102
0.214286	0.428571	0.11	0.570352	0.988406
0.285714	0.928571	0.91	0.739727	0.857971
0.357143	0.928571	0.92	0.790782	0.975362
0.428571	0.928571	0.87	0.797704	0.995652
0.5	0.785714	0.29	0.840527	0.989855
0.571429	0.642857	0.07	0.844695	0.998551
0.642857	0.642857	0.05	0.866476	0.998551
0.714286	0.714286	0.08	0.874757	0.994203
0.785714	0.785714	0.19	0.869417	0.998551
0.857143	0.785714	0.58	0.872464	1
0.928571	0.642857	0.05	0.872316	0.998551
1	0.785714	0.18	0.872136	0.997102

In Table 4, we give results of tests with granules of granular objects. In this case, classifiers depend on three parameters: granulation radii r_{gran}, catch radii r_{catch}, and ε's. Results are shown for granulation radii in terms of optimal values of r_{catch} and ε.

Table 5. CV-5; Australian credit; Algorithm 3_v2. r_catch=catch radius,ε_{opt}= optimal epsilon, acc= accuracy, cov= coverage. Best result for $r_{catch} = 0.1428$: $\varepsilon_{opt} = 0.35$, accuracy= 0.8681.

r_catch	optimal eps	acc	cov
nil	nil	0.845	1.0
0	0	0.555073	1.0
0.071428	0	0.83913	1.0
0.142857	0.35	0.868116	1.0
0.214286	0.5	0.863768	1.0
0.285714	0.52	0.831884	1.0
0.357143	0.93	0.801449	1.0
0.428571	1.0	0.514493	1.0
0.500000	1.0	0.465217	1.0
0.571429	1.0	0.115942	1.0

In Table 5, results are shown of tests with granules of rules from the training set for all granulation radii in terms of optimal value of ε.

6   Conclusions

Results of tests show that the tested method of building classifiers is very promising. Results of tests for optimal values of parameters exceed quality of classification by means of the standard exhaustive classifier in all cases and some of them, e.g., accuracy of 0. 868 in Table 5 or 0.8747 in Table 4 are among the best results obtained by rough set techniques, see [B].

Acknowledgement

The author wishes to thank Professor Lech Polkowski for his support and guidance.

References

[B]   Bazan, J.G.: A comparison of dynamic and non–dynamic rough set methods for extracting laws from decision tables. In: Polkowski, L., Skowron, A. (eds.) Rough Sets in Knowledge Discovery, vol. 1, pp. 321–365. Physica Verlag, Heidelberg (1998)

[H]   Nguyen, S.H.: Regularity analysis and its applications in Data Mining. In: Polkowski, L., Tsumoto, S., Lin, T.Y. (eds.) Rough Set Methods and Applications, pp. 289–378. Physica Verlag, Heidelberg (2000)

[Po1]   Polkowski, L.: Rough mereology in analysis of vagueness. In: RSKT 2008. LNCS (LNAI), vol. 5009, Springer, Heidelberg (in print, 2008)

[Po2]   Polkowski, L.: On the idea of using granular rough mereological structures in classification of data. In: RSKT 2008. LNCS (LNAI), vol. 5009, Springer, Heidelberg (in print, 2008)

[Po3]   Polkowski, L.: Formal granular calculi based on rough inclusions (a feature
        talk). In: Zhang, Y.-Q., Lin, T.Y. (eds.) IEEE GrC 2006, pp. 9–18. IEEE
        Press, Piscataway (2006)
[Po4]   Polkowski, L.: Formal granular calculi based on rough inclusions (a feature
        talk). In: Hu, X., Liu, Q., Skowron, A., Lin, T.Y., Yager, R.R., Zhang, B.
        (eds.) IEEE GrC 2005, pp. 57–62. IEEE Press, Piscataway (2005)
[RS]   Skowron, A., et al.: RSES, http://logic.mimuw.edu.pl/~rses/
[UCI]   UCI Repository, http://www.ics.uci.edu/~mlearn/databases/
[W]   Wróblewski, J.: Adaptive aspects of combining approximation spaces. In: Pal,
        S.K., Polkowski, L., Skowron, A. (eds.) Rough Neural Computing, pp. 139–156.
        Springer, Berlin (2004)

Rough Mereological Classifiers Obtained from Weak Variants of Rough Inclusions

Piotr Artiemjew

University of Warmia and Mazury
Zolnierska 14, 10-560 Olsztyn, Poland
artem@matman.uwm.edu.pl

Abstract. Granular reflections of data sets have turned out to be very effective in data classification. In this work we present results of classification of real data sets by means of an approach in which granules of objects or decision rules are built on the basis of weak variants of rough inclusions.

Keywords: rough sets, granulation of knowledge, rough inclusions, classification of data.

1 Introduction

An information system is a pair $I = (U, A)$ where U is a set of objects, and A is a set of attributes; a decision system is a triple (U, A, d) where $d \notin A$ is a decision. Objects in U are represented by means of information sets: $Inf(u) = \{(a, a(u)) : a \in A\}$ is the information set of the object u. Decision rules are expressions of the form $\bigwedge_{a \in A}(a, a(u)) \Rightarrow (d = d(u))$. The basic form of granulation in information/decision systems is partitioning of U into classes of the indiscernibility relation $IND(A) = \{(u, v) : a(u) = a(v) \text{ for each } a \in A\}$. Each class $[u]_A = \{v \in U : IND(v, u)\}$ is interpreted as an elementary granule and unions of elementary granules are granules of knowledge. Thus, granulation in this case means forming aggregates of objects indiscernible over sets of attributes.

Another approach to granulation, proposed by L.Polkowski, see [Po1], [Po3], [Po4], consists in using rough inclusions.

A rough inclusion is a relation $\mu \subseteq U \times U \times [0, 1]$ which can be regarded as a graded similarity relation extending the indiscernibility relation by relaxing restrictions on attribute values.

In this work we are using rough inclusions proposed by L. Polkowski, see [Po1], obtained from continuous t–norms by means of their residual implications.

For a continuous t–norm t, see, e.g., [Po5], the residual implication $x \Rightarrow_t y$ is a mapping from the square $[0, 1]^2$ into $[0, 1]$ defined as follows, see, e.g., [Po5],

$$x \Rightarrow_t y \geq z \text{ iff } t(x, z) \leq y; \tag{1}$$

thus, $x \Rightarrow_t y = max\{z : t(x, z) \leq y\}$.

G. Wang et al. (Eds.): RSKT 2008, LNAI 5009, pp. 229–236, 2008.

As defined in [Po1], $\Rightarrow_t$ does induce a rough inclusion on the interval $[0, 1]$:

$$\mu^{\Rightarrow t}(u, v, r) \text{ iff } x \Rightarrow_t y \geq r. \tag{2}$$

This rough inclusion can be transferred to the universe U of an information system as shown in [Po1]: for given objects u, v, and $\varepsilon \in [0, 1]$, factors: $dis_\varepsilon(u, v) = \frac{|\{a \in A: |a(u)-a(v)| \geq \varepsilon\}|}{|A|}$, and $ind_\varepsilon(u, v) = \frac{|\{a \in A: |a(u)-a(v)| < \varepsilon\}|}{|A|}$ are introduced.

The weak variant of rough inclusion $\mu^{\rightarrow t}$ is defined, see [Po1], as,

$$\mu_t * (u, v, r) \text{ iff } dis_\varepsilon(u, v) \Rightarrow_t ind_\varepsilon(u, v) \geq r. \tag{3}$$

Particular cases of this similarity measure induced by, respectively, t–norm min, t–norm $P(x, y)$, and t–norm L are, see [Po1],

For $t = min(x, y)$, $x \Rightarrow_{min} y$ is y in case $x > y$ and 1 otherwise hence $\mu_{min} * (u, v, r)$ iff $dis_\varepsilon(u, v) > ind_\varepsilon(u, v) \geq r$ with $r < 1$ and 1 otherwise.

For $t = P$ where $P(x, y) = x \cdot y$, $x \Rightarrow_P y = \frac{y}{x}$ when $x \neq 0$ and 1 when $x = 0$ hence $\mu_P * (u, v, r)$ iff $\frac{ind_\varepsilon(u,v)}{dis_\varepsilon(u,v)} \geq r$ with $r < 1$ and 1 otherwise.

For $t = L$, $x \Rightarrow_L y = min\{1, 1 - x + y\}$, hence $\mu_L * (u, v, r)$ iff $1 - dis_\varepsilon(u, v) + ind_\varepsilon(u, v) \geq r$ with $r < 1$ and 1 otherwise.

These similarity measures will be applied in building granules and then in data classification.

2   Granular Reflections of Data Sets

As introduced in [Po3], [Po4], granules defined by means of rough inclusions can be made into new data sets called granular reflections of the original data sets, see [Po1].

Granules of knowledge are defined by means of a rough inclusion μ as follows. For an object u and a granulation radius $r \in [0, 1]$, the granule $g(u, r, \mu)$ of the radius r about u relative to μ is defined as the set of all objects v such that $\mu(v, u, r)$ holds. This concerns as well weak variants of rough inclusions for which we keep the same defining formula,

$$g(u, r, \mu) = \{v : \mu(v, u, r)\}. \tag{4}$$

In particular, for the weak rough inclusion μ_t* (3), the granule $g(u, r, \mu)$ consists of objects v such that $dis_\varepsilon(u, v) \Rightarrow_t ind_\varepsilon(u, v) \geq r$.

The collection $G(r, \mu)$ of all granules of the given radius r relative to μ can be filtered through a strategy $\mathcal{G}$ in order to select a covering $C(r, \mu)$ of the set U of objects; in our approach, $\mathcal{G}$ is a random choice of granules with reduction of reduntant granules in each step. For each attribute a in the attribute set $A \cup \{d\}$, and each granule $g \in C(r, \mu)$, the factored attribute $\bar{a}$ takes on g the value,

$$\bar{a}(g) = \mathcal{S}(\{a(u) : u \in g\}) \tag{5}$$

where $\mathcal{S}$ is a strategy of choosing the value. In our work, $\mathcal{S}$ is always the majority voting with random tie resolution.

The granular reflection of the radius r defined up to a choice of $C(r, \mu)$, $\mathcal{G}$ and S is the data set $(C(r, \mu), \{\overline{a} : a \in A\}, \overline{d})$. This granular reflection was proposed in [Po4], see [Po1] to serve as a basis for classifier construction.

The validity of this proposal was confirmed by tests with real data reported, e.g., [Po2], [PoA] for granules computed with the rough inclusion based on the Lukasiewicz t–norm, see [Po1].

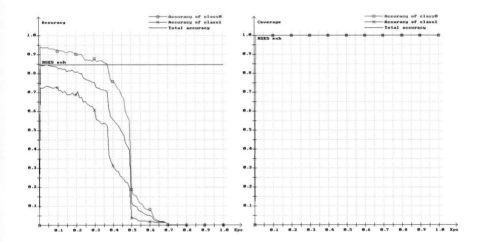

Fig. 1. Results for algorithm 5_v1, Best result for $\varepsilon = 0.04$: accuracy $= 0.847826$, coverage $= 1$

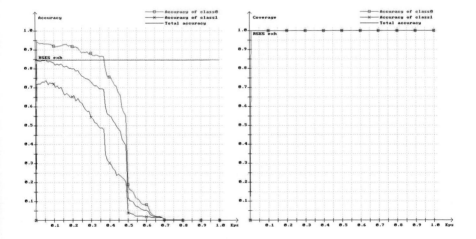

Fig. 2. Results for algorithm 6_v1, Best result for $\varepsilon = 0.06$: accuracy $= 0.847826$, coverage $= 1$

In this work, we construct and test classifiers built on the basis of granules computed by means of the weak rough inclusion μ_t*, given explicitly above.

3   Results of Experiments

Tests were carried out with Australian credit data set [UCI] and the method was CV-5 (the 5–fold cross validation). Results of classification have been judged

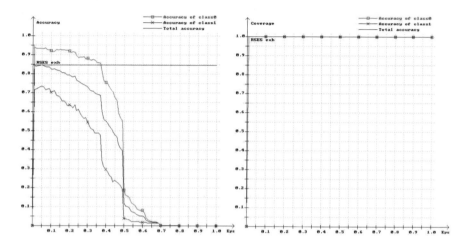

Fig. 3. Results for algorithm 7_v1, Best result for $\varepsilon = 0.05$: accuracy $= 0.846377$, coverage $= 1$

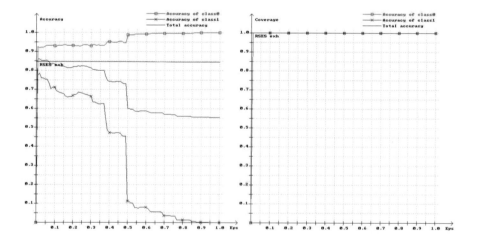

Fig. 4. Results for algorithm 5_v2, Best result for $\varepsilon = 0.02$: accuracy $= 0.86087$, coverage $= 1$

by accuracy and coverage factors, see [RSES]. The accuracy computed with the standard RSES exhaustive classifier [RSES] is for these data 0.845, and coverage is 1.0.

We have four cases for testing with: 1. granules of objects in the training set, 2. granules of rules from the training set, 3. granules of granular objects, for each of the three rough inclusions t=min, P, L.

In Case 1, training objects are made into granules for a given ε. Objects in each granule g about a test object u, vote for decision value at u as follows: for each decision class c, the value $p(c) = \dfrac{\Sigma \text{training object v in g falling in c}^{w(v,t)}}{\text{size of c in training set}}$ is

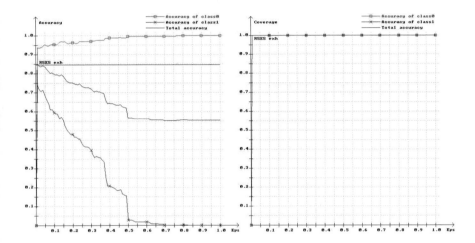

Fig. 5. Results for algorithm 6_v2, Best result for $\varepsilon = 0.01$: accuracy $= 0.850725$, coverage $= 1$

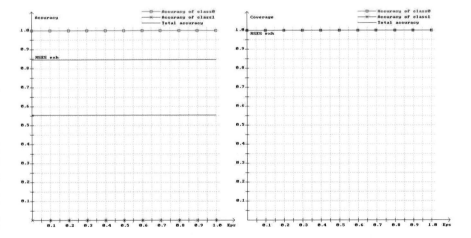

Fig. 6. Results for algorithm 7_v2, Best result for $\varepsilon = 0$, accuracy $=0.555073$, coverage$=1$

Table 1. CV-5; Australian credit; Algorithm 5_v3. r_gran=granulation radius, optimal_eps= optimal epsilon, acc= accuracy, cov=coverage, m_trn=mean training set

r_gran	optimal eps	acc	cov	m_trn
nil	nil	0.845	1.0	552
0.500000	0.03	0.834783	1.0	53.8
0.571429	0.02	0.791304	1.0	134.4
0.642857	0.01	0.798551	1.0	295.8
0.714286	0.02	0.83913	1.0	454.8
0.785714	0.05	0.855072	1.0	533.8
0.857143	0.05	0.847826	1.0	546.2
0.928571	0.04	0.847826	1.0	548
1.000000	0.04	0.847826	1.0	552

Table 2. CV-5; Australian credit; Algorithm 6_v3. r_gran=granulation radius, optimal_eps= optimal epsilon, acc= accuracy, cov= coverage, m_trn=mean training set

r_gran	optimal eps	acc	cov	m_trn
nil	nil	0.845	1.0	552
0	0.01	0.555073	1.0	1
0.500000	0.01	0.808696	1.0	54.8
0.571429	0.01	0.746377	1.0	131.8
0.642857	0.01	0.763768	1.0	295.2
0.714286	0.01	0.818841	1.0	454.4
0.785714	0.01	0.852174	1.0	533.2
0.857143	0.01	0.847826	1.0	546.2
0.928571	0.01	0.846377	1.0	548
1.000000	0.06	0.847826	1.0	552

Table 3. CV-5; Australian credit; Algorithm 7_v3. r_gran=granulation radius, optimal_eps= optimal epsilon, acc=Total accuracy, cov=Total coverage, m_trn=mean training set

r_gran	optimal eps	acc	cov	m_trn
nil	nil	0.845	1.0	552
0.500000	0.01	0.707247	1.0	53.2
0.571429	0.01	0.595652	1.0	132
0.642857	0.01	0.563768	1.0	292.2
0.714286	0.02	0.786956	1.0	457.6
0.785714	0.01	0.85942	1.0	533
0.857143	0.05	0.847826	1.0	546.2
0.928571	0.05	0.849275	1.0	548
1.000000	0.05	0.846377	1.0	552

computed where the weight $w(v,t)$ is computed for a given t–norm t as $w(v,t) = dis_\varepsilon(u,v) \rightarrow_t ind_\varepsilon(u,v)$. The class c* assigned to u is the one with the largest value of p. Results for the three chosen t–norms are given in Fig.1 (t=min), Fig.2 (t=P), Fig.3 (t=L).

In Case 2, weighted voting of rules in a given granule g for decision at test object u goes according to the formula d(u)= arg max p(c) where

$$p(c) = \frac{\sum \text{rule in g pointing to c } w(r,t) \cdot support(r)}{\text{size of c in training set}},$$

where weight $w(r,t)$ is computed as $dis_\varepsilon(u,r) \rightarrow_t ind_\varepsilon(u,r)$.
Results are shown in Fig. 4 (t=min), Fig. 5 (t=P), Fig.6 (t=L).

In Case 3, granular objects from granules vote for decision. The difference is in the fact that now we have two–parameter case with ε, r hence results are given in Table 1(t=min), Table 2 (t=P), Table 3 (t=L) in which for each row corresponding to the radius of granulation the best ε is given along with accuracy and coverage in that case.

4   Conclusions

Optimal results obtained with granules of training objects relative to the rules induced from the original training set for t=*min* and t=P (Figs. 4,5) are fully comparable with best results by rough set techniques see [B].

Acknowledgement

The author thanks Professor Lech Polkowski for his support and guidance.

References

[B]   Bazan, J.G.: A comparison of dynamic and non–dynamic rough set methods for extracting laws from decision tables. In: Polkowski, L., Skowron, A. (eds.) Rough Sets in Knowledge Discovery 1, pp. 321–365. Physica, Heidelberg (1998)

[Po1]   Polkowski, L.: On the idea of using granular rough mereological structures in classification of data. In: RSKT 2008. LNCS, vol. 5009, Springer, Heidelberg (in print 2008)

[Po2]   Polkowski, L.: The paradigm of granular rough computing. In: Zhang, D., Wang, Y., Kinsner, W. (eds.) ICCI 2007, pp. 145–163. IEEE Computer Society, Los Alamitos (2007)

[Po3]   Polkowski, L.: Formal granular calculi based on rough inclusions (a feature talk). In: Zhang, Y.-Q., Lin, T.Y. (eds.) IEEE GrC 2006, pp. 9–18. IEEE Press, Piscataway (2006)

[Po4]   Polkowski, L.: Formal granular calculi based on rough inclusions (a feature talk). In: Hu, X., Liu, Q., Skowron, A., Lin, T.Y., Yager, R.R., Zhang, B. (eds.) IEEE GrC 2005, pp. 57–62. IEEE Press, Piscataway (2005)

[Po5]   Polkowski, L.: Rough Sets. Mathematical Foundations. Physica, Heidelberg (2002)

[PoA]   Polkowski, L., Artiemjew, P.: On granular rough computing: Factoring clasi-
fiers through granular structures. In: Kryszkiewicz, M., Peters, J.F., Rybinski,
H., Skowron, A. (eds.) RSEISP 2007. LNCS (LNAI), vol. 4585, pp. 280–290.
Springer, Heidelberg (2007)
[RSES] RSES, http://logic.mimuw.edu.pl/rses
[UCI]   UCI Repository, http://www.ics.uci.edu/~mlearn/databases/

Color Image Interpolation Combined with Rough Sets Theory

Fengmei Liang[1] and Keming Xie[2]

[1] College of Information Engineering, Taiyuan University of Technology
Taiyuan, 030024, P.R. China
fm_liang@163.com
[2] College of Information Engineering, Taiyuan University of Technology
Taiyuan, 030024, P.R. China
kmxie@tyut.edu.cn

Abstract. A new realtime interpolation algorithm for color image is presented. The algorithm is based on the concept of indiscernibility relation in rough sets (RS) theory. By applying the concept of upper and lower approximation based on the continuity of images, the image is first divided into homogenous area, edge pixels and isolated pixels. Then *Bézier* surface interpolation is further achieved using the information of classification. Besides emulation, the technology has been applied to the visual presenter with low-resolution image sensor. Results demonstrate that the new algorithm improves substantially the subjective and objective quality of the interpolated images over conventional interpolation algorithms, and meets the requirements of real time image processing. The algorithm represents an attempt to incorporate RS in image processing.

Keywords: Rough sets, image interpolation, indiscernibility, upper approximation, lower approximation, *Bézier* surface.

1 Introduction

In many electronic imaging applications such as infrared imaging system, CCD and CMOS, the image resolution is limited to the array densities of the sensors. Moreover, the pixel difference of the optics, atmosphere and system noise will blur and warp the images. Therefore, it is important and economical to improve the image resolution by image interpolation technique. At present, common image interpolation methods include bilinear interpolation, B shape interpolation, SINC function, fractal interpolation, and so on. These interpolation algorithms merely consider relativity between image neighbor pixels and do not consider information degeneration after interpolation. Recently, many new image interpolation algorithms are put forward [1,2,3,4,5], however these algorithms are implemented in emulation mode. With greater learning capability, artificial neural network has been applied to image interpolation since 1995 [6,7,8,9]. Even so, tremendous nerve cell [6], huge calculation [8], or classification is desired, and

G. Wang et al. (Eds.): RSKT 2008, LNAI 5009, pp. 237–243, 2008.

these approaches have never better extensive effect on the lesser neural network scale [7].

RS theory, proposed by Pawlak in 1982 can be seen as a new mathematical approach to intelligent information processing. The RS approach seems to be of fundamental importance in artificial intelligence and cognitive sciences, especially in research areas such as machine learning, intelligent systems, inductive reasoning, pattern recognition, metrology, image processing, signal analysis, knowledge discovery, decision analysis, and expert systems. RS have been applied to image processing and comprehension by some researchers [10,11,12], which include image filtering, classification, segmentation, and so on. A new image interpolation algorithm is presented in this paper. The algorithm combined *Bézier* surface interpolation with RS theory. Based on the continuity of images, a RS criterion was put forward to distinguish noise pixels and edge pixels by applying the concept of upper and lower approximation. With the information of classification, then *Bézier* surface interpolation is further implemented. In order to confirm the above scheme's validity and feasibility, we apply the technology to the visual presenter with low-resolution image sensor. Results demonstrate that the new interpolation algorithm removes effectively noises and preserves the edge details during substantial resolution improvement.

2   RS Theory [13,14]

Suppose we are given a pair $\mathscr{A} = (U, A)$ of non-empty, finite sets U and A, where U is the universe of objects, and A – a set consisting of attributes, i.e. functions $a : U \longrightarrow Va$, where Va is the set of values of attribute a, called the domain of a. The pair $\mathscr{A} = (U, A)$ is called an information system. Any information system can be represented by a data table with rows labeled by objects and columns labeled by attributes. Any pair (x, a), where $x \in U$ and $a \in A$ defines the table entry consisting of the value $a(x)$.

Any subset B of A determines a binary relation $I(B)$ on U, called an indiscernibility relation, defined by $xI(B)y$ if and only if $a(x) = a(y)$ for every $a \in B$. where $a(x)$ denotes the value of attribute a for object x. If $(x, y) \in I(B)$ we will say that x and y are B-indiscernible. Equivalence classes of the relation I(B) are referred to as B-elementary sets or B-elementary granules. In the rough set approach the elementary sets are the basic building blocks (concepts) of our knowledge about reality. The unions of B-elementary sets are called B-definable sets.

For $B \subseteq A$ we denote by $Inf_B(x)$ the B-signature of $x \in U$, i.e., the set $(a, a(s)) : a \in A$. Let $Inf(B) = Inf_B(s) : s \in U$. Then for any objects $x, y \in U$, $xI(B)y$ if and only if $Inf_B(x) = Inf_B(y)$. The indiscernibility relation will be further used to define basic concepts of rough set theory. Let us define now the following two operations on sets $X \subseteq U$.

$$\begin{cases} B_* (X) = \{x \in U : B(x) \subseteq X\}. \\ B^* (X) = \{x \in U : B(x) \cap X \neq \Phi\}. \end{cases} \tag{1}$$

assigning to every subset X of the universe U two sets $B_*(X)$ and $B^*(X)$ called the $B - lower$ and the $B - upper\ approximation$ of X, respectively. The set

$BN_B(X) = B^*(X) - B_*(X)$ will be referred to as the $B - boundary region$ of X.

3   Image Interpolation Combined with RS Theory

3.1   *Bézier* Surface Interpolation

To consider the powerful advantages of *Bézier* surface method, which is very precise and can be implemented fast, the *Bézier* surface interpolation is selected.

Supposing a digital image as (i, j) , and its color value in *ith* row and *jth* column pixel is $Y_{i,j}$ $(i = 1, ..., m, j = 1, ..., n)$.

Firstly, $Y_1 = \{Y_{i,j} : (i = 1, ..., m, j = 1, ..., n)\}$ is extended to $Y_2 = \{Y_{i,j} : (i = 1, ..., m + 1, j = 1, ..., n + 1)\}$ with equation

$$\begin{cases} Y_{m+1,j} = 2Y_{m,j} - Y_{m-1,j} & (j = 1, ..., n) \\ Y_{i,n+1} = 2Y_{i,n} - Y_{i,n-1} & (i = 1, ..., m) \\ Y_{m+1,n+1} = 2Y_{m+1,n} + Y_{m,n+1} - Y_{m,n} \end{cases} \tag{2}$$

With similar method, Y_2 is extended to $Y_3 = \{Y_{i,j} : (i = 1, ..., m + 2, j = 1, ..., n + 2)\}$, so the *Bézier* surface is constructed as :

$$S_{i,j}(x, y) = \sum_{s=0}^{3} \sum_{t=0}^{3} Q_{s,t}^{i,j} B_s^3(x) B_t^3(y). \tag{3}$$

where

$$Q_{3\alpha,3\beta}^{i,j} = Y_{i+\alpha,j+\beta} \qquad (\alpha, \beta = 0, 1) \ \ (1 \le i \le m, 1 \le j \le n).$$

$$\begin{cases} Q_{1,3\beta}^{i,j} = Y_{i,j+\beta} + (Y_{i+1,j+\beta} - Y_{i-1,j+\beta})/6 \\ Q_{2,3\beta}^{i,j} = Y_{i+1,j+\beta} + (Y_{i,j+\beta} - Y_{i+2,j+\beta})/6 \\ Q_{3\alpha,1}^{i,j} = Y_{i+\alpha,j} + (Y_{i+\alpha,j+1} - Y_{i+\alpha,j-1})/6 \\ Q_{3\alpha,2}^{i,j} = Y_{i+\alpha,j+1} + (Y_{i+\alpha,j} - Y_{i+\alpha,j+2})/6 \end{cases} \qquad (\alpha, \beta = 0, 1). \tag{4}$$

$$\begin{cases} Q_{1,\beta}^{i,j} = Q_{0,\beta}^{i,j} + (Q_{3,\beta}^{i,j} - Q_{0,\beta}^{i,j-1})/6 \\ Q_{2,\beta}^{i,j} = Q_{3,\beta}^{i,j} + (Q_{0,\beta}^{i,j} - Q_{3,\beta}^{i,j+1})/6 \end{cases} \qquad (\beta = 1, 2).$$

Thus $Q_{\alpha,\beta}^{i,j}$ is the linear combination of sixteen pixels as $\{Y_{i+x,j+y}, -1 \le x, y \le 2\}$, and the whole *Bézier* surface is denoted as in (5), where[.]is integral form operation.

$$F(u, v) + S_{[u],[v]}(u - [u], v - [v]). \tag{5}$$

Compared to other conventional algorithms, the method has been proved to get the higher revolution image efficiently. Even now, there are some edge burring and noise remaining. So we propose the following scheme.

3.2 RS Criterion for Noise and Edge

An image can be supposed as the approximate universe of objects with $\mathscr{K} = (U, (B_1, B_2))$, where U is an image and (B_1, B_2) is the equivalence relation.

Here B_1 is defined as: if every one of two pixels is within selected noise parameter range, then two pixels B_1 correlation, and noise pixels can be got by $B_1 - lower\ approximation\ B_{1*}(X)$. And B_2 is defined as: if every one of two pixels is within selected edge parameter range, then two pixels B_2 correlation, and edge pixels can be got by $B_2 - lower\ approximation\ B_{2*}(X)$.

For a color image, every pixel is denoted by three primary colors, namely R, G and B. Supposing two pixels as M and N, certain color is denoted with x_M and x_N respectively, and the ratio of colors is denoted as $r(M : N) = x_M/x_N$. ($x_N = 1$ is supposed when $x_N = 0$, and the assumption couldn't influence image quality.). After practical test, eyes are unconscious of color variety while $0.9 \leq r(M : N) \leq 1.1$, and eyes are conscious of color variety while $r(M : N) < 0.9$ or $r(M : N) > 1.1$. Thereby, the ratio of colors can be served as RS criterion to judge isolated noise pixels and edge pixels.

Image can be divided into homogenous area, edge pixels and isolated pixels. Except for isolated pixels, there is great correlation between neighbor pixels in the natural image. If the color value is quite different from the neighbor region, the pixel is likely to be noise; If the color value is close to the neighbor region, the pixel is likely within the homogenous area; If the color value is close to two neighbor pixels, but is quite different from other neighbor pixels, the pixel is likely on the edge.

Supposing a digital image as matrix $[i, j]$, $(i, j$ expresses the pixel position), Then sub image with 3×3 rotation window is denoted as:

$$M(x_{i,j}) = \begin{bmatrix} x_{i-1,j-1} & x_{i-1,j} & x_{i-1,j+1} \\ x_{i,j-1} & x_{i,j} & x_{i,j+1} \\ x_{i+1,j-1} & x_{i+1,j} & x_{i+1,j+1} \end{bmatrix}. \tag{6}$$

By color ratio, the matrix is transformed into the following:

$$R(M(x_{i,j})) = \begin{bmatrix} r(x_{i-1,j-1} : x_{i,j}) & r(x_{i-1,j} : x_{i,j}) & r(x_{i-1,j+1} : x_{i,j}) \\ r(x_{i,j-1} : x_{i,j}) & r(x_{i,j} : x_{i,j}) & r(x_{i,j+1} : x_{i,j}) \\ r(x_{i+1,j-1} : x_{i,j}) & r(x_{i+1,j} : x_{i,j}) & r(x_{i+1,j+1} : x_{i,j}) \end{bmatrix}. \tag{7}$$

Then function $w(x_{m,n})$ and $C(x_{i,j})$ $(i - 1 \leq m \leq i + 1, j - 1 \leq n \leq j + 1)$ is denoted as:

$$w(x_{m,n}) = \begin{cases} 1, & r(x_{m,n} : x_{i,j}) \geqslant 1.1 \\ 1, & r(x_{m,n} : x_{i,j}) \leqslant 0.9 \\ 0, & others \end{cases} \tag{8}$$

$$C(x_{i,j}) = \sum_{m=i-1}^{i+1} \sum_{n=j-1}^{j+1} w(x_{m,n}). \tag{9}$$

Define equivalence relation $B_{1*}(X)$ and $B_{2*}(X)$: within 3×3 rotation window, ratios of center pixel to eight neighbor pixels are all greater than 1.1 or all lesser than 0.9, the pixel is noise. When ratios of center pixel to six neighbor pixels are greater than 1.1 or lesser than 0.9, the pixel is edge. Namely:

$$\begin{cases} B_{1*}(X) = \{x|C(x_{i,j}) = 8\}. \\ B_{2*}(X) = \{x|C(x_{i,j}) = 6\}. \end{cases} \tag{10}$$

3.3 RS Application on Image Interpolation

Before interpolation, image pixels are classified by RS criterion.

- while $x_{i,j}$ locates within the homogenous area, original interpolation is applied;
- while $x_{i,j}$ is noise, its color value is replaced with average of neighbor pixels;
- while $x_{i,j}$ is on the edge, its color value is replaced with average of two nearest pixels and itself.

3.4 Experiments

The standard image testing card is applied in experiments, and the visual results is shown as Fig. 1. Results showed that two conventional interpolations blur

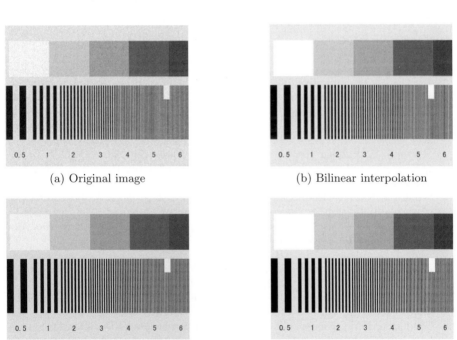

(a) Original image   (b) Bilinear interpolation

(c) *Bézier* surface interpolation   (d) Interpolation combined with RS

Fig. 1. Image interpolation experiment results

image edge and remain noise. The new interpolation algorithm enhanced details, meanwhile improved definition and vision effect by a long way.

From the following table, PSNR is increased compared with bilinear interpolation and *Bézier* surface interpolation.

Table 1. Interpolation Algorithm Comparison

Algorithm	Bilinear Interpolation	*Bézier* surface Interpolation	Interpolation with RS
PSNR(dB).	32.07	33.10	42.33

4 Conclusion

Due to classification by RS criterion, the new interpolation algorithm removes effectively noises and preserves edge details during substantial resolution improvement. The advantage of the algorithm is quite apparent.

Except for emulation, the above approach has been applied on our visual presenter with low-resolution image sensor, whose hardware platform core is DSC21CPU. The implement is excellent from both quality and computational complexity aspects obviously.

References

1. Darwish, A.M.: Adaptive Resampling Algorithm for Image Zooming. In: Proc. SPIE: Image and Video Processing, San Jose, CA, USA, vol. 2666, pp. 131–144 (1996)
2. Jensen, K., Anastassiou, D.: Subpixel Edge Localization and The Interpolation of Still Image. IEEE Transactions on Image Processing 4(3), 285–295 (1995)
3. Martucci, S.A.: Image Resizing in the Discrete Cosine Transform Domain. In: International Conference on Image Processing, Washington, DC, USA, vol. 2, pp. 244–247 (1995)
4. Schultz, R.R., Stevenson, R.L.: A Bayesian Approach to Image Expansion for Improved Definition. IEEE Transactions on Image Processing 3(3), 233–242 (1994)
5. Li, X., Michael, T.O.: New edge-directed interpolation. IEEE Transactions on Image Processing 10(10), 1521–1527 (2001)
6. Ahmed, F., Gustafsou, S.C.: High Fidelity Image Interpolation Using Radial Basis Function Neural Networks. In: Proc. IEEE National Aerospace and Electronics Conf., Dayton, OH, USA, vol. 2, pp. 588–592 (1995)
7. Plaziac, N.: Image Interpolation Using Neural Networks. IEEE Transactions on Image Processing 8(11), 1647–1651 (1999)
8. Candoncia, F.M., Principe, J.C.: Super Resolution of Images Based on Local Correlations. IEEE Transaction on Neural Networks 10(2), 372–380 (1999)
9. Behnke, S.: Hierarchical Neural Networks for Image Interpretation. LNCS, vol. 2766. Springer, Heidelberg (2003)
10. Wojcik, Z.M.: Rough Sets for Intelligent Image Filtering. In: Proceedings of the International Workshop on Rough Sets and Knowledge Discovery, Banff, Canada, pp. 399–410 (1993)

11. Shek, S., Lau, Y.: Image Segmentation Based on the Indiscernibility Relation. In: Proceedings of the International Workshop on Rough Sets and Knowledge Discovery, Banff, Canada, pp. 439–451 (1993)
12. Nguye, J.: Classication Based on Optimal Feature Extraction and the Theory of Rough Sets. In: SDSU, San Diego, CA, USA, pp. 439–451 (1995)
13. Pawlak, Z.: Rudiments of rough sets. International Jounal of Computing and Information Sciences 177, 3–27 (2007)
14. Pawlak, Z.: Rough sets: Some extensions. International Jounal of Computing and Information Sciences 177, 28–40 (2007)

Dominance-Based Rough Sets Using Indexed Blocks as Granules

Chien-Chung Chan[1,2] and Gwo-Hshiung Tzeng[3,4]

[1] Department of Computer Science, University of Akron, Akron, OH, 44325-4003,
USA
chan@uakron.edu
[2] Department of Information Communications, Kainan University, No. 1 Kainan
Road, Luchu, Taoyuan County 338, Taiwan
[3] Department of Business and Entrepreneurial Administration, Kainan University,
No. 1 Kainan Road, Luchu, Taoyuan County 338, Taiwan
ghtzeng@mail.knu.edu.tw
[4] Institute of Management of Technology, National Chiao Tung University, 1001
Ta-Hsueh Road, Hsinchu 300, Taiwan
ghtzeng@cc.nctu.edu.tw

Abstract. Dominance-based rough set introduced by Greco et al. is
an extension of Pawlak's classical rough set theory by using dominance
relations in place of equivalence relations for approximating sets of preference ordered decision classes satisfying upward and downward union
properties. This paper introduces a formulation of approximation spaces
based on multiple criteria decision tables by using the concept of indexed
blocks, which are sets of objects indexed by pairs of decision values. The
approximations of sets of decision classes are formulated in terms of exclusive neighborhoods of indexed blocks. We show that the set of indexed
blocks with exclusive neighborhoods forms a partition on the universe of
objects when approximating preference ordered decision classes with upward and downward union properties. Examples are given to illustrate
presented concepts.

Keywords: Rough sets, Dominance-based rough sets, Multiple criteria decision analysis (MCDA), Classification, sorting, Indexed blocks,
Granule.

1   Introduction

Dominance-based rough sets (DBRS) introduced by Greco, Matarazzo and
Slowinski [1, 2, 3] extend Pawlak's classical rough sets (CRS) [8, 9, 10] by considering attributes, called criteria, with preference-ordered domains and by substituting the indiscernibility relation in CRS with a dominance relation that is
reflexive and transitive. The DBRS approach was motivated by representing preference models for multiple criteria decision analysis (MCDA) problems, where
preference orderings on domains of attributes are quite typical in exemplary
based decision-making. It is also assumed that decision classes are ordered by

G. Wang et al. (Eds.): RSKT 2008, LNAI 5009, pp. 244–251, 2008.
© Springer-Verlag Berlin Heidelberg 2008

some preference ordering. More precisely, let $Cl = \{Cl_t | t \in T\}, T = \{1, 2, ..., n\}$, be a set of decision classes such that for each x in the universe U, x belongs to one and only one $Cl_t \in Cl$ and for all r, $s \in T$, if $r > s$, the decision from Cl_r are preferred to the decision from Cl_s. A consistent preference model is taken to be one that respects the dominance principle when assigning actions (objects) to the preference ordered decision classes. Action x is said to dominate action y if x is at least as good as y under all considered criteria. The dominance principle requires that if action x dominates action y, then x should be assigned to a class not worse than y. Given a total ordering on decision classes, in DBRS, the sets to be approximated are the upward union and downward union of decision classes [5]. The DBRS approach has been shown to be an effective tool for MCDA [12] and has been applied to solve multi-criteria sorting problems [4, 5]. Algorithms for inducing decision rules consistent with dominance principle were introduced in [6, 7].

This work is motivated by trying to study the relationship between the structures of approximation spaces based on dominance relations and the structures of preference ordered decision classes satisfying upward and downward union property. The basic idea is to consider relations on decision classes and the representation of objects related by pairs of decisions in a multi-criteria decision table. In addition, we are interested in computing the reduction of inconsistency when criteria are aggregated one by one incrementally. Here inconsistency is as a result of violating the dominance principle. In this study, we consider only decision tables with multiple criteria which are quantitative and totally ordered.

The remainder of this paper is organized as follows. In Section 2, after reviewing related concepts, the concept of indexed blocks is defined. In Section 3, we consider the combination of criteria and how to update indexed blocks. The concept of exclusive neighborhoods is introduced, and three rules for combining criteria are presented. An example is given to illustrate the concepts and rules. In Section 4, we formulate approximations of sets of decision classes in terms of indexed blocks and exclusive neighborhoods. We show that the set of indexed blocks forms a partition on universe when all neighborhoods are exclusive. Finally, conclusions are given in Section 5.

2   Related Concepts

2.1   Information Systems, Rough Sets, and Dominance Based Rough Sets

In rough sets theory [8, 9, 10], information of objects in a domain is represented by an information system $IS = (U, A, V, f)$, where U is a finite set of objects, A is a finite set of attributes, $V = \cup_{q \in A} V_q$ and V_q is the domain of attribute q, and $f : U \times A \rightarrow V$ is a total information function such that $f(x, q) \in V_q$ for every $q \in A$ and $x \in U$. In many applications, we use a special case of information systems called *decision tables* to represent data sets. In a decision table $(U, C \cup D = \{d\})$, there is a designated attribute $\{d\}$ called *decision attribute*, and attributes in C are called *condition attributes*. Each attribute q in $C \cup D$ is associated with an equivalence relation R_q on the set U of objects such that for each x and $y \in U$, xR_qy means $f(x,q) = f(y,q)$. For each x and $y \in U$, we say that x and y are *indiscernible* on attributes $P \subseteq C$ if and only if xR_qy for all $q \in P$.

In dominance-based rough sets, attributes with totally ordered domains are called *criteria*. More precisely, each criterion q in C is associated with an outranking relation [11] S_q on U such that for each x and $y \in U, xS_qy$ means $f(x,q) \geq f(y,q)$. For each x and $y \in U$, we say that x *dominates* y on criteria $P \subseteq C$ if and only if xS_qy for all $q \in P$. The dominance relations are taken to be total pre-ordered, i.e., strongly complete and transitive binary relations [5].

Dominance-based rough sets approach is capable of dealing with inconsistencies in MCDA problems based on the *principle of dominance*, namely: given two objects x and y, if x dominates y, then x should be assigned to a class not worse than y. Assignments of objects to decision classes are inconsistent if the dominance principle is violated. The sets of decision classes to be approximated are considered to have upward union and downward union properties. More precisely, let $Cl = \{Cl_t | t \in T\}, T = \{1, 2, ..., n\}$, be a set of decision classes such that for each $x \in U$, x belongs to one and only one $Cl_t \in Cl$ and for all r, s in T, if $r > s$, the decision from Cl_r is preferred to the decision from Cl_s. Based on this total ordering of decision classes, the upward union and downward union of decision classes are defined respectively as:

$$Cl_t^{\geq} = \cup_{s \geq t} Cl_s, \quad Cl_t^{\leq} = \cup_{s \leq t} Cl_s, \quad t = 1, 2, ..., n.$$

An object x is in $Cl_t^{\geq}$ means that x at least belongs to class Cl_t, and x is in $Cl_t^{\leq}$ means that x at most belongs to class Cl_t.

2.2   Indexed Blocks

To study the relationships between approximation spaces based on dominance relations and the sets of decision classes to be approximated, we introduce a new concept called *indexed blocks*, which are sets of objects indexed by pairs of decision values.

Let $(U, C \cup D = \{d\})$ be a multi-criteria decision table where condition attributes in C are criteria and decision attribute d is associated with a total preference ordering. For each condition criterion q and a decision value d_i of d, let $\min_q(d_i) = \min\{f(x, q) | f(x, d) = d_i\}$ and $\max_q(d_i) = \max\{f(x, q) | f(x, d) = d_i\}$. That is, $\min_q(d_i)$ denotes the minimum value of q among objects with decision value d_i in a multi-criteria table, and $\max_q(d_i)$ denotes the maximum value.

For each condition criterion q, the mapping $I_q(i, j) : D \times D \to \wp(V_q)$ is defined as

$$I_q(i,j) = \{f(x,q) = v | v \geqslant \min_q(d_j) \ and \ v \leqslant \max_q(d_i), \ for \ i < j; i, j = 1, ..., V_D\}$$

and $I_q(i, i) = \{f(x, q) | f(x, d) = i \ and \ f(x, q) \notin \cup_{i \neq j} I_q(i, j)\}$, where $\wp(V_q)$ denotes the power set of V_q.

Intuitively, $I_q(i, j)$ denotes the set of values of criterion q shared by objects of decision values i and j. We will denote the set of values as $[\min_q(j), \max_q(i)]$

or simply as $[\min_j, \max_i]$ for a decision value pair i and j with $i < j$. The set $I_q(i, i)$ denotes the values of criterion q where objects can be consistently labeled with decision value i. For $i < j$, values in $I_q(i, j)$ are conflicting or inconsistent in the sense that objects with higher values of criterion q are assigned to a lower decision class or vice versa, namely, the dominance principle is violated.

For each $I_q(i, j)$, the corresponding set of *ordered pairs* $[I_q(i, j)]$: $D \times D \rightarrow \wp(U \times U)$ is defined as $[I_q(i, j)] = \{(x, y) \in U \times U | f(x, d) = i, f(y, d) = j$ such that $f(x, q) \geq f(y, q)$ for $f(x, q), f(y, q) \in I_q \in (i, j)\}$.

For simplicity, we will take the set $[I_q(i, i)]$ to be reflexive.

For each $[I_q(i, j)]$, the restrictions of $[I_q(i, j)]$ to i and j are defined as:
$[I_q(i, j)]_i = \{x \in U |$ there exists $y \in U$ such that $(x, y) \in [I_q(i, j)]\}$ and
$[I_q(i, j)]_j = \{y \in U |$ there exists $x \in U$ such that $(x, y) \in [I_q(i, j)]\}$.

The corresponding indexed block $B_q(i, j) \subseteq U$ of $[I_q(i, j)]$ is defined as

$$B_q(i, j) = [I_q(i, j)]_i \cup [I_q(i, j)]_j.$$

For each criterion q, the union of its indexed blocks is a covering of U generally.

Example 1. The above concepts are illustrated using the following multi-criteria decision table, where U is the universe of objects, q_1 and q_2 are condition criteria and d is the decision with preference ordering $3 > 2 > 1$.

Table 1. Example of a multi-criteria decision table

U	q_1	q_2	d
1	1	2	2
2	1.5	1	1
3	2	2	1
4	1	1.5	1
5	2.5	3	2
6	3	2.5	3
7	2	2	3
8	3	3	3

In order to find out the minimum and maximum values for each decision class, we can apply sorting on the decision d first, followed by sorting on the criterion q_1. The result is shown in Table 2 where we can derive inconsistent intervals $I_{q_1}(i, j)$ with the following sets of ordered pairs:

$$[I_{q_1}(1, 1)] = [I_{q_1}(2, 2)] = \emptyset,$$

$$[I_{q_1}(1, 2)] = \{(4, 1), (2, 1), (3, 1)\},$$

$$[I_{q_1}(1, 3)] = \{(3, 7)\},$$

$$[I_{q_1}(2, 3)] = \{(5, 7)\},$$

$$[I_{q_1}(3, 3)] = \{(6, 6), (8, 8)\}.$$

The indexed blocks are shown in Table 3.

Table 2. Result after sorting in terms of $\{d\}$ followed by $\{q_1\}$

U	q_1	d
4	1	1
2	1.5	1
3	2	1
1	1	2
5	2.5	2
7	2	3
6	3	3
8	3	3

Table 3. Index blocks derived from $[I_{q_1}(i, j)]$

$D \times D$	1	2	3
1	$\emptyset$	$\{1, 2, 3, 4\}$	$\{3, 7\}$
2		$\emptyset$	$\{5, 7\}$
3			$\{6, 8\}$

3   Combination of Criteria

In this section, we consider the combination of two criteria using indexed blocks and the underlying sets of ordered pairs. For a criterion q and for each decision value $i \in V_d$, and for each indexed block $B_q(i, i)$ of $[I_q(i, i)]$, the neighborhood of $B_q(i, i)$ is defined as

$$NB(B_q(i, i)) = \{B_q(k, i) | k \geq 1 \text{ and } k < i\} \cup \{B_q(i, k) | k \geq 1 \text{ and } k > i\}.$$

Note that $B_q(i, i)$ is not part of its neighborhood, and the "exclusive" neighborhood of $B_q(i, i)$ corresponds to sets of objects which have inconsistent decision class assignments associated with decision i. Objects in $B_q(i, i)$ are assigned to decision i consistently. Blocks in the neighborhood of $B_q(i, i)$ are inconsistent blocks. Alternatively, we may take the neighborhood of $B_q(i, i)$ as a set of objects, i.e., $\cup NB(B_q(i, j))$. For x in U, we say x does not belong to the neighborhood of $B_q(i, i)$ iff $x \notin B$, for all B in $NB(B_q(i, i))$ iff $x \notin \cup NB(B_q(i, i))$.

When combining two criteria q_1 and q_2, the following three rules are used to update the sets of ordered pairs $[I_{\{q_1, q_2\}}(i, j)]$ and indexed blocks $B_{\{q_1, q_2\}}(i, j)$:

Rule 1: For decision pairs (i, i):

$$[I_{\{q_1, q_2\}}(i, i)] = [I_{q_1}(i, i)] \cup [I_{q_2}(i, i)].$$

Rule 2: For decision pairs (i, j) and $i < j$:

$$[I_{\{q_1, q_2\}}(i, j)] = [I_{q_1}(i, j)] \cap [I_{q_2}(i, j)].$$

Rule 3: For pairs (x, y) in $[I_{q_1}(i,j)] - [I_{\{q_1,q_2\}}(i,j)]$ or $[I_{q_2}(i,j)] - [I_{\{q_1,q_2\}}(i,j)]$:
  If $\{x\}$ does not belong to the neighborhood of $B_{\{q_1,q_2\}}(i,i)$,
  then add (x, x) to $[I_{\{q_1,q_2\}}(i,i)]$, which is the same as adding x to $B_{\{q_1,q_2\}}(i,i)$.
  If $\{y\}$ does not belong to the neighborhood of $B_{\{q_1,q_2\}}(j,j)$,
  then add (y,y) to $[I_{\{q_1,q_2\}}(j,j)]$,which is the same as adding y to $B_{\{q_1,q_2\}}(j,j)$.
  After applying the above three rules in sequence, we can obtain the updated indexed blocks accordingly. The working of the rules is illustrated in the following example.

Example 2. Consider the criterion q_2 in the multi-criteria decision table given in Table 1. The set of inconsistent intervals $I_{q_2}(i,j)$ with the following corresponding sets of ordered pairs: $[I_{q_2}(1,1)] = \{(2, 2), (4, 4)\}$, $[I_{q_2}(1,2)] = \{(3, 1)\}$, $[I_{q_2}(1,3)] = \{(3, 7)\}$, $[I_{q_2}(2,2)] = \emptyset$, $[I_{q_2}(2,3)] = \{(1, 7), (5, 7), (5, 6), (5, 8)\}$, $[I_{q_2}(3,3)] = \emptyset$.
  The indexed blocks $B_{q2}(i,j)$ are shown in Table 4.

Table 4. Indexed blocks $B_{q2}(i, j)$

$D \times D$	1	2	3
1	$\{2, 4\}$	$\{1, 3\}$	$\{3, 7\}$
2		$\emptyset$	$\{1, 5, 6, 7, 8\}$
3			$\emptyset$

Applying the three rules of combining q_1 and q_2 we have the combined indexed blocks shown in Table 5 with the following exclusive neighborhoods:

$NB(B_{\{q1,q2\}}(1, 1)) = \{B_{\{q1,q2\}}(1, 2), B_{\{q1,q2\}}(1, 3)\}$,
$NB(B_{\{q1,q2\}}(2, 2)) = \{B_{\{q1,q2\}}(1, 2), B_{\{q1,q2\}}(2, 3)\}$,
$NB(B_{\{q1,q2\}}(3, 3)) = \{B_{\{q1,q2\}}(1, 3), B_{\{q1,q2\}}(2, 3)\}$.

Table 5. Indexed blocks $B_{\{q1,q2\}}(i, j)$

$D \times D$	1	2	3
1	$\{2, 4\}$	$\{1, 3\}$	$\{3, 7\}$
2		$\emptyset$	$\{5, 7\}$
3			$\{6, 8\}$

4 Approximating Sets of Decision Classes

Let $(U, C \cup D = \{d\})$ be a multi-criteria decision table, $P \subseteq C$, and $\{B_P(i,j)|$ $(i, j) \in V_d \times V_d\}$ be the indexed blocks derived from P for $(i, j) \in V_d \times V_d$. For a decision class $Cl_i = \{x \in U | f(x, d) = d_i, d_i \in V_d\}$ with decision value d_i, the lower approximation of Cl_i by P is the indexed block $B_P(i,i)$, the boundary set of Cl_i is $\cup NB(B_P(i,i))$ the union of blocks in the neighborhood of $B_P(i, i)$, and the upper approximation of Cl_i is the union of $B_P(i,i)$ and the boundary set.

For a set of two decision values $D_2 = \{i, j\} \subseteq V_d$, the neighborhoods of indexed blocks $B_P(i,i)$ and $B_P(j,j)$ are reduced by removing objects associated with decision values $\{i, j\}$ only. This can be done by checking objects in the indexed block $B_P(i,j)$.

For x in $B_P(i,j)$, if $f(x,d) = i$ and x does not belong to $NB(B_P(i,i)) - B_P(i,j)$, then x is removed from the neighborhood $NB(B_P(i,i))$ and added to $B_P(i,i)$. Similarly, if $f(x,d) = j$ and x does not belong to $NB(B_P(j,j)) - B_P(i,j)$, then x is removed from the neighborhood $NB(B_P(j,j))$ and added to $B_P(j,j)$. More precisely, we define the updated indexed blocks as

$B_P(i,i)_{D2} = B_P(i,i) \cup \{x \in B_P(i,j) | f(x,d) = i$ and $x \notin NB(B_P(i,i) - B_P(i,j)\}$ and

$$B_P(j,j)_{D2} = B_P(j,j) \cup \{x \in B_P(i,j) | f(x,d)=j \text{ and } x \notin NB(B_P(j,j) - B_P(i,j)\}.$$

The neighborhood of $B_P(i, j)$ is defined as

$$NB(B_P(i,j)) = [NB(B_P(i,i)) \cup NB(B_P(j,j)) - B_P(i,j)] \cup [B_P(i,j)$$
$$\cap (NB(B_P(i,i)) - B_P(i,j))] \cup [B_P(i,j)$$
$$\cap (NB(B_P(j,j)) - B_P(i,j))].$$

In general, neighborhood of $B_P(i, j)$, for $i \neq j$, is not exclusive. It is partially inclusive, i.e., some objects of $B_P(i,j)$ may be associated with some blocks other than $B_P(i,j)$ in the neighborhood.

The lower approximation of $Cl_{D2} = Cl_i \cup Cl_j$ is defined as

$\underline{P}(Cl_{D2}) = B_P(i, i) \cup B_P(j, j) \cup [B_P(i, j) - \cup NB(B_P(i, j))]$
$= B_P(i, i)_{D2} \cup B_P(j, j)_{D2}$,

the boundary set of $Cl_i \cup Cl_j$ is defined as

$BN_P(Cl_{D2}) = \cup NB(B_P(i,j))$, and

the upper approximation of $Cl_i \cup Cl_j$ is defined as

$\bar{P}Cl_{D2} = \underline{P} Cl_{D2} \cup BN_P(Cl_{D2})$.

Now, for a set of k decision values $D_k = \{d_{i1}, \ldots, d_{ik}\} \subseteq V_d$, the lower approximation of the set $Cl_{Dk} = \cup \{Cl_i : i \in D_k\}$ of decision classes by the set P of criteria is defined as

$\underline{P} Cl_{Dk} = \cup \{\underline{P}(Cl_{D2}) : D_2 = (i, j) \text{ in } D_k \times D_k \text{ and } i < j\}$,

the boundary set of Cl_{Dk} is defined as

$BN_P(Cl_{Dk}) = \cup \{\cup NB(B_P(i, j)): (i, j) \text{ in } D_k \times D_k\}$, and

the upper approximation of Cl_{Dk} is defined as

$\bar{P}Cl_{Dk} = \underline{P} Cl_{Dk} \cup BN_P(Cl_{Dk})$.

Due to space limitation, examples and proof of the following proposition are omitted.

Proposition. Let $(U, C \cup D=\{d\})$ be a multi-criteria decision table where domains of all criteria $q \in C$ are totally ordered and the preference-ordered decision classes $Cl = \{Cl_t: t \in T\}$, $T = \{1, \ldots, n\}$ satisfying the upward and downward union properties. If the neighborhoods of indexed blocks $B_C(i, j)$ are exclusive for $(i,j) \in T \times T$, then the set of indexed blocks $B_C(i, j)$ is a partition on U.

5 Conclusion

In this paper we introduced the concept of indexed blocks to represent dominance based approximation spaces derived from multi-criteria decision tables. We used exclusive neighborhoods of indexed blocks to represent inconsistent information. It can be shown that for approximating preference-ordered decision classes with upward and downward union properties, the set of indexed blocks forms a partition on the universe when the neighborhoods of these indexed blocks are exclusive. These results demonstrate the efficacy of using indexed blocks as granules for representing dominance based approximation space. It provides new ways for understanding and studying dominance-based rough sets, and further development of decision rules induction algorithms.

References

1. Greco, S., Matarazzo, B., Slowinski, R.: Rough approximation of a preference relation by dominance relations, ICS Research Report 16/96, Warsaw University of Technology, Warsaw (1996); European Journal of Operational Research 117(1), 63–83 (1996)
2. Greco, S., Matarazzo, B., Slowinski, R.: A new rough set approach to evaluation of bankruptcy risk. In: Zopounidis, C. (ed.) Operational Tools in the Management of Financial Risks, pp. 121–136. Kluwer Academic Publishers, Dordrecht (1998)
3. Greco, S., Matarazzo, B., Slowinski, R.: The use of rough sets and fuzzy sets in MCDM. In: Gal, T., Stewart, T., Hanne, T. (eds.) Advances in Multiple Criteria Decisions Making, ch. 14, pp. 14.1–14.59. Kluwer Academic Publishers, Dordrecht (1999)
4. Greco, S., Matarazzo, B., Slowinski, R.: Rough sets theory for multicriteria decision analysis. European Journal of Operational Research 129(1), 1–47 (2001)
5. Greco, S., Matarazzo, B., Slowinski, R.: Rough sets methodology for sorting problems in presence of multiple attributes and criteria. European Journal of Operational Research 138(2), 247–259 (2002)
6. Greco, S., Matarazzo, B., Slowinski, R., Stefanowski, J.: An Algorithm for Induction of Decision Rules Consistent with the Dominance Principle. In: Ziarko, W., Yao, Y. (eds.) RSCTC 2000. LNCS (LNAI), vol. 2005, pp. 304–313. Springer, Heidelberg (2001)
7. Greco, S., Matarazzo, B., Slowinski, R., Stefanowski, J., Zurawski, M.: Incremental versus Non-incremental Rule Induction for Multicriteria Classification. In: Peters, J.F., Skowron, A., Dubois, D., Grzymała-Busse, J.W., Inuiguchi, M., Polkowski, L. (eds.) Transactions on Rough Sets II. LNCS, vol. 3135, pp. 33–53. Springer, Heidelberg (2004)
8. Pawlak, Z.: Rough sets: basic notion. International Journal of Computer and Information Science 11(15), 344–356 (1982)
9. Pawlak, Z.: Rough sets and decision tables. In: Skowron, A. (ed.) SCT 1984. LNCS, vol. 208, pp. 186–196. Springer, Heidelberg (1985)
10. Pawlak, Z., Grzymala-Busse, J., Slowinski, R., Ziarko, W.: Rough sets. Communication of ACM 38(11), 89–95 (1995)
11. Roy, B.: Methodologie Multicritere d'Aide a la Decision. Economica, Paris (1985)
12. Fan, T.F., Liu, D.R., Tzeng, G.H.: Rough set-based logics for multicriteria decision analysis. European Journal of Operational Research 182(1), 340–355 (2007)

Algebraic Structures
for Dominance-Based Rough Set Approach

Salvatore Greco[1], Benedetto Matarazzo[1], and Roman Słowiński[2]

[1] Faculty of Economics, University of Catania,
Corso Italia, 55, 95129 Catania, Italy
[2] Institute of Computing Science, Poznań University of Technology,
60-965 Poznań, and Institute for Systems Research,
Polish Academy of Sciences, 01-447 Warsaw, Poland

Abstract. Dominance-based Rough Set Approach (DRSA) has been proposed to generalize classical rough set approach when monotonicity between memberships to considered concepts has to be taken into account. This is typical for data describing various phenomena, e.g., "the larger the mass and the smaller the distance, the larger the gravity". These monotonicity relationships are fundamental in rough set approach to multiple criteria decision analysis. In this paper, we propose an algebraic structure for DRSA.

1 Introduction

Rough set theory has been proposed by Pawlak in the early 80s [6,7] as a tool for reasoning about data in terms of granules of knowledge. While the original rough set idea is very useful for classification support, it is not handling a background knowledge about monotonic relationship between evaluations of objects on condition attributes and their evaluations on decision attributes. Such a knowledge is typical for data describing various phenomena and for data describing multiple criteria decision problems, e.g., "the larger the mass and the smaller the distance, the larger the gravity", "the more a tomato is red, the more it is ripe" or "the better the school marks of a pupil, the better his overall classification". The monotonic relationships within multiple criteria decision problems follow from preferential ordering of value sets of attributes (scales of criteria), as well as preferential ordering of decision classes. In order to handle these monotonic relationships between conditions and decisions, Greco, Matarazzo and Słowiński [3,4,8] have proposed to substitute the indiscernibility relation with a dominance relation. Dominance-based Rough Set Approach (DRSA) permits approximation of ordered sets. When dealing with preferences, monotonicity is expressed through the following relationship: "the better is an object with respect to considered points of view (criteria), the more it is appreciated". While many algebraic characterizations of classical rough set approach have been presented, no result in this direction has been proposed for DRSA. In this paper we give an algebraic c haracterization of DRSA in terms of bipolar complemented de Morgan Brower-Zadeh distributive lattice, being a generalization of

G. Wang et al. (Eds.): RSKT 2008, LNAI 5009, pp. 252–259, 2008.
© Springer-Verlag Berlin Heidelberg 2008

the de Morgan Brower-Zadeh distributive lattice [1], already proposed to characterize classical rough set approach [2]. The paper is organized as follows. In the next section classical rough set approach is introduced. The third section presents the DRSA approximations. The fourth section recalls the de Morgan Brower-Zadeh distributive lattice and the characterization of the classical rough set approach in its terms. The fifth section introduces the bipolar complemented de Morgan Brower-Zadeh distributive lattice and the characterization of the DRSA approximations in its terms. The last sections contains conclusions.

2   Classical Rough Set as a Particular Case of the Monotonic Rough Approximation of a Fuzzy Set

In classical rough set approach [6,?], the original information is expressed by means of an *information system*, that is the 4-tuple $S = < U, Q, V, \phi >$, where U is a finite set of *objects* (universe), $Q=\{q_1, q_2, ..., q_m\}$ is a finite set of *attributes*, V_q is the set of values of the attribute q, $V = \bigcup_{q \in Q} V_q$ and $\phi : U \times Q \to V$ is a total function such that $\phi(x, q) \in V_q$ for each $q \in Q$, $x \in U$, called *information function*.

Therefore, each object x from U is described by a vector $Des_Q(x) = [\phi(x, q_1), \phi(x, q_2), ..., \phi(x, q_m)]$, called *description* of x in terms of the evaluations of the attributes from Q; it represents the available information about x. Obviously, $x \in U$ can be described in terms of any non-empty subset $P \subseteq Q$.

With every (non-empty) subset of attributes P there is associated an *indiscernibility relation* on U, denoted by I_P:

$$I_P = \{(x, y) \in U \times U : \phi(x, q) = \phi(y, q), \forall q \in P\}.$$

If $(x, y) \in I_P$, it is said that the objects x and y are P-indiscernible. Clearly, the indiscernibility relation thus defined is an equivalence relation (reflexive, symmetric and transitive). The family of all the equivalence classes of the relation I_P is denoted by $U|I_P$, and the equivalence class containing an element $x \in U$ by $I_P(x)$, i.e.

$$I_P(x) = \{y \in U : \phi(y, q) = \phi(x, q), \forall q \in P\}.$$

The equivalence classes of the relation I_P are called P-*elementary sets*.

Let S be an information system, X a non-empty subset of U and $\emptyset \neq P \subseteq Q$. The P-*lower approximation* and the P-*upper approximation* of X in S are defined, respectively, as:

$$\underline{P}(X) = \{x \in U : I_P(x) \subseteq X\},$$

$$\overline{P}(X) = \{x \in U : I_P(x) \cap X \neq \emptyset\}.$$

The elements of $\underline{P}(X)$ are all and only those objects $x \in U$ which belong to the equivalence classes generated by the indiscernibility relation I_P, *contained* in X; the elements of $\overline{P}(X)$ are all and only those objects $x \in U$ which belong to the equivalence classes generated by the indiscernibility relation I_P, *containing at least one* object x belonging to X. In other words, $\underline{P}(X)$ is the largest union of the P-elementary sets included in X, while $\overline{P}(X)$ is the smallest union of the P-elementary sets containing X.

3   Dominance Based-Rough Set Approach

In this section, we recall the Dominance-based Rough Set Approach [3], taking into account, without loss of generality, the case of rough approximation of fuzzy sets [5].

A *fuzzy information base* is the 3-tuple $\boldsymbol{B} =< U, F, \varphi >$, where U is a finite set of *objects* (universe), $F=\{f_1, f_2, ..., f_m\}$ is a finite set of *properties*, and $\varphi : U \times F \to [0, 1]$ is a function such that $\varphi(x, f_h) \in [0, 1]$ expresses the degree in which object x has property f_h. Therefore, each object x from U is described by a vector

$$Des_F(x) = [\varphi(x, f_1), \ldots, \varphi(x, f_m)]$$

called *description* of x in terms of the evaluations of the properties from F; it represents the available information about x. Obviously, $x \in U$ can be described in terms of any non-empty subset $E \subseteq F$ and in this case we have

$$Des_E(x) = [\varphi(x, f_h), f_h \in E].$$

Let us remark that the concept of fuzzy information base can be considered as a generalization of the concept of property system [9]. Indeed, in a property system an object may either possess a property or not, while in the fuzzy information base an object may possess a property in a given degree between 0 and 1.

With respect to any $E \subseteq F$, we can define the *dominance relation* D_E as follows: for any $x, y \in U$, x dominates y with respect to E (denoted as xD_Ey) if, for any $f_h \in E$,

$$\varphi(x, f_h) \geq \varphi(y, f_h).$$

For any $x \in U$ and for each non-empty $E \subseteq F$, let

$$D_E^+(x) = \{y \in U : yD_Ex\}, \quad D_E^-(x) = \{y \in U : xD_Ey\}.$$

Given $E \subseteq F$, for each $X \subseteq U$, we can define its *upward lower approximation* $\underline{E}^{(>)}(X)$ and its *upward upper approximation* $\overline{E}^{(>)}(X)$ as:

$$\underline{E}^{(>)}(X) = \left\{ x \in U : D_E^+(x) \subseteq X \right\},$$

$$\overline{E}^{(>)}(X) = \left\{ x \in U : D_E^-(x) \cap X \neq \emptyset \right\}.$$

Analogously, given $E \subseteq F$, for each $X \subseteq U$, we can define its *downward lower approximation* $\underline{E}^{(<)}(X)$ and its *downward upper approximation* $\overline{E}^{(<)}(X)$ as:

$$\underline{E}^{(<)}(X) = \left\{ x \in U : D_E^-(x) \subseteq X \right\},$$

$$\overline{E}^{(<)}(X) = \left\{ x \in U : D_E^+(x) \cap X \neq \emptyset \right\}.$$

Let us observe that in the above definition of rough approximations $\underline{E}^{(>)}(X)$, $\overline{E}^{(>)}(X)$, $\underline{E}^{(<)}(X)$, $\overline{E}^{(<)}(X)$, the elementary sets, which in the classical rough set theory are equivalence classes of the indiscernibility relation, are the sets $D_E^+(x)$ and $D_E^-(x)$, $x \in U$. Observe also that the rough approximations $\underline{E}^{(>)}(X)$, $\overline{E}^{(>)}(X)$, $\underline{E}^{(<)}(X)$, $\overline{E}^{(<)}(X)$ can be expressed as unions of the elementary sets [5] , i.e. for any $X \subseteq U$ and $E \subseteq F$

1. $\underline{E}^{(>)}(X) = \bigcup\limits_{x \in U} \left\{ D_E^+(x) : D_E^+(x) \subseteq X \right\}$,

2. $\overline{E}^{(>)}(X) = \bigcup\limits_{x \in U} \left\{ D_E^+(x) : D_E^-(x) \cap X \neq \emptyset \right\}$,

3. $\underline{E}^{(<)}(X) = \bigcup\limits_{x \in U} \left\{ D_E^-(x) : D_E^-(x) \subseteq X \right\}$,

4. $\overline{E}^{(<)}(X) = \bigcup\limits_{x \in U} \left\{ D_E^-(x) : D_E^+(x) \cap X \neq \emptyset \right\}.$

The rough approximations $\underline{E}^{(>)}(X)$, $\overline{E}^{(>)}(X)$, $\underline{E}^{(<)}(X)$, $\overline{E}^{(<)}(X)$ can be used to analyze data relative to gradual membership of objects to some concepts representing properties of objects and their assignment to decision classes. This analysis takes into account the following monotonicity principle: "the greater the degree to which an object has properties from $E \subseteq F$, the greater its degree of membership to a considered class". This principle can be formalized as follows. Let us consider a fuzzy set X in U, characterized by the membership function $\mu_X : U \to [0,1]$. This fuzzy set represents a class of interest, such that function μ specifies a graded membership of objects from U to considered class X. For each cutting level $\alpha \in [0,1]$, we can consider the following sets

– weak upward cut of fuzzy set X:

$$X^{\geq \alpha} = \left\{ x \in U : \mu(x) \geq \alpha \right\},$$

– strict upward cut of fuzzy set X:

$$X^{> \alpha} = \left\{ x \in U : \mu(x) > \alpha \right\},$$

– weak downward cut of fuzzy set X:

$$X^{\leq \alpha} = \left\{ x \in U : \mu(x) \leq \alpha \right\},$$

– strict upward cut of fuzzy set X:

$$X^{< \alpha} = \left\{ x \in U : \mu(x) < \alpha \right\}.$$

Let us remark that, for any fuzzy set X and for any $\alpha \in [0,1]$, we have that

$$U - X^{\geq \alpha} = X^{<\alpha}, \quad U - X^{\leq \alpha} = X^{>\alpha},$$

$$U - X^{>\alpha} = X^{\leq \alpha}, \quad U - X^{<\alpha} = X^{\geq \alpha}.$$

Given a family of fuzzy sets $\mathbf{X} = \{X_1, X_2, ..., X_p\}$ on U, whose respective membership functions are $\mu_1, \mu_2, ..., \mu_p$, let $P^>(\mathbf{X})$ be the set of all the sets obtained through unions and intersections of weak and strict upward cuts of component fuzzy sets. Analogously, let $P^<(\mathbf{X})$ be the set of all the sets obtained through unions and intersections of weak and strict downward cuts of component fuzzy sets.

$P^>(\mathbf{X})$ and $P^<(\mathbf{X})$ are closed under set union and set intersection operations, i.e. for all $Y_1, Y_2 \in P^>(\mathbf{X})$, $Y_1 \cup Y_2$ and $Y_1 \cap Y_2$ belong to $P^>(\mathbf{X})$, as well as for all $W_1, W_2 \in P^<(\mathbf{X})$, $W_1 \cup W_2$ and $W_1 \cap W_2$ belong to $P^<(\mathbf{X})$. Observe, moreover, that the universe U and the empty set $\emptyset$ belong both to $P^>(\mathbf{X})$ and to $P^<(\mathbf{X})$ because, for any fuzzy set $X_i \in \mathbf{X}$,

$$U = X_i^{\geq 0} = X_i^{\leq 1}$$

and

$$\emptyset = X_i^{>1} = X_i^{<0}.$$

4   Brower-Zadeh Distributive De Morgan Lattices

A system $\langle \Sigma, \wedge, \vee, ', ^\sim, 0, 1 \rangle$ is a quasi-Brower-Zadeh [1] distributive lattice if the following properties (1)-(4) hold:

(1) Σ is a distributive lattice with respect to the join and the meet operations $\vee, \wedge$ whose induced partial order relation is

$$a \leq b \text{ iff } a = a \wedge b \text{ (equivalently } b = a \vee b)$$

Moreover, it is required that Σ is bounded by the least element 0 and the greatest element 1:

$$\forall a \in \Sigma, \quad 0 \leq a \leq 1$$

(2) The unary operation $' : \Sigma \rightarrow \Sigma$ is a Kleene (also Zadeh or fuzzy) complementation. In other words, for arbitrary $a, b \in \Sigma$,
   (K1) $a'' = a$,
   (K2) $(a \vee b)' = a' \wedge b'$,
   (K3) $a \wedge a' \leq b \vee b'$.

(3) The unary operation $^\sim : \Sigma \rightarrow \Sigma$ is a Brower (or intuitionistic) complementation. In other words, for arbitrary $a, b \in \Sigma$,
   (B1) $a \wedge a^{\sim\sim} = a$,
   (B2) $(a \vee b)^\sim = a^\sim \wedge b^\sim$,
   (B3) $a \wedge a^\sim = 0$.

(4) The two complementations are linked by the interconnection rule which must hold for arbitrary $a \in \Sigma$:

(in) $a^{\sim} \leq a'$

A structure $\langle \Sigma, \wedge, \vee,',^{\sim}, 0, 1 \rangle$ is a Brower-Zadeh distributive lattice if it is a quasi-Brower-Zadeh distributive lattice satisfying the stronger interconnection rule:

$(s\text{-}in)$ $a^{\sim\sim} = a^{\sim\prime}$.

A Brower-Zadeh distributive lattice satisfying also the $\vee$ de Morgan property

(B2a) $(a \wedge b)^{\sim} = a^{\sim} \vee b^{\sim}$

is called a de Morgan Brower-Zadeh distributive lattice.

The de Morgan Brower-Zadeh distributive lattice is an algebraic structure which can be given to the collection of all rough approximations within the classical rough set approach as follows. Fixed $P \subseteq C$, for any $X \subseteq U$ let us consider the pair $\langle \underline{P}(X), U - \overline{P}(X) \rangle$ and

$$A = \left\{ (I, E) : \exists X \subseteq U \text{for which } I = \underline{P}(X) \text{and } E = U - \overline{P}(X) \right\}.$$

The following result holds.

Theorem 1. [2] The structure $\langle A, \sqcap, \sqcup,^{-},^{\approx}, \langle \emptyset, U \rangle, \langle U, \emptyset \rangle \rangle$ where, for any $\langle I_1, E_1 \rangle, \langle I_2, E_2 \rangle \in \mathbf{A}$,

$$\langle I_1, E_1 \rangle \sqcap \langle I_2, E_2 \rangle = \langle I_1 \cap I_2, E_1 \cup E_2 \rangle$$

$$\langle I_1, E_1 \rangle \sqcup \langle I_2, E_2 \rangle = \langle I_1 \cup I_2, E_1 \cap E_2 \rangle$$

$$\langle I_1, E_1 \rangle^{-} = \langle E_1, I_1 \rangle$$

$$\langle I_1, E_1 \rangle^{\approx} = \langle E_1, U - E_1 \rangle$$

is a de Morgan Brower-Zadeh distributive lattice.

5 Bipolar Complemented de Morgan Brower-Zadeh Distributive Lattices

A system $\langle \Sigma, \Sigma^+, \Sigma^-, \wedge, \vee,'^+,'^-,^{\sim+},^{\sim-}, 0, 1 \rangle$ is a *bipolar complemented quasi-Brower-Zadeh distributive lattice* if the following properties (1b)-(4b) hold:

(1b) Σ is a distributive lattice with respect to the join and the meet operations $\vee$ and $\wedge$

(1b') $\Sigma^+, \Sigma^- \subseteq \Sigma$ are distributive lattices with respect to the join and the meet operations $\vee$ and $\wedge$. Σ is bounded by the least element 0 and the greatest element 1, which implies that also Σ^+ and Σ^- are bounded.

(2b) The unary operations $'^+ : \Sigma^+ \to \Sigma^-$ and $'^- : \Sigma^- \to \Sigma^+$ are Kleene (also Zadeh or fuzzy) bipolar complementation, that is, for arbitrary $a, b \in \Sigma^+$ and $c, d \in \Sigma^-$,

(K1b) $a'^{+'-} = a$, $c'^{-'+} = c$,

(K2b) $(a \vee b)'^+ = a'^+ \wedge b'^+, (c \vee d)'^- = c'^- \wedge d'^-$,

(K3b) $a \wedge a'^+ \leq b \vee b'^+$, $c \wedge c'^- \leq d \vee d'^-$.

(3b) The unary operations $\sim^+ : \Sigma^+ \to \Sigma^-$ and $\sim^- : \Sigma^- \to \Sigma^+$ are Brower (or intuitionistic) bipolar complementations, that is, for arbitrary $a, b \in \Sigma^+$ and $c, d \in \Sigma^-$,

(B1b) $a \wedge a^{\sim^+ \sim^-} = a$, $c \wedge c^{\sim^- \sim^+} = c$

(B2b) $(a \vee b)^{\sim^+} = a^{\sim^+} \wedge b^{\sim^+}$, $(c \vee d)^{\sim^-} = c^{\sim^-} \wedge d^{\sim^-}$,

(B3b) $a \wedge a^{\sim^+} = 0$, $c \wedge c^{\sim^-} = 0$.

(4b) Complementation $'^+$ and complementation $\sim^+$ in one hand and complementation $'^-$ and complementation $\sim^-$ in the other hand are linked by the interconnection rule, that is, for arbitrary $a \in \Sigma^+$ and arbitrary $b \in \Sigma^-$: (in-b) $a^{\sim^+} \leq a'^+$, $b^{\sim^-} \leq b'^-$

A structure $\langle \Sigma, \Sigma^+, \Sigma^-, \wedge, \vee, '^+, '^-, \sim^+, \sim^-, 0, 1 \rangle$ is a *bipolar complemented Brower-Zadeh distributive lattice* if it is a quasi-Brower-Zadeh distributive lattice satisfying the stronger interconnection rule, that is, for arbitrary $a \in \Sigma^+$ and arbitrary $b \in \Sigma^-$:

$$(s\text{-}in\text{-}b) \; a^{\sim^+ \sim^-} = a^{\sim^+ '-}, \; b^{\sim^- \sim^+} = a^{\sim^- '+}.$$

A bipolar complemented Brower-Zadeh distributive lattice is a *bipolar complemented de Morgan Brower-Zadeh distributive lattice*, if it satisfies also the $\vee$ de Morgan property that is, for arbitrary $a, b \in \Sigma^+$ and $c, d \in \Sigma^-$

$$(B2a\text{-}b) \; (a \wedge b)^{\sim^+} = a^{\sim^+} \vee b^{\sim^+}, \; (c \wedge d)^{\sim^-} = c^{\sim^-} \vee d^{\sim^-}.$$

The bipolar complemented de Morgan Brower-Zadeh distributive lattice is an algebraic structure which can be given to the collection of all rough approximations within the DRSA as follows. Fixed $G \subseteq F$, for any $X \subseteq U$ let us consider the pairs $\left\langle \underline{G}^{(\leq)}(X), U - \overline{G}^{(\leq)}(X) \right\rangle$ and $\left\langle \underline{G}^{(\geq)}(X), U - \overline{G}^{(\geq)}(X) \right\rangle$ and the sets

$$B = \{(I, E) : I, E \subseteq U \text{ such that } I \cap E = \emptyset\},$$

$$B^- = \left\{(I, E) : \exists X \subseteq U \text{ for which } I = \underline{G}^{(\leq)}(X) \text{ and } E = U - \overline{G}^{(\leq)}(X)\right\},$$

$$B^+ = \left\{(I, E) : \exists X \subseteq U \text{ for which } I = \underline{G}^{(\geq)}(X) \text{ and } E = U - \overline{G}^{(\geq)}(X)\right\}.$$

The following result holds.

Theorem 2. The structure $\langle B, B^+, B^-, \sqcap, \sqcup, ^{--}, ^{-+}, ^{\approx-}, ^{\approx+}, \langle \emptyset, U \rangle, \langle U, \emptyset \rangle \rangle$ where, for any $\langle I_1, E_1 \rangle, \langle I_2, E_2 \rangle \in B$, $\langle I_3, E_3 \rangle \in B^-, \langle I_4, E_4 \rangle \in B^+$,

$$\langle I_1, E_1 \rangle \sqcap \langle I_2, E_2 \rangle = \langle I_1 \cap I_2, E_1 \cup E_2 \rangle$$

$$\langle I_1, E_1 \rangle \sqcup \langle I_2, E_2 \rangle = \langle I_1 \cup I_2, E_1 \cap E_2 \rangle$$

$$\langle I_3, E_3 \rangle^{--} = \langle E_3, I_3 \rangle, \; \langle I_4, E_4 \rangle^{-+} = \langle E_4, I_4 \rangle$$

$$\langle I_3, E_3 \rangle^{\approx-} = \langle E_3, U - E_3 \rangle, \; \langle I_4, E_4 \rangle^{\approx+} = \langle E_4, U - E_4 \rangle$$

is a bipolar complemented de Morgan Brower-Zadeh distributive lattice.

6 Conclusions

In this paper we take into account a new problem in rough set approach: the algebraic characterization of the DRSA. More precisely, we proposed a characterization of the DRSA approximations in terms of a specific algebraic structure: bipolar complemented de Morgan Brower-Zadeh distributive lattice. Future research will be oriented in investigating the formal properties of the bipolar complemented de Morgan Brower-Zadeh distributive lattice and in the characterization of DRSA in terms of other abstract algebras.

References

1. Cattaneo, G., Giuntini, R., Pilla, R.: BZMV^{dM} algebras and stonian MV-algebras (applications to fuzzy sets and rough approximations). Fuzzy Sets and Systems 108, 201–222 (1999)
2. Cattaneo, G., Ciucci, D.: Algebraic Structures for Rough Sets. In: Peters, J.F., Skowron, A., Dubois, D., Grzymała-Busse, J.W., Inuiguchi, M., Polkowski, L. (eds.) Transactions on Rough Sets II. LNCS, vol. 3135, pp. 208–252. Springer, Heidelberg (2004)
3. Greco, S., Matarazzo, B., Słowiński, R.: Rough set theory for multicriteria decision analysis. European Journal of Operational Research 129, 1–47 (2001)
4. Greco, S., Matarazzo, B., Słowiński, R.: Decision rule approach. In: Figueira, J., Greco, S., Erghott, M. (eds.) Multiple Criteria Decision Analysis: State of the Art Surveys, pp. 507–563. Springer, Berlin (2005)
5. Greco, S., Matarazzo, B., Słowiński, R.: Dominance-Based Rough Set Approach as a Proper Way of Handling Graduality in Rough Set Theory. In: Peters, J.F., Skowron, A., Marek, V.W., Orłowska, E., Słowiński, R., Ziarko, W. (eds.) Transactions on Rough Sets VII. LNCS, vol. 4400, pp. 36–52. Springer, Heidelberg (2007)
6. Pawlak, Z.: Rough Sets. International Journal of Computer and Information Sciences 11, 341–356 (1982)
7. Pawlak, Z.: Rough Sets. Kluwer, Dordrecht (1991)
8. Słowiński, R., Greco, S., Matarazzo, B.: Rough set based decision support. In: Burke, E., Kendall, G. (eds.) Search Methodologies: Introductory Tutorials in Optimization and Decision Support Techniques, ch. 16, pp. 475–527. Springer, New York (2005)
9. Vakarelov, D.: Consequence Relations and Information Systems. In: Słowiński, R. (ed.) Intelligent Decision Support. Handbook of Applications and Advances of the Rough Sets Theory, pp. 391–399. Kluwer, Dordrecht (1992)

Ensemble of Decision Rules for Ordinal Classification with Monotonicity Constraints

Krzysztof Dembczyński[1], Wojciech Kotłowski[1], and Roman Słowiński[1,2]

[1] Institute of Computing Science, Poznań University of Technology,
60-965 Poznań, Poland
{kdembczynski, wkotlowski, rslowinski}@cs.put.poznan.pl
[2] Systems Research Institute, Polish Academy of Sciences, 01-447 Warsaw, Poland

Abstract. Ordinal classification problems with monotonicity constraints (also referred to as multicriteria classification problems) often appear in real-life applications, however they are considered relatively less frequently in theoretical studies than regular classification problems. We introduce a rule induction algorithm based on forward stagewise additive modeling that is tailored for this type of problems. The algorithm monotonizes the dataset (excludes highly inconsistent objects) using Dominance-based Rough Set Approach and generates monotone rules. Experimental results indicate that taking into account the knowledge about order and monotonicity constraints in the classifier can improve the prediction accuracy.

1 Introduction

An *ordinal classification problem with monotonicity constraints* consists in assignment of objects to finite number of ordered *classes*. Objects are described by attributes with ordered value sets and *monotonicity constraints* are present in the data: a higher value of an object on an attribute, with other values being fixed, should not decrease its class assignment. Problems of ordinal classification in the presence of monotonicity constraints are commonly encountered in real-life applications. A typical representative is multiple-criteria classification considered within multiple-criteria decision analysis (MCDA) [11]. Moreover, in many other domains ordinal and monotone properties follow from the domain knowledge about the problem and should not be neglected. They are encountered in such problems as bankruptcy risk prediction [10], breast cancer diagnosis [18], house pricing [14], credit rating [6] and many others.

In order to solve ordinal classification problem with monotonicity constraints, one can apply two steps for improving the accuracy of the classifier. The first one consists in "monotonization" of the dataset, i.e. exclusion of objects highly violating the monotone relationships. The second one consists in imposing the constraints such that only monotone functions are taken into account.

Dominance-based Rough Set Approach (DRSA) [11] is one of the first approaches introduced to deal with this type of problems. By replacing indiscernibility relation, considered in classical rough sets [15], with a dominance relation,

G. Wang et al. (Eds.): RSKT 2008, LNAI 5009, pp. 260–267, 2008.

DRSA is able to handle inconsistencies following from violation of monotone relationships. In this context several specialized decision rule induction algorithms were proposed that were able to capture the ordinal nature of data and handle domain knowledge in the form of monotonicity constraints [12,5] (we will refer to rules consistent with monotonicity constraints as *monotone* rules). Among them, DOMLEM [12] seems to be the most popular one. It aims at finding a minimal set of monotone rules covering the dataset, using the well-known sequential covering procedure as a search heuristic.

We follow a different methodology for monotone rule induction that is based on *forward stagewise additive modeling* (FSAM) [7], i.e. greedy procedure for minimizing a loss function on the dataset. The algorithm introduced in this paper, called MORE (from MOnotone Rule Ensembles), treats a single rule as a subsidiary base classifier in the ensemble. The rules are added to the ensemble iteratively, one by one. Each rule is fitted by concentrating on the examples which were hardest to classify correctly by rules that have already been generated. The advantage of this approach is that we use a single measure only (value of the empirical risk) at all stages of learning procedure: setting the best cuts (conditions), stopping the rule's growth and determining the weight of the rule; no additional features (e.g. impurity measures, pruning procedures) are considered. Such an approach was already considered in ordinary classification problems and algorithms such as RuleFit [9], SLIPPER [2], LRI [21] or EDR [1] were introduced. The algorithm presented here can be seen as an extension of the last from the mentioned above methods. It monotonizes the dataset (excludes highly inconsistent objects) using DRSA and then generates monotone rules.

2   Problem Statement

In the *classification* problem, the aim is to predict the unknown class label $y \in Y = \{1, \ldots, K\}$ (decision value) of an object $\mathbf{x}$ using the description of the object in terms of p (condition) attributes, $\mathbf{x} = (x_1, x_2, \ldots, x_p) \in X$, where X is the *attribute space*. Here, we assume without loss of generality that value set of each attribute is a subset of $\mathbb{R}$, so that $X \subseteq \mathbb{R}^p$. In the *ordinal classification*, it is assumed that there is a meaningful order between classes which corresponds to the natural order between class labels. We also assume the presence of *monotonicity constraints* in the data.

In order to formalize the concept of monotonicity, we define the *dominance* relation as a binary relation on X in the following way: for any $\mathbf{x}, \mathbf{x}' \in X$ we say that $\mathbf{x}$ *dominates* $\mathbf{x}'$, denoted $\mathbf{x} \succeq \mathbf{x}'$, if on every attribute, $\mathbf{x}$ has value not smaller than $\mathbf{x}'$, $x_j \geq x'_j$, for all $j = 1, \ldots, p$. The dominance relation is a partial pre-order on X, i.e. it is reflexive and transitive. Having defined the dominance relation, we define the *monotone function* to be any function $f \colon X \to Y$ satisfying the monotonicity constraints:

$$\mathbf{x} \succeq \mathbf{x}' \to f(\mathbf{x}) \geq f(\mathbf{x}') \tag{1}$$

for any $\mathbf{x}, \mathbf{x}' \in X$.

Now, the problem of ordinal classification with monotonicity constraints can be stated as a problem of finding the *monotone classification function* $f(\mathbf{x})$ that predicts accurately values of y. The accuracy is measured in terms of the *loss function* $L(y, f(\mathbf{x}))$, which is the penalty for predicting $f(\mathbf{x})$ when the actual value is y. The overall accuracy of function $f(\mathbf{x})$ is defined as the expected loss (*risk*) according to the probability distribution of data to be predicted:

$$R(f) = E[L(y, f(\mathbf{x}))] \tag{2}$$

Since the data probability distribution is unknown, the function is learned from a set of n training examples $\{(\mathbf{x}_1, y_1), \ldots, (\mathbf{x}_n, y_n)\}$ (*training set*). In order to minimize the value of risk (2), the learning procedure usually performs minimization of the *empirical risk*:

$$R_{\mathrm{emp}}(f) = \frac{1}{n} \sum_{i=1}^{n} L(y_i, f(\mathbf{x}_i)), \tag{3}$$

which is the value of a loss function on the training set (training error). It is possible to use a variety of loss functions for measuring accuracy; here, for simplicity, we assume the loss function to be the *absolute error loss*,

$$L_{\mathrm{abs}}(y, f(\mathbf{x})) = |y - f(\mathbf{x})|. \tag{4}$$

Although in classification a 0-1 loss is often considered (defined as $L_{0-1}(y, f(\mathbf{x})) = 1$ if $y \neq f(\mathbf{x})$, 0 otherwise), absolute error loss has the advantage over 0-1 loss of being sensitive to the difference between predicted and actual class label, therefore taking the order between classes into account.

Solution to the ordinal classification problem with loss function (4) can be obtained by reducing the problem to $K - 1$ binary problems. Let us define for a given class label y, auxiliary class labels y_k equal to 1 if $y \geq k$, otherwise -1, for each $k = 2, \ldots, K$. Then, we have $y = 1 + \sum_{k=2}^{K} \frac{1}{2}(y_k + 1)$. Moreover, let $f_k(\mathbf{x}) \in \mathbb{R}$ to be a function such that if $f(\mathbf{x}) \geq k$, then $f_k(\mathbf{x}) > 0$, and $f_k(\mathbf{x}) < 0$ otherwise. Then, we have:

$$L_{\mathrm{abs}}(y, f(\mathbf{x})) = |y - f(\mathbf{x})| = \sum_{k=2}^{K} \left| \frac{1}{2}(y_k - \mathrm{sgn}(f_k(\mathbf{x}))) \right| = \sum_{k=2}^{K} L_{0-1}(y_k f_k(\mathbf{x}))$$

where $L_{0-1}(y_k f_k(\mathbf{x}))$ is so called *margin 0-1 loss* defined for binary problems as $L_{0-1}(y_k f_k(\mathbf{x})) = \theta(-y_k f_k(\mathbf{x}))$, where $\theta(a)$ is a step function, equal to 1 for $a \geq 0$, and 0 elsewhere. The only problem is to satisfy $f_k(\mathbf{x}) \geq f_{k-1}(\mathbf{x})$, for $k = 3, \ldots, K$. If this condition is violated, one can try to find $g_k(\mathbf{x})$, $k = 2, \ldots, K$ satisfying this condition and being as close to $f_k(\mathbf{x})$, $k = 2, \ldots, K$ as possible:

$$\min \sum_{k=2}^{K} (f_k(\mathbf{x}) - g_k(\mathbf{x}))^2$$

This is the problem of *isotonic regression* [17]. The final prediction is then $f(\mathbf{x}) = 1 + \sum_{k=2}^{K} \frac{1}{2}(\mathrm{sgn}(g_k(\mathbf{x})) + 1)$. However, one does not need to solve isotonic regression at all. It can be shown that simple voting procedure gives the same prediction. For a given object $\mathbf{x}$, if $f_k(\mathbf{x}) > 0$, then each class indicated by labels $k, \ldots, K$ gets vote $|f_k(\mathbf{x})|$; if $f_k(\mathbf{x}) < 0$, each class indicated by labels $1, \ldots, k-1$ gets vote $|f_k(\mathbf{x})|$. Votes are summed for each $k = 1, \ldots, K$ and $\mathbf{x}$ is classified to the class with the highest score.

From the monotonicity assumption *dominance principle* follows: for any two objects $\mathbf{x}_i, \mathbf{x}_j$ from the dataset, such that $\mathbf{x}_i \succeq \mathbf{x}_j$, it should hold $y_i \geq y_j$. However, it still may happen that in the dataset there exists an object $\mathbf{x}_i$, dominating another object $\mathbf{x}_j$, while it holds $y_i < y_j$. Such a situation violates the monotonicity assumption, so we shall call objects $\mathbf{x}_i$ and $\mathbf{x}_j$ *inconsistent*. Notice that no monotone function can approximate accurately inconsistent objects. Therefore, DRSA [11] and its stochastic extension [4] is applied in order to monotonize the data. Instead of using all data, we remove the inconsistent objects taking into account only stochastic lower approximations of decision classes:

$$\underline{Cl_k^{\geq}} = \{\mathbf{x}_i : \Pr(y \geq k | \mathbf{x}_i) \geq \alpha, i = 1, \ldots, n\},$$

$$\underline{Cl_k^{\leq}} = \{\mathbf{x}_i : \Pr(y \leq k | \mathbf{x}_i) \geq \alpha, i = 1, \ldots, n\}$$

where $\Pr(y \geq k | \mathbf{x}_i)$ $(\Pr(y \leq k | \mathbf{x}_i))$ is a probability, conditioned on $\mathbf{x}_i$, of class label at least (at most) k, and $\alpha \in (0.5, 1]$ is a chosen *consistency level*. The probabilities are obtained using maximum likelihood estimation taking into account monotonicity constraints [4].

3   Ensemble of Decision Rules

This section describes the general scheme for decision rule induction. Here we focus on the binary classification case and assume that $Y = \{-1, 1\}$, where a "positive" class is ranked higher (in the order) to a "negative" class. This algorithm can be used in to each of the $K - 1$ binary problems resulting from the reduction of the ordinal classification problem. One can also use lower approximations instead of whole dataset.

Decision rule is a logical statement of the form: *if [condition], then [decision]*. Let X_j be the set of all possible values for the j-th attribute. Condition part of the rule consist of elementary expressions of the form $x_j \geq s_j$ or $x_j \leq s_j$ for some $s_j \in X_j$. Let Φ denote the set of elementary expressions constituting the condition part of the rule, and let $\Phi(\mathbf{x})$ be an indicator function equal to 1 if an objects $\mathbf{x}$ satisfies the condition part of the rule (we also say that a rule *covers* an object), otherwise $\Phi(\mathbf{x}) = 0$. The decision is a single real value, denoted by α. Therefore, we define a decision rule as:

$$r(\mathbf{x}) = \alpha\Phi(\mathbf{x}). \tag{5}$$

Notice that the decision rule takes only two values, $r(\mathbf{x}) \in \{\alpha, 0\}$, depending whether $\mathbf{x}$ satisfies the conditions or not. In this paper, we assume the classification function is a linear combinations of M decision rules:

$$f(\mathbf{x}) = \alpha_0 + \sum_{m=1}^{M} r_m(\mathbf{x}), \tag{6}$$

Algorithm 1. Monotone Rule Ensemble – MORE

input : set of n training examples $\{(y_1, \mathbf{x}_1), \ldots, (y_n, \mathbf{x}_n)\}$,
$\qquad\quad$ M – number of decision rules to be generated.
output: default rule α_0, ensemble of decision rules $\{r_m(\mathbf{x})\}_1^M$.

$\alpha_0 = \arg\min_\alpha \sum_{i=1}^n \sigma(\alpha y_i)$;
$f_0(\mathbf{x}) = \alpha_0$;
for $m = 1$ *to* M **do**

$\qquad \Phi_m(\mathbf{x}) = \arg\max_\Phi \left| \sum_{\Phi(\mathbf{x}_i)=1} y_i \sigma'(y_i f_{m-1}(\mathbf{x}_i)) \right|$;
$\qquad \alpha_m = \arg\min_\alpha \sum_{\Phi_m(\mathbf{x}_i)=1} \sigma(y_i(f_{m-1}(\mathbf{x}_i) + \alpha))$;
$\qquad r_m(\mathbf{x}) = \alpha_m \Phi_m(\mathbf{x})$;
$\qquad f_m(\mathbf{x}) = f_{m-1}(\mathbf{x}) + r_m(\mathbf{x})$;
end

---

where α_0 is a constant value, which can be interpreted as a default rule, covering the whole X. Object $\mathbf{x}$ is classified to the class indicated by the sign of $f(\mathbf{x})$. The combination (6) has very simple interpretation as a voting procedure: rules with positive α vote for the positive class, while rules with negative α – for the negative class. Object $\mathbf{x}$ is classified to the class with higher vote (which is equivalent to the sign of $f(\mathbf{x})$). Notice that in order to maintain monotonicity of $f(\mathbf{x})$, it is necessary and sufficient that for the m-th rule, α_m is positive when all elementary expressions in Φ_m are of the form $x_j \geq s_j$; similarly, for negative α_m all the conditions must be of the form $x_j \leq s_j$.

Rule induction is performed by minimizing the margin 0-1 loss function (classification error) on the set of n training examples (empirical risk). Notice that this loss function, is neither smooth nor differentiable. Therefore, we approximate it with the sigmoid function:

$$\sigma(x) = \frac{1}{1 + e^x} \qquad (7)$$

Thus, we minimize the following empirical risk:

$$R_{\mathrm{emp}}(f) = \sum_{i=1}^n \sigma(y_i f(\mathbf{x}_i)) \qquad (8)$$

However, finding a set of rules minimizing (8) is computationally hard, that is why we follow here FSAM, i.e. the rules are added one by one, greedily minimizing (8). We start with the default rule defined as:

$$\alpha_0 = \arg\min_\alpha R_{\mathrm{emp}}(\alpha) = \arg\min_\alpha \sum_{i=1}^n \sigma(\alpha y_i). \qquad (9)$$

Let $f_{m-1}(\mathbf{x})$ be a classification function after $m-1$ iterations, consisting of first $m-1$ rules and the default rule. The m-th decision rule $r_m(\mathbf{x}) = \alpha_m \Phi_m(\mathbf{x})$ should be obtained from $r_m = \arg\min_r R_{\mathrm{emp}}(f_{m-1}+r)$, but in order to speed up computations, it is built in two steps. First, we obtain value of $\Phi_m(\mathbf{x})$. Then, we obtain value of α_m by solving the line-search problem with formerly determined $\Phi_m(\mathbf{x})$. To explain the procedure for determining $\Phi_m(\mathbf{x})$, let us expand the value of the loss function up to the first order using $g(x + \alpha) \simeq g(x) + \alpha \frac{dg(x)}{dx}$:

$$\sigma(y_i(f_{m-1}(\mathbf{x}_i) + \alpha)) = \sigma(y_i f_{m-1}(\mathbf{x}_i)) + \alpha y_i \sigma'(y_i f_{m-1}(\mathbf{x}_i)), \qquad (10)$$

where $\sigma'(x)$ is a derivative of sigmoid function $\sigma(x)$. Using (10) in (8) we approximate the empirical risk $R_{\text{emp}}(f_{m-1} + r)$ as:

$$\sum_{\Phi(\mathbf{x}_i)=1} \left[\sigma(y_i f_{m-1}(\mathbf{x}_i)) + \alpha y_i \sigma'(y_i f_{m-1}(\mathbf{x}_i)) \right] + \sum_{\Phi(\mathbf{x}_i)=0} \sigma(y_i f_{m-1}(\mathbf{x}_i)). \qquad (11)$$

However, minimizing (11) is equivalent to minimizing:

$$\mathcal{L}_m(\Phi) = \sum_{\Phi(\mathbf{x}_i)=1} y_i \sigma'(y_i f_{m-1}(\mathbf{x}_i)) \qquad (12)$$

for any positive value of α or maximizing (12) for any negative value of α. Thus, the general idea of algorithm for finding Φ_m is the following: first we search for Φ_m^+ with positive α by minimizing $\mathcal{L}_m(\Phi)$, next we search for Φ_m^- with negative α by maximizing $\mathcal{L}_m(\Phi)$, and we choose Φ_m with higher $|\mathcal{L}_m(\Phi)|$, $\Phi_m = \arg\max\{|\mathcal{L}_m(\Phi_m^+)|, |\mathcal{L}_m(\Phi_m^-)|\}$. The procedure for finding Φ_m^+ (Φ_m^-) resembles the way the decision trees are generated. Here, we look for only one branch instead of the whole decision tree. At the beginning, Φ_m^+ (Φ_m^-) is empty and in each next step an elementary expression $x_j \geq s_j$ ($x_j \leq s_j$) is added to Φ_m^+ (Φ_m^-) until $\mathcal{L}_m(\Phi_m^+)$ ($\mathcal{L}_m(\Phi_m^-)$) cannot be decreased. Let us underline that a minimal value of $\mathcal{L}_m(\Phi_m^+)$ ($\mathcal{L}_m(\Phi_m^-)$) is a natural stop criterion, what differs this procedure from those used for decision trees generation. After Φ_m has been determined, α_m can be obtained by simply using the line search procedure to:

$$\alpha_m = \arg\min_\alpha \sum_{\Phi_m(\mathbf{x}_i)=1} \sigma(y_i(f_{m-1} + \alpha)). \qquad (13)$$

In our implementation, to speed up the computations, instead of solving (9) and (13) we perform a gradient search with short step – we choose α_m to be a small, fixed value $\pm\gamma$ that corresponds to a learning rate.

4   Experimental Results

In order to test how our approach to rule induction behaves in practice, we selected eight datasets, for which it is known that monotonicity constraints make sense. Five datasets come from UCI repository [19]: Wisconsin Breast Cancer, CPU Performance, Ljubljana Breast Cancer, Boston House Pricing, Car. The other three were obtained from different sources: Den Bosch House Pricing [3], Bankruptcy Risk [10] and Windsor House Pricing [13]. Due to lack of space we omit the detailed characteristics of each dataset.

Table 1. Mean absolute error ± standard error. For each dataset, the best method and all methods within one standard error below the best are marked with bold.

Dataset	SVM	j48	Adaboost	MORE	MORE+
DenBosch	0.2055±.0042	0.1689±.0041	0.1294±.0025	**0.1181**±.0031	0.1303±.0035
CPU	0.4366±.0028	0.1261±.0037	0.5727±.0027	**0.0641**±.0023	**0.0641**±.0023
Wisconsin	**0.0324**±.0004	0.0536±.0015	0.0406±.0007	0.0359±.0007	0.0331±.0008
Bankruptcy	0.1692±.0039	0.1756±.0028	0.2692±.0090	**0.1256**±.0059	**0.1256**±.0059
Ljubljana	0.3203±.0035	**0.2437**±.0015	0.2727±.0024	0.2781±.0018	0.2510±.0028
Boston	0.3856±.0016	0.3813±.0042	0.7659±.0029	**0.3118**±.0019	**0.3101**±.0019
Windsor	0.5774±.0028	0.6440±.0042	0.9294±.0029	**0.5046**±.0025	**0.5040**±.0029
Car	0.6752±.0012	0.6517±.0016	0.4005±.0001	**0.0473**±.0007	0.0490±.0007

For each dataset we tested three regular classifiers which do not take order nor monotonicity into account: support vector machines (SVM) with linear kernel [20], j48 decision trees [16] and AdaBoost [8] with decision stump as a base learner. We used their implementations from Weka package [22]. Moreover, we also used two versions of our MORE algorithm. The first one induces decision rules from class unions. The second ("MORE+") employs Stochastic DRSA [4] and induces rules from lower approximations with consistency level 0.5. For SVM and j48, typical parameters from Weka were chosen; for AdaBoost we increased the number of iteration to 100 to make it more competitive; for MORE we have chosen $M = 100$ and $\gamma = 0.5$). For each dataset and for each algorithm, 10-fold cross validation was used and repeated 20 times to decrease the variance of the results. The measured error rate is mean absolute error, which is the value of absolute error loss on the testing set.

The results shown in Table 1 show a great improvement in accuracy when using monotone rule ensembles over the regular classifiers. This is probably due to the fact, that MORE utilizes the domain knowledge. Poor results of ordinary algorithms, e.g. AdaBoost for Windsor dataset, can be explained by the fact, that those algorithms are not adjusted to minimize absolute error. On the other hand, there is only a small improvement (if any) in using lower approximations in rule induction comparing to the rule induction from raw class unions.

5   Conclusions

We introduced a novel rule induction algorithm for ordinal classification problem in the presence of monotonicity constraints. The algorithm uses forward stage-wise additive modeling scheme for generating the ensemble of decision rules for binary problems. We show how to solve the ordinal classification problem with absolute error by solving binary subproblems with zero-one error. Due to specific nature of the problem, a syntax typical to monotone rules was used to find the statistically best ensemble. Moreover, we show how to use DRSA to deal with inconsistent objects. The experimental results show that incorporating such domain knowledge into classification algorithms can dramatically improve the prediction accuracy.

References

1. Błaszczyński, J., Dembczyński, K., Kotłowski, W., Słowiński, R., Szeląg, M.: Ensemble of decision rules. Foundations of Comp. and Decision Sc. (31), 3–4 (2006)
2. Cohen, W., Singer, Y.: A simple, fast, and effective rule learner. National Conference on Artificial Intelligence, pp. 335–342 (1999)
3. Daniels, H., Kamp, B.: Applications of MLP networks to bond rating and house pricing. Neural Computing and Applications 9, 226–234 (1999)
4. Dembczyński, K., Greco, S., Kotłowski, W., Słowiński, R.: Statistical model for rough set approach to multicriteria classification. In: Kok, J.N., Koronacki, J., Lopez de Mantaras, R., Matwin, S., Mladenič, D., Skowron, A. (eds.) PKDD 2007. LNCS (LNAI), vol. 4702, pp. 164–175. Springer, Heidelberg (2007)
5. Dembczyński, K., Pindur, R., Susmaga, R.: Generation of Exhaustive Set of Rules within Dominance-based Rough Set Approach. Electr. Notes Theor. Comp. Sc. 82 (2003)
6. Doumpos, M., Pasiouras, F.: Developing and testing models for replicating credit ratings: A multicriteria approach. Computational Economics 25, 327–341 (2005)
7. Hastie, T., Tibshirani, R., Friedman, J.: Elements of Statistical Learning: Data Mining, Inference, and Prediction. Springer, Heidelberg (2003)
8. Freund, Y., Schapire, R.: A decision-theoretic generalization of on-line learning and an application to boosting. J. of Comp. and System Sc. 55, 119–139 (1997)
9. Friedman, J., Popescu, B.: Predictive Learning via Rule Ensembles. Technical Report, Dept. of Statistics, Stanford University (2005)
10. Greco, S., Matarazzo, B., Słowiński, R.: A new rough set approach to evaluation of bankruptcy risk. In: Zopounidis, C. (ed.) Operational Tools in the Management of Financial Risks, pp. 121–136. Kluwer Academic Publishers, Dordrecht (1998)
11. Greco, S., Matarazzo, B., Słowiński, R.: Rough sets theory for multicriteria decision analysis. European Journal of Operational Research 129, 1–47 (2001)
12. Greco, S., Matarazzo, B., Słowiński, R., Stefanowski, J.: An Algorithm for Induction of Decision Rules Consistent with the Dominance Principle. In: Ziarko, W., Yao, Y. (eds.) RSCTC 2000. LNCS (LNAI), vol. 2005, pp. 304–313. Springer, London (2001)
13. Koop, G.: Analysis of Economic Data. John Wiley and Sons, Chichester (2000)
14. Potharst, R., Feelders, A.J.: Classification trees for problems with monotonicity constraints. SIGKDD Explorations 4, 1–10 (2002)
15. Pawlak, Z.: Rough Sets. In: Theoretical Aspects of Reasoning about Data, Kluwer Academic Publishers, Dordrecht (1991)
16. Quinlan, R.: C4.5: Programs for Machine Learning. Morgan Kaufmann Publishers, San Mateo (1993)
17. Robertson, T., Wright, F., Dykstra, R.: Order Restricted Statistical Inference. John Wiley & Sons, Chichester (1998)
18. Ryu, Y.U., Chandrasekaran, R., Jacob, V.: Data classification using the isotonic separation technique: Application to breast cancer prediction. European Journal of Operational Research 181, 842–854 (2007)
19. Asuncion, A., Newman, D.: (UCI) Repository of machine learning databases, University of California, Irvine, School of Information and Computer Sciences (1998), www.ics.uci.edu/~mlearn/MLRepository.html
20. Vapnik, V.: The Nature of Statistical Learning Theory, 2nd edn. Springer, Heidelberg (1998)
21. Weiss, S., Indurkhya, N.: Lightweight rule induction. In: International Conference on Machine Learning, pp. 1135–1142 (2000)
22. Witten, I., Frank, E.: Data Mining: Practical machine learning tools and techniques, 2nd edn. Morgan Kaufmann, San Francisco (2005)

Case-Based Reasoning Using Gradual Rules Induced from Dominance-Based Rough Approximations

Salvatore Greco[1], Benedetto Matarazzo[1], and Roman Słowiński[2]

[1] Faculty of Economics, University of Catania,
Corso Italia, 55, 95129 – Catania, Italy
[2] Institute of Computing Science, Poznań University of Technology,
60-965 Poznań, and Systems Research Institute,
Polish Academy of Sciences, 01-447 Warsaw, Poland

Abstract. Case-based reasoning (CBR) regards the inference of some proper conclusions related to a new situation by the analysis of similar cases from a memory of previous cases. We propose to represent similarity by gradual decision rules induced from rough approximations of fuzzy sets. Indeed, we are adopting the Dominance-based Rough Set Approach (DRSA) that is particularly appropriate in this context for its ability of handling monotonicity relationship of the type "the more similar is object y to object x, the more credible is that y belongs to the same set as x". At the level of marginal similarity concerning single features, we consider only ordinal properties of similarity, and for the aggregation of marginal similarities, we use a set of gradual decision rules based on the general monotonicity property of comprehensive similarity with respect to marginal similarities. We present formal properties of rough approximations used for CBR.

1 Introduction

The basic idea of case-based reasoning (CBR) (see e.g. [12]) can be found in the following sentence of Hume [11]: "From causes which appear *similar* we expect *similar* effects. This is the sum of all our experimental conclusions." Rephrasing Hume, one can say that "the more similar are the causes, the more similar one expects the effects". For this reason, measuring similarity is the essential point of all approaches to CBR and, in particular, of fuzzy set approach to CBR [1].

Problems with measuring similarity are encountered at two levels:

- at the level of single features: how to define a meaningful similarity measure with respect to a single feature?
- at the level of all features: how to properly aggregate the similarity measures with respect to single features in order to obtain a comprehensive similarity measure?

For the above reasons, we proposed in [8] a new way to deal with CBR, using the Dominance-based Rough Set Approach (DRSA). It tries to be possibly

G. Wang et al. (Eds.): RSKT 2008, LNAI 5009, pp. 268–275, 2008.

"neutral" and "objective" in approaching the problems of measuring similarity at the two levels mentioned above. At the level of similarity concerning single features, we consider only ordinal properties of similarity, and at the level of aggregation, we consider a set of decision rules based on the general monotonicity property of comprehensive similarity with respect to similarity of single features.

The Dominance-based Rough Set Approach (DRSA) has been proposed and characterized by the authors in [4,5,6,7,14]. It operates on a *decision table* composed of a set U of objects described by a set of *condition attributes* C and a set of *decision attributes* D. Decision attributes from set D (often reduced to a single decision attribute d) make a partition of set U into *decision classes*. DRSA takes into account background knowledge about *ordinal* properties of the considered attributes and *monotonic relationships* between condition and decision attributes, e.g.: "the larger the mass and the smaller the distance, the larger the gravity", "the more a tomato is red, the more it is ripe" or "the better the school marks of a pupil, the better his overall classification". Such monotonic relationships are typical for *ordinal classification problems* [10].

DRSA permits, moreover, a natural *hybridization of fuzzy set and rough set concepts*, without using any fuzzy connective [3,9,10]. *Gradual decision rules* following from this approach express monotonic relationships between membership values in conditions and in the decision. For example: "if a car is *speedy* with credibility at least 0.8 and it has *high fuel consumption* with credibility at most 0.7, then it is a *good* car with credibility at least 0.9".

The above gives the reasons of the ability of fuzzy rough approximations in handling monotonic relationships typical for CBR, i.e. monotonicity of the type: "the more similar is y to x, the more credible is that y belongs to the same set as x". Application of DRSA in this context leads to decision rules similar to the gradual decision rules:

"the more object z is similar to a referent object x w.r.t. condition attribute s, the more z is similar to a referent object x w.r.t. decision attribute d",

or, equivalently, but more technically, $s(z, x) \geq \alpha \Rightarrow d(z, x) \geq \alpha$, where functions s and d measure the credibility of similarity with respect to condition attribute and decision attribute, respectively. When there are multiple condition and decision attributes, functions s and d aggregate similarity with respect to these attributes.

The decision rules we propose do not need the aggregation of the similarity with respect to different attributes into one comprehensive similarity. This is important, because it permits to avoid using aggregation functions (involving operators, like weighted L_p norms, min, etc.) which are always arbitrary to some extent [2]. Moreover, the gradual decision rules we propose permit to consider different thresholds for degrees of credibility in the premise and in the conclusion, which is not the case for classical gradual rules.

The paper aims to state formal properties of DRSA applied to CBR presented in [8] and improved in [10]. The paper is organized as follows. Section 2 recalls DRSA applied to CBR. In section 3, we give relevant formal properties of DRSA applied to CBR. The final section contains conclusions.

2   DRSA for CBR

DRSA approximation of fuzzy sets for CBR is based on a *pairwise fuzzy information base* being the 3-tuple

$$\boldsymbol{B} = \; <U,F,\sigma>,$$

where U is a finite set of *objects* (universe), $F=\{f_1,f_2,...,f_m\}$ is a finite set of *features*, and $\sigma \; : \; U \times U \times F \rightarrow [0,1]$ is a function such that $\sigma(x,y,f_h) \in [0,1]$ expresses the credibility that object x is similar to object y w.r.t. feature f_h. The minimal requirement function σ must satisfy is that, for all $x \in U$ and for all $f_h \in F$, $\sigma(x,x,f_h)=1$. Therefore, each pair of objects $(x,y) \in U \times U$ is described by a vector

$$Des_F(x,y)=[\sigma(x,y,f_1), \; \ldots, \; \sigma(x,y,f_m)]$$

called *description* of (x,y) in terms of the credibilities of similarity with respect to features from F; it represents the available information about similarity between x and y. Obviously, similarity between x and y, $x,y \in U$, can be described in terms of any non-empty subset $E \subseteq F$, and in this case we have

$$Des_E(x,y)=[\sigma(x,y,f_h), \; f_h \in E].$$

With respect to any $E \subseteq F$, we can define the dominance relation D_E on $U \times U$ as follows: for any $x,y,w,z \in U$, (x,y) dominates (w,z) with respect to E (denotation $(x,y)D_E(w,z)$) if, for any $f_h \in E$,

$$\sigma(x,y,f_h) \geq \sigma(w,z,f_h).$$

Given $E \subseteq F$ and $x,y \in U$, let

$$D_E^+(y,x) = \{w \in U : (w,x)D_E(y,x)\},$$
$$D_E^-(y,x) = \{w \in U : (y,x)D_E(w,x)\}.$$

In the pair (y,x), x is considered to be a *reference object*, while y can be called a *limit object*, because it is conditioning the membership of w in $D_E^+(y,x)$ and in $D_E^-(y,x)$.

For each $X \subseteq U$, we can define its *upward lower approximation* $\underline{E}(x)_\sigma^{(>)}(X)$ and its *upward upper approximation* $\overline{E}(x)_\sigma^{(>)}(X)$, based on similarity σ with respect to $E \subseteq F$ and $x \in U$, as:

$$\underline{E}(x)_\sigma^{(>)}(X) = \left\{y \in U : D_E^+(y,x) \subseteq X\right\},$$
$$\overline{E}(x)_\sigma^{(>)}(X) = \left\{y \in U : D_E^-(y,x) \cap X \neq \emptyset\right\}.$$

Analogously, for each $X \subseteq U$, we can define its *downward lower approximation* $\underline{E}(x)_\sigma^{(<)}(X)$ and its *downward upper approximation* $\overline{E}(x)_\sigma^{(<)}(X)$, based on similarity σ with respect to $E \subseteq F$ and $x \in U$, as:

$$\underline{E}(x)_\sigma^{(<)}(X) = \left\{ y \in U : D_E^-(y, x) \subseteq X \right\},$$

$$\overline{E}(x)_\sigma^{(<)}(X) = \left\{ y \in U : D_E^+(y, x) \cap X \neq \emptyset \right\}.$$

Observe, that in the above definition of rough approximations $\underline{E}(x)_\sigma^{(>)}(X)$, $\overline{E}(x)_\sigma^{(>)}(X)$, $\underline{E}(x)_\sigma^{(<)}(X)$, $\overline{E}(x)_\sigma^{(<)}(X)$, the elementary sets are sets $D_E^-(y, x)$ and $D_E^+(y, x)$, $E \subseteq F$, $x, y \in U$, instead of equivalence classes of the indiscernibility relation in the classical rough set theory. Observe, moreover, that the rough approximations $\underline{E}(x)_\sigma^{(>)}(X)$, $\overline{E}(x)_\sigma^{(>)}(X)$, $\underline{E}(x)_\sigma^{(<)}(X)$, $\overline{E}(x)_\sigma^{(<)}(X)$ can be expressed as unions of the elementary sets, i.e. for any $X \subseteq U$ and $E \subseteq F$,

1) $\displaystyle \underline{E}(x)_\sigma^{(>)}(X) = \bigcup_{y \in U} \left\{ D_E^+(y, x) : D_E^+(y, x) \subseteq X \right\},$

2) $\displaystyle \overline{E}(x)_\sigma^{(>)}(X) = \bigcup_{y \in U} \left\{ D_E^+(y, x) : D_E^-(y, x) \cap X \neq \emptyset \right\},$

3) $\displaystyle \underline{E}(x)_\sigma^{(<)}(X) = \bigcup_{y \in U} \left\{ D_E^-(y, x) : D_E^-(y, x) \subseteq X \right\},$

4) $\displaystyle \overline{E}(x)_\sigma^{(<)}(X) = \bigcup_{y \in U} \left\{ D_E^-(y, x) : D_E^+(y, x) \cap X \neq \emptyset \right\}.$

The rough approximations $\underline{E}(x)_\sigma^{(>)}(X), \overline{E}(x)_\sigma^{(>)}(X), \underline{E}(x)_\sigma^{(<)}(X), \overline{E}(x)_\sigma^{(<)}(X)$ can be used to analyze data concerning *gradual membership* of objects to some concepts representing properties of objects and to decision classes. This analysis takes into account the following monotonicity principle: "the greater the degree to which an object w is similar to a reference object x with respect to features $E \subseteq F$, the greater its degree of membership to a considered decision class".

This principle can be formalized as follows. Let us consider a *fuzzy set* X in U, characterized by the membership function $\mu_X : U \to [0, 1]$. This fuzzy set represents a decision class of interest, such that function μ specifies a graded membership of objects from U to decision class X. For each cutting level $\alpha \in [0, 1]$, we can consider the following sets:

− weak upward cut of fuzzy set X:

$$X^{\geq \alpha} = \{x \in U : \mu(x) \geq \alpha\},$$

− strict upward cut of fuzzy set X:

$$X^{> \alpha} = \{x \in U : \mu(x) > \alpha\},$$

− weak downward cut of fuzzy set X:

$$X^{\leq \alpha} = \{x \in U : \mu(x) \leq \alpha\},$$

− strict downward cut of fuzzy set X:

$$X^{< \alpha} = \{x \in U : \mu(x) < \alpha\}.$$

Let us remark that, for any fuzzy set X and for any $\alpha \in [0,1]$, we have

$$U - X^{\geq \alpha} = X^{<\alpha}, \quad U - X^{\leq \alpha} = X^{>\alpha},$$

$$U - X^{>\alpha} = X^{\leq \alpha}, \quad U - X^{<\alpha} = X^{\geq \alpha}.$$

As the above cuts of fuzzy set X are crisp sets, the upward and downward rough approximations obviously apply to them. Let us denote by $Y \in U$ any such cut set, and rewrite the rough approximations $\underline{E}(x)_\sigma^{(>)}(Y)$, $\overline{E}(x)_\sigma^{(>)}(Y)$, $\underline{E}(x)_\sigma^{(<)}(Y)$, $\overline{E}(x)_\sigma^{(<)}(Y)$ as follows:

$$\underline{E}(x)_\sigma^{(>)}(Y) = \{y \in U : \forall w \in U, \ (w,x)D_E(y,x) \Rightarrow w \in Y\},$$

$$\overline{E}(x)_\sigma^{(>)}(Y) = \{y \in U : \exists w \in U \text{ such that } (y,x)D_E(w,x) \text{ and } w \in Y\},$$

$$\underline{E}(x)_\sigma^{(<)}(Y) = \{y \in U : \forall w \in U, \ (y,x)D_E(w,x) \Rightarrow w \in Y\},$$

$$\overline{E}(x)_\sigma^{(<)}(Y) = \{y \in U : \exists w \in U \text{ such that } (w,x)D_E(y,x) \text{ and } w \in Y\}.$$

This formulation of the rough approximation is concordant with the syntax of the decision rules induced by means of DRSA from a fuzzy pairwise information base. For example $\underline{E}(x)_\sigma^{(>)}(Y)$ is concordant with decision rules of the type:

"if object w is similar to object x w.r.t. feature f_{i1} to degree at least h_{i1} and w.r.t. feature f_{i2} to degree at least h_{i2} and ... and w.r.t. feature f_{ip} to degree at least h_{ip}, then object w belongs to set Y",

where $\{f_{i1}, \ldots, f_{ip}\} \subseteq E$ and $h_{i1}, \ldots, h_{ip} \in [0,1]$.

The above definitions of rough approximations and the syntax of decision rules are based on ordinal properties of similarity relations only. In fact, no algebraic operation, such as sum or product, involving cardinal properties of function σ measuring credibility of similarity relations is considered. This is an important characteristic of our approach in comparison with alternative approaches to CBR.

3 Formal Properties of DRSA for CBR

Given a family of fuzzy sets $\mathbf{X} = \{X_1, X_2, \ldots, X_p\}$ on U, whose respective membership functions are $\mu_1, \mu_2, \ldots, \mu_p$, let $P^>(\mathbf{X})$ be the set of all the sets obtained through unions and intersections of weak and strict upward cuts of component fuzzy sets. Analogously, let $P^<(\mathbf{X})$ be the set of all the sets obtained through unions and intersections of weak and strict downward cuts of component fuzzy sets.

$P^>(\mathbf{X})$ and $P^<(\mathbf{X})$ are closed under set union and set intersection operations, i.e. for all $Y_1, Y_2 \in P^>(\mathbf{X})$, $Y_1 \cup Y_2$ and $Y_1 \cap Y_2$ belong to $P^>(\mathbf{X})$, as well as for all $W_1, W_2 \in P^<(\mathbf{X})$, $W_1 \cup W_2$ and $W_1 \cap W_2$ belong to $P^<(\mathbf{X})$. Observe, moreover, that the universe U and the empty set $\emptyset$ belong both to $P^>(\mathbf{X})$ and to $P^<(\mathbf{X})$ because, for any fuzzy set $X_i \in \mathbf{X}$,

$$U = X_i^{\geq 0} = X_i^{\leq 1}$$

and
$$\emptyset = X_i^{>1} = X_i^{<0}.$$

The following theorem states some important properties of the dominance-based rough approximations for case based reasoning.

Theorem

1. For any $Y \in P^>(\mathbf{X})$ and for any $W \in P^<(\mathbf{X})$ and for any $E \subseteq F$ and for any $x \in U$,

$$\underline{E}(x)_\sigma^{(>)}(Y) \subseteq Y \subseteq \overline{E}(x)_\sigma^{(>)}(Y), \quad \underline{E}(x)_\sigma^{(<)}(W) \subseteq W \subseteq \overline{E}(x)_\sigma^{(<)}(W).$$

2. For any $E \subseteq F$ and for any $x \in U$,

$$\underline{E}(x)_\sigma^{(>)}(\emptyset) = \overline{E}(x)_\sigma^{(>)}(\emptyset) = \emptyset, \quad \underline{E}(x)_\sigma^{(<)}(\emptyset) = \overline{E}(x)_\sigma^{(<)}(\emptyset) = \emptyset,$$

$$\underline{E}(x)_\sigma^{(>)}(U) = \overline{E}(x)_\sigma^{(>)}(U) = U, \quad \underline{E}(x)_\sigma^{(<)}(U) = \overline{E}(x)_\sigma^{(<)}(U) = U.$$

3. For any $E \subseteq F$, for any $Y_1, Y_2 \in P^>(\mathbf{X})$ and for any $W_1, W_2 \in P^<(\mathbf{X})$, for any $x \in U$

$$\overline{E}(x)_\sigma^{(>)}(Y_1 \cup Y_2) = \overline{E}(x)_\sigma^{(>)}(Y_1) \cup \overline{E}(x)_\sigma^{(>)}(Y_2),$$

$$\overline{E}(x)_\sigma^{(<)}(W_1 \cup W_2) = \overline{E}(x)_\sigma^{(<)}(W_1) \cup \overline{E}(x)_\sigma^{(<)}(W_2).$$

4. For any $E \subseteq F$, for any $Y_1, Y_2 \in P^>(\mathbf{X})$ and for any $W_1, W_2 \in P^<(\mathbf{X})$ and for any $x \in U$,

$$\underline{E}(x)_\sigma^{(>)}(Y_1 \cap Y_2) = \underline{E}(x)_\sigma^{(>)}(Y_1) \cap \underline{E}(x)_\sigma^{(>)}(Y_2),$$

$$\underline{E}(x)_\sigma^{(<)}(W_1 \cap W_2) = \underline{E}(x)_\sigma^{(<)}(W_1) \cap \underline{E}(x)_\sigma^{(<)}(W_2).$$

5. For any $E \subseteq F$, for any $Y_1, Y_2 \in P^>(\mathbf{X})$ and for any $W_1, W_2 \in P^<(\mathbf{X})$ and for any $x \in U$,

$$Y_1 \subseteq Y_2 \Rightarrow \underline{E}(x)_\sigma^{(>)}(Y_1) \subseteq \underline{E}(x)_\sigma^{(>)}(Y_2),$$

$$W_1 \subseteq W_2 \Rightarrow \underline{E}(x)_\sigma^{(<)}(W_1) \subseteq \underline{E}(x)_\sigma^{(<)}(W_2).$$

6. For any $E \subseteq F$, for any $Y_1, Y_2 \in P^>(\mathbf{X})$ and for any $W_1, W_2 \in P^<(\mathbf{X})$ and for any $x \in U$,

$$Y_1 \subseteq Y_2 \Rightarrow \overline{E}(x)_\sigma^{(>)}(Y_1) \subseteq \overline{E}(x)_\sigma^{(>)}(Y_2),$$

$$W_1 \subseteq W_2 \Rightarrow \overline{E}(x)_\sigma^{(<)}(W_1) \subseteq \overline{E}(x)_\sigma^{(<)}(W_2).$$

7. For any $E \subseteq F$, for any $Y_1, Y_2 \in P^>(\mathbf{X})$ and for any $W_1, W_2 \in P^<(\mathbf{X})$ and for any $x \in U$,

$$\underline{E}(x)_\sigma^{(>)}(Y_1 \cup Y_2) \supseteq \underline{E}(x)_\sigma^{(>)}(Y_1) \cup \underline{E}(x)_\sigma^{(>)}(Y_2),$$

$$\underline{E}(x)_\sigma^{(<)}(W_1 \cup W_2) \supseteq \underline{E}(x)_\sigma^{(<)}(W_1) \cup \underline{E}(x)_\sigma^{(<)}(W_2).$$

8. For any $E \subseteq F$, for any $Y_1, Y_2 \in P^>(\mathbf{X})$ and for any $W_1, W_2 \in P^<(\mathbf{X})$ and for any $x \in U$,

$$\overline{E}(x)_\sigma^{\langle > \rangle}(Y_1 \cap Y_2) \subseteq \overline{E}(x)_\sigma^{\langle > \rangle}(Y_1) \cap \overline{E}(x)_\sigma^{\langle > \rangle}(Y_2),$$

$$\overline{E}(x)_\sigma^{\langle < \rangle}(W_1 \cap W_2) \subseteq \overline{E}(x)_\sigma^{\langle < \rangle}(W_1) \cap \overline{E}(x)_\sigma^{\langle < \rangle}(W_2).$$

9. For any $E \subseteq F$, for any $Y \in P^>(\mathbf{X})$ and for any $W \in P^<(\mathbf{X})$ and for any $x \in U$,

$$\underline{E}(x)_\sigma^{\langle > \rangle}(U - W) = U - \overline{E}(x)_\sigma^{\langle < \rangle}(W),$$

$$\underline{E}(x)_\sigma^{\langle < \rangle}(U - Y) = U - \overline{E}(x)_\sigma^{\langle > \rangle}(Y).$$

10. For any $E \subseteq F$, for any $Y \in P^>(\mathbf{X})$ and for any $W \in P^<(\mathbf{X})$ and for any $x \in U$,

$$\overline{E}(x)_\sigma^{\langle > \rangle}(U - W) = U - \underline{E}(x)_\sigma^{\langle < \rangle}(W),$$

$$\overline{E}(x)_\sigma^{\langle < \rangle}(U - Y) = U - \underline{E}(x)_\sigma^{\langle > \rangle}(Y).$$

11. For any $E \subseteq F$, for any $Y \in P^>(\mathbf{X})$ and for any $W \in P^<(\mathbf{X})$ and for any $x \in U$,

$$\underline{E}(x)_\sigma^{\langle > \rangle}[\underline{E}(x)_\sigma^{\langle > \rangle}(Y)] = \overline{E}(x)_\sigma^{\langle > \rangle}[\underline{E}(x)_\sigma^{\langle > \rangle}(Y)] = \underline{E}(x)_\sigma^{\langle > \rangle}(Y),$$

$$\underline{E}(x)_\sigma^{\langle < \rangle}[\underline{E}(x)_\sigma^{\langle < \rangle}(W)] = \overline{E}(x)_\sigma^{\langle < \rangle}[\underline{E}(x)_\sigma^{\langle < \rangle}(W)] = \underline{E}(x)_\sigma^{\langle < \rangle}(W).$$

12. For any $E \subseteq F$, for any $Y \in P^>(\mathbf{X})$ and for any $W \in P^<(\mathbf{X})$ and for any $x \in U$,

$$\overline{E}(x)_\sigma^{\langle > \rangle}[\overline{E}(x)_\sigma^{\langle > \rangle}(Y)] = \underline{E}(x)_\sigma^{\langle > \rangle}[\overline{E}(x)_\sigma^{\langle > \rangle}(Y)] = \overline{E}(x)_\sigma^{\langle > \rangle}(Y),$$

$$\overline{E}(x)_\sigma^{\langle < \rangle}[\overline{E}(x)_\sigma^{\langle < \rangle}(W)] = \underline{E}(x)_\sigma^{\langle < \rangle}[\overline{E}(x)_\sigma^{\langle < \rangle}(W)] = \overline{E}(x)_\sigma^{\langle < \rangle}(W). \qquad \square$$

The proof is omitted for the lack of space. The results given in the above Theorem correspond to well known properties of classical rough sets (see the original properties numbered in the same way in [13]), however, with the noticeable exception of properties 9 and 10 characterizing the specific nature of complementarity relations within DRSA.

4   Conclusions

DRSA applied to CBR exploits only ordinal character of similarity and avoids harmful aggregation of similarity measures with respect to single features into a real valued function. Instead, it proposes a very general aggregation using gradual decision rules induced from dominance-based rough approximations of fuzzy sets, corresponding to decision classes with degree of membership not smaller (or not greater) than some level α. The formal properties of the fuzzy-rough approximations prove that DRSA applied to CBR enjoys very good properties requiring relatively weak assumptions.

References

1. Dubois, D., Prade, H., Esteva, F., Garcia, P., Godo, L., Lopez de Mantara, R.: Fuzzy Set Modelling in Case-based Reasoning. International Journal of Intelligent Systems 13, 345–373 (1998)
2. Gilboa, I., Schmeidler, D.: A Theory of Case-Based Decisions. Cambridge University Press, Cambridge (2001)
3. Greco, S., Inuiguchi, M., Słowiński, R.: Fuzzy rough sets and multiple-premise gradual decision rules. International Journal of Approximate Reasoning 41, 179–211 (2006)
4. Greco, S., Matarazzo, B., Słowiński, R.: The use of rough sets and fuzzy sets in MCDM. In: Gal, T., Stewart, T., Hanne, T. (eds.) Advances in Multiple Criteria Decision Making, ch. 14, p. 14.1–14.59. Kluwer Academic Publishers, Boston (1999)
5. Greco, S., Matarazzo, B., Słowiński, R.: Rough sets theory for multicriteria decision analysis. European Journal of Operational Research 129, 1–47 (2001)
6. Greco, S., Matarazzo, B., Słowiński, R.: Dominance-Based Rough Set Approach to Knowledge Discovery (I) - General Perspective, ch. 20, and (II) - Extensions and Applications, ch. 21. In: Zhong, N., Liu, J. (eds.) Intelligent Technologies for Information Analysis, pp. 513–612. Springer, Berlin (2004)
7. Greco, S., Matarazzo, B., Słowiński, R.: Decision rule approach, ch. 13. In: Figueira, J., Greco, S., Ehrgott, M. (eds.) Multiple Criteria Decision Analysis: State of the Art Surveys, pp. 507–563. Springer, Berlin (2005)
8. Greco, S., Matarazzo, B., Słowiński, R.: Dominance-Based Rough Set Approach to Case-Based Reasoning. In: Torra, V., Narukawa, Y., Valls, A., Domingo-Ferrer, J. (eds.) MDAI 2006. LNCS (LNAI), vol. 3885, pp. 7–18. Springer, Heidelberg (2006)
9. Greco, S., Matarazzo, B., Słowiński, R.: Dominance-Based Rough Set Approach as a Proper Way of Handling Graduality in Rough Set Theory. In: Peters, J.F., Skowron, A., Marek, V.W., Orłowska, E., Słowiński, R., Ziarko, W. (eds.) Transactions on Rough Sets VII. LNCS, vol. 4400, pp. 36–52. Springer, Heidelberg (2007)
10. Greco, S., Matarazzo, B., Słowiński, R.: Granular computing for reasoning about ordered data: the dominance-based rough set approach. In: Pedrycz, W., Skowron, A., Kreinovich, V. (eds.) Handbook of Granular Computing, ch. 15, John Wiley & Sons, New York (2008)
11. Hume, D.: An Enquiry Concerning Human Understanding. Clarendon Press, Oxford (1748)
12. Kolodner, J.: Case-Based Reasoning. Morgan Kaufmann, San Mateo (1993)
13. Pawlak, Z.: Rough Sets. Kluwer, Dordrecht (1991)
14. Słowiński, R., Greco, S., Matarazzo, B.: Rough set based decision support. In: Burke, E.K., Kendall, G. (eds.) Search Methodologies: Introductory Tutorials in Optimization and Decision Support Techniques, ch. 16, pp. 475–527. Springer, New York (2005)

The Incremental Learning Methodology of VPRS Based on Complete Information System

Dun Liu[1], Pei Hu[2], and Chaozhe Jiang[3]

[1] Department of Economics and Management, Southwest Jiaotong University
chengdu, 610031, China
newton83@163.com
[2] Department of Economics and Management, Southwest Jiaotong University
chengdu, 610031, China
huhupei@126.com
[3] Department of Transportation, Southwest Jiaotong University
chengdu, 610031, China
jiangchaozhe@163.com

Abstract. By considering the inconsistent character in many information system, the variable precision rough set (VPRS) model is introduced
to solve decision-making problems in this paper. Firstly, the integrations
of the interesting and discernibility of knowledge based on VPRS model
are defined, and an approach for available knowledge is proposed. Then,
the incremental learning method of VPRS model in dynamic environment and the incremental updating for accuracy and coverage are also
studied. At last, a case is studied to validate the feasibility of our method.

Keywords: VPRS, Accuracy, Coverage, Incremental learning.

1 Introduction

The people would come up against many complicated and uncertain problems
which include some inaccuracy, inconsistent, incomplete information in today's
life. In generally, the probability in statistics and the membership function in
fuzzy mathematics were used to describe the uncertainty. Since Professor Pawlak
(1982) proposed the Rough sets Theory (RST) [1], the theory had developed very
fast in last 20 years to deal with the uncertain problems and had been used in
many fields such as business crisis prediction, database marketing and financial
investment, but the Pawlak RST had some shortages to deal with the inconsistent information system. Then Zarkio (1993) proposed the VPRS model to
extend the classical strict relation by introducing a probability value β (misclas-
sification parameter) [2], the new relation could induce some toleration rules,
which made the process of decision-making more reasonably and realism.

By considering the significance of VPRS model, many scholars had studied the
model in both theories and applications such as the methodologies of reduction
and classification [3-5]; the incremental learning methods of VPRS [6-7] and the
applications of VPRS for prediction [8-10].

G. Wang et al. (Eds.): RSKT 2008, LNAI 5009, pp. 276–283, 2008.

In this paper, some detailed and deeply works about VPRS model were discussed. Some basic concepts were reviewed in Section 2; the concept of available knowledge in our opinion was proposed in Section 3; the incremental learning method of VPRS in dynamic environment was studied in Section 4; the Section 5 showed a case to validate our model.

2   Preliminaries

The basic concepts, notations and results of rough sets as well as their extensions are briefly reviewed.

Definition 1. [11]: *A complete information system is defined as a pair* $S = U/R$, *where* U *is a non-empty finite set of objects. Let*:
$$P(X, Y) = \{1 - |X \bigcap Y|/|X|, |X| > 0; 0, |X| = 0\}$$

Where, $|\cdot|$ stands for the number of elements in sets, and we call $P(X, Y)$ as the relative misclassification parameter.

Definition 2. [11]: *Suppose* $S = U/R$ *is a complete information system. For* $\forall X \subseteq U$, *the* β-*lower approximation*, β-*upper approximation*, β-*boundary region are defined as follow*:
$$\underline{R}_\beta = \bigcup\{[x]_R | P([x]_R, Y) \leq \beta\};$$
$$\overline{R}^\beta = \bigcup\{[x]_R | P([x]_R, Y) < 1 - \beta\};$$
$$BNG_R = \bigcup\{[x]_R | \beta < P([x]_R, Y) < 1 - \beta\}.$$

In addition, we denote β-inclusion relation $\subseteq_\beta$ as: $X \subseteq_\beta Y \Leftrightarrow P(X, Y) < \beta$. In generally, we call β-inclusion relation as majority inclusion relation.

3   Interesting and Discernibility of Knowledge

Given an information system, the people usually want to obtain some available decision knowledge. On the one hand, it should be performed as high accuracy and high coverage simultaneously, which is called the interesting of the knowledge [11]; on the other hand, the knowledge should be discernibility [12].
Firstly, we consider the interesting of knowledge based on VPRS model.

Definition 3. [12]: *Suppose* $S = \{U, C \bigcup D, V, F\}$ *is a complete information system, we denote* $U/C = \{X_1, X_2, \cdots, X_m\}$, $U/D = \{D_1, D_2, \cdots, D_n\}$. *For* $\forall X_i \subseteq U/C$ *and* $\forall D_j \subseteq U/D$, *the support, accuracy and coverage of* $X_i \rightarrow D_j$ *are defined in the follow.*
Support of $X_i \rightarrow D_j$: $Sup(D_j | X_i) = |X_i \bigcap D_j|$;
Accuracy of $X_i \rightarrow D_j$: $Acc(D_j | X_i) = |X_i \bigcap D_j|/|X_i|$;
Coverage of $X_i \rightarrow D_j$: $Cov(D_j | X_i) = |X_i \bigcap D_j|/|D_j|$.

Based on the definition 3, we can construct the probability distributing matrix (also can be called the accuracy matrix) and the coverage matrix as follow:

$$Acc(D|X)=\begin{pmatrix} Acc(D_1|X_1) & Acc(D_1|X_2) & \cdots & Acc(D_1|X_m) \\ Acc(D_2|X_1) & Acc(D_2|X_2) & \cdots & Acc(D_2|X_m) \\ \vdots & \vdots & \vdots & \vdots \\ Acc(D_n|X_1) & Acc(D_n|X_2) & \ldots & Acc(D_n|X_m) \end{pmatrix} \quad (1)$$

$$Cov(D|X)=\begin{pmatrix} Cov(D_1|X_1) & Cov(D_1|X_2) & \cdots & Cov(D_1|X_m) \\ Cov(D_2|X_1) & Cov(D_2|X_2) & \cdots & Cov(D_2|X_m) \\ \vdots & \vdots & \vdots & \vdots \\ Cov(D_n|X_1) & Cov(D_n|X_2) & \ldots & Cov(D_n|X_m) \end{pmatrix} \quad (2)$$

Proposition 1. *For $\forall X_i \subseteq U/D (i = 1, 2, \cdots, m)$ in matrix (1), it satisfies :*
(1). $\sum_{j=1}^{n}(D_j|X_i) = 1$; (2). $0 \le (D_j|X_i) \le 1$.

Specially, If $\exists X_i$, it satisfies $Acc(D_j|X_i) = 1$, we have $X_i \subseteq D_j$. In addition, for $\forall X_i \subseteq U/C$, if $\max(D_j|X_i) \le 0.5$ holds, for $\forall \beta \subseteq [0, 0.5)$, we can get that $\overline{R}^\beta = \phi$. In this case, we can't find any interesting knowledge from the system.

Proposition 2. *For $\forall D_j \subseteq U/D (j = 1, 2, \cdots, n)$ in matrix (2), it satisfies :*
(1). $\sum_{i=1}^{m}(D_j|X_i) = 1$; (2). $0 \le (D_j|X_i) \le 1$.

In generally, we usually set two threshold value α and β when dealing with the practical problem. For $\forall X_i (i = 1, 2, \cdots, n)$ and $\forall D_j (j = 1, 2, \cdots, m)$, if $Acc(D_j|X_i) \ge \beta$ and $Cov(D_j|X_i) \ge \alpha$ hold, we call the rules induced by $X_i \to D_j$ are interesting knowledge.

After discussing the interesting knowledge acquisition, the affections of β to the discernibility of information system are discussed in the follow.

Definition 4. [11]: *Suppose $S = U/R$ is a complete information system, For $\forall X \subseteq U$, if $BNG_R = \phi(0 \le \beta < 0.5)$, we call X is β relative discernibility.*

Due to definition 3, for $\forall X \subseteq U$, we denote $\xi(X)$ as the minimum threshold to sustain the discernibility of the system, if it satisfies: $\xi(X) = \max(m_1, m_2)$. where, $m_1 = 1 - \min\{P(X_i)|\forall X_i \subseteq U/C, P(X_i, X) > 0.5\}$, $m_2 = \max\{P(X_i)| \forall X_i \subseteq U/C, P(X_i, X) < 0.5\}$. If $\exists \beta \in [\xi(X), 0.5)$, we can get $\underline{R}_\beta = \overline{R}_\beta$. In this case, the information system will be discernibility.

In addition, according to the definition 1 and definition 3, we can get: $P(X, D) + Acc(D|X) = 1$. So, as for the information system, we have $\beta = \max(\xi(D_1), \xi(D_2), \cdots, \xi(D_n))$. If $\beta \in [0, 0.5)$, the knowledge in the system will be discernibility.

Finally, if the information system is available (interesting and discernibility simultaneously), we have: $\xi(D) \in [0, 0.5)$ and for $\forall D_j \in D$, it satisfies $m_1 \le m_2$.

4   The Incremental Learning Method

For a complete information system $S = \{U, C \bigcup D, V, f\}$, we denote the partition based on condition attributes and decision attribute as $U^{(t)}/C = \{X_1, X_2, \cdots, X_m\}$ and $U^{(t)}/D = \{D_1, D_2, \cdots, D_n\}$ respectively in the tth time. $G^{(t)}$ is the element transfer function and $U^{(t+1)} = G^{(t)}(U^{(t)})$. In addition, $G^{(t)} = g^{(t)} \bigcup \overline{g}^{(t)}$, $g^{(t)}$ and $\overline{g}^{(t)}$ are the immigration and emigration functions respectively. It satisfies: $\exists x \overline{\in} U^{(t)}$, such that $g^{(t)} = u \in U^{(t+1)}$; $\exists x' \in U^{(t)}$, such that $\overline{g}(x') = u \overline{\in} U^{(t+1)}$.

Then, let's discuss the conditions when elements are entering and getting out of the information system.

By considering the phenomena that the process of element transformation can divide into individual, we just need to discuss the condition when one element enters or gets out of the system.

For $\forall x \in U^{(t)}$ and $\exists x \overline{\in} U^{(t+1)}$, x is an element to immigrate to the system, and there are four cases hold.

(1).$x \overline{\in} X_i (i = 1, 2, \cdots, n)$, x has the different antecedent with the element of $U^{(t)}$, and there are two cases hold.

(1.1). $x \overline{\in} D_j (j = 1, 2, \cdots, m)$the immigration of x is independent with the former system $U^{(t)}$, it generates a new conditional class X_{m+1} and a new decision class D_{n+1}, at this time, $Acc^{(t+1)}_{(m+1, n+1)} = 1$, $Cov^{(t+1)}_{(m+1, n+1)} = 1$.

(1.2). $\exists j \in [1, 2, \cdots, n]$, such that $x \in D_j$, which means the immigration of X add the cardinal number of D_j, and X also forms a new conditional class X_{m+1}.

For $X_{m+1} \to D_j$, $Acc^{(t+1)}_{(m+1, j)} = 1, Cov^{(t+1)}_{(m+1, j)} = 1/(|D_j| + 1)$;

For $X_u \to D_j (u \neq m + 1)$, $Acc^{(t+1)}_{(u, j)} = Acc^t_{(u, j)}$, $Cov^{(t+1)}_{(u, j)} = |X_u \bigcap D_j|/(|D_j| + 1)$. In addition,when $X_{m+1} \bigcap D_k (k \neq j)$, we have: $Acc^{(t+1)}_{(m+1, k)} = Cov^{(t+1)}_{(m+1, k)} = 0$.

(2). $\exists i \in [1, 2, \cdots, m]$, such that $x \in X_i$, X has the same antecedent with X_i, and there are two cases hold.

(2.1). $\exists j \in [1, 2, \cdots, n]$, such that $x \in D_j$, which means $(X_i \bigcup x) \bigcap D_j \neq \phi$, so the immigration of x support the rule of $X_i \to D_j$.

For $X_i \to D_j$,$Acc^{(t+1)}_{(i, j)} = (|X_i \bigcap D_j| + 1)/(|X_i| + 1)$; $Cov^{(t+1)}_{(i, j)} = (|X_i \bigcap D_j| + 1)/(|D_i| + 1)$.

For $X_i \to D_k (k \neq j)$, $Acc^{(t+1)}_{(i, k)} = (|X_i \bigcap D_k|)/(|X_i| + 1)$, $Cov^{(t+1)}_{(i, k)} = Cov^{(t)}_{(i, k)}$;

For $X_u \to D_j (u \neq m+1)$, $Acc^{(t+1)}_{(u, j)} = Acc^{(t)}_{(u, j)}$, $Cov^{(t+1)}_{(u, j)} = (X_u \bigcap D_j|)/(|D_j| + 1)$.

In particularly, if $X_i \subseteq D_j$, the immigration of X don't change the consistent of $X_i \to D_j$, it just support the rule; If $X_i \bigcap D_j = \phi$, for the condition equivalence class X_i, the immigration of x brings a inconsistent decision rule, we have: for $X_i \to D_j$,$Acc^{(t+1)}_{(i, j)} = (|X_i \bigcap D_j| + 1)/(|X_i| + 1) = 1/(|X_i| + 1)$, $Cov^{(t+1)}_{(i, j)} = (|X_i \bigcap D_j| + 1)/(|D_j| + 1) = 1/(|D_j| + 1)$.

(2.2). $x \overline{\in} D_j (j = 1, 2, \cdots, m)$the immigration of x brings a new inconsistent rule, and x forms a new decision class D_{n+1}.

For $X_i \to D_{n+1}$, $Acc^{(t+1)}_{(i, n+1)} = 1/(|X_i| + 1)$, $Cov^{(t+1)}_{(i, n+1)} = 1$;

For $X_i \to D_k (k \neq n + 1)$, $Acc^{(t+1)}_{(i, k)} = (|X_i \bigcap D_k|)/(|X_i| + 1)$, $Cov^{(t+1)}_{(i, k)} = Cov^{(t)}_{(i, k)}$;

For $X_u \to D_{n+1}$, due to $X_u \bigcap D_{n+1} = \phi$, then $Acc^{(t+1)}_{(u, n+1)} = Cov^{(t+1)}_{(u, n+1)} = 0$.

In the upper discussion, we are taking about the condition that the element enters to the system, then we are discussing the case that the element goes out of the system.

For $\forall x \overline{\in} U^{(t)}$ and $\exists x' \in U^{(t+1)}$, $\exists i \in [1, 2, \cdots, m], j \in [1, 2, \cdots, n]$, such that: $x' \in X_i$, $x' \in D_j$ in the tth time. Due to $x' \in X_i$, $x' \in D_j$, we have $X_i \bigcap D_j \neq \phi$, the new system is changed when x' is deleted:

For $X_i \rightarrow D_j$, $Acc_{(i,j)}^{(t+1)} = (|X_i \bigcap D_j| - 1)/(|X_i| - 1)$, $Cov_{(i,j)}^{(t+1)} = (|X_i \bigcap D_j| - 1)/(|D_i| - 1)$;

For $X_i \rightarrow D_k(k \neq j)$, $Acc_{(i,k)}^{(t+1)} = (|X_i \bigcap D_k|)/(|X_i| - 1)$, $Cov_{(i,k)}^{(t+1)} = Cov_{(i,k)}^{(t)}$;

For $X_u \rightarrow D_j(u \neq i)$, $Acc_{(u,j)}^{(t+1)} = Acc_{(u,j)}^{(t)}$, $Cov_{(u,j)}^{(t+1)} = (|X_u \bigcap D_j|)/(|D_j| - 1)$.

In particularly:

(1). If $X_u \subseteq D_j$, when x' is deleted, for $X_i \rightarrow D_j$: $Acc_{(i,j)}^{(t+1)} = (|X_i \bigcap D_j| - 1)/(|X_i| - 1) = (|X_i| - 1)/(|X_i| - 1) = 1 = Acc_{(i,j)}^{(t)}$;

For $X_i \rightarrow D_k(k \neq j)$, due to $X_i \bigcap D_k \neq \phi$, we have : $Acc_{(i,k)}^{(t+1)} = Cov_{(i,k)}^{(t+1)} = 0$.

(2). If $X_i \bigcap D_j = \{x\}$, when x' is deleted, $X_i \bigcap D_j - \{x\} = \phi$, for $X_i \rightarrow D_j$: $Acc_{(i,j)}^{(t+1)} = Cov_{(i,j)}^{(t+1)} = 0$

5   A Case Study

Suppose $S = \{U, C \bigcup D, V, f\}$ is a complete information system in the tth time, the condition attributes $C = \{a_1, a_2, a_3\}$, the decision attribute $D = \{d\}$. N stands for the cardinal number of elements, the detailed information are showed in Table 1.

Table 1. The data of a information system

U	a_1	a_2	a_3	d	N	U	a_1	a_2	a_3	d	N
x_1	0	0	0	0	10	x_7	1	2	2	0	3
x_2	0	1	0	0	15	x_8	1	2	2	1	20
x_3	0	1	0	1	5	x_9	1	2	2	2	47
x_4	0	1	1	1	25	x_{10}	2	2	2	1	5
x_5	0	1	1	3	3	x_{11}	2	2	2	2	15
x_6	1	1	2	2	42	x_{12}	2	2	2	3	30

According to the table, we denote $\{X_1\} = \{x_1\}, \{X_2\} = \{x_2, x_3\}$, $\{X_3\} = \{x_4, x_5\}, \{X_4\} = \{x_6\}$, $\{X_5\} = \{x_7, x_8, x_9\}$, $\{X_6\} = \{x_{10}, x_{11}, x_{12}\}$. Firstly, we construct the probability distributing matrix and the coverage matrix in the tth time as follow:

$$Acc^{(t)}(D|X) = \begin{pmatrix} 1 & 0.75 & 0 & 0 & 0.043 & 0 \\ 0 & 0.25 & 0.893 & 0 & 0.286 & 0.1 \\ 0 & 0 & 0 & 1 & 0.671 & 0.3 \\ 0 & 0 & 0.107 & 0 & 0 & 0.6 \end{pmatrix}$$

$$Cov^{(t)}(D|X)=\begin{pmatrix} 0.357 & 0.536 & 0 & 0 & 0.107 & 0 \\ 0 & 0.088 & 0.438 & 0 & 0.386 & 0.088 \\ 0 & 0 & 0 & 0.412 & 0.441 & 0.147 \\ 0 & 0 & 0.091 & 0 & 0 & 0.909 \end{pmatrix}$$

By calculating the two matrixes, we have $\xi\{D_1\}=0.25, \xi\{D_2\}=0.286, \xi\{D_3\}=0.358, \xi\{D_4\}=0.4$, So $\beta=\max\{\xi\{D_1\},\xi\{D_2\},\xi\{D_3\},\xi\{D_4\}\}=0.4$. Due to Section 3, the knowledge is discernibility when $\beta\in[0.4,0.5)$.

Then, we are considering about the case of element transformation in the $(t+1)$th time:

Suppose in the $(t+1)$th time: (1). 10 elements is going out of X_1; (2) 10 elements is entering to x_{12}; (3) a new sample set $x_{13}(a_1=1, a_2=1, a_3=1, d=1$ is immigrating to the system, the cardinal number of x_{13} is 20, we denoted $X_7=x_{13}$. So we can recalculate the new accuracy and coverage matrix according to the incremental learning method.

(1).10 elements is going out of X_1: We have $Acc^{(t+1)}_{(1,j)}=Cov^{(t+1)}_{(1,j)}=0$.

For $X_u\to D_1(u\neq 1)$, $Acc^{(t+1)}_{(u,1)}=Acc^{(t)}_{(u,1)}$, $Cov^{(t+1)}_{(u,1)}=(|X_u\cap D_1|)/(|D_1|-10)$.

(2).10 elements is entering to x_{12} in X_6:

For $X_6\to D_4$, $Acc^{(t+1)}_{(6,4)}=(|X_6\cap D_4|+10)/(|X_6|+10)$, $Cov^{(t+1)}_{(6,4)}=(|X_6\cap D_4|+10)/(|D_4|+10)$;

For $X_6\to D_k(k\neq 4)$, $Acc^{(t+1)}_{(6,k)}=(|X_6\cap D_k|)/(|X_6|+10)$, $Cov^{(t+1)}_{(6,k)}=Cov^{(t)}_{(6,k)}$;

For $X_u\to D_4(u\neq 6)$, $Acc^{(t+1)}_{(u,4)}=Acc^{(t)}_{(u,4)}$, $Cov^{(t+1)}_{(u,4)}=(|X_u\cap D_4|)/(|D_4|+10)$.

(3).a new sample set $x_{13}(a_1=1, a_2=1, a_3=1, d=1$ is immigrating to the system:

For $X_7\to D_2$, $Acc^{(t+1)}_{(7,2)}=1, Cov^{(t+1)}_{(7,2)}=20/(|D_2|+20)$,

For $X_7\to D_k(k\neq 2)$, due to $X_7\cap D_k=\phi$, we can get $Cov^{(t+1)}_{(7,k)}=Cov^{(t)}_{(7,k)}=0$,

For $X_u\to D_2(u\neq 7)$, $Acc^{(t+1)}_{(u,2)}=Acc^{(t)}_{(u,2)}=0$, $Cov^{(t+1)}_{(u,2)}=|X_u\cap D_2|/(|D_2|+20)$.

So, we can construct the probability distributing matrix and converge matrix in the $(t+1)$th time as follow:

$$Acc^{(t+1)}(D|X)=\begin{pmatrix} 1 & 0.75 & 0 & 0 & 0.043 & 0 & 0 \\ 0 & 0.25 & 0.893 & 0 & 0.286 & 0.083 & 1 \\ 0 & 0 & 0 & 1 & 0.671 & 0.25 & 0 \\ 0 & 0 & 0.107 & 0 & 0 & 0.667 & 0 \end{pmatrix}$$

$$Cov^{(t+1)}(D|X)=\begin{pmatrix} 0 & 0.833 & 0 & 0.167 & 0 & 0 & 0 \\ 0 & 0.067 & 0.334 & 0 & 0.226 & 0.067 & 0.226 \\ 0 & 0 & 0 & 0.412 & 0.441 & 0.147 & 0 \\ 0 & 0 & 0.07 & 0 & 0 & 0.93 & 0 \end{pmatrix}$$

By calculating the two matrixes, we have $\xi\{D_1\}=0.25, \xi\{D_2\}=0.286, \xi\{D_3\}=0.329, \xi\{D_4\}=0.333$, So $\beta=\max\{\xi\{D_1\},\xi\{D_2\},\xi\{D_3\},\xi\{D_4\}\}=0.333$. Due to Section 3, the new knowledge is discernibility when $\beta\in[0.333,0.5)$.

In addition, if we fix β and define $|R_\beta|/|U|$ as the measure of interesting knowledge inducing by accuracy. When β is changed, the relations between accuracy and the interesting of knowledge are showed in Fig. 1.

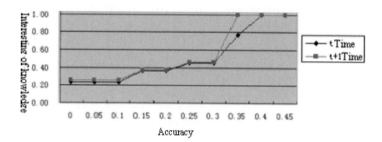

Fig. 1. The relations between accuracy and the interesting of knowledge

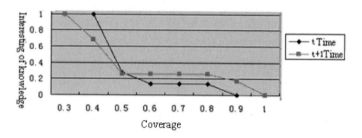

Fig. 2. The relations between coverage and the interesting of knowledge

In the same way, the relations between the coverage and the knowledge interesting can also be studied. For example, when the coverage threshold value is more than 0.8, we can easily get that: $X_i \to D_4$ is interesting in the tth time; $X_{i'} \to D_1$ and $X_{i'} \to D_4$ are interesting in the $(t+1)$th time. The relations between coverage and the interesting of knowledge are showed in Fig. 2.

6 Conclusion

In this paper, the integrations of the interesting and discernibility based on VPRS model are studied firstly. Then, we discuss the incremental learning method of VPRS model in dynamic environment. Then the mechanism of the knowledge acquiring based on the change of accuracy and coverage are also studied, which gives us a new thought to obtain the available knowledge. At last, a case is proposed to validated the rationality and validity of our method. But, the work of our study is only based on the complete information system and equivalence relation. Future research work of our study will be focus on the incomplete information system and extend the strict relation into dynamic environment, it also seems worthwhile to explore if the proposed approach can be extended to other generalized rough set models such as fuzzy rough set theory.

Acknowledgements

The authors would like to thank the MEPRC Doctoral Science Research Fund (20060613019) for the providing the research grant.

References

1. Pawlak, Z.: Rough sets. International Journal of Computer and Information Science 11, 341–356 (1982)
2. Ziarko, W.: Variable precision rough set model. Journal of Computer and System Sciences 46, 39–59 (1993)
3. Beynon, M.: Reducts within the variable precision rough sets model: a further investigation. European Journal of Operational Research 134, 592–605 (2001)
4. Mi, J.S.: Approaches to knowledge reduction based on variable precision rough sets model. Information Sciences 159, 255–272 (2004)
5. Mi, J.S.: Knowledge Reducts Based on Variable Precision Rough Set Theory. Systems Engineering-theory and Practice 24, 116–120 (2004)
6. Li., L.: Incremental learning based on Variable Precision Rough set Model. Computer Science 26, 55–61 (1999)
7. An, L.P.: Rough Set Approach to incremental acquisition of rules. Journal of Nankai University 36, 98–103 (2003)
8. Hong, T.P.: R Mining fuzzy β-certain and β-possible rules from quantitative data based on the variable precision rough-set model. Expert Systems with Applications 32, 223–232 (2007)
9. Beynon, M.: Variable precision rough set theory and data discretisation: an application to corporate failure prediction. The International Journal of Management Science 29, 576–581 (2001)
10. Su, C.T.: Precision parameter in the variable precision rough sets model: an application. The International Journal of Management Science 34, 149–157 (2006)
11. Zhang, W.X., Liang, J.Y.: Rough sets Theory and methodology. The Press of Science, Beijing (2001)
12. Han, J., Kamber, M.: Data Mining: Concepts and techniques. China Machine Press, Beijing (2006)
13. Cheng, Y.S., Zhang, Y.S.: Self-determining variable precision value approach based on variable precision rough sets model. Journal of System Simulation 19, 2555–2558 (2007)
14. Yao, Y.Y.: Decision-Theoretic Rough Set Models. In: Yao, J., Lingras, P., Wu, W.-Z., Szczuka, M., Cercone, N.J., Ślęzak, D. (eds.) RSKT 2007. LNCS (LNAI), vol. 4481, pp. 1–12. Springer, Heidelberg (2007)

Feature Selection with Fuzzy Decision Reducts

Chris Cornelis[1], Germán Hurtado Martín[1,2],
Richard Jensen[3], and Dominik Ślęzak[4]

[1] Dept. of Mathematics and Computer Science, Ghent University, Gent, Belgium
{Chris.Cornelis,German.HurtadoMartin}@UGent.be
[2] Dept. of Industrial Sciences, Hogeschool Gent, Gent, Belgium
[3] Dept. of Computer Science, The University of Wales, Aberystwyth, UK
rkj@aber.ac.uk
[4] Infobright Inc., Toronto, Canada
slezak@infobright.com

Abstract. In this paper, within the context of fuzzy rough set theory, we generalize the classical rough set framework for data-based attribute selection and reduction, based on the notion of fuzzy decision reducts. Experimental analysis confirms the potential of the approach.

Keywords: fuzzy sets, rough sets, decision reducts, classification.

1 Introduction

Rough set theory [5] is well-suited to semantics-preserving data dimensionality reduction, i.e.: to omit attributes (features) from decision systems without sacrificing the ability to *discern* between objects belonging to different concepts or classes. A minimal set of attributes that preserves the decision making power of the original system is called a *decision reduct*.

Traditionally, discernibility is modeled by an equivalence relation in the set of objects: objects are indiscernible w.r.t. a given set of attributes B if they have the same values for all attributes in B. This works well for qualitative data, in particular if the number of distinct values for each attribute is limited and there is no particular relationship among them. Quantitative data, however, involve *continuous* (i.e., real-valued) attributes like age, speed or length, and are tied to a natural scale of *closeness*, loosely expressing that the closer the attribute values of two objects are, the less discernible they are. While the standard methodology can be tailored to handle them by applying discretization, it is more natural to consider a notion of approximate equality between objects. Formally, such a notion can be modeled by means of a *fuzzy relation* [10] in the set of objects.

Guided by this principle, the original rough set framework for data-based attribute selection and reduction can be generalized (see e.g. [1,2,3,4,8]). This paper differs from previous research efforts by the introduction of the concept of a fuzzy decision reduct: conceptually, this is a weighted version of its classical counterpart that assigns to each attribute subset the degree to which it preserves the predictive ability of the original decision system. We consider alternative

G. Wang et al. (Eds.): RSKT 2008, LNAI 5009, pp. 284–291, 2008.

ways of defining fuzzy decision reducts, grouped along two directions: the first direction works with an extension of the well-known positive region, while the second one is based on an extension of the discernibility function from classical rough set analysis.

The remainder of this paper is organized as follows: we first recall preliminaries of rough sets, fuzzy sets and their hybridization in Section 2. In Section 3, we propose a general definition of a fuzzy decision reduct, and then develop a number of concrete instances of it. In Section 4, experiments are conducted to evaluate the effectiveness of these alternatives. Finally, in Section 5 we conclude.

2 Preliminaries

2.1 Rough Set Theory

Definitions. In rough set analysis, data is represented as an *information system* $(X, \mathcal{A})$, where $X = \{x_1, \ldots, x_n\}$ and $\mathcal{A} = \{a_1, \ldots, a_m\}$ are finite, non-empty sets of objects and attributes, respectively. Each a in $\mathcal{A}$ corresponds to an $X \rightarrow V_a$ mapping, in which V_a is the value set of a over X. For every subset B of $\mathcal{A}$, the B-indiscernibility relation R_B is defined as $R_B = \{(x, y) \in X^2$ and $(\forall a \in B)(a(x) = a(y))\}$. Clearly, R_B is an equivalence relation. Its equivalence classes $[x]_{R_B}$ can be used to approximate concepts, i.e., subsets of the universe X. Given $A \subseteq X$, its lower and upper approximation w.r.t. R_B are defined by $R_B{\downarrow}A = \{x \in X | [x]_{R_B} \subseteq A\}$ and $R_B{\uparrow}A = \{x \in X | [x]_{R_B} \cap A \neq \emptyset\}$.

A *decision system* $(X, \mathcal{A} \cup \{d\})$ is a special kind of information system, used in the context of classification, in which d ($d \notin \mathcal{A}$) is a designated attribute called decision. Based on the values v_k that d assumes (drawn from the finite[1] set V_d), X is partitioned into a number of decision classes X_k. Given $B \subseteq \mathcal{A}$, the B-positive region $POS_B = \bigcup_{v_k \in V_d} R_B{\downarrow}X_k$ contains the objects for which the values of B allow to predict the decision class unequivocally. The predictive ability w.r.t. d of the attributes in B is then measured by $\gamma_B = \frac{|POS_B|}{|X|}$ (degree of dependency of d on B). A subset B of $\mathcal{A}$ is called a *decision reduct* if $POS_B = POS_{\mathcal{A}}$, i.e., B preserves the decision making power of $\mathcal{A}$, and if it cannot be further reduced, i.e., there exists no proper subset B' of B such that $POS_{B'} = POS_{\mathcal{A}}$.

Example 1. Consider the decision system[2] in Table 1.a). There are two decision classes: X_0 contains all x for which $d(x) = 0$, while X_1 contains those with $d(x) = 1$. If we want to apply the standard rough set analysis approach, we first have to discretize the system; a possible discretization is given in Table 1.b). Then we can calculate the positive region. For example, for $B = \{a_4, a_5\}$, $POS_B = \{x_1, x_5, x_6, x_7\}$. Also, $POS_{\mathcal{A}} = \{x_1, x_2, x_3, x_4, x_5, x_6, x_7, x_8\}$.

---

[1] In this paper, we assume that decisions are always qualitative (discrete-valued).

[2] This is a sample taken from the Pima Indians Diabetes dataset, available at http://www.ics.uci.edu/~mlearn/MLRepository.html

Table 1. a) Original decision system b) Discretized decision system

	a_1	a_2	a_3	a_4	a_5	a_6	a_7	a_8	d
x_1	1	101	50	15	36	24.2	0.526	26	0
x_2	8	176	90	34	300	33.7	0.467	58	1
x_3	7	150	66	42	342	34.7	0.718	42	0
x_4	7	187	68	39	304	37.7	0.254	41	1
x_5	0	100	88	60	110	46.8	0.962	31	0
x_6	0	105	64	41	142	41.5	0.173	22	0
x_7	1	95	66	13	38	19.6	0.334	25	0

	a_1	a_2	a_3	a_4	a_5	a_6	a_7	a_8	d
x_1	0	0	0	0	0	0	2	0	0
x_2	1	2	2	1	1	1	1	1	1
x_3	1	1	1	1	1	2	2	1	0
x_4	1	2	1	1	1	2	0	1	1
x_5	0	0	2	1	0	3	2	1	0
x_6	0	0	1	1	0	3	0	0	0
x_7	0	0	1	0	0	0	1	0	0

Finding Decision Reducts. Decision reducts are used to synthesize minimal decision rules, which result from overlaying the reducts over the decision system and reading off the values. Below we recall a well-known approach to generate all reducts of a decision system based on its decision-relative discernibility matrix and function [7]. The decision-relative discernibility matrix of $(X, \mathcal{A} \cup \{d\})$ is the $n \times n$ matrix O, defined by, for i and j in $\{1, ..., n\}$, $O_{ij} = \emptyset$ if $d(x_i) = d(x_j)$ and $O_{ij} = \{a \in \mathcal{A} | a(x_i) \neq a(x_j)\}$ otherwise. On the other hand, the *discernibility function* of $(X, \mathcal{A} \cup \{d\})$ is the $\{0,1\}^m \to \{0,1\}$ mapping f, defined by $f(a_1^*, ..., a_m^*) = \bigwedge \{\bigvee O_{ij}^* | 1 \leq j < i \leq n$ and $O_{ij} \neq \emptyset\}$. in which $O_{ij}^* = \{a^* | a \in O_{ij}\}$. The boolean variables $a_1^*, ..., a_m^*$ correspond to the attributes from $\mathcal{A}$, and we denote $\mathcal{A}^* = \{a_1^*, ..., a_m^*\}$. If $B \subseteq \mathcal{A}$, then the valuation function $\mathcal{V}_B$ corresponding to B is defined by $\mathcal{V}_B(a^*) = 1$ iff $a \in B$. This valuation can be extended to arbitrary formulas, such that $\mathcal{V}_B(f(a_1^*, ..., a_m^*)) = f(\mathcal{V}_B(a_1^*), ..., \mathcal{V}_B(a_m^*))$. This expresses whether the attributes in B preserve the discernibility of $(X, \mathcal{A} \cup \{d\})$ (when its value is 1) or not (when it is 0). The discernibility function can be reduced to its disjunctive normal form, that is $f(a_1^*, ..., a_m^*) = \bigwedge A_1^* \vee ... \vee \bigwedge A_p^*$, in which $p \geq 1$, and for all i in $\{1, ..., p\}$ it holds that $A_i^* \subseteq \mathcal{A}^*$, and $A_i^* \not\subseteq A_j^*$ for $i \neq j$. If we define $a \in A_i$ iff $a^* \in A_i^*$, then it can be shown that $A_1, ..., A_p$ constitute exactly all decision reducts of $(X, \mathcal{A} \cup \{d\})$.

Example 2. The reduced discernibility function of the decision system in Table 1.b) is given by $f(a_1^*, ..., a_8^*) = a_2^* \vee (a_1^* \wedge a_7^*) \vee (a_5^* \wedge a_7^*) \vee (a_6^* \wedge a_7^*) \vee (a_7^* \wedge a_8^*)$. Hence, the decision reducts are $\{a_2\}$, $\{a_1, a_7\}$, $\{a_5, a_7\}$, $\{a_6, a_7\}$ and $\{a_7, a_8\}$.

2.2   Fuzzy Set Theory

Recall that a fuzzy set [10] in X is an $X \to [0, 1]$ mapping, while a fuzzy relation in X is a fuzzy set in $X \times X$. For all y in X, the R-foreset of y is the fuzzy set Ry defined by $Ry(x) = R(x, y)$ for all x in X. If R is reflexive and symmetric, i.e., $R(x, x) = 1$ and $R(x, y) = R(y, x)$ hold for all x and y in X, then R is called a fuzzy tolerance relation. For fuzzy sets A and B in X, $A \subseteq B \Leftrightarrow (\forall x \in X)(A(x) \leq B(x))$. If X is finite, the cardinality of A equals $|A| = \sum_{x \in X} A(x)$.

Fuzzy logic connectives play an important role in the development of fuzzy rough set theory. We therefore recall some important definitions. A triangular norm (t-norm for short) $\mathcal{T}$ is any increasing, commutative and associative

$[0,1]^2 \to [0,1]$ mapping satisfying $\mathcal{T}(1,x) = x$, for all x in $[0,1]$. In this paper, we consider $\mathcal{T}_M$ and $\mathcal{T}_L$, defined by $\mathcal{T}_M(x,y) = \min(x,y)$ and $\mathcal{T}_L(x,y) = \max(0, x+y-1)$ for x,y in $[0,1]$. An implicator is any $[0,1]^2 \to [0,1]$-mapping $\mathcal{I}$ satisfying $\mathcal{I}(0,0) = 1, \mathcal{I}(1,x) = x$, for all x in $[0,1]$. Moreover, we require $\mathcal{I}$ to be decreasing in its first, and increasing in its second component. In this paper, we consider $\mathcal{I}_M$ and $\mathcal{I}_L$, defined by, for x,y in $[0,1]$, $\mathcal{I}_M(x,y) = 1$ if $x \le y$ and $\mathcal{I}_M(x,y) = y$ otherwise, and $\mathcal{I}_L(x,y) = \min(1, 1 - x + y)$.

2.3   Fuzzy Rough Set Theory

Research on hybridizing fuzzy sets and rough sets has focused mainly on fuzzifying the formulas for lower and upper approximation. In this process, the set A is generalized to a fuzzy set in X, allowing that objects can belong to a concept to varying degrees. Also, rather than assessing objects' indiscernibility, we may measure their *closeness*, represented by a fuzzy tolerance relation R. For the lower and upper approximation of A by means of a fuzzy tolerance relation R, we adopt the definitions proposed in [6]: given an implicator $\mathcal{I}$ and a t-norm $\mathcal{T}$, $R{\downarrow}A$ and $R{\uparrow}A$ are defined by, for all y in X, $(R{\downarrow}A)(y) = \inf\limits_{x \in X} \mathcal{I}(R(x,y), A(x))$ and $(R{\uparrow}A)(y) = \sup\limits_{x \in X} \mathcal{T}(R(x,y), A(x))$.

In this paper, given a quantitative attribute a with range $l(a)$, we compute the approximate equality between two objects w.r.t. a, by the parametrized relation R_a, defined by, for x and y in X, $R_a(x,y) = \max\left(0, \min\left(1, \beta - \alpha\frac{|a(x)-a(y)|}{l(a)}\right)\right)$. The parameters α and β ($\alpha \ge \beta \ge 1$) determine the granularity of R_a.

Discernibility, or distance, of two objects x and y w.r.t. a can be computed as the complement of their closeness: $1 - R_a(x,y)$. Assuming that for a qualitative (i.e., nominal) attribute a, the classical way of discerning objects is used, i.e., $R_a(x,y) = 1$ if $a(x) = a(y)$ and $R_a(x,y) = 0$ otherwise, we can define, for any subset B of $\mathcal{A}$, the fuzzy B-indiscernibility relation by $R_B(x,y) = \min\limits_{a \in B} R_a(x,y)$. It can easily be seen that R_B is a fuzzy tolerance relation, and also that if only qualititative attributes (possibly stemming from discretization) are used, then the traditional concept of B-indiscernibility relation is recovered.

Example 3. For the non-dicretized decision system in Table 1a), assume that $\alpha = 5$ and $\beta = 1.2$ are used in R_a for each attribute a, and that the attributes' ranges are determined by the minimal and maximal occurring values in the decision system. It can be verified e.g. that $R_{a_1}(x_2, x_3) = 0.575, R_{a_2}(x_2, x_3) = 0$, $R_{a_4}(x_3, x_6) = 1$, and also that $R_{\{a_3, a_4\}}(x_3, x_4) = \min(0.95, 0.88) = 0.88$.

3   Fuzzy-Rough Attribute Reduction

In this section, we extend the framework for rough set analysis described in Section 2.1 using concepts of fuzzy set theory, to deal with quantitative attributes more appropriately. In order to do so, we introduce a number of increasing, $[0,1]$-valued measures to evaluate subsets of $\mathcal{A}$ w.r.t. their ability to maintain discernibility relative to the decision attribute. Once such a measure, say $\mathcal{M}$, is obtained, we can associate a notion of fuzzy decision reduct with it.

Definition 1. (Fuzzy $\mathcal{M}$-decision reduct) *Let $\mathcal{M}$ be a monotonic $\mathcal{P}(\mathcal{A}) \to [0,1]$ mapping, $B \subseteq \mathcal{A}$ and $0 < \alpha \leq 1$. B is called a fuzzy $\mathcal{M}$-decision reduct to degree α if $\mathcal{M}(B) \geq \alpha$ and for all $B' \subset B$, $\mathcal{M}(B') < \alpha$.*

3.1 Fuzzy Positive Region

Using fuzzy B-indiscernibility relations, we can define, for y in U, $POS_B(y) = \left(\bigcup_{v_k \in V_d} R_B \downarrow X_k \right)(y)$. Hence, POS_B is a fuzzy set in X, to which y belongs to the extent that its R_B-foreset is included into *at least one* of the decision classes. However, only the decision class y belongs to needs to be inspected:

Proposition 1. *For y in X, $POS_B(y) = (R_B \downarrow X_{k^*})(y)$ with $X_{k^*}(y) = 1$.*

Example 4. Let us come back to the decision system in Table 1a). Using the same indiscernibility relations as in Ex. 3, and $\mathcal{I} = \mathcal{I}_L$, we can calculate the fuzzy positive region for $B = \{a_4, a_5\}$. For instance, $POS_B(x_3) = 0.42$. The complete result is $POS_B = \{(x_1, 1), (x_2, 0.65), (x_3, 0.42), (x_4, 0.42), (x_5, 1), (x_6, 1), (x_7, 1)\}$. Compare this with Ex. 1, where POS_B was computed for the discretized system: the fuzzy positive region allows gradual membership values, and hence is able to express that e.g. x_2 is a less problematic object than x_3 and x_4. Finally, it can also be verified that, with these parameters, still $POS_\mathcal{A} = X$.

Once we have fixed the fuzzy positive region, we can define an increasing $[0,1]$-valued measure to obtain fuzzy decision reducts. We may extend the degree of dependency, as proposed by Jensen and Shen in [3,4], or, rather than considering an average, it is also possible to focus on the most problematic element. These alternatives are reflected by the following normalized[3] measures:

$$\gamma_B = \frac{|POS_B|}{|POS_\mathcal{A}|} \qquad \delta_B = \frac{\min\limits_{x \in X} POS_B(x)}{\min\limits_{x \in X} POS_\mathcal{A}(x)}$$

Proposition 2. *If $B_1 \subseteq B_2 \subseteq \mathcal{A}$, then $\gamma_{B_1} \leq \gamma_{B_2}$ and $\delta_{B_1} \leq \delta_{B_2}$.*

Example 5. For B as in Ex. 4, $\gamma_B = 5.49/7 = 0.78$ and $\delta_B = 0.42$. Also, B is a fuzzy γ-decision reduct to degree 0.77, since for $B' \subset B$, $\gamma_{B'} < 0.77$.

3.2 Fuzzy Discernibility Function

The closeness relation R_B can be used to redefine the discernibility function as an $\{0,1\}^m \to [0,1]$ mapping, such that, for each combination of conditional attributes, a value between 0 and 1 is obtained indicating how well these attributes maintain discernibility, relative to the decision attribute, between all objects.

---

[3] In this paper, we assume $POS_\mathcal{A}(x) > 0$ for every x in X.

A faithful extension of the decision-relative discernibility matrix, in which O_{ij} (i, j in $\{1, \ldots, n\}$) is a fuzzy set in $\mathcal{A}$, is obtained by defining, for any attribute a in $\mathcal{A}$, $O_{ij}(a) = 0$ if $d(x_i) = d(x_j)$ and $O_{ij}(a) = 1 - R_a(x_i, x_j)$ otherwise. Accordingly, we can define O_{ij}^* as the fuzzy set in $\mathcal{A}^*$, such that $O_{ij}^*(a^*) = O_{ij}(a)$. Interpreting the connectives in the crisp discernibility function by the minimum and the maximum, we can then extend it to a $\{0,1\}^m \rightarrow [0,1]$ mapping:

$$f(a_1^*, \ldots, a_m^*) = \min_{1 \le i < j \le n} c_{ij}(a_1^*, \ldots, a_m^*) \tag{1}$$

$$c_{ij}(a_1^*, \ldots, a_m^*) = \begin{cases} 1 & \text{if } O_{ij} = \emptyset \\ \frac{\max(O_{ij}^*(a_1^*)a_1^*, \ldots, O_{ij}^*(a_m^*)a_m^*)}{1 - R_{\mathcal{A}}(x_i, x_j)} & \text{otherwise} \end{cases} \tag{2}$$

Referring again to the valuation $\mathcal{V}_B$ corresponding to a subset B of $\mathcal{A}$, $\mathcal{V}_B(f(a_1^*, \ldots, a_m^*))$ is now a value between 0 and 1 that expresses the degree to which, for all object pairs, different values in attributes of B correspond to different values of d. Rather than taking a minimum operation in (1), which is rather strict, one can also consider the average over all object pairs:

$$g(a_1^*, \ldots, a_m^*) = \frac{2 \cdot \sum_{1 \le i < j \le n} c_{ij}(a_1^*, \ldots, a_m^*)}{n(n-1)} \tag{3}$$

The following two propositions express that f and g are monotonic, and that they assume the value 1 when all the attributes are considered.

Proposition 3. *If $B_1 \subseteq B_2 \subseteq \mathcal{A}$, then $\mathcal{V}_{B_1}(f(a_1^*, \ldots, a_m^*)) \le \mathcal{V}_{B_2}(f(a_1^*, \ldots, a_m^*))$ and $\mathcal{V}_{B_1}(g(a_1^*, \ldots, a_m^*)) \le \mathcal{V}_{B_2}(g(a_1^*, \ldots, a_m^*))$.*

Proposition 4. $\mathcal{V}_{\mathcal{A}}(f(a_1^*, \ldots, a_m^*)) = \mathcal{V}_{\mathcal{A}}(g(a_1^*, \ldots, a_m^*)) = 1$.

Example 6. For B as in Ex. 4, it can be verified that $V_B(f(a_1^*, \ldots, a_m^*)) = f(0, 0, 0, 1, 1, 0, 0, 0) = 0.42$, and that $V_B(g(a_1^*, \ldots, a_m^*)) = g(0, 0, 0, 1, 1, 0, 0, 0) = 0.96$. Here, it holds e.g. that B is a fuzzy g-decision reduct to degree 0.95.

4   Experimental Analysis

In this section, we evaluate the performance of our measures in classification, and compare the results to the approaches from [3,4]; the latter have already been shown to outperform other state-of-the-art feature selection techniques in terms of accuracy. In order to select suitable attribute subsets of a decision system $(X, \mathcal{A} \cup \{d\})$ according to a given measure $\mathcal{M}$ and threshold α, we used a heuristic algorithm called ReverseReduct, adapted from [3]. ReverseReduct starts off with $B = \mathcal{A}$, and progressively eliminates attributes from B as long as $\mathcal{M}(B) \ge \alpha$; at each step, the attribute yielding the smallest decrease in $\mathcal{M}$ is omitted. By construction, when the algorithm finishes, B is a fuzzy $\mathcal{M}$-reduct of $(X, \mathcal{A} \cup \{d\})$ to degree α. After feature selection, the decision system is reduced and classified. In our experiments, we used JRip for classification, implemented

Table 2. Classification accuracy (%) and reduct size

Dataset	γ-$\mathcal{I}_L$	γ-$\mathcal{I}_M$	δ-$\mathcal{I}_L$	δ-$\mathcal{I}_M$	f	g	Unred.	[4]
Pima	**77.0** (7)	76.0 (8)	76.8 (6)	76.0 (8)	77.0 (7)	**77.6** (2)	76.0 (8)	76.0 (8)
Cleveland	**54.2** (9)	54.5 (9)	53.2 (8)	53.9 (9)	53.2 (8)	53.9 (2)	52.2 (13)	54.6 (8)
Glass	65.4 (6)	67.8 (6)	63.1 (5)	**71.5** (9)	65.9 (8)	55.1 (3)	71.5 (9)	71.5 (9)
Heart	80.7 (7)	**81.9** (8)	73.7 (8)	73.7 (8)	73.7 (8)	75.2 (2)	77.4 (13)	78.5 (10)
Olitos	67.5 (8)	65.0 (12)	**68.3** (5)	60.8 (6)	**68.3** (5)	64.2 (2)	70.8 (25)	71.7 (5)
Water 2	82.8 (11)	81.0 (17)	83.1 (8)	83.1 (8)	83.1 (8)	**83.3** (1)	83.9 (38)	85.6 (6)
Water 3	83.3 (11)	83.6 (17)	82.8 (7)	82.8 (7)	82.8 (7)	**85.9** (2)	82.8 (38)	82.8 (11)
Wine	**92.7** (6)	87.6 (8)	84.3 (5)	84.3 (5)	84.3 (5)	88.2 (2)	92.7 (13)	95.5 (5)

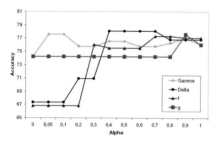
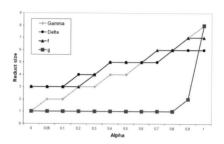

Fig. 1. Accuracy results and reduct size for varying values of threshold parameter α

in WEKA [9]. The benchmark datasets come from [4], and also include the full version of the Pima dataset used in our running example.

In a first experiment, we fixed α to 0.9 (ReverseReduct looks for a fuzzy $\mathcal{M}$-decision reduct to degree 0.9). Table 2 records the results obtained with γ, δ, and the min- and average-based variants of the fuzzy discernibility function; for the measures based on the positive region, we worked with $\mathcal{I} = \mathcal{I}_L$ and $\mathcal{I} = \mathcal{I}_M$ as implicators. To compute approximate equality, we used R_B as defined in Section 2.3, with $\alpha = 5$ and $\beta = 1.2$ for the average-based approaches, and $\alpha = 15$ and $\beta = 1$ for the min-based approaches[4]. The one but last column contains accuracy and size for the unreduced dataset, and the last one records the best accuracy obtained in [4], along with the size of the corresponding attribute set.

The results show that, on the whole, our methods are competitive with those from [4]. Moreover, for three of the datasets, strictly better accuracy results can be obtained with at least one of the new approaches. Also, in many cases shorter attribute subsets are produced. In particular, note that g generates very short subsets that have reasonable, and sometimes even excellent, accuracy.

We also investigated the influence of α on the quality of the fuzzy decision reducts; Fig. 1 plots the results obtained for Pima with the four approaches[5] as

---

[4] Since the min-based approaches are stricter, they require crisper definitions of approximate equality to perform well.

[5] For γ and δ, $\mathcal{I}_L$ was used as implicator.

a function of α. All approaches reach their optimum for $\alpha < 1$, which clearly endorses using fuzzy decision reducts. For the average-based measures, $\alpha = 0.9$ seems a good compromise value, while the min-based approaches generally require smaller values[6]. The corresponding reduct size decreases gradually for most approaches, except for g which is sensitive to small changes when α is large.

5 Conclusion

We have introduced a framework for fuzzy-rough set based feature selection, built up around the formal notion of a fuzzy reduct. By expressing that an attribute subset should retain the quality of the full feature set to a certain extent only, we are able to generate shorter attribute subsets, without paying a price in accuracy. For the future, we plan to further investigate the role of the various parameters. We also hope to extend the approach to deal with quantitative decisions.

Acknowledgment

Chris Cornelis' research is funded by the Research Foundation—Flanders.

References

1. Chen, D., Tsang, E., Zhao, S.: An approach of attributes reduction based on fuzzy rough sets. In: IEEE Int. Conf. on Systems, Man, and Cybernetics, pp. 486–491 (2007)
2. Chen, D., Tsang, E., Zhao, S.: Attribute Reduction Based on Fuzzy Rough Sets. In: Int. Conf. on Rough Sets and Intelligent Systems Paradigms, pp. 381–390 (2007)
3. Jensen, R., Shen, Q.: Fuzzy-rough sets assisted attribute selection. IEEE Transactions on Fuzzy Systems 15(1), 73–89 (2007)
4. Jensen, R., Shen, Q.: New approaches to fuzzy-rough feature selection. IEEE Transactions on Fuzzy Systems (to appear)
5. Pawlak, Z.: Rough sets. International Journal of Computer and Information Sciences 11(5), 341–356 (1982)
6. Radzikowska, A., Kerre, E.: A comparative study of fuzzy rough sets. Fuzzy Sets and Systems 126, 137–156 (2002)
7. Skowron, A., Rauszer, C.: The Discernibility Matrices and Functions in Information Systems. In: Słowiński, R. (ed.) Intelligent Decision Support: Handbook of Applications and Advances of the Rough Sets Theory, pp. 331–362. Kluwer Academic Publishers, Dordrecht (1992)
8. Wang, X., Tsang, E., Zhao, S., Chen, D., Yeung, D.: Learning fuzzy rules from fuzzy samples based on rough set technique. Information Sciences 177, 4493–4514 (2007)
9. Witten, I., Frank, E.: Data Mining: Practical machine learning tools and techniques, 2nd edn. Morgan Kaufmann, San Francisco (2005)
10. Zadeh, L.: Fuzzy sets. Information and Control 8, 338–353 (1965)

---

[6] Incidentally, the best overall accuracy, 78.1%, was obtained for δ with $\alpha \in [0.4, 0.7]$.

Rough-Fuzzy Relational Clustering Algorithm for Biological Sequence Mining

Pradipta Maji and Sankar K. Pal

Center for Soft Computing Research
Machine Intelligence Unit, Indian Statistical Institute, India
{pmaji,sankar}@isical.ac.in

Abstract. This paper presents a hybrid relational clustering algorithm, termed as rough-fuzzy c-medoids, to cluster biological sequences. It comprises a judicious integration of the principles of rough sets, fuzzy sets, c-medoids algorithm, and amino acid mutation matrix used in biology. The concept of crisp lower bound and fuzzy boundary of a class, introduced in rough-fuzzy c-medoids, enables efficient selection of cluster prototypes. The effectiveness of the algorithm, along with a comparison with other algorithms, is demonstrated on different protein data sets.

1 Introduction

Cluster analysis is a technique for finding natural groups present in the data. It divides a given data set into a set of clusters in such a way that two objects from the same cluster are as similar as possible and the objects from different clusters are as dissimilar as possible. In biological sequences, the only available information is the numerical values that represent the degrees to which pairs of sequences in the data set are related. Algorithms that generate partitions of that type of relational data are usually referred to as relational or pair-wise clustering algorithms. An well-known relational clustering algorithm is c-medoids due to Kaufman and Rousseeuw [1].

One of the main problems with biological sequence is the uncertainty. Some of the sources of this uncertainty include incompleteness and vagueness in class definitions of biological data. In this background, fuzzy sets theory [2] and rough sets theory [3], have gained popularity in modeling and propagating uncertainty. Both fuzzy and rough sets provide a mathematical framework to capture uncertainties associated with the data [3]. A recent fuzzy relational clustering algorithm is Krishnapuram's fuzzy c-medoids [4]. It offers the opportunity to deal with the data that belong to more than one cluster at the same time. Also, it can handle with the uncertainties arising from overlapping cluster boundaries. However, it is very sensitive to noise and outliers. The possibilistic c-medoids [4] is an extension of fuzzy c-medoids, which handles efficiently data sets containing noise and outliers. But, it sometimes generates coincident clusters.

In this paper, we present a relational clustering algorithm, termed as rough-fuzzy c-medoids algorithm, based on rough sets and fuzzy sets to cluster biological sequences. While the membership function of fuzzy sets enables efficient

G. Wang et al. (Eds.): RSKT 2008, LNAI 5009, pp. 292–299, 2008.
© Springer-Verlag Berlin Heidelberg 2008

handling of overlapping partitions, the concept of lower and upper approximations of rough sets deals with uncertainty, vagueness, and incompleteness in class
definition. Each partition is represented by a medoid, a crisp lower approximation, and a fuzzy boundary. The medoid depends on the weighting average of
the crisp lower approximation and fuzzy boundary. The similarity between two
sequences is computed with reference to a biological similarity matrix (amino
acid mutation matrix). In effect, the biological content in the sequences can be
maximally utilized for accurate clustering. Some quantitative measures are used
to evaluate the quality of the relational clustering algorithm. The effectiveness
of the algorithm, along with a comparison with hard c-medoids [1] and fuzzy
c-medoids [4], has been demonstrated on different protein data sets.

2   Rough-Fuzzy C-Medoids Algorithm

In this section, we first describe hard c-medoids [1] and fuzzy c-medoids [4], for
clustering biological sequences. Next, we describe a novel relational clustering
algorithm, termed as rough-fuzzy c-medoids.

2.1   Hard C-Medoids and Fuzzy C-Medoids

The hard c-medoids algorithm [1] uses the most centrally located object in a
cluster, which is termed as the medoid. A medoid is essentially an existing data
from the cluster, which is closest to the mean of the cluster.

Let $\mathbb{A}$ be the set of 20 amino acids, $X = \{x_1, \cdots, x_j, \cdots, x_n\}$ be the set of
n sequences with m residues, and $V = \{v_1, \cdots, v_i, \cdots, v_c\} \subset X$ be the set of
c medoids such that $v_{ik}, x_{jk} \in \mathbb{A}$, $\forall_{i=1}^{c}, \forall_{j=1}^{n}, \forall_{k=1}^{m}$. The non-gapped pair-wise
homology alignment score is considered to compute the similarity between two
sequences, which can be calculated using an amino acid mutation matrix [5].
The pair-wise alignment score between x_j and v_i is then defined as

$$\mathcal{S}(x_j, v_i) = \sum_{k=1}^{m} \mathcal{M}(x_{jk}, v_{ik}) \qquad (1)$$

where $\mathcal{M}(x_{jk}, v_{ik})$ can be obtained from an amino acid mutation matrix through
a table look-up method. The function value is high if two sequences are similar
or close to each other, and small if two sequences are distinct.

The objective of the hard c-medoids algorithm for clustering biological sequences is to assign n sequences to c clusters. Each of the clusters β_i is represented by a medoid v_i for that cluster. The process begins by randomly choosing
c sequences as the medoids. The sequences are assigned to one of the c clusters
based on the maximum value of the non-gapped pair-wise homology alignment
score $\mathcal{S}(x_j, v_i)$ between the sequence x_j and the medoid v_i. After the assignment
of all the sequences to various clusters, the new medoid are calculated as follows:

$$v_i = x_q \qquad \text{where} \ \ q = \arg\max\{\mathcal{S}(x_k, x_j)\}; \ \ x_j \in \beta_i; \ \ x_k \in \beta_i \qquad (2)$$

and $\mathcal{S}(x_k, x_j)$ can be calculated as per (1).

The fuzzy c-medoids provides a fuzzification of the hard c-medoids algorithm [4]. For relational clustering of biological sequences, it maximizes

$$J = \sum_{j=1}^{n} \sum_{i=1}^{c} (\mu_{ij})^{\acute{m}} \{\mathcal{S}(x_j, v_i)\} \tag{3}$$

where $1 \leq \acute{m} < \infty$ is the fuzzifier, $\mu_{ij} \in [0, 1]$ is the fuzzy membership of the sequence x_j in cluster β_i, such that

$$\mu_{ij} = \sum_{l=1}^{c} \left\{ \frac{\mathcal{S}(x_j, v_i)}{\mathcal{S}(x_j, v_l)} \right\}^{\frac{1}{\acute{m}-1}} \text{ subject to } \sum_{i=1}^{c} \mu_{ij} = 1, \forall j, \ 0 < \sum_{j=1}^{n} \mu_{ij} < n, \forall i. \tag{4}$$

The new medoids are calculated as follows:

$$v_i = x_q \quad \text{where} \quad q = \arg \max \sum_{k=1}^{n} (\mu_{ik})^{\acute{m}} \{\mathcal{S}(x_k, x_j)\}; \ \ 1 \leq j \leq n. \tag{5}$$

2.2   Rough-Fuzzy C-Medoids

Let $\underline{A}(\beta_i)$ and $\overline{A}(\beta_i)$ be the lower and upper approximations of cluster β_i, and $B(\beta_i) = \overline{A}(\beta_i) - \underline{A}(\beta_i)$ denotes the boundary region of cluster β_i. However, it is possible to define a pair of lower and upper bounds $[\underline{A}(\beta_i), \overline{A}(\beta_i)]$ or a rough set for every set $\beta_i \subseteq U$, U be the set of objects of concern [3]. The family of upper and lower bounds are required to follow some of the basic rough set properties: (i) an object x_j can be part of at most one lower bound; (ii) $x_j \in \underline{A}(\beta_i) \Rightarrow x_j \in \overline{A}(\beta_i)$; and (iii) an object x_j is not part of any lower bound $\Rightarrow x_j$ belongs to two or more upper bounds. Incorporating both fuzzy and rough sets, next we describe the rough-fuzzy c-medoids algorithm. It adds the concept of fuzzy membership of fuzzy sets, and lower and upper approximations of rough sets into c-medoids algorithm. While the lower and upper bounds of rough sets deal with uncertainty, vagueness, and incompleteness in class definition, the membership of fuzzy sets enables efficient handling of overlapping partitions.

In fuzzy c-medoids, the medoid depends on the fuzzy membership values of different sequences. Whereas in rough-fuzzy c-medoids, after computing the memberships for c clusters and n sequences, the membership values of each sequence are sorted and the difference of two highest memberships is compared with a threshold value δ. Let μ_{ij} and μ_{kj} be the highest and second highest memberships of sequence x_j. If $(\mu_{ij} - \mu_{kj}) > \delta$, then $x_j \in \underline{A}(\beta_i)$ as well as $x_j \in \overline{A}(\beta_i)$ and $x_j \notin \underline{A}(\beta_k)$, otherwise $x_j \in B(\beta_i)$ and $x_j \in B(\beta_k)$. That is, the algorithm first separates the "core" and overlapping portions of each cluster β_i based on the threshold value δ. The "core" portion of the cluster β_i is represented by its lower approximation $\underline{A}(\beta_i)$, while the boundary region $B(\beta_i)$ represents the overlapping portion. In effect, it minimizes the vagueness and incompleteness in cluster definition. According to the definitions of lower approximations and boundary of rough sets, if a sequence $x_j \in \underline{A}(\beta_i)$, then $x_j \notin \underline{A}(\beta_k), \forall k \neq i$, and

$x_j \notin \underline{B}(\beta_i), \forall i$. That is, the sequence x_j is contained in β_i definitely. Thus, the weights of the sequences in lower approximation of a cluster should be independent of other medoids and clusters, and should not be coupled with their similarity with respect to other medoids. Also, the sequences in lower approximation of a cluster should have similar influence on the corresponding medoid and cluster. Whereas, if $x_j \in B(\beta_i)$, then the sequence x_j possibly belongs to β_i and potentially belongs to another cluster. Hence, the sequences in boundary regions should have different influence on the medoids and clusters.

So, in rough-fuzzy c-medoids, after assigning each sequence in lower approximations and boundary regions of different clusters based on δ, the memberships μ_{ij} of the sequences are modified. The membership values of the sequences in lower approximation are set to 1, while those in boundary regions are remain unchanged. In other word, the proposed c-medoids first partitions the data into two classes - lower approximation and boundary. The concept of fuzzy memberships is applied only to the sequences of boundary region, which enables the algorithm to handle overlapping clusters. Thus, in rough-fuzzy c-medoids, each cluster is represented by a medoid, a crisp lower approximation, and a fuzzy boundary. The lower approximation influences the fuzziness of final partition. The fuzzy c-medoids can be reduced from rough-fuzzy c-medoids when $\underline{A}(\beta_i) = \emptyset, \forall i$. Thus, the proposed algorithm is the generalization of existing fuzzy c-medoids. The new medoids are calculated based on the weighting average of the crisp lower approximation and fuzzy boundary. The medoids calculation is given by:

$$
v_i = x_q \quad \text{where} \quad q = \arg\max
\begin{cases}
w \times \mathcal{A} + \tilde{w} \times \mathcal{B} & \text{if } \underline{A}(\beta_i) \neq \emptyset, \, B(\beta_i) \neq \emptyset \\
\mathcal{A} & \text{if } \underline{A}(\beta_i) \neq \emptyset, \, B(\beta_i) = \emptyset \\
\mathcal{B} & \text{if } \underline{A}(\beta_i) = \emptyset, \, B(\beta_i) \neq \emptyset
\end{cases} \quad (6)
$$

$$
\mathcal{A} = \sum_{x_k \in \underline{A}(\beta_i)} \mathcal{S}(x_k, x_j); \qquad \mathcal{B} = \sum_{x_k \in B(\beta_i)} (\mu_{ik})^{\acute{m}} \mathcal{S}(x_k, x_j)
$$

The parameters w and $\tilde{w}$ $(= 1 - w)$ correspond to the relative importance of lower bound and boundary region. Since the sequences lying in lower approximation definitely belong to a cluster, they are assigned a higher weight w compared to $\tilde{w}$ of the sequences lying in boundary region. That is, $0 < \tilde{w} < w < 1$.

3 Quantitative Measure

In this section we present some quantitative indices to evaluate the quality of relational clustering for biological sequences.

β **Index:** It is defined as

$$
\beta = \frac{1}{c} \sum_{i=1}^{c} \frac{1}{n_i} \sum_{x_j \in \beta_i} \frac{\mathcal{S}(x_j, v_i)}{\mathcal{S}(v_i, v_i)} \tag{7}
$$

where n_i is the number of sequences in the ith cluster β_i and $\mathcal{S}(x_j, v_i)$ is the non-gapped pair-wise homology alignment scores between sequence x_j and medoid

v_i. The β index is the average normalized homology alignment scores of input sequences with respect to their corresponding medoids. The β index increases with increase in homology alignment scores within a cluster. The value of β also increases with c. In an extreme case when the number of clusters is maximum, i.e., $c = n$, the total number of sequences, we have $\beta = 1$. Thus, $0 < \beta \leq 1$.

γ **Index:** It can be defined as

$$\gamma = \max_{i,j} \frac{1}{2} \left\{ \frac{\mathcal{S}(v_j, v_i)}{\mathcal{S}(v_i, v_i)} + \frac{\mathcal{S}(v_i, v_j)}{\mathcal{S}(v_j, v_j)} \right\} \tag{8}$$

$0 < \gamma < 1$. The γ index calculates the maximum normalized homology alignment score between medoids. A good clustering procedure for medoids selection should make the homology alignment score between all medoids as low as possible. The γ index minimizes the between-cluster homology alignment score.

Based on the mutual information, the β index would be as follows:

$$\overline{\beta} = \frac{1}{c} \sum_{i=1}^{c} \frac{1}{n_i} \sum_{x_j \in \beta_i} \frac{\mathrm{MI}(x_j, v_i)}{\mathrm{MI}(v_i, v_i)}; \qquad \mathrm{MI}(x_i, x_j) = \mathrm{H}(x_i) + \mathrm{H}(x_j) - \mathrm{H}(x_i, x_j) \tag{9}$$

$\mathrm{MI}(x_i, x_j)$ is the mutual information between sequences x_i and x_j with $\mathrm{H}(x_i)$ and $\mathrm{H}(x_j)$ being the entropy of sequences x_i and x_j respectively, and $\mathrm{H}(x_i, x_j)$ their joint entropy. $\mathrm{H}(x_i)$ and $\mathrm{H}(x_i, x_j)$ are defined as

$$\mathrm{H}(x_i) = -\mathrm{p}(x_i)\mathrm{lnp}(x_i) \qquad \mathrm{H}(x_i, x_j) = -\mathrm{p}(x_i, x_j)\mathrm{lnp}(x_i, x_j) \tag{10}$$

$\mathrm{p}(x_i)$ and $\mathrm{p}(x_i, x_j)$ are the a priori probability of x_i and joint probability of x_i and x_j respectively. Similarly, γ index would be

$$\overline{\gamma} = \max_{i,j} \frac{1}{2} \left\{ \frac{\mathrm{MI}(v_i, v_j)}{\mathrm{MI}(v_i, v_i)} + \frac{\mathrm{MI}(v_i, v_j)}{\mathrm{MI}(v_j, v_j)} \right\} \tag{11}$$

4   Experimental Results

The performance of rough-fuzzy c-medoids (RFCMdd) is compared extensively with that of hard c-medoids (HCMdd) [1] and fuzzy c-medoids (FCMdd) [4]. To analyze the performance of the RFCMdd, we use Cai-Chou HIV data set [6] and caspase cleavage protein sequences downloaded from the NCBI (www.ncbi.nih.gov). The Dayhoff amino acid mutation matrix [5] is used to calculate the non-gapped pair-wise homology score between two sequences.

4.1   Optimum Values of Parameters $\acute{m}$, w, and δ

Tables 1-3 report the performance of different c-medoids for different values of $\acute{m}$, w, and δ respectively. The results and subsequent discussions are presented here with respect to β, γ, $\overline{\beta}$, and $\overline{\gamma}$. The fuzzifier $\acute{m}$ controls the extent of membership sharing between fuzzy clusters. From Table 1, it is seen that as the value of $\acute{m}$

Table 1. Performance of RFCMdd and FCMdd for Different Values of $\acute{m}$

Value of $\acute{m}$	Algorithms	Cai-Chou HIV Data Set				Caspase Cleavage Proteins			
		β	γ	$\bar{\beta}$	$\bar{\gamma}$	β	γ	$\bar{\beta}$	$\bar{\gamma}$
1.7	RFCMdd	0.794	0.677	0.895	0.950	0.785	0.647	0.907	0.977
	FCMdd	0.750	0.728	0.868	0.973	0.772	0.671	0.883	0.978
1.8	RFCMdd	0.818	0.639	0.907	0.932	0.803	0.628	0.923	0.972
	FCMdd	0.764	0.695	0.890	0.954	0.795	0.671	0.890	0.978
1.9	RFCMdd	0.829	0.618	0.911	0.927	0.814	0.611	0.937	0.965
	FCMdd	0.809	0.656	0.903	0.941	0.808	0.668	0.898	0.962
2.0	RFCMdd	0.829	0.618	0.911	0.927	0.839	0.608	0.942	0.944
	FCMdd	0.809	0.656	0.903	0.941	0.816	0.662	0.901	0.953
2.1	RFCMdd	0.811	0.622	0.908	0.945	0.826	0.617	0.935	0.949
	FCMdd	0.802	0.671	0.901	0.948	0.801	0.665	0.899	0.973
2.2	RFCMdd	0.802	0.640	0.903	0.958	0.817	0.639	0.928	0.954
	FCMdd	0.767	0.692	0.892	0.977	0.798	0.665	0.895	0.973
2.3	RFCMdd	0.791	0.658	0.882	0.961	0.801	0.641	0.901	0.961
	FCMdd	0.760	0.703	0.877	0.982	0.784	0.668	0.886	0.979

Table 2. Performance of RFCMdd for Different Values of w $(= 1 - \tilde{w})$

Value of w	Cai-Chou HIV Data Set				Caspase Cleavage Proteins			
	β	γ	$\bar{\beta}$	$\bar{\gamma}$	β	γ	$\bar{\beta}$	$\bar{\gamma}$
0.51	0.684	0.827	0.806	1.000	0.683	0.714	0.808	1.000
0.60	0.788	0.708	0.883	0.991	0.779	0.649	0.883	0.983
0.70	0.829	0.618	0.911	0.927	0.839	0.608	0.942	0.944
0.80	0.793	0.651	0.874	0.978	0.817	0.622	0.914	0.964
0.90	0.748	0.711	0.829	1.000	0.761	0.682	0.825	1.000
0.99	0.671	0.813	0.802	1.000	0.675	0.762	0.798	1.000

increases, the values of β and $\bar{\beta}$ increase, while γ and $\bar{\gamma}$ decrease. The RFCMdd and FCMdd achieve their best performance with $\acute{m} = 1.9$ and 2.0 for Cai-Chou HIV data set and $\acute{m} = 2.0$ for caspase cleavage protein sequences respectively. But, for $\acute{m} > 2.0$, the performance of both algorithms decreases with the increase in $\acute{m}$. That is, the best performance of RFCMdd and FCMdd is achieved when the fuzzy membership value of a sequence in a cluster is equal to its normalized homology alignment score with respect to all the medoids.

The parameter w has an influence on the performance of RFCMdd. Since the sequences lying in lower approximation definitely belong to a cluster, they are assigned a higher weight w compared to $\tilde{w}$ of the sequences lying in boundary regions. Hence, for RFCMdd, $0 < \tilde{w} < w < 1$. Table 2 presents the performance of RFCMdd for different values w considering $\acute{m} = 2.0$ and $\delta = 0.20$. When the sequences of both lower approximation and boundary region are assigned approximately equal weights, the performance of RFCMdd is significantly poorer than HCMdd. As the value of w increases, the values of β and $\bar{\beta}$ increase, while γ and $\bar{\gamma}$ decrease. The best performance of both algorithms is achieved with

$w = 0.70$. The performance significantly reduces with $w \simeq 1.00$. In this case, since the clusters cannot see the sequences of boundary regions, the mobility of the clusters and the medoids reduces. As a result, some medoids get stuck in local optimum. On the other hand, when $w = 0.70$, the sequences of lower approximations are assigned a higher weight compared to that of boundary regions as well as the clusters and the medoids have a greater degree of freedom to move. In effect, the quality of generated clusters is better compared to other values of w.

Table 3. Performance of RFCMdd for Different Values of δ

Value of δ	Cai-Chou HIV Data Set				Caspase Cleavage Proteins			
	β	γ	$\overline{\beta}$	$\overline{\gamma}$	β	γ	$\overline{\beta}$	$\overline{\gamma}$
0.00	0.713	0.782	0.817	1.000	0.707	0.698	0.862	1.000
0.05	0.753	0.707	0.868	1.000	0.766	0.683	0.881	1.000
0.10	0.794	0.683	0.882	0.991	0.801	0.641	0.907	0.995
0.15	0.806	0.629	0.902	0.964	0.819	0.622	0.928	0.973
0.20	0.829	0.618	0.911	0.927	0.839	0.608	0.942	0.944
0.25	0.811	0.638	0.907	0.952	0.814	0.631	0.932	0.980
0.30	0.805	0.681	0.894	0.988	0.791	0.667	0.908	0.995
0.35	0.784	0.704	0.875	1.000	0.772	0.671	0.881	1.000

The performance of RFCMdd also depends on the value of δ, which determines the class labels of all the sequences. In other word, the RFCMdd partitions the data set of a cluster into two classes - lower approximation and boundary, based on the value of δ. Table 3 presents the performance of RFCMdd for different values of δ considering $\acute{m} = 2.0$ and $w = 0.70$. For $\delta = 0.0$, all the sequences will be in lower approximations of different clusters and $B(\beta_i) = \emptyset, \forall i$. In effect, the RFCMdd reduces to conventional HCMdd. On the other hand, for $\delta = 1.0$, $\underline{A}(\beta_i) = \emptyset, \forall i$ and all the sequences will be in the boundary regions of different clusters. That is, the RFCMdd boils down to FCMdd. The best performance of RFCMdd with respect to β, $\overline{\beta}$, γ, and $\overline{\gamma}$ is achieved with $\delta = 0.2$. This is approximately equal to the average difference of highest and second highest fuzzy membership values of all the sequences.

4.2   Comparative Performance of Different Relational Algorithms

Finally, Table 4 provides the comparative results of different algorithms. It is seen that the RFCMdd produces medoids having the highest β and $\overline{\beta}$ values and lowest γ and $\overline{\gamma}$ values for all the cases. Table 4 also provides execution time of different algorithms for two data sets. The execution time required for RFCMdd is comparable to FCMdd. For the HCMdd, although the execution time is less, the performance is significantly poorer than that of FCMdd and RFCMdd. Use of rough and fuzzy sets adds a small computational load to the HCMdd; however the corresponding integrated methods (FCMdd and RFCMdd) show a definite increase in β and $\overline{\beta}$ values and decrease in γ and $\overline{\gamma}$ values. Integration of rough

Table 4. Comparative Performance of Different Methods

Data Set	Algorithms	β	γ	$\overline{\beta}$	$\overline{\gamma}$	Time (milli sec.)
Cai-	RFCMdd	0.829	0.618	0.911	0.927	6217
Chou	FCMdd	0.809	0.656	0.903	0.941	4083
HIV	HCMdd	0.713	0.782	0.817	1.000	718
Caspase	RFCMdd	0.839	0.608	0.942	0.944	513704
Cleavage	FCMdd	0.816	0.662	0.901	0.953	510961
Protein	HCMdd	0.707	0.698	0.862	1.000	8326

sets, fuzzy sets, and c-medoids, in the RFCMdd algorithm produces a set of most informative medoids in the comparable computation time.

5   Conclusion

The main contribution of the paper is to develop a methodology integrating the merits of rough sets, fuzzy sets, c-medoids algorithm, and amino acid mutation matrix for clustering biological sequences. Although the methodology has been efficiently demonstrated for biological sequence analysis, the concept can be applied to other relational unsupervised classification problems.

Acknowledgement.The authors would like to thank the DST, Govt. of India for funding the CSCR under its IRHPA scheme. The work was done when one of the authors, S.K. Pal, was a J. C. Bose Fellow of the Govt. of India.

References

1. Kaufman, L., Rousseeuw, P.J.: Finding Groups in Data, An Introduction to Cluster Analysis. JohnWiley & Sons, Brussels, Belgium (1990)
2. Zadeh, L.A.: Fuzzy Sets. Information and Control 8, 338–353 (1965)
3. Pawlak, Z.: Rough Sets, Theoretical Aspects of Resoning About Data. Kluwer, Dordrecht, The Netherlands (1991)
4. Krishnapuram, R., Joshi, A., Nasraoui, O., Yi, L.: Low Complexity Fuzzy Relational Clustering Algorithms for Web Mining. IEEE Transactions on Fuzzy System 9, 595–607 (2001)
5. Johnson, M.S., Overington, J.P.: A Structural Basis for Sequence Comparisons: An Evaluation of Scoring Methodologies. Journal of Molecular Biology 233, 716–738 (1993)
6. Cai, Y.D., Chou, K.C.: Artificial Neural Network Model for Predicting HIV Protease Cleavage Sites in Protein. Advances in Engineering Software 29(2), 119–128 (1998)

Rough-Fuzzy Clustering: An Application to Medical Imagery

Sushmita Mitra[1] and Bishal Barman[2]

[1] Center for Soft Computing Research
Indian Statistical Institute, Kolkata - 700 108, India
`sushmita@isical.ac.in`
[2] Electrical Engineering Department
S. V. National Institute of Technology, Surat - 395 007, Gujarat, India
`bishalbarman@gmail.com`

Abstract. A novel application of rough-fuzzy clustering is presented for synthetic as well as CT scan images of the brain. It is observed that the algorithm generates good prototypes even in the presence of outliers. The rough-fuzzy clustering simultaneously handles overlap of clusters and uncertainty involved in class boundary, thereby yielding the best approximation of a given structure in unlabeled data. The number of clusters is automatically optimized in terms of various validity indices. A comparative study is made with related partitive algorithms. Experimental results demonstrate the diagnosis of the extent of brain infarction in CT scan images, and is validated by medical experts.

Keywords: Rough-fuzzy clustering, cluster validation, image segmentation, CT scan imaging.

1 Introduction

Soft computing consists of methodologies that work synergistically and provides flexible information processing capability for handling real life ambiguity [1]. The main constituents of soft computing, at this juncture, include fuzzy logic, neural networks, genetic algorithms and rough sets. Rough set theory, proposed by Pawlak [2], is a paradigm to handle vagueness, uncertainty and incompleteness in information systems. The theory of rough sets arises from the notion of approximation spaces. The use of rough sets in clustering has been reported in literature [3]. Fuzzy sets and rough sets were incorporated in the c-means framework to develop the fuzzy c-means (FCM) [4] and rough c-means (RCM) [3] algorithms. While membership in FCM enables efficient handling of overlapping partitions, the concept of rough sets [2] deals with uncertainty, vagueness and incompleteness in data in terms of upper and lower approximations.

Rough-fuzzy clustering was developed [5] [6] to incorporate the merits of both rough and fuzzy sets in a hybridized framework. In this paper, we investigate the application of rough-fuzzy clustering for segmentation of CT scan imagery and for determining prototypes in the presence of outliers. The use of rough set

G. Wang et al. (Eds.): RSKT 2008, LNAI 5009, pp. 300–307, 2008.

helps in automatically controlling the effect of uncertainty among patterns lying between the upper and lower approximation. Patterns within the lower approximation play a central role during clustering. The incorporation of membership in the RCM framework enhances the robustness of the algorithm. The Davies Bouldin [7] and Xie-Beni [8] indices and the Silhouette statistic [9] are employed to automatically determine the optimum number of clusters.

The article is organized into five sections. The rough and the rough-fuzzy algorithms are discussed in Section 2. Cluster validation is concisely presented in Section 3 while the results on synthetic data and CT scan imagery are presented in Section 4. The application demonstrates an effective Computer Aided Diagnostic (CAD) methodology to provide a second opinion to radiologists. Finally, Section 5 concludes the paper.

2 Rough Clustering Algorithms

Rough sets [2] are used to model clusters in terms of upper and lower approximations, that are weighted by a pair of parameters while computing cluster prototypes [3]. We observe that the rough set theory assigns objects into two distinct regions, *viz.*, lower and upper approximations, such that objects in lower approximation indicate definite inclusion in the concept under discussion while those in the upper approximation correspond to possible inclusion in it [2]. Since there is no concept of membership involved in rough clustering, therefore any measure of closeness of patterns to the clusters cannot be determined. Rough-fuzzy hybridized clustering [6] attempts to overcome this limitation.

2.1 Rough C-Means (RCM)

In RCM [3], the concept of c-means is extended by viewing each cluster as an interval or rough set X. It is characterized by the lower and upper approximations $\underline{B}X$ and $\overline{B}X$ respectively, with the following properties. (i) An object $\mathbf{x}_k$ can be part of at most *one* lower approximation. (ii) If $\mathbf{x}_k \in \underline{B}X$ of cluster X, then simultaneously $\mathbf{x}_k \in \overline{B}X$. (iii) If $\mathbf{x}_k$ is not a part of any lower approximation, then it belongs to two or more upper approximations. This permits overlaps between clusters. The centroid $\mathbf{v}_i$ of cluster U_i is evaluated as

$$\mathbf{v}_i = \begin{cases} w_{low}\frac{\sum_{\mathbf{x}_k \in \underline{B}U_i} \mathbf{x}_k}{|\underline{B}U_i|} + w_{up}\frac{\sum_{\mathbf{x}_k \in (\overline{B}U_i - \underline{B}U_i)} \mathbf{x}_k}{|\overline{B}U_i - \underline{B}U_i|} & \text{if } \underline{B}U_i \neq \emptyset \wedge \overline{B}U_i - \underline{B}U_i \neq \emptyset, \\ \frac{\sum_{\mathbf{x}_k \in (\overline{B}U_i - \underline{B}U_i)} \mathbf{x}_k}{|\overline{B}U_i - \underline{B}U_i|} & \text{if } \underline{B}U_i = \emptyset \wedge \overline{B}U_i - \underline{B}U_i \neq \emptyset, \\ \frac{\sum_{\mathbf{x}_k \in \underline{B}U_i} \mathbf{x}_k}{|\underline{B}U_i|} & \text{otherwise,} \end{cases} \tag{1}$$

where the parameters w_{low} and w_{up} correspond to the relative importance of the lower and upper approximations respectively. Here $|\underline{B}U_i|$ indicates the number of patterns in the lower approximation of cluster U_i, while $|\overline{B}U_i - \underline{B}U_i|$ is the number of patterns in the rough boundary lying between the two approximations. The algorithm is outlined as follows.

1. Assign initial means $\mathbf{v}_i$ for the c clusters.
2. Assign each data object (pattern) $\mathbf{x}_k$ to the lower approximation $\underline{B}U_i$ or upper approximation $\overline{B}U_i$, $\overline{B}U_j$ of cluster pairs U_i, U_j by computing the difference in its distance $d_{ik} - d_{jk}$ from cluster centroid pairs $\mathbf{v}_i$ and $\mathbf{v}_j$.
3. Let d_{ik} be minimum and d_{jk} be the next to minimum.
   If $d_{jk} - d_{ik}$ is less than some *threshold*
   then $\mathbf{x}_k \in \overline{B}U_i$ and $\mathbf{x}_k \in \overline{B}U_j$ and $\mathbf{x}_k$ cannot be a member of any lower approximation,
   else $\mathbf{x}_k \in \underline{B}U_i$ such that distance d_{ik} is minimum over the c clusters.
4. Compute new mean for each cluster U_i using eqn. (1).
5. **Repeat** Steps 2-4 **until** convergence, *i.e.*, there are no more new assignments of objects.

The parameter *threshold* measures the relative distance of an object $\mathbf{x}_k$ from a pair of clusters having centroids $\mathbf{v}_i$ and $\mathbf{v}_j$. The parameter w_{low} controls the importance of the objects lying within the lower approximation of a cluster in determining its centroid. Hence an optimal selection of these parameters is an issue of reasonable interest. We allowed $w_{up} = 1 - w_{low}$, $0.5 < w_{low} < 1$ and $0 < threshold < 0.5$.

2.2   Rough-Fuzzy C-Means (RFCM)

A rough-fuzzy c-means algorithm, involving an integration of fuzzy and rough sets, has been developed [5] [6]. This allows one to incorporate fuzzy membership value u_{ik} of a sample $\mathbf{x}_k$ to a cluster mean $\mathbf{v}_i$, relative to all other means $\mathbf{v}_j$ $\forall$ $j \neq i$, instead of the absolute individual distance d_{ik} from the centroid. This sort of relativistic measure enhances the robustness of the clustering with respect to different choices of parameters. The major steps of the algorithm are provided below.

1. Assign initial means $\mathbf{v}_i$ for the c clusters.
2. Compute membership u_{ik} for c clusters and N data objects as

$$u_{ik} = \frac{1}{\sum_{j=1}^{c} \left(\frac{d_{ik}}{d_{jk}}\right)^{\frac{2}{m-1}}}. \tag{2}$$

3. Assign each data object (pattern) $\mathbf{x}_k$ to the lower approximation $\underline{B}U_i$ or upper approximation $\overline{B}U_i$, $\overline{B}U_j$ of cluster pairs U_i, U_j by computing the difference in its membership $u_{ik} - u_{jk}$ to cluster centroid pairs $\mathbf{v}_i$ and $\mathbf{v}_j$.
4. Let u_{ik} be maximum and u_{jk} be the next to maximum.
   If $u_{ik} - u_{jk}$ is less than some *threshold*
   then $\mathbf{x}_k \in \overline{B}U_i$ and $\mathbf{x}_k \in \overline{B}U_j$ and $\mathbf{x}_k$ cannot be a member of any lower approximation,
   else $\mathbf{x}_k \in \underline{B}U_i$ such that membership u_{ik} is maximum over the c clusters.

5. Compute new mean for each cluster U_i as

$$
\mathbf{v}_i = \begin{cases} w_{low} \frac{\sum_{\mathbf{x}_k \in \underline{B}U_i} u_{ik}^m \mathbf{x}_k}{\sum_{\mathbf{x}_k \in \underline{B}U_i} u_{ik}^m} + w_{up} \frac{\sum_{\mathbf{x}_k \in (\overline{B}U_i - \underline{B}U_i)} u_{ik}^m \mathbf{x}_k}{\sum_{\mathbf{x}_k \in (\overline{B}U_i - \underline{B}U_i)} u_{ik}^m} & \text{if } \underline{B}U_i \neq \emptyset \wedge \overline{B}U_i - \underline{B}U_i \neq \emptyset, \\[3mm] \frac{\sum_{\mathbf{x}_k \in (\overline{B}U_i - \underline{B}U_i)} u_{ik}^m \mathbf{x}_k}{\sum_{\mathbf{x}_k \in (\overline{B}U_i - \underline{B}U_i)} u_{ik}^m} & \text{if } \underline{B}U_i = \emptyset \wedge \overline{B}U_i - \underline{B}U_i \neq \emptyset, \\[3mm] \frac{\sum_{\mathbf{x}_k \in \underline{B}U_i} u_{ik}^m \mathbf{x}_k}{\sum_{\mathbf{x}_k \in \underline{B}U_i} u_{ik}^m} & \text{otherwise.} \end{cases}
\tag{3}
$$

6. **Repeat** Steps 2-5 **until** convergence, *i.e.*, there are no more new assignments.

As in the case of RCM, we use $w_{up} = 1 - w_{low}$, $0.5 < w_{low} < 1$, $m = 2$, and $0 < threshold < 0.5$.

3   Cluster Validation

Partitive clustering algorithms typically require prespecification of the number of clusters. Hence the results are dependent on the choice of c. However there exist validity indices to evaluate the goodness of clustering, corresponding to a given value of c. In this article we compute the optimal number of clusters c_0 in terms of the Davies-Bouldin cluster validity index [7], Xie-Beni index [8] and Silhouette index [9].

3.1   Davies-Bouldin Index

The Davies-Bouldin index is a function of the ratio of the sum of within-cluster distance to between-cluster separation. The optimal clustering, for $c = c_0$, minimizes

$$
DB = \frac{1}{c} \sum_{k=1}^{c} \max_{l \neq k} \left\{ \frac{d_w(U_k) + d_w(U_l)}{d_b(U_k, U_l)} \right\},
\tag{4}
$$

for $1 \leq k, l \leq c$. In this process, the within-cluster distance $d_w(U_k)$ is minimized and the between-cluster separation $d_b(U_k, U_l)$ is maximized. The distance can be chosen as the traditional Euclidean metric for numeric features.

3.2   Xie-Beni Index

The Xie-Beni index [8] presents a fuzzy-validity criterion based on a validity function which identifies overall compact and separate fuzzy c-partitions. This function depends upon the data set, geometric distance measure, distance between cluster centroids and on the fuzzy partition, irrespective of any fuzzy algorithm used. We define χ as a fuzzy clustering validity function

$$
\chi = \frac{\sum_{i=1}^{c} \sum_{j=1}^{N} u_{ij}^2 \|\mathbf{v_i} - \mathbf{x_j}\|^2}{N \min_{i,j} \|\mathbf{v_i} - \mathbf{v_j}\|^2}.
\tag{5}
$$

In case of FCM and RFCM algorithms, with $m = 2$, eqn. (5) reduces to

$$\chi = \frac{J_2}{N * (d_{min})^2},$$ (6)

where J_2 is the fuzzy objective function with Euclidean norm and $d_{min} = \min_{i,j} ||\mathbf{v_i} - \mathbf{v_j}||$. The more separate the clusters, the larger $(d_{min})^2$ and the smaller χ. Thus the smallest χ, corresponding to $c = c_0$, indeed indicates a valid optimal partition.

3.3   Silhouette Statistic

The Silhouette statistic [9], though computationally more intensive, is another way of estimating the number of clusters in a distribution. The Silhouette index, S, computes for each point a width depending on its membership in any cluster. This silhouette width is then an average over all observations. We define S_k as

$$S_k = \frac{1}{N} \sum_{i=1}^{N} \frac{b_i - a_i}{max(a_i, b_i)},$$ (7)

where N is the total number of points, a_i is the average distance between point i and all other points in its own cluster and b_i is the minimum of the average dissimilarities between i and points in other clusters. Finally, the global silhouette index, S, of the clustering is given by

$$S = \frac{1}{c} \sum_{k=1}^{c} S_k.$$ (8)

The partition with highest S is taken to be optimal.

4   Results

The utility of rough sets in rough-fuzzy clustering is demonstrated on synthetic and real life CT scan imagery. The *synthetic* data of Fig. 1 consists of 32 points with two clusters. Three points shown in the upper part of the all the scatter plot in Fig. 1, are outliers, which have been purposely inserted to test the ability of the algorithms to resist a bias in the estimation of cluster prototypes.

The CT scan images were acquired by the Siemens Emotion-Duo model. The images were of size 512 x 512 pixels with 16-bit gray levels. The brain images were taken of patients in an age-range of 30-65 years, and exhibit different cases of brain infarction. Fig. 2.(a) illustrates a sample image for patient P45 indicating fresh vascular insult.

4.1   Synthetic Data

Fig. 1 represents the original data set and the centroids generated using different clustering algorithms. The centroids are marked by rectangles on the figure. It

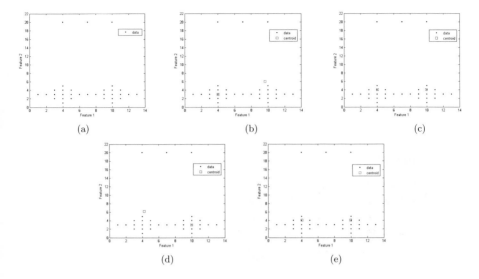

(a)   (b)   (c)

(d)   (e)

Fig. 1. *Synthetic* data set. (a) Original, and after clustering with (b) HCM (c) FCM (d) RCM (e) RFCM algorithms.

can be easily seen that the noise or outlier has maximum effect on HCM and RCM, while FCM and RFCM show reasonable improvement. Moreover, RFCM gives the best estimation of the centroids in the presence of noise. The hybridization of fuzzy membership with rough approximation leads to the better modeling. All the cluster validity indices concur to generate two optimal partitions. The results are tabulated below to further illustrate this central idea. It is observed that the last column corresponds to the best results by RFCM, viz. minimum for Davies-Bouldin and Xie-Beni, and maximum for Silhouette. It should be noted that all the algorithms were randomly initialized and 9 runs were taken. The indices were averaged over the 9 runs to compensate for possibly very bad starting points.

4.2   CT Scan Imagery

Segmentation is a process of partitioning an image into some non-overlapping meaningful regions [10]. Pixel clustering is one of the popular techniques of constituting a homogeneous region for segmentation. In this subsection, we present sample results of different members of the family of c-means algorithms on segmentation of the infarcted region in CT scan images of the brain.

The patient under study, P45, is suffering from fresh vascular insult. The infarction is observable on the left, with the left side compressing the right side such that the third ventricle is not visible due to this severe edema. Dilation of the blood ventricles is the main cause of the edema here. The problem at hand is modeled as the task of segmenting six regions comprising the gray matter (GM), the white matter (WM), the infarcted region, the skull and the background.

Table 1. Cluster Validity Indices, for two clusters, on *Synthetic* data

Index	HCM	FCM	RCM	RFCM
Davies-Bouldin	0.6550	0.5973	0.6034	**0.5971**
Xie-Beni	0.5210	0.4560	0.4572	**0.4477**
Silhouette	-0.6087	-0.2384	-0.2411	**-0.2233**

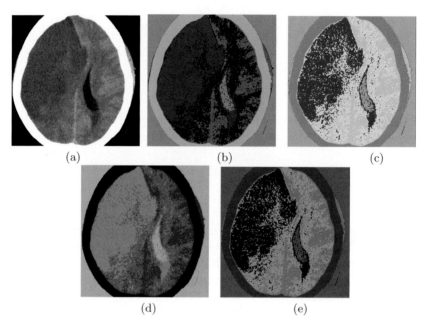

(a)   (b)   (c)

(d)   (e)

Fig. 2. Sample case of Fresh Infarction for patient, P45. (a) Original CT scan image, and the corresponding segmented versions for (b) HCM (c) FCM (d) RCM (e) RFCM clustering.

Fig. 2 shows the results of segmentation under HCM, FCM, RCM and RFCM. In the absence of an accurate index to test the accuracy of segmentation in CT scan images, we resorted to expert domain knowledge. In all, 36 frames of the patient P45 were studied, and the ground-truth regrading the best segmentation was validated my an experienced radiologist. As before, the RFCM algorithm produced the best results as verified by expert radiologists. It could be noted that RFCM not only performed better for case of fresh infarction, but also in cases of chronic infarction and subtle cases of infarction.

5   Conclusion

It is easily observed that the rough-fuzzy hybridization in RFCM enables better performance, as compared to other algorithms of the *c*-means family. Although simultaneous use of fuzzy and rough sets increases the computation complexity, but there is a marked improvement in the three validity indices, viz. Davies-Bouldin, Xie-Beni and Silhouette. The hybrid approach aims to maximize the utility of both fuzzy sets and rough sets with aim of estimating the best representative of a structure, viz. the cluster prototype. The effectiveness of the approach in knowledge discovery in case of both synthetic and real-life data is suitably illustrated. The results describe a novel way of segmenting CT scan imagery. They also promise to provide a helpful second opinion to radiologists in case of Computer Aided Diagnostic (CAD).

Acknowledgement

Mr. Bishal Barman would like to thank the Indian National Academy of Engineering (INAE) for a fellowship to carry out this work at the Center for Soft Computing Research, Indian Statistical Institute, Kolkata, India.

References

1. Zadeh, L.A.: Fuzzy logic, neural networks, and soft computing. Communications of the ACM 37, 77–84 (1994)
2. Pawlak, Z.: Rough Sets, Theoretical Aspects of Reasoning about Data. Kluwer Academic, Dordrecht (1991)
3. Lingras, P., West, C.: Interval set clustering of Web users with rough k-means. Journal of Intelligent Information Systems 23, 5–16 (2004)
4. Bezdek, J.C.: Pattern Recognition with Fuzzy Objective Function Algorithms. Plenum Press, New York (1981)
5. Dubois, D., Prade, H.: Rough fuzzy sets and fuzzy rough sets. International Journal of General Systems 17, 191–209 (1990)
6. Mitra, S., Banka, H., Pedrycz, W.: Rough-fuzzy collaborative clustering. IEEE Transactions on Systems, Man, and Cybernetics, Part-B 36, 795–805 (2006)
7. Bezdek, J.C., Pal, N.R.: Some new indexes for cluster validity. IEEE Transactions on Systems, Man, and Cybernetics, Part-B 28, 301–315 (1998)
8. Xie, X.L., Beni, G.: A validity measure for fuzzy clustering. IEEE Transactions on Pattern Analysis and Machine Intelligence 13, 841–847 (1991)
9. Rousseeuw, P.: Silhouette: A graphical aid to the interpretation and validation of cluster analysis. Journal of Computational and Applied Mathematics 20, 53–65 (1987)
10. Gonzalez, R.C., Woods, R.E.: Digital Image Processing. Prentice Hall, Upper Saddle River (2002)

Roughness Measures of Intuitionistic Fuzzy Sets

Lei Zhou[1,2], Wen-Xiu Zhang[1], and Wei-Zhi Wu[2]

[1] Institute for Information and System Sciences, Faculty of Science,
Xi'an Jiaotong University, Xi'an, Shaan'xi 710049, P.R. China
[2] School of Mathematics, Physics and Information Science,
Zhejiang Ocean University, Zhoushan, Zhejiang 316004, P.R. China
zhl_2021@126.com, wxzhang@mail.xjtu.edu.cn, wuwz@zjou.edu.cn

Abstract. In this paper, rough approximations of intuitionistic fuzzy sets with respect to an intuitionistic fuzzy approximation space are first introduced. Basic properties of intuitionistic fuzzy rough sets are then examined. Finally, roughness measures of intuitionistic fuzzy sets are defined and their properties are explored.

Keywords: Intuitionistic fuzzy rough sets, intuitionistic fuzzy sets, roughness measures, rough sets.

1 Introduction

The theory of rough sets, proposed by Pawlak [14], is an extension of classical set theory for the study of intelligent systems characterized by insufficient and incomplete information. In Pawlak's original rough set theory, a partition or an equivalence (indiscernibility) relation is an important and primitive concept. However, an equivalence relation is so stringent that it may limit the application domain of the rough set model. The generalization of Pawlak's rough set model is thus one of the main directions for the study of rough set theory (see e.g. [4,9,13,15,16,17,20,22,23,24,25]. Another interesting direction in the research of rough set theory is to aim at defining uncertainty measures such as similarity measures, roughness measures and fuzziness measures for rough sets which may be useful in application to pattern recognition and image analysis problems etc. [3,10].

As a more general case of fuzzy sets, the concept of intuitionistic fuzzy sets (IF sets for short), which was originated by Atanassov [1], has played an important role in the analysis of uncertainty of data [8,12,21]. Unlike a fuzzy set which gives a degree to which an element belongs to a set, an IF set gives both a membership degree and a nonmembership degree and is thus more precise to describe the vagueness and uncertainty than a fuzzy set. More recently, rough set approximations had been introduced into IF sets [6,7,11,18,19].

In this paper, we will investigate a generalized IF rough set model by employing the "max − min" operations. We will also define roughness measures of IF rough sets to describe the rough approximation qualities of IF sets and explore their properties.

G. Wang et al. (Eds.): RSKT 2008, LNAI 5009, pp. 308–315, 2008.

2   Preliminaries

Let U be a nonempty set called the universe of discourse. The class of all subsets (fuzzy subsets, respectively) of U will be denoted by $\mathcal{P}(U)$ (by $\mathcal{F}(U)$, respectively).

Definition 1. [1] *Let a set U be fixed. An IF set A in U is an object having the form*
$$A = \{\langle x, \mu_A(x), \gamma_A(x)\rangle \mid x \in U\},$$
where $\mu_A : U \to [0, 1]$ and $\gamma_A : U \to [0, 1]$ satisfy $0 \leq \mu_A(x) + \gamma_A(x) \leq 1$ for all $x \in U$, and $\mu_A(x)$ and $\gamma_A(x)$ are called the degree of membership and the degree of non-membership of the element $x \in U$ to A respectively. The family of all IF subsets of U is denoted by $\mathcal{IF}(U)$. The complement of an IF set A is denoted by $\sim A = \{\langle x, \gamma_A(x), \mu_A(x)\rangle \mid x \in U\}$.

The basic set operations on $\mathcal{IF}(U)$ are defined as follows [1,2]: $\forall A, B \in \mathcal{IF}(U)$,

$A \subseteq B$ if and only if (iff) $\mu_A(x) \leq \mu_B(x)$ and $\gamma_A(x) \geq \gamma_B(x)$ for all $x \in U$,

$A \supseteq B$ iff $B \subseteq A$,

$A = B$ iff $A \subseteq B$ and $B \subseteq A$,

$A \cap B = \{\langle x, \min(\mu_A(x), \mu_B(x)), \max(\gamma_A(x), \gamma_B(x))\rangle \mid x \in U\}$,

$A \cup B = \{\langle x, \max(\mu_A(x), \mu_B(x)), \min(\gamma_A(x), \gamma_B(x))\rangle \mid x \in U\}$.

Here we define a constant IF set $\widehat{(\alpha, \beta)} = \{\langle x, \alpha, \beta\rangle \mid x \in U\}$, where $0 \leq \alpha, \beta \leq 1$, and $\alpha + \beta \leq 1$. The IF universe set is $1_{\sim} = U = \{\langle x, 1, 0\rangle \mid x \in U\}$ and the IF empty set is $0_{\sim} = \emptyset = \{\langle x, 0, 1\rangle \mid x \in U\}$.

An IF relation R on U is an IF subset of $U \times U$, that is,
$$R = \{\langle (x, y), \mu_R(x, y), \gamma_R(x, y)\rangle \mid x, y \in U\},$$
where $\mu_R, \gamma_R : U \times U \to [0, 1]$ satisfy $0 \leq \mu_R(x, y) + \gamma_R(x, y) \leq 1$ for all $(x, y) \in U \times U$. We denote the family of all IF relations on U by $IFR(U \times U)$.

Definition 2. [5] *Let $R \in IFR(U \times U)$, we say that R is*

(1) *reflexive if $\mu_R(x, x) = 1$ and $\gamma_R(x, x) = 0$ for all $x \in U$,*

(2) *symmetric if for any $(x, y) \in U \times U$,*
$$\mu_R(x, y) = \mu_R(y, x) \text{ and } \gamma_R(x, y) = \gamma_R(y, x).$$

(3) *transitive if $R \geq R \overset{\vee, \wedge}{\odot} R$, i.e., for any $(x, z) \in U \times U$,*
$$\mu_R(x, z) \geq \vee_{y \in U}[\mu_R(x, y) \overset{\wedge, \vee}{\wedge} \mu_R(y, z)] \text{ and } \gamma_R(x, z) \leq \wedge_{y \in U}[\gamma_R(x, y) \vee \gamma_R(y, z)].$$

3   Intuitionistic Fuzzy Rough Sets

In this section, we introduce IF rough approximation operators by employing the "$\max - \min$" fuzzy logic operators and investigate their properties.

Definition 3. *Let U be a nonempty and finite universe of discourse and $R \in IFR(U \times U)$, the pair (U, R) is called an IF approximation space. For any $A \in \mathcal{IF}(U)$, the upper and lower approximations of A w.r.t. (U, R), denoted by $\overline{R}(A)$ and $\underline{R}(A)$, are two IF sets and are defined respectively as follows:*
$$\overline{R}(A) = \{\langle x, \mu_{\overline{R}(A)}(x), \gamma_{\overline{R}(A)}(x)\rangle \mid x \in U\},$$

$$\underline{R}(A) = \left\{ \langle x, \mu_{\underline{R}(A)}(x), \gamma_{\underline{R}(A)}(x) \rangle \mid x \in U \right\},$$

where

$$\mu_{\overline{R}(A)}(x) = \vee_{y \in U}[\mu_R(x,y) \wedge \mu_A(y)], \ \gamma_{\overline{R}(A)}(x) = \wedge_{y \in U}[\gamma_R(x,y) \vee \gamma_A(y)];$$

$$\mu_{\underline{R}(A)}(x) = \wedge_{y \in U}[\gamma_R(x,y) \vee \mu_A(y)], \ \gamma_{\underline{R}(A)}(x) = \vee_{y \in U}[\mu_R(x,y) \wedge \gamma_A(y)].$$

$\overline{R}(A)$ and $\underline{R}(A)$ are called the upper and lower IF rough approximations of A w.r.t. (U, R), respectively. The pair $(\underline{R}(A), \overline{R}(A))$ is called the IF rough set of A w.r.t. (U, R), and $\overline{R}$, $\underline{R} : \mathcal{IF}(U) \to \mathcal{IF}(U)$ are referred to as upper and lower IF rough approximation operators respectively.

Theorem 1. *Let (U, R) be an IF approximation space. Then the upper and lower IF rough approximation operators in Definition 3 satisfy the following properties:* $\forall A, B \in \mathcal{IF}(U), \alpha, \beta \in [0, 1]$ *with* $\alpha + \beta \leq 1$,

(IL1) $\underline{R}(A) = \sim \overline{R}(\sim A)$,　　　　(IU1) $\overline{R}(A) = \sim \underline{R}(\sim A)$;

(IL2) $\underline{R}(1_\sim) = 1_\sim$,　　　　　　　　(IU2) $\overline{R}(0_\sim) = 0_\sim$;

(IL3) $\underline{R}(A \cap B) = \underline{R}(A) \cap \underline{R}(B)$,　(IU3) $\overline{R}(A \cup B) = \overline{R}(A) \cup \overline{R}(B)$;

(IL4) $A \subseteq B \Longrightarrow \underline{R}(A) \subseteq \underline{R}(B)$,　(IU4) $A \subseteq B \Longrightarrow \overline{R}(A) \subseteq \overline{R}(B)$;

(IL5) $\underline{R}(A \cup B) \supseteq \underline{R}(A) \cup \underline{R}(B)$,　(IU5) $\overline{R}(A \cap B) \subseteq \overline{R}(A) \cap \overline{R}(B)$;

(IL6) $\underline{R}(A \cup \widehat{(\alpha, \beta)}) = \underline{R}(A) \cup \widehat{(\alpha, \beta)}$, (IU6) $\overline{R}(A \cap \widehat{(\alpha, \beta)}) = \overline{R}(A) \cap \widehat{(\alpha, \beta)}$.

Moreover, if R is an IF reflexive relation, then

(IRL7) $\underline{R}(A) \subseteq A$,　　(IRU7) $A \subseteq \overline{R}(A)$.

Proof. (1) Let $A = \{\langle x, \mu_A(x), \gamma_A(x) \rangle \mid x \in U\}$, then $\sim A = \{\langle x, \gamma_A(x), \mu_A(x) \rangle \mid x \in U\}$. On one hand, for any $x \in U$, by definition we have

$$\mu_{\underline{R}(\sim A)}(x) = \wedge_{y \in U}[\gamma_R(x,y) \vee \mu_{\sim A}(y)] = \wedge_{y \in U}[\gamma_R(x,y) \vee \gamma_A(y)] = \gamma_{\overline{R}(A)}(x),$$

$$\gamma_{\underline{R}(\sim A)}(x) = \vee_{y \in U}[\mu_R(x,y) \wedge \gamma_{\sim A}(y)] = \vee_{y \in U}[\mu_R(x,y) \vee \mu_A(y)] = \mu_{\overline{R}(A)}(x).$$

On the other hand, from $\overline{R}(A) = \{\langle x, \mu_{\overline{R}(A)}(x), \gamma_{\overline{R}(A)}(x) \rangle \mid x \in U\}$, we have $\sim \overline{R}(A) = \{\langle x, \gamma_{\overline{R}(A)}(x), \mu_{\overline{R}(A)}(x) \rangle \mid x \in U\}$. Hence, $\underline{R}(\sim A) = \sim \overline{R}(A)$, thus we conclude (IL1). From (IL1) we can easily deduce (IU1).

(2) (IL2) and (IU2) follow directly from Definition 3.

(3) For any $A = \{\langle x, \mu_A(x), \gamma_A(x) \rangle \mid x \in U\}$ and $B = \{\langle x, \mu_B(x), \gamma_B(x) \rangle \mid x \in U\}$. Note that

$$A \cap B = \{\langle x, \min(\mu_A(x), \mu_B(x)), \max(\gamma_A(x), \gamma_B(x)) \rangle \mid x \in U\}.$$

Then, $\forall x \in U$,

$$\begin{aligned}
\mu_{\underline{R}(A \cap B)}(x) &= \wedge_{y \in U}\big[\gamma_R(x,y) \vee \min\big(\mu_A(y), \mu_B(y)\big)\big] \\
&= \wedge_{y \in U}\big[\min\big(\gamma_R(x,y) \vee \mu_A(y), \gamma_R(x,y) \vee \mu_B(y)\big)\big] \\
&= \min\big[\wedge_{y \in U}(\gamma_R(x,y) \vee \mu_A(y)), \wedge_{y \in U}(\gamma_R(x,y) \vee \mu_B(y))\big] \\
&= \min\big(\mu_{\underline{R}(A)}(x), \mu_{\underline{R}(B)}(x)\big),
\end{aligned}$$

and

$$\begin{aligned}
\gamma_{\underline{R}(A \cap B)}(x) &= \vee_{y \in U}\big[\mu_R(x,y) \wedge \max\big(\gamma_A(y), \gamma_B(y)\big)\big] \\
&= \vee_{y \in U}\big[\max\big(\mu_R(x,y) \wedge \gamma_A(y), \mu_R(x,y) \wedge \gamma_B(y)\big)\big] \\
&= \max\big[\vee_{y \in U}(\mu_R(x,y) \wedge \gamma_A(y)), \vee_{y \in U}(\mu_R(x,y) \wedge \gamma_B(y))\big] \\
&= \max\big(\gamma_{\underline{R}(A)}(x), \gamma_{\underline{R}(B)}(x)\big).
\end{aligned}$$

Thus (IL3) holds. Similarly, we can conclude (IU3).

(4), (5) Properties (IL4) and (IL5) can be directly derived from (IL3), and (IU4) and (IU5) can be induced from (IU3).

(6) Since $A \cup (\widehat{\alpha, \beta}) = \{\langle x, \mu_A(x) \vee \alpha, \gamma_A(x) \wedge \beta \rangle \mid x \in U\}$, we have, $\forall x \in U$,

$$\mu_{\underline{R}(A \cup (\widehat{\alpha, \beta}))}(x) = \wedge_{y \in U} \left[\gamma_R(x, y) \vee (\mu_A(y) \vee \alpha) \right] = \wedge_{y \in U} \left[(\gamma_R(x, y) \vee \mu_A(y)) \vee \alpha \right]$$
$$= \wedge_{y \in U} [\gamma_R(x, y) \vee \mu_A(y)] \vee \alpha = \mu_{\underline{R}(A)}(x) \vee \alpha = \mu_{\underline{R}(A) \cup (\widehat{\alpha, \beta})}(x),$$

and

$$\gamma_{\underline{R}(A \cup (\widehat{\alpha, \beta}))}(x) = \vee_{y \in U} \left[\mu_R(x, y) \wedge (\gamma_A(y) \wedge \beta) \right] = \vee_{y \in U} \left[(\mu_R(x, y) \wedge \gamma_A(y)) \wedge \beta \right]$$
$$= \vee_{y \in U} [\mu_R(x, y) \wedge \gamma_A(y)] \wedge \beta = \gamma_{\underline{R}(A)}(x) \wedge \beta = \gamma_{\underline{R}(A) \cup (\widehat{\alpha, \beta})}(x).$$

Therefore, $\underline{R}(A \cup (\widehat{\alpha, \beta})) = \underline{R}(A) \cup (\widehat{\alpha, \beta})$, i.e., (IL6) holds. Likewise, we can conclude (IU6).

(7) Assume that R is an IF reflexive relation and $A \in \mathcal{IF}(U)$. For any $x \in U$, by the reflexivity of R we have $\mu_R(x, x) = 1$ and $\gamma_R(x, x) = 0$. Then

$$\mu_{\overline{R}(A)}(x) = \vee_{y \in U} [\mu_R(x, y) \wedge \mu_A(y)] \geq \mu_R(x, x) \wedge \mu_A(x) = \mu_A(x),$$

and

$$\gamma_{\overline{R}(A)}(x) = \wedge_{y \in U} [\gamma_R(x, y) \vee \gamma_A(y)] \leq \gamma_R(x, x) \vee \gamma_A(x) = \gamma_A(x).$$

Thus we obtain (IRU7). Likewise, we can conclude (IRL7).

Definition 4. *Let (U, R) be an IF approximation space and $A, B \in \mathcal{IF}(U)$, A and B are called IF lower $R-$equal if $\underline{R}(A) = \underline{R}(B)$, denote by $A \approx B$; A and B are called IF upper $R-$equal if $\overline{R}(A) = \overline{R}(B)$, denote by $A \simeq B$; A and B are called IF $R-$equal if $\underline{R}(A) = \underline{R}(B)$ and $\overline{R}(A) = \overline{R}(B)$, denote by $A \approx B$.*

Theorem 2. *Let (U, R) be an IF approximation space. Then for sets in $\mathcal{IF}(U)$ we have*
(1) *$A \approx B$ iff $A \cap B \approx A$ and $A \cap B \approx B$;*
(2) *$A \simeq B$ iff $A \cup B \simeq A$ and $A \cup B \simeq B$;*
(3) *If $A \approx A'$ and $B \approx B'$, then $A \cap B \approx A' \cap B'$;*
(4) *If $A \simeq A'$ and $B \simeq B'$, then $A \cup B \simeq A' \cup B'$;*
(5) *If $A \subseteq B$ and $A \approx U$, then $B \approx U$;*
(6) *If $A \subseteq B$ and $B \simeq \emptyset$, then $A \simeq \emptyset$;*
(7) *If $A \approx \emptyset$ or $B \approx \emptyset$, then $A \cap B \approx \emptyset$;*
(8) *If $A \simeq U$ or $B \simeq U$, then $A \cup B \simeq U$.*

4 Roughness Measures of IF sets

In this section, we will investigate roughness measures of IF sets. Throughout this section, we always assume that (U, R) is an IF reflexive approximation space, i.e., R is an IF reflexive relation on U.

Definition 5. *Let $A \in \mathcal{IF}(U)$ and $\alpha, \beta \in [0, 1]$, the $\alpha-$level bottom cut set of A, denoted by A_α, and the $\beta-$level top cut set of A, denoted by A^β, are defined respectively as follows:*

$$A_\alpha = \{x \in U | \mu_A(x) \geq \alpha\}, \quad A^\beta = \{x \in U | \gamma_A(x) \leq \beta\}.$$

In terms of above definition, we introduce the definitions of roughness measures of an IF set.

Definition 6. *Let (U, R) be an IF approximation space. $A \in \mathcal{IF}(U), 0 < \beta \leq \alpha \leq 1$, an absolute roughness measure of A w.r.t. parameters α, β and the approximation space (U, R) is defined by*

$$\rho_R^{\alpha,\beta}(A) = 1 - \frac{|[\underline{R}(A)]_\alpha|}{|[\overline{R}(A)]_\beta|},$$

meanwhile, a relative roughness measure of A w.r.t. parameters α, β and the approximation space (U, R) is defined by

$$\varrho_R^{\alpha,\beta}(A) = 1 - \frac{|[\underline{R}(A)]^{1-\alpha}|}{|[\overline{R}(A)]^{1-\beta}|}.$$

In special, we think that $\rho_R^{\alpha,\beta}(A) = 0$ when $|[\overline{R}(A)]_\beta| = 0$, and $\varrho_R^{\alpha,\beta}(A) = 0$ when $|[\overline{R}(A)]^{1-\beta}| = 0$, where $|X|$ is the cardinality of the crisp set X.

Remark 1. If (U, R) is a fuzzy approximation space and $A \in \mathcal{F}(U)$, it can be easily checked that $[\underline{R}(A)]_\alpha = [\underline{R}(A)]^{1-\alpha}$ and $[\overline{R}(A)]_\beta = [\overline{R}(A)]^{1-\beta}$, then we have $\rho_R^{\alpha,\beta}(A) = \varrho_R^{\alpha,\beta}(A)$, thus the roughness measures in Definition 6 degenerate to roughness measure of fuzzy set A defined by Banerjee and Pal [3].

Remark 2. If $\underline{R}(A) = \overline{R}(A)$, i.e., A is a definable IF set on (U, R), and $\alpha = \beta$, then $\rho_R^{\alpha,\beta}(A) = 0$ and $\varrho_R^{\alpha,\beta}(A) = 0$.

Absolute roughness measure and relative roughness measure express the same essence from two different aspects—membership and non-membership. These measures may help us to analyze rough approximation qualities of IF sets in fuzzy data processing. According to Definition 6, it immediately follows

Theorem 3. *Let (U, R) be an IF approximation space, $A \in \mathcal{IF}(U)$, and $0 < \beta \leq \alpha \leq 1$. Then*

(1) $0 \leq \rho_R^{\alpha,\beta}(A) \leq 1$, $0 \leq \varrho_R^{\alpha,\beta}(A) \leq 1$;

(2) *If β is fixed, then $\rho_R^{\alpha,\beta}(A)$ and $\varrho_R^{\alpha,\beta}(A)$ increase with α increasing; If α is fixed, then $\rho_R^{\alpha,\beta}(A)$ and $\varrho_R^{\alpha,\beta}(A)$ decrease with β increasing.*

Theorem 4. *Let (U, R) be an IF approximation space and $0 < \beta \leq \alpha \leq 1$. If A is a constant IF set, i.e., there exist two numbers $\delta, \theta \in [0,1]$ such that $A = \widehat{(\delta, \theta)}$, then*

(1) *if $0 < \beta < \delta < \alpha \leq 1$, then $\rho_R^{\alpha,\beta}(A) = 1$; and for $0 < \delta < \beta \leq \alpha \leq 1$ or $0 < \beta \leq \alpha \leq \delta \leq 1$, we have $\rho_R^{\alpha,\beta}(A) = 0$*

(2) *if $0 < 1 - \alpha < \theta < 1 - \beta$, then $\varrho_R^{\alpha,\beta}(A) = 1$; for $0 < \theta \leq 1 - \alpha \leq 1 - \beta \leq 1$ or $0 < 1 - \alpha \leq 1 - \beta < \theta \leq 1$, we have $\varrho_R^{\alpha,\beta}(A) = 0$.*

Proof. (1) Notice that R is reflexive, by Theorem 1 (6) and (7), it can be easily concluded that $\underline{R}(A) = \overline{R}(A) = \widehat{(\delta, \theta)}$. Then for $0 < \beta < \delta < \alpha \leq 1$, we have $[\underline{R}(A)]_\alpha = \emptyset$ and $[\overline{R}(A)]_\beta \neq \emptyset$. Hence $\rho_R^{\alpha,\beta}(A) = 1$. While for $0 < \delta < \beta \leq \alpha \leq 1$, $[\overline{R}(A)]_\beta = \emptyset$, thus $\varrho_R^{\alpha,\beta}(A) = 0$. For $0 < \beta \leq \alpha \leq \delta \leq 1$, $[\underline{R}(A)]_\alpha = [\overline{R}(A)]_\beta = U$, then $\rho_R^{\alpha,\beta}(A) = 0$. Similarly, we can prove (2).

Theorem 5. *Let (U, R) be an IF approximation space, $A, B \in \mathcal{IF}(U)$ and $0 < \beta \leq \alpha \leq 1$. If $A \subseteq B$, then*

(1) $\rho_R^{\alpha,\beta}(A) \geq \rho_R^{\alpha,\beta}(B)$ *when* $[\overline{R}(A)]_\beta = [\overline{R}(B)]_\beta$, *and* $\rho_R^{\alpha,\beta}(A) \leq \rho_R^{\alpha,\beta}(B)$ *when* $[\underline{R}(A)]_\alpha = [\underline{R}(B)]_\alpha$;

(2) $\varrho_R^{\alpha,\beta}(A) \geq \varrho_R^{\alpha,\beta}(B)$ *when* $[\overline{R}(A)]^{1-\beta} = [\overline{R}(B)]^{1-\beta}$, *and* $\varrho_R^{\alpha,\beta}(A) \leq \varrho_R^{\alpha,\beta}(B)$ *when* $[\underline{R}(A)]^{1-\alpha} = [\underline{R}(B)]^{1-\alpha}$.

Proof. From $A \subseteq B$ we have $[\underline{R}(A)]_\alpha \subseteq [\underline{R}(B)]_\alpha$ and $[\overline{R}(A)]_\beta \subseteq [\overline{R}(B)]_\beta$, meanwhile, $[\underline{R}(A)]^{1-\alpha} \subseteq [\underline{R}(B)]^{1-\alpha}$ and $[\overline{R}(A)]^{1-\beta} \subseteq [\overline{R}(B)]^{1-\beta}$. So it is easy to conclude the results of (1) and (2).

Theorem 6. *Let (U, R) be an IF approximation space, $A, B \in \mathcal{IF}(U)$, and $0 < \beta \leq \alpha \leq 1$. Then*

(1) $A \approx B \Longrightarrow \rho_R^{\alpha,\beta}(A \cap B) \leq \rho_R^{\alpha,\beta}(A)$, $\rho_R^{\alpha,\beta}(A \cap B) \leq \rho_R^{\alpha,\beta}(B)$, $\varrho_R^{\alpha,\beta}(A \cap B) \leq \varrho_R^{\alpha,\beta}(A)$, *and* $\varrho_R^{\alpha,\beta}(A \cap B) \leq \varrho_R^{\alpha,\beta}(B)$;

(2) $A \simeq B \Longrightarrow \rho_R^{\alpha,\beta}(A \cup B) \leq \rho_R^{\alpha,\beta}(A)$, $\rho_R^{\alpha,\beta}(A \cup B) \leq \rho_R^{\alpha,\beta}(B)$, $\varrho_R^{\alpha,\beta}(A \cup B) \leq \varrho_R^{\alpha,\beta}(A)$, *and* $\varrho_R^{\alpha,\beta}(A \cup B) \leq \varrho_R^{\alpha,\beta}(B)$;

(3) $A \approx B \Longrightarrow \rho_R^{\alpha,\beta}(A) = \rho_R^{\alpha,\beta}(B)$, $\varrho_R^{\alpha,\beta}(A) = \varrho_R^{\alpha,\beta}(B)$.

Proof. (1) Since $A \approx B$, by definition we have $\underline{R}(A) = \underline{R}(B)$. Then in terms of Theorem 1, we obtain $\underline{R}(A \cap B) = \underline{R}(A) \cap \underline{R}(B) = \underline{R}(A) = \underline{R}(B)$, and $\overline{R}(A \cap B) \subseteq \overline{R}(A)$ and $\overline{R}(A \cap B) \subseteq \overline{R}(B)$. So $[\underline{R}(A \cap B)]_\alpha = [\underline{R}(A)]_\alpha = [\underline{R}(B)]_\alpha$ and $[\underline{R}(A \cap B)]^{1-\alpha} = [\underline{R}(A)]^{1-\alpha} = [\underline{R}(B)]^{1-\alpha}$, $[\overline{R}(A \cap B)]_\beta \subseteq [\overline{R}(A)]_\beta$ and $[\overline{R}(A \cap B)]_\beta \subseteq [\overline{R}(B)]_\beta$, $[\overline{R}(A \cap B)]^{1-\beta} \subseteq [\overline{R}(A)]^{1-\beta}$ and $[\overline{R}(A \cap B)]^{1-\beta} \subseteq [\overline{R}(B)]^{1-\beta}$. Thus,

$$\rho_R^{\alpha,\beta}(A \cap B) = 1 - \frac{|[\underline{R}(A \cap B)]_\alpha|}{|[\overline{R}(A \cap B)]_\beta|} = 1 - \frac{|[\underline{R}(A)]_\alpha|}{|[\overline{R}(A \cap B)]_\beta|} \leq 1 - \frac{|[\underline{R}(A)]_\alpha|}{|[\overline{R}(A)]_\beta|} = \rho_R^{\alpha,\beta}(A).$$

Likewise, we can prove $\rho_R^{\alpha,\beta}(A \cap B) \leq \rho_R^{\alpha,\beta}(B)$. Meanwhile

$$\begin{aligned} \varrho_R^{\alpha,\beta}(A \cap B) &= 1 - \frac{|[\underline{R}(A \cap B)]^{1-\alpha}|}{|[\overline{R}(A \cap B)]^{1-\beta}|} = 1 - \frac{|[\underline{R}(A)]^{1-\alpha}|}{|[\overline{R}(A \cap B)]^{1-\beta}|} \\ &\leq 1 - \frac{|[\underline{R}(A)]^{1-\alpha}|}{|[\overline{R}(A)]^{1-\beta}|} = \varrho_R^{\alpha,\beta}(A). \end{aligned}$$

Similarly, we can conclude $\varrho_R^{\alpha,\beta}(A \cap B) \leq \varrho_R^{\alpha,\beta}(B)$.

(2) It is similar to the proof of (1).

(3) Since $A \approx B$, we have $\underline{R}(A) = \underline{R}(B)$ and $\overline{R}(A) = \overline{R}(B)$. Thus by Definition 6 we conclude $\rho_R^{\alpha,\beta}(A) = \rho_R^{\alpha,\beta}(B)$ and $\varrho_R^{\alpha,\beta}(A) = \varrho_R^{\alpha,\beta}(B)$.

Example 1. Let $U = \{x_1, x_2\}, \alpha = 0.3, \beta = 0.3$, and R is an IF relation on U defined as follows:

$R = \{\langle (x_1, x_1), 1, 0 \rangle, \langle (x_1, x_2), 0.7, 0.2 \rangle, \langle (x_2, x_1), 0.4, 0.4 \rangle, \langle (x_2, x_2), 1, 0 \rangle\}.$

If $A = \{\langle x_1, 0.1, 0.5 \rangle, \langle x_2, 0.6, 0.4 \rangle\}$ and $B = \{\langle x_1, 0.3, 0.4 \rangle, \langle x_2, 0.6, 0.4 \rangle\}$, then according to Definition 3, we have

$\underline{R}(A) = \{\langle x_1, 0.1, 0.5 \rangle, \langle x_2, 0.4, 0.4 \rangle\}, \overline{R}(A) = \{\langle x_1, 0.6, 0.4 \rangle, \langle x_2, 0.6, 0.4 \rangle\},$
$\underline{R}(B) = \{\langle x_1, 0.3, 0.4 \rangle, \langle x_2, 0.4, 0.4 \rangle\}, \overline{R}(B) = \{\langle x_1, 0.6, 0.4 \rangle, \langle x_2, 0.6, 0.4 \rangle\}.$

Thus $A \simeq B$, and

$$\rho_R^{\alpha,\beta}(A) = 1 - \frac{|[\underline{R}(A)]_{0.3}|}{|[\overline{R}(A)]_{0.3}|} = 1 - \frac{1}{2} = 0.5, \quad \varrho_R^{\alpha,\beta}(A) = 1 - \frac{|[\underline{R}(A)]^{1-0.3}|}{|[\overline{R}(A)]^{1-0.3}|} = 1 - 1 = 0,$$

$$\rho_R^{\alpha,\beta}(B) = 1 - \frac{|[\underline{R}(B)]_{0.3}|}{|[\overline{R}(B)]_{0.3}|} = 1 - 1 = 0, \quad \varrho_R^{\alpha,\beta}(B) = 1 - \frac{|[\underline{R}(B)]^{1-0.3}|}{|[\overline{R}(B)]^{1-0.3}|} = 1 - 1 = 0.$$

Theorem 7. *Let (U, R) and (U, S) be two IF approximation spaces, $A \in \mathcal{IF}(U)$, and $0 < \beta \leq \alpha \leq 1$. If $S \subseteq R$, then $\rho_S^{\alpha,\beta}(A) \leq \rho_R^{\alpha,\beta}(A)$ and $\varrho_S^{\alpha,\beta}(A) \leq \varrho_R^{\alpha,\beta}(A)$.*

Proof. Since $S \subseteq R$, for any $(x, y) \in U \times U$, we have
$$\mu_S(x, y) \leq \mu_R(x, y) \text{ and } \gamma_S(x, y) \geq \gamma_R(x, y).$$
Then according to Definition 3, it can be concluded that $\underline{S}(A) \supseteq \underline{R}(A)$ and $\overline{S}(A) \subseteq \overline{R}(A)$. Hence, $[\underline{S}(A)]_\alpha \supseteq [\underline{R}(A)]_\alpha$, $[\underline{S}(A)]^{1-\alpha} \supseteq [\underline{R}(A)]^{1-\alpha}$ and $[\overline{S}(A)]_\beta \subseteq [\overline{R}(A)]_\beta$, $[\overline{S}(A)]^{1-\beta} \subseteq [\overline{R}(A)]^{1-\beta}$. Therefore, by Definition 6 we obtain $\rho_S^{\alpha,\beta}(A) \leq \rho_R^{\alpha,\beta}(A)$ and $\varrho_S^{\alpha,\beta}(A) \leq \varrho_R^{\alpha,\beta}(A)$.

5   Conclusion

In this paper we have introduced rough approximations of IF sets with respect to an IF approximation space and explored basic properties of IF rough approximation operators. We have also defined absolute roughness measures and relative roughness measures of IF sets to describe roughness of the IF sets from membership degree and non-membership degree. For further study, the roughness measures of IF sets in rough data analysis are needed.

Acknowledgement

This work was supported by grants from the National Natural Science Foundation of China (No. 60673096, No. 60773174, and No. 60703117) and the Natural Science Foundation of Zhejiang Province in China (No. Y107262).

References

1. Atanassov, K.: Intuitionistic Fuzzy Sets. Fuzzy Set and Systems 20, 87–96 (1986)
2. Atanassov, K.: Intuitionistic Fuzzy Sets. Physica-Verlag, Heidelberg (1999)
3. Banerjee, M., Pal, S.K.: Roughness of a Fuzzy Set. Inform. Sci. 93, 235–246 (1996)
4. Boixader, D., Jacas, J., Recasens, J.: Upper and Lower Approximations of Fuzzy Sets. Int. J. Gen. Syst. 29, 555–568 (2000)
5. Bustince, H., Burillo, P.: Structures on Intuitionistic Fuzzy Relations. Fuzzy Sets and Systems 78, 293–303 (1996)
6. Chakrabarty, K., Gedeon, T., Koczy, L.: Intuitionistic Fuzzy Rough Sets. In: Proceedings of 4th Joint Conference on Information Sciences (JCIS), Durham, NC, pp. 211–214 (1998)
7. Cornelis, C., De Cock, M., Kerre, E.E.: Intuitionistic Fuzzy Rough Sets: at the Crossroads of Imperfect Knowledge. Expert Systems 20, 260–270 (2003)

8. Deschrijver, G., Kerre, E.E.: On the Position of Intuitionistic Fuzzy Set Theory in the Framework of Theories Modelling Imprecision. Inform. Sci. 177, 1860–1866 (2007)

9. Dubois, D., Prade, H.: Rough Fuzzy Sets and Fuzzy Rough Sets. Int. J. Gen. Syst. 17, 191–209 (1990)

10. Huynh, V.-N., Nakamori, Y.: A Roughness Measure for Fuzzy Sets. Inform. Sci. 173, 255–275 (2005)

11. Jena, S.P., Ghosh, S.K.: Intuitionistic Fuzzy Rough Sets. Notes Intuit. Fuzzy Sets 8, 1–18 (2002)

12. Lin, L., Yuan, X.-H., Xia, Z.-Q.: Multicriteria Fuzzy Decision-Making Methods Based on Intuitionistic Fuzzy Sets. J. Comput. Syst. Sci. 73, 84–88 (2007)

13. Nanda, S., Majumda, S.: Fuzzy Rough Sets. Fuzzy Sets and Systems 45, 157–160 (1992)

14. Pawlak, Z.: Rough Sets. Int. J. Comput. Inform. Sci. 11, 341–356 (1982)

15. Pawlak, Z., Skowron, A.: Rudiments of Rough Sets. Inform. Sci. 177, 3–27 (2007)

16. Pawlak, Z., Skowron, A.: Rough Sets: Some Extensions. Inform. Sci. 177, 28–40 (2007)

17. Radzikowska, A.M., Kerre, E.E.: A Comparative Study of Fuzzy Rough Sets. Fuzzy Sets and Systems 126, 137–155 (2002)

18. Rizvi, S., Naqvi, H.J., Nadeem, D.: Rough Intuitionistic Fuzzy Sets. In: Proceedings of the 6th Joint Conference on Information Sciences (JCIS), Durham, NC, pp. 101–104 (2002)

19. Samanta, S.K., Mondal, T.K.: Intuitionistic Fuzzy Rough Sets and Rough Intuitionistic Fuzzy Sets. J. Fuzzy Math. 9, 561–582 (2001)

20. Slowinski, R., Vanderpooten, D.: A Generalized Definition of Rough Approximations Based on Similarity. IEEE Trans. Knowledge and Data Eng. 12, 331–336 (2000)

21. Vlachos, L.K., Sergiadis, G.D.: Intuitionistic Fuzzy Information—Applications to Pattern Recognition. Pattern Recognition Letters 28, 197–206 (2007)

22. Wu, W.-Z., Mi, J.-S., Zhang, W.-X.: Generalized Fuzzy Rough Sets. Inform. Sci. 151, 263–282 (2003)

23. Wu, W.-Z., Zhang, W.-X.: Constructive and Axiomatic Approaches of Fuzzy Approximation Operators. Inform. Sci. 159, 233–254 (2004)

24. Yao, Y.Y.: Constructive and Algebraic Methods of the Theory of Rough Sets. J. Inform. Sci. 109, 21–47 (1998)

25. Yao, Y.Y.: Generalized Rough Set Model. In: Polkowski, L., Skowron, A. (eds.) Rough Sets in Knowledge Discovery 1. Methodology and Applications, pp. 286–318. Physica-Verlag, Heidelberg (1998)

Intra-cluster Similarity Index Based on Fuzzy Rough Sets for Fuzzy C-Means Algorithm

Fan Li, Fan Min, and Qihe Liu

School of Computer Science and Engineering
University of Electronic Science and Technology of China
Chengdu, 610051, P.R. China
{lifan,minfan,qiheliu}@uestc.edu.cn

Abstract. Cluster validity indices have been used to evaluate the quality of fuzzy partitions. In this paper, we propose a new index, which uses concepts of Fuzzy Rough sets to evaluate the average intra-cluster similarity of fuzzy clusters produced by the fuzzy c-means algorithm. Experimental results show that contrasted with several well-known cluster validity indices, the proposed index can yield more desirable cluster number estimation.

Keywords: Fuzzy c-means algorithm, Fuzzy Rough sets, Intra-cluster similarity, Cluster validity index.

1 Introduction

Cluster analysis for revealing the structure existing in a given data (patterns) set can be viewed as the problem of dividing the data set into a few compact subsets. The fuzzy c-means (FCM) algorithm [1] for cluster analysis has been the dominant approach in both theoretical and practical applications of fuzzy techniques for the last two decades. The aim of FCM is to partition a given set of data points (patterns) $X = \{\mathbf{x}_1, \mathbf{x}_2, \cdots, \mathbf{x}_n\} \subset \mathcal{R}^p$ into c clusters represented as fuzzy sets $F_1, F_2, \cdots, F_c$. The FCM objective function has the form of

$$J_m(\mathbf{U}, V) = \sum_{i=1}^{c} \sum_{j=1}^{n} u_{ij}^m \parallel \mathbf{x}_j - \mathbf{v}_i \parallel^2, \tag{1}$$

where $\mathbf{v}_i$ is the centroid of the fuzzy cluster F_i, $\parallel \cdot \parallel$ is a certain distance function, the exponent $m > 1$ is a fuzzifier, $u_{ij} = F_i(\mathbf{x}_j)$ is the membership value of $\mathbf{x}_j$ belonging to F_i satisfying $\sum_{i=1}^{c} u_{ij} = 1$ ($j = 1, 2, \cdots, n$) and $0 < \sum_{j=1}^{n} u_{ij} < n$ ($i = 1, 2, \cdots, c$), $\mathbf{U} = [u_{ij}]$ is the partition matrix, and $V = \{\mathbf{v}_1, \mathbf{v}_2, \cdots, \mathbf{v}_c\}$ is the set of all cluster centroids. FCM iteratively updates $\mathbf{U}$ and V to minimize $J_m(\mathbf{U}, V)$ until a certain termination criterion has been satisfied. In FCM, a fuzzy partition is denoted as $(\mathbf{U}, V)$.

In FCM, if c is not known a priori, a cluster validity index must be used to evaluate the quality of fuzzy partitions for different values of c to find out the optimal cluster number. In most cited indices, e.g. the Xie-Beni index [2] and the

G. Wang et al. (Eds.): RSKT 2008, LNAI 5009, pp. 316–323, 2008.

Fukuyama-Sugeno index [3], the intra-cluster similarity of a fuzzy partition is estimated by using distances between data points and cluster centroids. But this approach is not effective for large values of c, because $\lim_{c \to n} \parallel \mathbf{x}_j - \mathbf{v}_i \parallel^2 = 0$ (see [4,5]). To overcome this shortcoming, the Kwon index [5] is proposed, and another kind of index has been proposed in recent years [6,7]. This kind of index only considers the inter-cluster proximity, which is evaluated by the membership values of each data point belonging to all fuzzy clusters whereas the distance function is not taken into account.

In this paper, we propose a new method to assess the intra-cluster similarity of a fuzzy cluster by using the concepts of Fuzzy Rough sets. And the intra-cluster similarity index of a fuzzy partition obtained from FCM is defined as the average intra-cluster similarity of all fuzzy clusters. Experimental results indicate that the proposed index can find the correct cluster number and is reliable in comparison with several well-known cluster validity indices.

2   Basic Concepts

The concepts of Fuzzy Rough sets, which were proposed by Dubois and Prade [8,9], aim at extending the classical Rough sets theory [10,11] to fuzzy information systems. Let U be a nonempty set of objects. A fuzzy binary relation R on U is called a T-similarity relation if R satisfies:
(1) Reflectivity: $R(x, x) = 1, \forall x \in U$;
(2) Symmetry: $R(x, y) = R(y, x), \forall x, y \in U$; and
(3) T-transitivity: $R(x, z) \geq T(R(x, y), R(y, z)), \forall x, y, z \in U$,
where T is a t-norm.

Definition 1. *Let F be a fuzzy subset of U, and R a T-similarity relation, the R-lower approximation and R-upper approximation of F, denoted by two fuzzy sets $\underline{R}(F)$ and $\overline{R}(F)$ respectively, are defined as:*

$$\underline{R}(F)(x) = \inf_{y \in U} max\{1 - R(x, y), F(y)\}, \tag{2}$$

$$\overline{R}(F)(x) = \sup_{y \in U} min\{R(x, y), F(y)\}. \tag{3}$$

The pair $(\underline{R}(F), \overline{R}(F))$ is called a Fuzzy Rough set.

The above definitions were generalized in [12]. The R-lower approximation and R-upper approximation of F are defined as:

$$\underline{R}(F)(x) = \inf_{y \in U} I_T\{R(x, y), F(y)\}, \tag{4}$$

$$\overline{R}(F)(x) = \sup_{y \in U} T\{R(x, y), F(y)\}, \tag{5}$$

where I_T is the residuation implication of T, i.e. $I_T(a, b) = sup\{c \in [0, 1] : T(a, c) \leq b\}$ for every $a, b \in [0, 1]$.

In general, the distance between two data points can qualify their similarity. The longer distance indicates the less degree of similarity, and vice versa. This intuition can be used to construct a fuzzy binary relation, which reflects the whole structure of the given data set. Thus, based on it, we can construct the lower approximations of $F_1, F_2, \cdots, F_c$, and use these approximations to estimate the quality of the corresponding fuzzy partition.

3   Proposed Intra-cluster Similarity Index

Definition 2. *Let* $X = \{\mathbf{x}_1, \mathbf{x}_2, \cdots, \mathbf{x}_n\} \subset \mathcal{R}^p$ *be a given set of data points. A fuzzy binary relation S on X is defined as:* $\forall \mathbf{x}_i, \mathbf{x}_j \in X$,

$$S(\mathbf{x}_i, \mathbf{x}_j) = 1 - \frac{\| \mathbf{x}_i - \mathbf{x}_j \|}{d_{max}}, \tag{6}$$

where $d_{max} = max_{i,j}\{\| \mathbf{x}_i - \mathbf{x}_j \|\}$.

Proposition 1. *S is a T_L-similarity relation, where T_L is the Lukasiewicz t-norm:* $T_L(a, b) = \max\{0, a + b - 1\}$ *for every* $a, b \in [0, 1]$.

For the Lukasiewicz t-norm T_L, $I_{T_L}(a, b) = min\{1, 1 - a + b\}$ for every $a, b \in [0, 1]$. Let F_i be a fuzzy cluster of X. By Eq. 4, we have:

$$\underline{S}(F_i)(\mathbf{x}_i) = \inf_{\mathbf{x}_j \in X} \min\{1, 1 - S(\mathbf{x}_i, \mathbf{x}_j) + u_{ij}\}. \tag{7}$$

$\underline{S}(F_i)(\mathbf{x}_i)$ can be seen as the certainty degree of the event that a data point in X belongs to the fuzzy cluster F_i according to the similarity between this data point and $\mathbf{x}_i$. Intuitively, since S reflects the structure of the data set, we can estimate the intra-cluster similarity of each fuzzy cluster based on this concept.

Definition 3. *Let* $X = \{\mathbf{x}_1, \mathbf{x}_2, \cdots, \mathbf{x}_n\} \subset \mathcal{R}^p$ *be a given set of data points, and* $F = \{F_1, F_2, \cdots, F_c\}$ *a fuzzy partition of X.* $\forall F_i \in F$, *the intra-cluster similarity of F_i is defined as:*

$$IS(F_i) = \frac{1}{|\underline{S}(F_i)|} \sum_{\mathbf{x} \in B_i} \underline{S}(F_i)(\mathbf{x}), \tag{8}$$

where $B_i = \{\mathbf{x} \in X | F_i(\mathbf{x}) \geq l\}$ *is the l-level set of F_i,* $\frac{1}{c} \leq l < 1$, *and* $| A | = \sum_i A(x_i)$ *is the cardinality of a fuzzy set A.*

Generally speaking, B_i contains "important" data points of the fuzzy cluster F_i. $IS(F_i)$ reflects the proportion of the sum of those "important" data points' membership values belonging to the S-lower approximation of F_i to the cardinality of the S-lower approximation of F_i.

Definition 4. *Let* $X = \{\mathbf{x}_1, \mathbf{x}_2, \cdots, \mathbf{x}_n\} \subset \mathcal{R}^p$ *be a given set of data points, and* $F = \{F_1, F_2, \cdots, F_c\}$ *a fuzzy partition of X. The intra-cluster similarity index (IS) of F is defined as:*

$$IS(F) = \frac{1}{c} \sum_{i=1}^{c} IS(F_i). \tag{9}$$

Table 1. Three existing validity indices for FCM

Index	Functional description
XB	$\frac{J_m(\mathbf{U},V)}{n\,\min_{i\neq j}\|\mathbf{v}_i-\mathbf{v}_j\|^2}$
FS	$J_m(\mathbf{U},V)-\Sigma_{i=1}^c\Sigma_{j=1}^n u_{ij}^m \parallel \mathbf{v}_i-\overline{\mathbf{v}}_1 \parallel^2$
K	$\frac{J_m(\mathbf{U},V)+\frac{1}{c}\Sigma_{i=1}^c\|\mathbf{v}_i-\overline{\mathbf{v}}_2\|^2}{\min_{i\neq j}\|\mathbf{v}_i-\mathbf{v}_j\|^2}$
$\overline{\mathbf{v}}_1=\frac{1}{c}\Sigma_{i=1}^c\mathbf{v}_i,\ \overline{\mathbf{v}}_2=\frac{1}{n}\Sigma_{i=1}^n x_i$	

$IS(F)$ is the average intra-cluster similarity of all fuzzy clusters in the fuzzy partition F. A large value of $IS(F)$ indicates a good intra-cluster similarity of the fuzzy partition F.

In [13], two validity indices, DB_r and D_r, are also defined in rough-fuzzy framework. The two indices extend the traditional Davies-Bouldin index and Dunn index (see [14]), which are used for crisp clustering, respectively. The main differences between the two indices and $IS(F)$ are as follows. Firstly, DB_r and D_r are used for Rough-Fuzzy c-means algorithm (a variation of Rough c-means algorithm [15]), whereas $IS(F)$ is used for FCM. Since a crisp set can be viewed as a special case of fuzzy sets, $IS(F)$ can be used for crisp clustering either. Secondly, DB_r and D_r use the distance between each data point and the corresponding cluster center to evaluate the intra-cluster similarity, whereas $IS(F)$ uses the concept of the lower approximation to do so. This concept can be interpreted based on Zadeh's possibility theory [9]. Finally, two parameters, w_{low} and w_{up}, must be assigned in DB_r and D_r. These two parameters correspond to the relative importance of the lower and upper approximations respectively. In $IS(F)$, the threshold value l must be assigned. But in general, the task of deciding the value of l is easier than that of w_{low} and w_{up}.

4   Experimental Results

In order to evaluate the performance of the proposed index (IS), we applied IS and several well-known cluster validity indices, including the extended Xie-Beni index (XB) [2], the Fukuyama-Sugeno index (FS) [3] and the extended Kown index (K) [5], to fuzzy partitions obtained from FCM for two data sets. The functional description of the above three index is listed in Table 1.

The first data set is a synthetic data set, which is shown in Fig. 1. It consists of five clusters with 10 data points per cluster. The second one is the IRIS data set from the UCI repository of machine learning databases [16], which represents different categories of irises with four features. There are three classes in this data set: Setosa, Versicolor and Virginica, with 50 samples per class. It is known that two classes Versicolor and Virginica have a substantial overlap while the class Setosa is linearly separable from the other two. Thus, the most suitable cluster number is two or three.

For the mentioned data sets, we ran FCM for different values of c (c=2–9). For a particular c and data set, FCM started from the same initial partition and

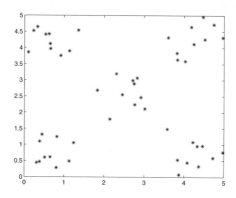

Fig. 1. Synthetic data set

Table 2. Preferable values of c for the Synthetic data set chosen by each index $(c = 2-9)$

m	XB	FS	K	IS
1.5	5	9	5	5
1.6	5	9	5	5
1.7	5	7	5	5
1.8	5	5	5	5
1.9	5	5	5	5
2.0	5	5	5	5
2.1	5	5	5	5
2.2	5	5	5	5
2.3	5	5	4	5
2.4	5	5	4	5
2.5	5	4	4	5

Table 3. Preferable values of c for the IRIS data set chosen by each index $(c = 2-9)$

m	XB	FS	K	IS
1.5	2	9	2	2
1.6	2	4	2	2
1.7	2	5	2	2
1.8	2	5	2	2
1.9	2	5	2	2
2.0	2	5	2	2
2.1	2	5	2	2
2.2	2	5	2	2
2.3	2	5	2	2
2.4	2	5	2	2
2.5	2	5	2	2

Table 4. Values of the four indices on the Synthetic data set for c =2–9

c	XB	FS	K	IS	c	XB	FS	K	IS
2	0.362	55.647	18.366	0.864	2	0.371	63.161	18.809	0.793
3	0.161	-79.486	8.388	0.817	3	0.152	-67.492	7.954	0.748
4	0.079	-192.881	4.508	0.894	4	0.063	-193.353	3.685	0.832
5	**0.045**	-231.212	**3.029**	**0.957**	5	**0.043**	-226.713	**2.914**	**0.893**
6	0.528	-232.163	36.637	0.874	6	0.365	-225.784	24.984	0.814
7	0.466	-235.497	32.527	0.833	7	0.436	**-226.840**	31.159	0.758
8	0.498	-231.379	34.406	0.807	8	0.461	-221.323	32.741	0.713
9	0.392	**-246.722**	29.577	0.824	9	0.368	-221.530	26.066	0.671
	(a) m=1.5					(b) m=1.7			

c	XB	FS	K	IS	c	XB	FS	K	IS
2	0.337	59.517	17.101	0.709	2	0.275	49.525	14.012	0.619
3	0.124	-56.816	6.563	0.670	3	0.087	-39.217	4.716	0.609
4	0.047	-182.982	2.844	0.750	4	0.028	**-136.591**	**1.937**	0.611
5	**0.037**	**-200.951**	**2.625**	**0.753**	5	**0.024**	-136.372	2.011	**0.643**
6	0.342	-193.478	24.817	0.686	6	0.230	-122.279	20.219	0.564
7	0.353	-188.133	27.196	0.625	7	0.202	-109.660	18.412	0.523
8	0.335	-179.649	25.975	0.609	8	0.164	-102.443	16.733	0.500
9	0.289	-175.988	24.622	0.593	9	0.168	-104.169	20.062	0.466
	(c) m=2.0					(d) m=2.5			

Table 5. Values of the four indices on the IRIS data set for c =2–9

c	XB	FS	K	IS	c	XB	FS	K	IS
2	**0.062**	-431.455	**9.553**	**0.959**	2	**0.059**	-424.416	**9.153**	**0.927**
3	0.156	-515.259	24.812	0.893	3	0.150	-496.773	23.899	0.846
4	0.183	-568.406	29.390	0.840	4	0.174	-542.094	28.183	0.752
5	0.507	-533.600	83.792	0.801	5	0.262	**-628.713**	43.251	0.679
6	0.228	-554.874	39.040	0.773	6	0.211	-497.527	36.547	0.658
7	0.345	-559.269	59.150	0.737	7	0.314	-498.966	54.866	0.664
8	0.548	-546.903	96.007	0.722	8	0.307	-509.351	54.274	0.651
9	0.392	**-577.570**	67.790	0.737	9	0.338	-510.541	59.708	0.629
	(a) m=1.5					(b) m=1.7			

c	XB	FS	K	IS	c	XB	FS	K	IS
2	**0.054**	-401.801	**8.376**	**0.877**	2	**0.044**	-341.890	**6.865**	**0.824**
3	0.137	-450.495	21.955	0.777	3	0.108	-344.731	17.709	0.693
4	0.159	-476.000	26.020	0.663	4	0.124	-332.390	21.073	0.585
5	0.228	**-544.972**	38.241	0.578	5	0.162	**-348.558**	28.637	0.512
6	0.175	-389.115	31.386	0.577	6	0.114	-213.866	22.871	0.525
7	0.537	-344.699	99.514	0.549	7	0.433	-162.868	93.035	0.554
8	0.254	-389.922	47.002	0.568	8	0.197	-201.132	42.886	0.524
9	0.308	-384.021	57.666	0.556	9	0.288	-149.523	69.877	0.537
	(c) m=2.0					(d) m=2.5			

ran for different values of m (m=1.5–2.5). After the fuzzy partition is obtained, the four indices were computed. In the computation of IS, $l = \frac{1}{c}$.

The results are shown in Tables 2– 5. As indicated in Tables 2 and 3, only IS and XB correctly recognize correct cluster numbers of the two data sets for all values of m. Furthermore, a cluster validity index is considered as a reliable index when it is insensitive to changes in m [4,6]. From this point of view, IS provides more reliable results compared to other indices, as shown in Tables 4 and 5, where the optimal value of each index is marked by boldface.

Thus we can conclude that the proposed index provides the best cluster number estimation for all test data sets.

5   Conclusions

The fuzzy c-means (FCM) algorithm is an effective tool for cluster analysis. In FCM, if the cluster number c is not known a priori, a validation index must be used to find out the optimal number of clusters. By using the concepts of Fuzzy Rough sets, this paper presents a new intra-cluster similarity index to assess the intra-cluster similarity of fuzzy partitions obtained from FCM. Experimental results show that contrasted with some existing cluster validity indices, the proposed index yields the correct cluster number and is reliable in comparison with several well-known cluster validity indices.

In future works, we plan to carry out extensive experiments and theoretical analysis to firmly establish the utility of the proposed index. We also plan to apply the basic ideas described in this paper to the evaluation of the inter-cluster proximity as well as to the cluster validity analysis for crisp clustering algorithms.

Acknowledgement

This work was supported by an information distribution project under grant No. 9140A06060106DZ223, Program for New Century Excellent Talents in University (NCET-06-0811), and Young Foundation of UESTC, grant No. L080106015X0748.

References

1. Bezdek, J.C.: Pattern Recognition with Fuzzy Objective Function Algorithms. Plenum, New York (1981)
2. Xie, X.L., Beni, G.: A validity measure for fuzzy clustering. IEEE Trans. Pattern Anal. Mach. Intell 13(8), 841–847 (1991)
3. Fukuyama, Y., Sugeno, M.: A new method of choosing the number of clusters for the fuzzy c-mean method. In: 5th Fuzzy Systems Symposium, Japan, pp. 247–250 (1989)
4. Pal, N.R., Bezdek, J.C.: On cluster validity for the fuzzy c-means model. IEEE Trans. Fuzzy Syst. 13(3), 370–379 (1995)

5. Kwon, S.H.: Cluster validity index for fuzzy clustering. Electron. Lett 34(22), 2176–2177 (1998)
6. Kim, D., Lee, K.H., Lee, D.: On cluster validity index for estimation of the optimal number of fuzzy clusters. Pattern Recognition 37(10), 2561–2574 (2004)
7. Kim, Y., Kim, D., et al.: A cluster validation index for GK cluster analysis based on relative degree of sharing. Information Science 168(1-4), 225–242 (2004)
8. Dubois, D., Prade, H.: Rough fuzzy sets and fuzzy rough sets. Internat. J. General Systems 17(2–3), 191–209 (1990)
9. Dubois, D., Prade, H.: Putting rough sets and fuzzy sets together. In: Slowinski, R. (ed.) Intelligent Decision Support: Handbook of Applications and Advances of the Rough Sets Theory, pp. 203–222. Kluwer, The Netherlands (1992)
10. Pawlak, Z.: Rough sets. International J. Comp. Inform. Science 11, 341–356 (1982)
11. Pawlak, Z.: Some Issues on Rough Sets. In: Peters, J.F., Skowron, A., Grzymała-Busse, J.W., Kostek, B.z., Świniarski, R.W., Szczuka, M. (eds.) Transactions on Rough Sets I. LNCS, vol. 3100, pp. 1–58. Springer, Heidelberg (2004)
12. Morsi, N.N., Yakout, M.M.: Axiomatics for fuzzy rough set. Fuzzy Sets Syst. 100(1-3), 327–342 (1998)
13. Mitra, S., Banka, H., Pedrycz, W.: Rough-fuzzy collaborative clustering. IEEE Trans. Syst., Man, Cybern. B, Cybern. 36(4), 795–805 (2006)
14. Bezdek, J.C., Pal, N.R.: Some New Indexes of Cluster Validity. IEEE Trans. Syst., Man, Cybern. B, Cybern. 28(3), 301–315 (1998)
15. Lingras, P., Yan, R., West, C.: Comparison of conventional and rough k-means clustering. In: Wang, G., et al. (eds.) RSFDGrC 2003. LNCS (LNAI), vol. 2639, pp. 130–137. Springer, Heidelberg (2003)
16. UCI Repository of machine learning databases, http://www.ics.uci.edu/~mlearn/MLRepository.html

A New SVM-Based Decision Fusion Method Using Multiple Granular Windows for Protein Secondary Structure Prediction

Anjum Reyaz-Ahmed and Yanqing Zhang

Department of Computer Science
Georgia State University
Atlanta, GA 30302-3994
USA
anjumrahmed@gmail.com zhang@taichi.cs.gsu

Abstract. Support vector machines (SVM) have shown strong generalization ability in a number of application areas, including protein structure prediction. Bioinformatics techniques to protein secondary structure prediction mostly depend on the information available in amino acid sequence. In this study, a new sliding window scheme is introduced with multiple granular windows to form the protein data for training and testing SVM. Orthogonal encoding scheme coupled with BLOSUM62 matrix is used to make the prediction. The prediction of binary classifiers using multiple windows is compared with single window scheme, the results shows single window not to be good in all cases. New classifier is introduced for effective tertiary classification. The accuracy level of the new architectures are determined and compared with other studies. The tertiary architecture is better than most available techniques.

Keywords: Binary classifier, BLOSUM62, encoding scheme, granular computing, orthogonal profile, support vector machine (SVM), tertiary Classifier.

1 Introduction

Protein secondary structure is closely related to the protein tertiary structure, which determines the functional character of proteins. The success of genome sequencing program resulted in massive amounts of protein sequence data (that are produced by DNA sequencing) [HUMAN GENOME PROJECT]. This means the output of experimentally determined protein structure is lagging far behind the output of protein sequence. It is therefore becoming increasingly important to predict protein structure from its amino acid sequence, using insight obtained from already known structures.

The SVM method is a comparatively new learning system which is mostly used in pattern recognition problems. This machines uses hypothesis space of linear functions in a high-dimensional feature space, and it is trained with a learning algorithm based on optimization theory. To compare the results of this study

G. Wang et al. (Eds.): RSKT 2008, LNAI 5009, pp. 324–331, 2008.
© Springer-Verlag Berlin Heidelberg 2008

with previous results RS126 data set is used. RS 126 data set [1] is proposed by Rost and Sander. Among neural networks Chandonia and Karplus [2] introduced a novel method for processing and decoding the protein sequence with NNs by using large training data set such as 681 non homologous proteins. And with the use of jury method, this scheme records 74.8% percentage accuracy. Some of the recent studies adopting this SVM learning machine for secondary structure prediction are the one which used frequent profiles with evolutionary information as an encoding schemes for SVM [3], the one which adopted two layers of SVM, with a weighted cost function [4], the one which applied PSI-BLAST PSSM profiles [5] as an input vector and the sliding window scheme with SVM_Representative architecture [6]. This paper is part of a research done on protein secondary structure prediction the other section contains details about another new tertiary architecture which combines both the svm and the neural networks and uses genetic algorithms for optimization [13].

Granular computing is a study in which the details of data are seen from different dimensions. It is an approach in which data are seen at different levels/scales of granularity [14,15]. It is noted that the same data seen at different levels of resolution depicts different patterns and regularities. The difference between multiple windows and single window method can also be considered as different approach for scaling of protein sequence data to predict its structure. The different course employed in considering granules in the encoding scheme reflects on the accuracy of the method.

2   Multiple Granular Windows Encoding Scheme

The single window technique is challenged with multiple windows encoding scheme in this study. In the case of multiple windows scheme, instead of using a single sliding window multiple sliding windows are used. The center element of the middle window becomes the target and all other windows are used as feature values to train and test the SVM. Only the elements/residues/granules inside the window forms the training/testing data, some residues in the middle are skipped. Sliding window technique is applied to move to the next residue. In this study windows of equal sizes are considered. Windows of different sizes will be studied as future technique. In the case of different size windows, the window in the middle will have more residues than windows at each side. In all the multiple windows cases consider have three windows with different lengths. Initially in this research BLOSUM_62 matrix [7] coupled with orthogonal encoding scheme is used. The single window technique is compared with multiple windows encoding scheme. In both cases same parameter values are used.

The comparison of the two techniques reveals single window scheme not to be good in all cases. For window of size 15 the simulation results show the multiple windows to be better than single window for all the six binary classifiers. The results are shown in the Table 1.In this case the single window is of length 15 and in the multiple windows case, 3 windows each of size 5 with gaps between the

Table 1. Comparing Single Window and Multiple Windows

Binary Classifier	Multiple Windows	Single Window
H/ H	73.59%	73.52%
E/ E	78.39%	78.39%
C/ C	69.69%	69.62%
H/E	72.94%	72.33%
E/C	75.9%	75.59%
C/H	71.93%	71.74%
Average	73.74%	73.53%

Table 2. Simulation II: Single Window vs. Multiple Windows

Binary Classifier	Accuracy of Multiple Windows	Accuracy of Single Window
H/ H	72.37%	74.67%
E/ E	78.41%	78.34%
C/ C	70.00%	69.63%
H/E	72.24%	73.7%
E/C	75.45%	74.3%
C/H	71.43%	72.9%
Average	73.32%	73.92%

windows is used. In both the cases RBF kernel is used with the same parameter values (gamma γ and cost co-efficient C).

In another simulation single window of size 21 is compared with 3 windows, each of size 5 and a gap of 3 residues (gap means these three residue was not considered to form the data for SVM) between the windows. The results of this simulation are shown in Table 2. This indicates single window not be good in all cases and multiple windows has less information to process (as it has only 15 residues to consider where as single window have 21 residues in each set). Considering all the points multiple windows shows scope of performance. This study was conducted to determine if single window scheme is solely the best method to do protein secondary structure prediction, empirically there is scope for other methods too. The optimal window length and other optimal values of the parameters are selected to be the same as those used in previous studies. As the previous studies have already run simulations and have obtained the optimal values for all the parameters, further research is avoided.

3   Support Vector Machines

Support Vector Machines (SVM) are learning systems that use a hypothesis space of linear function in a high dimensional feature space, trained with a learning algorithm from optimization theory that implements a learning bias derived from statistical learning theory. This learning strategy introduced by Vapnik [8] and co-workers is a principled and very powerful method that in the few years since its introduction has already outperformed most other systems in a wide variety of applications [9].

Binary classifier is frequently implemented by using a real-valued function $f : X \subseteq \Re^2 \to \Re$ in the following way: the input $x = (x_1, x_2, ..., x_n)$ is assigned to the positive class, if $f(x) \geq 0$, and otherwise to the negative class. If we consider the case where $f(x)$ is a linear function of $x \in X$, so that it can be written as

$$f(x) = w \bullet x + b. \tag{1}$$

$$= \sum_{i=1}^{n} w_i x_i + b. \tag{2}$$

The above algorithm for separable data, when applied to non-separable data, will find no feasible solution: this will be evidenced by the objective function(i.e. the dual Langrangian) growing arbitrarily large. So to extend these ideas to handle non-separable data, the constraints (1) and (2) are relaxed, but only when necessary, that is, by introducing further cost (i.e. an increase in the primal objective function). This can be done by introducing positive slack variables ξ_i, $i = 1, .., n$ in the constraints, which then become:

$$x_i \bullet w + b \geq +1 - \xi_i \; for \; y_i = +1. \tag{3}$$

$$x_i \bullet w + b \leq -1 + \xi_i \; for \; y_i = -1. \tag{4}$$

4   SVM_Complete: New Tertiary Classifier

In this method all the six binary classifiers are used to form the tertiary classifier. In SVM_Represnt. scheme [6], no matter what the distance values are positive or negative, the classifier with the absolute maximum distance is chosen as the representative classifier for the final decision of the class. In this paper, we consider that fact that among the three one-versus-one classifier, two classifier try to identify the same class, for example H/E and C/H tries to classify H (only difference is in H/E H is the positive class and in C/H H is the negative class). So we add up the values of one-versus-one classifier which classifies the same class. Then we also add the value of one-versus-rest classifier, to sum up the total strength of the specific class.

For example, for calculating the strength of H, we have to:

Step 1: Check if SVM (H/E) is positive, if true
        H = absolute value of SVM (H/E)
Step2: Check if SVM (C/H) is negative, if true
        H = H + absolute value of SVM (C/H)
Step 3: Add one-versus-rest prediction value
        H = H + value of SVM (H/H̃). *
* Note here we add the actual value not absolute, since we want to determine H's total strength.

Similarly strength of E and C are calculated and final result is produced depending upon which class has the highest value. Here SVM (H/E) means the exact output the support vector machine gives after classifying the given data.

In SVM_VOTE [3], all six binary classifiers are combined by using a simple voting scheme in which the testing sample is predicted to be state i (i is among H, E and C) if the largest number of the six binary classifiers classify it as state i. In case the testing samples have two classifications in each state, it is considered to be a coil. Though all six binary classifiers are considered for tertiary classification, only one constitutes the results. In SVM_Complete all six classifiers are used to calculate individual strengths of each class and finally the one with highest strength is considered as the predicted secondary structure.

5   Training and Testing Datasets

For comparing the results of this study with previously published results [6], RS 126 data set is used. The RS 126 data set is proposed by Rost & Sander [1] and according to their definition, it is non-homologous set. They used percentage identity to measure the homology and defines non-homologous as no two proteins in the set share more than 25% sequence identity over a length of more than 80 residues.

Table 3. 8-to-3 state reduction method in secondary structure assignment

DSSP Class	8 − state symbol	3 − state symbol	Class Name
3_{10}- helix	G		
α-helix	H	H	Helix
π-helix	I		
β-strand	E	E	Sheet
isolated β-bridge	B		
Bend	S		
Turn	T	C	loop
Rest (connection region)	−		

For each data set, the seven fold cross validation is done [1,3,6]. In the seven-fold cross validation test, one subset is chosen for testing and remaining 6 subsets are used for training and this process is repeated until all the subsets are chosen for the testing. The secondary structure is converted from the experimentally determined tertiary structure by DSSP [8], STRIDE [7] or DEFINE. In this study, the DSSP scheme is used since it is the most generally used secondary structure prediction method. In this study, these eight classes are reduced into three regular classes based on the following Table 3.

6 Accuracy Metrics

There are several standard evaluation methods of secondary structure prediction. Among them, Q_3, Matthew's Correlation Coefficient and Segment Overlap Measure (SOV) are widely used assessing methods. We have simulated results comparing the Q_3 percentage value of different tertiary classifier. Q_3 is one of the most commonly used performance measures in the protein secondary structure prediction and it refers to the three-state overall percentage of correctly predicted residues. This measure is defined as,

$$Q_3 = \frac{\sum_{i \in \{H,E,C\}} \# \ of \ residues \ correctly \ predicted_i}{\sum_{i \in \{H,E,C\}} \# \ number \ of \ residues \ in \ class \ i} \times 100 \qquad (5)$$

$$Q_I = \frac{\# \ of \ residues \ correctly \ predicted \ in \ state \ I}{\# \ number \ of \ residues \ in \ state \ I} \times 100 \qquad (6)$$

$I \in \{H, E, C\}$

The new tertiary classifier is compared with other tertiary architectures of former studies. The 7 fold test cases have been performed for a valid comparison of the new tertiary classifier with that of the SVM_Represnt. [6] contributed classifier. The accuracy level of the tertiary classifier is important from research point of view, as the main objective in this study is to accurately determine the secondary structure of the protein sequence. The Table 4 gives the accuracy level of all the methods [6] and also the accuracy levels the new SVM_Complete (tertiary classifier of this research). As shown in the table SVM_Complete is better than other available methods.

The table is adopted from [6].

Table 4. Accuracy of tertiary Classifiers on the RS 126 data set. Combined results of 7-fold cross validation are shown.

Tertiary Classifier	$Q_3(\%)$	$Q_H(\%)$	$Q_E(\%)$	$Q_C(\%)$
TREE_HEC [3]	63.2	51.0	45.2	79.9
TREE_ECH [3]	62.3	62.4	26.2	79.0
TREE_CHE [3]	61.2	64.8	47.3	65.2
SVM_VOTE [3]	62.0	73.5	34.7	65
SVM_MAX_D [3]	63.2	61.0	40.1	75.5
DAG [3]	63.2	59.2	41.6	76.0
SVM_REPRESNT. [6]	63.2	70.6	35.4	70.5
SVM_Complete	66.7	64.0	40.8	80.3

First the accuracy levels of single window encoding scheme was compared with different former methods. As seen in Table 4 the average accuracy of the SVM_Complete (a new classifier of this study) is better than other available methods. The table 4 is the accuracy level for window of size 15. The results in the table are obtained after 7-fold cross validation for window of size 15. The accuracies are compared with other classifiers that use single window encoding scheme. In table 4 accuracy levels of SVM_Represnt. [6] And SVM_VOTE [3] are obtained by simulation after 7-fold cross validation. All other former classifiers accuracies are adopted from [6]. The simulations were performed for many window sizes. The accuracy levels of SVM_Complete and SVM_Represnt. are shown in table 5 for window sizes 15, 13 and 11.

Table 5. Q_3 % for Different Window Sizes

Window Size	SVM_Complete(%)	SVM_Represnt.
15	66.7%	63.15%
13	64.1%	62.43%
11	56.82%	55.93%

The are many researches that show accuracy greater than 75% but all this studies use PSSM (Position Specific Scoring Matrix) as their encoding scheme for binary classifier. The reason for higher tertiary classifier accuracy is due to the fact that their binary classifiers have over 85 % of prediction accuracy. Accuracy levels of tertiary classifier using multiple windows encoding scheme are shown in Table 6.

Table 6. Accuracy of Tertiary Classifier Using Multiple Windows Scheme

Tertiary Classifier	Q_3(%)	Q_H(%)	Q_E(%)	Q_C(%)
SVM_VOTE [3]	62.6	78.9	39.7	62.4
SVM_REPRESNT. [6]	64.8	72.1	41.8	72.0
SVM_Complete	68.4	69.1	45.0	78.8

7   Conclusion

After many demonstrations, it is now established that single window scheme is not the only best method to encode while considering BLOSUM62 and orthogonal matrix. Multiple windows scheme performed better in some cases where the data given to the learning machine (SVM) was less informative than that given in single window scheme. When both were encoded with equal amount of information, multiple window schemes' performance is better than single window scheme in every case.

All the tertiary classifiers discussed in this study have less accuracy when directly compared with new tertiary classifiers introduces in this study. Though the study is not better when compared directly to the claimed accuracy levels of the former methods, the encoding scheme of binary classifiers used in those methods is different and better than the one used in this study.

References

1. Rost, B., Sander, C.: Improved prediction of protein secondary structure by use of sequence profile and neural networks. Proc. Natl. Acad. Sci. U S A 90, 7558–7562 (1993)
2. Chandonia, J.M., Karplus, M.: New Method for accuracy prediction of protein secondary structure. Proteins 35, 293–306 (1999)
3. Hua, S., Sun, Z.: A Novel Method of Protein Secondary Structure Prediction with High Segment Overlap Measure: Support Vector Machine Approach. J. Mol. Biol. 308, 397–407 (2001)
4. Casbon, J.: Protein Secondary Structure Prediction with Support Vector Machines (2002)
5. Jones, D.T.: Protein Secondary Structure Prediction Based on Position-Specific-Scoring Matrices. J. Mol. Biol. 292, 195–202 (1999)
6. Hu, H., Yi, P.: Improved Secondary Structure Prediction Using Support Vector Machines with a New Encoding Scheme and an Advanced Tertiary Classifier. IEEE - Transaction on Nanobioscience 3(4) (2004)
7. Henikoff, S., Henikoff, J.G.: Amino acid substitution matrices from protein blocks. PNAS 89, 10915–10919 (1992)
8. Vapnik, V., Corter, C.: Support vector networks. Machine Learning 20, 273–293 (1995)
9. Christianini, N., Shawe-Taylor, J.: An introduction to Support Vector Machines. Cambridge University Press, Cambridge (2000)
10. Burges, C.J.C.: A Tutorial on Support Vector Machines for Pattern Recognition (1998), http://www.kernel-machines.org/papers/Burges98.ps.gz
11. Joachims, T.: SVM light (2002), http://www.cs.cornell.edu/People/tj/svm_light
12. Kim, H., Park, H.: Protein Secondary Structure Prediction Based on an Improved Support Vector Machines Approach. Protein Eng. 16, 553–560 (2003)
13. Reyaz-Ahmed, A., Zhang, Y.-Q.: Protein Secondary Structure Prediction Using Genetic Neural Support Vector Machines. In: Proc. of IEEE 7th International Conference on BioInformatics and BioEngineering, Boston, October 14-17, pp. 1355–1359 (2007)
14. Tang, Y.C., Jin, B., Zhang, Y.-Q.: Granular Support Vector Machines with Association Rules Mining for Protein Homology Prediction, Artificial Intelligence in Medicine. Special Issue on Computational Intelligence Techniques in Bioinformatics 35(1-2), 121–134 (2005)
15. Jin, B., Zhang, Y.-Q., Wang, B.H.: Granular Kernel Trees with Parallel Genetic Algorithms for Drug Activity Comparisons. International Journal of Data Mining and Bioinformatics 1(3), 270–285 (2007)

Hierarchical Clustering of Non-Euclidean Relational Data Using Indiscernibility-Level

Shoji Hirano and Shusaku Tsumoto

Department of Medical Informatics, Shimane University, School of Medicine
89-1 Enya-cho, Izumo, Shimane 693-8501, Japan
hirano@ieee.org, tsumoto@computer.org

Abstract. In this paper, we present a clustering method for non-Euclidean relational data based on the combination of indiscernibility level and linkage algorithm. Indiscernibility level quantifys the level of global agreement for classifying two objects into the same category as indiscernible objects. Single-linkage grouping is then used to merge objects according to the indiscernibility level from bottom to top and construct the dendrogram. This scheme enables users to examine the hierarchy of data granularity and obtain the set of indiscernible objects that meets the given level of granularity. Additionally, since indiscernibility level is derived based on the binary classifications assigned independently to each object, it can be applied to non-Euclidean, asymmetric relational data.

1 Introduction

In some application areas where the relationships between instances are of main concern, *non-Euclidean (non-metric) relational data* can be collected and provided as a subject for analysis. Non-Euclidean relational data involves the following properties: (1) objects are not represented in a usual feature vector space but their relationships (usually similarity or dissimilarity) are measured and stored in a relational data matrix. (2) The dissimilarity can be non-Euclidean; that means the dissimilarity may not satisfy the triangular inequality nor symmetry. Examples in social sciences include subjectively judged relations between students and input/output of the persons between countries [1].

Clustering of such non-Euclidean relational data has attracted much interests as a method for discovering interesting groups of objects based on their pairwise relationships. However, since attribute vectors do not exist and dissimilarities are non-metric, the choice of clustering methods is limited compared to the metric and/or non-relational cases. For example, methods such as k-means may not be directly applied to this type of data as they assume the existence of data vectors. Conventional hierarchical clusterings are capable of dealing with relative or subjective measures. However, they involve other problems such as erosion or expansion of data space by intermediate objects between large clusters, and in some cases the results may change according to the order of processing objects [2]. The NERF c-means proposed by Hathaway et al. [3] is an extension of fuzzy

G. Wang et al. (Eds.): RSKT 2008, LNAI 5009, pp. 332–339, 2008.

c-means and capable of handling the non-Euclidean relational data. However, as it is a sort of partitional clustering method, it is still difficult to examine the structure of the data, namely, the hierarchy of data groups. Additionally, most of these methods are not designed to deal with asymmetric dissimilarity.

In this paper, we present a clustering method for non-Euclidean relational data based on the combination of indiscernibility level and linkage algorithm. Instead of using (dis-)similarity of objects, we use indiscernibility of objects as proximity. The indiscernibility level represents the degree of global agreement for classifying a pair of objects as indiscernible objects, and is calculated based on the binary classifications determined independently to each object. Then the simple nearest neighbor hierarchical clustering is used to construct a dendrogram of objects, which represents the hierarchy of indiscernibility. This scheme allows us to control the granularity of resultant object groups, by interactively selecting the threshold level of indiscernibility. The benefits of this method also include that the dissimilarity of objects for forming the binary classifications does not need to satisfy symmetry nor triangular inequality; thus it could be applied to various kind of datasets including relational data.

2   Preliminaries

This section provides basic definitions about indiscernibility, mostly come from Rough Sets [4]. Let $U \neq \phi$ be a universe of discourse and X be a subset of U. An equivalence relation R classifies U into a set of subsets $U/R = \{X_1, X_2, ...X_N\}$ that satisfies the following conditions: (1) $X_i \subseteq U$, $X_i \neq \phi$ for any i, (2)$X_i \cap X_j = \phi$ for any $i, j, i \neq j$, (3)$\cup_{i=1,2,...N} X_i = U$. Any subset X_i is called a category and represents an equivalence class of R. A category in R containing an object $x \in U$ is denoted by $[x]_R$. Objects x_i and x_j in U are *indiscernible on R* if $(x_i, x_j) \in P$ where $P \in U/R$. For a family of equivalence relations $\mathbf{P} \subseteq \mathbf{R}$, an indiscernibility relation over $\mathbf{P}$ is defined as the intersection of individual relations $Q \in \mathbf{P}$.

3   Method

The proposed method consists of three steps:

1. Assign a binary classification to each object.
2. Compute the indiscernibility level for each pair of objects according to the binary classifications. Then construct a symmetric square matrix of indiscernibility level.
3. Construct a dendrogram from the indiscernibility matrix using the single linkage (nearest-neighbor) hierarchical clustering.

3.1   Binary Classifications

First, each object independently classifies the entire set of objects U into two disjoint sets P and $U - P$. This binary classification is formalized using an equivalence relation. Let $U = \{x_1, x_2, ..., x_N\}$ be the set of objects we are

Table 1. An example of asymmetric, non-Euclidean dissimilarity matrix

	x_1	x_2	x_3	x_4	x_5
x_1	0.0	0.1	0.1	0.7	0.9
x_2	0.2	0.0	0.1	0.6	0.8
x_3	0.7	0.1	0.0	0.2	0.8
x_4	0.2	0.3	0.2	0.0	0.6
x_5	0.7	0.6	0.9	0.1	0.0

interested in and let R_i be an equivalence relation defined for object x_i. Then binary classification for x_i is defined by

$$U/R_i = \{P_i,\ U - P_i\}, \tag{1}$$

where P_i contains objects that are indiscernible to x_i, and $U - P_i$ contains objects that are discernible to x_i.

Method for determining the binary classification is arbitrary. R_i should provide some criteria to form P_i; however, it may not necessarily be defined explicitly. For example, one may simply form P_i according to the proximity between objects as

$$P_i = \{x_j|\ d(x_i, x_j) \leq Th_{di}\},\ \ \forall x_j \in U. \tag{2}$$

where $d(x_i, x_j)$ denotes dissimilarity between objects x_i and x_j, and Th_{di} denotes a threshold value of dissimilarity for object x_i. Other methods can be used as alternatives if they are appropriate with respect to the property of the data. We have introduced a method for constructing binary grouping based on the denseness of the objects in [5]; however, one may use any method, including the choice of proximity measure, under the condition that it has the ability of performing binary classification on U.

An important point to note is that U/R_i can be defined locally and independently to each object x_i, $i = 1, 2, ..., N$. For example, U/R_1 can be defined according only to the relationships between x_1 and other $N - 1$ objects, without taking into account other information such as relationships between x_2 and x_3. Similarly, U/R_i can be defined according only to the relationships between x_i and other $N - 1$ objects, without taking into account the relationships between x_j and x_k, where $j, k \neq i$. This property enables us to employ an asymmetric, non-Euclidean proximity matrix as input data.

[Example 1:] Binary Classification
Let us assume $U = \{x_1, x_2, x_3, x_4, x_5\}$ and consider an asymmetric, non-Euclidean dissimilarity matrix shown in Table 1. Suppose we define binary classifications U/R_i as

$$U/R_i = \{P_i,\ U - P_i\},$$
$$P_i = \{x_j|\ d(x_i, x_j) \leq 0.5\},\ \ \forall x_j \in U. \tag{3}$$

Then we obtain the following five binary classifications.

$$U/R_1 = \{\{x_1, x_2, x_3\}, \{x_4, x_5\}\},$$

$$U/R_2 = \{\{x_1, x_2, x_3\}, \{x_4, x_5\}\},$$
$$U/R_3 = \{\{x_2, x_3, x_4\}, \{x_1, x_5\}\},$$
$$U/R_4 = \{\{x_1, x_2, x_3, x_4\}, \{x_5\}\},$$
$$U/R_5 = \{\{x_4, x_5\}, \{x_1, x_2, x_3\}\}. \tag{4}$$

Note that these five classifications are derived independently. Objects such as x_1 and x_2 are classified as indiscernible in U/R_1 and U/R_2, but classified as discernible in U/R_3. □

3.2 Indiscernibility Level

The family of binary classifications $U/\mathbf{R}$, where $\mathbf{R} = \{R_1, R_2, \ldots, R_N\}$, produces the finest sets of objects by taking intersection of all binary classifications. In this scheme, objects fall into the same category in $U/\mathbf{R}$ only when all of the N relations agree to classify them as indiscernible. If there is at least one relation that discriminate them, they are regarded as discernible in $U/\mathbf{R}$ even when other $N-1$ relations agree to classify them as indiscernible.

Now recall the example shown in Eq. (4). This example contains three types of binary classifications: $U/R_1 (= U/R_2 = U/R_5)$, U/R_3 and U/R_4. Since they are slightly different, classification of U by the family of binary classifications $\mathbf{R}$, $U/\mathbf{R}$, results in producing four very small, almost independent categories.

$$U/\mathbf{R} = \{\{x_1\}, \{x_2, x_3\}, \{x_4\}, \{x_5\}\}. \tag{5}$$

This simple example demonstrates the following problems that prevent us from observing data with appropriate granularity.

1. Binary classifications are defined independently; thus global relationships between each classification is not taken into account.
2. Binary representation of indiscernible/discernible makes it difficult to reflect the global agreement for classifying objects.

We here introduce *indiscernibility level*, a novel measure that solves the above problems and makes it possible to represent the granularity of objects while keeping the use of independently defined binary classifications. The indiscernibility level, $\gamma(x_i, x_j)$, defined for a pair of objects x_i and x_j, quantifies the ratio of binary classifications that agree to classify x_i and x_j as indiscernible. The higher level of indiscernibility implies that although there is small number of counterview, they are likely to be treated as indiscernible, and vise versa.

The *indiscernibility level* $\gamma(x_i, x_j)$ for objects x_i and x_j is defined as follows.

$$\gamma(x_i, x_j) = \frac{\sum_{k=1}^{|U|} \delta_k^{indis}(x_i, x_j)}{\sum_{k=1}^{|U|} \delta_k^{indis}(x_i, x_j) + \sum_{k=1}^{|U|} \delta_k^{dis}(x_i, x_j)}, \tag{6}$$

where

$$\delta_k^{indis}(x_i, x_j) = \begin{cases} 1, & \text{if } (x_i \in [x_k]_{R_k} \wedge x_j \in [x_k]_{R_k}) \\ 0, & \text{otherwise.} \end{cases} \tag{7}$$

and

$$\delta_k^{dis}(x_i, x_j) = \begin{cases} 1, \text{ if } (x_i \in [x_k]_{R_k} \wedge x_j \notin [x_k]_{R_k}) \text{ or} \\ \quad \text{if } (x_i \notin [x_k]_{R_k} \wedge x_j \in [x_k]_{R_k}) \\ 0, \text{ otherwise.} \end{cases} \quad (8)$$

Equation (7) means that $\delta_k^{indis}(x_i, x_j) = 1$ holds only when x_i and x_j are *indiscernible* on U/R_k under the condition that they are also indiscernible with x_k. Equation (8) means that $\delta_k^{dis}(x_i, x_j) = 1$ holds only when x_i and x_j are *discernible* on U/R_k, under the condition that either of them is indiscernible with x_k. By taking the sum of $\delta_k^{indis}(x_i, x_j)$ and $\delta_k^{dis}(x_i, x_j)$ for all $k(1 \leq k \leq |U|)$ as in Equation (6), we obtain the ratio of binary classifications that agree to treat x_i and x_j as indiscernible objects. Note that in Equation (7), we excluded the case when x_i and x_j are indiscernible but not indiscernible with x_k. This is to exclude the case where R_k does not significantly put weight on discerning x_i and x_j. P_k for U/R_k is often determined by focusing on similar objects rather than dissimilar objects. This means that when both of x_i and x_j are highly dissimilar to x_k, their dissimilarity is not significant for x_k. Thus we only count the number of binary classifications that certainly evaluate the dissimilarity of x_i and x_j.

[**Example 2:**] **Indiscernibility Level**
The indiscernibility level $\gamma(x_1, x_2)$ of objects x_1 and x_2 in Example 1 is calculated as follows.

$$\begin{aligned} \gamma(x_1, x_2) &= \frac{\sum_{k=1}^{5} \delta_k^{indis}(x_1, x_2)}{\sum_{k=1}^{5} \delta_k^{indis}(x_1, x_2) + \sum_{k=1}^{5} \delta_k^{dis}(x_1, x_2)} \\ &= \frac{1+1+0+1+0}{(1+1+0+1+0)+(0+0+1+0+0)} = \frac{3}{4}. \end{aligned} \quad (9)$$

Let us explain this example with the calculation of the numerator $(1+1+0+1+0)$. The first value 1 is for $\delta_1^{indis}(x_1, x_2)$. Since x_1 and x_2 are indiscernible on U/R_1 and obviously they are in the same class to x_1, $\delta_1^{indis}(x_1, x_2) = 1$ holds. The second value is for $\delta_2^{indis}(x_1, x_2)$, and analogously, it equals 1. The third value is for $\delta_3^{indis}(x_1, x_2)$. Since x_1 and x_2 are discernible on U/R_3, it becomes 0. The fourth value is for $\delta_4^{indis}(x_1, x_2)$ and it obviously, becomes 1. The last value is for $\delta_5^{indis}(x_1, x_2)$. Although x_1 and x_2 are indiscernible on U/R_5, their class is different to that of x_5. Thus $\delta_5^{indis}(x_1, x_2)$ becomes 0.

Indiscernibility levels for all pairs in U are tabulated in Table 2. Note that the indiscernibility level of object x_i to itself, $\gamma(x_i, x_i)$, will always be 1. □

3.3   Hierarchy of Indiscernibility Level

The indiscernibility level can be used with thresholding to control the granularity of data. According to the definition of indiscernibility level, any objects (x_i, x_j) whose indiscernibility level exceeds the threshold value Th_γ, namely, if $\gamma(x_i, x_j) \geq Th_\gamma$ holds, they should be treated as *indiscernible*. Since treating two objects as indiscernible is equal to merge the two objects, and it is a stepwise abstraction process that goes hierarchically from bottom to top according to the

Table 2. Indiscernibility level γ for objects in Eq. (4)

	x_1	x_2	x_3	x_4	x_5
x_1	3/3	3/4	3/4	1/5	0/4
x_2		4/4	4/4	2/5	0/5
x_3			4/4	2/5	0/5
x_4				3/3	1/3
x_5					1/1

Table 3. Indiscernibility level γ for objects in Eq. (4)(recalculated)

	x_1	x_2	x_3	x_4	x_5
x_1	1.0	0.75	0.75	0.2	0.0
x_2		1.0	1.0	0.4	0.0
x_3			1.0	0.4	0.0
x_4				1.0	0.33
x_5					1.0

Table 4. Hierarchical merge process

Step	pairs	γ	clusters
1	x_2, x_3	1.0	$\{x_1\}, \{x_2, x_3\}, \{x_4\}, \{x_5\}$
2	x_1, x_2	0.75	$\{x_1, x_2, x_3\}, \{x_4\}, \{x_5\}$
3	x_2, x_4	0.4	$\{x_1, x_2, x_3, x_4\}, \{x_5\}$
4	x_4, x_5	0.33	$\{x_1, x_2, x_3, x_4, x_5\}$

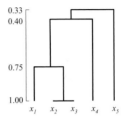

Fig. 1. Dendrogram for Example 3

indiscernibility level, it is possible to construct a dendrogram that represents the hierarchy of indiscernibility by using conventional single-linkage hierarchical grouping method. By setting an appropriate threshold on the dendrogram, one can obtain abstracted groups of objects that meet the given level of indiscernibility. Namely, one can interactively change the granularity of data. The lowest threshold produces the finest groups of objects (granules) and the highest threshold produces the coarsest groups.

[Example 3:] Hierarchy of Indiscernibility Level
Let us recall the case in Example 2. The matrix of indiscernibility levels is provided in Table 2. For the sake of easy understandings, we provide in Table 3 recalculated values. Here we treat the indiscernibility level as similarity because the mergence should proceed in decreasing order of $\gamma(x_i, x_j)$. If one prefers to treat it as dissimilarity, simply use $1 - \gamma(x_i, x_j)$ instead of $\gamma(x_i, x_j)$.

Table 4 and Figure 1 provide the detail of merging process and the dendrogram respectively. Since $\gamma(x_2, x_3) = 1.0$, these objects are indiscernible at the lowest level; thus $\{x_1\}, \{x_2, x_3\}, \{x_4\}, \{x_5\}$ constitute the finest sets of objects (granules). At $\gamma = 0.75$, x_1 becomes indiscernible with x_2. Since x_2 and x_3 are also indiscernible, $\{x_1, x_2, x_3\}, \{x_4\}, \{x_5\}$ constitute an abstracted sets of objects. Similarly, at $\gamma = 0.40$, x_4 becomes indiscernible with x_2 and $\{x_1, x_2, x_3, x_4\}, \{x_5\}$ constitute the more abstracted sets of objects. Finally, at $\gamma = 0.33$, all objects are considered to be indiscernible and the most abstracted set is obtained. The level of abstraction can be interactively set by changing the threshold value Th_γ on the dendrogram. For example, in Figure 1, one can set $Th_\gamma = 0.5$ as a reasonable level since the difference of γ between steps is relatively large.   □

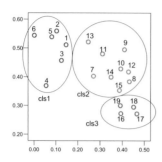

Fig. 2. 2D plot of the test data

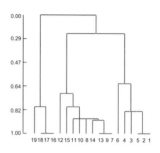

Fig. 3. Dendrogram for the Test data

4   Experimental Results

We applied our method to a synthetic dataset. The dataset contained 19 objects in two-dimensional space as shown in Figure 2. The dataset was generated by Neyman-Scott method [6] with cluster number = 3. The label 'cls 1' to 'cls 3' shows the original class that each object belongs to.

The proposed method starts with determining a binary classification, U/R_i, for each object x_i, $i = 1, 2, \ldots, 19$. In order to seclude the inference of methods/parameters for determining U/R_i, we used the following perfect binary classifications, which were generated based on the class labels of data.

$$U/R_i = \{P_i,\ U - P_i\},$$
$$P_i = \{x_j|\ c[x_i] = c[x_j]\},\quad \forall x_j \in U. \tag{10}$$

Then, in order to simulate the non-Euclidean properties, we applied random disturbance to the perfect binary classifications. Taking the randomly disturbed perfect classifications as input, we calculated the indiscernibility levels and constructed a dendrogram. Table 5 provides all the disturbed binary classifications ($U - P_i$ omitted for simplicity. x of x_i also omitted in P_i for simplicity).

Using the binary classifications in Table 5, we calculated indiscernibility level for each pair of objects. Then we generated the dendrogram shown in Figure 3.

At the lowest level of indiscernibility, 13 sets of objects were generated as the finest granules because the randomly disturbed binary classifications were slightly different each other. However, their disturbance affected locally; therefore the binary classifications retrained the ability of classifying most of the object pairs correctly. In other words, although there exist a few counterviews, object pairs in the same class retained higher level of agreement among binary classifications to be classified as indiscernible objects. Therefore, if we changed the threshold of indiscernibility level to a slightly lower value, for example to 0.8, most of the object pairs could recover their original indiscernibility. The shapes of dendrogram around the bottom part visualizes this characteristics. As the threshold level decreases (goes toward upper direction on the dendrogram), the granularity of the data quickly became coarser, and then became stable for

Table 5. Binary classifications for the test data

x_i	P_i of U/R_i	x_i	P_i of U/R_i
x_1	1,2,4,5,6,15	x_{11}	7,8,9,10,11,12,13,14,15,12
x_2	1,2,3,4,5,4	x_{12}	7,8,9,10,11,13,14,15
x_3	1,2,3,4,5,6,6	x_{13}	7,8,9,10,11,12,13,14,15,6
x_4	1,2,3,4,5,6,12	x_{14}	7,8,9,10,12,13,14,15,15
x_5	1,2,3,4,5,6,19	x_{15}	7,8,9,10,11,12,13,14,15,6
x_6	1,2,3,5,6,14	x_{16}	16,17,18,19
x_7	7,8,9,10,11,13,14,15	x_{17}	16,17,18,19
x_8	7,9,10,11,12,13,14,15	x_{18}	16,17,18,19
x_9	7,8,9,11,12,13,14,15	x_{19}	16,17,18,19
x_{10}	7,8,9,10,11,12,13,14		

$Th_\gamma = 0.63$ to 0.25. For these values, the method generated correct clusters, which corresponded to the appropriate level of object granularity. If we further set Th_γ to lower value, objects with very low indiscernibility became merged and excessively abstracted sets would be obtained.

The above results demonstrated that (1)the proposed method could visualize the hierarchy of indiscernibility using dendrogram, (2) by changing the threshold level on the dendrogram, users could interactively change the granularity of objects defined based on the indiscernibility level, and (3) the method could handle non-Euclidean relational data in which asymmetry and local disturbance of the triangular inequality could occur.

5 Conclusions

In this paper, we presented a clustering method for non-Euclidean relational data based on the combination of indiscernibility level and single-linkage AHC. Using a simple synthetic dataset, we have demonstrated that the method could produce clusters that meet the user-specified level of granularity, and could handle asymmetric dissimilarity using independently determined binary classifications. It remains as a future work to apply this method to other real-world data, and to compare the performance with other methods such as NERFCM [3].

References

1. Romesburg, H.C.: Cluster Analysis for Researchers. Krieger Publishing Inc. (1989)
2. Everitt, B.S., Landau, S., Leese, M.: Cluster Analysis, 4th edn. Arnold Publishers (2001)
3. Hathaway, R.J., Bezdek, J.C.: NERF c-means: Non-Euclidean relational fuzzy clustering. Pattern Recognition 27(3), 429–437 (1994)
4. Pawlak, Z.: Rough Sets. In: Theoretical Aspects of Reasoning about Data, Kluwer Academic Publishers, Dordrecht (1991)
5. Hirano, S., Tsumoto, S.: An indiscernibility-based clustering method with iterative refinement of equivalence relations - rough clustering. Journal of Advanced Computational Intelligence and Intelligent Informatics 7, 169–177 (2003)
6. Neyman, J., Scott, E.L.: Statistical Approach to Problems of Cosmology. Journal of the Royal Statistical Society, Series B20, 1–43 (1958)

Online Granular Prediction Model for Web Prefetching

Zhimin Gu, Zhijie Ban, Hongli Zhang, Zhaolei Duan, and Xiaojin Ren

School of Computer Science and Technology,
Beijing Institute of Technology, P.R. China
zmgu@x263.net

Abstract. Web prefetching is a primary means to reduce user access latency. The PPM was used to predict user request patterns in traditional literature. However the existing PPM models are usually constructed in offline case, they could not be updated incrementally for user coming new request, such models are only suitable for the relatively stable user access patterns. In this paper, we present an online PPM granular prediction model to capture the changing patterns and the limitation of memory, its implementation is based on a noncompact suffix tree and a sliding window W, the results show that our granular prediction model gives the best result comparing with existing PPM prediction models.

Keywords: Granular Computing, Data Mining, Algorithms.

1 Introduction

Although high-speed network research has never stopped, users are experiencing Web access latency more often than ever [1]. To tackle this latency problem, Web caching proxies have been widely deployed as a means to reduce network traffic and improve response time for Web accesses [2]. But its benefit has been significantly limited by the rapid change of objects in the web [3,4]. Web prefetching is characterized as one of the efficient schemes to reduce the user access latency. Web prefetching fetches objects and stores them in advance, the prefetched objects are likely to be accessed in the near future, and such accesses would be satisfied from the cache rather than by retrieving the objects from the Web server[5]. The core of Web prefetching is to build a prediction model that describes user request patterns and makes prediction. Prediction by Partial Match (PPM) [6-9] is a commonly used technique, where prefetching decisions are made based on historical user requests in a Markov prediction tree. The web content is often dynamic changed, new user request may be available at any instant time, user request patterns may change too. On the other hand, a PPM model may soon become too big to size of its memory, so it is highly desirable to perform the online update of the model incrementally. Therefore, the prediction model should have the important online property. The online characteristic requires that user requests are incrementally inserted and deleted to an existing PPM. But existing PPM prediction models have the common limitations that they

G. Wang et al. (Eds.): RSKT 2008, LNAI 5009, pp. 340–347, 2008.

aren't online. They are normally constructed in off-line case. Such models are only suitable for the relatively stable user access patterns. In past to reduce size of PPM and higher prediction precision, the longest repeated subsequence (LRS PPM [8]) and error-pruned (selective PPM [6]) been given, but the models could not be updated in online too.

In this paper, we use some methods of granular computing[11] and propose an online PPM granular prediction model which attempts to capture the changing of user request patterns under limitation of the memory. To construct an real prefetching system, we discuss some offline prediction granular models too, they are important complementarity of our online granular model. The contributions of this paper are the following: 1). Offline granular Models based on granular computing is presented, they are important complementarity of our online granular model. 2). Our online PPM granular model based on noncompact suffix tree is presented, it could incrementally insert the newest user request and deletes the oldest user request. The detail of Larsson's suffix tree construction algorithm is seen in [10]. 3). A sliding window is proposed to control the number of user requests in the granular model. Thus our online model could capture changing of user patterns under size limitation of memory.

2 Offline Granular Prediction Models

2.1 Prediction Granula

A proxy log is viewed as a sequence L=r1,r2,...,ri,...,rn of page references, ri is the five-tuples $(r_{id}$, IP, s_{id}, p_{id},t), where r_{id} is an unique identifier, IP is an request IP, s_{id} is a being requested server identifier, p_{id} is a being requested page, t is the time of the request. Here,U=the set of all ri in L.

Definition 1. *IP Sequence: An IP Sequence $L_{IP} \subseteq L$, if a) $\forall r \in L_{IP}$, then r.IP = IP; b) $\forall r \in L$ and r.IP = IP, then $r \in L_{IP}$. We have an equivalence relation L_{IP1},..., L_{IPk} in U. An IP sequence is often refered to an user sessions as they assume that each IP address represents an individual web user.*

Definition 2. *t Sequence: An t Sequence $L_t \subseteq L$, if a)$\forall r \in L_t$, then r.t=t; b) $\forall r \in L$ and r.t=t, then $r \in L_t$.We have an equivalence relation L_{t1},..., L_{tk} in U, where ti is a day or some days or a month or other times.*

Definition 3. *IP-Based Prediction Granula: If U is the set of L_{IP1}, L_{IP2},..., L_{IPk}, we use a subset Gi to be prediction granula, here, Gi is a set of L_{IP} and IP in IP-SET, the subsets could be an equivalence partition R in U, then we have U/R={G1, G2, ..., Gi, ..., Gm}, $\forall X \subseteq U$, then $X_* = \bigcup \{Gi \in U/R | Gi \subseteq X\}$, $X^* = \bigcup \{Gi \in U/R | Gi \bigcap X \neq \Phi\}$. For example, G1=the set of L_{IP1} and L_{IP2}, G2=the set of L_{IP3}, L_{IP4} and L_{IP5},..., Gm=the set of L_{IPk-2}, L_{IPk-1} and L_{IPk}.*

Definition 4. *t-Based Prediction Granula: If U is the set of L_{t1}, L_{t2},..., L_{tk}, we use a subset Gi to be prediction granula, here, Gi is the set of L_t and t in T, the subsets could be an equivalence partition R in U, then we have*

U/R={$G1, G2, ..., Gi, ..., Gm$}, $\forall X \subseteq U$, then $X_* = \bigcup$ {$Gi \in U/R | Gi \subseteq X$}, $X^* = \bigcup$ {$Gi \in U/R | Gi \cap X \neq \Phi$}. *For example, G1= the set of t1 and t2, G2= the set of t3, t4, t5 and t6,..., Gm=the set of t_{k-1} and tk. Here, Gi could be a time-based partition group.*

2.2 PPM-Based Prediction Granular Model

PPM model belongs to the context models. The context is a finite sequence of symbols preceding the current symbol. The length of the sequence is called the order of context. The context model keeps information about count of symbols' appearances for the context. All context relations is used to build the PPM model. PPM describes user request patterns in a Markov prediction tree. A k-order PPM model maintains the Markov prediction tree with height k+1, which corresponds to the contexts of length from 0 to k. Each node represents the request sequence of Web pages that can be found by traversing the tree from the root to the node. Figure 1 shows the prediction tree structure of the 3-order PPM model for the request sequences Granula G1 of L_{IP1}={ACD} and L_{IP2}={ACC}.It records the counts of request sequence occuring in the path. For example, the notation C/2 indicates that request sequence{AC}was accessed twice. Here, for $G1 = \{Lt1, Lt2\}, Lt1 = \{ACD\}, Lt2 = \{ACC\}$,we have same as figure 1 in some conditions too.

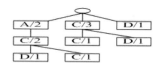

Fig. 1. Tree Structure of PPM for Request Sequences of {ACD} and {ACC}

3   Online t-Based Prediction Granular Model

Our idea of online prediction Granular model is based on two reasons: 1) the proxy log will write dynamically down user being requested new page, a user session in the log is an ordered user request sequence when the time difference of consecutive request pages from same IP is less than some period, and the U in definition 4 need be changed because of online needs. For example, the boundary Ltk is changed into Ltk+Δtk; 2) because the size of online prediction Granula Gm is usually limited, Gm need be change into Gm=a offline granula Gm_offline $\bigcup$ online granula Gm_online , and the oldest r element of Gm is added to Gm_offline, the new r is added to Gm_online. Our p rediction granula always incrementally inserts the newest user request and deletes the oldest user request based on a noncompact suffix tree. To adapt the memory size and the change of user request patterns, it keeps the recent requests by a sliding window W and deletes the oldest requests as they drop out of the sliding window. A head pointer points to the newest user request and a tail pointer points to the oldest user request. When new user request arrives, the head pointer is changed to the request. The new request is directly inserted into if the difference between the head pointer and the tail pointer is less than W, otherwise the request which the tail pointer points to is deleted, then the new request is inserted and the tail pointer is changed to next oldest request, so we have the following definition.

Definition 5. *t-Based Prediction Granula with Δtk* :
If $U=\{Lt1, Lt2, ..., Ltk, L\Delta tk\}$, we use a subset $Gi=\{Lt|t \in T\}$ of U to be prediction granula, and the subsets could be an equivalence partition R in U, then we have $U/R=\{G1, G2, ..., Gi, ..., Gm = Gm\_offline \bigcup Gm\_online\}$, $\forall X \subseteq U$, then $X_* =\bigcup\{Gi \in U/R|Gi \subseteq X\}, X^* = \bigcup\{Gi \in U/R|Gi \bigcap X \neq \Phi\}$.

3.1   Model's Online Insertion

We use the noncompact suffix tree to implement the online prediction Granular model. The noncompact suffix tree is an efficient data structure for string matching. A noncompact suffix tree indexes all substrings of a given string and can be fast constructed. In the noncompact suffix tree, every node represents a request sequence from the root to the node and includes a suffix pointer. In order to fast update the model and predict requests, an additional structure keeps track of the current longest context of active user sessions. A new user request A is added to the model in the following fashion: 1). If A belongs to a new user session S, S becomes active user session and the corresponding longest current context points to the root node. 2). For the longest current context of A's active user session and all its suffixes, to check if any of child nodes represent the new request A. 3). If such a node exists, its context count increments one. Otherwise, create a new child node and set 1 as its context count. The longest current context points to the new one. 4). If the new node is created, its suffix pointer points to its suffix node.

The online insertion algorithm of our model is described as follows:
Algorithms ModelOnlineInsertion(A, T)
Input:new user request A, prediction tree T; Output:prediction tree T
Begin
    i=Get_Active_Session(A ,T); p=Longest_Current_Context[i];
    flag=0; Temp_Suffix=0;
    While (p)
    Begin   If (q is a child node of p and it represents A) q.count=q.count+1
    Else Begin q is created ; q.count=1 End;
    If (flag equals 0) Longest_Current_Context[i]=q; flag=1; p=p.Suffix_Pointer;
    If(Temp_Suffix!=0) Temp_Suffix.Suffix_Pointer=q; Temp_Suffix=q;
    End; Return T;
End
Suppose user request sequence is $\{AABABC\}$ and also is described as $\{121321\}$ for corresponding users. The label 1, the label 2 and label 3 represents user 1, user 2 and user 3 respectively. The sequence is converted into user sessions. The

session of user 1 is $\{ABC\}$, the session of user 2 is $\{AB\}$ and the session of user 3 is $\{A\}$. The size of the sliding window is 6. Figure 2 shows that the request sequence is inserted into the PPM model. The arrow lines stand for suffix pointers. Every step orderly inserts a request of $\{AABABC\}$. We use (6) in Figure 2 to explain insertion principles. It inserts the sequence's sixth request. The current active contexts are $\{AB\}$, $\{B\}$ and the root node, thus new nodes are created and the corresponding context is $\{ABC\}$, $\{BC\}$ and $\{C\}$ respectively.

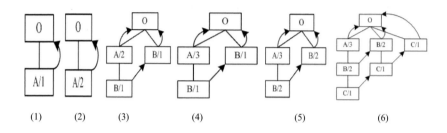

Fig. 2. Insertion Sequence $\{AABABC\}$ Into the PPM

3.2 Model's Online Deletion

We orderly delete these requests when they drop out of the sliding window W. The oldest user request A is deleted from the model in the following fashion: 1). The node p represents the request sequence A. Find p among the root's child nodes. 2). The node's count reduces one from p to the corresponding longest context node. The node is deleted when its count equals 0.

The online deletion algorithm is described as follows:

Algorithm ModelOnlineDeletion (T)
Input:prediction tree T; Output:prediction tree T
Begin
    While (exceeds W)
    Begin A=Tail.Request; Tail=Tail.Next; S=AX1X2X3...Xn; Find the node p representing the request sequence A among the child nodes of the root; j=1;
    While (p) Begin push(p); Find the node q which represents the request sequence S[1]S[j+1] among the child nodes of p; If (q!=NULL) Begin push(q); p=q; End; j++; End;
    While (the stack is not empty) Begin p=pop(); p.count=p.count-1; If (p.count equals 0) delete p; End;
    End; Return T;
End.

Here, S represents the request sequence from A to the end request of the A's user session. Assume the size of the sliding window is adjusted to 4. Figure 3 shows the result, which orderly deletes the leftmost two requests of $\{AABABC\}$ from (6) in Figure 2. We use (1) in Figure 3 to explain deletion principles. It deletes the first request of $\{AABABC\}$ and the request belongs to user 1. Therefore,

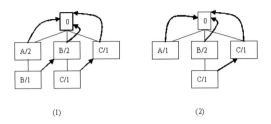

Fig. 3. Orderly Deletion the Leftmost Two Requests of $\{AABABC\}$

the counts of the contexts $\{ABC\}$, $\{AB\}$ and $\{A\}$ all reduce one. The relative node of the context $\{ABC\}$ is deleted because its count has been changed to 0.

4   Experiments

We make the trace-driven simulation by an real trace file. The file is from a Chinese University's proxy server log that contains HTTP requests of half a month. In this period there are totally 9.39×10^6 requests. There are a total of 477 unique IP addresses and 1,579,206 unique pages. We remove all dynamically generated files. These files can be in types of .asp, .php ,.cgi and so on. We also filter out the embedded image files such as .gif and .jpg. Finally, we also remove requests with unsuccessful HTTP response code. The experiment takes 1/2 of the log as training set and the remaining 1/2 as prediction set.

Precision is the ratio between the number of correct predictions and the number of total predictions. If the predicted page is accessed in a subsequent request, this prediction is considered to be correct, otherwise incorrect. Hit rate (HR) is the ratio between the number of correct predictions and the number of total user requests. Traffic incremental rate (TIR) is the ratio between the traffic from incorrect prefetching and the traffic from the total user requests.

We compare the performance of our online granular prediction model (OLPPM) with the standard PPM (STPPM) and the longest repeated sequence (LRS PPM). Each PPM model predicts at most an request according to a user's current request every time. In our model, the sliding windows' size is the recent 7 days' data and we create the first model using the training data and incrementally update it using the test data. In each test, 2-order PPM model is employed. Figure 4.(1) shows the precisions of three PPM models as the probability threshold increases, the precision of our prediction model is consistently higher than the other two prediction models when probability threshold varies from 0.5 to 0.9. There are several reasons to explain this. First, our prediction model keeps the most recent user requests using a sliding window and captures the changing user request patterns by online updating. Second, we consider the conditional probability distribution of every node and predict the next request using the node with low entropy. The precision of the LRS PPM model is higher than that of the standard PPM model because it uses the longest repeated sequence

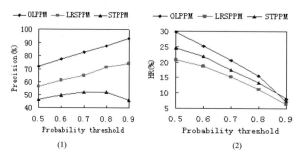

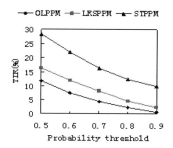

Fig. 4. Precision and Hit Rate

Fig. 5. Traffic Incremental Rate

to predict the next request. Figure 4.(2) compares the hit rate of three PPM models when the probability threshold varies from 0.5 to 0.9. The hit rates of three PPM models decrease as the probability threshold increases, the hit rate of our prediction model outperforms the LRS PPM and the standard PPM as the probability threshold increases from 0.5 to 0.8. The hit rate of our model is lower than that of the standard PPM model when the probability threshold is 0.9. The hit rate of LRS PPM is the lowest because it only uses longest repeated subsequence to predict the users' future requests. Figure 5 compares traffic incremental rates among the three prediction models. The traffic increment rates of three prediction models decrease as the probability threshold increases. Our prediction model has the lowest traffic incremental rate. The main reason is that our prediction model captures the changing web environment and predicts more beneficial web request every time. The standard PPM model has the highest traffic incremental rate because its prediction precision is lowest and the prediction number is high.

5   Conclusions

In this paper, we propose an online PPM granular prediction model based on the noncompact suffix tree, it uses a sliding window of memory and could capture the changing user request patterns,and could insert the newest user request and delete the oldest request that drops out of the sliding window. It predicts user requests based on the longest match. The results show that our online PPM model gives the best result comparing with existing PPM prediction models. However, seeing from definition 3 and 4, our online PPM granular prediction model is only an online prediction granula, other Gi is an offline PPM-based prediction granula, how an real prediction system to be constructed will be our future goal.

References

1. Xu, C.Z., Ibrahim, T.I.: A keyword-based semantic prefetching approach in Internet news services. IEEE Transactions on Knowledge and Data Engineering 16, 5601–5611 (2004)
2. Aniket, M., Anirban, M., Williamson, C.: Locality characteristics of web streams revisited. In: The SCS Symposium on Performance Evaluation of Computer and Telecommunication Systems, pp. 795–803 (2005)
3. Padmanabhan, V.N., Mogul, J.C.: Using Predictive Prefetching to Improve World Wide Web Latency. Computer Communication Review 26(3), 22–36 (1996)
4. Griffioen, J., Appleton, R.: Reducing File System Latency Using a Predictive Approach. In: Summer USENIX Conference, pp. 197–207 (1994)
5. Wu, B., Kshemkalyani, A.D.: Objective-Optimal Algorithms for Long-Term Web Prefetching. IEEE Transactions on Computers 5(1), 2–17 (2006)
6. Deshpande, M., Karypis, G.: Selective Markov Models for Predicting Web Page Accesses. ACM Transactions on Internet Technology 4(2), 163–184 (2004)
7. Palpanas, T., Mendelzon, A.: Web Prefetching Using Partial Match Prediction. In: The Fourth Web Caching Workshop (WCW 1999) (1999)
8. Pitko, J., Pirolli, P.: Mining Longest Repeating Subsequences to Predict World Wide Web Surfing. In: Second USENIX Symposium on Internet Technologies and Systems, pp. 139–150 (1999)
9. Fan, L., Cao, P., Jacobson, Q.: Web Prefetching Between Low-Bandwidth Clients and Proxies: Potential and Performance. In: The ACM SIGMETRICS 1999, Atlanta, Georgia (May 1999)
10. Larsson, N.J.: Extended application of suffix trees to data compression. In: The Conference on Data Compression, pp. 190–199 (1996)
11. Pawlak, Z.: Rough Sets. In: Theoretical Aspects of Reasoning about Data. Kluwer Academic Publishers, Dordrecht (1991)

Axiomatic Definition of Knowledge Granularity and Its Constructive Method

Ming-qing Zhao, Qiang Yang, and De-zhi Gao

College of Information Science and Engineering,
Shandong University of Science and Technology,
Qingdao 266510, Shandong, P.R. China
zhaomq64@163.com, yqiang725@163.com

Abstract. Granular computing has impact on development methods in such areas as knowledge discovery and data mining. This paper firstly presents a new axiomatic definition of knowledge granularity, then gives a series of methods of measuring knowledge granularity including concrete measurements lacking parameters and general measurements with parameters. Furthermore, several combinatorial forms of different granularity formulas are described. The principal results seem to have some theoretic and applied value to build granularity computation in information system.

Keywords: Granular computing, Axiomation, Information system.

1 Introduction

Granularity has its origin in physics, where it refers to the average metric of the size of particles, in fact, physics granularity means subdivision that makes physical object finer. When we regard knowledge or information as the research object, the measurement the knowledge granularity or information granularity. In an information system, the granularity means actually average measure of refinement degree of knowledge and information. In general knowledge granularity can represent discernibility ability of knowledge in rough set theory, in fact, the smaller the knowledge granularity is, the stronger its discernibility ability is.

Granularity computation is a very new active direction in the research of artificial intelligence. This is a kind of new concept of information processing and calculation normal form, which has extensive application prospects in numerous fields such as artificial intelligence, knowledge discovery, image compression, semantic Web service etc. The basic idea of granular computing is reflected in a lot of fields including rough set theory, cluster analysis, machine learning, the database and information retrieval, etc [1,2]. At present, there are several kinds of models in granular computing based on: (1) rough sets[3]; (2) word computing model [4]; (3) quotient space model [5].

In 1979, the problem of fuzzy information granularity was introduced by Zadeh in [6]. Especially, several measures in an information system closely associated with granular computing such as granulation measure, information entropy,

G. Wang et al. (Eds.): RSKT 2008, LNAI 5009, pp. 348–354, 2008.

rough entropy and knowledge granulation and their relationships were discussed in [7,8]. In [9,10], the constructive method is used to the research of knowledge granularity. The current status of granular computiong is summariuzed in [11]. The paper [12,13] has proposed a kind of axiomatic definition of knowledge granularity, which defines granularity as one mapping satisfying a certain condition, aiming at establishing the corresponding relation between attribute subset and non-negative number, but this definition has some deficiencies as follows: firstly, if we take one constant function as the mapping, all the conditions in the definition are still satisfied, it is obviously unreasonable to weigh different knowledge granularity with it; in addition, the value of granularity is a non-negative number, the upper boundary is indeterminate, that is, the granularity might not be 0 when it is the finest division, and the granularity is not be 1 while the dividing is the most coarse, which does not accord with people's habit of thought greatly . For overcoming the deficiencies in axiomatic definition of knowledge granularity in [12,13], here we put forward a new axiomatic definition of knowledge granularity that is more succinct, and provides a series of methods of measuring knowledge granularity on what we have researched above. Furthermore, we have also discussed the combinatorial forms of the knowledge granularity.

2   Axiomatic Definition of Knowledge Granularity

In the following, an axiom definition of knowledge granularity is given.

Definition 2.1 Let $S = (U, A)$ be an information system, G be a mapping from the power set of A to the set of real numbers. We say that G is a knowledge granularity in an information system if G satisfies the following conditions:

(1)$\forall P, Q \subseteq A$,if there exists a bijection $f : U/ind(P) \to U/ind(Q)$,which satisfies $|P_i| = |f(P_i)|$, $P_i \in U/ind(P)$, then $G(P) = G(Q)$ (invariance);
(2)$\forall P, Q \subseteq A$, if $P \prec Q$, then $G(P) < G(Q)$ (strictly monotone);
(3)$\forall P \subseteq A$, if $U/ind(P) = \{\{u_i\}|u_i \in U\}$, then $G(P) = 0$; if $U/ind(P) = \{\{U\}\}$, then $G(P) = 1$ (boundedness).

We can get $\forall P \subseteq A$, $0 \leq G(P) \leq 1$ from above definition.

Remark. The definition 2.1 is also established in incomplete information system.

3   Measure of Knowledge Granularity

Set out from definition 2.1, on the basis of the "average"thought in statistics, including arithmetic average and harmonic average, we construct in our article some new knowledge granulations have. These granulations are all significative from the average meaning. It is to be noted here that the concept of geometric average can't be used in the structure of knowledge granularity.

Definition 3.1 Let $S = (U, A)$ be an information system, $\forall R \subseteq A$, $U/ind(R) = \{R_1, R_2, \cdots, R_m\}$, then knowledge granularity can be defined as follows:

$$G(R) = \frac{1}{|U|^2 - 1}\left(\frac{|U|}{\sum\limits_{i=1}^{m} 1/|R_i|} - 1\right).$$

Theorem 3.1 Let $S = (U, A)$ be an information system, $\forall R \subseteq A$, $U/ind(R) = \{R_1, R_2, \cdots, R_m\}$, then G in definition 3.1 is knowledge granularity under definition 2.1.

Proof. G in definition 3.1 satisfies condition (1),(3) of definition 2.1 obviously, we prove that the condition (2) holds too.

Let $P, Q \subseteq A$ two subsets on A, such that $P \prec Q$, and let $U/ind(P) = \{P_1, P_2, \cdots, P_m\}$ and $U/ind(Q) = \{Q_1, Q_2, \cdots, Q_n\}$. Because $P \prec Q$, so $m > n$, and there exists a partition $C = \{C_1, C_2, \cdots, C_n\}$ of $\{1, 2, \cdots, m\}$, such that

$$Q_j = \bigcup_{i \in C_j} P_i, j \leq n.$$

Hence,

$$G(Q) = \frac{1}{|U|^2 - 1}\left(\frac{|U|}{\sum\limits_{j=1}^{n} 1/|Q_j|} - 1\right)$$

$$= \frac{1}{|U|^2 - 1}\left(\frac{|U|}{\sum\limits_{j=1}^{n} 1/|\bigcup\limits_{i \in C_j} P_i|} - 1\right)$$

$$= \frac{1}{|U|^2 - 1}\left(\frac{|U|}{\sum\limits_{j=1}^{n} 1/\sum\limits_{i \in C_j} |P_i|} - 1\right).$$

Since $m > n$, so there exists $C_{j_0} \in C$, satisfying $|C_{j_0}| > 1$. Consequently,

$$\frac{1}{\sum\limits_{i \in C_{j_0}} |P_i|} < \sum\limits_{i \in C_{j_0}} \frac{1}{|P_i|}, \frac{1}{\sum\limits_{i \in C_j, j \neq j_0} |P_i|} \leq \sum\limits_{i \in C_j, j \neq j_0} \frac{1}{|P_i|}.$$

That is,

$$\sum\limits_{i=1}^{m} \frac{1}{|P_i|} > \sum\limits_{j=1}^{n} \frac{1}{|Q_j|}.$$

So we obtain that:

$$G(Q) = \frac{1}{|U|^2 - 1}\left(\frac{|U|}{\sum\limits_{j=1}^{n} 1/|Q_j|} - 1\right)$$

$$> \frac{1}{|U|^2 - 1}\left(\frac{|U|}{\sum\limits_{i=1}^{m} 1/|P_i|} - 1\right) = G(P).$$

Thus, $G(R)$ in definition 3.1 is the knowledge granulation under definition 2.1.

Next, we will give the generalization form of theorem 3.1:

Theorem 3.2 Let $S = (U, A)$ be an information system, $\forall R \subseteq A$, $U/R = \{R_1, R_2, \cdots, R_m\}$, then

$$G_\alpha(P) = \frac{1}{|U|^{\alpha+1} - 1}\left(\frac{|U|}{\sum\limits_{i=1}^{m} 1/|R_i|^\alpha} - 1\right), \alpha \geq 0$$

is knowledge granularity under definition 2.1.

The proof of this theorem is similar to the certification process of theorem 3.1, only need to pay attention to the following inequality:

$$\sum_{i=1}^{n} \frac{1}{x_i^\alpha} > \frac{1}{\left(\sum\limits_{i=1}^{n} x_i\right)^\alpha}, x_i > 0, \alpha \geq 0, n > 1 (integer).$$

From $G_\alpha(R)$ defined in theorem 3.2, $G_0(R) = \frac{1}{|U|-1}(\frac{|U|}{m} - 1)$ is a particular case of $G_\alpha(R)$ when $\alpha = 0$, this is the most basic calculational formula of knowledge granularity.

Definition 3.2 Let $S = (U, A)$ be an information system, $\forall R \subseteq A$, $U/ind(R) = \{R_1, R_2, \cdots, R_m\}$, the knowledge granularity of R can be defined as follows:

$$GK(R) = \frac{1}{|U| - 1}\left(\frac{1}{|U|}\sum_{i=1}^{m}|R_i|^2 - 1\right).$$

Theorem 3.3 Let $S = (U, A)$ be an information system, $\forall R \subseteq A$, $U/ind(R) = \{R_1, R_2, \cdots, R_m\}$, then GK in definition 3.2 is knowledge granularity under definition 2.1 .

The proof of the theorem 3.3 can be finished by imitating certification process of theorem 3.1, its generalization form of GK in definition 3.2 is the following theorem:

Theorem 3.4 Let $S = (U, A)$ be an information system, $\forall R \subseteq A$, $U/ind(R) = \{R_1, R_2, \cdots, R_m\}$, then

$$GK_\alpha(R) = \frac{1}{|U|^{\alpha-1} - 1}\left(\frac{1}{|U|}\sum_{i=1}^{m}|R_i|^\alpha - 1\right), \alpha > 1$$

is knowledge granularity under definition 2.1 .

The proof of this theorem is similar to the certification process of theorem 3.1, only need to pay attention to the following inequality:

$$\left(\sum_{i=1}^{n} x_i\right)^{\alpha} > \sum_{i=1}^{n} x_i^{\alpha}, x_i > 0, \alpha > 1, n > 1 (integer).$$

That is, $\sum_{i=1}^{n} (x_i / \sum_{i=1}^{n} x_i)^{\alpha} < 1$, omitting detailed certification process.

Next, we will research the limit of $GK_{\alpha}(R)$ when $\alpha \to 1^+$:

$$\lim_{\alpha \to 1^+} GK_{\alpha}(R) = \lim_{\alpha \to 1^+} \frac{1}{|U|^{\alpha-1} - 1} \left(\frac{1}{|U|} \sum_{i=1}^{m} |R_i|^{\alpha} - 1\right)$$

$$= \lim_{\alpha \to 1^+} \frac{1}{|U|^{\alpha-1} ln|U|} \left(\frac{1}{|U|} \sum_{i=1}^{m} |R_i|^{\alpha} ln|R_i|\right)$$

$$= \frac{1}{|U| ln|U|} \sum_{i=1}^{m} |R_i| ln|R_i|.$$

Is this a knowledge granularity formula? We can verify that it satisfies every condition of definition 2.1, so it is really a knowledge granularity.

4   Combination Forms of Knowledge Granularity

The knowledge granularity formulas can be combined depending on the below conclusion:

Theorem 4.1 Suppose n-ary function $f(x_1, x_2, \cdots, x^n) (0 \le x_i \le 1, i = 1, 2, \cdots, n)$ satisfies the following conditions:

(1)the function is strictly increasing on every variable;
(2)$f(0, 0, \cdots, 0) = 0, f(1, 1, \cdots, 1) = 1$.

Let $S = (U, A)$ be an information system, $\forall R \subseteq A$, if $G_1, G_2, \cdots, G_n$ are n knowledge granularity formulas of R , so are $f(G_1(R), G_2(R), \cdots, G_n(R))$. The theorem can be verified directly. We can obtain the following deduction:

Deduction 4.1 Let $S = (U, A)$ be an information system, $\forall R \subseteq A$, if $G_1, G_2, \cdots, G_n$ are n knowledge granularity formulas of R , then, the next four formulas are also knowledge granularity of R:

(1)$\sum_{i=1}^{n} \alpha_i G_i, 0 < \alpha_i < 1, \sum_{i=1}^{n} \alpha_i = 1$;

(2)$\prod_{i=1}^{n} G_i^{\alpha_i}, 0 < \alpha_i < 1, \sum_{i=1}^{n} \alpha_i = 1$;

(3)$\bigvee_{i=1}^{n} G_i$;

(4)$\bigwedge_{i=1}^{n} G_i$.

5   Conclusions

This paper presents a new definition of knowledge granularity, then gives a series of methods of measuring knowledge granularity. Especially, several combination forms for different granularity are described. These results can explain the variety of knowledge granularity, we can choose one to calculate knowledge granularity of information system depending on the detailed conditions, even consider the combination of different knowledge granularity formulas, to balance different attitudes to understanding of knowledge granularity, or to emphasize a certain attitude, which can achieve better results. The presented results seem to have a theoretic and applied value to build granularity computation in information system. They can be used in knowledge reduction, measurement of attribute significance and rules acquisition. The practical application of the knowledge granularity this text provides is the subject that we further want to study. We will discuss in another paper examples of such applications.

Acknowledgements

This work was supported by the national natural science foundation of China (No. 60603090).

References

1. Zadeh, L.A.: Toward a theory of fuzzy information granulation and its centrality in human reasonding and fuzzy logic. Fuzzy Sets and System 19(1), 111–127 (1997)
2. Zadeh, L.A.: Fuzzy sets and information granularity. In: Gupta, N., Ragade, R., Yager, R. (eds.) Advances in fuzzy set theory and application, pp. 111–127. North-Holland, Amsterdam (1979)
3. Pawlak, Z.: Granularity of knowledge, indiscernibility and rough sets. In: Proceedings of 1998 IEEE International Conference on Fuzzy System, pp. 106–110 (1998)
4. Zadeh, L.A.: Fuzzy logic=computing with words. IEEE Transactions on Fuzzy Systems 4(1), 103–111 (1996)
5. Zhang, L., Zhang, B.: Theory of Fuzzy Quotient Space (Methods of Fuzzy Granular Computing). Journal of Software 14(4), 770–776 (2003)
6. Zadeh, L.A.: Fuzzy sets and information granularity. In: Gupta, M., Yager, R. (eds.) Advances in Fuzzy Set Theory and Application, pp. 3–18. North-Holland, Amsterdam (1979)
7. Liang, J.Y., Shi, Z.Z., Li, D.Y., Wierman, M.J.: The information entropy,rough entropy and knowledge granulation in incomplete information system. International Journal of General Systems (to appear)
8. Liang, J.Y., Shi, Z.Z.: The information entropy, rough entropy and knowledge granulation in rough set theory. International Journal of Uncertainty, Fuzziness and Knowledge-Based Systems 12(1), 37–46 (2004)
9. Wierman, M.J.: Measuring uncertainty in rough set theory[J]. International Journal of General Systems 28(4), 283–297 (1999)

10. Qian, Y.H., Liang, J.Y.: Combination Entropy and Combination Granulation in Incomplete Information System. In: Wang, G.-Y., Peters, J.F., Skowron, A., Yao, Y. (eds.) RSKT 2006. LNCS (LNAI), vol. 4062, pp. 184–190. Springer, Heidelberg (2006)
11. Pedrycz, W., Skowron, A., Kreinovich, V. (eds.): Handbook of Granular Computing. John Wiley & Sons, New York (in press)
12. Liang, J.Y., Qian, Y.H.: Knowledge, knowledge granulation in information system.Internation Journal of Intelligent System (to appear)
13. Liang, J.Y., Qian, Y.H.: Axiomatic Approach of Knowledge Granulation in Information System. In: Sattar, A., Kang, B.-h. (eds.) AI 2006. LNCS (LNAI), vol. 4304, pp. 1074–1078. Springer, Heidelberg (2006)

Intuitionistic Fuzzy Approximations and Intuitionistic Fuzzy Sigma-Algebras

Wei-Zhi Wu[1] and Lei Zhou[2]

[1] School of Mathematics, Physics and Information Science,
Zhejiang Ocean University, Zhoushan, Zhejiang 316004, P.R. China
[2] Institute for Information and System Sciences, Faculty of Science,
Xi'an Jiaotong University, Xi'an, Shaan'xi 710049, P.R. China
`wuwz@zjou.edu.cn`, `zhl_2021@126.com`

Abstract. In this paper, concepts of intuitionistic fuzzy measurable spaces and intuitionistic fuzzy σ-algebras are first introduced. Relationships between intuitionistic fuzzy rough approximations and intuitionistic fuzzy measurable spaces are then discussed. It is proved that the family of all intuitionistic fuzzy definable sets induced from an intuitionistic fuzzy serial approximation space forms an intuitionistic fuzzy σ-algebra. Conversely, for an intuitionistic fuzzy σ-algebra generated by a crisp algebra in a finite universe, there must exist an approximation space such that the family of all intuitionistic fuzzy definable sets is the class of all measurable sets in the given intuitionistic fuzzy measurable space.

Keywords: Approximation spaces, intuitionistic fuzzy rough sets, intuitionistic fuzzy sets, measurable spaces, rough sets, σ-algebras.

1 Introduction

Approximation spaces in rough set theory and measurable spaces in measure theory are two important structures to represent knowledge. An approximation space in Pawlak's rough set theory consists of a universe of discourse and an equivalence relation imposed on it [6]. Based on the approximation space, the notions of lower and upper approximation operators are induced. A set is said to be definable if its lower and upper approximations are the same, and undefinable otherwise [6,12]. The notion of definable sets in rough set theory plays an important role. A measurable space in measure theory contains a universe of discourse and a family of measurable sets called a σ-algebra [4]. Based on the measurable space, uncertainty of knowledge can be analyzed.

An equivalence relation in Pawlak's original rough set model is a very restrictive condition which may affect the application of rough set theory. The generalization of Pawlak's rough set model is thus one of the main directions for the study of rough set theory. Many authors have developed Pawlak's rough set model by using nonequivalence relations in crisp and/or fuzzy environments (see e.g. literature cited in [7,8]). Other researchers have also defined rough approximations of intuitionistic fuzzy sets [2,3,5,9,10].

G. Wang et al. (Eds.): RSKT 2008, LNAI 5009, pp. 355–362, 2008.

It seems useful to investigate connections of definability in rough set theory and measurability in measure theory. In [11], Wu and Zhang explored the connections between rough set algebras and measurable spaces in both crisp and fuzzy environments. In this paper, we will further study relationship between the two knowledge representation structures in intuitionistic fuzzy environment.

2   Concepts Related to Intuitionistic Fuzzy Sets

Let U be a nonempty set called the universe of discourse. The class of all subsets (fuzzy subsets, respectively) of U will be denoted by $\mathcal{P}(U)$ (by $\mathcal{F}(U)$, respectively). For $A \in \mathcal{P}(U)$, 1_A will denote the characteristic function of A, i.e., $1_A(x) = 1$ for $x \in A$ and 0 otherwise. $\mathbf{N}$ will denote the set of all positive integers.

Definition 1. [1] *Let a set U be fixed. An intuitionistic fuzzy (IF for short) set A in U is an object having the form*
$$A = \{\langle x, \mu_A(x), \gamma_A(x) \rangle \mid x \in U\},$$
where $\mu_A : U \to [0,1]$ and $\gamma_A : U \to [0,1]$ satisfy $0 \leq \mu_A(x) + \gamma_A(x) \leq 1$ for all $x \in U$, and $\mu_A(x)$ and $\gamma_A(x)$ are called the degree of membership and the degree of non-membership of the element $x \in U$ to A, respectively. The family of all IF subsets in U is denoted by $\mathcal{IF}(U)$. The complement of an IF set A is denoted by $\sim A = \{\langle x, \gamma_A(x), \mu_A(x) \rangle \mid x \in U\}.$

Obviously, a fuzzy set $A = \{\langle x, \mu_A(x) \rangle \mid x \in U\}$ can be identified with the IF set of the form $\{\langle x, \mu_A(x), 1 - \mu_A(x) \rangle \mid x \in U\}$. We can observe that an IF set A in U associates with two fuzzy sets μ_A and γ_A, we will simply write $A = (\mu_A, \gamma_A)$ and $A(x) = (\mu_A(x), \gamma_A(x))$ for $x \in U$.

The basic set operations on $\mathcal{IF}(U)$ are defined as follows [1]: $\forall A, B \in \mathcal{IF}(U)$,

- $A \subseteq B$ if and only if (iff) $\mu_A(x) \leq \mu_B(x)$ and $\gamma_A(x) \geq \gamma_B(x)$ for all $x \in U$,
- $A \supseteq B$ iff $B \subseteq A$,
- $A = B$ iff $A \subseteq B$ and $B \subseteq A$,
- $A \cap B = \{\langle x, \min(\mu_A(x), \mu_B(x)), \max(\gamma_A(x), \gamma_B(x)) \rangle \mid x \in U\}$,
- $A \cup B = \{\langle x, \max(\mu_A(x), \mu_B(x)), \min(\gamma_A(x), \gamma_B(x)) \rangle \mid x \in U\}$,
- $\bigcap\limits_{i \in J} A_i = \{\langle x, \bigwedge\limits_{i \in J} \mu_{A_i}(x), \bigvee\limits_{i \in J} \gamma_{A_i}(x) \rangle \mid x \in U\}$, $A_i \in \mathcal{IF}(U)$, $i \in J$, J is an index set,
- $\bigcup\limits_{i \in J} A_i = \{\langle x, \bigvee\limits_{i \in J} \mu_{A_i}(x), \bigwedge\limits_{i \in J} \gamma_{A_i}(x) \rangle \mid x \in U\}$, $A_i \in \mathcal{IF}(U)$, $i \in J$, J is an index set.

We define a constant IF set $\widehat{(\alpha, \beta)} = \{\langle x, \alpha, \beta \rangle \mid x \in U\}$, where $0 \leq \alpha, \beta \leq 1, \alpha + \beta \leq 1$. A special IF set (IF singleton set) $1_y = (\mu_{1_y}, \gamma_{1_y})$ for $y \in U$ is defined as follows:

$$\mu_{1_y}(x) = \begin{cases} 1, & \text{if} \quad x = y, \\ 0, & \text{if} \quad x \neq y. \end{cases} \qquad \gamma_{1_y}(x) = \begin{cases} 0, & \text{if} \quad x = y, \\ 1, & \text{if} \quad x \neq y. \end{cases}$$

The IF universe set is $1_\sim = U = \{\langle x, 1, 0 \rangle \mid x \in U\}$ and the IF empty set is $0_\sim = \emptyset = \{\langle x, 0, 1 \rangle \mid x \in U\}$.

Definition 2. *An IF relation R on U is an IF subset of $U \times U$, that is,*
$$R = \{\langle (x,y), \mu_R(x,y), \gamma_R(x,y) \rangle \mid x, y \in U \},$$
where $\mu_R, \gamma_R : U \times U \to [0,1]$ satisfy $0 \le \mu_R(x,y) + \gamma_R(x,y) \le 1$ for all $(x,y) \in U \times U$. An IF relation R on U is referred to as serial if for each $x \in U$ there exists an $y \in U$ such that $\mu_R(x,y) = 1$ and $\gamma_R(x,y) = 0$. A crisp relation R on U is referred to as serial if for each $x \in U$ there exists an $y \in U$ such that $(x,y) \in R$. A crisp relation R on U is referred to as equivalent if it is reflexive, symmetric and transitive.

3 Intuitionistic Fuzzy Measurable Spaces

Definition 3. *([4]) Let U be a nonempty set. A subset $\mathcal{A}$ of $\mathcal{P}(U)$ is referred to as a σ-algebra iff it satisfies following axioms:*
  (A1) $U \in \mathcal{A}$,
  (A2) $\{X_n \mid n \in \mathbf{N}\} \subset \mathcal{A} \Longrightarrow \bigcup_{n \in \mathbf{N}} X_n \in \mathcal{A}$,
  (A3) $X \in \mathcal{A} \Longrightarrow \sim X \in \mathcal{A}$.
The sets in $\mathcal{A}$ are called measurable sets (also called observable sets) and the pair $(U, \mathcal{A})$ a measurable space.

With the definition we can see that $\emptyset \in \mathcal{A}$ and
  (A2)$'$ $\{X_n \mid n \in \mathbf{N}\} \subset \mathcal{A} \Longrightarrow \bigcap_{n \in \mathbf{N}} X_n \in \mathcal{A}$
  If U is a finite universe of discourse, then axiom (A2) in Definition 3 can be replaced by axiom (A2)$''$:
  (A2)$''$ $X, Y \in \mathcal{A} \Longrightarrow X \cup Y \in \mathcal{A}$.
In such a case, $\mathcal{A}$ is called an algebra, alternatively, in a finite universe of discourse a family of sets of U is a $\sigma-$algebra if and only if it is an algebra. Moreover, if we denote
$$[x]_{\mathcal{A}} = \cap \{X \in \mathcal{A} \mid x \in X\}, \quad x \in U,$$
then by (A2)$'$ we have $[x]_{\mathcal{A}} \in \mathcal{A}$, it can be checked that $\{[x]_{\mathcal{A}} \mid x \in U\}$ forms a partition of U and $[x]_{\mathcal{A}}$ is called the atomic set of $\mathcal{A}$ containing x.

Definition 4. *Let U be a nonempty set. A subset $\mathcal{F}$ of $\mathcal{IF}(U)$ is referred to as an IF σ-algebra iff it satisfies following axioms:*
  (IF1) $(\widehat{\alpha, \beta}) \in \mathcal{IF}$ for all $\alpha, \beta \in [0,1]$ with $\alpha + \beta \le 1$,
  (IF2) $\{A_n \mid n \in \mathbf{N}\} \subset \mathcal{F} \Longrightarrow \bigcup_{n \in \mathbf{N}} A_n \in \mathcal{F}$,
  (IF3) $A \in \mathcal{F} \Longrightarrow \sim A \in \mathcal{F}$.
The sets in $\mathcal{F}$ are called IF measurable sets and the pair $(U, \mathcal{F})$ an IF measurable space.

With the definition we can see that $0_\sim, 1_\sim \in \mathcal{F}$ and
  (IF2)$'$ $\{A_n \mid n \in \mathbf{N}\} \subset \mathcal{F} \Longrightarrow \bigcap_{n \in \mathbf{N}} A_n \in \mathcal{F}$.
  If U is a finite universe of discourse, then axiom (IF2) in Definition 4 can be replaced by axiom (IF2)$''$:
  (IF2)$''$ $A, B \in \mathcal{F} \Longrightarrow A \cup B \in \mathcal{F}$.
In such a case, $\mathcal{F}$ is also called an IF algebra.

Definition 5. *Let U be a finite universe of discourse. An IF algebra $\mathcal{F}$ on U is said to be generated by a crisp algebra $\mathcal{A}$ iff for each $A \in \mathcal{F}$ there exist $a_i, b_i \in [0,1], i = 1, 2, \ldots, k$, such that*

$$\mu_A(x) = \sum_{i=1}^{k} a_i 1_{C_i}(x), \quad \gamma_A(x) = \sum_{i=1}^{k} b_i 1_{C_i}(x), \quad \forall x \in U,$$

where $\{C_1, C_2, \ldots, C_k\} = \{[x]_{\mathcal{A}} | x \in U\}$ is the atomic sets of $\mathcal{A}$.

Remark 1. If an IF algebra $\mathcal{F}$ on U is generated by a crisp algebra $\mathcal{A}$, we can see that, for each $A \in \mathcal{F}$, $\mu_A, \gamma_A : U \to [0,1]$ are measurable with respect to (w.r.t.) $\mathcal{A}$-$\mathcal{B}$, where $\mathcal{B}$ is the Borel subsets of $[0,1]$, alternatively, μ_A and γ_A are measurable in the sense of Zadeh [13].

4   Intuitionistic Fuzzy Rough Sets

In this section, we introduce IF rough approximation operators induced from an IF approximation space and present their properties.

Definition 6. *Let U be a nonempty universe of discourse and R an IF relation on U, the pair (U, R) is called an IF approximation space. For any $A \in \mathcal{IF}(U)$, the upper and lower approximations of A w.r.t. (U, R), denoted by $\overline{R}(A)$ and $\underline{R}(A)$, are two IF sets and are defined respectively as follows:*

$$\overline{R}(A) = \big\{ \langle x, \mu_{\overline{R}(A)}(x), \gamma_{\overline{R}(A)}(x) \rangle \mid x \in U \big\},$$
$$\underline{R}(A) = \big\{ \langle x, \mu_{\underline{R}(A)}(x), \gamma_{\underline{R}(A)}(x) \rangle \mid x \in U \big\},$$
where

$$\mu_{\overline{R}(A)}(x) = \vee_{y \in U} [\mu_R(x, y) \wedge \mu_A(y)], \quad \gamma_{\overline{R}(A)}(x) = \wedge_{y \in U} [\gamma_R(x, y) \vee \gamma_A(y)];$$

$$\mu_{\underline{R}(A)}(x) = \wedge_{y \in U} [\gamma_R(x, y) \vee \mu_A(y)], \quad \gamma_{\underline{R}(A)}(x) = \vee_{y \in U} [\mu_R(x, y) \wedge \gamma_A(y)].$$

The pair $(\underline{R}(A), \overline{R}(A))$ is called the IF rough set of A w.r.t. (U, R), $\overline{R}, \underline{R} : \mathcal{IF}(U) \to \mathcal{IF}(U)$ are referred to as upper and lower IF rough approximation operators respectively and the system $(\mathcal{IF}(U), \cap, \cup, \sim, \underline{R}, \overline{R})$ an IF rough set algebra. If $\underline{R}(A) = A = \overline{R}(A)$, then A is referred to as definable, otherwise it is undefinable.

Remark 2. If (U, R) is a Pawlak approximation space, i.e., R is an equivalence (crisp) relation on U, then it can be checked that

$$\mu_{\overline{R}(A)}(x) = \vee_{y \in [x]_R} \mu_A(y), \; \gamma_{\overline{R}(A)}(x) = \wedge_{y \in [x]_R} \gamma_A(y);$$

$$\mu_{\underline{R}(A)}(x) = \wedge_{y \in [x]_R} \mu_A(y), \; \gamma_{\underline{R}(A)}(x) = \vee_{y \in [x]_R} \gamma_A(y),$$
where $[x]_R$ is the R-equivalent class containing x.

The following Theorem 1 can be easily concluded from Definition 6.

Theorem 1. *Let (U, R) be an IF approximation space. Then the upper and lower approximation operators in Definition 6 satisfy the following properties:*

$A, A_i \in \mathcal{IF}(U)$, $i \in J$, J is an index set, $\alpha, \beta \in [0,1]$ with $\alpha + \beta \leq 1$,

(IL1) $\underline{R}(A) = \sim \overline{R}(\sim A)$,

(IU1) $\overline{R}(A) = \sim \underline{R}(\sim A)$;

(IL2) $\underline{R}(A \cup \widehat{(\alpha, \beta)}) = \underline{R}(A) \cup \widehat{(\alpha, \beta)}$, (IU2) $\overline{R}(A \cap \widehat{(\alpha, \beta)}) = \overline{R}(A) \cap \widehat{(\alpha, \beta)}$,

(IL3) $\underline{R}(\bigcap\limits_{i \in J} A_i) = \bigcap\limits_{i \in J} \underline{R}(A_i)$, (IU3) $\overline{R}(\bigcup\limits_{i \in J} A_i) = \bigcup\limits_{i \in J} \overline{R}(A_i)$.

Properties (IL1) and (IU1) show that the IF rough approximation operators $\underline{R}$ and $\overline{R}$ are dual to each other. Properties with the same number may be also considered as dual ones. It can be observed that properties (IL2) and (IU2) respectively imply following (IL4) and (IU4):

$$(\text{IL4}) \quad \underline{R}(1_\sim) = 1_\sim, \quad (\text{IU4}) \quad \overline{R}(0_\sim) = 0_\sim.$$

The following properties can be induced from (IL3) and (IU3):

(IL5) $\underline{R}(\bigcup\limits_{i \in J} A_i) \supseteq \bigcup\limits_{i \in J} \underline{R}(A_i)$, (IU5) $\overline{R}(\bigcap\limits_{i \in J} A_i) \subseteq \bigcap\limits_{i \in J} \overline{R}(A_i)$;

(IL6) $A \subseteq B \Longrightarrow \underline{R}(A) \subseteq \underline{R}(B)$, (IU6) $A \subseteq B \Longrightarrow \overline{R}(A) \subseteq \overline{R}(B)$.

For IF rough approximation operators induced from an IF serial approximation space in which the IF relation is serial, we can conclude the following

Theorem 2. *Let (U, R) be an IF approximation space, then*

R is serial $\Longleftrightarrow$ *(IL0)* $\underline{R}(\widehat{(\alpha, \beta)}) = \widehat{(\alpha, \beta)}$, $\forall \alpha, \beta \in [0,1], \alpha + \beta \leq 1$,

$\Longleftrightarrow$ *(IU0)* $\overline{R}(\widehat{(\alpha, \beta)}) = \widehat{(\alpha, \beta)}$, $\forall \alpha, \beta \in [0,1], \alpha + \beta \leq 1$,

$\Longleftrightarrow$ *(IL0)'* $\underline{R}(0_\sim) = 0_\sim$,

$\Longleftrightarrow$ *(IU0)'* $\overline{R}(1_\sim) = 1_\sim$,

$\Longleftrightarrow$ *(ILU)* $\underline{R}(A) \subseteq \overline{R}(A)$, $\forall A \in \mathcal{IF}(U)$.

5   Relationships between IF Approximation Spaces and IF Measurable Spaces

In this section, we examine under which conditions an IF approximation space can associate with an IF measurable space such that the class of all IF measurable sets in the IF measurable space is the family of all definable sets induced from the IF approximation space.

Theorem 3. *Assume that (U, R) is an IF serial approximation space and $\underline{R}$ and $\overline{R}$ are IF rough approximation operators defined in Definition 6. Denote*

$$\mathcal{F} = \{A \in \mathcal{IF}(U) | \underline{R}(A) = A = \overline{R}(A)\}.$$

Then $\mathcal{F}$ is an IF σ-algebra on U.

Proof. (IF1) For any $\alpha, \beta \in [0,1]$ with $\alpha + \beta \leq 1$, since R is an IF serial relation on U, by Theorem 2 we have $\underline{R}(\widehat{(\alpha, \beta)}) = \widehat{(\alpha, \beta)} = \overline{R}(\widehat{(\alpha, \beta)})$. Thus $\widehat{(\alpha, \beta)} \in \mathcal{F}$.

(IF2) Assume that $A_i \in \mathcal{F}$, $i \in \mathbf{N}$, that is,

$$\underline{R}(A_i) = A_i = \overline{R}(A_i), \quad \forall i \in \mathbf{N}.$$

Since R is an IF serial relation on U, by property (ILU) in Theorem 2 and (IU3) in Theorem 1 we have

$$\underline{R}(\bigcup_{i\in\mathbf{N}} A_i) \subseteq \overline{R}(\bigcup_{i\in\mathbf{N}} A_i) = \bigcup_{i\in\mathbf{N}} \overline{R}(A_i) = \bigcup_{i\in\mathbf{N}} A_i.$$

On the other hand, by the assumption and in terms of (IL5) we obtain

$$\bigcup_{i\in\mathbf{N}} A_i = \bigcup_{i\in\mathbf{N}} \underline{R}(A_i) \subseteq \underline{R}(\bigcup_{i\in\mathbf{N}} A_i).$$

Hence

$$\underline{R}(\bigcup_{i\in\mathbf{N}} A_i) = \bigcup_{i\in\mathbf{N}} A_i = \overline{R}(\bigcup_{i\in\mathbf{N}} A_i).$$

Thus $\bigcup_{i\in\mathbf{N}} A_i \in \mathcal{F}$.

(IF3) Assume that $A \in \mathcal{F}$, i.e., $\underline{R}(A) = A = \overline{R}(A)$. Then by using the dual properties (IL1) and (IU1) we have

$$\underline{R}(\sim A) =\sim \overline{R}(A) =\sim A =\sim \underline{R}(A) = \overline{R}(\sim A).$$

Thus $\sim A \in \mathcal{F}$.

Therefore we have proved that $\mathcal{F}$ is an IF algebra.

Theorem 3 shows that the family of all definable sets induced from an IF serial approximation space forms an IF $\sigma-$algebra.

Theorem 4. *Let U be a nonempty finite universe of discourse and $\mathcal{F}$ an IF algebra on U. If $\mathcal{F}$ is generated by a crisp algebra $\mathcal{A}$, then there exists an approximation space (U, R) such that*

$$\mathcal{F} = \{A \in \mathcal{IF}(U)|\underline{R}(A) = A = \overline{R}(A)\}.$$

Proof. For any $x \in U$, let $[x]_{\mathcal{A}} = \cap\{C \in \mathcal{A}|x \in C\}$. It is easy to see that $\{[x]_{\mathcal{A}}|x \in U\}$ is the family of atomic sets of $\mathcal{A}$ and it forms a partition of U. In such a case we can find an equivalence binary relation R on U such that $[x]_R = [x]_{\mathcal{A}}$ for all $x \in U$. With no loss of generality, we write the partition $\{[x]_{\mathcal{A}}|x \in U\}$ as $\{C_1, C_2, \ldots, C_k\}$. We now define two IF set operators $\underline{R}, \overline{R} : \mathcal{IF}(U) \to \mathcal{IF}(U)$ as follows:

$$\underline{R}(A)(x) = (\bigwedge_{y\in[x]_{\mathcal{A}}} \mu_A(y), \bigvee_{y\in[x]_{\mathcal{A}}} \gamma_A(y)), \quad x \in U, \quad A \in \mathcal{IF}(U),$$

$$\overline{R}(A)(x) = (\bigvee_{y\in[x]_{\mathcal{A}}} \mu_A(y), \bigwedge_{y\in[x]_{\mathcal{A}}} \gamma_A(y)), \quad x \in U, \quad A \in \mathcal{IF}(U).$$

Since R is an equivalence crisp relation, (U, R) can be regarded as a special IF serial approximation space, by Theorem 3 we see that $\{A \in \mathcal{IF}(U)|\underline{R}(A) = A = \overline{R}(A)\}$

is an IF algebra. We only need to prove that $\mathcal{F} = \{A \in \mathcal{IF}(U) | \underline{R}(A) = A = \overline{R}(A)\}$.

For any $A \in \mathcal{F}$, notice that $\mathcal{F}$ is generated by the crisp algebra $\mathcal{A}$, then there exist $a_i, b_i \in [0,1], a_i + b_i \leq 1, i = 1, 2, \ldots, k$, such that

$$A(x) = (\sum_{i=1}^{k} a_i 1_{C_i}(x), \sum_{i=1}^{k} b_i 1_{C_i}(x)), \quad \forall x \in U.$$

For any $x \in U$, if $x \in C_i$, of course $[x]_A = [x]_R = C_i$, then

$$\underline{R}(A)(x) = (\bigwedge_{y \in [x]_R} \mu_A(y), \bigvee_{y \in [x]_R} \gamma_A(y)) = (\bigwedge_{y \in C_i} \mu_A(y), \bigvee_{y \in C_i} \gamma_A(y)) = (a_i, b_i),$$

$$\overline{R}(A)(x) = (\bigvee_{y \in [x]_R} \mu_A(y), \bigwedge_{y \in [x]_R} \gamma_A(y)) = (\bigvee_{y \in C_i} \mu_A(y), \bigwedge_{y \in C_i} \gamma_A(y)) = (a_i, b_i).$$

Thus $\underline{R}(A)(x) = \overline{R}(A)(x) = A(x) = (a_i, b_i)$. Hence, $\underline{R}(A) = \overline{R}(A) = A$, from which it follows that

$$\mathcal{F} \subseteq \{A \in \mathcal{IF}(U) | \underline{R}(A) = A = \overline{R}(A)\}.$$

Conversely, assume that $B \in \{A \in \mathcal{IF}(U) | \underline{R}(A) = A = \overline{R}(A)\}$. For any $C_i \in \{C_1, C_2, \ldots, C_k\}$ and $x \in C_i$, clearly, $x \in C_i = [x]_R$. Then

$$\underline{R}(A)(x) = (\bigwedge_{y \in C_i} \mu_A(y), \bigvee_{y \in C_i} \gamma_A(y)),$$

$$\overline{R}(A)(x) = (\bigvee_{y \in C_i} \mu_A(y), \bigwedge_{y \in C_i} \gamma_A(y)).$$

Since $\underline{R}(A)(x) = A(x) = \overline{R}(A)(x)$, we have

$$\bigwedge_{y \in C_i} \mu_B(y) = \bigvee_{y \in C_i} \mu_B(y) = \mu_B(x),$$

$$\bigvee_{y \in C_i} \gamma_B(y) = \bigwedge_{y \in C_i} \gamma_B(y)) = \gamma_B(x).$$

Then

$$\mu_B(y) = \mu_B(x), \quad \gamma_B(y) = \gamma_B(x), \quad \forall y \in C_i = [x]_R.$$

Let $\mu_B(x) = \alpha_i, \gamma_B(x) = \beta_i$, obviously, $\alpha_i + \beta_i \leq 1$, we then obtain

$$\mu_B(y) = \alpha_i, \ \gamma_B(y) = \beta_i, \forall y \in C_i, i = 1, 2, \ldots, k.$$

Hence

$$B(x) = (\sum_{i=1}^{k} \alpha_i 1_{C_i}(x), \sum_{i=1}^{k} \beta_i 1_{C_i}(x)), \quad \forall x \in U.$$

Therefore, $B \in \mathcal{F}$, from which it follows that

$$\{A \in \mathcal{IF}(U) | \underline{R}(A) = A = \overline{R}(A)\} \subseteq \mathcal{F}.$$

Thus we have proved that

$$\{A \in \mathcal{IF}(U) | \underline{R}(A) = A = \overline{R}(A)\} = \mathcal{F}.$$

6   Conclusion

In this paper, we have introduced the concepts of IF measurable spaces and IF σ–algebras. We have also discussed the relationships between IF rough set approximations and IF measurable spaces. We have proved that the family of all IF definable sets induced from an IF serial approximation space forms an IF σ–algebra. On the other hand, in a finite universe, for an IF σ–fuzzy algebra generated by a crisp algebra there must exist an approximation space such that the family of all IF definable sets is the class of all measurable sets in the given IF measurable space. Based on this observation, we hope to gain more insights into the two structures of knowledge representation——definable sets in theory set theory and measurable sets in measure theory.

Acknowledgement

This work was supported by grants from the National Natural Science Foundation of China (No. 60673096 and No. 60773174) and the Natural Science Foundation of Zhejiang Province in China (No. Y107262).

References

1. Atanassov, K.: Intuitionistic Fuzzy Sets. Physica-Verlag, Heidelberg (1999)
2. Chakrabarty, K., Gedeon, T., Koczy, L.: Intuitionistic Fuzzy Rough Sets. In: Proceedings of 4th Joint Conference on Information Sciences (JCIS), Durham, NC, pp. 211–214 (1998)
3. Cornelis, C., Cock, D.M., Kerre, E.E.: Intuitionistic Fuzzy Rough Sets: at the Crossroads of Imperfect Knowledge. Expert Systems 20, 260–270 (2003)
4. Halmos, P.R.: Measure Theory. Van Nostrand-Reinhold, New York (1950)
5. Jena, S.P., Ghosh, S.K.: Intuitionistic Fuzzy Rough Sets. Notes Intuit. Fuzzy Sets 8, 1–18 (2002)
6. Pawlak, Z.: Rough Sets: Theoretical Aspects of Reasoning about Data. Kluwer Academic Publishers, Boston (1991)
7. Pawlak, Z., Skowron, A.: Rudiments of Rough Sets. Inform. Sci. 177, 3–27 (2007)
8. Pawlak, Z., Skowron, A.: Rough Sets: Some Extensions. Inform. Sci. 177, 28–40 (2007)
9. Rizvi, S., Naqvi, H.J., Nadeem, D.: Rough Intuitionistic Fuzzy Sets. In: Proceedings of the 6th Joint Conference on Information Sciences (JCIS), Durham, NC, pp. 101–104 (2002)
10. Samanta, S.K., Mondal, T.K.: Intuitionistic Fuzzy Rough Sets and Rough Intuitionistic Fuzzy Sets. J. Fuzzy Math. 9, 561–582 (2001)
11. Wu, W.-Z., Zhang, W.-X.: Rough Set Approximations vs. Measurable Spaces. In: IEEE International Conference on Granular Computing, pp. 329–332. IEEE Press, New York (2006)
12. Yao, Y.Y.: A Note on Definability and Approximations. In: Peters, J.F., Skowron, A., Marek, V.W., Orłowska, E., Słowiński, R., Ziarko, W. (eds.) Transactions on Rough Sets VII. LNCS, vol. 4400, pp. 274–282. Springer, Heidelberg (2007)
13. Zadeh, L.A.: Probability Measures of Fuzzy Events. J. Math. Anal. Appl. 23, 421–427 (1968)

A Hierarchical Lattice Closure Approach to Abstract Rough Approximation Spaces*

Gianpiero Cattaneo and Davide Ciucci

Dipartimento di Informatica, Sistemistica e Comunicazione
Università di Milano – Bicocca, Viale Sarca 336–U14, I–20126 Milano, Italia
{cattang,ciucci}@disco.unimib.it

Abstract. A hierarchy of closure operators on the abstract context of lattice structures is investigated, and compared to the abstract approach to rough approximation spaces. In particular, the Tarski, the Kuratowski and the Halmos closures are treated, with the corresponding models of covering, topological and partition spaces.

1 Closure Operation in Lattice and Rough Approximation Spaces

In [3,4] a first investigation about closure operations in a lattice context has been done, based on the proof of the equivalence of (weak) closure with the notion of rough approximation space according to the definition given in [2]. In this paper we give a further in depth analysis of this argument. First of all, let us recall the notion of closure, developed by Tarski in his study in *consequences* in logic [16].

Definition 1. *A* Tarski closure lattice *is a structure* $\langle \Sigma, \wedge, \vee, ', *, 0, 1 \rangle$ *where:*

(Cl-dM1) The substructure $\langle \Sigma, \wedge, \vee, ', 0, 1 \rangle$ *is a* de Morgan *lattice, i.e., a bounded lattice, equipped with a de Morgan complementation mapping* $' : \Sigma \mapsto \Sigma$ *that satisfies, for arbitrary* $a, b \in \Sigma$, *the conditions: (dM1)* $a = a''$; *(dM2)* $(a \vee b)' = a' \wedge b'$.

(Cl-dM2) The mapping $* : \Sigma \to \Sigma$ *is a* Tarski closure operation, *that is, it satisfies:*

(C1)	$0^* = 0$	(normalized)
(C2)	$a \leq a^*$	(increasing)
(C3)	$a^* = a^{**}$	(idempotent)
(C4)	$a^* \vee b^* \leq (a \vee b)^*$	(sub–additive)

The subset of *closed elements* is defined as the collection of elements which are equal to their closure: formally, $\mathcal{C}(\Sigma) = \{a \in \Sigma : a = a^*\}$. This set is not

---

* The author's work has been supported by MIUR\PRIN project "Automata and Formal languages: mathematical and application driven studies".
   For lack of space, this is an extended abstract of a forthcoming larger version.

G. Wang et al. (Eds.): RSKT 2008, LNAI 5009, pp. 363–370, 2008.
© Springer-Verlag Berlin Heidelberg 2008

empty since $0, 1$ are closed elements. Condition (C3) says that for any element $a \in \Sigma$ the corresponding a^* is closed; this element is called the *closure* of a. With respect to the notion of closure in a de Morgan lattice, the dual notion is the following one.

Theorem 1. *Suppose a Tarski closure (de Morgan) lattice* $\mathcal{T} = \langle \Sigma, \wedge, \vee, ', *, 0, 1 \rangle$. *The mapping* $^o : \Sigma \to \Sigma$ *defined by the law*

$$\forall a \in \Sigma, \ a^o := ((a')^*)' \tag{1}$$

is a (Tarski) interior operation, i.e., it satisfies the followings:

(I1)	$1^o = 1$	(normalized)
(I2)	$a^o \leq a$	(decreasing)
(I3)	$a^o = a^{oo}$	(idempotent)
(I4)	$(a \wedge b)^o \leq a^o \wedge b^o$	(sub–multiplicative)

Analogously to the closure case, the structure $\mathcal{T}^i = \langle \Sigma, \wedge, \vee, ', {}^o, 0, 1 \rangle$ *is said to be a* Tarski interior lattice.

The subset of *open elements* is defined as the collection of elements which are equal to their interior. Formally, $\mathcal{O}(\Sigma) = \{a \in \Sigma : a = a^o\}$. Also this set is not empty since the two elements $0, 1$ are open too. Condition (I3) says that for every $a \in \Sigma$ the element a^o, called the *interior* of a, is open.

Other important concepts, which is possible to introduce are: the set of *clopen* elements: $\mathcal{CO}(\Sigma) := \mathcal{C}(\Sigma) \cap \mathcal{O}(\Sigma)$, which is not empty and, in general, do not coincide with the set of open or closed elements; the so–called *exterior* of an element $a \in \Sigma$ defined as the open element $\mathbf{e}(a) := (a^*)' \in \mathcal{O}(\Sigma)$; the *boundary* defined as the closed element $\mathbf{b}(a) := (a^o)' \wedge a^* \in \mathcal{C}(\Sigma)$.

In order to grasp the intuitive aspects of the abstract notion of rough approximation space, as introduced in [2], we have to recall the following result.

Theorem 2. *Suppose a (de Morgan) lattice equipped with a Tarski interior and a Tarski closure operations* $\mathcal{A} = \langle \Sigma, {}^o, {}^* \rangle$. *Then, the structure* $\mathcal{A}^{\blacktriangledown} := \langle \Sigma, \mathcal{O}(\Sigma), \mathcal{C}(\Sigma) \rangle$ *is a (de Morgan) rough approximation space in the sense that, defined for any arbitrary* approximable *element* $a \in \Sigma$ *its* lower approximation *as* $l(a) := a^o$ *(the interior of a) and its* upper approximation *as* $u(a) := a^*$ *(the closure of a), the following hold:*

(In1) $l(a) \in \mathcal{O}(\Sigma)$ *(In2)* $l(a) \leq a$ *(In3)* $\forall \beta \in \mathcal{O}(\Sigma), \ \beta \leq a \Rightarrow \beta \leq l(a)$;
(Up1) $u(a) \in \mathcal{C}(\Sigma)$ *(Up2)* $a \leq u(a)$ *(Up3)* $\forall \gamma \in \mathcal{C}(\Sigma), \ a \leq \gamma \Rightarrow u(a) \leq \gamma$.

Any structure $\langle \Sigma, \mathcal{C}(\Sigma), u \rangle$ (resp., $\langle \Sigma, \mathcal{O}(\Sigma), l \rangle$) satisfying the conditions (Up1)–(Up3) (resp., (In1)–(In3)) is called an *upper* (resp., a *lower*) *rough approximation* space, and the mapping $l : \Sigma \mapsto \mathcal{O}(\Sigma)$ (resp., $u : \Sigma \mapsto \mathcal{C}(\Sigma)$) the *lower* (resp., *upper*) *approximation map*. Relatively to these last structures, the element of $\mathcal{C}(\Sigma)$ (resp., $\mathcal{O}(\Sigma)$) are also called *upper* (resp., *lower*) *crisp* elements.

In general, the above outlined structures of Tarski closure and induced interior lattices allow one to introduce the important notion of *rough approximation mapping* $r : \Sigma \mapsto \mathcal{C}(\Sigma) \times \mathcal{O}(\Sigma)$ which associates to any element $a \in \Sigma$ its *open–closed rough approximation* $r(a) = (a^o, a^*)$, with $a^o \in \mathcal{O}(\Sigma)$, $a^* \in \mathcal{C}(\Sigma)$, and $a^0 \leq a \leq a^*$ (see *I2* and *C2*). In this context, an element e is said to be *crisp* (or *exact*, also *sharp*) iff $r(e) = (e, e)$, and this is equivalent to ask that $e \in \mathcal{CO}(\Sigma)$.

Some remarks.

(i) The non–equational conditions of Theorem 2 capture the intuitive aspects of an expected rough approximation space: the lower (resp., upper) approximations $l(a)$ (resp., $u(a)$) is the best approximation of the approximable element a from the bottom (resp., top) by open (resp., closed) elements.

(ii) Theorem 2 is only a part of a more complete version discussed in [3] which can be summarized by the statement that the concrete category of de Morgan lattice with Tarski closure (and induced interior) operator and the one of de Morgan lattice with upper (and induced inner) approximation map are categorical equivalent (isomorphic) between them.

(iii) Thus, if one works in the context of a generalization of the closure operator conditions, for instance weakening (or suppressing) some of the above conditions (C1)–(C4) of definition 1, then some of the "intuitive" conditions expressed by (Up1)–(Up3) are correspondingly lost, or modified.

2   A Hierarchy of Closure Operators

Starting from the original Tarski closure given in definition 1, it is possible to introduce a hierarchy of closure operators with increasingly strong behaviour.

Definition 2. *A closure operation on a de Morgan lattice is said to be* topological *iff the sub–additive condition (C4) is substituted by the stronger* additive *property: (C4T)* $a^* \vee b^* = (a \vee b)^*$.

Citing from Rasiowa–Sikorski [14]: "The closure operation * satisfies the conditions (C1)–(C3) and (C4T). These Axioms are due to Kuratowski [7]. For a detailed exposition of the theory of topological spaces see e.g. Kelley [8], Kuratowski [9,10]." In agreement with this quotation, in the sequel we shall call *Kuratowski closure operation* any topological closure on a de Morgan lattice. A de Morgan lattice equipped with a Kuratowski closure operations is called a *Kuratowski closure lattice*.

The *Kuratowski interior operator* dually defined according to the equation (1) satisfies the *multiplicative* property: (I4T) $a^o \wedge b^o = (a \wedge b)^o$.

Definition 3. *A Kuratowski closure operation on a de Morgan lattice is a* Halmos closure *iff the idempotent condition (C3) is substituted by the stronger one: (sC3)* $a^{*\prime*} = a^{*\prime}$. *A Halmos closure lattice is any de Morgan lattice equipped with a Halmos closure operation.*

This definition has been introduced by Halmos in [5] (collected in [6]) on the base of a Boolean algebra structure as the algebraic formalization of the *existential quantifier*. In the present context, we generalize this structure to the case of de Morgan lattices. The following result assures that the Halmos closure is a Kuratowski closure too.

Proposition 1. *Let Σ be a de Morgan lattice with an operation* $* : \Sigma \mapsto \Sigma$ *which satisfies conditions* (C1), (C2), *and* (C4T). *Then the following implication holds:* (sC3) $\forall a \in \Sigma, \ a^{*\prime *} = a^{*\prime}$ *implies*   (C3) $\forall a \in \Sigma, \ a^{*} = a^{**}$.

The following property characterizes de Morgan lattices with Halmos closure.

Proposition 2. *Let Σ be a de Morgan lattice with a topological (i.e., Kuratowski) closure operation. Then the following are equivalent:*

1. *the topological closure operation satisfies condition (sC3), i.e., it is Halmos;*
2. *the collection of closed elements and the collection of open elements coincide:* $\mathcal{C}(\Sigma) = \mathcal{O}(\Sigma)$.

3   Tarski, Kuratowski and Halmos Closure Models on a Concrete Universe

In this section we investigate three *concrete* examples of the three abstract notions of closure, all based on a concrete nonempty set, the *universe* of the discourse X, and its power set $\mathcal{P}(X)$, the Boolean lattice of all subsets of X.

3.1   Coverings of a Universe as Models of Tarski Closure

First of all let us consider a *covering* of the universe X, i.e., a family $\gamma = \{C_i \in \mathcal{P}(X) : i \in I\}$ of nonempty subsets of X (indexed by the index set I) which satisfies the *covering* condition: $X = \cup\{C_i \in \gamma : i \in I\}$.

Open sets are defined as the set theoretic union of subsets from the covering plus the empty set $\emptyset$. Denoting by $\mathcal{O}_\gamma(X)$ their collection, we have that

$$L \in \mathcal{O}_\gamma(X) \quad \text{iff} \quad \exists\{C_j \in \gamma : j \in I\} : L = \bigcup_{j \in I} C_j \tag{2}$$

The following properties of an open family hold:

(PO1) $\mathcal{O}_\gamma(X)$ contains both the empty set and the whole universe, and
(PO2)   it is closed with respect to arbitrary set theoretic union.

To any subset A of the universe X it is possible to assign the open set $l_\gamma(A) = \cup\{L \in \mathcal{O}_\gamma(X) : L \subseteq A\}$. The main result is the following.

Proposition 3. *The mapping* $l_\gamma : \mathcal{P}(X) \mapsto \mathcal{P}(X)$, $A \to l_\gamma(A)$ *is a Tarski interior operator, or from an equivalent point, it is a lower approximation map.*

The *lower rough approximation space* with respect to the covering γ of X is then $\mathfrak{LR}_\gamma = \langle \mathcal{P}(X), \mathcal{O}_\gamma(X), l_\gamma \rangle$, where the *lower (open) approximation* of A is the *interior* $A^o := l_\gamma(A)$. In particular, from the point of view of a rough approximation space: the role of the bounded de Morgan lattice Σ of approximable elements is played by the Boolean lattice $\mathcal{P}(X)$; the set of open elements $\mathcal{O}_\gamma(X)$ is the lattice of *lower crisp elements*; for any subset A of X (the approximable element) the corresponding lower approximation is its *interior*, i.e., the greatest open set contained in A.

The dual of the covering γ is the *anti–covering* $\gamma_d := \{D \in \mathcal{P}(X) : \exists C \in \gamma \ s.t \ D = C^c\}$, which satisfies the *disjointness* condition: $\emptyset = \cap\{D : D \in \gamma_d\}$. A *closed* subset is then the set theoretic intersection of elements from the dual covering. The collection of all closed sets induced by the covering γ will be denoted by $\mathcal{C}_\gamma(X)$ and so

$$U \in \mathcal{C}_\gamma(X) \quad \text{iff} \quad \exists \{D_s \in \gamma_d : s \in S\} : U = \bigcap_{s \in S} D_s \qquad (3)$$

with $U \in \mathcal{C}_\gamma(X)$ iff $\exists L \in \mathcal{O}_\gamma(X)$ s.t. $U = L^c$. Of course, the following properties of the closed family holds:

(PC1) $\mathcal{C}_\gamma(X)$ contains both the empty set and the whole universe, and
(PC2)   it is closed with respect to arbitrary set theoretic intersection.

These properties qualify $\mathcal{C}_\gamma(X)$ as a *Moore family* of subsets of X (see [1, p. 111]). The *upper (closed) approximation* of any subset A of X is $u_\gamma(A) = \cap\{U \in \mathcal{C}_\gamma(X) : A \subseteq U\}$ and we have the following result.

Proposition 4. *The mapping* $u_\gamma : \mathcal{P}(X) \mapsto \mathcal{P}(X)$, $A \to u_\gamma(A)$ *is a* Tarski closure operator *or, equivalently, it is an* upper approximation map.

In this closure framework the corresponding *upper approximation space* with respect to the covering γ is then $\mathfrak{CR}_\gamma = \langle \mathcal{P}(X), \mathcal{C}_\gamma(X), u_\gamma \rangle$, where $\mathcal{P}(X)$ is the collection of approximable elements; the collection $\mathcal{C}_\gamma(X)$ of all closed sets is the lattice of *upper crisp elements*; for any $A \subseteq X$ (the approximable element) the corresponding upper approximation is its *closure*, i.e., the smallest closed set containing A usually denoted by $A^* := u(A)$.

The *rough approximation space* induced by the covering γ is then the structure $\mathfrak{R}_T = \langle \mathcal{P}(X), \mathcal{O}_\gamma(X), \mathcal{C}_\gamma(X), l_\gamma, u_\gamma \rangle$ where the *rough approximation mapping* $r_\gamma : \mathcal{P}(X) \mapsto \mathcal{O}_\gamma(X) \times \mathcal{C}_\gamma(X)$ assigns to any subset A of the universe X its *rough approximation* as the open–closed pair $r_\gamma(A) = (A^o, A^*)$, with $A^o \subseteq A \subseteq A^*$. Note that *crisp elements* are just the clopen $\mathcal{C}_\gamma \mathcal{O}_\gamma(\Sigma) = \mathcal{C}_\gamma(\Sigma) \cap \mathcal{O}_\gamma(\Sigma)$.

Borrowing some definitions from the (discussed in the sequel) partition case, in the present covering context it is also possible to introduce the *two open approximations* of any subset A as

$$l_\gamma^{(op)}(A) := \cup\{C \in \gamma : C \subseteq A\} = l_\gamma(A) \qquad (4a)$$

$$u_\gamma^{(op)}(A) := \cup\{D \in \gamma : D \cap A \neq \emptyset\} \neq u_\gamma(A) \qquad (4b)$$

This second definition, in general does not give the same result, since $u_\gamma^{(op)}(A) \in \mathcal{O}(X)$ is an open set (as set theoretic union of open sets), whereas $u_\gamma(A) = A^* \in \mathcal{C}(X)$ is a closed set (as set theoretic intersection of closed sets). From a terminological point of view, $u_\gamma(A)$ can be called the *closed upper approximation* and $u_\gamma^{(op)}(A)$ the *open upper approximation* of A.

Proposition 5. *In any covering space the following inequality between upper approximations holds:* $\forall A \in \mathcal{P}(X), \quad u_\gamma(A) \subseteq u_\gamma^{(op)}(A)$.

3.2   Topologies of a Universe as Models of Kuratowski Closure

Let us recall that a topological space can be introduced as a pair $\tau = (X, \mathcal{B})$ consisting of a set X and a base for a topology (simply *open base*), i.e., a collection $\mathcal{B} = \{B_i \in \mathcal{P}(X) : i \in I\}$ of subsets of X, each of which is an *open granule* (or adopting the topological terminology, see [15, p. 99], a *basic open set*), such that the following hold:

 (C) *covering* condition: $X = \cup \{B_i \in \mathcal{B} : i \in I\}$;
(OB) *open base* condition: for any pair B_i, B_j of subsets of the base $\mathcal{B}$ for which $B_i \cap B_j \neq \emptyset$ a collection $\{\widehat{B}_{i_h} \in \mathcal{B} : h \in H\}$ of elements from the base exists such that $B_i \cap B_j = \cup \{\widehat{B}_{i_h} \in \mathcal{B} : h \in H\}$.

This constitutes a strengthening of the notion of covering owing to the further condition (OB). Thus, similarly to the covering case, see equation (2), we introduce the notion of *topological open* sets as any subset of X which is either a set theoretic union of elements from the open base $\mathcal{B}$ or the empty set $\emptyset$. In this way we have induced a topological space defined as a pair $(X, \mathcal{O}(X))$ consisting of a nonempty set X equipped with a family of *open subsets* $\mathcal{O}(X)$In particular any element of the open basis is an open set, that is $\mathcal{B} \subseteq \mathcal{O}(X)$.

Further, we can construct the dual structure $\tau_d = (X, \mathcal{K})$, where $\mathcal{K} := \{K \in \mathcal{P}(X) : \exists B \in \mathcal{B} \text{ s.t. } K = B^c\}$ is the dual base of *closed granules*. This collection $\mathcal{K}$ is a base of closed sets (simply, *closed base*, see [15, p. 112]) for a topology on X in the sense that the following hold:

(DB) *disjointness* condition: $\emptyset = \cap \{K \in \mathcal{P}(X) : K \in \mathcal{K}\}$;
(CB) *closed base* condition: for any pair K_i, K_j of subsets of the closed base $\mathcal{K}$ for which $K_i \cup K_j \neq X$ a collection $\{\widehat{K}_{i_v} \in \mathcal{K} : v \in V\}$ of elements from the closed base exists such that $K_i \cup K_j = \cap \{\widehat{K}_{i_v} \in \mathcal{K} : v \in V\}$.

Similarly to the covering case, see equation (3), a *topological closed* set is either the set theoretic intersection of elements from the closed base or the whole space X. Trivially, a subset of X is *closed* iff it is the set theoretic complement of an open set: $C \in \mathcal{C}(X)$ iff $\exists O \in \mathcal{O}(X) : C = O^c$. All the elements of the closed base are also closed, that is $\mathcal{K} \subseteq \mathcal{C}(X)$.

Note that, as a particular case of closure, also in the now treated case of a topology on a universe the notions of lower, upper, and rough approximation spaces can be introduced inheriting all the properties discussed in the general covering context. In particular, the inclusion $u_\gamma(A) \subset u_\gamma^{(op)}(A)$ continues to hold in this context. Moreover, the lower (resp., upper) approximation is a Kuratowski interior (resp., closure) operation.

3.3   Partitions of a Universe as Models of Halmos Closure

A *partition* of X is a (non necessarily finite) collection $\pi = \{G_i : i \in I\}$ of nonempty subsets of X such that the following hold: *covering* condition: $X = \cup\{G_i \in \pi : i \in I\}$; *disjointness* condition: for any pair G_h, G_k of different elements, $G_h \neq G_k$, of the partition π it is $G_h \cap G_k = \emptyset$. The usual approach to rough set theory as introduced by Pawlak [11,12,13] is based on a concrete *partition space*, that is a pair (X, π) consisting of a nonempty set X, the *universe* of discourse (with corresponding power set $\mathcal{P}(X)$, the collection of all subsets of X, which are the *approximable* sets), and a partition $\pi := \{G_i \in \mathcal{P}(X) : i \in I\}$ of X whose elements are the *elementary sets*. The partition π can be characterized by the induced equivalence relation $\mathcal{R} \subseteq X \times X$, defined as $(x, y) \in \mathcal{R}$ iff $\exists G \in \pi : x, y \in G$. In this case x, y are said to be *indistinguishable* with respect to $\mathcal{R}$ and the equivalence relation $\mathcal{R}$ is called an *indistinguishability* relation.

Of course, a partition space (X, π) generates a topological space whose open base is just the family π, i.e., $\mathcal{B} = \pi$; the corresponding closed base is defined in the usual way. A partition gives rise to a topology of clopen, denoted by $\mathcal{E}_\pi(X)$, formally: $\mathcal{E}_\pi(X) = \mathcal{O}_\pi(X) = \mathcal{C}_\pi(X)$. Thus $\mathcal{E}_\pi(X)$ is the collection of all crisp sets (i.e., if in this model the collection of all approximable elements is $\Sigma = \mathcal{P}(X)$, then the corresponding collection of all crisp set is $\Sigma_c = \mathcal{E}_\pi(X)$).

Thus, the concrete *rough approximation space* generated by the partition π consists of the structure $\mathfrak{R}_\pi := \langle \mathcal{P}(X), \mathcal{E}_\pi(X), l_\pi, u_\pi \rangle$ where: $\mathcal{P}(X)$ is the Boolean atomic (complete) lattice of all *approximable* subsets A of the universe X; the two sets of lower crisp elements and of upper crisp elements coincide with the set of *clopen* (exact) elements $\mathcal{E}_\pi(X)$; the map $l_\pi : \mathcal{P}(X) \mapsto \mathcal{E}_\pi(X)$ associates with any subset A of X its lower approximation:

$$l_\pi(A) := \cup\{O \in \mathcal{E}_\pi(X) : O \subseteq A\} = \cup\{G \in \pi : G \subseteq A\}; \tag{5a}$$

the map $u_\pi : \mathcal{P}(X) \mapsto \mathcal{E}_\pi(X)$ associates with any subset A of X its upper approximation

$$u_\pi(A) := \cap\{C \in \mathcal{E}_\pi(X) : A \subseteq C\} = \cup\{H \in \pi : H \cap A \neq \emptyset\}. \tag{5b}$$

Note that for any subset A, the universe turns out to be the set theoretic union of the mutually disjoint sets $X = l_\pi(A) \cup b_\pi(A) \cup e_\pi(A)$, where $b_\pi(A) := u_\pi(A) \setminus l_\pi(A)$ is the *boundary* and $e_\pi(A) := X \setminus u_\pi(A)$ is the *exterior* of A.

Let us stress that in the case of a covering approximation space, as generalization of the partition approximation space, the equations (4a) and (4b) are generalizations of the corresponding partition versions (5a) and (5b). Further, the lower (resp., upper) approximation l_π (resp., u_π) is a Halmos interior (resp., closure) operation.

4   Conclusion

In the present work, we discussed three different notions of closure on a De Morgan lattice: Tarski, Kuratowski and Halmos, giving rise to a hierarchy of operators. For each of them a model of sets is discussed: coverings, topologies and partitions, respectively. Further, to any closure operator it is possible to associate a dual interior operator such that the pair interior-closure is a rough approximation.

References

1. Birkhoff, G.: Lattice theory, 3 ed., American Mathematical Society Colloquium Publication, vol. XXV, American Mathematical Society, Providence, Rhode Island (1967) (first edition 1940, second (revisited) edition 1948)
2. Cattaneo, G.: Abstract approximation spaces for rough theories. In: Polkowski, L., Skowron, A. (eds.) Rough Sets in Knowledge Discovery, New York, pp. 59–98. Physica–Verlag, Heidelberg (1998)
3. Cattaneo, G., Ciucci, D.: Some Methodological Remarks About Categorical Equivalences in the Abstract Approach to Roughness – Part I. In: Wang, G.-Y., Peters, J.F., Skowron, A., Yao, Y. (eds.) RSKT 2006. LNCS (LNAI), vol. 4062, pp. 277–283. Springer, Heidelberg (2006)
4. Cattaneo, G., Ciucci, D.: Some Methodological Remarks About Categorical Equivalences in the Abstract Approach to Roughness – Part II. In: Wang, G.-Y., Peters, J.F., Skowron, A., Yao, Y. (eds.) RSKT 2006. LNCS (LNAI), vol. 4062, pp. 284–289. Springer, Heidelberg (2006)
5. Halmos, P.R.: The basic concepts of algebraic logic. American Mathematical Monthly 53, 363–387 (1956)
6. Halmos, P.R.: Algebraic logic. Chelsea Pub. Co., New York (1962)
7. Kuratowski, C.: Sur l'opération $\overline{A}$ de l'analysis situs. Fundamenta Mathematicae 3, 182–199 (1922)
8. Kelley, J.L.: General topology. Springer, New York (1955) (second edition 1957 by Van Nostrand)
9. Kelley, J.L.: Topologie. I, Monografie Matematyczne, Warszawa (1933) (Other editions 1948 and 1958)
10. Kelley, J.L.: Topologie. II, Monografie Matematyczne, Warszawa, (1961)
11. Pawlak, Z.: Information systems - theoretical foundations. Information Systems 6, 205–218 (1981)
12. Pawlak, Z.: Rough sets. Int. J. of Computer and Information Sciences 11, 341–356 (1982)
13. Pawlak, Z.: Rough sets: A new approach to vagueness. In: Zadeh, L.A., Kacprzyc, J. (eds.) Fuzzy Logic for the Management of Uncertainty, pp. 105–118. J. Wiley and Sons, New York (1992)
14. Rasiowa, H., Sikorski, R.: The mathematics of metamathematics. In: Rasiowa, H., Sikorski, R. (eds.) Monografie Matematyczne, 3rd edn., vol. 41, Polish Scientific Publishers, Warszawa (1970)
15. Simmons, G.F.: Topology and modern analysis. McGraw-Hill Book Company Inc., New York (1963)
16. Tarski, A.: Fundamentale Begriffe der Methodologie der deduktiven Wissennschaften. I, Monathshefte fur Mathematik und Physik 37, 361–404 (English version in [17]) (1930)
17. Tarski, A.: Logic, semantics, metamathematics, Hackett, Indianapolis (1983) (Second Edition–First Edition, 1956 by Oxford)

A Unifying Abstract Approach for Rough Models

Davide Ciucci*

Dip. di Informatica Sistemistica e Comunicazione – Università di Milano Bicocca
ciucci@disco.unimib.it

Abstract. *Rough approximation algebra* is defined with the aim to give a general abstract approach to all rough sets models, based either on Boolean or Fuzzy sets. Further, a *rough approximations framework* is a structure which is intended as an abstraction of all those cases where several approximations are possibile on the same set. Some properties and models of these structures are given.

1 Introduction

Since the first study by Pawlak [14] there has been a growing interest in rough sets, with thousands of publications both from the theoretical and application standpoint. As a result, several new models and generalizations of the original Pawlak approach have been defined [17]. Then, new models mixing fuzzy and rough sets have been introduced [10,18] and generalization to more abstract structures performed [3,12] Of course, this kind of research is still going on, with new paradigms being defined. Now, the question is what all these models have in common and why they are all referred to as *rough*. Here, we try to give an answer by looking at their properties and their algebraic structure.

Both in the case of boolean rough sets and fuzzy rough sets, the properties of the upper and lower approximations of several models have been studied (for instance, [11,21]). If we consider the classic Pawlak model we have that the lower and upper ($\mathbf{L}$ and $\mathbf{U}$, respectively) approximations satisfy very strong properties, which are listed below.

(U_1)	$\mathbf{U}(H) = (\mathbf{L}(H^c))^c$	(L_1)	$\mathbf{L}(H) = (\mathbf{U}(H^c))^c$
(U_2)	$\emptyset = \mathbf{U}(\emptyset)$	(L_2)	$X = \mathbf{L}(X)$
(U_3)	$\mathbf{U}(H \cup K) = \mathbf{U}(H) \cup \mathbf{U}(K)$	(L_3)	$\mathbf{L}(H \cap K) = \mathbf{L}(H) \cap \mathbf{L}(K)$
(U_4)	$\mathbf{U}(H \cap K) \subseteq \mathbf{U}(H) \cap \mathbf{U}(K)$	(L_4)	$\mathbf{L}(H \cup K) \supseteq \mathbf{L}(H) \cup \mathbf{L}(K)$
(U_5)	$H \subseteq K \;\Rightarrow\; \mathbf{U}(H) \subseteq \mathbf{U}(K)$	(L_5)	$H \subseteq K \;\Rightarrow\; \mathbf{L}(H) \subseteq \mathbf{L}(K)$
(U_6)	$\mathbf{U}(X) = X$	(L_6)	$\mathbf{L}(\emptyset) = \emptyset$
(U_7)	$H \subseteq \mathbf{U}(H)$	(L_7)	$\mathbf{L}(H) \subseteq H$

* The author's work has been supported by MIUR\COFIN project "Formal Languages and Automata: Methods, Models and Applications".

G. Wang et al. (Eds.): RSKT 2008, LNAI 5009, pp. 371–378, 2008.

(U_8) $\mathbf{U}(\mathbf{U}(H)) \subseteq \mathbf{U}(H)$ (L_8) $\mathbf{L}(H) \subseteq \mathbf{L}(\mathbf{L}(H))$

(U_9) $\mathbf{U}(\mathbf{L}(H)) \subseteq \mathbf{L}(H)$ (L_9) $\mathbf{U}(H) \subseteq \mathbf{L}(\mathbf{U}(H))$

When analyzing generalized models, we loose some of these properties. For instance, in tolerance rough sets [20] properties 8 and 9 do not hold. In VPRS [29], instead of U_3 and L_3 we have the weaker forms $\mathbf{U}(H) \cup \mathbf{U}(K) \subseteq \mathbf{U}(H \cup K)$ and $\mathbf{L}(H \cap K) \subseteq \mathbf{L}(H) \cap \mathbf{L}(K)$. and also U_7 is not satisfied. Moreover, when considering extended VPRS [13] or fuzzy rough sets [18], U_1 and L_1 are generally not satisfied. Now, only properties 2, 4 and 6 are left. We could also go on and define even weaker models. At the end, we could consider just two mappings l, u with any further requirement. However, it will then be questionable whether to still call them *rough* approximations. On the other hand, if we adopt an axiomatic approach, we could assume these properties as defining a basic lower and upper approximation mappings.

In the next section, we will define an algebraic structure with two mappings whose intended meaning is exactly of lower and upper approximations and whose behaviour is characterized by some basic properties, including the above quoted ones. Let us note that this approach has been already followed in literature but with stronger requirements and so less generality (see for instance [23,3]). Finally, in several rough set models more than a pair of approximation mappings is defined on the same domain. In general, these approximations are linked by an order relation such that one is better, i.e., closer to the object under approximation, than another. Here, we give some examples and define a general structure to cope with them.

2   Rough Approximation Algebra

In this section, we give the main definitions: rough approximation algebra and framework, together with some properties.

Definition 1.   *A rough approximation algebra is a structure* $\langle A, \wedge, \vee, ', l, u, 0, 1 \rangle$ *where* $\langle A, \wedge, \vee, 0, 1 \rangle$ *is a bounded lattice;* $'$ *is a de Morgan (or involutive) negation, i.e., for all* a, b *it holds (N1)* $a = a''$ *and (N2)* $(a \wedge b)' = a' \vee b'$; *the lower and* upper *approximation mappings* $l, u : A \mapsto A$ *satisfy the following properties*

(R1) $\forall a \in A$, $l(a) \leq u(a)$ *(approximation property);*
(R2) $l(0) = u(0) = 0$, $l(1) = u(1) = 1$ *(boundary conditions);*
(R3) $a \leq b$ *implies* $l(a) \leq l(b)$ *and* $u(a) \leq u(b)$ *(monotonicity).*

The negative *mapping* $n : A \mapsto A$ *is defined as* $n(a) := (u(a))'$.

All the above properties (R1)–(R3) seem reasonable to us. In particular, (R1) just says that the lower approximation must not be greater than the upper one and this is coherent with the semantic we want to give to "lower" and "upper". It is even questionable if there should not hold the stronger property $l(a) \leq a \leq u(a)$. We preferred not to include this stronger version, since it does not hold in several models. Boundary conditions (R2) require that 0 and 1 are known exactly.

Now, given an element $a \in A$ a *rough approximation* of a is just the pair lower-upper approximation $\langle l(a), u(a) \rangle$, which due to the fact that $'$ is involutive, is equivalent to the lower-negative pair $\langle l(a), n(a) \rangle$. Further, we note that from conditions (N1) and (N2) it follows the dual de Morgan law $(a \vee b)' = a' \wedge b'$ and the contraposition law "if $a \leq b$ then $b' \leq a'$".

Let us observe that in [24] two mappings l, u are called lower and upper approximations if they are dual, i.e., satisfy property U_1, and $l(x) \leq u(x)$. Here, we consider operators which are not necessarily dual, a condition which is not always guaranteed to hold (for instance, in fuzzy rough sets), but require also monotonicity and boundary conditions.

Definition 2. *A* rough approximations framework *is a collection of rough approximation algebras, i.e., for a family of indexes $i \in I$, we consider a collection of rough approximations (l_i, u_i) on the algebra $\langle A, \wedge, \vee, ', 0, 1 \rangle$ such that (l_i, u_i) satisfy conditions (R1)–(R3). A rough approximation framework is said to be* regular *if $\forall a \in A$, $l_i(a) \leq l_{i+1}(a)$ and $u_{i+1}(a) \leq u_i(a)$.*

Let us remark that we do not tackle the problem of defining what an exact set is [26]. Thus, l and u are mappings from A to A and not to the collection of "exact" elements $A_e \subseteq A$. Indeed, in this general case, it is not well understood what we mean by exact or definable. If we define exact elements the ones such that $l(a) = a$ (or $u(a) = a$), then we should have that $l(l(a)) = l(a)$ (resp., $u(u(a)) = u(a)$) for all elements in A. But we did not consider this property as a fundamental one. Further, in [4] a pair of fuzzy sets is used to approximate a Boolean set, that is, even the condition $A_e \subseteq A$ does not hold in that particular model. Now, we prove some properties satisfied by l and u.

Proposition 1. *Let $\langle A, \wedge, \vee, ', l, u, 0, 1 \rangle$ a rough approximation algebra. The following hold for any $a, b \in A$.*

$(u1)$	$u(a) \vee u(b) \leq u(a \vee b)$	$(l1)$	$l(a \wedge b) \leq l(a) \wedge l(b)$
$(u2)$	$u(a \wedge b) \leq u(a) \wedge u(b)$	$(l2)$	$l(a) \vee l(b) \leq l(a \wedge b)$
(neg)	$n(a) \leq l'(a)$		

In the next sections we will give some paradigmatic examples of rough approximation algebras and frameworks. We will analyze the wide two categories of Boolean and Fuzzy Rough Sets, showing that our rough approximation algebra can be considered as an abstraction of all these models.

3   Boolean Rough Sets

We consider, now, all the rough approximations of Boolean sets through Boolean sets. In other words, the collection of elements A of our algebra is the Boolean algebra $\mathcal{P}(X)$, powerset of a given universe X. Of course, all the models known in literature simply as rough sets (the term Boolean is not specified) belong to this category. In this context the lattice operators are the intersection and union of sets, and the de Morgan negation, is the usual complementation of sets. In this case, we will use the usual notation $\langle \mathcal{P}(X), \cap, \cup, ^c, \emptyset, X \rangle$.

Since we are dealing with a Boolean algebra, if we define the *boundary* mapping $bnd : A \mapsto A$ as $bnd(a) := u(a) \wedge (l(a))'$, then, it easily follows that $\forall a \in A$, $l(a) \vee bnd(a) \vee n(a) = 1$. That is, the "union" of lower, boundary and negative elements give the maximum 1.

We now consider two main streams of research about (Boolean) rough sets: decision theoretic rough sets, which are based on a probabilistic setting, and rough sets obtained by a general (not necessarily equivalence) binary relation.

3.1 Decision Theoretic Rough Sets

All the models of this field are based on an equivalence relation $\mathcal{R}$ on the universe X, which partitions X in equivalence classes $[x] := \{y \in X : x\mathcal{R}y\}$.

Now, let us define the conditional probability $P(A|[x])$ as the probability that an element y is in A due to the fact that $y \in [x]$. Often, when dealing with finite universes, it is assumed that the conditional probability $P(A|[x])$ is given by the rough membership function [15,30]: $\frac{|A \cap [x]|}{|[x]|}$.

Now, using this probability it is possible to define the decision theoretic rough set model [28,27]. In this model, the lower and upper approximations are defined according to the following equations:

$$l_\alpha(A) = \{x \in X | P(A|[x]) \geq \alpha\} \qquad u_\beta(A) = \{x \in X | P(A|[x]) > \beta\} \qquad (1)$$

where $\alpha, \beta \in [0,1]$ are two parameters under the condition that $\alpha > \beta$. Let us note that this "reasonable" condition ensures that "the cost of classifying an element into the boundary region is closer to the cost of a correct classification than to the cost of an incorrect classification" [27]. Also in our approach this condition is fundamental since it assures that for all $A \subseteq X$, $l_\alpha(A) \subseteq u_\beta(A)$.

Proposition 2. *For $\alpha, \beta \in [0,1]$, $\alpha > \beta$, let l_α, u_β be defined as in equation 1. Then, $\langle \mathcal{P}(X), \cap, \cup, ^c, l_\alpha, u_\beta, \emptyset, X \rangle$ is a (Boolean) rough approximation algebra.*

Now, if we consider the approximations given by different αs and βs, the following properties hold [27]: $\alpha_2 \geq \alpha_1$ implies $l_{\alpha_2}(A) \subseteq l_{\alpha_1}(A)$ and $\beta_2 \leq \beta_1$ implies $u_{\beta_1}(A) \subseteq u_{\beta_2}(A)$. Thus, for a set of parameters $\alpha_i > \beta_i$ depending on the index $i \in I$ and such that $\alpha_i \leq \alpha_{i+1}$ and $\beta_{i+1} \leq \beta_i$, we have that $\{i \in I : \langle A, \wedge, \vee, ', l_{\alpha_i}, u_{\beta_i}, 0, 1 \rangle\}$ is a regular rough approximations framework.

Finally, we underline that Variable Precision Rough Sets and hence Pawlak rough sets can be obtained by an adequate choice of the parameters α and β [27]. Thus, as expected, one deduces that they are both a model of Boolean rough approximation algebra.

3.2 Rough Sets under General Binary Relations

Pawlak rough sets are based on an equivalence relation $\mathcal{R}$ which partitions the underlying universe in equivalence classes. By relaxing the properties that

$\mathcal{R}$ must satisfy, several generalized approaches have been defined and studied [17,25,11]. Here, we recall the definition of rough sets based on a general binary relation and see under which conditions we obtain a rough approximation algebra.

Let $\mathcal{R}$ be a binary relation on a universe U, $\mathcal{R}$ is said *serial* iff $\forall x \in U$, $\exists y \in U : x\mathcal{R}y$. For any element x of the universe, the *granule* (generalization of the equivalence classes of the classical model) generated by $\mathcal{R}$ is the set $g_{\mathcal{R}}(x) := \{y \in U : x\mathcal{R}y\}$. Given a subset $A \subseteq U$ the lower and upper approximations given by $\mathcal{R}$ are respectively defined as:

$$\mathcal{L}_{\mathcal{R}}(A) := \{x \in U : g_{\mathcal{R}}(x) \subseteq A\} \qquad \mathcal{U}_{\mathcal{R}}(A) := \{x \in U : g_{\mathcal{R}}(x) \cap A \neq \emptyset\} \qquad (2)$$

Now, the following properties old for any relation $\mathcal{R}$.

Lemma 1. *[24] Let $\mathcal{R}$ be a binary relation on U. Then,*

- *$\mathcal{L}_{\mathcal{R}}$ and $\mathcal{U}_{\mathcal{R}}$ are monotonic and $\mathcal{L}_{\mathcal{R}}(U) = U$, $\mathcal{U}_{\mathcal{R}}(\emptyset) = \emptyset$.*
- *The relation $\mathcal{R}$ is serial iff for all $A \subseteq U$ it holds $\mathcal{L}_{\mathcal{R}}(A) \subseteq \mathcal{U}_{\mathcal{R}}(A)$*
- *If $\mathcal{R}$ is serial then $\mathcal{U}_{\mathcal{R}}(U) = U$ and $\mathcal{L}_{\mathcal{R}}(\emptyset) = \emptyset$*

Thus, by these properties we can conclude that

Proposition 3. *If $\mathcal{R}$ is a serial binary relation then $\langle U, \cap, \cup, \mathcal{L}_{\mathcal{R}}, \mathcal{U}_{\mathcal{R}}, \emptyset, U \rangle$ is a (Boolean) rough approximation algebra.*

Thus, according to our definition 1 we do consider as *rough* approximations only the ones generated by a relation which is at least serial. That is, we exclude the cases where there are objects which are not in relation with any object including themselves.

Finally, let us note that this is not the only one way to define a rough approximation using a binary relation. Here, we just quote the one given in [2] which gives rise to a regular rough approximations framework which can also be generalized to the fuzzy case [9].

4   Fuzzy Rough Sets

In this section, rough approximations made of a pair of fuzzy sets are considered, showing that they give rise to rough approximations algebras and framework. First of all, we give some basic notions about fuzzy sets then define and analyze fuzzy rough approximation spaces.

4.1   Preliminary Notions on Fuzzy Sets

Let X be the universe of investigation, a fuzzy set on X is represented by its membership function $f : X \mapsto [0,1]$. The collection of all fuzzy sets on X will be denoted as $[0,1]^X$ and a constant fuzzy set as $\underline{a}$, for $a \in [0,1]$. A fuzzy binary relation is a mapping $\mathcal{R} : X \times X \mapsto [0,1]$, $\mathcal{R}$ is serial if $\forall x$, $\exists y$ such that $\mathcal{R}(x,y) = 1$; reflexive if $\forall x \in X$ $\mathcal{R}(x,x) = 1$ and symmetric if $\forall x, y \in X$ $\mathcal{R}(x,y) = \mathcal{R}(y,x)$. For a fuzzy binary relation $\mathcal{R}$ on X and an element $y \in X$, the $\mathcal{R}$-*foreset* of y is defined as the fuzzy set $R_y : X \mapsto [0,1]$, $R_y(x) := R(x,y)$.

Definition 3. *A* triangular norm *(t–norm) is a mapping* $* : [0,1] \times [0,1] \mapsto [0,1]$*, which is commutative, associative, monotonic and such that* $a * 1 = a$*.*

A function $\to: [0,1]^2 \mapsto [0,1]$ *is an* implicator *(or implication) if* $1 \to 0 = 0$ *and* $1 \to 1 = 0 \to 1 = 0 \to 0 = 1$*. Further, it is said a* border implicator *if* $\forall x \in [0,1]$, $1 \to x = x$*.*

A typical example of border implicator is the residual of a left-continuous t-norm $*$, defined as $a \to_* b := \sup\{c \in [0,1] : a * c \leq b\}$.

In the sequel we will consider the fuzzy algebra $\langle [0,1]^X, \wedge, \vee, ', \underline{0}, \underline{1} \rangle$ where $\wedge, \vee$ are the usual Gödel t–norm : $a \wedge b := \min\{a,b\}$, $a \vee b := \max\{a,b\}$ and $'$ the standard involutive negation: $x' = 1 - x$. The order relation in this lattice is the usual pointwise one: $f \leq g$ iff $\forall x \in X : f(x) \leq g(x)$.

4.2   Fuzzy Rough Approximations

Now, we have all the instruments to define a fuzzy-rough approximation.

Definition 4. *A* fuzzy approximation space *is a pair* $(X, \mathcal{R})$ *with* $\mathcal{R}$ *a binary fuzzy relation on* X*. Given an implicator* $\to$ *and a t-norm* $*$*, a* fuzzy rough approximation *for a fuzzy set* $f : X \mapsto [0,1]$ *is given by the pair of fuzzy sets* $\langle L_{\mathcal{R}}(f), U_{\mathcal{R}}(f) \rangle$ *defined pointwise as*

$$L_{\mathcal{R}}(f)(x) := \inf_{y \in X}\{\mathcal{R}(x,y) \to f(y)\} \qquad U_{\mathcal{R}}(f)(x) := \sup_{y \in X}\{\mathcal{R}(x,y) * f(y)\}$$

In [21] an axiomatic approach to $L_{\mathcal{R}}$ and $U_{\mathcal{R}}$ is given and the properties arising from different kinds of relation $\mathcal{R}$ analyzed. We give some results which are relevant to the present work.

Proposition 4. *Let* $(X, \mathcal{R})$ *be a fuzzy approximation space. Then,* $L_{\mathcal{R}}$ *and* $U_{\mathcal{R}}$ *satisfy the following properties*

(F1) $L_{\mathcal{R}}(\underline{1}) = \underline{1}$, $U_{\mathcal{R}}(\underline{0}) = \underline{0}$
(F2) $f \leq g$ *implies* $L_{\mathcal{R}}(f) \leq L_{\mathcal{R}}(g)$ *and* $U_{\mathcal{R}}(f) \leq U_{\mathcal{R}}(g)$
(F3) $\mathcal{R}$ *is serial iff* $U_{\mathcal{R}}(\underline{1}) = \underline{1}$
(F4) *If* $\mathcal{R}$ *is serial then* $L_{\mathcal{R}}(\underline{0}) = \underline{0}$
(F5) *If* $\to$ *is a border implication and* R *is serial, then* $L(f) \leq U(f)$*.*

Proof. Properties (F1)–(F3) are proved in [21].
(F4) If $\mathcal{R}$ is serial than forall x, there exists y such that $\mathcal{R}(x,y) = 1$. For this element y we have $\mathcal{R}(x,y) \to \underline{0}(y) = 1 \to 0 = 0$.
(F5) The required property is satisfied if for all x there exists (at least) an element y such that $R(x,y) \to f(y) \leq R(x,y) * f(y)$. If R is serial, then forall x, there exists y such that $R(x,y) = 1$. Thus, by hypothesis we get $1 \to f(y) = f(y) \leq f(y) = 1 * f(y)$.

Proposition 5. *Let* $(X, \mathcal{R})$ *be a fuzzy approximation space. If* $\to$ *is a border implication and* R *is serial, then* $\langle [0,1]^X, L_{\mathcal{R}}, U_{\mathcal{R}}, \wedge, \vee, ', \underline{0}, \underline{1} \rangle$ *is a rough approximation algebra.*

Due to property (F3) we know that the requirement that $\mathcal{R}$ is serial is a necessary condition. However, we do not know if this is also sufficient, since in order to prove (F5) we also used the fact that $\rightarrow$ is a border implicator. Thus, a deeper study is needed to understand which conditions are necessary and sufficient to obtain a rough approximation algebra based on fuzzy rough sets.

Also in the case of fuzzy rough sets, there is the possibility to define a set of rough approximations linked by an order relation, i.e., what we called a regular rough approximations framework. This is mainly based on the work [9], but for a lack of space, we cannot enter here into details.

Finally, we only mention the work [22] the only one case, to the best of our knowledge, where the approximation of type-2 fuzzy sets is treated. Also in this case, the lower and upper maps give rise to a rough approximation algebra.

5   Conclusions

Two new algebraic structures have been defined: rough approximation algebra and rough approximations framework, in order to give an abstract approach to rough sets models. Several paradigms have been investigated, showing under which conditions they are a model of the introduced algebras. For the sake of brevity, we did not consider the class of lattice-based approximations, which are also a model of the new algebras [12,19,6,8,5,7]. This class will be surely considered in a following study. Another open problem is the analysis of what an "exact" set is in this abstract context.

We underline that, to the best of our knowledge, rough approximations framework is the first tentative to study, in a general way, the situations where several approximations are possible on the same element.

References

1. Abd El-Monsef, M.M., Kilany, N.M.: Decision analysis via granulation based on general binary relation. International Journal of Mathematics and Mathematical Sciences (2007) doi:10.1155/2007/12714
2. Cattaneo, G.: Generalized rough sets (preclusivity fuzzy-intuitionistic BZ lattices). Studia Logica 58, 47–77 (1997)
3. Cattaneo, G.: Abstract approximation spaces for rough theories. In: Polkowski, Skowron (eds.) [16], pp. 59–98 (1998)
4. Cornelis, C., De Cock, M., Radzikowska, A.M.: Vaguely Quantified Rough Sets. In: An, A., Stefanowski, J., Ramanna, S., Butz, C.J., Pedrycz, W., Wang, G. (eds.) RSFDGrC 2007. LNCS (LNAI), vol. 4482, pp. 87–94. Springer, Heidelberg (2007)
5. Ciucci, D., Flaminio, T.: Generalized rough approximations in $\Pi\frac{1}{2}$. International Journal of Approximate Reasoning (in press, 2007)
6. Ciucci, D.: On the axioms of residuated structures: Independence, dependencies and rough approximations. Fundamenta Informaticae 69, 359–387 (2006)
7. Chen, X., Li, Q.: Construction of rough approximations in fuzzy setting. Fuzzy sets and Systems 158, 2641–2653 (2007)
8. Chen, D., Zhang, W.X., Yeung, D., Tsang, E.C.C.: Rough approximations on a complete completely distributive lattice with applications to generalized rough sets. Information Sciences 176, 1829–1848 (2006)

9. De Cock, M., Cornelis, C., Kerre, E.E.: Fuzzy rough sets: The forgotten step. IEEE Transactions on Fuzzy Systems 15, 121–130 (2007)
10. Dubois, D., Prade, H.: Putting rough sets and fuzzy sets together. In: SLowinski, R. (ed.) Intelligent Decision Support – Handbook of Applications and Advances of the Rough Sets Thepry, pp. 203–232. Kluwer Academic Publishers, Dordrecht (1992)
11. Inuiguchi, M.: Generalizations of Rough Sets and Rule Extraction. In: Peters, J.F., Skowron, A., Grzymała-Busse, J.W., Kostek, B.z., Świniarski, R.W., Szczuka, M. (eds.) Transactions on Rough Sets I. LNCS, vol. 3100, pp. 96–119. Springer, Heidelberg (2004)
12. Jarvinen, J.: On the structure of rough approximations. Fundamenta Informaticae 53, 135–153 (2002)
13. Katzberg, J.D., Ziarko, W.: Variable precision extension of rough sets. Fundamenta Informaticae 27(2,3), 155–168 (1996)
14. Pawlak, Z.: Rough sets. Int. J. of Computer and Information Sciences 11, 341–356 (1982)
15. Pawlak, Z., Skowron, A.: Rough membership functions, pp. 251–271. Wiley, Chichester (1994)
16. Polkowski, L., Skowron, A. (eds.): Rough Sets in Knowledge Discovery 1. Physica–Verlag, Heidelberg, New York (1998)
17. Pawlak, Z., Skowron, A.: Rough sets: some extensions. Information Sciences 177, 28–40 (2007)
18. Radzikowska, A.M., Kerre, E.E.: A comparative study of fuzzy rough sets. Fuzzy Sets and Systems 126, 137–155 (2002)
19. Radzikowska, A.M., Kerre, E.E.: Fuzzy Rough Sets Based on Residuated Lattices. In: Peters, J.F., Skowron, A., Dubois, D., Grzymała-Busse, J.W., Inuiguchi, M., Polkowski, L. (eds.) Transactions on Rough Sets II. LNCS, vol. 3135, pp. 278–296. Springer, Heidelberg (2004)
20. Skowron, A., Stepaniuk, J.: Tolerance approximation spaces. Fundamenta Informaticae 27, 245–253 (1996)
21. Wu, W.Z., Leung, Y., Mi, J.S.: On characterizations of $(\rangle, \sqcup)$ - fuzzy rough approximation operators. Fuzzy Sets and Systems 154, 76–102 (2005)
22. Wu, H., Wu, Y., Liu, H., Zhang, H.: Roughness of type 2 fuzzy sets based on similarity relations. International Journal of Uncertainty, Fuzziness and Knowledge Based System 15, 513–517 (2007)
23. Yao, Y.Y.: Two view of the theory of rough sets in finite universes. International Journal of Approximate Reasoning 15, 291–317 (1996)
24. Yao, Y.Y.: Constructive and algebraic methods of the theory of rough sets. Journal of Information Sciences 109, 21–47 (1998)
25. Yao, Y.Y.: Generalized rough set models. In: Polkowski, Skowron (eds.) [16], pp. 286–318 (1998)
26. Yao, Y.Y.: A Note on Definability and Approximations. In: Peters, J.F., Skowron, A., Marek, V.W., Orłowska, E., Słowiński, R., Ziarko, W. (eds.) Transactions on Rough Sets VII. LNCS, vol. 4400, pp. 274–282. Springer, Heidelberg (2007)
27. Yao, Y.Y.: Probabilistic rough set approximations. International Journal of Approximate Reasoning (in press, 2007)
28. Yao, Y.Y., Wong, S.K.M.: A decision theoretic framework for approximating concepts. International Journal of Man–Machine Studies 37, 793–809 (1992)
29. Ziarko, W.: Variable precision rough sets model. Journal of Computer and Systems Sciences 43(1), 39–59 (1993)
30. Ziarko, W.: Probabilistic approach to rough sets. In: International Journal of Approximate Reasoning (in press, 2007)

Quorum Based Data Replication in Grid Environment

Rohaya Latip, Hamidah Ibrahim, Mohamed Othman, Md Nasir Sulaiman,
and Azizol Abdullah

Faculty of Computer Science and Information Technology,
Universiti Putra Malaysia
{rohaya,hamidah,mothman,nasir,azizol}@fsktm.upm.edu.my

Abstract. Replication is a useful technique for distributed database systems and can be implemented in a grid computation environment to provide a high availability, fault tolerant, and enhance the performance of the system. This paper discusses a new protocol named Diagonal Data Replication in 2D Mesh structure (DR2M) protocol where the performance addressed are data availability which is compared with the previous replication protocols, Read-One Write-All (ROWA), Voting (VT), Tree Quorum (TQ), Grid Configuration (GC), and Neighbor Replication on Grid (NRG). DR2M protocol is organized in a logical 2D mesh structure and by using quorums and voting techniques to improve the performance and availability of the replication protocol where it reduce the number of copies of data replication for read or write operations. The data file is copied at the selected node of the diagonal site in a quorum. The selection of a replica depends on the diagonal location of the structured 2D mesh network where the middle node is selected because it is the best location to get a copy of the data if every node has the equal number of request and data accessing in the network. The algorithm in this paper also calculates the best number of nodes in each quorum and how many quorums are needed for N number of nodes in a network. DR2M protocol also ensures that the data for read and write operations is consistency, by proofing the quorum must not have a nonempty intersection quorum. To evaluate DR2M protocol, we developed a simulation model in Java. Our results prove that DR2M protocol improves the performance of the data availability compare to the previous data replication protocol, ROWA, VT, TQ, GC and NRG.

Keywords: Data replication, Grid, Data management, Availability, Replica control protocol.

1 Introduction

A grid is a distributed network computing system, a virtual computer formed by a networked set of heterogeneous machines that agree to share their local resources with each other. A grid is a very large scale, generalized distributed network computing system that can scale to internet size environment with machines distributed across multiple organizations and administrative domains [1, 2]. Ensuring efficient access to such a huge network and widely distributed data is a challenge to

G. Wang et al. (Eds.): RSKT 2008, LNAI 5009, pp. 379–386, 2008.
© Springer-Verlag Berlin Heidelberg 2008

those who design, maintain, and manage the grid network. The availability of a data on a large network is an issue [3, 4, 5, 6] because geographically it is distributed and has different database management to share across the grid network whereas replicating data can become expensive if the number of operations such as read or write operations is high. In our work, we investigate the use of replication on a grid network to improve its ability to access data efficiently.

Distributed computing manages thousands of computer systems and this has limited its memory and processing power. On the other hand, grid computing has some extra characteristics. It is concerned to efficient utilization of a pool of heterogeneous systems with optimal workload management utilizing an enterprise's entire computational resources (servers, networks, storage, and information) acting together to create one or more large pools of computing resources. There is no limitation of users or originations in grid computing. Even though grid sometime can be as minimum one node but for our protocol the best number of nodes should be more than five nodes to implement the protocol. This protocol is suitable for large network such as grid environment.

There are some research been done for replica control protocol in distributed database and grid such as Read-One Write-All (ROWA) [7], Voting (VT) [8], Tree Quorum (TQ) [9, 10], Grid Configuration (GC) [11, 12], and the latest research in year 2007 is Neighbor Replication on Grid (NRG) [13, 14, 15]. Each protocol has its own way of optimizing the data availability. The usage of replica is to get an optimize data accessing in a large and complex grid. Fig. 1 illustrates the usage of replica protocol in grid where it is located in the replica optimization component [16].

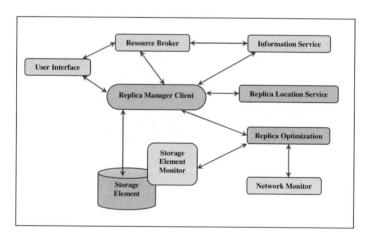

Fig. 1. Grid Components Interact with Replica Manager in Grid

In this paper, we present a new quorum-based protocol for data replication that provides both high data availability and low response time [17]. The proposed protocol imposes a logical two dimensional mesh structure to produce the best number of quorums and obtain good performance of data availability.

Quorums improved the performance of fault tolerant and availability of replication protocols [18]. Quorums reduce the number of copies involved in reading or writing data. To address the availability, we replicate data on the selected node from the diagonal site of the 2D mesh structure, which has been organized in quorums. This is because it is easy to access the database because of its middle location in the quorum. We use Java to run this replication protocol.

The paper is organized as follows. Section two introduces Diagonal Replication in 2D Mesh (DR2M) protocol as a new replica control protocol. Section three, illustrates the DR2M algorithm. Section four discusses the results by comparing DR2M with the existing protocols and the paper ends with a conclusion.

2 Diagonal Replication in 2D Mesh (DR2M) Protocol

In our protocol, all nodes are logically organized into two dimensional mesh structures. We assume that the replica copies are in the form of data files and all nodes are operational meaning that the copy at the nodes is available. The data file is replicated to only one middle node at the diagonal site of each quorum.

This protocol uses quorum to arrange nodes in cluster. Voting approach assigned every copy of replicated data object a certain number of votes and a transaction has to collect a read quorum of r votes to read a data object, and a write quorum of w votes to write the data object. In this approach the quorum must satisfy two constraints which are $r + w$ must be larger than the total number of votes, v assigned to the copies of the data object and $w > v/2$ [8].

Quorum is grouping the nodes or databases as shown in Fig. 2. This figure illustrates how the quorums for network size of 81 nodes are grouped by nodes of 5 x 5 in each quorum. Nodes which are formed in a quorum intersect with other quorums. This is to ensure that these quorums can read or write other data from other nodes which are in another quorum. The number of nodes grouped in a quorum, q must be odd so that only one middle node from the diagonal site can be selected such as site $s(3,3)$ colored in black circles in Fig. 2. Site $s(3,3)$ has the copy of the data file for read and write operation to be executed.

Definition 2.1. Assume that a database system consists of n x n nodes that are logically organized in the form of two dimensional grid structures. All sites are labeled $s(i,j)$, $1 \leq i \leq n$, $1 \leq j \leq n$. The diagonal site of $s(i,j)$ is $s(n,n)$, where $n = 1, 2, ..., n$.

From Fig. 2, for $q1$, the nodes of the diagonal site, $D(s)$ is $\{s(1,1), s(2,2), s(3,3), s(4,4), s(5,5)\}$ in each quorum and the middle node $s(3,3)$ has the copy of the data file. This figure shows that 81 nodes have four quorums where each quorum actually intersects with each other. Node e in $q1$ is actually node a in $q2$ and node e in $q3$ is actually node a in $q4$ and etc.

Since the data file is replicated only on one node for each quorum, thus it minimizes the number of database operations. The selected node in the diagonal sites is assigned with vote one or vote zero. A vote assignment on grid, B, is a function such that, $B(s(i,j)) \in \{0, 1\}$, $1 \leq i \leq n$, $1 \leq j \leq n$ where $B(s(i,j))$ is the vote assigned to site $s(i,j)$. This assignment is treated as an allocation of replicated copies and a vote assigned to the site results in a copy allocated at the diagonal site. That is,

1 vote ≡ 1 copy.   Let $L_B = \sum B(s(i,j))$, $s(i,j) \in D(s)$

where L_B is the total number of votes assigned to the selected node as a primary replica in each quorum. Thus $L_B = 1$ in each quorum.

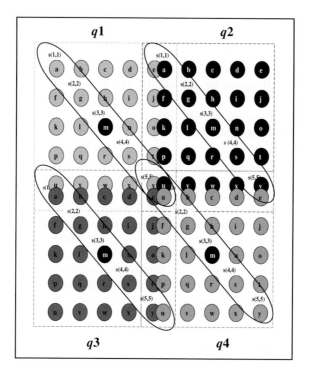

Fig. 2. A grid organization with 81 nodes, each of the node has a data file a, b,..., and y, respectively

Let r and w denote the read quorum and write quorum, respectively. To ensure that read operation always gets the updated data, $r + w$ must be greater than the total number of votes assigned to all sites. To make sure the consistency is obtained, the following conditions must be fulfilled [8].

i. $1 \leq r \leq L_B, 1 \leq w \leq L_B$
ii. $r + w = L_B + 1$.

These two conditions ensure that there is a nonempty intersection of copies between read and write quorums. Thus, these conditions ensure that a read operation accesses the most recently updated copy of the replicated data.

Let $S(B) = \{s(i,j)| B(s(i,j)) = 1, 1 \leq i \leq n, 1 \leq j \leq n\}$ where $i = j$

Definition 2.2. For a quorum q, a quorum group is any subset of $S(B)$ where the size is greater than or equal to q. The collection of quorum group is defined as the quorum set.

Let $Q(B_m,q)$ be the quorum set with respect to assignment B and quorum q, then

$Q(B_m,q) = \{G| G \subseteq S(B) \text{ and } |G| \geq q\}$.

For example, from Fig. 2, let site $s(3,3)$ be the primary database of the master data file m. Its diagonal sites are $s(1,1)$, $s(2,2)$, $s(3,3)$, $s(4,4)$, and $s(5,5)$. Consider an assignment B for the data file m, where $B_m(s(i,j))$ is the vote assigned to site $s(i,j)$ and $L_{B,m}$ is the total number of votes assigned to primary database in each quorum which is $(s(3,3))$ for data file m, such that $B_m(s(1,1)) = B_m(s(2,2)) = B_m(s(3,3)) = B_m(s(4,4)) = B_m(s(5,5)) = 1$ and $L_{B,m} = B_m(s(3,3))$. Therefore, $S(B) = \{s(3,3)\}$.

For simplicity, a read quorum for the data file m, is equal to write quorum. The quorum sets for read and write operations are $Q(B_m,q1)$, $Q(B_m,q2)$, $Q(B_m,q3)$, and $Q(B_m,q4)$, respectively, where $Q(B_m,q1) = \{s(3,3)\}$, $Q(B_m,q2) = \{s(3,3)\}$, $Q(B_m,q3) = \{s(3,3)\}$, and $Q(B_m,q4) = \{s(3,3)\}$. Therefore, the number of replicated data file m is four.

3   DR2M Algorithm

DR2M is developed using Java. The algorithm of the model is as in Fig. 3, where it shows the flow of the protocol. For the development of the simulation some assumptions were made such as the network has no failure and nodes are accessible.

To analyze the performance of read and write availability, below are the equations where n is the 2D mesh column or row size, example n is 7, thus 7 x 7 nodes of grid is the network size and p is the probability of data available which is between 0 to 1 whereas, q is the number of quorum for a certain operation such as read or write operation. Eq. (1) is to calculate the data availability.

$$\text{Av,}_q = \sum_{i=q}^{n} \binom{n}{i} (p^i(1-p)^{n-i}) \tag{1}$$

Main
Input number of row or column, N

If $\sqrt{N}$ is odd integer then
  Find the number of quorum, Q

  $Q = \left\lfloor \sqrt{n} - \dfrac{n}{10} \right\rfloor$

  Find number of nodes in each quorum, X

  $X = {}^{n}\!/_{Q}$

  Get the next odd integer after X
  Select the middle replica, R
  Copy the data file at R
Else add one virtual column and row, $\text{Col}_{new} + \text{Row}_{New}$
  $N = \text{Col}_{new} * \text{Row}_{New}$ then
  Return N to Main.

Fig. 3. Algorithm of DR2M protocol

A larger network has more quorums. The number of columns and rows in each quorum must be odd, to get the middle replica. DR2M protocol assumes all nodes have the same data access level and request number. Fig. 3 is the algorithm for DR2M protocol. It illustrates how the algorithm is designed and implemented.

For Fig. 3, the number of nodes in the network must be odd and if it is an even number then a virtual column and row is added meaning an empty node of rows and column are located to make the selection of the middle node easier. To find the best number of quorum, Q, for the whole network size, n x n, Eq. (2) is used where n is the number of nodes for row or column.

$$Q = \left\lfloor \sqrt{n} - \frac{n}{10} \right\rfloor \qquad (2)$$

For obtaining the best number of nodes for each quorum, X, Eq. (3) is used where the number of nodes must be odd to make the selection of the middle node easy and if X is not odd then X will be the next odd number after n/Q.

$$X = \frac{n}{Q} \qquad (3)$$

After the selected node is chosen, the data file is copied where it acts as a primary database for that particular quorum. Some protocol depends on the frequent usage of the data to select that particular node as the primary database. But for this protocol, an assumption is made where every node has the same level of data access and number of request.

4   Results and Discussion

In this section, DR2M protocol is compared with the results of read and write availability of the existing protocols, namely: ROWA, VT, TQ, GC, and NRG. Fig. 4 shows the results of read availability in 81 nodes of network size. ROWA protocol has the higher read availability about average of 2.425% for probability of data accessing between 0.1 to 0.9 even when the number of nodes is increased. This is because only one replica is accessed by a read operation for all n nodes of network size but ROWA has the lowest write availability.

Fig. 5 proves that the DR2M protocol has 4.163% higher of write availability for all probabilities of data accessing. This is due to the fact that replicas are selected from the middle location of all nodes in each quorum and by using quorum approach helps to reduce the number of copies in a large network.

Fig. 5 illustrates the write availability for 81 numbers of nodes, where the probability is from 0.1 to 0.9.

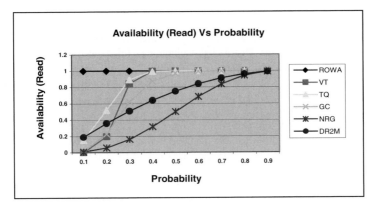

Fig. 4. Read availability results for existing protocols

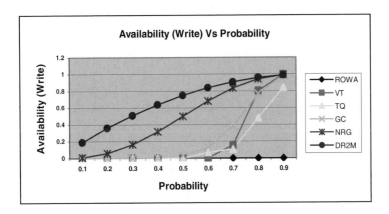

Fig. 5. Write availability results for existing protocols

6   Conclusions

In this paper, DR2M protocol selects a primary database from the middle location of the diagonal site where the nodes are organized in structure of 2D Mesh. By getting the best number of quorum and using the voting techniques have improved the availability for write operation compared to all protocols and has higher read availability compared to the latest technique, NRG. In the future, we will investigate the response time to access a range of data size in the grid environment and this investigation will be used to evaluate the performance of our DR2M protocol.

Acknowledgements. We would like to thank the Malaysian Ministry of Science, Technology and Innovation (MOSTI) for supporting our project under the Fundamental Grant No: 02-01-07-269FR.

References

1. Krauter, K., Buyya, R., Maheswaran, M.: A Taxanomy and Survey of Grid Resource Management Systems for Distributed Computing. International Journal of Software Practice and Experience 32(2), 135–164 (2002)
2. Foster, I., Kesselman, C., Nick, J., Tuecke, S.: Grid Services for Distributed System Integration. Computer 35(6), 37–46 (2002)
3. Ranganathan, K., Foster, I.: Identifying Dynamic Replication Strategies for a High Performance Data Grid. In: International Workshop on Grid Computing, Denver (2001)
4. Lamehamedi, H., Szymanski, B., Shentu, Z., Deelman, E.: Data Replication Strategies in Grid Environment. In: ICAP 2003, pp. 378–383. IEEE Computer Science Press, Los Alamitos (2002)
5. Lamehamedi, H., Shentu, Z., Szymanski, B.: Simulation of Dynamic Data Replication Strategies in Data Grids. In: The 17th International Symposium on Parallel and Distributed Processing, pp. 22–26 (2003)
6. Lamehamedi, H.: Decentralized Data Management Framework for Data Grids. Ph.D. thesis, Rensselaer Polytechnic Institute Troy, New York (2005)
7. Agrawal, D., El Abbadi, A.: Using Reconfiguration for Efficient Management of Replicated Data. IEEE Transactions on Knowledge and Data Engineering 8(5), 786–801 (1996)
8. Mat Deris, M.: Efficient Access of Replication Data in Distributed Database Systems. Thesis PhD, Universiti Putra Malaysia (2001)
9. Agrawal, D., El Abbadi, A.: The Generalized Tree Quorum Protocol: An Efficient Approach for Managing Replicated Data. ACM Transactions Database System 17(4), 689–717 (1992)
10. Agrawal, D., El Abbadi, A.: The Tree Quorum Protocol: An Efficient Approach for Managing Replicated Data. In: 16th International Conference on Very Large databases, pp. 243–254 (1990)
11. Maekawa, M.: A $\sqrt{n}$ Algorithm for Mutual Exclusion in Decentralized Systems. ACM Transactions Computer System 3(2), 145–159 (1992)
12. Cheung, S.Y., Ammar, M.H., Ahmad, M.: The Grid Protocol: A High Performance Schema for Maintaining Replicated Data. IEEE Transactions on Knowledge and Data Engineering 4(6), 582–592 (1992)
13. Mat Deris, M., Evans, D.J., Saman, M.Y., Ahmad, N.: Binary Vote Assignment on Grid For Efficient Access of Replicated Data. Int'l Journal of Computer Mathematics, 1489–1498 (2003)
14. Mat Deris, M., Abawajy, J.H., Suzuri, H.M.: An Efficient Replicated Data Access Approach for Large Scale Distributed Systems. In: IEEE/ACM Conf. on Cluster Computing and Grid, Chicago, USA (2004)
15. Ahamad, N., Mat Deris, M.: Managing Neighbor Replication Transactions in Distributed Systems. In: DCABES 2006, China (2006)
16. Kunszt, P., Laure, E., Stockinger, H., Stockinger, K.: Advanced Replica Management with Reptor. In: CERN, European Organization for Nuclear Research, CH-1211 Geneva 23, Switzerland (2000)
17. Mat Deris, M., Abawajy, J.H., Mamat, A.: An Efficient Replicated Data Access Approach for Large-scale Distributed Systems. Future Generation Computer Systems 24, 1–9 (2007)
18. Jimenez-Peris, R., et al.: Are Quorums an Alternative for Data Replication? ACM Transactions on Database Systems 28(3) (2003)

On Non–pointwise Entropies of Coverings: Relationship with Anti–monotonicity*

Daniela Bianucci and Gianpiero Cattaneo

Dipartimento di Informatica, Sistemistica e Comunicazione
Università di Milano – Bicocca
Viale Sarca 336/U14, I–20126 Milano, Italia
{bianucci,cattang}@disco.unimib.it

Abstract. Some approaches to the entropy of coverings, generalizing the case of the partition entropy, and some definitions of orderings and quasi–orderings of coverings as extensions to the covering context of various formulations of the standard order on partitions. Unfortunately, we here show that the problem of anti–monotonicity of *non–pointwise* entropy of coverings is still open. On the other side, we will illustrate an approach to generate a partition from a covering; we will then show how we can indirectly compare the entropies of two coverings, that are in a certain order relation, by comparing the entropies of the induced partitions.

Keywords: Partitions, Coverings, Partial order, Quasi–order, Entropy, Co–entropy, Monotonicity, Anti–monotonicity, Induced partition.

1 Introduction to Information Systems and Partition Entropy

An information system, according to the original definition given by Pawlak [11], is formalized as a triple $\langle X, Att, F \rangle$, where X, called the "universe of discourse", is a finite collection of objects, Att a finite family of attributes, and F a mapping defined on $X \times Att$ which assigns a value $F(x, a)$ to each pair $(x, a) \in X \times Att$ which is the value assumed by the object x relatively to the attribute a. An information system can be represented by a table in which each row represents an object x of X and each column represents an attribute a of Att [8, 13, 17]. An information system is said to be *complete* if it does not contain undefined values, i.e., if for each pair (x, a) the mapping F assumes a well known value. The information system is said to be *incomplete* if the "information" about some pair (x, a) is missing, denoted by $F(x, a) = *$.

---

* The author's work has been supported by MIUR\PRIN project "Automata and Formal languages: mathematical and application driven studies"and by "Funds of *Sovvenzione Globale INGENIO* allocated by *Fondo Sociale Europeo, Ministero del Lavoro e della Previdenza Sociale*, and *Regione Lombardia*".

For lack of space, this is an extended abstract of a forthcoming larger version.

G. Wang et al. (Eds.): RSKT 2008, LNAI 5009, pp. 387–394, 2008.

On a complete information system we can define an equivalence relation, called "*indiscernibility*" [12], which states that, chosen a subfamily $\mathcal{A} \subseteq Att$ of attributes, two objects x and y are *indiscernible* with respect to $\mathcal{A}$ iff $\forall\, a \in \mathcal{A}$ they assume the same value, $F(x, a) = F(y, a)$. Through this equivalence relation we obtain a partition of the universe X. In the case of an incomplete information system in which some values are missing we can use a generalization of the indiscernibility relation, called "*similarity*" relation [9], for which we say that, given a subset of attributes $\mathcal{A} \subseteq Att$, two objects x and y are *similar* with respect to $\mathcal{A}$ (denoted by $x \sim_{\mathcal{A}} y$) iff $\forall\, a \in \mathcal{A}$ either they assume the same value or one of the two objects (either x or y) assumes the undefined $*$ value. Each object $x \in X$ generates a similarity class through the similarity relation, $s_{\mathcal{A}}(x) = \{y \in X : x \sim_{\mathcal{A}} y\}$, and the collection of all the similarity classes thus induced constitutes a covering of the universe X.

We will first recall the main aspects and definitions involved in the partition situation. A partition of a finite universe X is a finite collection $\pi = \{A_1, A_2, \ldots, A_N\}$ of nonempty pairwise disjoint measurable subsets A_i of X whose set theoretic union is X. Let us observe that, as previously briefly illustrated, if we deal with a complete information system, given a set of attributes $\mathcal{A}$, through the indiscernibility relation we obtain a partition of X depending from $\mathcal{A}$. On the collection $\Pi(X)$ of all partitions of X we can introduce the following binary relations [3, 1, 4, 6] defined for any pair of partitions $\pi_1, \pi_2 \in \Pi(X)$ of the universe X as follows:

$$\pi_1 \preceq \pi_2 \quad \text{iff} \quad \forall\, A \in \pi_1,\ \exists\, B \in \pi_2 : A \subseteq B; \tag{1}$$

$$\pi_1 \ll \pi_2 \quad \text{iff} \quad \forall\, B \in \pi_2,\ \exists\, \{A_{i_1}, \ldots, A_{i_h}\} \subseteq \pi_1 : B = A_{i_1} \cup \ldots \cup A_{i_h}. \tag{2}$$

$$\pi_1 \leq_W \pi_2 \quad \text{iff} \quad \forall\, A_i \in \pi_1,\ \forall\, B_j \in \pi_2,\ A_i \cap B_j \neq \emptyset \quad \text{implies} \quad A_i \subseteq B_j \tag{3}$$

These three formulations are mutually equivalent on $\Pi(X)$ and define the same partial order, which in the partition context will be denoted by $\preceq$ in the sequel. This does not happen in their extensions to the covering case, where they correspond to three different relations: the first two are quasi–order (reflexive, transitive, but not anti–symmetric) relations, the latter is a standard ordering.

Let us now describe the main details and definitions regarding the entropy and co–entropy for partitions. In the sequel we consider measurable universes with a non-trivial finite measure, i.e., $0 \neq m(X) < +\infty$. From the monotonicity of m [16] we always have that for every measurable set $A \subseteq X$ the inequality $m(A) \leq m(X)$ holds . Hence $p(A) = \frac{m(A)}{m(X)}$ define a probability measure representing the probability of occurrence of each elementary event A. In the case of a finite measurable partition π the elementary events are the equivalence classes $A_1, A_2, \ldots, A_N$. Let us observe that the vector $\boldsymbol{p}(\pi) = (p(A_1), p(A_2), \ldots, p(A_N))$ is a *probability distribution* since each $p(A_i) > 0$ and $\sum_{i=1}^{N} p(A_i) = 1$. The vector $I(\pi) = (-\log p(A_1), \ldots, -\log p(A_N))$ is the discrete random variable of the partition *uncertainty*. The resulting average uncertainty related to a partition π of X is the information entropy defined, according to Shannon [15], as $H(\pi) = -\sum_{i=1}^{N} p(A_i) \log p(A_i)$. When dealing with partitions the entropy can also be equivalently expressed by $H(\pi) = \log m(X) - \frac{1}{m(X)} \sum_{i=1}^{N} m(A_i) \log m(A_i)$. If we consider the second term, i.e., $E(\pi) = \frac{1}{m(X)} \sum_{i=1}^{N} m(A_i) \log m(A_i)$, we can observe that it could be an average by itself of a discrete random variable $G(\pi) = (\log m(A_1), \ldots, \log m(A_N))$ furnishing a granularity measure of π. Thus $E(\pi)$ does not regard the average *uncertainty* of the events A_i, but it is the average *granularity* related to π. This measure, complements the entropy with respect to the fixed quantity $\log m(X)$ and for this reason it is here called *co–entropy* of the partition π [3, 2, 4, 1, 6].

In the partition context the entropy (resp., co–entropy) behaves anti–monotonically (resp., monotonically) with respect to the previously described partial order on $\Pi(X)$. Formally:

$$\pi_1 \preceq \pi_2 \quad \text{implies} \quad H(\pi_2) \leq H(\pi_1) \quad \text{and} \quad E(\pi_1) \leq E(\pi_2). \tag{4}$$

This is of great interest and importance in the context of rough sets when dealing with a complete information system and a partition induced by it through the equivalence relation of *indiscernibility* [6]. Indeed, in a complete information system given two sets of attributes $\mathcal{B}, \mathcal{A} \subseteq Att$, $\mathcal{B} \subseteq \mathcal{A}$ implies $\pi(\mathcal{A}) \preceq \pi(\mathcal{B})$, and so the required and expected monotonic condition is satisfied. Thus, making use of (4) the further monotonicities are true: $\mathcal{B} \subseteq \mathcal{A}$ implies $H(\pi(\mathcal{B})) \leq H(\pi(\mathcal{A}))$ and $E(\pi(\mathcal{A})) \leq E(\pi(\mathcal{B}))$.

2   The Covering Case

In the concrete applications, the real information systems are often incomplete and this is one of the reasons for which it is important to study the covering case. As briefly illustrated in the introduction, given a family of attributes, through the similarity relation we obtain a covering $\gamma(\mathcal{A})$ of X constituted by the similarity classes $s_\mathcal{A}(x)$ generated by all the objects x of X. In the sequel we present a survey of the state of the art about orderings and quasi–orderings on coverings and about *non–pointwise* entropies of coverings, not making distinctions between coverings induced by an incomplete information system via a similarity relation, and generic coverings of a universe X, unless explicitly said. We will then illustrate an important result: how to compare two generic coverings which are in a certain order relation, by generating a partition from each of them and computing their entropies.

Orderings and Quasi–Orderings of Coverings. In the previous section we described some different formulations (1), (2) and (3) of the same partial order relation on $\Pi(X)$. The extension to coverings of these formulations generates two different quasi–orderings, i.e., reflexive and transitive, but in general non anti–symmetric relations. In the class $\Gamma(X)$ of all coverings of the universe X, (1) and (2) become respectively:

$$\gamma \preceq \delta \quad \text{iff} \quad \forall C_i \in \gamma \quad \exists D_j \in \delta : C_i \subseteq D_j \tag{5}$$

$$\gamma \ll \delta \quad \text{iff} \quad \forall D \in \delta \quad \exists \{C_1, C_2, \ldots, C_p\} \subseteq \gamma : D = C_1 \cup C_2 \cup \ldots \cup C_p \tag{6}$$

where γ and δ are coverings from $\Gamma(X)$. These are both quasi–order relations in $\Gamma(X)$. If we consider the class of *genuine* coverings $\Gamma_g(X)$ consisting of all those coverings which do not contain redundant elements (formally: for any $C_i, C_j \in \gamma$, $C_i \subseteq C_j$ implies $C_i = C_j$, i.e., for any set $C_i \in \gamma$ we have that there is no $C_j \in \gamma$ such that $C_i \subset C_j$ - see [3] for more details and examples) and in which there is no element equal to the whole X, then we have that (5) is a partial order relation.

In the coverings case, in general these two quasi–orderings are different and no general relationship exists between them (see example 5.1 in [3]), even when we consider the restricted class of genuine coverings. Let us now illustrate a binary relation introduced by Wierman (in the covering context in an unpublished work which he kindly sent to us):

$$\gamma \lhd_W \delta \quad \text{iff} \quad \forall C_i \in \gamma \;\; \forall D_j \in \delta \;\; C_i \cap D_j \neq \emptyset \quad \text{implies} \quad C_i \subseteq D_j \qquad (7)$$

This binary relation, that is the extension to coverings of (3), has the advantage of being anti–symmetric on the whole $\Gamma(X)$; but it presents the drawback (as explained by Wierman himself) that it is not reflexive in the covering context (it is easy to find examples). For this reason, in order to define an ordering on coverings, Wierman added the further condition $\gamma = \delta$ in the following way:

$$\gamma \leq_W \delta \quad \text{iff} \quad \gamma = \delta \quad \text{or} \quad \gamma \lhd_W \delta \qquad (8)$$

So, we can see that in the covering case it is difficult to maintain the three properties of reflexivity, transitivity and anti–symmetry at the same time unless one adds more conditions in the definition of the binary relation, or restricts the applicability on a subclass of coverings, such as the class of all genuine ones. Another advantage of (8) (as illustrated by Wierman himself) is that the pair $(\Gamma(X), \leq_W)$ is a poset lower bounded by the discrete partition $\pi_d = \{\{x_1\}, \{x_2\}, \ldots, \{x_{m(X)}\}\}$, which is the least element, and upper bounded by the trivial partition $\pi_t = X$ which is the greatest element. Moreover, it is a lattice.

Let us now illustrate how one can extract a partition $\pi(\gamma)$ from a covering γ. We thought of a method consisting in two main steps (see [5]): first we create the covering completion γ_c, which consists of all the sets C_i of γ and of all the corresponding complements C_i^c; then, for each $x \in X$, we generate the granule $gr(x) = \bigcap(C \in \gamma_c : x \in C)$. The collection of all granules $gr(x)$ is a partition. The Wierman approach to generate a partition from a covering, presents a different formulation, i.e., $gr(x) = \bigcap_{x \in C} C \setminus \bigcup_{x \notin C} C$, which is equal to the just introduced granule. The following proposition is easy to prove:

Proposition 1. *For any two coverings γ and δ the following holds:*

$$\gamma \leq_W \delta \quad \text{implies} \quad \pi(\gamma) \preceq \pi(\delta)$$

This property, not described by Wierman, is important because it allows us to compare two coverings through the entropies of their induced partitions, which behave anti–monotonically with respect to the standard order relation (1) on $\Pi(X)$. Hence, the following can be proved:

Proposition 2. *For any two coverings γ and δ the following holds:*

$$\gamma \leq_W \delta \quad implies \quad H(\pi(\delta)) \leq H(\pi(\gamma)) \quad and \quad E(\pi(\gamma)) \leq E(\pi(\delta))$$

As we can observe, by simply expressing in a third way the same order relation on partitions, we have obtained an *order relation* on coverings with respect to which the entropy (resp. co–entropy) of the partition induced by a covering behaves anti–monotonically (resp. monotonically).

Entropies and Co–Entropies for Coverings. In [3] we introduced some definitions of entropies with corresponding co–entropies for coverings, whose restrictions to partitions induce the standard entropy and co–entropy. For lack of space, we will here only recall the definitions of the entropies. More details and considerations will be soon illustrated in a further paper.

Let us consider a covering $\gamma = \{B_1, B_2, \ldots, B_N\}$ of the universe X. We started from an entropy based on a probability distribution in which the probability of an elementary event was represented by $p(B_i) = \frac{\eta(B_i)}{m(X)}$, where $\eta(B_i) = \sum_{x \in X} \frac{1}{\sum_{i=1}^{N} \chi_{B_i}(x)} \chi_{B_i}(x)$ ($\chi(B_i)$ being the characteristic functional of the set B_i for any point $x \in X$) (see also [3, 1, 4, 6]). The resulting entropy is then:

$$H(\gamma) = - \sum_{i=1}^{N} p(B_i) \log p(B_i) \tag{9}$$

We then described a second approach to entropy and co–entropy for coverings. We defined the *total outer measure* of X induced from γ as $m^*(\gamma) := \sum_{i=1}^{N} m(B_i) \geq m(X) > 0$. We illustrated an alternative probability of occurrence of the elementary event B_i from the covering γ as $p^*(B_i) := \frac{m(B_i)}{m^*(\gamma)}$ obtaining that the vector $\boldsymbol{p^*}(\gamma) := (p^*(B_1), p^*(B_2), \ldots, p^*(B_N))$ is a probability distribution since trivially: (1) every $p^*(B_i) \geq 0$; (2) $\sum_{i=1}^{N} p^*(B_i) = 1$. Hence we defined a second entropy of a covering as

$$H^*(\gamma) = \log m^*(\gamma) - \frac{1}{m^*(\gamma)} \sum_{i=1}^{N} m(B_i) \log m(B_i) \tag{10}$$

In the same work we illustrated a third approach to entropy and co–entropy for coverings starting from the probability of the elementary event B_i defined as $p_{LX}(B_i) := \frac{m(B_i)}{m(X)}$ (see [3, 1]). In this case we observed that the probability vector $\boldsymbol{p_{LX}}(\gamma) := (p_{LX}(B_1), p_{LX}(B_2), \ldots, p_{LX}(B_N))$ does *not* define a probability distribution since in general $\sum_{i=1}^{N} p_{LX}(B_i) \geq 1$. Keeping in mind this characteristic, we defined the following *pseudo–entropy* (originally denoted by $H_{LX}^{(g)}$):

$$H^{(g)}(\gamma) = m^*(\gamma) \frac{\log m(X)}{m(X)} - \frac{1}{m(X)} \sum_{i=1}^{N} m(B_i) \log m(B_i) \tag{11}$$

Anti–Monotonic Behavior. As for the behavior of the here illustrated co–entropies with respect to the quasi–orderings for coverings (5) and (6), the reader can find a deep analysis and various examples in [3, 1]. We here only recall that the entropies H, H^* and $H^{(g)}$ for coverings behave neither monotonically nor anti–monotonically with respect to these two quasi–order relations, even in the more favorable context of *genuine* coverings.

For what concerns the order relation $\leq_W$ on $\Gamma(X)$ described in (8), we have found examples in which the entropy H shows both behaviors, i.e., monotonic in some cases and anti–monotonic in others, and so it does not present a general behavior. We could find counterexamples to the behavior of anti–monotonicity of the two entropies (10) and (11) with respect to the partial order relation (8) in which the finest of the two coverings was a partition (for instance, the induced partition).

The Case of Incomplete Information Systems. Let us now see what happens considering the partial order relation $\leq_W$ (8) in the case of coverings induced from an incomplete information system. Unfortunately, it is easy to find examples in which two coverings, $\gamma(\mathcal{A})$ and $\gamma(\mathcal{B})$, obtained via similarity relation depending on two subsets of attributes, $\mathcal{A}$ and $\mathcal{B}$, with $\mathcal{B} \subseteq \mathcal{A}$, are not in $\leq_W$ relation. This means that given two families of attribute sets $\mathcal{A}$ and $\mathcal{B}$, condition $\mathcal{B} \subseteq \mathcal{A}$ in general does not imply $\gamma(\mathcal{A}) \leq_W \gamma(\mathcal{B})$. Let us recall that with respect to the quasi–ordering (5) we always have that $\mathcal{B} \subseteq \mathcal{A}$ implies $\gamma(\mathcal{A}) \preceq \gamma(\mathcal{B})$. On the coverings generated from an incomplete information system we can use the following *pointwise* binary relation [10]; let us consider $\mathcal{A}, \mathcal{B} \subseteq Att$, we define:

$$\gamma(\mathcal{A}) \leq_s \gamma(\mathcal{B}) \quad \text{iff} \quad \forall x \in X \quad s_{\mathcal{A}}(x) \subseteq s_{\mathcal{B}}(x) \tag{12}$$

This is a partial order relation, and we have that $\mathcal{B} \subseteq \mathcal{A}$ always implies that $\gamma(\mathcal{A}) \leq_s \gamma(\mathcal{B})$. On the other hand, in this case we have that in general $\gamma(\mathcal{A}) \leq_s \gamma(\mathcal{B})$ does not imply $\pi(\gamma(\mathcal{A})) \preceq \pi(\gamma(\mathcal{B}))$. The same happens when considering the quasi–orderings (5), (6): for instance, either $\gamma(\mathcal{A}) \preceq \gamma(\mathcal{B})$ or $\gamma(\mathcal{A}) \ll \gamma(\mathcal{B})$ in general do not imply $\pi(\gamma(\mathcal{A})) \preceq \pi(\gamma(\mathcal{B}))$.

On the other side, for a covering induced by a similarity relation from an incomplete information system there exist some *pointwise* entropies as: $H_{LSLW} = -\sum_{x \in X} \frac{1}{m(X)} \log \frac{m(s_{\mathcal{A}}(x))}{m(X)}$ [7], or also $H_{LX} = -\sum_{x \in X} \frac{m(s_{\mathcal{A}}(x))}{m(X)} \log \frac{m(s_{\mathcal{A}}(x))}{m(X)}$ [10, 3], in which each class generated by each $x \in X$ is considered in the computation, even in the case that some of the classes are the same. Moreover, we also have an entropy for incomplete information systems based on pseudo–probability partitions (see [2] for details). All these entropies behave monotonically with respect to the set inclusion of subfamilies of attributes of an incomplete information system. The drawback is that they are *not directly* applicable to generic coverings, i.e., to those coverings not necessarily generated from an incomplete information system through a similarity relation.

3   Conclusions and Open Problems

We have analyzed some orderings and quasi–orderings on coverings resulting as extensions to the covering case of different (but equivalent) formulations of the standard order relation on $\Pi(X)$. In particular, we have analyzed the partial order relation $\leq_W$ of equation (8) and we have shown that if a pair of coverings is in this relation, the pair of the induced partitions is in the partial order relation on partitions $\preceq$ of equation (1); hence we can compare the two coverings through the entropies of their induced partitions. This result is important for the context of generic coverings, since at the moment there is no non–pointwise entropy which allows a comparison between coverings.

We have observed that in the case of coverings induced by a similarity relation on an incomplete information system, the set inclusion of subfamilies of attributes implies the quasi–order relation $\preceq$ of equation (5) (or better the order relation $\leq_s$ of equation (12)), but in general it does not imply the order relation $\leq_W$ of equation (8). This means that it is not possible in general to compare the resulting coverings through the entropies of the partitions generated by the coverings.

Then we analyzed some *non–pointwise* entropies of coverings defined in previous works and we have observed that in general they all seem to behave neither monotonically, nor anti-monotonically with respect to the partial order relation $\leq_W$ of equation (8).

We have to stress that the only example of monotonic behavior for the entropies H^* of equation (10) and $H^{(g)}$ of equation (11) concerns the comparison of a covering with its induced partition.

For what concerns the case of incomplete information systems, there exist some entropies (one based on pseudo–probability partitions, and two pointwise entropies) which behave anti–monotonically with respect to some orderings (for instance, $\leq_s$) and quasi–orderings ($\preceq$ and $\ll$). Although they behave as expected, in our opinion the pointwise entropies and co–entropies present a drawback: for their computation it is necessary to consider all the similarity classes generated by all the objects $x \in X$, and this can be seen as an excess of information, apart from a possible problem of computational complexity. The entropy and co–entropy based on the approach of pseudo–probability partitions (see [2]) do not have this drawback.

References

[1] Bianucci, D., Cattaneo, G.: Monotonic Behavior of Entropies and Co–entropies for Coverings with Respect to Different Quasi–orderings. In: Kryszkiewicz, M., Peters, J.F., Rybinski, H., Skowron, A. (eds.) RSEISP 2007. LNCS (LNAI), vol. 4585, pp. 584–593. Springer, Heidelberg (2007)
[2] Bianucci, D., Cattaneo, G., Ciucci, D.: Entropies and Co–entropies for Incomplete Information Systems. In: Yao, J., Lingras, P., Wu, W.-Z., Szczuka, M., Cercone, N.J., Ślęzak, D. (eds.) RSKT 2007. LNCS (LNAI), vol. 4481, pp. 84–92. Springer, Heidelberg (2007)

[3] Bianucci, D., Cattaneo, G., Ciucci, D.: Entropies and co–entropies of coverings with application to incomplete information systems. Fundam. Inform. 75, 77–105 (2007)

[4] Cattaneo, G., Ciucci, D., Bianucci, D.: Information Entropy and Co–entropy of Crisp and Fuzzy Granulations. In: Masulli, F., Mitra, S., Pasi, G. (eds.) WILF 2007. LNCS (LNAI), vol. 4578, pp. 9–19. Springer, Heidelberg (2007)

[5] Bianucci, D.: Rough entropies for complete and incomplete information systems, Ph.D. thesis (2007)

[6] Cattaneo, G., Ciucci, D., Bianucci, D.: Entropy and co-entropy of partitions and coverings with applications to roughness theory, Granular Computing: at the Junction of Fuzzy Sets and Rough Sets. In: Bello, R., Falcon, R., Pedrycz, W., Kacprzyk, J. (eds.) Studies in Fuzziness and Soft Computing, vol. 224, pp. 55–77. Springer, Heidelberg (2008)

[7] Liang, J.Y., Shi, Z.Z., Li, D.Y., Wierman, M.J.: Information entropy, rough entropy and knowledge granulation in incomplete information systems. Int. J. Gen. Sys. 35(6), 641–654 (2006)

[8] Komorowski, J., Pawlak, Z., Polkowski, L., Skowron, A.: Rough sets: A tutorial. In: Pal, S., Skowron, A. (eds.) Rough Fuzzy Hybridization, pp. 3–98. Springer, Singapore (1999)

[9] Kryszkiewicz, M.: Rough set approach to incomplete information systems. Inf. Sci. 112, 39–49 (1998)

[10] Liang, J., Xu, Z.: Uncertainty measure of randomness of knowledge and rough sets in incomplete information systems, Intelligent Control and Automata. In: Proc. of the 3rd World Congress on Intelligent Control and Automata, vol. 4, pp. 526–2529 (2000)

[11] Pawlak, Z.: Information systems - theoretical foundations. Inf. Sys. 6, 205–218 (1981)

[12] Pawlak, Z.: Rough sets. Int. J. Inform. Comput. Sci. 11, 341–356 (1982)

[13] Pawlak, Z.: Rough sets: Theoretical aspects of reasoning about data. Kluwer Academic Publishers, Dordrecht (1991)

[14] Polkowski, L., Skowron, A. (eds.): Rough sets in knowledge discovery 1. Physica–Verlag, Heidelberg, New York (1998)

[15] Shannon, C.E.: A mathematical theory of communication. Bell Sys. Tech. Journal 27, 379–423, 623–656 (1948)

[16] Taylor, A.E.: General theory of functions and integration. Dover Publications, New York (1985)

[17] Vakarelov, D.: A modal logic for similarity relations in Pawlak knowledge representation systems. Fundam. Inform. XV, 61–79 (1991)

A Family of Fuzzy Description Logics with Comparison Expressions*

Jianjiang Lu[1], Dazhou Kang[2], Yafei Zhang[1], Yanhui Li[2], and Bo Zhou[1]

[1] Institute of Command Automation, PLA University of Science and Technology,
Nanjing 210007, China
[2] School of Computer Science and Engineering, Southeast University,
Nanjing 210096, China

Abstract. The fuzzy knowledge plays an important role in many applications on the semantic web which faces imprecise and vague information. The current ontology languages on the semantic web use description logics as their logic foundation, which are insufficient to deal with fuzzy knowledge. Comparisons expressions between fuzzy membership degrees are frequently used in fuzzy knowledge systems. However, the current fuzzy extensions of description logics are not support the expression of such comparisons. This paper defines fuzzy comparison cuts to represent comparison expressions, extends fuzzy description logics by importing fuzzy comparison cuts and introducing new constructors. Furthermore, the reasoning algorithm is proposed. It enables representation and reasoning for fuzzy knowledge on the semantic web.

Keywords: Ontology, Distributed, Fuzzy, Description logic.

1 Introduction

Description logics (DLs) [1] are a family of knowledge representation languages widely used in the semantic web as a logic foundation for knowledge representation and reasoning. It is often necessary to represent fuzzy knowledge in real-life applications [2]. The fuzzy knowledge plays an important role in many domains which faces a huge amount of imprecise and vague knowledge and information, such as text mining, multimedia information system, medical informatics, machine learning, human natural language processing. However, classical DLs are insufficient to representing fuzzy knowledge [3]. Fuzzy DLs import the fuzzy set theory to enable the capability of dealing with fuzzy knowledge.

Many research work on fuzzy DLs have been carried out. Yen provided a fuzzy extension of DL $\mathcal{FL}^-$ [4]. Tresp presents a fuzzy extension of $\mathcal{ALC}$, $\mathcal{ALC}_{F_M}$ [3]. Straccia presented fuzzy $\mathcal{ALC}$ and an algorithm for assertional reasoning [8]. There are many extensions of fuzzy $\mathcal{ALC}$. Höldobler introduced the membership manipulator constructor to present $\mathcal{ALC}_{FH}$ [9]. Sanchez generalized the quantification in fuzzy $\mathcal{ALCQ}$ [10]. Stoilos provided pure ABoxes reasoning algorithms

* This work was supported in part by the National High Technology Research and Development Program of China (No. 2007AA01Z126, 863 Program).

G. Wang et al. (Eds.): RSKT 2008, LNAI 5009, pp. 395–402, 2008.

for the fuzzy extensions of $\mathcal{SHIN}$ [11]. There are several works using the idea of fuzzy cuts. Straccia transformed fuzzy $\mathcal{ALC}$ into classical $\mathcal{ALCH}$ [5]. Li presented a family of extended fuzzy DLs ($\mathcal{EF}$DLs) [6]. Calegari showed the fuzzy OWL [12] and Straccia presented a fuzzy $\mathcal{SHOIN}(D)$[13].

It is a familiar description that "Tom is taller than Mike," which can be seemed as a comparison between two fuzzy membership degrees. We call such descriptions *comparison expressions* on fuzzy membership degrees. However, the current fuzzy DLs do not support the expression of comparisons between fuzzy membership degrees. So it is necessary to extend fuzzy DLs with the ability of expressing comparison expressions.

This paper defines *fuzzy comparison cuts* (*cuts* for short) to represent comparison expressions on fuzzy membership degrees. The reasoning algorithm is proposed. It enables representation and reasoning for expressive fuzzy knowledge on the semantic web.

2   Fuzzy DLs with Comparison Expressions

2.1   Represent Comparison Expressions

For an individual a and a fuzzy concept C, let $a : C$ be the degree to which a is an instance of C. Similarly, $(a, b) : R$ is the degree to which two individuals a and b has a role R. In fuzzy DLs, the degrees can have their values in $[0, 1]$. We can show ranges of degrees in a set of *fuzzy assertions* of the form $\langle \alpha \bowtie n \rangle$, where α is a degree, $n \in [0, 1]$ is a constant and $\bowtie \in \{=, \neq, <, \leq, \geq, >\}$.

It is often necessary to compare the fuzzy membership degrees. There can be different kinds of comparisons between fuzzy membership degrees:

- A *numerical comparison* compares a degree to a constant. $Tom : Tall > 0.8$ means "Tom is quite tall." $Mike : Tall \leq 0.9$ means "Mike is not very tall."
- An *abstract comparison* compares degrees of the same individual. $Tom : Absolutist < Tom : Liberalist$ means "Tom prefers liberalism to absolutism."
- A *relative comparison* compares degrees between different individuals. $Tom : Tall < Mike : Tall$ means "Tom is taller than Mike."
- A *complex comparison* is constructed from the above kinds of simple comparisons. If for any person x such that $(Tom, x) : hasFriend > 0.9$, it holds $Tom : Tall > x : Tall$ or $Tom : Strong > x : Strong$, then we can say "No close friend (the degree of friendship is greater than 0.9) of Tom is taller and stronger than him."

Our idea is to define new elements to express the above kinds of comparisons, and integrate them into the current fuzzy DLs. We call the new elements *fuzzy comparison cuts*. In the fuzzy set theory[2], the cut sets are indeed classical sets, but facilitate a normative theory for formalizing fuzzy set theory. The idea of fuzzy cuts can also be used for fuzzy DLs. [5] use the idea of cut sets of fuzzy concepts to transform fuzzy DL $\mathcal{ALC}$ to classical DL $\mathcal{ALCH}$. [6] defined cuts of fuzzy concepts for more expressive ability.

2.2   Syntax and Semantics

The new languages with comparison cuts is called FCDLs. The syntax of FCDLs starts from three disjoint sets: N_I, N_C and N_R: N_I is a set of individual names, N_C is a set of fuzzy concept names, and N_R is a set of fuzzy role names. Complex fuzzy descriptions can be built from them inductively with fuzzy concept constructors and fuzzy role constructors.

Definition 1. *The set of* fuzzy role descriptions *(or* fuzzy roles *for short) is defined as: every fuzzy role name $R \in N_R$ is a fuzzy role; and for any fuzzy role R, R^- is also a fuzzy role (let $(R^-)^- := R$). For two fuzzy role R, S, $R \sqsubseteq S$ is called a* fuzzy role inclusion axiom. *A finite set of role inclusions is called a* role hierarchy. *For individual names $a, b \in N_I$ and a constant $n \in [0,1]$, $\langle (a,b) : R \bowtie n \rangle$ is called a* fuzzy role assertion.

If for any a, b, $(a,b) : R = n$ iff $(b,a) : S = n$, then S is an inverse role of R, written R^-. The role inclusion is transitive and $R \sqsubseteq S$ implies $R^- \sqsubseteq S^-$. For a role hierarchy $\mathcal{R}$, let $\sqsubseteq_\mathcal{R}$ be the transitive reflexive closure of $\sqsubseteq$ on $\mathcal{R} \cup \{R^- \sqsubseteq S^- | R \sqsubseteq S \in \mathcal{R}\}$. Beside the definition, N_R consists both transitive and normal fuzzy role names $N_R = N_{R+} \cup N_{R^P}$, where $N_{R+} \cap N_{R^P} = \emptyset$. For a transitive fuzzy role R, if $(a,b) : R \geq n$ and $(b,c) : R \geq n$, then it must have $(a,c) : R \geq n$. Let $\mathrm{trans}(S, \mathcal{R})$ be true if for some R with $R = S$ or $R \equiv_\mathcal{R} S$ such that $R \in N_{R+}$ or $R^- \in N_{R+}$, where $R \equiv_\mathcal{R} S$ is an abbreviation for $R \sqsubseteq_\mathcal{R} S$ and $S \sqsubseteq_\mathcal{R} R$. A role R is *simple* w.r.t. $\mathcal{R}$ iff $\mathrm{trans}(S, \mathcal{R})$ is not true for all $S \sqsubseteq_\mathcal{R} R$. Simple roles is required in order to avoid undecidable logics [1].

Definition 2. *The set of* fuzzy concepts *is that*

1. *every fuzzy concept name $A \in N_C$ is a fuzzy concept, $\top$ and $\bot$ are fuzzy concepts,*
2. *if C, D are fuzzy concepts, $o \in N_I$ is an individual name, R is a fuzzy role, S is a simple fuzzy role, and $q \in \mathbb{N}$, then $\neg C, C \sqcap D, C \sqcup D, \forall R.C, \exists R.C, \geq qS.C, \leq qS.C, \{o\}$ are also fuzzy concepts,*
3. *if R is a fuzzy role, S is a simple fuzzy role, P is a cut, and $q \in \mathbb{N}$, then $\exists R.P, \forall R.P, \geq qS.P, \leq qS.P$ are also fuzzy concepts.*

For two fuzzy concept C, D, $C \sqsubseteq D$ is called a fuzzy concept inclusion. *For $a \in N_I$ and $n \in [0,1]$, $\langle a : C \bowtie n \rangle$ is called a* fuzzy concept assertion.

Definition 3. *The set of* fuzzy comparison cuts *(or* cuts *for short) is defined as: if C, D are fuzzy concepts, $n \in [0,1]$ and $\bowtie \in \{=, \neq, >, \geq, <, \leq\}$, then $[C \bowtie n]$, $[C \bowtie D]$ and $[C \bowtie D^\uparrow]$ are cuts (and $[C \bowtie]$ is an abbreviation of $[C \bowtie C^\uparrow]$); if P, Q are cuts, then $\neg P$, $P \sqcap Q$ and $P \sqcup Q$ are also cuts. For any cut P and $a \in N_I$, $P(a)$ is called an* absolute cut. *If a cut P contains no $\uparrow$, then P itself is an absolute cut, and we do not distinguish P and $P(a)$ for any a. For two absolute cuts P, Q, $P \sqsubseteq Q$ is called a* cut inclusion. *For an absolute cut $P(b)$ and $a \in N_I$, $\langle a : P(b) \rangle$ is called a* cut assertion.

Definition 4. *A fuzzy interpretation* $\mathcal{I} = \langle \Delta^{\mathcal{I}}, \cdot^{\mathcal{I}} \rangle$ *consists a nonempty set* $\Delta^{\mathcal{I}}$ *as its domain, and a function* $\cdot^{\mathcal{I}}$ *maps every individual* $a \in N_I$ *to an element* $a^{\mathcal{I}} \in \Delta^{\mathcal{I}}$, *maps every fuzzy concept name* $A \in N_C$ *to a function* $A^{\mathcal{I}} : \Delta^{\mathcal{I}} \to [0,1]$, *and maps every fuzzy role name* $R \in N_R$ *to a function* $R^{\mathcal{I}} : \Delta^{\mathcal{I}} \times \Delta^{\mathcal{I}} \to [0,1]$.

The interpretation function is also extended to complex descriptions. It maps every fuzzy concept C *to a function* $C^{\mathcal{I}} : \Delta^{\mathcal{I}} \to [0,1]$, *maps every fuzzy role* R *to a function* $R^{\mathcal{I}} : \Delta^{\mathcal{I}} \times \Delta^{\mathcal{I}} \to [0,1]$, *maps every cut* P *to a function* $P^{\mathcal{I}} : \Delta^{\mathcal{I}} \to 2^{\Delta^{\mathcal{I}}}$, *and maps every absolute cut* $P(a)$ *to a set* $P^{\mathcal{I}}(a^{\mathcal{I}}) \subseteq \Delta^{\mathcal{I}}$.

The syntax and semantics of FCDLs are showed in Table 1. Table 1 does not list all available constructors, but only selects the most common ones.

From the semantics, it is clear that the interpretation of $[C \bowtie n]^{\mathcal{I}}(s)$ and $[C \bowtie D]^{\mathcal{I}}(s)$ do not depend on s. For any cut P and individual name a, $P(a)$ is an absolute cut, and $(P(a))^{\mathcal{I}} = P^{\mathcal{I}}(a^{\mathcal{I}})$. If a cut P contains no $\uparrow$, then $P^{\mathcal{I}}(s)$ is independent of s. So P itself is an absolute cut.

With the cuts and new constructors, FCDLs are more expressive than the current fuzzy DLs. FCDLs support all kinds of comparison expressions. They enable representation of expressive fuzzy knowledge on the semantic web.

2.3 Knowledge Bases and Reasoning

Definition 5. *A* knowledge base (KB) *of FCDLs is consists of an ABox, a TBox and a RBox:*

An ABox *is a finite set of concept assertions of the form* $\langle a : C \bowtie n \rangle$, *role assertions of the form* $\langle (a,b) : R \bowtie n \rangle$, *and cut asserions of the form* $\langle a : P(b) \rangle$. *An interpretation* $\mathcal{I}$ *satisfies an ABox* $\mathcal{A}$ *iff* $\mathcal{I}$ *satisfies each assertion in* $\mathcal{A}$; *such* $\mathcal{I}$ *is called a model of* $\mathcal{A}$. $\mathcal{I}$ *satisfies* $\langle a : C \bowtie n \rangle$ *iff* $C^{\mathcal{I}}(a^{\mathcal{I}}) \bowtie n$, $\mathcal{I}$ *satisfies* $\langle (a,b) : R \bowtie n \rangle$ *iff* $R^{\mathcal{I}}(a^{\mathcal{I}}, b^{\mathcal{I}}) \bowtie n$, $\mathcal{I}$ *satisfies* $\langle a : P(b) \rangle$ *iff* $a^{\mathcal{I}} \in P^{\mathcal{I}}(b^{\mathcal{I}})$.

An TBox *is a finite set of concept inclusions of the form* $C \sqsubseteq D$, *and cut inclusions of the form* $P \sqsubseteq Q$. $\mathcal{I}$ *satisfies a TBox* $\mathcal{T}$ *iff* $\mathcal{I}$ *satisfies each inclusion in* $\mathcal{T}$; *such* $\mathcal{I}$ *is called a model of* $\mathcal{T}$. $\mathcal{I}$ *satisfies* $C \sqsubseteq D$ *iff for any* $s \in \Delta^{\mathcal{I}}$, $C^{\mathcal{I}}(s) \leq D^{\mathcal{I}}(s)$, $\mathcal{I}$ *satisfies* $P \sqsubseteq Q$ *iff* $P^{\mathcal{I}} \subseteq Q^{\mathcal{I}}(s)$.

An RBox *(or called role hierarchy) is a finite set of role inclusions of the form* $R \sqsubseteq S$. $\mathcal{I}$ *satisfies an RBox* $\mathcal{R}$ *iff* $R^{\mathcal{I}}(s,t) \leq S^{\mathcal{I}}(s,t)$ *for each* $R \sqsubseteq S \in \mathcal{R}$; *such* $\mathcal{I}$ *is called a model of* $\mathcal{R}$.

For a knowledge base $\mathcal{K} = \langle \mathcal{A}, \mathcal{T}, \mathcal{R} \rangle$, *if* $\mathcal{I}$ *is a model of* $\mathcal{T}$, $\mathcal{R}$ *and* $\mathcal{A}$, *then* $\mathcal{I}$ *is called a model of* $\mathcal{K}$

Definition 6. *The basic inference problems of FCDLs include*

- Satisfiability of concepts: *a concept* C *is satisfiable w.r.t. a TBox* $\mathcal{T}$ *and a RBox* $\mathcal{R}$ *to a given degree* n, *iff there exists a model* $\mathcal{I}$ *of* $\mathcal{T}$ *and* $\mathcal{R}$ *with* $\exists s \in \Delta^{\mathcal{I}}, C^{\mathcal{I}}(s) \geq n$.
- Consistency of ABoxes: *an ABox* $\mathcal{A}$ *is consistent w.r.t.* $\mathcal{T}$ *and* $\mathcal{R}$, *iff there exists a model of* $\mathcal{T}$, $\mathcal{R}$ *and* $\mathcal{A}$.

$\mathcal{ALC}_{fc}$ is the most basic FCDL. For any $\mathcal{ALC}_{fc}$-role R, $R \in N_{R^P}$; and $\mathcal{ALC}_{fc}$-concepts are $C, D ::= A|\top|\bot|\neg C|C \sqcap D|C \sqcup D|\exists R.C|\forall R.C|\exists R.P|\forall R.P$. The $\mathcal{ALC}_{fc}$-cuts are $P, Q ::= [C \bowtie n]|[C \bowtie D]|[C \bowtie D^{\uparrow}]|\neg P|P \sqcap Q|P \sqcup Q$.

Table 1. Syntax and semantics of FCDLs

Syntax	Semantics	Symbol
R	$R^{\mathcal{I}}(s,t) \in [0,1]$	
R^-	$(R^-)^{\mathcal{I}}(s,t) = R^{\mathcal{I}}(t,s)$	$\mathcal{I}$
$R \in N_{R^+}$	$\sup_{t_i \in \Delta^{\mathcal{I}}}\{R^{\mathcal{I}}(s_1,t_i) \wedge R^{\mathcal{I}}(t_i,s_2)\} \le R^{\mathcal{I}}(s_1,s_2)$	$\mathcal{R}^+(\mathcal{S})$
A	$A^{\mathcal{I}}(s) \in [0,1]$	
$\top$	$\top^{\mathcal{I}}(s) = 1$	
$\bot$	$\bot^{\mathcal{I}}(s) = 0$	
$\neg A$	$(\neg A)^{\mathcal{I}}(s) = 1 - A^{\mathcal{I}}(s)$	$\mathcal{AL}(\mathcal{S})$
$C \sqcap D$	$(C \sqcap D)^{\mathcal{I}}(s) = \min(C^{\mathcal{I}}(s), D^{\mathcal{I}}(s))$	$\mathcal{AL}(\mathcal{S})$
$C \sqcup D$	$(C \sqcup D)^{\mathcal{I}}(s) = \max(C^{\mathcal{I}}(s), D^{\mathcal{I}}(s))$	$\mathcal{U}(\mathcal{S})$
$\neg C$	$(\neg C)^{\mathcal{I}}(s) = 1 - C^{\mathcal{I}}(s)$	$\mathcal{C}(\mathcal{S})$
$\forall R.C$	$(\forall R.C)^{\mathcal{I}}(s) = \inf_{t \in \Delta^{\mathcal{I}}}\{\max(1 - R^{\mathcal{I}}(s,t), C^{\mathcal{I}}(t))\}$	$\mathcal{AL}(\mathcal{S})$
$\exists R.C$	$(\exists R.C)^{\mathcal{I}}(s) = \sup_{t \in \Delta^{\mathcal{I}}}\{\min(R^{\mathcal{I}}(s,t), C^{\mathcal{I}}(t))\}$	$\mathcal{E}(\mathcal{S})$
$\ge qR$	$(\ge qR)^{\mathcal{I}}(s) = \sup_{t_1,\ldots,t_q \in \Delta^{\mathcal{I}}} \min_{i=1}^{q}\{R^{\mathcal{I}}(s,t_i)\}$	$\mathcal{N}$
$\le qR$	$(\le qR)^{\mathcal{I}}(s) = \inf_{t_1,\ldots,t_{q+1} \in \Delta^{\mathcal{I}}} \max_{i=1}^{q+1}\{1 - R^{\mathcal{I}}(s,t_i)\}$	
$\ge qR.C$	$(\ge qR.C)^{\mathcal{I}}(s) = \sup_{t_1,\ldots,t_q \in \Delta^{\mathcal{I}}} \min_{i=1}^{q}\{R^{\mathcal{I}}(s,t_i), C^{\mathcal{I}}(t_i)\}$	$\mathcal{Q}$
$\le qR.C$	$(\le qR.C)^{\mathcal{I}}(s) = \inf_{t_1,\ldots,t_{q+1} \in \Delta^{\mathcal{I}}} \max_{i=1}^{q+1}\{1 - R^{\mathcal{I}}(s,t_i), C^{\mathcal{I}}(t_i)\}$	
$\{o\}$	$\{o\}^{\mathcal{I}}(s) = \begin{cases} 1 \text{ if } s=o^{\mathcal{I}} \\ 0 \text{ if } s \neq o^{\mathcal{I}} \end{cases}$	$\mathcal{O}$
$\forall R.P$	$(\forall R.P)^{\mathcal{I}}(s) = \inf_{t \in P^{\mathcal{I}}(x)}\{1 - R^{\mathcal{I}}(s,t)\}$	$\mathcal{AL}(\mathcal{S})$
$\exists R.P$	$(\exists R.P)^{\mathcal{I}}(s) = \sup_{t \in P^{\mathcal{I}}(x)}\{R^{\mathcal{I}}(s,t)\}$	$\mathcal{E}(\mathcal{S})$
$\ge qR.P$	$(\ge qR.P)^{\mathcal{I}}(s) = \sup_{t_1,\ldots,t_q) \in P^{\mathcal{I}}(x)} \min_{i=1}^{q}\{R^{\mathcal{I}}(s,t_i)\}$	$\mathcal{Q}$
$\le qR.P$	$(\le qR.P)^{\mathcal{I}}(s) = \inf_{t_1,\ldots,t_{q+1} \in P^{\mathcal{I}}(x)} \max_{i=1}^{q+1}\{1 - R^{\mathcal{I}}(s,t_i)\}$	$\mathcal{Q}$
$[C \bowtie n]$	$[C \bowtie n]^{\mathcal{I}}(s) = \{t | C^{\mathcal{I}}(t) \bowtie n\}$	
$[C \bowtie D]$	$[C \bowtie D]^{\mathcal{I}}(s) = \{t | C^{\mathcal{I}}(t) \bowtie D^{\mathcal{I}}(t)\}$	
$[C \bowtie D^{\dagger}]$	$[C \bowtie D^{\dagger}]^{\mathcal{I}}(s) = \{t | C^{\mathcal{I}}(t) \bowtie D^{\mathcal{I}}(s)\}$	
$\neg P$	$(\neg P)^{\mathcal{I}}(s) = \Delta^{\mathcal{I}} \backslash P^{\mathcal{I}}(s)$	
$P \sqcap Q$	$(P \sqcap Q)^{\mathcal{I}}(s) = P^{\mathcal{I}}(s) \cap Q^{\mathcal{I}}(s)$	$\mathcal{AL}(\mathcal{S})$
$P \sqcup Q$	$(P \sqcup Q)^{\mathcal{I}}(s) = P^{\mathcal{I}}(s) \cup Q^{\mathcal{I}}(s)$	$\mathcal{U}(\mathcal{S})$
$R \sqsubseteq S$	$\forall s,t \in \Delta^{\mathcal{I}}, R^{\mathcal{I}}(s,t) \le S^{\mathcal{I}}(s,t)$	$\mathcal{H}$
$(a,b):R \bowtie n$	$R^{\mathcal{I}}(a^{\mathcal{I}},b^{\mathcal{I}}) \bowtie n$	
$C \sqsubseteq D$	$\forall s \in \Delta^{\mathcal{I}}, C^{\mathcal{I}}(s) \le D^{\mathcal{I}}(s)$	
$a:C \bowtie n$	$C^{\mathcal{I}}(a^{\mathcal{I}}) \bowtie n$	
$P \sqsubseteq Q$	$P^{\mathcal{I}} \subseteq Q^{\mathcal{I}}$ where P,Q are absolute cuts	
$a:P(b)$	$a^{\mathcal{I}} \in P^{\mathcal{I}}(b^{\mathcal{I}})$	

3   Reasoning Algorithm

Here presents an algorithm to decide the consistency for $\mathcal{ALC}_{fc}$ ABoxes by constructing completion graphs.

Definition 7. *A completion graph is* $T = \langle S, E, L, \delta \rangle$, *where S is a set of nodes in the graph. E is a set of edges (pairs of nodes) in the graph. L is a function:*

R1	if	$\neg C \in L(x)$, and not $C(x) =_\delta 1 - (\neg C)(x)$
		then $L(x) \to L(x) \cup \{C\}$, and $C(x) =_\delta 1 - (\neg C)(x)$
R2	if	$C \sqcap D \in L(x)$, and not $\min(C(x), D(x)) =_\delta (C \sqcap D)(x)$
		then $L(x) \to L(x) \cup \{C, D\}$, and $\min(C(x), D(x)) =_\delta (C \sqcap D)(x)$
R3	if	$C \sqcup D \in L(x)$, and not $(C \sqcup D)(x) =_\delta 1 - (\neg C \sqcap \neg D)(x)$
		then $L(x) \to L(x) \cup \{C \sqcup D\}$, and $(C \sqcup D)(x) =_\delta 1 - (\neg C \sqcap \neg D)(x)$
R4	if	$\exists R.C \in L(x)$, and there is y with $R \in L(x, y)$
		but not $X \leq_\delta (\exists R.C)(x)$ for some $X \in \{R(x,y), C(x)\}$
		then $L(y) \to L(y) \cup \{C\}$, and $X \leq_\delta (\exists R.C)(x)$
R5	if	$\exists R.C \in L(x)$, and there is no y with $X =_\delta (\exists R.C)(x)$
		or $X <_\delta (\exists R.C)(x)$, for some $X \in \{R(x,y), C(x)\}$
		then add a new node y with $L(x,y) = \{R\}, L(y) = \{C\}$,
		and $X =_\delta (\exists R.C)(x)$ or $X <_\delta (\exists R.C)(x)$
R6	if	$\forall R.C \in L(x)$, and not $(\forall R.C)(x)_\delta = 1 - (\exists R.\neg C)(x)$
		then $L(x) \to L(x) \cup \{\exists R.\neg C\}$, and $(\forall R.C)(x) =_\delta 1 - (\exists R.\neg C)(x)$
R7	if	$\exists R.P \in L(x)$, and there is y with $R \in L(x, y)$
		but not $R(x,y)_\delta \leq (\exists R.C)(x)$ nor $\neg P(x)_\delta \in L(y)$
		then $R(x,y) \leq_\delta (\exists R.C)(x)$, or $L(y) \to L(y) \cup \{\neg P(x)\}$
R8	if	$\exists R.P \in L(x)$, and there is no y with $P(x) \in L(y)$,
		$R(x,y) =_\delta (\exists R.C)(x)$ or $R(x,y) <_\delta (\exists R.C)(x)$
		then add a new node y with $L(x,y) = \{R\}, L(y) = \{P(x)\}$,
		and $R(x,y) =_\delta (\exists R.C)(x)$ or $R(x,y) <_\delta (\exists R.C)(x)$
R9	if	$\forall R.P \in L(x)$, and not $(\forall R.P)(x) =_\delta 1 - (\exists R.\neg P)(x)$
		then $L(x) \to L(x) \cup \{\exists R.\neg P\}$, and $(\forall R.P)(x) =_\delta 1 - (\exists R.\neg P)(x)$
R10	if	$[C \bowtie n] \in L(x)$, and not $C(x) \bowtie_\delta n$
		then $L(x) \to L(x) \cup \{C\}$, and $C(x) \bowtie_\delta n$
R11	if	$[C \bowtie D] \in L(x)$, and not $C(x) \bowtie_\delta D(x)$
		then $L(x) \to L(x) \cup \{C, D\}$, and $C(x) \bowtie_\delta D(x)$
R12	if	$[C \bowtie D^\dagger](y) \in L(x)$, and not $C(x) \bowtie_\delta D(y)$
		then $L(x) \to L(x) \cup \{C\}$, $L(y) \to L(y) \cup \{D\}$, and $C(x) \bowtie_\delta D(y)$
R13	if	$(P \sqcap Q)(y) \in L(x)$, and not $\{P(y), Q(y)\} \subseteq L(x)$
		then $L(x) \to L(x) \cup \{P(y), Q(y)\}$
R14	if	$(P \sqcup Q)(y) \in L(x)$, and $\{P(y), Q(y)\} \cap L(x) = \emptyset$
		then $L(x) \to L(x) \cup \{X\}$ for some $X \in \{P(y), Q(y)\}$
R15	if	$C \sqsubseteq D \in \mathcal{T}$, and there is x with no $C(x) \leq_\delta D(x)$
		then $L(x) \to L(x) \cup \{C, D\}$, and $C(x) \leq_\delta D(x)$
R16	if	$C \sqsubseteq D \in \mathcal{T}$, and there is x with no $C(x) <_\delta D(x)$
		then $L(x) \to L(x) \cup \{C, D\}$, and $C(x) <_\delta D(x)$
R17	if	$C \in L(x)$ or $R \in L(x,y)$, and let $X = C(x)$ or $R(x,y)$
		there is no i such that $v_i <_\delta X <_\delta v_{i+1}$, or $X =_\delta v_i$
		then $v_i <_\delta X <_\delta v_{i+1}$ for some v_i, v_{i+1}, or $X =_\delta v_i$ for some v_i

Fig. 1. Expansion rules for $\mathcal{ALC}_{fc}$

for every node $x \in S$, $L(x)$ is a set of concepts or absolute cuts; for every edge $(x, y) \in E$, $L(x, y)$ is a set of roles. δ is a set of formulas of the form $X \leq Y$, $X \neq Y$ or $X \lessdot Y$, where $X, Y ::= n|C(x)|R(x, y)|1 - X$ such that $n \in [0, 1]$, C is a concept, R is a role, $x, y \in S$, and for any X, $1 - (1 - X) = X$.

The completion graph T of an ABox $\mathcal{A}$ w.r.t. a TBox $\mathcal{T}$ initializes with: $S = \{a \in N_I | a$ occurs in $\mathcal{A}\}$; for any $\langle a : P(b) \rangle \in \mathcal{A}$, $P(b) \in L(a)$; for any $\langle (a, b) : R \bowtie n \rangle \in \mathcal{A}$, $R \in L(a, b)$ and $R \bowtie_\delta n$. Let $V_0 = \{v_1, v_2, \ldots, v_k\} = \{0, 1, 0.5\} \cup \{n \in [0, 1] | n$ or $1 - n$ occurs in $\mathcal{A}$ or $\mathcal{T}\}$, where $0 = v_1 < v_2 < \cdots < v_k = 1$. For any $v_i < v_j$, let $v_i <_\delta v_j$. Then the graph grows up by applying the *expansion rules* showed in Fig. 1. If a rule applied to x creates a new node y, then y is a *successor* of x. Let *descendant* be transitive closure of successor. Several abbreviations are defined below:

$$X \leq_\delta Y =_{def} X \leq Y \in \delta, \text{ or } X \leq_\delta Y, Y \leq_\delta Z, \text{ or } 1 - Y \leq_\delta 1 - X$$

$$\min(X, Y) =_\delta Z =_{def} Z \leq_\delta X, Z \leq_\delta Y, W \leq_\delta Z \text{ for some } W \in \{X, Y\};$$

$$X \neq_\delta Y =_{def} X \neq Y \in \delta; \qquad X \lessdot_\delta Y =_{def} X \lessdot Y \in \delta;$$

$$X \geq_\delta Y =_{def} Y \leq_\delta X; \qquad X =_\delta Y =_{def} X \leq_\delta Y, Y \leq_\delta X;$$

$$X <_\delta Y =_{def} X \leq_\delta Y, X \neq_\delta Y; \qquad X_\delta > Y =_{def} Y \leq_\delta X, X \neq_\delta Y.$$

The $\lessdot_\delta$ relation is used to simulate the infinite supreme. For any $a \in N_I$, $\mathrm{lev}(a) = 1$. If $\mathrm{lev}(x) = i$, y is a successor of x by updating $\lessdot_\delta$, then $\mathrm{lev}(y) = i+1$. For any X of the form $C(x)$, $1 - C(x)$, $R(y, x)$, or $1 - R(y, x)$, if $\mathrm{lev}(x) = i$, then $X \in V_i$. If $X \lessdot_\delta Y$ and $Y \in V_i$, then for any $Z \in V_j$ such that $j \leq i$, $Z <_\delta Y \rightarrow Z <_\delta X$ and $Z >_\delta X \rightarrow Z \geq_\delta Y$. So $X \lessdot_\delta Y$ means X is greater than any $Z < Y$ such that $Z \in V_0 \cup V_1 \cup \cdots \cup V_i$ and $Y \in V_i$. It ensures that for any given constant ε, we can assign values to the variables in V such that $X - Y < \varepsilon$ without inducing any conflict.

Since there are variables, the blocking condition in $\mathcal{ALC}_{fc}$ is different from classical DLs. It has to consider the comparisons between degrees. For any x, let $\delta(x) = \{X \bowtie Y | X \bowtie_\delta Y$, X, Y are of the form $C(x)$, $1 - C(x)$, or v_i. A node x is *blocked* by y, iff x is an descendant of y, and $\delta(x) = [x/y]\delta(y)$, where $[x/y]\delta(y)$ means to replace any y in $\delta(y)$ by x. Then we call y *blocks* x. When x is blocked, all descendants of x is also blocked. No rules in Fig. 1 can be applied to blocked nodes. T is said to contain a *clash* if $\{X \neq_\delta Y, X =_\delta Y\} \subseteq \delta$, or $X >_\delta 1$, or $X <_\delta 0$. T is said to be *clash-free* if it contains no clash. If none of the expansion rules can be applied to T, then T is said to be *complete*.

From the blocking condition and the number of concepts in any $L(x)$ is finite, the algorithm terminates. An $\mathcal{ALC}_{fc}$ ABox $\mathcal{A}$ is consistent w.r.t. TBox $\mathcal{T}$ iff a complete and clash-free completion graph can be constructed from $\mathcal{A}$ w.r.t. $\mathcal{T}$.

4 Conclusions

It is important to compare fuzzy membership degrees in representation of expressive fuzzy knowledge. This paper defines fuzzy comparison cuts to represent comparison expressions on fuzzy membership degrees, and extends fuzzy DLs by importing them and introducing new constructors. They enable representation and reasoning for expressive fuzzy knowledge on the semantic web. The future work is to design reasoning algorithms for more expressive FCDLs, implement reasoners for FCDLs and construct fuzzy knowledge systems based on FCDLs.

References

1. Baader, F., Calvanese, D., McGuinness, D., Nardi, D., Patel-Schneider, P.F. (eds.): The description logic handbook: theory, implementation, and applications. Cambridge University Press, Cambridge (2003)
2. Zadeh, L.A., Kacprzyk, J. (eds.): Fuzzy logic for the management of uncertainty. John Wiley & Sons, Inc., New York (1992)
3. Tresp, C.B., Molitor, R.: A description logic for vague knowledge. In: Proceedings of the 13th Biennial European Conference on Artificial Intelligence (ECAI 1998), Brighton, UK, pp. 361–365. J. Wiley and Sons, Chichester (1998)
4. Yen, J.: Generalizing term subsumption languages to fuzzy logic. In: Proceedings of the 12th International Joint Conference on Artificial Intelligence, Sidney, Australia, pp. 472–477 (1991)
5. Straccia, U.: Transforming fuzzy description logics into classical description logics. In: Alferes, J.J., Leite, J.A. (eds.) JELIA 2004. LNCS (LNAI), vol. 3229, pp. 385–399. Springer, Heidelberg (2004)
6. Li, Y., Xu, B., Lu, J., Kang, D., Wang, P.: A family of extended fuzzy description logics. In: Proceedings of the IEEE 29th Annual International Computer Software and Applications Conference, Edinburgh, Scotland, pp. 221–226 (2005)
7. Kang, D., Xu, B., Lu, J., Li, Y.: Reasoning for a fuzzy description logic with comparison expressions. In: The 2006 International Workshop on Description Logics (DL 2006) (2006)
8. Straccia, U.: Reasoning within fuzzy description logics. Journal of Artificial Intelligence Research 14, 137–166 (2001)
9. Hölldobler, S., Störr, H.P., Tran, D.K.: The fuzzy description logic alcfh with hedge algebras as concept modifiers. Journal of Advanced Computational Intelligence and Intelligent Informatics 7(3), 294–305 (2003)
10. Sánchez, D., Tettamanzi, A.G.: Generalizing quantification in fuzzy description logic. In: Proceedings of the 8th Fuzzy Days in Dortmund: Computational Intelligence, Theory and Applications. Advances in Soft Computing Series, Springer, Berlin (2004)
11. Stoilos, G., Stamou, G., Tzouvaras, V., Pan, J.Z., Horrocks, I.: The fuzzy description logic f-shin. In: Proceedings of the International Workshop on Uncertainty Reasoning for the Semantic Web, Galway, Ireland (2005)
12. Calegari, S., Ciucci, D.: Fuzzy ontology, fuzzy description logics and fuzzy-OWL. In: Masulli, F., Mitra, S., Pasi, G. (eds.) WILF 2007. LNCS (LNAI), vol. 4578, pp. 118–126. Springer, Heidelberg (2007)
13. Straccia, U.: A fuzzy description logic for the semantic web. In: Sanchez (ed.) Fuzzy Logic and the semantic web, Elsevier, Amsterdam (2006)

Soft Sets and Soft Modules

Qiu-Mei Sun, Zi-Long Zhang, and Jing Liu

College of Mathematics and Information Science,
Hebei Normal University, Shijiazhuang, Hebei, 050016, P.R. China
sunqiumei666@tom.com, zhaozhang1982@yahoo.com.cn

Abstract. Molodtsov introduced the concept of soft sets. Recently, Aktaş et al. generalized soft sets by defining the concept of soft groups. In this paper, we present the definition of soft modules and construct some basic properties using modules and Molodtsov's definition of soft sets.

Keywords: soft sets, soft modules, soft submodules, soft homomorphism.

1 Introduction

At present, the complexities of modeling uncertain data in economics, engineering, environmental science and many other fields cannot be dealt with by classical methods. Probability theory, fuzzy sets [9], rough sets [8], and other mathematical tools have their inherent difficulties. Consequently, Molodtsov [5] proposed a new completely approach, which is so-called soft set theory.

Now, work on the soft set theory is progressing rapidly. Maji et al. described the application of soft set theory [6] and have published a detailed theoretical study on soft sets [4]; Molodtsov [7] demonstrated lots of potential applications in many different fields; Chen et al. [3] present a new definition of soft set parametrization reduction; Aktaş et al. [1] introduced the basic properties of soft sets, and compared soft sets with the related concepts of fuzzy sets and rough sets. At the same time, Aktaş gave a definition of soft groups, and derived their basic properties, using Molodtsov's definition of the soft sets.

This paper begin with the basic concepts of soft set theory, and some simple theories of modules.

The main purpose of this paper is to introduce a basic version of soft module theory, which extends the notion of module by including some algebraic structures in soft sets. A soft module defined in this paper is actually a parameterized family of submodules, and has some properties similar to those of modules.

2 Soft Sets

First let us recall some basic concepts of soft set theory ([1], [4]). Throughout this section U refers to an initial universe, E is a set of parameters, $P(U)$ is the power set of U, and $A \subseteq E$.

Definition 1. *A pair (F, A) is called a soft set over U where F is a mapping given by $F : A \to P(U)$.*

G. Wang et al. (Eds.): RSKT 2008, LNAI 5009, pp. 403–409, 2008.

Definition 2. *For two soft sets (F, A) and (G, B) over U, (F, A) is called a soft subset of (G, B) if*
(1) *$A \subset B$ and*
(2) *$\forall \varepsilon \in A, F(\varepsilon)$ and $G(\varepsilon)$ are identical approximations.*

This relationship is denoted by $(F, A)\widetilde{\subset}(G, B)$. Similarly, (F, A) is called a soft superset of (G, B) if (G, B) is a soft subset of (F, A). This relationship is denoted by $(F, A)\widetilde{\supset}(G, B)$.

Tow soft sets (F, A) and (G, B) over U are called soft equal if (F, A) is a soft subset of (G, B), and (G, B) is a soft subset of (F, A).

Definition 3. *Let (F, A) and (G, B) be two soft sets over U*
(1) *The intersection of (F, A) and (G, B) is the soft set (H, C), where $C = A \cap B$ and $\forall \varepsilon \in C, H(\varepsilon) = F(\varepsilon)$ or $G(\varepsilon)$ (as both are same set). This relationship is denoted by $(F, A)\widetilde{\cap}(G, B) = (H, C)$.*
(2) *The union of (F, A) and (G, B) is the soft set (H, C), where $C = A \cup B$ and $\forall \varepsilon \in C$*

$$H(\varepsilon) = \begin{cases} F(\varepsilon) & \varepsilon \in A - B. \\ G(\varepsilon) & \varepsilon \in B - A. \\ F(\varepsilon) \cup G(\varepsilon) & \varepsilon \in A \cap B. \end{cases}$$

This relationship is denoted by $(F, A)\widetilde{\cup}(G, B) = (H, C)$.

Example 1. suppose that $U = \{h_1, h_2, h_3, h_4, h_5, h_6, h_7\}$, $A = \{expensive, middle, cheap\}$ and $B = \{beautiful, modern, cheap\}$. Let $F(expensive) = \{h_2, h_4\}$, $F(middle) = \{h_1, h_3, h_5\}, F(cheap) = \{h_6, h_7\}, G(beautiful) = \{h_2, h_3, h_4\}$, $G(modern) = \{h_1, h_5, h_6\}, G(cheap) = \{h_6, h_7\}$.Then we have:
$(F, A)\widetilde{\cap}(G, B) = (H, C), where C = A \cap B, H(cheap) = \{h_6, h_7\}$.
$(F, A)\widetilde{\cup}(G, B) = (H, C), where C = A \cup B$,
$H(expensive) = \{h_2, h_4\}, H(middle) = \{h_1, h_3, h_5\}, H(cheap) = \{h_6, h_7\}$,
$H(beautiful) = \{h_2, h_3, h_4\}, H(modern) = \{h_1, h_5, h_6\}$.

3   Module

In this section we introduce some theories of modules ([2], [10]).

Definition 4. *Let R be a ring with identity. M is said to be a left R-module if left scalar multiplication $\lambda : R \times M \to M$ via $(a, x) \mapsto ax$ satisfying the axioms $\forall r, r_1, r_2, 1 \in R; m, m_1, m_2 \in M$:*
(1) *M is an abelian group (which we shall write additively);*
(2) *$r(m_1 + m_2) = rm_1 + rm_2, (r_1 + r_2)m = r_1m + r_2m$;*
(3) *$(r_1r_2)m = r_1(r_2m)$;*
(4) *$1m = m$.*

Left R-module is denoted by $_R M$ or M for short. Similarly we can define right R-module and denote it by M_R.

For $_RM$ and M_S, M is said to be S-R-bimodule if the following conditions are satisfied :

$$s(mr) = (sm)r \qquad \forall\ s \in S, m \in M, r \in R.$$

Let M be a left R-module, then an abelian subgroup N of M is a left R-submodule of M in case it is closed under scalar multiplication by R. This relationship will be denoted by $N < M$.

Proposition 1. A is a subset of M, then the following statements are equivalent
   (1) $N < M$;
   (2) A is an abelian subgroup of M and $ar \in A$ for all $a \in A$, $r \in R$;
   (3) $a_1 + a_2 \in A$, $ar \in A$ for all a, a_1, $a_2 \in A$, $r \in R$.

Definition 5. N is a non-trivial submodule of M
   (1)if there is no submodule of M which contain N, we call N the maximal submodules.
   (2) if there is no non-zero submodule of M which contained in N, we call N the minimal submodules.

Definition 6. Let $\{M_i \mid i \in I\}$ be a nonempty family of submodules
   (1) if $\{M_i \mid i \in I\}$ is a family of maximal submodules, then $\bigcap\limits_{i \in I} M_i$ is a submodule of M called Jacobson radical of module. This is denoted by $radM$.
   (2) if $\{M_i \mid i \in I\}$ is a family of minimal submodules, then $\sum\limits_{i \in I} M_i$ is a submodule of M called socle of module. This is denoted by $socM$.

Definition 7. Let $\{M_i \mid i \in I\}$ be a nonempty family of R-modules, $P = \prod\limits_{i \in I} M_i = \{(x_i) \mid x_i \in M_i\}$ is a direct product set, if the operations on the product are given by

$$(x_i) + (y_i) = (x_i + y_i) \qquad r(x_i) = (rx_i).$$

then P do induce a left R-module structure called direct product of $\{M_i \mid i \in I\}$, which will be denoted by $\prod\limits_{i \in I} M_i$.

Proposition 2. Let $\{M_i \mid i \in I\}$ be a family of submodules of M, then $\bigcap\limits_{i \in I} M_i$ and $\sum\limits_{i \in I} M_i$ are all submodules of M .

Definition 8. All the elements (x_i) in the direct product $\prod\limits_{i \in I} M_i$, where x_i is zero for almost all $i \in I$ except finite one, establish a submodule of $\prod\limits_{i \in I} M_i$ which called direct sum of $\{M_i \mid i \in I\}$, will be denoted by $\coprod\limits_{i \in I} M_i$ or $\bigoplus\limits_{i \in I} M_i$.

Definition 9. The homomorphism sequence of R-modules $\cdots \to M_{n-1} \xrightarrow{f_{n-1}} M_n \xrightarrow{f_n} M_{n+1} \to \cdots$ is called exact sequence of modules if $Imf_{n-1} = Kerf_n$

for all $n \in N$, and we call the exact sequence of modules form as $0 \to M' \xrightarrow{f} M \xrightarrow{g} M'' \to 0$ the short exact sequence of modules.

4   Soft Modules

Throughout this section, let M be a left R-module, A be any nonempty set. $F : A \to P(M)$ refer to a set-valued function and the pair (F, A) is a soft set over M.

Definition 10. *Let (F, A) be a soft set over M. (F, A) is said to be a soft module over M if and only if $F(x) < M$ for all $x \in A$.*

Proposition 3. *Let (F, A) and (G, B) be two soft modules over M.*
(1) *$(F, A) \widetilde{\cap} (G, B)$ is a soft module over M.*
(2) *$(F, A) \widetilde{\cup} (G, B)$ is a soft module over M if $A \cap B = \emptyset$.*

Proof. (1) From Definition 3 we know that $(F, A) \widetilde{\cap} (G, B) = (H, C)$ is a soft set over M, where $C = A \cap B$ and $H(x) = F(x) < M$ or $H(x) = G(x) < M$ for all $x \in C$ since (F, A) and (G, B) are soft module over M, so $(F, A) \widetilde{\cap} (G, B)$ is a soft module over M.

(2) We know $(F, A) \widetilde{\cup} (G, B) = (H, C)$ is a soft set, where $C = A \cup B$ and

$$H(x) = \begin{cases} F(x) & x \in A - B. \\ G(x) & x \in B - A. \\ F(x) \cup G(x) & x \in A \cap B. \end{cases}$$

$x \in A - B$ or $x \in B - A$ as $A \cap B = \emptyset$, thus (H, C) is a soft module over M by (F, A) and (G, B) are soft modules.

Definition 11. *Let (F, A) and (G, B) be two soft modules over M. Then $(F, A) + (G, B)$ is defined as $(H, A \times B)$, where $H(x, y) = F(x) + G(y) \forall (x, y) \in A \times B$.*

Proposition 4. *Let (F, A) and (G, B) be two soft modules over M. Then $(F, A) + (G, B)$ is soft module over M.*

Proof. This is easily obtained by Proposition 2.

Definition 12. *Let (F, A) and (G, B) be two soft modules over M and N respectively. Then $(F, A) \times (G, B) = (H, A \times B)$ is defined as $H(x, y) = F(x) \times G(y)$ for all $(x, y) \in A \times B$.*

Proposition 5. *Let (F, A) and (G, B) be two soft modules over M and N respectively. Then $(F, A) \times (G, B)$ is soft module over $M \times N$.*

Proof. This is easily obtained by Definition 8.

Direct product and direct sum are the same when the dimension is finite, so "$\oplus$" can instead of "$\times$" in the above proposition.

Definition 13. *Let (F, A) and (G, B) be two soft modules over M. Then (G, B) is soft submodule of (F, A) if*
  (1) $B \subset A$ *and*
  (2) $G(x) < F(x), \forall\ x \in B$.
This is denoted by $(G, B)\widetilde{<}(F, A)$.

Proposition 6. *Let (F, A) and (G, B) be two soft modules over M. (G, B) is soft submodule of (F, A) if $G(x) \subseteq F(x), \forall\ x \in A$.*

Proof. straight forward.

Definition 14. *Let $E = \{e\}$, where e is unit of A. Then every soft module (F, A) over M at least have two soft submodules (F, A) and (F, E) called trivial soft submodule.*

Proposition 7. *Let (F, A) be a soft module over M, and $\{(G_i, B_i) \mid i \in I\}$ be a nonempty family of soft submodules of (F, A). Then*
  (1) $\sum\limits_{i \in I}(G_i, B_i)$ *is soft submodule of (F, A).*
  (2) $\bigcap\limits_{i \in I}(G_i, B_i)$ *is soft submodule of (F, A).*
  (3) $\bigcup\limits_{i \in I}(G_i, B_i)$ *is soft submodule of (F, A), if $B_i \cap B_j = \emptyset$ for all $i, j \in I$.*

Proof. straight forward.

Proposition 8. *Let (F, A) and (G, B) be two soft modules over M, and (G, B) be soft submodule of (F, A). If $f : M \to N$ is homomorphism of module, then $(f(F), A)$ and $(f(G), B)$ are all soft modules over N, and $(f(G), B)$ is soft submodule of $(f(F), A)$.*

Proof. This is easy to proof as the image of a homomorphism of submodule is a submodule.

Definition 15. *Let (F, A) and (G, B) be two soft modules over M and N respectively, $f : M \to N$, $g : A \to B$ be two functions. Then we say that (f, g) is a soft homomorphism if the following conditions are satisfied:*
  (1) $f : M \to N$ *is homomorphism of module;*
  (2) $g : A \to B$ *is a mapping;*
  (3) $f(F(x)) = G(g(x)), \forall\ x \in A$.

At the same time, we say (F, A) is soft homomorphic to (G, B), which denoted by $(F, A) \simeq (G, B)$.

In this definition, if f is an isomorphism from M to N and g is a one-to-one mapping from A onto B, then we say that (F, A) is a soft isomorphism and that (F, A) is a soft isomorphic to (G, B), this is denoted by $(F, A) \cong (G, B)$.

Definition 16. *Let (F, A) and (G, B) be two soft modules over M, and (G, B) be soft submodule of (F, A). We say (G, B) is maximal soft submodule of (F, A) if $G(x)$ is maximal submodule of $F(x)$ for all $x \in B$. We say (G, B) is minimal soft submodule of (F, A) if $G(x)$ is minimal submodule of $F(x)$ for all $x \in B$.*

Proposition 9. *Let (F, A) be a soft module over M*

(1) *if $\{(G_i, B_i) \mid i \in I\}$ is a nonempty family of maximal soft submodules of (F, A), then $\bigcap_{i \in I}(G_i, B_i)$ is maximal soft submodule of (F, A).*

(2) *if $\{(G_i, B_i) \mid i \in I\}$ is a nonempty family of minimal soft submodules of (F, A), then $\sum_{i \in I}(G_i, B_i)$ is minimal soft submodule of (F, A).*

Proof. It is easy to prove.

Definition 17. *Let (F, A) be a soft module over M, then*

(1) *(F, A) is said to be a null soft module over M if $F(x) = 0$ for all $x \in A$, where 0 is zero element of M.*

(2) *(F, A) is said to be an absolute soft module over M if $F(x) = M$ for all $x \in A$.*

Proposition 10. (1) *Let (F, A) be a soft module over M and $f : M \to N$ be a homomorphism. if $F(x) = Kerf$ for all $x \in A$, then $(f(F), A)$ is the null soft module over N.*

(2) *Let (F, A) be an absolute soft module over M and let $f : M \to N$ be an epimorphism, then $(f(F), A)$ is an absolute soft module over N.*

Proof. straight forward.

Proposition 11. *Let (F, A) be a null soft module over module P and (G, B) be an absolute soft module over module Q. if $0 \to P \xrightarrow{f} M \xrightarrow{g} Q \to 0$ is a short exact sequence, then $0 \to F(x) \xrightarrow{\tilde{f}} M \xrightarrow{\tilde{g}} G(y) \to 0$ is a short exact sequence for all $x \in A, y \in B$.*

Proof. $F(x) = 0$, $\forall\, x \in A$ since (F, A) is a null soft module over P, so $\tilde{f}$ is a monomorphism.

$G(y) = Q$, $\forall\, y \in B$ since (G, B) is an absolute soft module over Q. $g : M \to Q$ is an epimorphism as $0 \to P \xrightarrow{f} M \xrightarrow{g} Q \to 0$ is a short exact sequence, so $\tilde{g}$ is an epimorphism.

Proposition 12. *Let (F, A) be a null soft module over module P and (G, B) be an absolute soft module over module M. if $0 \to P \xrightarrow{f} M \xrightarrow{g} Q \to 0$ is a short exact sequence, then $0 \to f(F)(x) \xrightarrow{\tilde{f}} M \xrightarrow{\tilde{g}} g(G)(y) \to 0$ is a short exact sequence for all $x \in A$, $y \in B$.*

Proof. $F(x) = 0$, $\forall\, x \in A$ since (F, A) is a null soft module over P. $Kerf = 0$ so $Kerf = F(x)$, $\forall\, x \in A$, consequently $(f(F), A)$ is null soft module over M .

(G, B) is an absolute soft module over M and $g : M \to Q$ is an epimorphism, so $(g(G), B)$ is an absolute soft module over Q, thus $0 \to f(F)(x) \xrightarrow{\tilde{f}} M \xrightarrow{\tilde{g}} g(G)(y) \to 0$ is a short exact sequence for all $x \in A$, $y \in B$ by Proposition 11.

5   Conclusion

This paper summarized the basic concepts of soft sets and modules. Then studied the algebraic properties of soft sets in module structure. This work focused on soft modules, soft submodules, and soft homomorphism, which extend soft sets.

Acknowledgements

This paper is supported by the Natural Science Foundation of China (10671053 10771051).

References

1. Aktas, H., Cagman, N.: Soft Sets and Soft Groups. Inform. Sci. 177, 2726–2735 (2007)
2. Anderson, F.W., Fuller, K.R.: Rings and Categories of Modules. Springer, Heidelberg (1992)
3. Chen, D.: The Parameterization Reduction of Soft Sets and Its Applications. Comput. Math. Appl. 49, 757–763 (2005)
4. Maji, P.K., Bismas, R., Roy, A.R.: Soft Set Theory. Comput. Math. Appl. 45, 555–562 (2003)
5. Molodtsov, D.: The Theory of Soft Sets. URSS Publishers, Moscow (2004)
6. Maji, P.K.: An Application of Soft Set in Decision Making Problem. Comput. Math. Appl. 44, 1077–1083 (2002)
7. Molodtsov, D.: Soft Set Theory: First Results. Comput. Math. Appl. 37, 19–31 (1999)
8. Pawlak, Z.: Rough Sets. Int. J. Inform. Comput. Sci. 11, 341–356 (1982)
9. Zadeh, L.A.: Fuzzy Sets. Inform. Control. 8, 338–353 (1965)
10. Kassel, C.: Quantum Groups. Springer, Berlin (1991)

The Separation of Duty with Privilege Calculus*

Chenggong Lv, Jun Wang, Lu Liu, and Weijia You

Beihang University, Beijing 100083, P.R. China
lcgong@gmail.com, king.wang@buaa.edu.cn,
liulu@buaa.edu.cn, weijiawx@gmail.com

Abstract. This paper presents Privilege Calculus (PC) as a new approach of knowledge representation for Separation of Duty (SD) in the view of process and intents to improve the reconfigurability and traceability of SD. PC presumes that the structure of SD should be reduced to the structure of privilege and then the regulation of system should be analyzed with the help of forms of privilege.

1 Introduction

The Separation of Duty (SD) is a security principle that is used to formulate multi-person control policies, which requires that two or more different people be responsible for completion of a task or a set of related tasks [1]. The Role-Based Access Control (RBAC) system is defined by a state machine model and characterized by the fact that a user's rights to access objects are defined by the user's membership to a "role" and by the roles' permissions to perform operations on those objects [2]. Hence, the role is a semantic referent of duty representation and the structure of role is a division of rights in cross-organization systems. With the help of assignment operation, the user-role assignment can be handled by one while permission-role assignment is handled by another [3].

Because the permission assignment on role hierarchy is static, Sandhu [4] introduced the Role Activation Hierarchy (RAH). RAH extends the permission-usage hierarchy and makes the role activation governed by an activation hierarchy. Sandhu argued that the administration of RBAC must itself be decentralized and managed by administrative roles. Moreover, Ferraiolo [5] argued that static separation of duty enforces constraints on the assignment of users to roles, and dynamic separation duty places constraints on roles that can be activated within or across a user's session.

Although the delegation model [6] is helpful to resolve the temporal permission assignment problem by the delivery of duty in trust, the permission delegated has to crosscut two or more roles in RAH and the definition working to map between them is not easy. Also, for the constraints in RBAC, there is an inconsistency between the access control policy and the constraints that are specified to limit this policy. One transform limit may preclude, by a constraint, the change in

---

* We are grateful for the supporting of the National Natural Science Foundation of China (NSFC, Project No. 70401001).

G. Wang et al. (Eds.): RSKT 2008, LNAI 5009, pp. 410–417, 2008.
© Springer-Verlag Berlin Heidelberg 2008

another transform limit even though the rights that embody the conflict have not been assigned yet [7]. So extra mechanisms were integrated to detect [8] and resolve [9] the conflict. Jaeger has argued that since fail-safety is often a goal of secure systems, some form of conflict resolution may not be unreasonable, but the trade-off is not clear-cut [7].

It is the question that how to keep change of condition predictable and how control exists after reconfiguration in dynamic way, for which the essential challenge is, we believe , the representation of SD still. Our approach is enlightened by π-calculus that makes process reconfigurable [10], and assumes that the duty is composed of the interaction commitment of process, i.e. privilege(see section 3.3), and the result of SD is a collection of interaction commitments, i.e. regulation(see section 2). The examples in section 5 show the flexibility and usefulness of our approach.

2   Regulation

There are two synchronized complementary actions in an interaction [10]. The guarded action is an action with one preceding action that has not been reduced. We have two processors that execute these actions respectively. These actions represent the semantics of this interaction of the two processors.

A component is featured with the composition of distinct functions and consists of corresponding processors. One function features one processor in design, and one processor runs one action in one process (runtime). The sequence of *observed* action represents a process and reflects the implementation of function intention. So the sequence of *programmed* action represents an interaction commitment. Moreover, the intersection of interaction commitment involved in an interaction are not empty.

Although component is neutral, system works in a conservative way. The framework of system is a guarding processor and guards each interaction of two managed components. The guarding interaction of framework precedes the guarded interaction of component.

Regulation of system is a collection of interaction commitments, including the interaction commitments of framework and of component. For the systems based on privilege calculus, the result of separation of duty is regulation, i.e. a collection of privilege.

3   Structure of Privilege

In this section, we give the structure of privilege with the help of notions, employment and condition. The notion of employment is the refined structure of function intention.

3.1   Employment

Definition 1. *The function-entity employment f/e means that function f is employed on entity e.*

Proposition 1. *There are employments,* f_1/e_1 *and* f_2/e_2,

$$f_1/e_1 + f_2/e_2 = \emptyset \iff f_1/e_1 = \emptyset \wedge f_2/e_2 = \emptyset$$

Then we introduce the left employment mergence of function-entity.

Proposition 2. *There are employments,* $f_1/e_1 \neq \emptyset$ *and* $f_2/e_2 \neq \emptyset$.

$$(f_1/e_1) * (f_2/e_2) = \begin{cases} f/e, & \text{if } f = f_1 = f_2 \neq \emptyset \text{ and } e = e_1 = e_2 \neq \emptyset; \\ \emptyset, & \text{otherwise.} \end{cases}$$

Definition 2. *F is a collection of functions, and E is a collection of entities. The employment F/E is a set $\{f/e | f \in F, e \in E\}$.*

Let F, F_1, and F_2 be respectively a collection of functions, and let E, E_1, and E_2 be a collection of entities. We have $f_1 \in F_1$, $f_2 \in F_2$, $e_1 \in E_1$, and $e_2 \in E_2$. The mergence of employment is

$$F_1/E_1 * F_2/E_2 = \{f_1/e_1 * f_2/e_2 \neq \emptyset\} \ . \tag{1}$$

The composition of employment is

$$F_1/E_1 + F_2/E_2 = \{f_1/e_1 \neq \emptyset \vee f_2/e_2 \neq \emptyset\} \ . \tag{2}$$

For the convenience of computation, we give $F/\emptyset = \emptyset$, $\emptyset/E = \emptyset$ and $\emptyset/\emptyset = \emptyset$. If no confusion arises, these expressions, f/e, $\{f\}/e$ and $f/\{e\}$, are the same as $\{f\}/\{e\}$. With definition 2 and equations 1 and 2, we prove that the employment are associative, commutative and distributive.

3.2 Condition

Regulation is different from process, which we have discussed in section 2. The condition acts as the connection with the state of "process world". In this subsection, we propose the definition of condition.

Definition 3. *The fact set T is a collection of subsets of statement collection S. The fact set T on S has the following properties:*

1. *$\emptyset$ and S are in T.*
2. *The union of the elements of any sub-collection of T is in T.*
3. *The intersection of the elements of any finite sub-collection of T in T.*

Definition 4. *Fact set T on S, condition r is a function $r : T_s \rightarrow \{1, 0\}$ with the property: $\forall x_1, x_2 \in T$ and $x_1 \cap x_2 = \emptyset$, $r(x_1 \cup x_2) = r(x_1) \vee r(x_2)$.*

The $\{1, 0\}$ is the true value. If the fact $x \in T$, we call that the condition r is supported on the fact x, or the fact x supports the condition r.

Proposition 3. *For fact set T on S, $\forall x_1, x_2 \in T$ and $x_1 \subset x_2$, $r(x_1) \rightarrow r(x_2)$.*

Definition 5. *For fact set T on S and condition r, if $r(x)$ is true, the fact $x \in T$ is the evidence to r.*

Definition 6. *For fact set T on S, $\exists x^* \in T$ and such that x^* is the evidence to the condition r, if $\nexists x \subset x^*$ and such that x is the evidence to r, then the x^* is the minimum evidence to r.*

3.3   Privilege

Definition 7. *For a collection of functions F, a collection of entities E and a collection of conditions R, the privilege is $(F/E, R)$.*

For convenience, we define, $(\emptyset, r) = \emptyset$.

Definition 8. *The privilege space $\mathcal{P}$ is a collection of subsets of P with the following properties:*

1. (Privilege Mergence) *For all privilege, $u, v \in \mathcal{P}$, $u = (f_1/E_1, R_1)$, and $v = (f_2/E_2, R_2)$,*
$$u * v = \{(f_1 * f_2/(E_1 \cap E_2), R_1 \cap R_2)\} \; ;$$

2. (Privilege Composition) *For all privilege, $u, v \in \mathcal{P}$, $u = (f_1/E_1, R_1)$, and $v = (f_2/E_2, R_2)$,*

$$u + v = \{(f_1/E_1, R_1) \cup (f_2/E_2, R_2)\} \; ;$$

3. *For all privilege, $u, v \in \mathcal{P}$, $u * v = v * u$;*
4. *For all privilege, $u, v \in \mathcal{P}$, $u + v = v + u$;*
5. *For all privilege, $u, v, w \in \mathcal{P}$, $(u * v) * w = v * (u * w)$;*
6. *For all privilege, $u, v, w \in \mathcal{P}$, $(u + v) + w = v + (u + w)$;*
7. *For all privilege, $u, v, w \in \mathcal{P}$, $u * (v + w) = u * v + u * w$.*

4   Normal Form of Privilege

Definition 9. *The employment arrangement M is a finite collection of employment and such that $\forall m, n \in M, \quad m \neq n \wedge m * n = \emptyset$.*

Definition 10. *To employment arrangement M, the normal form of privilege p is*

$$\mathrm{nfm}_M(p) = \sum_i^M m_i = \sum_i^M (f_i/E_i, c_i) \; ,$$

where f_i/E_i is an element of M and c_i is a condition.

Proposition 4. *To employment arrangement M, every privilege is structurally equal to its normal form.*

Definition 11. *To employment arrangement M, the privileges are structural equivalence, if and only if they have the same normal form,*

$$u \overset{M}{=} v \iff \mathrm{nfm}_M(u) = \mathrm{nfm}_M(v) \; .$$

When one condition has an evidence, these privileges that involve the condition are pulsed. Corresponding to normal form of privilege, there is the pulsed form.

Definition 12. *To employment arrangement M, on the fact $t \in T$, the pulsed form of privilege p is*

$$\text{pfm}_M(p, t) = \sum_i^M (f_i/E_i, c_i(t)) \ ,$$

where f_i/E_i is an element of M and c_i is a condition.

We have a sequence of fact $Q = (t_0, t_1, \ldots, t_j, \ldots)$. We get the sequence of pulse to privilege t,

$$\text{pfm}_M(p, Q) = (\text{pfm}_M(p, t_0), \text{pfm}_M(p, t_1), \ldots, \text{pfm}_M(p, t_j), \ldots) \ .$$

This sequence of pulsed form describes the trace of process about privilege p. The trace matrix $(c_{i,j})$ of privilege p is made from this sequence, where $c_{i,j} \in \{1, 0\}$.

	t_0	t_1	$\cdots$	t_j	$\cdots$
f_0/E_0	$c_{0,0}$	$c_{0,1}$	$\cdots$	$c_{0,j}$	$\cdots$
f_1/E_1	$c_{1,0}$	$c_{1,1}$	$\cdots$	$c_{1,j}$	$\cdots$
$\vdots$	$\vdots$	$\vdots$	$\ddots$	$\vdots$	$\ddots$
f_i/E_i	$c_{i,0}$	$c_{i,1}$	$\cdots$	$c_{i,j}$	$\cdots$
$\vdots$	$\vdots$	$\vdots$	$\ddots$	$\vdots$	$\ddots$
f_n/E_n	$c_{n,0}$	$c_{n,1}$	$\cdots$	$c_{n,j}$	$\cdots$

For example, we have two operations (privileges) op_1 and op_2, and three people (privileges) u_1, u_2 and u_3. We want to know what will happen at time (facts) t_0 and t_1. So we define a gauging privilege, $g = (u_1 + u_2 + u_3) * (op_1 + op_2)$. And the sequence of pulse is $(\text{pfm}_M(g, t_0), \text{pfm}_M(g, t_1))$.

Definition 13. *To employment arrangement M, privileges, u and v, are congruent on fact $t \in T$, $a \overset{t}{\sim} b$, if and only if u and v have the same pulsed form.*

Definition 14. *To employment arrangement M, on fact $t \in T$, privilege p is compliant to privilege q, $p \overset{t}{\looparrowright} q$, if and only if $(p * q) \overset{t}{\sim} q$.*

The congruence $\sim$ and the compliance $\looparrowright$ are a function $P \times P \times T \rightarrow \{1, 0\}$. So they can be a condition in one high-order privilege. For a compliance example, we have the privileges, g, p and q, and such that $g = [p \looparrowright q]$. We call that the privilege g is a high-order privilege of p and q.

5   Discussion

In general, the role-based models, such as RBAC reference model [11,5], AR-BAC [12], and T-RBAC [13], have constructs, such as, USERS, ROLES, OPS (operations), and OBJS (objects), and relations, such as UA(user-to-role assignment), PA(permission-to-role assignment), PRMS (set of permission), and RH (role inheritance relation). These constructs are able to be defined with privilege and these relation with privileges. And these privileges are glued by privilege's operations, such as privilege mergence and privilege composition.

The following code is a demonstration written in PAL(Privilege Analysis Language) that is a reference implementation based on privilege calculus. With this demonstration we discuss cases about privilege representation.

```
namespace "example" {

    let doc1 is TechDoc

    reader := (read + list)/TechDoc
    manager := (reader + write + remove)/TechDoc

    bob := reader + write/TechDoc
    may := manager

    phone := read + list
    officepc := read + list + write + remove
}
```

Shown by the above code, we have four operations, *read, list, write,* and *remove,* two roles, *reader* and *manager,* two users, *bob* and *may,* and two terminals, *officepc* and *phone.* The statement *"let"* declares that *doc1* is a document in the category $TechDoc$. The role *reader* can read any documents in $TechDoc$ and *list* entries of those, and the role *manager* can *write* and *remove* any one in $TechDoc$ and *manager* inherits all of *reader*'s privileges that are limited in $TechDoc$. User *bob* plays the role *reader* and User *may* has the role *manager.* The mobile *phone,* a terminal device, has a limitation to access, *read* and *list.*

So far, we have defined these privileges: *read, list, write, remove, reader, manager, bob, may, officepc, phone, doc1,* and $TechDoc$.

While user *bob* has logged in system at his *officepc,* and the system creates his session, $session_1 = bob * officepc$. In $session_1$, *bob* is able to *read, list* and *write* any one in $TechDoc$.

Later *bob* uses his personal *phone* to navigate the system, the $session_2$ is created automatically, $session_2 = bob * phone$. The $session_2$'s privileges are different from $session_1$'s. We set an employment arrangement, $M = read + list + write + remove$. Thus,

$$session_1 \overset{M}{=} bob * officepc$$
$$\overset{M}{=} (reader + write/TechDoc) * (read + list + write + remove)$$
$$\overset{M}{=} read/TechDoc + list/TechDoc + write/TechDoc ,$$

$$session_2 \overset{M}{=} bob * phone$$

$$\overset{M}{=} ((read + list)/TechDoc + write) * (read + list)$$

$$\overset{M}{=} read/TechDoc + list/TechDoc .$$

With the above computation, we know the $session_2$ lacks the employment '*write*' on $TechDoc$. It is interesting that the session in system can be created as a privilege and these constructs, such as session, user, role, permission, group, location etc., could be represented by privilege.

We continue the story. User *bob* wants to read the document *doc1* that is a $TechDoc$. The guard *readguard* to the action *read* is

$$readguard = read * [session_1 \nleftrightarrow (read/doc1)] .$$

The *readguard* is the high-order privilege of $session_1$ and $read/doc1$. The pulse of *readguard* depends on the $session_1$'s compliance to $read/doc1$.

User *may* has logged in, and her session is $session_3$. She wants to write the document *doc1*. The regulation does concern not only *may*'s privilege but also the *doc1*'s. So the privilege *doc1* is redefined, $doc1 = readable + writable$. Because the *doc1*'s "writable" action and the *may*'s "write" action are complementary in this synchronized interaction, *writeguard* and *writableguard* are defined,

$$writegurad = write * [session_3 \nleftrightarrow (write/doc1)] ,$$
$$writableguard = writable * [doc1 \nleftrightarrow (writable)] .$$

Thus, we have the interaction guard *interactionguard*,

$$interactionguard = writeguard + writableguard .$$

Finally, the $session_3$'s compliance and the *doc1*'s compliance consistently make the pulse of *interactionguard*.

6   Conclusion

Separation of duty is critical not only in security control but also in modeling and monitoring of business logic. For improving reconfigurability of representation of duty, we propose *privilege calculus*. With the help of privilege's normal form and pulsed form, we are able to analyze the structure of privilege and to monitor the change in process. We also have demonstrated that the access control model based on privilege calculus is compatible with RBAC, ACL.

So far, we have only begun to explore the computation of privilege and representation of regulation in access control logic. But we have little knowledge about the relationship among regulation, business process and business rule. On all accounts, we hope that the paper will throw some light on the knowledge representation in separation of duty domain to facilitate the analysis of business rules and business processes.

References

1. Simon, R., Zurko, M.: Separation of duty in role-based environments. In: Proceedings of the 10th Computer Security Foundations Workshop, pp. 183–194. IEEE Press, New York (1997)
2. Gligor, V., Gavrila, S., Ferraiolo, D.: On the formal definition of separation-of-duty policies and their composition. In: Proceedings of Symposium on Security and Privacy, pp. 172–183. IEEE Press, New York (1998)
3. Sandhu, R.: Future Directions in Role-Based Access Control Models. In: Gorodetski, V.I., Skormin, V.A., Popyack, L.J. (eds.) MMM-ACNS 2001. LNCS, vol. 2052, pp. 22–26. Springer, Heidelberg (2001)
4. Sandhu, R.: Role activation hierarchies. In: Proceedings of the third ACM workshop on role-based access control, pp. 33–40. ACM Press, New York (1998)
5. Ferraiolo, D.F., Sandhu, R., Gavrila, S., Kuhn, D.R., Chandramouli, R.: Proposed NIST standard for role-based access control. ACM Transactions on Information and System Security 4(3), 224–274 (2001)
6. Barka, E., Sandhu, R.: Framework for role-based delegation models. In: Proceedings of the 16th Annual Computer Security Applications Conference, pp. 168–176. IEEE Press, New York (2000)
7. Jaeger, T.: On the increasing importance of constraints. In: Proceedings of the fourth ACM workshop on role-based access control, pp. 33–42. ACM Press, New York (1999)
8. Schaad, A.: Detecting Conflicts in a Role-based Delegation Model. In: Proceedings of the 17th Annual Computer Security Applications Conference, pp. 117–126. IEEE Press, New York (2001)
9. Jaeger, T., Sailer, R., Zhang, X.: Resolving constraint conflicts. In: Proceedings of the 9th ACM symposium on Access control models and technologies, pp. 105–114. ACM Press, New York (2004)
10. Milner, R.: Communicating and Mobile Systems: the π-Calculus. Cambridge University Press, Cambridge (1999)
11. Sandhu, R., Coyne, E., Feinstein, H., Youman, C.: Role-Based Access Control Models. Computer 29(2), 38–47 (1996)
12. Sandhu, R., Bhamidipati, V., Munawer, Q.: The ARBAC97 model for role-based administration of roles. ACM Transactions on Information and System Security 2(1), 105–135 (1999)
13. Oh, S., Park, S.: Task-role-based access control model. Information Systems 28(6), 533–562 (2003)

Maximal Invariable Update Operator Based on Splitting*

Maonian Wu[1] and Mingyi Zhang[2]

[1] College of Computer Science & Technology, Guizhou University, P.R. China
gzu_wu@yahoo.com
[2] Guizhou Academy of Sciences, P.R. China
Zhangmingyi045@yahoo.com.cn

Abstract. There are a very large number of beliefs in an agent generally. Now many researchers think that efficient belief revision should be
performed only in the part of its relevant states at a time. So Parikh
proposed relevance criterion and showed that AGM belief change operations do not always satisfy it. By introducing the notion of maximal
invariable partial meet contraction into AGM model, we obtain a class
of partial meet contractions that satisfy Parikh's relevance criterion. It
benefits computing an update operator by a local update one. Together
with the notion of essential letter, an open problem which proposed by
Makinson was resolved.

Keywords: Belief change, Splitting language, Maximal invariable.

1 Introduction

Belief revision is a topic of much interest in theoretical computer science and
logic, and it forms a central problem in research into artificial intelligence (infer
to [5,10]). In the logic of belief revision, a belief state (or database) is often represented by a set of formulae. Its notable methodology is AGM model, formulated
by Alchourron, Gardenfors and Makinson in 1985 [1].

An agent in the world usually holds a very large number of beliefs and receives
new information from the exterior world. Thus it may be inconsistent when new
information is added into its beliefs. In order to deal with the contradictions,
some original beliefs must be given up accordingly. What was given up is the
motivation about why we introduce belief revision. Many works about belief
revision have proposed many useful methods to deal with it. The AGM model
provides a milestone work in belief revision. But it is possible to give up almost
all original beliefs in AGM model. Recently some researches formulated that
relevant beliefs should be affected only when beliefs are changed. For example,
Parikh proposed that a belief change operator which respects his relevance criterion should protect any irrelevant formulae [8]. In that case, belief revision only
changes locally in original belief.

---

* Supported partially NSFC 60573009 and 90718009 and Science Research Fund For
Talents Recruiting of Guizhou University(X065018).

G. Wang et al. (Eds.): RSKT 2008, LNAI 5009, pp. 418–425, 2008.

In order to deal with the local change, Parikh defined splitting of any set of formulates and irrelevant formulae in belief revision over a finite language [8]. Kourousias and Makinson extended these definitions to an infinite language and showed that any set K of formulae has a unique finest splitting [4]. Parikh still formulated a relevance criterion in belief update and showed that AGM's operator dissatisfies it generally [8]. Many researches have been conducted to attack this issue by exploring some postulates to ensure the relevance criterion. These works are carried out by [2-4,8-9,11].

From the works above, more and more people think that belief revision should be performed only in a local belief subset when belief was changed. By introducing the notion of maximal invariable partial meet contraction into AGM model, we gain a subset $M(K,x)$ of entire set of partial meet contraction over K by x such that its each element satisfies parikh's relevance criterion. In other words, the set satisfies AGM postulates and relevance criterion. In general, there are many belief change operators in AGM model for given K and x. We propose the maximal invariable partial meet contraction to reduce the set of belief change operators. Inducting belief revision into decision-making model, our method reduces the selective set and respects the relevance criterion. It benefits computing an update operator by a local update one even if the language is infinite when every element in the finest splitting of belief set is finite. Meanwhile essential letter play a important role in the partial meet contraction based on splitting.

In the classic logic and some other logics, the essential letter of a set of formulae is very important. For example, it reduces solver's spaces in SAT generally. Schlechta has observed that it is possible to generalize the notion of an essential letter, making it relative to a set of valuations rather than to a set of formulae. Makinson proposed a theorem(lemma 2.10) about the essential letter of a set of formulae. He wants to formulate a theorem about the essential letter of a set of valuations as lemma 2.10. But his idea does not finished by a counterexample in [7]. He left it as an open problem. An results, no letter is essential to nonempty set V of valuations on infinitely numberable letter set P iff V consists of some union of $\{v : |\{p \in X : v(p) = 1\}| < \omega \ \& \ |\{p \in P \backslash X : v(p) = 0\}| < \omega\}$ where $X \subseteq P$, was formulated in the paper. So the open problem was resolved. Though just now people start the research in essential letter of a set of valuations, I belief the result will a base of future study in essential letter.

The rest of the paper is organized as follows: In section 2, we recall many preliminaries. We discuss maximal invariable partial meet contraction in section 3. In the next section, we answer an open problem. We compare our approach with some related methods and discuss conclusion and future work in section 5.

2   Preliminaries

We review the preliminaries for the paper. We always assume a propositional logic with set of infinite or finite letters including the zero-ary truth $\top$ among the primitive connectives. We use lower case letters $a, b, ..., x, y, z, \alpha, \beta, ...$ to range over formulae of classical propositional logic. Sets of formulae are denoted by upper case letters $A, B, ..., X, Y, ...$, reserving L for the set of all formulae, E for the set of all elementary letters (alias propositional variables) of language. For any formula α, we write $E(\alpha)$ to mean the set of the elementary letters occurring in α; similarly for sets of formulae. Let $F \subseteq E$, $L(F)$ stands for the sub-language generated by F, i.e. the set of all formulae x with $E(x) \subseteq F$. Classical consequence is written as $\vdash$ when treated as a relation over $2^L \times L$, classical consequence operation is written as Cn when treated as an operation on 2^L into itself. The relation of classical equivalence is written $\dashv\vdash$. Set K of formulae is a belief set if $K = Cn(K)$. To lighten notation, $v(A) = 1$ is short for: $v(\alpha) = 1$ for all $\alpha \in A$, while $v(A) = 0$ abbreviates: $v(\alpha) = 0$ for some $\alpha \in A$.

The following notations are needed in section 3. We write $\underline{Cn}(X)$ with an underline under Cn, classical consequence over the entire language, is classical consequence over the restricted language $L(E(X))$, i.e. $\underline{Cn}(X) = Cn(X) \cap L(E(X))$. Let us recall the definition of the essential letter of a set of formulae and valuations in [7]. Let F be an arbitrary set of formulae. We say that an elementary letter p is *essential* to F iff there are two valuations that agree on all letters other than p, but disagree in the value they give to F. Let W be an arbitrary set of valuations. We say that an elementary letter p is *essential* to W iff there are two valuations that agree on all letters other than p, but one in and the other outside W. A formula y is an essential formula of x iff $x \dashv\vdash y$ and for every formula z with $z \dashv\vdash x$ satisfy $E(y) \subseteq E(z)$. It is clear that the essential formula of a formula is not unique. But the essential formulae of a formula has the same elementary letter by the least letter-set theorem [7].

For any belief set K, $K \perp x$ is the set of all maximal subsets A of K such that $A \nvdash x$. In other words, $A \in K \perp x$ iff

(1) $A \subseteq K$,
(2) $A \nvdash x$,
(3) any $\varphi \in K \setminus A$, $A, \varphi \vdash x$.

And γ is any function such that for every formula x, $\gamma(K \perp x)$ is a nonempty subset of $K \perp x$, if the latter is nonempty, and $\gamma(K \perp x) = \{K\}$ otherwise. Such a function is called a selection function for K. We say that γ is transitively relational over K iff there is a transitive relation $\leq$ over 2^K such that for all $x \notin Cn(\emptyset)$, $\leq$ marks off $\gamma(K \perp x)$ in the sense the following identity, which we call the marking off identity, holds:

$$\gamma(K \perp x) = \{B \in K \perp x : B' \leq B \text{ for all } B' \in K \perp x\}.$$

The operation $\div$ defined by putting $K \div x = \bigcap \gamma(K \perp x)$ for all x is called the partial meet contraction over K determined by γ.

Note that the concept of partial meet contraction includes, as special cases, those of maxichoice contraction and full meet contraction. The former is a partial meet contraction with $\gamma(K \perp x)$ a singleton; the latter is the partial meet contraction with $\gamma(K \perp x)$ the entire set $(K \perp x)$.

Lemma 2.1 ([1]). Let $K \div x$ be a function defined for belief set K and a formula x. $K \div x$ is a partial meet contraction operation over K iff $K \div x$ satisfies AGM postulates $(\div 1) - (\div 6)$ for contraction over K.

Lemma 2.2 ([1]). Let K be any belief set, and $K \div x$ a partial meet contraction function over K, determined by a selection function γ. Then $K \div x$ is a transitive relation over K if and only if $K \div x$ satisfies AGM postulates $(\div 1) - (\div 8)$.

Many matters relate to each other in the world, but only partial relations between matters are essential. Parikh proposed the concept of splitting for set of formulate. It allows us carve up a set of formulae into disjoint pieces about different subject matters.

Definition 2.3 (Splitting [4]). Let $\mathbf{E} = \{E_i\}_{i \in I}$ be any partition of the set E of all elementary letters of the language. we say that $\mathbf{E}$ is a splitting of set K of formulae iff $\bigcup\{Cn(K) \cap L(E_i)\}_{i \in I} \vdash K$, equivalently, iff there is a family $\{B_i\}_{i \in I}$ with each $E(B_i) \subseteq E_i$ such that $\bigcup\{B_i\}_{i \in I} \dashv\vdash K$.

Generally, people often hope to split a belief set as fine as possible such that essential relation matters in same piece possibly.

Definition 2.4 (Fineness of a Partition[4]). Following customary terminology, we say that a partition $\mathbf{E} = \{E_i\}_{i \in I}$ of all elementary letters set E of the language is at least as fine as another partition $\mathbf{F} = \{F_j\}_{j \in J}$ of E, and we write $\mathbf{E} \leqslant \mathbf{F}$, iff every cell of $\mathbf{F}$ is the union of cells of $\mathbf{E}$. Equivalently, $R_{\mathbf{E}} \subseteq R_{\mathbf{F}}$, where $R_{\mathbf{E}}$ (resp. $R_{\mathbf{F}}$) is the equivalence relation over E associated with $\mathbf{E}$ (resp. $\mathbf{F}$).

Parikh showed that there was a unique finest splitting of K for finite language and Kourousias and Makinson proved the result for any language.

Lemma 2.5 ([4]). Every set K of formulae has a unique finest splitting.

The lemma says that there is a unique way to think of K as being composed of disjoint information about certain subject matters.

Lemma 2.6 (Parallel interpolation theorem[4]). Let $A = \bigcup\{A_i\}_{i \in I}$ where the letter sets $E(A_i)$ are pairwise disjoint, and suppose $\bigcup\{A_i\}_{i \in I} \vdash x$. Then there are formulae b_i such that each $E(b_i) \subseteq E(A_i) \cap E(x)$, $A_i \vdash b_i$, and $\bigcup\{b_i\}_{i \in I} \vdash x$.

Parikh defined the irrelevant formulae in a belief change over a finite language. Kourousias and Makinson extended the definition for an infinite language.

Definition 2.7 (Irrelevant formulae in a belief change [4]). Let K be any consistent set of formulae, with x a formula that is a candidate for contracting from K or integrating into K by a process of revision. Let $\mathbf{E} = \bigcup\{E_i\}_{i \in I}$ be the unique finest splitting of K. We say that a formula y is irrelevant to the contraction or revision of K by x (briefly: $y \in K$ is irrelevant to x modulo K) iff $E(y)$ is disjoint from $\bigcup\{E_j\}_{j \in J}$, where $\bigcup\{E_j\}_{j \in J}$ is the subfamily of cells in $\mathbf{E}$ that share some elementary letter with $E(x)$. We denote the set of the irrelevant formulae to x modulo K by $I_{K,x}$ and simply as I_x in contexts where the identity of K is clear. Formally,

$I_x = \{y \in K | E(y) \cap \bigcup_{j \in J}\{E_j\} = \emptyset$ where $E_j \cap E(x) \neq \emptyset$ for all $j \in J\}$.

The *relevance criterion* may be put as follows: whenever an element $y \in K$ is irrelevant to x modulo K, then it remains an element of the result of contracting or revising K by x. In other words, $I_x \subseteq K \div x$ or $I_x \subseteq K * x$.

Lemma 2.8 ([11]). Let $K \div^* x$ be a function defined for a belief set K and a formula x. The partial meet contraction $K \div^* x$ determined by a transitive relation γ^* over K iff $K \dot{-}^* x$ satisfies AGM postulates $(\dot{-}1)-(\dot{-}8)$ for contraction and also satisfies the relevance criterion.

The lemma formulated a representation theorem for the relevance criterion and AGM postulates for contraction.

In the following, we will consider a special operator in belief contraction.

Lemma 2.9 ([12]). Let $K \dot{-}^* x$ be any partial meet contraction function based on splitting over belief set K. Then $K \dot{-}^* x$ satisfies fullness postulate iff $K \dot{-}^* x$ is a maxichoice contraction function.

The following lemma discusses the essential letters of a set of formulae.

Lemma 2.10([7]). Let A be any set of formulae. Then A is contingent (neither a tautology nor a contradiction) iff at least one of its elementary letters is essential to it.

3   Maximal Invariable Partial Meet Contraction

In belief revision, some original beliefs are given up when a new belief enter the belief set generally. People often care for the change in entire set of original beliefs over a sub-language L_1 with set E of letters. Specially, the original beliefs in L_1 unchange when belief revision, i.e. $K \cap L_1 \subseteq K \dot{-} x$ or $K \cap L(E) \subseteq K \dot{-} x$.

Naturally, we can define the invariable letter E' of a partial meet contraction $K \dot{-} x$ as $(K \dot{-} x) \cap L(E') = K \cap L(E')$. But it seems not to be well-defined. For example(Ex1), let $K = Cn\{p \leftrightarrow q, r \leftrightarrow s, t\}$ and $x = (p \rightarrow q) \wedge (r \rightarrow s) \vee (p \leftarrow q) \wedge (r \leftarrow s)$. Let a partial meet contraction $K \dot{-} x = Cn\{(p \rightarrow q) \wedge ((r \rightarrow s), t\}$. Then $\{p\}$ is an invariable letter set of $K \dot{-} x$ since $K \cap L(p) = \underline{Cn}(\emptyset)$. But it is clear that letter p occurring in $q \rightarrow p$ and $q \rightarrow p \notin K \dot{-} x$.

The splitting of a belief set carves up the belief set into many disjoint pieces about different subject matters. Using the property of splitting, we definite an invariable letter of a partial meet contraction based on splitting as follows.

Definition 3.1(Invariable letter set). Let K be a set of formulae and x a formula in a language with set E of letters. Given any partial meet contraction $K \dot{-} x$, we say that a set $E' \subseteq E$ is an invariable letter set of $K \dot{-} x$ if $\{E', E \backslash E'\}$ is a splitting of K and $K \cap L(E') = (K \dot{-} x) \cap L(E')$.

For example(Ex2), let $K = Cn\{p \leftrightarrow q, r \leftrightarrow s, t\}$ and $x = (p \rightarrow q) \wedge (r \rightarrow s) \vee (p \leftarrow q) \wedge (r \leftarrow s)$. Let $K \dot{-}_1 x = Cn\{p \leftrightarrow q, t\}$, $K \dot{-}_2 x = Cn\{r \leftrightarrow s, t\}$ and $K \dot{-}_3 x = Cn\{(p \rightarrow q) \wedge ((r \rightarrow s), t\}$. In the example, It is clear that $\emptyset$ and $\{t\}$ are the invariable letter set of $K \dot{-}_3 x$; $\emptyset$, $\{t\}$ and $\{r, s, t\}$ are the invariable letter set of $K \dot{-}_2 x$; $\emptyset$, $\{t\}$ and $\{p, q, t\}$ are the invariable letter set of $K \dot{-}_1 x$.

Let K be a set of formulae and x a formula. Let $\mathbf{E} = \{E_i\}_{i \in I}$ be the finest splitting of K. Given any partial meet contraction $K \dot{-} x$, let $MIL(K \dot{-} x) = \bigcup\{E_i \in \mathbf{E} : K \cap L(E_i) = (K \dot{-} x) \cap L(E_i)\}$. given partial meet contraction $K \dot{-} x$, we prove that $MIL(K \dot{-} x)$ is the maximal invariable letter set of $K \dot{-} x$.

Lemma 3.2 Let K be a set of formulae and x a formula in a language with set E of letters. Given any partial meet contraction $K \dot{-} x$, $MIL(K \dot{-} x)$ is invariable letter set of $K \dot{-} x$.

Lemma 3.3 Let K be a set of formulae and x a formula in a language with set E of letters. Given any partial meet contraction $K \div x$, $MIL(K \div x)$ is maximal invariable letter set of $K \div x$.

By lemmas 3.2, 3.3, the following theorem is clear.

Theorem 3.4 Let K be a set of formulae and x a formula. Given any partial meet contraction $K \div x$, $MIL(K \div x)$ is the unique maximal invariable letter set of $K \div x$.

From the above discussion, we can compute invariable letter set for any partial meet contraction. Now we consider maximal element in the entire invariable letter set for all partial meet contractions over set K of formulae by formula x.

Definition 3.5(Maximal invariable contraction). Let K be a set of formulae and x a formula in a language with set E of letters. Given any partial meet contraction $K \div x$, we say that $K \div x$ is a maximal invariable partial meet contraction if there is no partial meet contraction $K \div_1 x$ such that $MIL(K \div x) \subsetneq MIL(K \div_1 x)$. Equivalently, $K \div x$ is maximal invariable partial meet contraction iff $MIL(K \div x)$ is an maximal element in set $\{MIL(K \div x) : K \div x \text{ is a partial meet contraction }\}$.

For the example in this section, $K \div_1 x$ and $K \div_2 x$ are maximal invariable partial meet contraction. But $K \div_3 x$ is not the one even if it satisfies the relevance criterion. Now let set M(K,x) only contains all maximal invariable partial meet contraction over K by x. In other words, $M(K, x) = \{K \div x : K \div x \text{ is a maximal invariable partial meet contraction over K by x }\}$.

Theorem 3.6 Let K be a set of formulae and x a formula. If $K \div x$ is a maximal invariable partial meet contraction then $K \div x$ satisfies relevance criterion.

Proof: Let $E_x = E(I_{\overline{x}})$. It is clear that $E_x = \bigcup \{E_i \in \mathbf{E} : E_i \cap E(\overline{x}) = \emptyset\}$ by definition of $I_{\overline{x}}$ where $\mathbf{E} = \{E_i\}_{i \in I}$ is the finest splitting if K and $\overline{x}$ is some essential formula of x. Now we suppose the theorem is dissatisfies. Then $I_{\overline{x}} \subsetneq K \div x$ for some maximal invariable partial meet contraction $K \div x$ over K and x. In other word, $E_x \subsetneq MIL(K \div x)$. Let $E_1 = E_x \setminus MIL(K \div x)$ and $E_2 = MIL(K \div x) \setminus E_x$. It is evident that $E_1 \neq \emptyset$ and $\{E_1, E_2, E \setminus (E_1 \cup E_2)\}$ is a splitting of K. By $K \div x \nvdash x$ and definition of $MIL(K \div x)$, we have $(K \div x) \cap L(MIL(K \div x)) \nvdash x$, $(K \div x) \cap L(MIL(K \div x)) \nvdash \overline{x}$ and $K \cap L(MIL(K \div x)) \nvdash \overline{x}$ where $\overline{x}$ is some essential formula of x. So $K \cap L(MIL(K \div x)) \bigcup (L(E_1) \cap K) \nvdash \overline{x}$ by parallel interpolation theorem and splitting $\{E_1, E_2, E \setminus (E_1 \cup E_2)\}$ of K. Hence there is a belief contraction $\div'$ over K and x such that $MIL(K \div' \overline{x})$ includes E_1 and E_2. It is evident that $MIL(K \div' x)$ includes E_1 and E_2. This is a contradiction to $K \div x$ is a maximal invariable partial meet contraction.

We form a subset $M(K, x)$ of entire set of partial meet contraction over K by x such that every element in it satisfies relevance criterion. It benefits computing an update operator by a local update one even if the language is infinite when every element in the finest splitting of belief set is finite.

4   Answer to an Open Problem

In [7], Makinson proposed lemma 2.10 about the essential letter of a set of formulae. He wanted to formulate a theorem about the essential letter of a set of valuations as the lemma 2.10. But it did not holds by a counterexample in [7]. He left it as an open problem. The theorem, no letter is essential to nonempty set V of valuations on infinitely numberable letter set P iff V consists of some unions of $\{v : |\{p \in X : v(p) = 1\}| < \omega \ \& \ |\{p \in P\backslash X : v(p) = 0\}| < \omega\}$ where $X \subseteq P$, was proved in the section. So we resolve the open problem.

In the section, let V be a set of valuations and V' entire sets of valuations on a language with infinite numberable set $P = \{p_1, p_2, ...\}$ of letters. Let $X_u^1 = \{p \in P : u(p) = 1\}$ and $X_u^0 = \{p \in P : u(p) = 0\}$ when $u \in V'$.

Lemma 4.1 If no letter is essential to V and $u \in V$ then $\{v : |\{p \in X_u^0 : v(p) = 1\}| < \omega \ \& \ |\{p \in X_u^1 : v(p) = 0\}| < \omega\} \subseteq V$.

Proof: Suppose the contrary. We assume that there is a valuation $x \in \{v : |\{p \in X_u^0 : v(p) = 1\}| < \omega \ \& \ |\{p \in X_u^1 : v(p) = 0\}| < \omega\}$ and $x \notin V$. Let $u - x = \{p \in P : u(p) \neq x(p)\}$. By $u, x \in \{v : |\{p \in X_u^0 : v(p) = 1\}| < \omega \ \& \ |\{p \in X_u^1 : v(p) = 0\}| < \omega\}$, we show that $|u - x| < \omega$. Without loss of generality (abbreviated to WLOG), let $u - x = \{r_1, r_2, ..., r_n\}$. WLOG, assume $u(r_1) = 0$. Then $u(r_1/1) \in V$ since no letter is essential to V. WLOG, assume $u(r_1/1)(r_2) = 0$. Then $u(r_1/1, r_2/1) \in V$ since no letter is essential to V. Similarly, we have $u(r_1/1, r_2/1, ..., r_n/1) = x \in V$. It is a contradiction to $x \notin V$.

Lemma 4.2 Let $u \in V$. No letter is essential to V iff no letter is essential to $V\backslash\{v : |\{p \in X_u^0 : v(p) = 1\}| < \omega \ \& \ |\{p \in X_u^1 : v(p) = 0\}| < \omega\}$.

Theorem 4.3 No letter is essential to $V(\neq \emptyset)$ iff V consists of some unions of $\{v : |\{p \in X : v(p) = 1\}| < \omega \ \& \ |\{p \in P\backslash X : v(p) = 0\}| < \omega\}$ where $X \subseteq P$.

Though just now people start the research in essential letter of a set of valuations, I belief the result will a base of future study in essential letter.

5   Comparison and Conclusion

Firstly, we compare maximal invariable partial meet contraction with the partial meet contraction based on relevance [11].

By lemma 2.8, $M(K, x) \subseteq K \div^* x$ when K is a belief set. But $K \div^* x \not\subseteq M(K, x)$ even when K is belief set by the example in section 3. $K \div_3 x$ is a partial meet contraction which satisfies relevance criterion. But it is not a maximal invariable contraction since $MIL(K \div_3 x) \subsetneq MIL(K \div_1 x)$ or $MIL(K \div_3 x) \subsetneq MIL(K \div_2 x)$. So $K \div^* x \not\subseteq M(K, x)$ even when K is belief set.

Secondly, we compare maximal invariable partial meet contraction over belief set with maxichoice contraction base on splitting [12].

The following examples show that there exists a maximal invariable partial meet contraction $K \div x$ is not maxichoice contraction. And there exists maxichoice contraction $K \div x$ which satisfies relevance criterion such that it is not a maximal invariable partial meet contraction.

Example Ex3: Let p, q, r are distinct letters. Let $K = Cn\{p \leftrightarrow q, r\}$ and $x = p \leftrightarrow q$. Then $K \div x = Cn\{p \rightarrow q, r\} \cap Cn\{q \rightarrow p, r\} = Cn(r)$ is a maximal invariable partial meet contraction. But it is not maxichoice contraction.

Ex4: Let p, q, r are distinct letters. Let $K = Cn\{p, q, r\}$ and $x = p \wedge q$. Then $K \div x = Cn\{p \leftrightarrow q, r\}$ is maxichoice contraction which satisfies relevance criterion. But it is not a maximal invariable partial meet contraction.

Finally, introducing the notion of maximal invariable update operation into AGM model, we obtain a class of partial meet contraction operations that respect Parikh's relevance criterion. It benefits computing an update operator by a local update one. Together with the notion of essential letter, we resolve an open problem proposed by Makinson in 2005. In future, we want to give the syntax of the maximal invariable partial meet contraction over belief K by new information x and apply the essential letter of a set of valuations in other areas.

Acknowledgements

We are grateful to David Makinson, PC chairmans and two unknown referees for their insightful suggestions and objective argumentation.

References

1. Alchourron, C.A., Gardenfors, P., Makinson, D.: On the Logic of Theory Change: Partial Meet Contraction and Revision Functions. J. Symbolic Logic 50, 510–530 (1985)
2. Chopra, S., Georgatos, K., Parikh, R.: Relevance Sensitive Non-monotonic Inference on Belief Sequences, CoRR. cs.AI/0003021 (2000)
3. Chopra, S., Parikh, R.: Relevance Sensitive Belief Structures. Ann. Math. Artif. Intell. 28, 259–285 (2000)
4. Kourousias, G., Makinson, D.: Parallel Interpolation, Splitting, and Relevance in Belief Change. J. Symbolic Logic 72, 994–1002 (2007)
5. Hansson, S.O.: A Textbook of Belief Dynamics: Theory Change and Database Updating. Kluwer Academic Publishers, Dordrecht (1999)
6. Hansson, S.O., Wassermann, R.: Local Change. Studia Logica 70, 49–76 (2002)
7. Makinson, D.: Friendliness for Logicians. We will Show Them! essays in honour of Gabbay, D. In: Artemov, S.N., Barringer, H., Garcez, A.S.A., Lamb, L.C., Woods, W. (eds.), vol. 2, pp. 259–292. College Publications (2005)
8. Parikh, R.: Beliefs, Belief revision, and Splitting Languages. Logic, language, and Computation. In: Moss, L., Ginzburg, J., Rijke, M. (eds.), vol. 2, pp. 266–278. CSLI Publications, Stanford (1999)
9. Peppas, P., Chopra, S., Foo, N.: Distance Semantics for Relevance- Sensitive Bbelief Revision. Principles of Knowledge Representation and Rreasoning. In: Dubois, D., Welty, C.A., Williams, M. (eds.) KR 2004, pp. 319–328. AAAI Press, Menlo Park (2004)
10. Rott, H.: Change, Choice and Inference: A Study of Belief Revision and Non-monotonic Reasoning. Oxford University Press, Oxford (2001)
11. Wu, M., Zhu, Z., Zhang, M.: Partial Meet Contraction Based on Relevance Criterion. In: IMECS (to appear, 2008)
12. Wu, M., Zhang, M.: Partial Meet Contraction Based on Splitting (submitted)

A Novel Approach to Attribute Reduction in Formal Concept Lattices

Jing Liu and Ju-Sheng Mi

College of Mathematics and Information Science,
Hebei Normal University, Shijiazhuang, Hebei, 050016, P.R. China
lj9068@163.com, mijsh@263.net

Abstract. The theory of concept lattice is an efficient tool for knowledge representation and knowledge discovery. One of the key problems of knowledge discovery is attribute reduction. This paper presents a novel approach to attribute reduction in formal concept lattices. The approach employs all the extents of the meet-irreducible elements in the lattice. Each of them determines a family of attribute sets common to the objects in the extent. By various combinations of minimal elements from each family we can produce reducts of the formal context. Furthermore, a related algorithm is developed, and an illustration example is employed to perform the reduction process of the proposed method.

Keywords: Formal context, Concept lattice, Meet-irreducible element, Attribute reduction.

1 Introduction

The theory of concept lattice [12], proposed by Wille in 1982, has emerged as an important mathematical tool for dealing with uncertain information. Many researches on concept lattices concentrate on the construction and pruning algorithm of concept lattices, the relationship between concept lattices and rough sets, knowledge reduction and acquisition [3,5,7,9,10,12,13,15]. As an efficient tool for data analysis, the theory of concept lattice has potential applications in many fields, such as decision making, information retrieval, data mining, knowledge discovery and so on [1,2,4,6,11].

Attribute reduction is one of the key problems of the theory of concept lattice. It is to search for a minimal attribute set which can determine all the formal concepts and the structures of a concept lattice. That is, an optimal attribute set should be able to replace the whole attribute set but still maintains the same amount of information in the original concept lattice.

Recently, work on the attribute reduction of a concept lattice is progressing rapidly. Zhang et al. [14,16] proposed the reduction theory of concept lattices, obtained many judgment theorems of consistent sets, and analyzed the characteristics of different attributes. They also presented an approach to attribute reduction by using a discernibility matrix. Another version of attribute reduction was introduced in [10] which determines the same set of all meet-irreducible

G. Wang et al. (Eds.): RSKT 2008, LNAI 5009, pp. 426–433, 2008.
© Springer-Verlag Berlin Heidelberg 2008

elements of the concept lattice as the one determined by all attributes. Shao [9] studied the attribute reduction of two kinds of generalized concept lattice.

In this paper, we proceed this topic, and propose a new method of attribute reduction in a formal concept lattice. The approach first employs all the extents of the meet-irreducible elements in the concept lattice. Each of them determines a family of attribute sets common to the objects in the extent. A combination of the minimal elements from every family can produce a reduct of the concept lattice. Finally, we develop a related algorithm and provide an illustration example is employed to perform the reduction process of the proposed approach.

2 Preliminaries

To make this paper self-contained, the involved notions in formal concept analysis and reduction of concept lattices are introduced in this section [5,16].

Definition 1. (see [5]) *An element a of a lattice L is called* $meet-irreducible$, *if*

$$a = b \wedge c \Longrightarrow a = b \ \ or \ \ a = c$$

holds for any b, $c \in L$.

Proposition 1. (see [5]) *Every element a of a finite lattice L can be represented as the meet of some meet-irreducible elements.*

A *formal context* is an ordered triplet $T = (U, A, I)$, where U, A are finite nonempty sets and $I \subseteq U \times A$ is a correspondence from U to A. The elements in U are interpreted to be *objects*, elements in A are said to be *attributes*. If $(x, a) \in U \times A$ is such that $(x, a) \in I$, then the object x is said to have the attribute a. The correspondence I can be naturally represented by an incidence table: the rows of the table are labelled by objects, columns by attributes; if $(x, a) \in I$, the intersection of the row labelled by x and the column labelled by a contains 1; otherwise it contains 0.

For a set $B \subseteq A$ of attributes we define

$$\psi(B) = \{x \in U \mid xIa \ for \ all \ a \in B\}.$$

Correspondingly, for a set $X \subseteq U$ of objects we define

$$\phi(X) = \{a \in A \mid xIa \ for \ all \ x \in X\}.$$

Evidently, $\psi(B) = \bigcap_{a \in X} \psi(a)$, and $\phi(X) = \bigcap_{x \in X} \phi(x)$. Where $\phi(x) = \phi(\{x\})$ and $\psi(a) = \psi(\{a\})$ for short.

Definition 2. (see [5]) *A formal concept of the context (U, A, I) is a pair (X, B) with $X \subseteq U$, $B \subseteq A$, $\phi(X) = B$ and $\psi(B) = X$. We call X the extent and B the intent of the concept (X, B). $\mathcal{L}(U, A, I)$ denotes the set of all concepts of the context (U, A, I).*

From the above definition, the following proposition can be easily proved [16].

Proposition 2. *Let (U, A, I) be a formal context, X, X_1, $X_2 \subseteq U$ and B, B_1, $B_2 \subseteq A$, then*

(1) $X_1 \subseteq X_2 \Longrightarrow \phi(X_2) \subseteq \phi(X_1)$, $B_1 \subseteq B_2 \Longrightarrow \psi(B_2) \subseteq \psi(B_1)$.
(2) $X \subseteq \psi(\phi(X))$, $B \subseteq \phi(\psi(B))$.
(3) $\phi(X) = \phi(\psi(\phi(X)))$, $\psi(B) = \psi(\phi(\psi(B)))$.
(4) $X \subseteq \psi(B) \Longleftrightarrow B \subseteq \phi(X)$.
(5) $\phi(X_1 \cup X_2) = \phi(X_1) \cap \phi(X_2)$, $\psi(B_1 \cup B_2) = \psi(B_1) \cap \psi(B_2)$.
(6) $\phi(X_1 \cap X_2) \supseteq \phi(X_1) \cup \phi(X_2)$, $\psi(B_1 \cap B_2) \supseteq \psi(B_1) \cup \psi(B_2)$.
(7) $(\psi(\phi(X)), \phi(X))$ and $(\psi(B), \phi(\psi(B)))$ are formal concepts.

The concepts of a given formal context are naturally ordered by the subconcept-superconcept relation defined by

$$(X_1, B_1) \leq (X_2, B_2) \Leftrightarrow X_1 \subseteq X_2 (\Leftrightarrow B_2 \subseteq B_1).$$

The ordered set of all concepts in (U, A, I) is denoted by $\mathcal{L}(U, A, I)$ and called the concept lattice of (U, A, I). In fact, $\mathcal{L}(U, A, I)$ is a complete lattice in which the infimum and supremum are given by

$$(X_1, B_1) \wedge (X_2, B_2) = (X_1 \cap X_2, \phi(\psi(B_1 \cup B_2))),$$

and

$$(X_1, B_1) \vee (X_2, B_2) = (\psi(\phi(X_1 \cup X_2)), B_1 \cap B_2).$$

Definition 3. (see [16]) *Let $\mathcal{L}(U, A_1, I_1)$ and $\mathcal{L}(U, A_2, I_2)$ be two concept lattices. If for any concept $(X, B) \in \mathcal{L}(U, A_2, I_2)$, there exists $(X^{'}, B^{'}) \in \mathcal{L}(U, A_1, I_1)$ such that $X = X^{'}$, then $\mathcal{L}(U, A_1, I_1)$ is said to be finer than $\mathcal{L}(U, A_2, I_2)$, denoted by $\mathcal{L}(U, A_1, I_1) \leq \mathcal{L}(U, A_2, I_2)$. If $\mathcal{L}(U, A_1, I_1) \leq \mathcal{L}(U, A_2, I_2)$ and $\mathcal{L}(U, A_2, I_2) \leq \mathcal{L}(U, A_1, I_1)$, then the two concept lattices are said to be isomorphic to each other, and denoted by $\mathcal{L}(U, A_1, I_1) \cong \mathcal{L}(U, A_2, I_2)$.*

Let (U, A, I) be a formal context, for any set $D \subseteq A$ of attributes, denote by $I_D = I \cap (U \times D)$, then (U, D, I_D) is also a formal context. For any $X \subseteq U$, $\phi(X)$ is represented by $\phi_D(X)$ in (U, D, I_D). It is evident that $\phi_D(X) = \phi(X) \cap D$.

Proposition 3. (see [16]) *Let (U, A, I) be a formal context, then*

$$\mathcal{L}(U, A, I) \leq \mathcal{L}(U, D, I_D)$$

holds for any attribute set $\emptyset \neq D \subseteq A$.

Definition 4. (see [16]) *$B \subseteq A$ is called a consistent set of formal context (U, A, I) if $\mathcal{L}(U, B, I_B) \cong \mathcal{L}(U, A, I)$ holds. Furthermore, $\forall b \in B$, $\mathcal{L}(U, B \setminus \{b\}, I_{B \setminus \{b\}}) \ncong \mathcal{L}(U, A, I)$, then B is called a reduct of the formal context (U, A, I).*

Proposition 4. (see [16]) *Let (U, A, I) be a formal context, $\emptyset \neq D \subseteq A$. Then*

$$D \text{ is a consistent set of } (U, A, I) \Longleftrightarrow \mathcal{L}(U, D, I_D) \leq \mathcal{L}(U, A, I).$$

3   A Novel Approach to Attribute Reduction in Concept Lattices

After analyzing the theory of attribute reduction in concept lattices presented in [16], we propose in this section a novel approach to attribute reduction. Dislike the approach in [16] which is based on the discernibility matrix, the proposed method employs all the meet-irreducible elements in the lattice.

In what follows, let $S = (U, A, I)$ be a formal context. Denote by $\mathcal{F} = \{ \bigcap\limits_{a \in B} \psi(a) : B \subseteq A\}$ all the extents of $\mathcal{L}(U, A, I)$, and $\mathcal{F}_0 = \{X \in \mathcal{F} | \ (X, \phi(X)) \in \mathcal{L}(U, A, I)$ is an meet-irreducible element of $\mathcal{L}(U, A, I)\} = \{X_1, \cdots, X_m\}$. Let $[X_p] = \{B \subseteq A \mid \psi(B) = X_p\}$.

Proposition 5. $\forall X \in \mathcal{F}_0$, every minimal element of $[X]$, in the sense of set-inclusion, is a singleton.

Proof. If E is a minimal element of $[X]$ with $|E| > 1$, suppose $b_1, b_2 \in E$. Then we have $X = \psi(E) = \psi(\bigcup\limits_{b \in E} b) = \bigcap\limits_{b \in E} \psi(b) = (\bigcap\limits_{b \in \{b_1, b_2\} \subseteq E} \psi(b)) \cap (\bigcap\limits_{b \in E \setminus \{b_1, b_2\}} \psi(b))$. If $|E \setminus \{b_1, b_2\}| = 0$, then $X = (\bigcap\limits_{b \in \{b_1, b_2\} \subseteq E} \psi(b)) \cap U = \psi(b_1) \cap \psi(b_2)$. Thus $(X, \phi(X)) = (\psi(b_1), \phi(\psi(b_1))) \wedge (\psi(b_2), \phi(\psi(b_2)))$. Since $(X, \phi(X))$ is an meet-irreducible element, $X = \psi(b_1)$ or $X = \psi(b_2)$ holds. Hence, $\{b_1\} \in [X]$ or $\{b_2\} \in [X]$. Which means $\{b_1\}$ or $\{b_2\}$ is a minimal element of $[X]$, which is a contradiction to the hypothesis. If $|E \setminus \{b_1, b_2\}| = 1$, suppose $E = \{b_1, b_2, b\}$, then $X = (\psi(b_1) \cap \psi(b_2)) \cap \psi(b)$. Since $X \in \mathcal{F}_0$, similar to the above proof we conclude that $X = \psi(b_1) \cap \psi(b_2)$ or $X = \psi(b)$. Clearly, $X = \psi(b)$ means that $\{b\}$ is a minimal element of $[X]$, which is a contradiction to the hypothesis. Therefore, E is a singleton. Since A is a finite set, we conclude the statement inductively.

Proposition 6. For every $X_p \in \mathcal{F}_0$, $1 \leq p \leq m$, choose an arbitrary minimal element B_p of $[X_p]$. Then $\bigcup\limits_{p=1}^{m} B_p$ is a reduct of the formal context S.

Proof. Let $C = \bigcup\limits_{p=1}^{m} B_p$, first we prove that C is a consistent set of S.

$\forall (X, B) \in \mathcal{L}(U, A, I)$, if $X \in \mathcal{F}_0$, then $B \in [X]$. If B is a minimal element of $[X]$, then $B \subseteq C$, and $(X, B) \in \mathcal{L}(U, C, I_C)$. If B is not a minimal element of $[X]$, then there exists a minimal element B_k of $[X]$ such that $B_k \subseteq B$. Which implies that $B \cap C = B \cap (\bigcup\limits_{p=1}^{m} B_p) = B_k \cup (\bigcup\limits_{p \neq k} (B \cap B_p))$. As a result, $\psi(B \cap C) = \psi(B_k) \cap (\bigcap\limits_{p \neq k} \psi(B \cap B_p)) = X \cap (\bigcap\limits_{p \neq k} \psi(B \cap B_p)) \subseteq X$. Clearly $\psi(B \cap C) \supseteq X$, we then have $\psi(B \cap C) = X$. But $\phi_C(X) = \phi(X) \cap C = B \cap C$, we conclude that $(X, B \cap C) \in \mathcal{L}(U, C, I_C)$.

If $X \notin \mathcal{F}_0$, then $(X, B) \in \mathcal{L}(U, A, I)$ is not a meet-irreducible element of $\mathcal{F}$. By virtue of Proposition 2, there exist meet-irreducible elements $(X_{i1}, \phi(X_{i1}))$, $\cdots, (X_{il}, \phi(X_{il}))$ $(l \leq m)$ of $\mathcal{L}(U, A, I)$ such that $(X, B) = (X_{i1}, \phi(X_{i1})) \wedge \cdots \wedge (X_{il}, \phi(X_{il}))$, thus we have $X = X_{i1} \cap \cdots \cap X_{il}$. Since each X_{ij} $(1 \leq j \leq l)$

belongs to $\mathcal{F}_0$, there exist a minimal element B_{ij} of $[X_{ij}]$ such that $B_{ij} \subseteq C$. Which implies that

$$X = \bigcap_{j=1}^{l} X_{ij} = \psi(\bigcup_{j=1}^{l} B_{ij}) = \psi(B).$$

Therefore,

$$\bigcup_{j=1}^{l} B_{ij} = \bigcup_{j=1}^{l} \phi(X_{ij}) \subseteq \phi(\bigcap_{j=1}^{l} X_{ij}) = B.$$

Clearly,

$$\phi_C(X) = \phi(X) \cap C = B \cap C = B \cap (\bigcup_{p=1}^{m} B_P) = \bigcup_{j=1}^{l} B_{ij} \cup (\bigcup_{p \neq i1,\cdots,il} (B \cap B_p)).$$

Thus,

$$X \subseteq \psi(\phi_C(X)) = \psi(\bigcup_{j=1}^{l} B_{ij}) \cap \psi(\bigcup_{p \neq i1,\cdots,il} (B \cap B_p)) = X \cap (\bigcup_{p \neq i1,\cdots,il} \psi(B \cap B_p)) \subseteq X.$$

Consequently, $\psi(\phi_C(X)) = X$ and $(X, \phi_C(X)) \in \mathcal{L}(U, C, I_C)$.

Combining the above, we conclude that C is a consistent set of S.

If there exists an element $c_0 \in C$ such that $C \setminus \{c_0\}$ is a consistent set of S, then $\mathcal{L}(U, A, I) \cong \mathcal{L}(U, C \setminus \{c_0\}, I_{C \setminus \{c_0\}})$, and thus $\mathcal{L}(U, C \setminus \{c_0\}, I_{C \setminus \{c_0\}}) \leq \mathcal{L}(U, C, I_C)$. For $c_0 \in C$, by virtue of Proposition 5 and the definition of C, there exists an element $X_0 \in \mathcal{F}_0$ such that $\{c_0\}$ is a minimal element of $[X_0]$. Since $(\psi(c_0), \phi_C(\psi(c_0))) \in \mathcal{L}(U, C, I_C)$, there exists an attribute set $E \subseteq C \setminus \{c_0\}$ such that $(\psi(c_0), E) \in \mathcal{L}(U, C \setminus \{c_0\}, I_{C \setminus \{c_0\}})$, which implies that $\psi(E) = \psi(c_0)$ and $E \in [X_0]$. Noticing that $c_0 \notin E$, it must have a minimal element F of $[X_0]$ such that $F \subseteq E$. Of course, $F \not\subseteq C$, and thus $E \not\subseteq C$, a contradiction. Therefore, $C \setminus \{c\}$ is not a consistent set of S for any $c \in C$. That is to say, C is a reduct of the formal context S.

By using Proposition 6, we can naturally develop an algorithm to calculate reducts of a formal context.

Algorithm for attribute reduction

Input: formal context $S = (U, A, I)$.
Output: reduct $RED(A)$ of S.

(1) For all $a \in A$, compute $\psi(a)$.
(2) Compute $\mathcal{F}$: $\forall B \subseteq A$, compute $\bigcap_{a \in B} \psi(a)$, then $\mathcal{F}$ is composed of all $\bigcap_{a \in B} \psi(a)$, $B \subseteq A$.

(3) Compute $\mathcal{F}_0$: let $X \in \mathcal{F}$, if $X \neq Y \cap Z$ holds for any $Y, Z \in \mathcal{F}$, then $X \in \mathcal{F}_0$. $\mathcal{F}_0$ is composed of all these X.
(4) $\forall X_i \in \mathcal{F}_0 (1 \leq i \leq |\mathcal{F}_0|)$, compute $[X_i]$.
(5) $\forall X_i \in \mathcal{F}_0$, compute all the minimal elements of $[X_i]$.
(6) Output $RED(A) = \bigcup_{i=1}^{p} B_i$, where $p = |\mathcal{F}_0|$, B_i is a minimal element of $[X_i]$.

The following example (quote from [16]) will illustrate our algorithm of attribute reduction described above.

Example 1. A formal context $T = (U, A, I)$ is shown in Table 1, where the object set $U = \{1, 2, 3, 4\}$, the attribute set $A = \{a, b, c, d, e\}$.

Table 1. A formal context (U, A, I)

U	a	b	c	d	e
1	1	1	0	1	1
2	1	1	1	0	0
3	0	0	0	1	0
4	1	1	1	0	0

Step1. For all $x \in A$, compute $\psi(x)$.

$$\psi(a) = \{1, 2, 4\},$$
$$\psi(b) = \{1, 2, 4\},$$
$$\psi(c) = \{2, 4\},$$
$$\psi(d) = \{1, 3\},$$
$$\psi(e) = \{1\}.$$

Step2. Compute $\mathcal{F}$.

$$\mathcal{F} = \{\{1, 2, 4\}, \{2, 4\}, \{1\}, \{1, 3\}, \emptyset\}.$$

Step3. Compute $\mathcal{F}_0$.

$$\mathcal{F}_0 = \{\{1, 2, 4\}, \{2, 4\}, \{1, 3\}\}.$$

Step4. For any $X \in \mathcal{F}_0$, compute $[X]$.

$$[\{1, 2, 4\}] = \{\{a\}, \{b\}, \{a, b\}\},$$
$$[\{2, 4\}] = \{\{c\}, \{a, c\}, \{b, c\}, \{a, b, c\}\},$$
$$[\{1, 3\}] = \{\{d\}\}.$$

Step5. Compute the minimal elements of $[X]$, for any $X \in \mathcal{F}_0$.

$\{a\}$ and $\{b\}$ are two minimal elements of $[\{1, 2, 4\}]$,
$\{c\}$ is the only minimal element of $[\{2, 4\}]$,
$\{d\}$ is the minimal element of $[\{1, 3\}]$.

Step6. $C_1 = \{a, c, d\}$ and $C_2 = \{b, c, d\}$ are reducts of S.

By using the approach proposed in [16], we obtain the discernibility matrix of formal context of $S = (U, A, I)$ as shown in Table 2.

Table 2. The discernibility matrix

FC	$FC1$	$FC2$	$FC3$	$FC4$	$FC5$	$FC6$
$FC1$	ϕ	$\{c, d, e\}$	$\{a, b, e\}$	$\{d, e\}$	$\{a, b, d, e\}$	$\{c\}$
$FC2$	$\{c, d, e\}$	ϕ	$\{a, b, c, d\}$	$\{c\}$	$\{a, b, c\}$	$\{d, e\}$
$FC3$	$\{a, b, e\}$	$\{a, b, c, d\}$	ϕ	$\{a, b, d\}$	$\{d\}$	$\{a, b, c, e\}$
$FC4$	$\{d, e\}$	$\{c\}$	$\{a, b, d\}$	ϕ	$\{a, b\}$	$\{c, d, e\}$
$FC5$	$\{a, b, d, e\}$	$\{a, b, c\}$	$\{d\}$	$\{a, b\}$	ϕ	A
$FC6$	$\{c\}$	$\{d, e\}$	$\{a, b, c, e\}$	$\{c, d, e\}$	A	ϕ

Where $FC1 = (\{1\}, \{a, b, d, e\})$, $FC2 = (\{2, 4\}, \{a, b, c\})$, $FC3 = (\{1, 3\}, \{d\})$, $FC4 = (\{1, 2, 4\}, \{a, b\})$, $FC5 = (U, \phi)$, $FC6 = (\phi, A)$.

By discernibility formula in [15], we can easily obtain that $C_1 = \{a, c, d\}$ and $C_2 = \{b, c, d\}$ are two minimal sets satisfying the condition $C_i \cap D \neq \phi$ $(i = 1, 2)$, where D is arbitrary nonempty element in the discernibility matrix, therefore C_1 and C_2 are two reducts of the formal context.

We can see that the results obtained by the two methods are the same.

4   Conclusion

The theory of attribute reduction in concept lattice is one of the important issues in data mining. It makes the hidden knowledge more clear and precise, and promotes wide-ranging application of concept lattice. In this paper, we have proposed a new method of attribute reduction based on all the meet-irreducible elements of the concept lattice. It is proved that the proposed method is equivalent to the one in [15]. The related algorithm of attribute reduction has also been provided.

Acknowledgements

This paper is supported by the Natural Science Foundation of China (60773174, 60673096, 60703117), the Natural Science Foundation of Hebei Province (A2006-000129), and the Science Foundation of Hebei Normal University (L2005Z01).

References

1. Carpineto, C., Romano, G.: A Lattice Conceptual Clustering System and Its Application to Browsing Retrieval. Mach. Learn. 10, 95–122 (1996)
2. Faid, M., Missaoi, R., Godin, R.: Knowledge Discovery in Complex Objects. Comput. Intell. 15, 28–49 (1999)

3. Godin, R.: Incremental Concept Formation Algorithm Based on Calois (Concept) Lattice. Comput. Intell. 11, 246–267 (1995)
4. Godin, R., Missaoi, R.: An Incremental Concept Formation Approach for Learning from Databases. Theor. Comp. Sci. 133, 387–419 (1994)
5. Ganter, B., Wille, R.: Formal Concept Analysis: Mathematical Foundations. Springer, Berlin (1999)
6. Harms, S.K., Deogum, J.S.: Sequential Association Rule Mining With Time Lags. J. Intell. Inform. Sys. 22, 7–22 (2004)
7. Kent, R.E.: Rough Concept Analysis: A Synthesis of Rough Set and Formal Concept Analysis. Fund. Inform. 27, 169–181 (1996)
8. Oosthuizen, G.D.: The Application of Concept Lattice to Machine Learning. Technical Report, University of Pretoria, South Africa (1996)
9. Shao, M.W.: The Reduction for Two Kinds of Generalized Concept Lattice. In: Proceedings of the Fourth International Conference on Machine Learning and Cybernetics, pp. 2217–2222. IEEE Press, New York (2005)
10. Wang, X., Ma, J.M.: A Novel Approach to Attribute Reduction in Concept Lattices. In: Wang, G.-Y., Peters, J.F., Skowron, A., Yao, Y. (eds.) RSKT 2006. LNCS (LNAI), vol. 4062, pp. 522–529. Springer, Heidelberg (2006)
11. Wille, R.: Knowledge Acquisition by Methods of Formal Concept Analysis. In: Diday, E. (ed.) Data Analysis, Learning Symbolic and Numeric Knowledge, pp. 365–380. Nova Science, New York (1989)
12. Wille, R.: Restructuring Lattice Theory: An Approach Based on Hierarchies of Concepts. In: Rival, I. (ed.) Ordered sets, pp. 445–470. Dordrecht-Boston, Reidel (1982)
13. Yao, Y.Y.: Concept Lattices in Rough Set Theory. In: Dick, S., Kurgan, L., Pedrycz, W., Reformat, M. (eds.) Proceedings of 2004 Annual Meeting of the North American Fuzzy Information Processing Society (NAFIPS 2004), pp. 796–801. IEEE Press, Now York (2004)
14. Zhang, W.X., Wei, L., Qi, J.J.: Attribute Reduction in Concept Lattice Based on Discernibility Matrix. In: Ślęzak, D., Yao, J., Peters, J.F., Ziarko, W., Hu, X. (eds.) RSFDGrC 2005. LNCS (LNAI), vol. 3642, pp. 157–165. Springer, Heidelberg (2005)
15. Zhang, W.X., Qiu, G.F.: Uncertain Decision Making Based on Rough Sets. Tsinghua University Press, Beijing (2005)
16. Zhang, W.X., Wei, L., Qi, J.J.: Attribute Reduction Theory and Approach to Concept Lattice. Sci. China (F) 48, 713–726 (2005)

A Novel Deformation Framework for Face Modeling from a Few Control Points

Xun Gong[1,2] and Guoyin Wang[1,2]

[1] Institute of Computer Science and Technology,
Chongqing University of Posts and Telecommunications,
Chongqing 400065, P.R. China
[2] School of Information Science and Technology,
Southwest Jiaotong University,
Chengdu 600031, P.R. China
gongxun@foxmail.com

Abstract. This paper presents a novel method to reconstruct realistic 3D faces from a set of control points: (i) a local deformation approach is proposed, which could effectively preserve the information when only a few features are known; (ii) a Ternary Deformation Framework (TDF), combining the strengths of both local modification and global calculation, is developed to accurately recover the face shape. Simulation results show that TDF outperforms the conventional methods with respect to the modeling precision and could generate realistic face models.

1 Introduction

Realistic face synthesis is one of the most fundamental and difficult problems in computer graphics. The aim of this work is to provide an accurate solution of modeling face shape from a few control points.

Actually, studies in modeling realistic faces and animating it date back to the early 1970's [1], with hundreds of research papers published. Current commercial systems are commonly equipped with laser-based scanners, or project a pattern on the subject's face. The major drawback of such systems is the high cost of hardware required. On the other hand, reconstruction from multiple view images and video has produced a number of shape-from-motion techniques [2,3,4,5]. The basic idea of these methods is to adjust a general model to match the key features in the images (video frames). Although some of these methods [4,5] could create 3D faces realistically to a certain extent, but their computing cost is always high and they are not stable enough. A 3D morphable model (3DMM) [6,7] is another representative approach that could reconstruct human face automatically from a single image. 3DMM has demonstrated the advantage of using linear class model. However, the time-consuming procedure and instability are the main shortcomings of this analysis-by-synthesis method.

Hence, modeling a 3D face from a few feature points on one image is potentially providing a tradeoff between quality and speed. Some works have been done in [8,9]. However, these approaches are always converged to a solution

G. Wang et al. (Eds.): RSKT 2008, LNAI 5009, pp. 434–441, 2008.

very close to the initial value, resulting in a reconstruction which resembles the generic model rather than the particular face. The work presented here is inspired by Knothe et. al. [10] who use Local Feature Analysis (LFA) to characterize the coarse fitted results of a holistic method. However, the reshaped surface is rugged due to the localization strategy. To overcome this shortcoming, we propose a new localization strategy to reinforce the local property, and thus a novel Weighted LFA (WLFA) is proposed. Consequently, we develop a Ternary Deformation Framework (TDF), a combination of the strengths of both local modifications and global calculations, to generate more precise models.

2 Localization Strategy

Addressing the problem of shape deforming by a set of control vertices, a localization strategy has to guarantee that the interaction between controls is null. And, each control attaches more influence to near-neighboring vertices than to those far away, e.g. moving the nose tip should have less to do with vertices on forehead. Deformations based on one control point are illustrated in Fig. 1.

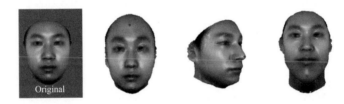

Fig. 1. Deformations modeled by one control point using LFA

Voronoi Tessellation (VT) was used in [10] to compute the influence for each control point. This is achieved by setting a weight value, decreasing linearly in each tessellation, to 1 at the control point while to 0 at its boundary, as shown in Fig. 2(a,b). However, the influence mask of this type encounters some problems in deformation and even would lead to unexpected results. Three cases of curve deformation by two control points are simulated in 2D cases (see Fig. 3(a)), where bold fine lines are the original curves, and dashed ones are their reshaped results, a and b are control points, c is an arbitrary point in T_1, which under the control of a, spline is fitted from these three points. We can see that the deformations using this localization strategy are poor. The first case, especially, has totally changed derivative of the original shape.

The main cause to VT's problem is that tessellation divisions make each point be controlled by only one control point (as shown in Fig. 2(a,b)), ignoring others which even may be near around. To compute the mask properly, a simple Distance Measurement (DM) approach, here, is developed. Given k controls $C = \{c_t, 1 \leq t \leq k\}$ on a surface, a vertex v_j of the rest is adjusted by all

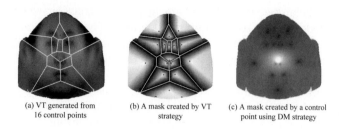

(a) VT generated from   (b) A mask created by VT   (c) A mask created by a control
16 control points   strategy   point using DM strategy

Fig. 2. Localization effects of VT and DM

controls at different weights $w_j = (w_{j,1}, ..., w_{j,k})$ which are calculated by the distance between c_t and p, let $r_t = 1/d(c_t, p)$ be the reciprocal distance, thus

$$w_{j,t} = r_{j,t} \Big/ \sum r_{j,t}, \quad 1 \leq t \leq k. \tag{1}$$

Eq. (1) tells that the new position of a vertex is determined by all controls, the weights of which are changed gradually resulting in a smooth mask, as shown in Fig. 2(c). An illustration shown in Fig. 3(b) has validated this point, too.

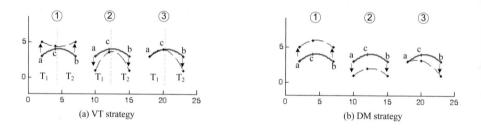

(a) VT strategy   (b) DM strategy

Fig. 3. Curves' deformation based on two control points a, b

3   Face Reconstruction

In this section, we introduce the linear class model of human face at first. Then, a local deformation method Weighted LFA (WLFA) is proposed on the basis of the localization strategy. After that, we combine the global method DCDM (Dynamic Component Deforming Model) and local WLFA with a smooth function RBFs (Radial Based Functions) to form a Ternary Deformation Framework.

3.1   Linear Class of Human Face

A face shape s_i can be denoted by a vector:

$$s_i = (x_1, y_1, z_1, \cdots, x_j, y_j, z_j, \cdots, x_n, y_n, z_n)^T \in \Re^{3n}, 1 \leq j \leq n, \tag{2}$$

where (x_j, y_j, z_j) is the coordinates of the j-th vertex v_j, n is vertex number. A group of m faces can be regarded as a linear class $\boldsymbol{S} = (\boldsymbol{s}_1, \cdots, \boldsymbol{s}_m) \in \Re^{3n \times m}$, and then a novel face $\boldsymbol{s}_{new}$ can be represented as a linear combination:

$$\boldsymbol{s}_{new} = \boldsymbol{S} \cdot \boldsymbol{\alpha} \tag{3}$$

where $\boldsymbol{\alpha} = (a_1, \cdots a_i, \cdots, a_m)^T, a_i \in [0, 1]$ and $\sum_{i=1}^{m} a_i = 1$.

PCA is applied to get a compact representation: let $\boldsymbol{Q} = (\boldsymbol{q}_1, \cdots, \boldsymbol{q}_{m'}) \in \Re^{3n \times m'}$ be the eigenmatrix by concatenating prior $m'(\leq m - 1)$ eigenvectors, the corresponding eigenvalues are $\boldsymbol{\sigma} = (\sigma_1^2, \cdots, \sigma_{m'}^2)$, where $\sigma_1^2 \geq \cdots \geq \sigma_{m'}^2$. With the scaled eigenmatrix [8], Eq. (3) can be rewritten as:

$$\boldsymbol{s}_{new} = \bar{\boldsymbol{s}} + \boldsymbol{Q}_s \cdot \boldsymbol{\beta} = \bar{\boldsymbol{s}} + \Delta \boldsymbol{s} \tag{4}$$

where $\boldsymbol{Q}_s = (\sigma_1 \boldsymbol{q}_1, \cdots, \sigma_{m'} \boldsymbol{q}_{m'})$, $\bar{\boldsymbol{s}} = \frac{1}{m} \sum_{i=1}^{m} \boldsymbol{s}_i$, and $\boldsymbol{\beta} = (\beta_1, \cdots, \beta_{m'})^T \in \Re^{m'}$. Eq. (4) is a Reconstruction Function (RF) which indicates that a special face can be deformed from an average face with a certain amount of changes $\Delta \boldsymbol{s}$.

3.2 Local Modification

The traditional statistic methods, like PCA [8], is holistic and typically is not topographic. To get a personalized face shape, as mentioned in section 2, a local modification is needed to depict the personality trails. Based on DM, here, we propose a method termed Weighted Local Feature Analysis (WLFA) that is extended from another powerful statistic tool — Local Feature Analysis (LFA) [11], which has local property and preserves topography.

Starting from an average face $\bar{\boldsymbol{s}}$, let $\boldsymbol{v}_t'$ be a new position of a control point $\boldsymbol{v}_t$, and $\Delta \boldsymbol{v}_t = \boldsymbol{v}_t' - \boldsymbol{v}_t$. The corresponding rows of $\boldsymbol{Q}$ are denoted by $\boldsymbol{Q}_t \in \Re^{l \times m'}(l=2$ or $l=3$ depends on whether 2D or 3D coordinate is used). And,

$$\boldsymbol{\Omega} = diag(1/\sigma_i) \in \Re^{m' \times m'}, \quad 1 \leq i \leq m', \tag{5}$$

is a normalizing matrix of LFA kernel $\boldsymbol{K} = \boldsymbol{Q} \cdot \boldsymbol{\Omega} \cdot \boldsymbol{Q}^T$, where, σ_i is standard deviation of the i-th principal component, $diag(.)$ means creating a diagonal matrix for a vector. Then, the LFA-coefficients $\boldsymbol{c}_t$ of the t-th control is

$$\boldsymbol{c}_t = \left(\boldsymbol{Q}_t \cdot \boldsymbol{\Omega}^{-1} \cdot \boldsymbol{Q}_t^T\right)^{-1} \cdot \Delta \boldsymbol{v}_t. \tag{6}$$

Thus, the total displacements $\Delta \boldsymbol{s}^t$ created by $\Delta \boldsymbol{v}_t$ are

$$\Delta \boldsymbol{s}^t = \boldsymbol{Q} \cdot \left(\boldsymbol{\Omega}^{-1} \cdot \boldsymbol{Q}_t^T \cdot \boldsymbol{c}_t\right) \in \Re^{3n}. \tag{7}$$

WLFA uses the localization strategy DM when matching an average shape to k controls. First, we construct a displacement matrix for all control points:

$$\boldsymbol{D} = \left(\Delta \boldsymbol{s}^1, \cdots, \Delta \boldsymbol{s}^t, \cdots, \Delta \boldsymbol{s}^k\right) \in \Re^{3n \times k}. \tag{8}$$

Then, a weight map W for all vertices is created by:

$$W = \left(w_1^T, w_1^T, w_1^T, \cdots, w_j^T, w_j^T, w_j^T, \cdots, w_n^T, w_n^T, w_n^T\right) \in \Re^{k \times 3n}, \qquad (9)$$

where, w_j is computed by Eq. (1). Thus, the WLFA based displacements are

$$L\left(\Delta s\right) = diag\left(D \cdot W\right), \qquad (10)$$

with $diag(.)$ forms a vector by extracting the diagonal elements of a matrix.

3.3   Ternary Deformation Framework

TDF is a combination of the global method DCDM, local WLFA and a smooth function RBFs. In the following, we introduce DCDM and RBFs briefly at first, and then use TDF to formulate the final deformation.

Different from the traditional PCA, DCDM is an effective tool [12] to alleviate the over-fitting problem by eliminating useless information from training set. Given k features on a novel face s_{new}, let $s_{new}^f \in \Re^l$ ($l = 2k$ or $l = 3k$ depends on if 2D or 3D coordinate is used) be a sparse version of s_{new} by concatenating the features' position. Likewise, the corresponding sparse eigenmatrix $Q_s^f = (\sigma_1 q_1^f, \cdots, \sigma_{m'} q_{m'}^f) \in \Re^{l \times m'}$. Hence, a sparse RF (see Eq. (4)) can be:

$$s_{new}^f = \bar{s} + \Delta s^f = \bar{s}^f + Q_s^f \cdot \beta. \qquad (11)$$

In this way, a coarse fitting coefficient can be achieved by using the approximation with regularization [8]:

$$\beta^0 = V \cdot \left(\frac{\lambda_i}{\lambda_i^2 + \eta}\right) \cdot U^T \cdot \Delta s^f, \qquad (12)$$

with $\eta \geq 0$ a weight factor controls a tradeoff between the precision of feature point matching and prior probability, and, $Q_s^f = U \Lambda V^T$, $\Lambda = diag(\lambda_i) \in \Re^{l \times m'}$, $U \in \Re^{l \times l}$, $V \in \Re^{m' \times m'}$. The t-test is used to determine the significance of each component according to the characteristics of a novel face. It concatenates the most correlative ones to form a Personal EigenMatrix (PEM) and updates β^0 to β^1, hence the DCDM based displacements are calculated by

$$D\left(\Delta s\right) = Q_s \cdot \beta^1. \qquad (13)$$

Radial Based Functions (RBFs), known as the most accurate as well as stable interpolation method [13] that has been widely used in surface reconstruction and modeling, is utilized to smooth the surface modeled by DCDM. The core of RBFs is a smooth function $f(v)$, defined as:

$$f(v) = p(v) + \sum_{t=1}^{k} \lambda_t \cdot \varphi\left(d\left(v, v_t\right)\right), \qquad (14)$$

where v_t is the t-th feature, k is the number of features, $d(\cdot, \cdot)$ is the Euclidean distance between two vertices. Polynomial $p(v) = M \cdot v + t$ represents the part of affine transformation, $\varphi(\|r\|) = e^{-r/40}$. Substituting all feature points in Eq. (14) to construct a linear system, which is then solved to get the parameters M, t, $\boldsymbol{\lambda} = (\lambda_1, \cdots, \lambda_k)$, thus, $f(\boldsymbol{v})$ is fixed. So, the RBFs based displacements are

$$R(\Delta s) = \left(f(\boldsymbol{v}_1)^T, \cdots, f(\boldsymbol{v}_j)^T, \cdots, f(\boldsymbol{v}_n)^T \right)^T. \tag{15}$$

In TDF, we use DCDM and RBFs to obtain an approximate and smooth estimation of a face shape, and WLFA is utilized to depict the local details. Hence, there is no need of a weight factor for WLFA displacements. Then, based on Eq. (10), (13), (15), TDF is formulated as a weighted sum:

$$T(\Delta s) = \mu_1 \cdot D(\Delta s) + \mu_2 \cdot R(\Delta s) + L(\Delta s), \quad \sum \mu_i = 1. \tag{16}$$

4 Experiments and Discussion

400 laser-scanned faces are selected from the BJUT-3D Face Database [14], which are then split into a training set and a testing set of m=200 faces each. From training set, we compute 199 components which serve afterward as basic space. Excessive uses of components may lead to over-fitting [12], we just use prior 50 of them in modeling. To evaluate reconstruction, we use the average Euclidean distance [8] to measure two surfaces s_r and s_t.

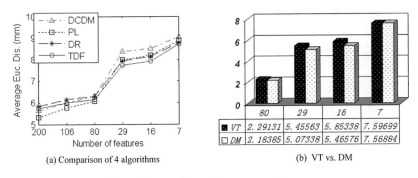

(a) Comparison of 4 algorithms   (b) VT vs. DM

Fig. 4. Comparison of reconstruction errors

The advantage of the proposed TDF is evaluated on the testing set (use 2D coordinate), in which algorithms PCA+LFA (PL) [10], DCDM [12], and DCDM+RBFs (DR) are assessed as counterparts. For PCA, the regularization $\eta = 0.0001$. The significance level α used in DCDM is 0.05. For DR and TDF, $\mu_1 = \mu_2 = 0.5$. The results (see Fig. 4(a)) reveal that the merit of TDF is more evident when the features' number is small. In previous work [12], we have also argued that DCDM outperforms Sparse Deforming Model [9] and SRSD (Surface Reconstruction from Sparse Data) [8]. Hence, TDF could be deemed as the state-of-arts among the relative methods. Fig. 5 shows that TDF could create a more realistic and smooth surface than that of PL method.

To find out the contribution of DM for modeling improvements, we run the algorithm PL twice with DM and VT respectively (use 3D coordinate). The results shown in Fig. 4(b) reveals that DM performs better than VT. A visual comparison is demonstrated in 2nd and 3rd images of Fig. 5.

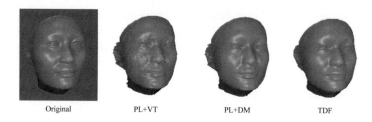

Original   PL+VT   PL+DM   TDF

Fig. 5. Demonstrations of reconstruction on testing face by various algorithm

To illustrate the performance of TDF in modeling face shapes from 2D photographs. Here, we apply it to a group of real photographs take from our lab. The picture is projected orthogonally to create the textures. Using 9 salient features, Fig. 6 shows that the generated 3D faces are realistic within a certain rotations.

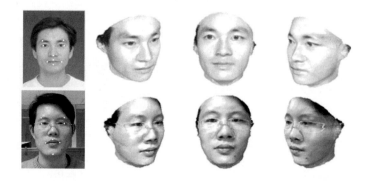

Fig. 6. Reconstruction on real photographs by TDF

5   Conclusion

We propose a novel system to reconstruct 3D face shape from a few control points. At first, a novel Weighted LFA is proposed to recover the local details of a special face. Then, we develop a TDF to combine the strengths of local modification with global calculations to generate more accurate results. Experiment results show that our method is quite effective and the 3D faces generated are rather realistic.

Acknowledgements

This paper is partially supported by the Program for New Century Excellent Talents in University (NCET), Open Funds of Chongqing Key lab of Computer Network and Communication, Natural Science Foundation of Chongqing under Grant (No. CSTC2007BB2445). Portions of the research in this paper use the BJUT-3D Face Database collected under the joint sponsor of National Natural Science Foundation of China.

References

1. Parke, F.I.: Computer generated animation of faces. In: ACM National Conference, pp. 451–457. ACM Press, New York (1972)
2. Lee, W.S., Magnenat, T.N.: Fast head modeling for animation. Image and Vision Computing 4, 355–364 (2000)
3. Wang, K., Zheng, N.N.: 3D face modeling based on SFM algorithm. Chinese Journal of Computers 6, 1048–1053 (2005)
4. Dimitrijevic, M., Ilic, S., Fua, P.: Accurate face models from uncalibrated and ill-lit video sequence. In: IEEE Conference on Computer Vision and Pattern Recognition, pp. 188–202. IEEE press, New York (2004)
5. Pighin, F., Hecker, J.: Synthesizing realistic facial expressions from photographs. In: 25th Annual Conference on Computer Graphics, pp. 75–84. ACM press, New York (1998)
6. Blanz, V., Vetter, T.: A morphable model for the synthesis of 3D-faces. In: SIG-GRAPH 1999 Conference Proceedings, pp. 187–194. ACM press, New York (1999)
7. Wang, C.Z., Yin, B.C., Sun, Y.F., Hu, Y.L.: An improved 3D face modeling method based on morphable model. Acta Automatica Sinica 3, 232–239 (2007)
8. Blanz, V., Vetter, T.: Reconstructing the complete 3D shape of faces from partial information. Informationstechnik und Technische Informatik 6, 295–302 (2002)
9. Chai, X.J., Shan, S.G., Qing, L.Y.: Pose and illumination invariant face recognition based on 3D face reconstruction. Journal of Software 3, 525–534 (2006)
10. Knothe, R., Romdhani, S., Vetter, T.: Combining PCA and LFA for surface reconstruction from a sparse set of control points. In: IEEE Conference on Automatic Face and Gesture Recognition, pp. 637–644. IEEE press, New York (2006)
11. Penev, P.S., Atick, J.J.: Local Feature Analysis: A general statistical theory for object representation. Network: Computation in Neural Systems 3, 477–500 (1996)
12. Gong, X., Wang, G.Y.: A Dynamic Component Deforming Model for Face Shape Reconstruction. In: Bebis, G., Boyle, R., Parvin, B., Koracin, D., Paragios, N., Tanveer, S.-M., Ju, T., Liu, Z., Coquillart, S., Cruz-Neira, C., Müller, T., Malzbender, T. (eds.) ISVC 2007, Part I. LNCS, vol. 4841, pp. 488–497. Springer, Heidelberg (2007)
13. Carr, J.C., Fright, W.R., Beatson, R.K.: Surface interpolation with radial basis functions for medical imaging. IEEE Trans. Medical Imaging 1, 96–107 (1997)
14. Multimedia Tech & Graphics Lab, Beijing University of Technology: The BJUT-3D Large-Scale Chinese Face Database. Technical report (2005)

Ranking with Partial Orders and Pairwise Comparisons

Ryszard Janicki

Department of Computing and Software,
McMaster University,
Hamilton, ON, L8S 4K1 Canada
janicki@mcmaster.ca

Abstract. A new approach to *Pairwise Comparisons based Ranking* is presented. An abstract model based on partial orders instead of numerical scales is introduced and analysed. The importance of the concept of *indifference* and the power of *weak order extensions* are discussed.

1 Introduction

The *Pairwise Comparisons* method is based on the observation that while ranking the importance of *several* objects is often problematic, it is much easier when to do restricted to *two* objects. The problem is then reduced to constructing a global ranking from the set of partially ordered pairs. The method could be traced to the Marquis de Condorcet's 1785 paper (see [1]), was explicitly mentioned and analysed by Fechner in 1860 [3], made popular by Thurstone in 1927 [13], and was transformed into a kind of semi-formal methodology by Saaty in 1977 (called *AHP*, Analytic Hierarchy Process, see [2,6,11]).

At present Pairwise Comparisons are practically identified with the controversial Saaty's AHP. On one hand AHP has respected practical applications, on the other it is still considered by many (see [2,7,10]) as a flawed procedure that produces arbitrary rankings. Due to a lack of space, the sources of the strange and contradictory examples of rankings obtained by AHP will not be discussed. An interesting reader is referred to [2,7,9,10]. However we believe that most of the problems mentioned in [2,7,9,10] and others, stem mainly from the following two sources:

1. The final outcome is always expected to be totally ordered (i.e. for all a, b, either $a < b$ or $b > a$),
2. Numbers are used to calculate the final outcome.

Non-numerical solutions were proposed and discussed in [7,8,9]. The model presented in this paper stems from [8], was highly influenced by [4], and is orthogonal to that of [7]. The concept of "consistency", crucial in [7] is not discussed in this paper at all. Algorithms for automatic construction of a final ranking are the essence of this paper, but they are not discussed in [7]. The model presented below uses no numbers and is entirely based on the concept of partial orders.

G. Wang et al. (Eds.): RSKT 2008, LNAI 5009, pp. 442–451, 2008.

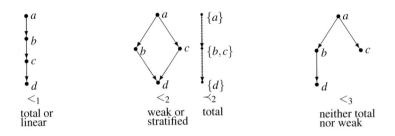

Fig. 1. Various types of partial orders (represented as Hasse diagrams). The total order $\prec_2$ represents the weak order $<_2$., The partial order $<_3$ is nether total nor weak.

2   Total, Weak and Partial Orders

Let X be a *finite* set. A relation $\lhd \subseteq X \times X$ is a (*sharp*) *partial order* if it is irreflexive and transitive, i.e. if $a \lhd b \Rightarrow \neg(b \lhd a)$ and $a \lhd b \lhd c \Rightarrow a \lhd c$, for all $a, b, c \in X$. A pair $(X, \lhd)$ is called a *partially ordered set*. We will often identify $(X, \lhd)$ with $\lhd$, when X is known.

We write $a \sim_\lhd b$ if $\neg(a \lhd b) \wedge \neg(b \lhd a)$, that is if a and b are either *distinctly incompatible* (w.r.t. $\lhd$) or *identical* elements of X. We also write

$$a \approx_\lhd b \iff \{x \mid x \sim_\lhd a\} = \{x \mid x \sim_\lhd b\}.$$

The relation $\approx_\lhd$ is an equivalence relation (i.e. it is reflexive, symmetric and transitive) and it is called *the equivalence with respect to* $\lhd$, since if $a \approx_\lhd b$, there is nothing in $\lhd$ that can distinquish between a and b (see [4] for details). We always have $a \approx_\lhd b \Rightarrow a \sim_\lhd b$, and one can show that [4]:

$$a \approx_\lhd b \iff \{x \mid a \lhd x\} = \{x \mid b \lhd x\} \wedge \{x \mid x \lhd a\} = \{x \mid x \lhd b\}$$

A partial order is [4]

- *total* or *linear*, if $\sim_\lhd$ is empty, i.e., for all $a, b \in X$. $a \lhd b \vee b \lhd a$,
- *weak* or *stratified*, if $a \sim_\lhd b \sim_\lhd c \Rightarrow a \sim_\lhd c$, i.e. if $\sim_\lhd$ is an equivalence relation,

Evidently, every total order is weak. Weak orders are often defined in an alternative way, namely [4],

- a partial order $(X, \lhd)$ is a weak order iff there exists a total order $(Y, \prec)$ and a mapping $\phi : X \to Y$ such that $\forall x, y \in X.\ x \lhd y \iff \phi(x) \prec \phi(y)$.

This definition is illustrated in Figure 1, let $\phi : \{a, b, c, d\} \to \{\{a\}, \{b, c\}, \{d\}\}$ and $\phi(a) = \{a\}$, $\phi(b) = \phi(c) = \{b, c\}$, $\phi(d) = \{d\}$. Note that for all $x, y \in \{a, b, c, d\}$ we have $x <_2 y \iff \phi(x) \prec_2 \phi(y)$.

Following [4], in this paper $a \lhd b$ is interpreted as "*a is less preferred than b*", and $a \approx_\lhd b$ is interpreted as "*a and b are indifferent*".

The preferred outcome of any ranking is a total order. For any total order $\lhd$, both $\sim_\lhd$ and $\approx_\lhd$ are just the equality relation. A total order has two natural models, both deeply embedded in the human perception of reality, namely: *time* and *numbers*.

Unfortunately in many cases it is not reasonable to insist that everything can or should be totally ordered. We may not have sufficient knowledge or such a perfect ranking may not even exist [1]. Quite often insisting on a totally ordered ranking results in an artificial and misleading "global index".

Weak (stratified) orders are a very natural generalization of total orders. They allow the modelling of some regular indifference, their interpretation is very simple and intuitive, and they are reluctantly accepted by decision makers. Although not as much as one might expect given the huge theory of such orders (see [4,6]).

If $\lhd$ is a weak order then $a \approx_\lhd b \iff a \sim_\lhd b$, so indifference means distinct incomparability or identity, and the relation $\lhd$ can be interpreted as a sequence of equivalence classes of $\sim_\lhd$. For the weak order $<_2$ from Figure 3, the equivalence classes of $\sim_{<_2}$ are $\{a\}$, $\{b,c\}$, and $\{d\}$. There are, however, cases where insisting on weak orders may not be reasonable. Those cases will not be discusses in this paper, the reader is referred to [4,6] for more details.

3   Pairwise Comparisons Ranking, Ranking Problem and Data

A *ranking* is just a partial order $Rank = (X, <^{rank})$, where X is the set of objects to be ranked and $<^{rank}$ is a ranking relation. We assume that $<^{rank}$ is a weak or total order. The ranking relation $<^{rank}$ is unknown and the *ranking problem* is to construct $<^{rank}$ on the basis of *ranking data*.

A *pairwise comparisons ranking data* is a pair $PCRD = (X, \mathcal{R})$, where $\mathcal{R}$ is a total function $\mathcal{R} : X \times X \to RV$. The elements of set $RV = \{v_0, v_1, ..., v_k\}$, $k \geq 1$ are called *ranking values*. The value v_0 is interpreted as *indifference*, so we assume $\mathcal{R}(x,x) = v_0$ for all $x \in X$. The values $v_1, ..., v_k$ are interpreted as *preferences*. We assume preferences are totally ordered and $v_k \hookleftarrow v_{k-1} \hookleftarrow ... \hookleftarrow v_1$. The total order $\hookleftarrow$ describes the degree of preference represented by the elements of *RS*. If $v_i \hookleftarrow v_j$ then v_i represents stronger preference than v_j (for example v_i represents *better* and v_j represents *slightly in favour*). Usually we will write $a\, v_i\, b$ instaed of $\mathcal{R}(a,b) = v_i$, $i = 0, ..., k$.

The function $\mathcal{R}$ is constructed using the *Pairwise Comparisons* paradigm. For each pair $x, y \in X$ the value $\mathcal{R}(x,y)$ is decided based on the analysis of x and y only, independently of the rest of X.

For example we may define *RS* (see [7]) as $RV = \{\approx, \sqsubset, \subset, <, \prec\}$, with the following interpretation $a \approx b$: a and b are *indifferent*, $a \sqsubset b$: *slightly in favour* of b, $a \subset b$: *in favour* of b, $a < b$: b is *better*, $a \prec b$: b is *much better*. The list $\sqsubset, \subset, <, \prec$ may be shorter or longer, but not empty and not much longer (due to limitations of the human mind[12]). In this case we assume: $a \prec b \implies a < b \implies a \subset b \implies a \sqsubset c$, i.e. $\prec \hookleftarrow < \hookleftarrow \subset \hookleftarrow \sqsubset$.

Given pairwise comparisons ranking data $PCRD = (X, \mathcal{R})$, with $RV = \{v_0, v_1, ..., v_k\}$, we may define the relations $R_i \subseteq X \times X$, $i = 0, ..., k$ in the following manner:
$$x R_0 y \iff \mathcal{R}(x,y) = v_0,$$

$$xR_ky \iff \mathcal{R}(x,y) = v_k,$$
$$xR_iy \iff \mathcal{R}(x,y) \in \{v_i, v_{i+1}, ..., v_k\}, i = 1, ..., k-1.$$

Corollary 1. *1. $R_0 \cup R_1 \cup ... \cup R_k = X \times X$*
2. $R_k \subseteq R_{k-1} \subseteq ... \subseteq R_1$
3. $v_i \hookleftarrow v_j \iff R_i \subseteq R_j, i, j = 1, ..., k$ □

It is often useful to represent pairwise comparison ranking data *PCRD* as a tuple $(X, R_0, R_1, ..., R_k)$ (see [7]) instead of a pair $(X, \mathcal{R})$. Usually we use the same symbol to denote both v_i and R_i. For example $(X, \approx, \sqsubset, \subset, <, \prec)$ is a pairwise comparison ranking data (with the interpretation described above).

We may now describe the *ranking problem* as follows: "*derive the ranking relation $<^{rank}$ from given pairwise comparison ranking data PCRD*". Note that in the general case, *none* of the relations R_i, $i = 1, ..., k$, might be even a partial order. The problem is that X is believed to be partially or weakly ordered by the ranking relation $<^{rank}$ but the data acquisition process may be so influenced by informational noise, imprecision, randomness, or expert ignorance that the collected data $R_1, R_2, ..., R_k$ are only some relations on X. We may say that they give a fuzzy picture of ranking, and to focus it, we must do some pruning and/or extension.

- For a given pairwise comparison ranking data $PCRD = (X, \mathcal{R})$, the ranking relation derived from *PCRD* will be denoted by $<^{rank}_{PCRD}$, or $<^{rank}_{(X,\mathcal{R})}$.

The tools needed to solve the ranking problem will be presented in the next section.

4 Partial and Weak Order Approximations

Let X be a set, R and $\lhd$ be two relations on X such that $\lhd$ is a partial order and $\lhd \subseteq R$. The relation R may or may not be a partial order. Our goal is to find a relation $<_{(R,\lhd)}$ on X which could be interpreted as the "best" partial order approximation of R satisfying $\lhd \subseteq <_{(R,\lhd)}$. If R is a partial order then obviously $<_{(R,\lhd)}$ equals R.

Suppose R is not a partial order. Without loss of generality we may assume that R is irreflexive, i.e. $(x,x) \notin R$, which means that R is not transitive. The "best" transitive approximation of R can be defined as the smallest transitive relation containing R. Such a relation is called *transitive closure* of R, and it is defined as $R^+ = \bigcup_{i=1}^{\infty} R^i$, where $R^{i+1} = R^i \circ R$ (c.f. [4]). Evidently $R \subseteq R^+$ and R^+ is transitive. Even though R is irreflexive, the relation R^+ may not be irreflexive, but in such a case we can use the following classical result.

Lemma 1 (Schröder 1895, see [8]). *Let $Q \subseteq X \times X$ be a transitive relation. Define:*

$$x <_Q y \iff xQy \land \neg yQx.$$

The relation $<_Q$ is a partial order.   □
The following properties of $<_{R^+}$ can easily be proved.

Proposition 1. *1. The relation $<_{R^+}$ is the biggest partial order $<$ that satisifes:*
(a) $x < y \implies xR^+y$
(b) $x < y \implies \neg(yRx)$.

2. *If R is a partial order then $<_{R^+}$ equals R.*
3. *If R^+ is a partial order then $<_{R^+}$ equals R^+.*   □

From Proposition 1 it follows that the partial order $<_{R^+}$ is most likely the "best" *partial order approximation* of R. If $\lhd \subseteq <_{R^+}$ we set $<_{(R,\lhd)}$ as $<_{R^+}$. It may happen however that $a \lhd b$ and $\neg(a <_{R^+} b)$ for some $a, b \in X$, i.e. it is not true that $\lhd \subseteq <_{R^+}$ (see Figure 1 in [8]).

Let us define the relation $<_{(R,\lhd)}$ as follows:

$$<_{(R,\lhd)} = (\lhd \cup <_{R^+})^+.$$

The below result suggests that $<_{(R,\lhd)}$ could be regarded as the "best" approximation of R that contains $\lhd$.

Proposition 2.   *1. $<_{(R,\lhd)}$ is the smallest partial order $<$ satisfying: $\lhd \cup <_{R^+} \subseteq <$,*
2. $\lhd \subseteq <_{(R,\lhd)}$,
3. $x <_{(R,\lhd)} y \implies \neg(yRx)$,
4. *if R is a partial order then $<_{(R,\lhd)}$ equals R,*
5. *if R^+ is a partial order then $<_{(R,\lhd)}$ equals R^+,*
6. *if $\lhd \subseteq <_{R^+}$ then $<_{(R,\lhd)}$ equals $<_{R^+}$ (this includes the case $\lhd$ equal to $\emptyset$).*   □

The relations $<_{R^+}$ and $<_{(R,\lhd)}$ are usually not weak orders.

Let X be a set and let $\lhd$ be a partial order on X. The relation $\lhd$ may or may not be a weak order. We are looking for the "best" weak order extension of $\lhd$. It appears that in this case the solution may not be unique.

Note that weak order extensions reflect the fact that if $x \approx_\lhd y$ than *all reasonable methods* for extending $\lhd$ will have x equivalent to y in the extension since there is nothing in the data that distinguishes between them (for details see [4]), which leads to the definition below (for both weak an total orders).

A weak (or total) order $\lhd^w \subseteq X \times X$ is a *proper weak (or total) order extension of* $\lhd$ if and only if : $(x \lhd y \Rightarrow x \lhd^w y)$ and $(x \approx_\lhd y \Rightarrow x \sim_{\lhd^w} y)$.

If X is finite then for every partial order $\lhd$ its proper weak extension always exists. If $\lhd$ is weak, than its only proper weak extension is $\lhd^w = \lhd$. If $\lhd$ if not weak, there are usually multiple such extensions. Various methods were proposed and discussed in [4] and especially in [5]. For our purposes, the best seems to be the method based on the concept of a *global score function* [4], which is defined as (for every finite set X, $\| X \|$ denotes its number of elements):

$$g_\lhd(x) = \| \{z \mid z \lhd x\} \| - \| \{z \mid x \lhd z\} \| .$$

Given the global score function $g_\lhd(x)$, we define the relation $\lhd_g^w \subseteq X \times X$ as

$$a \lhd_g^w b \iff g_\lhd(a) < g_\lhd(b).$$

Lemma 2 ([4]). *The relation $\lhd_g^w$ is a proper weak extension of a partial order $\lhd$.*   □

Some other variations of $g_\lhd$ and their interpretations were analyzed in [8]. From Lemma 2 it follows that every finite partial order has a proper weak extension. The well known procedure "topological sorting", popular in scheduling problems, guarantees that every finite partial order has a total extension (the Szpilrain Theorem guarantees it for all partial orders [4]), but even finite partial orders usually *do not have* proper total extensions. Note that the total order $\lhd_t$ is a proper total extension of $\lhd$ if and only if the relation $\approx_\lhd$ equals the identity, i.e $a \approx_\lhd b \iff a = b$. For example no weak order has a proper total extension unless it is also already total. This indicates that while *expecting a final ordering to be weak may be reasonable, expecting a final total ordering is often unreasonable*. It may however happen, and often does, that a proper weak extension is a total order, which suggests that *we should stop seeking a priori total orderings since weak orders appear to be more natural models of preferences than total orders.*

5 Some Solutions to Ranking Problem

CASE 1. We start with the simplest and most likely the most common case, $k = 1$. In this case $\mathcal{R}(a,b) = v_0$ means a and b are indifferent, and $\mathcal{R}(a,b) = v_1$ means b is preferred over a. Since $R_0 \cup R_1 = X \times X$, the case is reduced to finding the best weak order approximation of the relation R_1. This case was analysed in the context of Social Choice and Arrows' axioms (see [1,6]) in [8] and in the context of traditional numerical pairwise comparisons approach (but with Koczkodaj's consistency [10], not the more popular Saaty's consistency [11]) in [9].

Using terminology and notation from the previous section, we need to calculate first $<_{(R_1, \emptyset)}$, which equals $<_{R_1^+}$, and then to find a proper weak extension of $<_{R_1^+}$, preferably $\left(<_{R_1^+}\right)_g^w$.

- We may then set the ranking relation $<_{(X,\mathcal{R})}^{rank}$ as $\left(<_{R_1^+}\right)_g^w$.

In this case we will often write $<_{R_1}^{rank}$ instead of $<_{(X,\mathcal{R})}^{rank}$. The outcome is a weak order, and it may or may not be a total order.

The shape of the function $\mathcal{R}$ that is starting point in the process of creating $<_{R_1}^{rank}$ depends on what kind of preference is used. A stronger preference (for instance *much better* instead of *slightly better*) results in a smaller relation R_1 and bigger relation R_0 (which represents indifference). On the other hand, since the data acquisition process is imprecise(due to informational noise, imprecision, randomness, expert ignorance, etc.), weaker preferences reflect smaller confidence and allow a greater chance for an incorrect assement. That is, the chance that one has an assessment of "a is slightly better than b" when in fact b is better than a or they are indifferent is much larger than the chance that one has an assessment of "a is much better than b", as if there is any doubt one gets an indifferent assessment.

In other words, for stronger preferences we may expect that aR_1b implies $a <^{rank} b$ for all $a,b \in X$, and that R_1 is also a partial order; while for weak preferences we should rather be expecting aR_1b and $\neg(a <^{rank} b)$ for some $a,b \in X$. Which approach is better? Should we insist on finding a data acquisition process with strong discriminatory power? This is usually expensive and the confidence level for the results is rather low. Or, should we apply a discriminatory power for which we have a high confidence level (but which might yield a relatively big indifference relation R_0) and assume that the

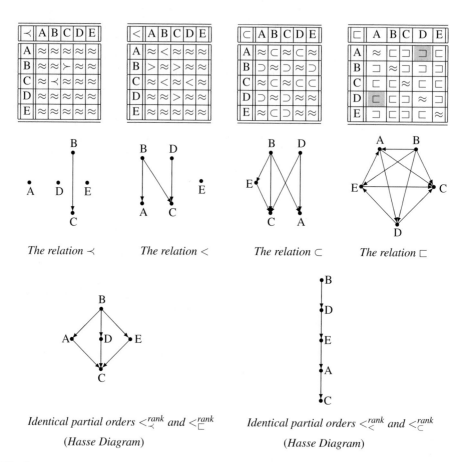

Fig. 2. Four pairwise comparisons ranking data $(X, \approx, \prec)$, $(X, \approx, <)$, $(X, \approx, \sqsubset)$ and $(X, \approx, \sqsubseteq)$, acquired for the same set of objects $X = \{A, B, C, D, E\}$, and the ranking relations they generate. Results of one experiment conducted to justify some claims of [7]. The relations $\approx$, $\prec$, $<$, $\sqsubset$, $\sqsubseteq$ are interpreted as described in Section 3. The wrong judgement of $\sqsubseteq$ is in two grey cells.

correction process (i.e. calculating the relation $(<_{R_1^+})_g^w$) presented above, will correctly identify the relation $<^{rank}$?

We were unable to find much in the literature on this subject for partial orders. However, to justify some of the claims presented in [7] the following experiment has been conducted. A blindfolded person compared the weights of stones. The person put one stone in their left hand and another in their right, and then decided which of the relations $\approx$, $\sqsubseteq$, $\sqsubset$, $<$, or $\prec$ (interpreted as described in Section 3) held. The experiment was repeated for the same set of stones by various people; and then again for different stones and different number of stones; and again for various subsets of $\{\sqsubseteq, \sqsubset, <, \prec\}$. The results of one such experiment are presented in Figure 2. There were many similar results but with more stones involved, so we presented only the smallest case.

The stones were weighed and their weights created an increasing total order C, A, E, D, B, exactly the same as $<_{<}^{rank} = <_{\sqsubset}^{rank}$, but different than $<_{\sqsubset}^{rank}$ - the result of using the finest preference. In fact in this case, the most discrimatory preference $\sqsubset$, and the least discrimatory and very crude preference $\prec$ produced the same outcome, different than actual ordering. On the other hand, the medium discrimatory preference $\sqsubset$ and the relatively low preference $<$ produced the correct ranking. The relations $\prec$, $<$, $\sqsubset$ were partial orders included in the correct total ranking, but none of them was even a weak order. The relation $\sqsubset$ was not a partial order and it was not transitive.

The weights difference among A, D, E were relatively small, so different persons provided different relations $\sqsubset$. For one person the $\sqsubset$-preference beetween A and D depended on which stone was put in which hand. On the other hand the outcomes for $\prec$, $<$ and $\sqsubset$ were the same for all persons.

The experiments conducted to justify some claims of [7] have most likely prehistoric roots. Our ancestors probably used this technique to decide which stone is better to kill an enemy or an animal. However the experiments were rather one-sided and by no means they covered all aspects of pairwise comparison techniques. No statistical data were also provided. More experiments covering various different aspects of pairwise comparison techniques are planned to be conducted soon. Nevertheless we believe the general results will be close to those for "weighting with hands only".

Summing up the case for $k = 1$, we conclude the following:

1. The order identification power of weak extension procedures is substantial and *vastly underestimated.*

2. If the ranked set of objects is, by its nature, expected to be totally ordered, the weak extension can detect it, even if the pairwise comparison process is not very precise, and often results in "indifference".

3. It is a *serious error* to attempt to find a *total extension* without going through a weak extension process.

4. In general, *admitting incomparability on the level of pairwise comparisons is better than insisting on an order at any cost.* The latter approach leads to an arbitrary and often incorrect total ordering.

5. Using less fine but more certain preferences is better than finer but uncertain preferences.

CASE 2. For a moment consider again the case $k = 1$ and the pairwise comparison ranking data (X, R_0, R_1). Even if R_1 may in general be imprecise, in most cases *some parts* of R_1 describe the precise ranking. For instance if R_1 is the result of expert voting, if all experts agree that aR_1b, than we may assume that $a <^{rank} b$ (see *Pereto's principle* [6]). Similarly if a person a is both taller and heavier than b, we would rather say that a is bigger than b, where "bigger" is a calculated ranking relation.

This leads us to *Case 2*, where $k = 2$, and the pairwise comparison ranking data $PCRD = (X, R_0, R_1, R_2)$, where R_2 is a partial order and $R_2 \subseteq <^{rank}$. To be consistent with section 3, we will write R instead of R_1 and $\lhd$ instead of R_2.

- In this case we set $<_{PCRD}^{rank}$ as $(<_{(R,\lhd)})_g^w$, and denote it as $<_{(R,\lhd)}^{rank}$.

An experiment illustrating this case is in Figure 3. The stone were weighted and their weights created and increasing total order C, A, E, D, B, G, F. This order was not detected by neither the fine preference $\sqsubset$, nor by "certain" preference $\lhd$, but was correctly detected by combining $\sqsubset$ and $\lhd$, i.e. by the pairwise comparisons ranking data $(X, \approx, \sqsubset, \lhd)$, where $X = \{A, B, C, D, E, F, G\}$.

- Note that in most cases deriving some $\lhd$ from R is rather easy and natural process. Therefore our final comment for *Case 2* is, *transform first the case* $(X, \approx, \sqsubset)$ *into the case* $(X, \approx, \sqsubset, \lhd)$ *and deal with the latter one.*

CASE 3. For arbitrary k, pairwise comparisons ranking data $(X, \mathcal{R})$ can be defined as a tuple $(X, R_0, R_1, ..., R_k)$ with $R_k \subseteq R_{k-1} \subseteq ... \subseteq R_1$. Without any loss of generality we

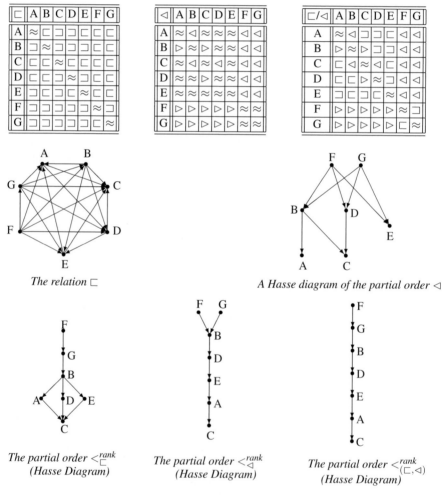

Fig. 3. Three pairwise comparisons ranking data $(X, \approx, \sqsubset)$, $(X, \approx, \lhd)$ and $(X, \approx, \sqsubset, \lhd)$, acquired for the same set of objects $X = \{A, B, C, D, E, F, G\}$, and the ranking relations they generate. Results of one experiment conducted to justify some claims of [7]. The relation $\approx$ is indifference, $\sqsubset$ is interpreted as "slightly in favour" and $\lhd$ as "strongly better". We assume $a \lhd b$ implies $a \sqsubset b$.

may assume that R_k is a partial order. If it is not we may construct $<_{R_k^+}$, set new R_k as $<_{R_k^+}$, and modify respectively $R_0, ..., R_{k-1}$.

Let j be the smallest number such that for all $i \geq j$, R_i are partial orders. If $k = 1$, we just set $<_{(X,\mathcal{R})}^{rank}$ as $(R_1)_g^w$. Otherwise define

$$\widehat{R}_{j-1} = <_{(R_{j-1},R_j)}, \widehat{R}_{j-2} = <_{(R_{j-2},\widehat{R}_j)}, ... , \widehat{R}_1 = <_{(R_1,\widehat{R}_2)}.$$

- We now set $<_{(X,\mathcal{R})}^{rank} = (\widehat{R}_1)_g^w$.

The algorithm presented above is orthogonal to that from [7]. It worked well for the "weighting with hands" experiment. In general the tuple $(X, \widehat{R}_0, \widehat{R}_1, ..., \widehat{R}_k)$, where $x\widehat{R}_0 y \iff \neg(\exists i \geq 1. x\widehat{R}_i y)$, may not satisfy the consistency rules proposed in [7], even though for "weighting with hands" experiments it usually does. The algorithm presented above is easy to program, while the method presented in [7] requires human intervention (changing of preferences).

6 Final Comments

The concepts of ranking, ranking problem and pairwise comparisons ranking methods have been defined and analysed in the partial orders setting. Some solutions have been presented. No numbers were used whatsoever, which we believe is more fair and objective approach. The importance of the indifference relation and the power of the weak order extension procedure have been emphasised.

References

1. Arrow, K.J.: Social Choice and Individual Values. Wiley, J. New York (1951)
2. Dyer, J.S.: Remarks on the Analytic Hierarchy Process. Management Science 36, 244–258 (1990)
3. Fechner, G.T.: Elemente der Psychophysik. Breitkopf und Härtel, Leipzig (1860)
4. Fishburn, P.C.: Interval Orders and Interval Graphs. Wiley, J. New York (1985)
5. Fishburn, P.C., Gehrlein, W.V.: A comparative analysis of methods for constructing weak orders from partial orders. J. Math. Sociol. 4, 93–102 (1975)
6. French, S.: Decision Theory. Ellis Horwood, New York (1986)
7. Janicki, R.: Pairwise Comparisons, Incomparability and Partial Orders. In: Proc. of ICEIS 2007 Int. Conference on Enterprise Information Systems. Artificial Intelligence and Decision Support System, vol. 2, pp. 396–397. Funchal, Portugal (2007)
8. Janicki, R., Koczkodaj, W.W.: Weak Order Approach to Group Ranking. Computers Math. Applic. 32(2), 51–59 (1996)
9. Janicki, R., Koczkodaj, W.W.: A Weak Order Solution to a Group Ranking and Consistency-Driven Pairwise Comparisons. Applied Mathematics and Computation 94, 227–241 (1998)
10. Koczkodaj, W.W.: A new definition of consistency of pairwise comparisons. Mathematical and Computer Modelling 18, 79–84 (1993)
11. Saaty, T.L.: A Scaling Methods for Priorities in Hierarchical Structure. Journal of Mathematical Psychology 15, 234–281 (1977)
12. Saaty, T.L.: Axiomatic Foundations of the Analytic Hierarchy Process. Management Science 32(7), 841–855 (1986)
13. Thurstone, L.L.: A Law of Comparative Judgments. Psychol. Reviews 34, 273–286 (1927)

Combination and Decomposition Theories of Formal Contexts Based on Same Attribute Set

Ling Wei[1] and Jian-Jun Qi[2]

[1] Department of Mathematics, Northwest University, Xi'an 710069, P.R. China
wl@nwu.edu.cn
[2] School of Computer Science & Technology,
Xidian University, Xi'an 710071, P.R. China
qijj@mail.xidian.edu.cn

Abstract. The theory of concept lattices is an efficient tool for knowledge discovery. To make the knowledge discovery of a larger formal context easier, this paper discusses the combination and decomposition of formal contexts when the attribute set is the same, and obtains the corresponding concept lattices. First, the methods how to obtain the whole concept lattice from sub-contexts of a larger formal context is proposed. Then, the converse situation is also studied, and the method how to obtain a sub-lattice from the original formal context is given. At the same time, the relationship between intension set of original context and that of its sub-context is analyzed. Finally, the combination and decomposition theories are generalized when there are multiple sub-contexts.

1 Introduction

The theory of concept lattices was proposed by Wille R in 1982 [14]. A concept lattice is an ordered hierarchical structure of formal concepts that are defined by a binary relation between an object set and an attribute set. As an efficient tool for data analysis and knowledge processing, the theory of concept lattices has been applied to various fields, such as data mining, information retrieval, and software engineering. Most of the researches on concept lattices focus on such topics as: construction of concept lattices [2,7,9], pruning of concept lattices [10], acquisition of rules [2,7,18], relationship with rough set [3,4,11,12,13,16], reduction of concept lattices [13,15,17], and applications [1,5,8,10,18].

As we have known, information acquisition of a larger formal context is difficult. In this paper, we discuss the combination and decomposition of formal contexts in the case of the same attribute set. It makes the discovery of implicit knowledge in larger contexts easier.

The paper is organized as follows. Section 2 recalls basic definitions in formal concept analysis. Section 3 proposes the combination and decomposition theories of formal contexts based on the same attribute set, and also give corresponding theorems. Section 4 discusses the relationship between intension set of original context and that of its sub-context. Section 5 generalizes combination and decomposition theories in the case of multiple sub-contexts. Section 6 gives an example. Finally, Section 7 concludes the paper.

G. Wang et al. (Eds.): RSKT 2008, LNAI 5009, pp. 452–459, 2008.

2   Basic Definitions in Formal Concept Analysis

To make this paper self-contained, we introduce the involved notions in formal concept analysis in this section [6].

Definition 1. *A triple (U, A, I) is called a formal context, if U and A are sets and $I \subseteq U \times A$ is a binary relation between U and A. $U = \{x_1, \dots, x_n\}$, each $x_i(i = 1, \dots n)$ is called an object. $A = \{a_1, \dots, a_m\}$, each $a_j(j = 1, \dots m)$ is called an attribute.*

In this paper, $(x, a) \in I$ is denoted by 1, and $(x, a) \notin I$ is denoted by 0.
A pair of dual operators are defined in (U, A, I) for $X \subseteq U$ and $B \subseteq A$ by:
$$X^* = \{a \in A | (x, a) \in I \text{ for all } x \in X\} \ ,$$
$$B^\star = \{x \in U | (x, a) \in I \text{ for all } a \in B\} \ .$$

Definition 2. *Let (U, A, I) be a formal context. A pair (X, B) is called a formal concept, for short, a concept, of (U, A, I), if and only if,*
$$X \subseteq U, \quad B \subseteq A, \quad X^* = B, \quad \text{and} \quad X = B^\star.$$
X is called the extension and B is called the intension of the concept (X, B).

The concepts of a formal context (U, A, I) are ordered by
$$(X_1, B_1) \leq (X_2, B_2) \Leftrightarrow X_1 \subseteq X_2 (\Leftrightarrow B_1 \supseteq B_2) \ .$$
The set of all concepts can form a complete lattice, which is called the concept lattice of (U, A, I), and is denoted by $L(U, A, I)$. The infimum and supremum are given by:
$$(X_1, B_1) \wedge (X_2, B_2) = (X_1 \cap X_2, (B_1 \cup B_2)^{\star\star}) \ ,$$
$$(X_1, B_1) \vee (X_2, B_2) = ((X_1 \cup X_2)^{\star\star}, B_1 \cap B_2) \ .$$
With respect to a formal context (U, A, I), the following properties hold: for all $X_1, X_2, X \subseteq U$ and all $B_1, B_2, B \subseteq A$,

1. $X_1 \subseteq X_2 \Rightarrow X_2^* \subseteq X_1^*, \quad B_1 \subseteq B_2 \Rightarrow B_2^\star \subseteq B_1^\star.$
2. $X \subseteq X^{**}, \quad B \subseteq B^{\star\star}.$
3. $X^* = X^{***}, \quad B^\star = B^{\star\star\star}.$
4. $X \subseteq B^\star \Leftrightarrow B \subseteq X^*.$
5. $(X_1 \cup X_2)^* = X_1^* \cap X_2^*, \quad (B_1 \cup B_2)^\star = B_1^\star \cap B_2^\star.$
6. $(X_1 \cap X_2)^* \supseteq X_1^* \cup X_2^*, \quad (B_1 \cap B_2)^\star \supseteq B_1^\star \cup B_2^\star.$
7. (X^{**}, X^*) and $(B^\star, B^{\star\star})$ both are concepts.

3   Combination and Decomposition Theories of Formal Contexts Based on the Same Attribute Set

In this section, we propose the combination and decomposition theories of formal contexts based on the same attribute set.

If a formal context (U, A, I) can be combined by two contexts (U_1, A, I_1) and (U_2, A, I_2) where $U_1 \cup U_2 = U$, $U_1 \cap U_2 = \emptyset$, and $I = I_1 \cup I_2$, then we can obtain the combination theorem to get all the concepts of the combined one from the two small contexts. Conversely, if a formal context (U, A, I) can be decomposed into two sub-contexts (U_1, A, I_1) and (U_2, A, I_2), where $U_1 \cup U_2 = U$, $U_1 \cap U_2 = \emptyset$, $I_1 = I \cap (U_1 \times A)$, $I_2 = I \cap (U_2 \times A)$, we can obtain the decomposition theorem. Whatever above cases, we say (U, A, I) is the direct sum of (U_1, A, I_1) and (U_2, A, I_2), (U_1, A, I_1) and (U_2, A, I_2) are complementary. Which is denoted by $(U, A, I) = (U_1, A, I_1) + (U_2, A, I_2)$.

The lattices of (U, A, I), (U_1, A, I_1) and (U_2, A, I_2) are denoted by L, L_1, and L_2 respectively, and we use $*i$ $(\star i)$ in $L_i (i = 1, 2)$ and $*$ $(\star)$ in L respectively.

The following lemma can be obtained naturally.

Lemma 1. *If $(U, A, I) = (U_1, A, I_1) + (U_2, A, I_2)$, then, for $i = 1, 2$,*
1). $\forall B \subseteq A$, $B^\star = B^{\star 1} \cup B^{\star 2}$; $B^{\star i} = B^\star \cap U_i$.
2). $\forall X \subseteq U$, $X^* = (X \cap U_1)^{*1} \cap (X \cap U_2)^{*2}$. Especially, $\forall X \subseteq U_i$, $X^{*i} = X^*$.

Definition 3. *Let (U_1, A, I_1) and (U_2, A, I_2) be two formal contexts, and L_{1A}, L_{2A} be their intension sets respectively, that is, $L_{iA} = \{B_i | (X_i, B_i) \in L_i\}$, $i = 1, 2$. Then we define their inner intersection of L_{1A} and L_{2A} as follows.*

$$L_{1A} \sqcap L_{2A} = \{B_1 \cap B_2 | (X_1, B_1) \in L_1, (X_2, B_2) \in L_2\} . \tag{1}$$

Theorem 1. *(**Combination theorem**) Suppose $(U, A, I) = (U_1, A, I_1) + (U_2, A, I_2)$. Then, we have the following statements.*
1). $\forall (X_1, B_1) \in L_1, (X_2, B_2) \in L_2, ((B_1 \cap B_2)^\star, B_1 \cap B_2) \in L$.
2). $L_A = L_{1A} \sqcap L_{2A}$.

Proof. 1). We only need to prove that $(B_1 \cap B_2)^{\star\star} = B_1 \cap B_2$.

Since $(B_1 \cap B_2)^{\star\star} = ((X_1)^{*1} \cap (X_2)^{*2})^{\star\star} = ((X_1)^* \cap (X_2)^*)^{\star\star} = (X_1 \cup X_2)^{\star\star\star} = (X_1 \cup X_2)^\star = (X_1)^* \cap (X_2)^* = (X_1)^{*1} \cap (X_2)^{*2} = B_1 \cap B_2$. Thus, the proposition 1) is concluded.

2). From the proof of proposition 1), we have $L_{1A} \sqcap L_{2A} \subseteq L_A$. Thus, we only need to prove its converse, i.e., $L_A \subseteq L_{1A} \sqcap L_{2A}$.

Suppose $C \in L_A$, that is, there exists $(Y, C) \in L$ such that $Y^* = C$, $C^\star = Y$. We prove the result from 3 cases.

If $Y \subseteq U_1$, then, $C = Y^* = Y^{*1} = Y^{*1} \cap A \in L_{1A} \sqcap L_{2A}$.

If $Y \subseteq U_2$, then, $C = Y^* = Y^{*2} = Y^{*2} \cap A \in L_{2A} \sqcap L_{1A} = L_{1A} \sqcap L_{2A}$.

If $Y \subseteq U$, $Y \nsubseteq U_1$, $Y \nsubseteq U_2$, then we assume $Y = Y_1 \cup Y_2$, $Y_1 \subseteq U_1$, $Y_2 \subseteq U_2$. Thus, $C = Y^* = (Y_1 \cup Y_2)^* = Y_1^* \cap Y_2^* = Y_1^{*1} \cap Y_2^{*2} \in L_{1A} \sqcap L_{2A}$.

Therefore, $L_A \subseteq L_{1A} \sqcap L_{2A}$ holds. Thus, proposition 2) is proved.

This theorem shows that all the concepts in the combined lattice $L = L(U, A, I)$ can be obtained by its two complementary sub-contexts.

Theorem 2. *(**Decomposition theorem**) Suppose $(U, A, I) = (U_1, A, I_1) + (U_2, A, I_2)$. Then, we have the following statements.*
1). $\forall (X, B) \in L$, there must be $(X \cap U_i, (X \cap U_i)^{*i}) \in L_i, i = 1, 2$.
2). There must exist $(X, B) \in L$ such that any concept in L_1 and L_2 can be described as 1).

Proof. Suppose $(X, B) \in L$.

1). We only prove the conclusion when $i = 1$, that is, $(X \cap U_1, (X \cap U_1)^{*1}) \in L_1$. The proof of the case $i = 2$ is similar.

It needs to be proved $(X \cap U_1)^{*1*1} = X \cap U_1$. In fact, $(X \cap U_1)^{*1*1} = (B^\star \cap U_1)^{*1*1} = ((B^{*1} \cup B^{*2}) \cap U_1)^{*1*1} = ((B^{*1} \cap U_1) \cup (B^{*2} \cap U_1))^{*1*1} = (B^{*1} \cup \emptyset)^{*1*1} = B^{*1*1*1} = B^{*1} = B^\star \cap U_1 = X \cap U_1$.

2). We also prove it only in L_1. That is, we need to show, $\forall (X_1, B_1) \in L_1$, there exists $(X, B) \in L$ such that $X_1 = X \cap U_1$ and $B_1 = (X \cap U_1)^{*1}$.

In fact, the existed $(X, B) \in L$ is $(X, B) = (X_1 \cup B_1^{*2}, B_1)$. Therefore, we need to prove $(X_1 \cup B_1^{*2}, B_1) \in L$, and $X_1 = (X_1 \cup B_1^{*2}) \cap U_1$.

Firstly, we have $X_1 \cup B_1^{*2} = B_1^{*1} \cup B_1^{*2} = B_1^\star$, and $(X_1 \cup B_1^{*2})^\star = X_1^* \cap B_1^{*2*} = X_1^{*1} \cap B_1^{*2*2} = B_1 \cap B_1^{*2*2} = B_1$, so, $(X_1 \cup B_1^{*2}, B_1) \in L$ holds.

Secondly, it is obvious that $(X_1 \cup B_1^{*2}) \cap U_1 = (X_1 \cap U_1) \cup (B_1^{*2} \cap U_1) = X_1 \cup \emptyset = X_1$.

Thus, the proof is completed.

This theorem says that if a large context can be decomposed to two sub-contexts satisfying some conditions described in this theorem, then, any concept in the sub-context can be obtained from the original context. Furthermore, concepts obtained by this method are all of its concepts.

4   The Relationship between the Intension Set of a Formal Context and That of Its Sub-contexts

This section discusses the relationship between the intention set of a formal context and that of its sub-contexts.

Theorem 3. Let (U, A, I) be a formal context, and (U_1, A, I_1) be one of its sub-context. Let L, L_A be the lattice and intention set of (U, A, I), and L_1, L_{1A} be the lattice and intention set of (U_1, A, I_1). Then, we have $L_{1A} \subseteq L_A$.

Proof. Suppose $B \in L_{1A}$, that is, there exists $(X, B) \in L_1$ satisfying $X^{*1} = B, B^{*1} = X$. It needs to be proved that $(B^\star, B) \in L$, i.e., $B^{**} = B$. In fact, $B^{**} = (X^{*1})^{**} = X^{***} = X^* = X^{*1} = B$. Therefore, $L_{1A} \subseteq L_A$ is concluded.

Corollary 1. Let (U, A, I) be a formal context. If there exists one sub-context (U_1, A, I_1) such that $L_A \subseteq L_{1A}$, then, $L_{1A} = L_A$.

Definition 4. Let $L_1 = L(U_1, A, I_1)$ and $L_2 = L(U_2, A, I_2)$ be two concept lattices. If $\forall (X_2, B_2) \in L_2$, there exists $(X_1, B_1) \in L_1$ such that $B_1 = B_2$, then we say L_1 is finer than L_2 with respect to the attribute set A, and is denoted by $L_1 \leq_A L_2$.

Theorem 4. Let (U, A, I) be a formal context, and $(U, A, I) = (U_1, A, I_1) + (U_2, A, I_2)$, L, L_1, L_2 and L_A, L_{1A}, L_{2A} be their lattices and intention sets respectively. Then, we have

$$L_A = L_{1A} \Leftrightarrow L_1 \leq_A L_2. \tag{2}$$

Proof. Necessity. Suppose $L_A = L_{1A}$. It should be proved that $\forall(X_2, B_2) \in L_2$, there exists $(X_1, B_1) \in L_1$ such that $B_1 = B_2$. Since $L_A = L_{1A}$, we only need to prove that there exists $(X_0, B_0) \in L$ such that $B_0 = B_2$. In fact, we can obtain from Theorem 2 that there exists $(X, B) \in L$ such that $B_2 = (X \cap U_2)^*$, and it is obvious that $((X \cap U_2)^{**}, (X \cap U_2)^*) \in L$, that is, the existed $B_0 = (X \cap U_2)^* = B_2$. Necessity is concluded.

Sufficiency. Since Corollary 1, we only need to prove that $L_1 \leq_A L_2 \Rightarrow L_A \subseteq L_{1A}$. From Theorem 1, we know that $\forall(X, B) \in L$, there exist $(X_1, B_1) \in L_1$ and $(X_2, B_2) \in L_2$ such that $X = (B_1 \cap B_2)^\star = (X_1 \cup X_2)^{**}, B = B_1 \cap B_2$. Since $L_1 \leq_A L_2$, for the existed $(X_2, B_2) \in L_2$, there must exist $(Y, C) \in L_1$ such that $B_2 = C$. Thus, $B = B_1 \cap B_2 = B_1 \cap C$. Then, $(B_1 \cap C)^{\star 1 \star 1} = (X_1^{*1} \cap Y^{*1})^{\star 1 \star 1} = (X_1 \cup Y)^{*1 \star 1 \star 1} = (X_1 \cup Y)^{*1} = X_1^{*1} \cap Y^{*1} = B_1 \cap C$. Therefore, $((B_1 \cap C)^{\star 1}, B_1 \cap C) \in L_1$, i.e., $B = B_1 \cap B_2 = B_1 \cap C \in L_{1A}$. Sufficiency is proved.

Theorem 5. *If $(U, A, I) = (U_1, A, I_1) + (U_2, A, I_2)$, and $L_A = L_{1A}$, then any concept $(X, B) \in L$ can be obtained by*

$$X = X_1 \cup (\underset{t \in \tau}{\cup} X_{2t}), \qquad B = B_1.$$

Where, $(X_1, B_1) \in L_1, B_1 \subseteq B_{2t}, (X_{2t}, B_{2t}) \in L_2$.

Proof. We need to show $B^\star = X$ and $X^* = B$, that is, $(X_1 \cup (\underset{t \in \tau}{\cup} X_{2t}))^\star = B_1$, and $B_1^* = X_1 \cup (\underset{t \in \tau}{\cup} X_{2t})$.

Firstly, we show $(X_1 \cup (\underset{t \in \tau}{\cup} X_{2t}))^\star = B_1$. Which is easy to be obtained since $(X_1 \cup (\underset{t \in \tau}{\cup} X_{2t}))^* = X_1^* \cap (\underset{t \in \tau}{\cup} X_{2t})^* = X_1^{*1} \cap (\underset{t \in \tau}{\cup} X_{2t})^{*2} = B_1 \cap (\underset{t \in \tau}{\cap} X_{2t}^{*2}) = B_1 \cap (\underset{t \in \tau}{\cap} B_{2t}) = B_1$.

Secondly, we show $B_1^\star = X_1 \cup (\underset{t \in \tau}{\cup} X_{2t})$. We have $B_1^\star = B_1^{\star 1} \cup B_1^{\star 2} = X_1 \cup B_1^{\star 2} = X_1 \cup (\underset{t \in \tau}{\cup} X_{2t})$ since $B_1^{\star 2} = (\underset{t \in \tau}{\cup} X_{2t})$. In fact, $B_1 \subseteq B_{2t} \Rightarrow B_1^{\star 2} \supseteq B_{2t}^{\star 2} = X_{2t} \Rightarrow B_1^{\star 2} \supseteq \underset{t \in \tau}{\cup} X_{2t}$, and we can also prove $B_1^{\star 2} \subseteq \underset{t \in \tau}{\cup} X_{2t}$. Assume that there exists $y \in B_1^{\star 2}, y \notin \underset{t \in \tau}{\cup} X_{2t}$, i.e., $\forall t \in \tau, y \notin X_{2t}$. Since $X_{2t} \subseteq B_1^{\star 2}, y \notin B_1^{\star 2}$ is concluded, which is contradiction to the assumption. Therefore, $B_1^{\star 2} \subseteq \underset{t \in \tau}{\cup} X_{2t}$. Thus, $B_1^{\star 2} = \underset{t \in \tau}{\cup} X_{2t}$ is obtained.

5 Generalization of Combination and Decomposition Theories

If $(U, A, I) = (U_1, A, I_1) + (U_2, A, I_2) + ... + (U_l, A, I_l)$ satisfying $U_i \cap U_j = \emptyset (\forall i \neq j), \overset{l}{\underset{i=1}{\cup}} U_i = U, I_i = I \cap (U_i \times A)$(or $I = I_1 \cup ... \cup I_l$), that is, (U, A, I) is the direct sum of $(U_i, A, I_i), i = 1, ...l$, then we can also obtain the following generalized combination theorem and decomposition theorem. This theorem can be concluded by using Theorem 1 and Theorem 2 repeatedly.

Theorem 6. *Suppose a formal context (U, A, I) can be described as the direct sum of $(U_i, A, I_i), i = 1, ...l$. Let L and $L_i(i = 1, ...l)$ be their corresponding lattices respectively, L_A and L_{iA} $(i = 1, ...l)$ be their intention sets respectively. Then we have the following propositions.*

1) $L_A = L_{1A} \sqcap L_{2A} \sqcap ... \sqcap L_{lA}$.

2) $\forall(X_i, B_i) \in L_i, i = 1, 2, ...l, ((\bigcap_{i=1}^{l} B_i)^\star, \bigcap_{i=1}^{l} B_i) \in L$.

3) $\forall(X, B) \in L, (X \cap U_i, (X \cap U_i)^{*i}) \in L_i, i = 1, ...l$.

6   An Example

Table 1 shows a formal context (U, A, I). In which, $U = \{1, 2, 3, 4, 5, 6, 7, 8\}, A = \{a, b, c, d, e\}$. In order to obtain its lattice, we take it as $(U, A, I) = (U_1, A, I_1) + (U_2, A, I_2)$, where, $U_1 = \{1, 2, 3, 4\}, I_1 = I \cap (U_1 \times A), U_2 = \{5, 6, 7, 8\}, I_2 = I \cap (U_2 \times A)$. Here, we use Theorem 1 to obtain the original lattice.

Table 1. A formal context (U, A, I)

	a	b	c	d	e
1	1	1	0	1	1
2	1	1	1	0	0
3	0	0	0	1	0
4	1	1	1	0	0
5	1	0	1	0	1
6	1	1	0	0	1
7	0	1	1	1	0
8	0	1	1	1	0

The lattices of $L(U_1, A, I_1)$ and $L(U_2, A, I_2)$ are shown in Fig.1 and Fig.2 respectively. The combined one obtained by Theorem 1 is shown in Fig.3.

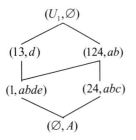
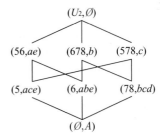

Fig. 1. The concept lattice of (U_1, A, I_1) **Fig. 2.** The concept lattice of (U_2, A, I_2)

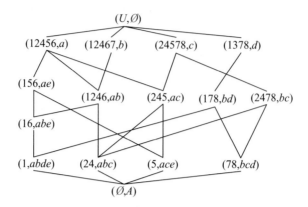

Fig. 3. The concept lattice of (U, A, I)

It is easy to see that obtaining the original lattice using the combination method is easier than compute it directly. Similarly, if we have known the original lattice $L = L(U, A, I)$, then we can also obtain the lattices of any sub-context using Theorem 2.

7   Conclusion

This paper has discussed combination and decomposition theories of formal contexts based on the same attribute set to obtain concept lattices. The theory proposed in this paper can be applied to knowledge discovery when data updating occurs. In parallel, Our next work will focus on the combination and decomposition theories of formal contexts based on the same object set. Furthermore, based on this theory, rules acquisition of whole lattice can be studied through that of sub-lattices. All of these significance topics will be researched in the future work.

Acknowledgements

The authors gratefully acknowledge the support of the National Natural Science Foundation of China(No.60703117), and the Doctor Research Fund of Northwest University in China.

References

1. Boucher-Ryan, P., Bridge, D.: Collaborative Recommending using Formal Concept Analysis. Knowledge-Based Systems 19(5), 309–315 (2006)
2. Carpineto, C., Romano, G.G.: An order-theoretic approach to conceptual clustering. In: Utgoff, P. (ed.) Proceedings of ICML 293, pp. 33–40. Elsevier, Amherst (1993)

3. Deogun, J.S., Saquer, J.: Concept approximations for formal concept analysis. In: Stumme, G. (ed.) Working with Conceptual Structures. Contributions to ICCS 2000, pp. 73–83. Verlag Shaker Aachen (2000)
4. Düntsch, I., Gediga, G.: Algebraic aspects of attribute dependencies in information systems. Fundamenta Informaticae 29(1-2), 119–133 (1997)
5. Ganter, B., Stumme, G., Wille, R.: Formal Concept Analysis, Foundations and Applications. Springer, Berlin (2005)
6. Ganter, B., Wille, R.: Formal Concept Analysis, Mathematical Foundations. Springer, Berlin (1999)
7. Godin, R.: Incremental concept formation algorithm based on Galois (concept) lattices. Computational Intelligence 11(2), 246–267 (1995)
8. Grigoriev, P.A., Yevtushenko, S.A.: Elements of an Agile Discovery Environment. In: Grieser, G., Tanaka, Y., Yamamoto, A. (eds.) DS 2003. LNCS (LNAI), vol. 2843, pp. 309–316. Springer, Heidelberg (2003)
9. Ho, T.B.: An approach to concept formation based on formal concept analysis. IEICE Trans. Information and Systems E782D(5), 553–559 (1995)
10. Oosthuizen, G.D.: The Application of Concept Lattice to Machine Learning. Technical Report, University of Pretoria, South Africa (1996)
11. Oosthuizen, G.D.: Rough sets and concept lattices. In: Ziarko, W.P. (ed.) Rough Sets, and Fuzzy Sets and Knowledge Discovery (RSKD 1993), pp. 24–31. Springer, Heidelberg (1994)
12. Pagliani, P.: From concept lattices to approximation spaces: Algebraic structures of some spaces of partial objects. Fundamenta Informaticae 18(1), 1–25 (1993)
13. Wei, L.: Reduction theory and approach to rough set and concept lattice, PhD Thesis, Xi'an Jiaotong University. Xi'an Jiaotong University Press, Xi'an (2005)
14. Wille, R.: Restructuring lattice theory: an approach based on hierarchies of concepts. In: Rival, I. (ed.) Ordered sets, pp. 445–470. Reidel, Dordrecht-Boston (1982)
15. Wu, W.Z.: Attribute Granules in Formal Contexts. In: An, A., Stefanowski, J., Ramanna, S., Butz, C.J., Pedrycz, W., Wang, G. (eds.) RSFDGrC 2007. LNCS (LNAI), vol. 4482, pp. 395–402. Springer, Heidelberg (2007)
16. Yao, Y.Y.: Concept lattices in rough set theory. In: Dick, S., Kurgan, L., Pedrycz, W., Reformat, M. (eds.) Proceedings of 2004 Annual Meeting of the North American Fuzzy Information Processing Society (NAFIPS 2004), IEEE Catalog Number: 04TH8736, June 27-30, pp. 796–801 (2004)
17. Zhang, W.X., Wei, L., Qi, J.J.: Attribute Reduction in Concept Lattice Based on Discernibility Matrix. In: Proceedings of the 10th International Conference on Rough Sets, Fuzzy Sets, Data Mining, and Granular Computing, Regina, Canada, August 31-September 3, pp. 157–165 (2005)
18. Zhao, Y., Yao, Y.Y.: Classification based on logical concept analysis. In: Proceedings of 19th Conference on the Canadian Society for Computational Studies of Intelligence (Canadian AI 2006), Québec City, Québec, Canada, June 7-9, pp. 419–430 (2006)

An Approach for Constructing Concept Lattices

Yutian Zhong[1], Keyun Qin[2], and Zhengjiang Wu[2]

[1] Computer and information technology College, Henan Normal University,
Xinxiang, Henan 610031, China
zhongyutian@126.com
[2] Department of mathematics,
Southwest Jiaotong University , Chengdu , Sichuan, 610031, China
keyunqin@263.net

Abstract. Galois (concept) lattices and formal concept analysis have been proved useful in the resolution of many problems of theoretical and practical interest. Recent studies have put the emphasis on the need for both efficient and flexible algorithms to construct the lattice. In this paper, some equivalent conditions for an attributes subset to be a reduction of a formal concept are presented. Further more, the structure of concept lattice was analyzed and it is proved that each concept is the meet of some single attribute generalized concepts. Based on the above research, reduction-based approaches towards constructing concept lattice was presented.

1 Introduction

Formal concept analysis(FCA) is a discipline that studies the hierarchical structures induced by a binary relation between a pair of sets. The structure, made up of the closed subsets ordered by set-theoretical inclusion, satisfies the properties of a complete lattice and has been firstly mentioned in the work of Birkhoff[1]. The term concept lattice and formal concept analysis are due to Wille[2], [3], [4]. Later on, it has been the subject of an extensive study with many interesting results. As a classification tool, FCA has been used in several areas such as data mining, knowledge discovery, and software engineering.

One of the important challenges in FCA is to get efficient and flexible algorithms to construct the concept lattice from the formal context. The algorithms can be mainly divided into two groups: algorithms which extract the set of concepts[5], [7] only, and algorithms for constructing the entire lattice[8], [9], [10] i.e., concepts together with lattice order. An efficient algorithm has been suggested by Bordat[8] which generates both the concept set and the Hasse diagram of the lattice. The obvious drawback of the method is that a concept is generated several times. The design of flexible algorithms was pioneered by Godin et al.[9] who designed an incremental method for constructing the concept lattices. The lattice is constructed starting from a single object and gradually incorporating new objects. Nourine and Raynaud[10] suggested a general approach

G. Wang et al. (Eds.): RSKT 2008, LNAI 5009, pp. 460–466, 2008.
© Springer-Verlag Berlin Heidelberg 2008

towards the computation of closure structures and showed how it could be used to construct concept lattices. Valtchev et al.[11] presents a novel approach for concept lattice construction based on the apposition of binary relation fragments. In[12], the concept of attribute reduction of formal concept was proposed with its properties being discussed. The $CL-$Axiom and some equivalent conditions for an attributes subset to be a reduction of a formal concept are presented.

In this paper, based on[12], some equivalent conditions for an attributes subset to be a reduction of a formal concept are presented. Further more, the structure of concept lattice was analyzed and it is proved that each concept is the meet of some single attribute generalized concepts. Based on the above research, reduction-based approaches towards constructing concept lattice was presented.

2   Fundamentals of FCA

Definition 1. *A formal context is an ordered triple* $T = (G, M, I)$ *where* G, M *are finite nonempty sets and* $I \subseteq G \times M$ *is an incidence relation. The elements in* G *are interpreted to be objects, elements in* M *are said to be attributes. If* $(g, m) \in G \times M$ *is such that* $(g, m) \in I$, *then the object* g *is said to have the attribute* m.

The incidence relation of a formal context can be naturally represented by an incidence table.

Example 1. [6] $T = (G, M, I)$ is a formal context, where $G = \{1, 2, 3, 4, 5, 6, 7, 8\}$, $M = \{a, b, c, d, e, f, g, h, i\}$ and the table below describes incidence relation:

Table 1. The incidence relation of the formal context

	a	b	c	d	e	f	g	h	i
1	x	x					x		
2	x	x					x	x	
3	x	x	x				x	x	
4	x		x				x	x	x
5	x	x		x		x			
6	x	x	x	x		x			
7	x		x	x	x				
8	x		x	x		x			

To introduce the definition of the formal concept, Wille used two set-valued functions, $\uparrow$ and $\downarrow$, given by the expressions:

$$\uparrow: P(G) \to P(M), X^{\uparrow} = \{m \in M; \forall g \in X, (g, m) \in I\},$$

$$\downarrow: P(M) \to P(G), Y^{\downarrow} = \{g \in G; \forall m \in Y, (g, m) \in I\}.$$

Definition 2. *A formal concept of a context* $T = (G, M, I)$ *is a pair* $(A, B) \in P(G) \times P(M)$ *such that* $A^{\uparrow} = B$ *and* $B^{\downarrow} = A$. *The set* A *is called its extent, the set* B *its intent.*

The subset $L(G, M, I)$ of $P(G) \times P(M)$ formed by all the concepts of the context is a complete lattice with the order relation:

$(A, B) \leq (C, D)$ if and only if $A \subseteq C$ (or equivalently $B \supseteq D$).

This relation shows the hierarchy between the concepts of the context. The lattice $(L(G, M, I), \leq)$ is said to be the formal concept lattice of the context (G, M, I) with LUB and GLB are given as follows:

$$\bigvee_{i=1}^{n} (A_i, B_i) = ((\bigcup_{i=1}^{n} A_i)^{\uparrow\downarrow}, \bigcap_{i=1}^{n} B_i),$$

$$\bigwedge_{i=1}^{n} (A_i, B_i) = (\bigcap_{i=1}^{n} A_i, (\bigcup_{i=1}^{n} B_i)^{\downarrow\uparrow}).$$

For convenience reasons, we simplify the standard set notation by dropping out all the separators (e.g., 124 will stand for the set of objects $\{1, 2, 4\}$ and cd for the set of attributes $\{c,d\}$). The concept lattice of Example 1 is showed in Fig. 1.

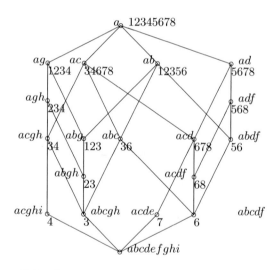

Fig. 1. Galois/concept lattice corresponding to Table 1

3 The Attribute Reduction of Formal Concepts

Let $(G, M.I)$ be a formal context and $(A, B) \in P(G) \times P(M)$ a formal concept. We introduce the notation $\varepsilon_{(A,B)}$ by

$$\varepsilon_{(A,B)} = \{Y \subseteq M; Y^{\downarrow} = A\}.$$

For any $Y \in P(M)$, $(Y^{\downarrow}, Y^{\downarrow\uparrow})$ is a concept, it is said to be the concept generated by the set Y of attributes. It follows that $\varepsilon_{(A,B)}$ is the family of subsets of attributes which generate same concept as B does.

Theorem 1. *[12] Let $(G, M.I)$ be a formal context and $(A, B) \in P(G) \times P(M)$ a formal concept.*
(1) B is the greatest element in the poset $(\varepsilon_{(A,B)}, \subseteq)$.
(2) If $Y_1 \in \varepsilon_{(A,B)}$ and $Y_1 \subseteq Y_2 \subseteq B$, then $Y_2 \in \varepsilon_{(A,B)}$.

Definition 3. *Let $(G, M.I)$ be a formal context and $(A, B) \in P(G) \times P(M)$ a formal concept.*
(1) A minimal element in $(\varepsilon_{(A,B)}, \subseteq)$ is said to be an attribute reduction of (A, B).
(2) If $a \in B$ is such that $(B - \{a\})^{\downarrow} \supset A$, then a is said to be a core attribute of (A, B).

We denote by $Core(A, B)$ the set of all core attributes of (A, B) and by $Red(A, B)$ the set of all attribute reductions of (A, B), that is

$$Core(A, B) = \{a \in B; (B - \{a\})^{\downarrow} \supset A\}, \tag{1}$$

$$Red(A, B) = \{Y; Y \text{ is an attribute reduction of } (A, B)\}. \tag{2}$$

Theorem 2. *[12] $\cap Red(A, B) = Core(A, B)$.*

Example 2. For the formal context in Example 1, $(23, abgh)$ is a concept. It is trivial to verify that

$$\varepsilon_{(23,abgh)} = \{bh, abh, bgh, abgh\},$$

bh is the unique attribute reduction of $(23, abgh)$ and $Core(23, abgh) = \{b, h\}$. For the concept $(6, abcdf)$,

$$\varepsilon_{(6,abcdf)} = \{bcd, bcf, bcdf, abcdf\}.$$

It follows that $Red(6, abcdf) = \{bcd, bcf\}$ and $Core(6, abcdf) = \{b, c\}$.

Theorem 3. *[12] Let (G, M, I) be a formal context, $Y \subseteq M, Y \neq \emptyset$. Y is an attribute reduction of a concept if and only if*

$$\triangle(b) = \{x \in G; I(x, b) = 0, \prod_{a \in Y \setminus \{b\}} I(x, a) = 1\} \neq \emptyset$$

for each $b \in Y$.

Based on the above Theorem, the CL−Axiom for attribute subset $Y \subseteq M$ is introduced as follows:
CL-Axiom: $\sum_{b \in Y} \delta(\triangle(b)) = |Y|$, where

$$\delta(\triangle(b)) = \begin{cases} 1, & \text{if} \triangle(b) \neq \emptyset, \\ 0, & \text{if} \triangle(b) = \emptyset. \end{cases} \tag{3}$$

Theorem 4. *[12] Let (G, M, I) be a formal context, $Y \subseteq M, Y \neq \emptyset$. Y is an attribute reduction of a concept if and only if Y satisfies $CL-Axiom$.*

For each concept, there exist its attribute reduction. From Theorem 7, the concept lattice $L(G, M, I)$ with respect to formal context (G, M, I) can be constructed as follows:

(1) For each $Y \subseteq M$, check Y satisfies $CL-$Axiom or not and form the set: $\lambda(T) = \{Y \subseteq M; Y \, satisfy \, \text{CL-}Axiom\}$;

(2) For each $Y \in \lambda(T)$, compute the concept $(Y^{\downarrow}, Y^{\downarrow\uparrow})$ and form the concept lattice

$$L(G, M, I) = \{(Y^{\downarrow}, Y^{\downarrow\uparrow}); Y \in \lambda(T)\}.$$

Theorem 5. *Let (G, M, I) be a formal context, $Y \subseteq M, Y \neq \emptyset$. Y does not satisfy $CL-Axiom$ if and only if there exist formal concept (A, B) and its attribute reduction Z such that $Z \subset Y \subseteq B$.*

Proof 6. *Assume that (A, B) is a concept and Z its attribute reduction such that $Z \subset Y \subseteq B$. Let $b \in Y - Z$. If there exist $x \in G$ such that $x \in \triangle(b)$, then $\prod_{a \in Y - \{b\}} I(x, a) = 1$ and hence $\prod_{a \in Z} I(x, a) = 1$, that is $x \in Z^{\downarrow}$. Consequently $I(x, b) = 1$ by $b \in B = Z^{\downarrow\uparrow}$, a contradiction with $I(x, b) = 0$. It follows that $\triangle(b) = \emptyset$ and Y does not satisfy $CL-Axiom$.*

Conversely, assume that Y does not satisfy $CL-Axiom$. It follows that there exist $b \in Y$ such that $\triangle(b) = \emptyset$. Consequently, for each $x \in G$, if $\prod_{a \in Y - \{b\}} I(x, a) = 1$, then $I(x, b) = 1$. It follows that $Y^{\downarrow} = (Y - \{b\})^{\downarrow}$. We consider the concept $((Y - \{b\})^{\downarrow}, (Y - \{b\})^{\downarrow\uparrow})$. It follows that there exist attribute reduction Z of $((Y - \{b\})^{\downarrow}, (Y - \{b\})^{\downarrow\uparrow})$ such that $Z \subseteq Y - \{b\}$. Consequently,

$$Z \subseteq Y - \{b\} \subset Y \subseteq Y^{\downarrow\uparrow} = (Y - \{b\})^{\downarrow\uparrow}. \tag{4}$$

Theorem 7. *Let (G, M, I) be a formal context. If $Y \subseteq M, Y \neq \emptyset$ is not attribute reduction of any concept, then Z is not attribute reduction of any concept for each $Y \subseteq Z \subseteq M$.*

4   The Structure of Concept Lattice

In this section, we discuss the structure of concept lattice.

Let $T = (G, M, I)$ be a formal context. It is trivial to verify that $\{a\}$ satisfies $CL-$Axiom for each $a \in M - G^{\uparrow}$. For each $(A, B) \in L(G, M, I)$, $M^{\uparrow} \subseteq B$ and $B - M^{\uparrow} \neq \emptyset$ if (A, B) is not the greatest element of $L(G, M, I)$.

Theorem 8. *Let $T = (G, M, I)$ be a formal context and $(A, B) \in L(G, M, I)$. If $B - M^{\uparrow} \neq \emptyset$, then*

$$(A, B) = \bigwedge_{a \in B - G^{\uparrow}} (\{a\}^{\downarrow}, \{a\}^{\downarrow\uparrow}) = (\bigcap_{a \in B - G^{\uparrow}} \{a\}^{\downarrow}, (\bigcap_{a \in B - G^{\uparrow}} \{a\}^{\downarrow})^{\uparrow}).$$

Proof 9. *For each* $a \in B - G^{\uparrow}$, $\{a\} \subseteq B$ *and it follows that* $\{a\}^{\downarrow\uparrow} \subseteq B^{\downarrow\uparrow} = B$. *Consequently* $(A, B) \leq (\{a\}^{\downarrow}, \{a\}^{\downarrow\uparrow})$ *and hence*

$$(A, B) \leq \bigwedge_{a \in B - G^{\uparrow}} (\{a\}^{\downarrow}, \{a\}^{\downarrow\uparrow}) = (\bigcap_{a \in B - G^{\uparrow}} \{a\}^{\downarrow}, (\bigcap_{a \in B - G^{\uparrow}} \{a\}^{\downarrow})^{\uparrow}).$$

If $(C, D) \in L(G, M, I)$ *is such that* $(C, D) \leq (\{a\}^{\downarrow}, \{a\}^{\downarrow\uparrow})$ *for each* $a \in B - G^{\uparrow}$, *then* $D \supseteq \{a\}^{\downarrow\uparrow}$ *and hence* $D \supseteq \bigcup_{a \in B - G^{\uparrow}} \{a\}^{\downarrow\uparrow}$.

For each $b \in B$, *if* $b \in B - G^{\uparrow}$, *then* $b \in \{b\}^{\downarrow\uparrow} \subseteq \bigcup_{a \in B - G^{\uparrow}} \{a\}^{\downarrow\uparrow}$. *If* $b \in G^{\uparrow}$, *we suppose that* $a_0 \in B - G^{\uparrow}$, *it follows that* $b \in \{a_0\}^{\downarrow\uparrow} \subseteq \bigcup_{a \in B - G^{\uparrow}} \{a\}^{\downarrow\uparrow}$. *Consequently,* $B \subseteq \bigcup_{a \in B - G^{\uparrow}} \{a\}^{\downarrow\uparrow}$.

By $D \supseteq \bigcup_{a \in B - G^{\uparrow}} \{a\}^{\downarrow\uparrow}$ *and* $B \subseteq \bigcup_{a \in B - G^{\uparrow}} \{a\}^{\downarrow\uparrow}$, *it follows that* $D \supseteq B$ *and* $(C, D) \leq (A, B)$.

By this theorem, each element of $L(G, M, I)$ is the meet of a subset of $\{(\{a\}^{\downarrow}, \{a\}^{\downarrow\uparrow}); a \in M - G^{\uparrow}\}$ (we make the appointment that the greatest element $(G, G^{\uparrow})$ of $L(G, M, I)$ satisfies $(G, G^{\uparrow}) = \cap\emptyset$). It follows that the concept lattice $L(G, M, I)$ with respect to formal context $T = (G, M, I)$ can be constructed as follows:

(1) Compute $G^{\uparrow}$;
(2) Compute $\alpha(T) = \{\{a\}^{\downarrow}; a \in M - G^{\uparrow}\}$;
(3) Compute $\beta(T) = \{\bigcap A; A \subseteq \alpha(T)\}$;
(4) Compute $\gamma(T) = \{(A, A^{\uparrow}); A \in \beta(T)\}$;
(5) $\gamma(T) \cup \{(G, G^{\uparrow})\} = L(G, M, I)$ is the concept lattice with respect to T.

It is worth noting that for $A \subseteq \alpha(T)$, $\bigcap A$ is just set-theoretical intersection of elements in A. That is to say, $\beta(T)$ is the set of elements which is the intersection of some elements in $\alpha(T)$.

Example 3. For the formal context in Example 1,
(1) $G^{\uparrow} = \{a\}$;
(2) $M - G^{\uparrow} = \{b, c, d, e, f, g, h, i\}$ and

$$\alpha(T) = \{12356, 34678, 5678, 7, 568, 1234, 234, 4\};$$

(3) $\beta(T) = \alpha(T) \cup \{36, 56, \emptyset, 123, 23, 678, 68, 34, 3, 6\}$;
(4) For each $A \in \beta(T)$, compute $(A, A^{\uparrow})$ and form the set $\gamma(T)$:
$\gamma(T) = \{(12356, ab), (34678, ac), (5678, ad), (7, acde), (568, adf), (1234, ag),$
$(234, agh), (4, acghi), (36, abc), (56, abdf), (\emptyset, abcdefghi), (123, abg),$
$(23, abgh), (678, acd), (68, acdf), (34, acgh), (3, abcgh), (6, abcdf)\}$;
(5) $\gamma(T) \cup \{(12345678, a)\}$ is the concept lattice with respect to T.

5   Conclusions

This paper is devoted to the discussion of constructing approach for concept lattice. Based on[12], some equivalent conditions for an attributes subset to be a reduction of a formal concept are presented. Further more, the structure of concept lattice was analyzed and it is proved that each concept is the meet of some single attribute generalized concepts. Based on the above research, reduction-based approaches towards constructing concept lattice was presented. We will study the algorithm for constructing concept lattice based on the theory presented in this paper in our future works.

Acknowledgements

The authors are grateful to the referees for their valuable comments and suggestions. This work has been supported by the National Natural Science Foundation of China (Grant No. 60474022).

References

1. Birkhoff, B.: (revised edition): Lattice Theory, vol. 25. American Mathematical Society Colloquium Publ., Providence, RI (1973)
2. Wille, R.: Restructuring the lattice theory: an approach based on hierarchies of concepts. In: Rival, I. (ed.) Ordered Sets, pp. 445–470. Reidel, Boston (1982)
3. Wille, R.: Lattices in data analysis: how to draw them with a computer. In: Algorithms and order, pp. 33–58. Kluwer Acad. Publ., Dordrecht (1989)
4. Wille, R.: Concepe lattices and conceptual knowledge systems. Comput. Math. Apll. 23(6-9), 493–515 (1992)
5. Ganter, B. (ed.): Two basic algorithms in concept analysis. Technische Hochschule, Darmstadt (1984)
6. Ganter, B., Wille, R. (eds.): Formal Concept Analysis. Mathematical Foundations. Springer, Berlin (1999)
7. Norris, E.M.: An algorithm for computing the maximal rectangles in a binary relation. Rev. Roumaine Math. Pures Appl. 23(2), 243–250 (1978)
8. Bordat, J.P.: Calcul pratique du treillis de Galois d'une correspondance. Mat. Sci. Hum. 96, 31–47 (1986)
9. Godin, R., Missaoui, R., Alaoui, H.: Incremental concept formation algorithms based on Galois (concept) lattices. Computational Intelligence 11(2), 246–267 (1995)
10. Nourine, L., Raynaud, O.: A fast algorithm for building lattices. Inform. Process. Lett. 71, 199–204 (1999)
11. Valtchev, P., Missaoui, R., Lebrun, P.: A partition-based approach towards constructing Galois (concept) lattices. Discrete Mathematics 256, 801–829 (2002)
12. Jingyu, J., Keyun, Q., Zheng, P.: Reduction-based approaches towards constructing Galois (concept) lattices. In: Wang, G.-Y., Peters, J.F., Skowron, A., Yao, Y. (eds.) RSKT 2006. LNCS (LNAI), vol. 4062, pp. 107–113. Springer, Heidelberg (2006)

Analyzing Correlation Coefficients of Objective Rule Evaluation Indices on Classification Rules

Hidenao Abe and Shusaku Tsumoto

Shimane University
89-1 Enya-cho Izumo Shimane, 6938501, Japan
abe@med.shimane-u.ac.jp, tsumoto@computer.org

Abstract. In data mining post-processing, which is one of important procedures in a data mining process, at least 39 metrics have been proposed to find out valuable knowledge. However, their functional properties have never been clearly articulated under the same condition. Therefore, we carried out a correlation analysis of functional properties between each objective rule evaluation indices on classification rule sets using correlation coefficients between each index. In this analysis, we calculated average values of each index using bootstrap method on 34 classification rule sets learned based on information gain ratio. Then, we found the following relationships based on correlation coefficient values: similar pairs, discrepant pairs, and independent indices. With regarding to this result, we discuss about relative functional relationships between each group of objective indices.

Keywords: Data Mining, Post-processing, Rule Evaluation Index, Correlation Analysis.

1 Introduction

In recent years, enormous amounts of data are stored on information systems in natural science, social science, and business domains. People have been able to obtain valuable knowledge due to the development of information technology. Besides, data mining techniques combine different kinds of technologies such as database technologies, statistical methods, and machine learning methods. Then, data mining has been well known for utilizing data stored on database systems. In particular, if-then rules, which are produced by rule induction algorithms, are considered as one of the highly usable and readable outputs of data mining. However, to large datasets with hundreds of attributes including noise, the process often obtains many thousands of rules. From such a large rule set, it is difficult for human experts to find out valuable knowledge, which are rarely included in the rule set.

To support such a rule selection, many efforts have done using objective rule evaluation indices such as recall, precision, and other interestingness measurements [1,2,3] (Hereafter, we refer to these indices as "objective indices"). Although their properties are identified with their definition, their functional properties are not investigated with any promising method.

G. Wang et al. (Eds.): RSKT 2008, LNAI 5009, pp. 467–474, 2008.
© Springer-Verlag Berlin Heidelberg 2008

With regard to the above-mentioned issues, we present an correlation analysis method to identify the functional properties of objective indices in Section 3. Then, with the 39 objective indices and classification rule sets from 34 UCI datasets, we identified the following relationships based on the correlation analysis method: similar pairs of indices, contradict pairs of indices, and independent indices. Based on the result in Section 4, we discuss about these relationships and differences between functional properties and original definitions.

2   Interestingness Measures and Related Work

Many studies have investigated the selection of valuable rules from a large mined rule set based on objective rule evaluation indices. Some of these works suggested the indices to discover interesting rules from such a large number of rules [1,2,3]. These interestingness measures are based on two different approaches[4]: the objective (data-driven) approach and the subjective approach.

To avoid confusing real human interest, the objective index, and the subjective index, we clearly define these three items as follows: **Objective Index :** features such as the correctness, uniqueness, and strength of a rule, which are calculated mathematically. An objective index does not include any human evaluation criteria. **Subjective Index :** The similarity or difference between the information on interestingness given beforehand by a human expert and that obtained from a rule. Although some human criteria are included in its initial state, the similarity or difference is mainly calculated mathematically.

Focusing on interesting rule selection with objective indices, researchers have developed more than forty objective indices based on number of instances, probability, statistics values, information quantity, distance or attributes of rules, and complexity of rules. The behavior of each of these indices with respect to their functional natures has been investigated in a number of studies[5,6,7].

However, there has been not yet done to analyze some functional relationships among objective indices on any actually obtained classification rule set totally.

3   Correlation Analysis for the Objective Rule Evaluation Indices

In this section, we describe a correlation analysis method to identify functional properties of objective indices. To analyze functional relationships between objective indices, we should gather the following materials: values of objective indices of each classification rule set learned from each dataset, and correlation coefficients between objective indices with the values. The process of the analysis is shown in Figure 1.

First, we obtain multiple rule sets from some datasets to get values of objective indices. When gathering these values, we should care the statistical correctness of each value. Therefore, the values are averaged adequately large number (> 100) of values from bootstrap samples.

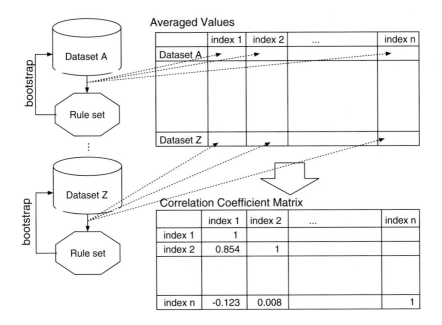

Fig. 1. An overview of the correlation analysis method

Then, the following correlation coefficients r between indices, x and y, are calculated for n datasets.

$$r = \frac{\frac{1}{n}\sum_{i=1}^{n}(x_i - \overline{x})(y_i - \overline{y})}{\sqrt{\frac{1}{n}\sum_{i=1}^{n}(x_i - \overline{x})^2}\sqrt{\frac{1}{n}\sum_{i=1}^{n}(y_i - \overline{y})^2}}$$

With these coefficient values, we identified similar pairs, contradict pairs, and independent indices.

4   Analyzing the Objective Rule Evaluation Indices on UCI Datasets

In this section, we describe the correlation analysis of the 39 objective indices with twelve UCI datasets. Table 1 shows the 39 objective indices investigated and reformulated for classification rules by Osaki et. al.[8].

As for datasets, we have taken the following 34 datasets from UCI machine learning repository[20]: anneal, audiology, autos, balance-scale, breast-cancer, breast-w, colic, credit-a, credit-g, diabetes, glass, heart-c, heart-h, heart-statlog, hepatitis, hypothyroid, iris, ionosphere, kr-vs-kp, labor, letter, lymph, mushroom, primary-tumor, segment, sick, sonar, soybean, splice, vehicle, vote, vowel, waveform-5000 and zoo.

Table 1. Objective rule evaluation indices for classification rules used in this research. **P:** Probability of the antecedent and/or consequent of a rule. **S:** Statistical variable based on P. **I:** Information of the antecedent and/or consequent of a rule. **N:** Number of instances included in the antecedent and/or consequent of a rule. **D:** Distance of a rule from the others based on rule attributes.

Theory	Index Name (**Abbreviation**) [Reference Number of Literature]
P	Coverage (**Coverage**), Prevalence (**Prevalence**)
	Precision (**Precision**), Recall (**Recall**)
	Support (**Support**), Specificity (**Specificity**)
	Accuracy (**Accuracy**), Lift (**Lift**)
	Leverage (**Leverage**), Added Value (**Added Value**)[2]
	Klösgen's Interestingness (**KI**)[9], Relative Risk (**RR**)[10]
	Brin's Interest (**BI**)[11], Brin's Conviction (**BC**)[11]
	Certainty Factor (**CF**)[2], Jaccard Coefficient (**Jaccard**)[2]
	F-Measure (**F-M**)[12], Odds Ratio (**OR**)[2]
	Yule's Q (**YuleQ**)[2], Yule's Y (**YuleY**)[2]
	Kappa (**Kappa**)[2], Collective Strength (**CST**)[2]
	Gray and Orlowska's Interestingness weighting Dependency (**GOI**)[13]
	Gini Gain (**Gini**)[2], Credibility (**Credibility**)[14]
S	χ^2 Measure for One Quadrant (χ^2-**M1**)[15]
	χ^2 Measure for Four Quadrant (χ^2-**M4**)[15]
I	J-Measure (**J-M**)[16], K-Measure (**K-M**)[8]
	Mutual Information (**MI**)[2]
	Yao and Liu's Interestingness 1 based on one-way support (**YLI1**)[3]
	Yao and Liu's Interestingness 2 based on two-way support (**YLI2**)[3]
	Yao and Zhong's Interestingness (**YZI**)[3]
N	Cosine Similarity (**CSI**)[2], Laplace Correction (**LC**)[2]
	ϕ Coefficient (ϕ)[2], Piatetsky-Shapiro's Interestingness (**PSI**)[17]
D	Gago and Bento's Interestingness (**GBI**)[18]
	Peculiarity (**Peculiarity**)[19]

For the above datasets, we obtained rule sets with PART[21] implemented in Weka[22]. PART constructs a rule set based on information gain ratio. This means the obtained rule sets are biased with the correctness of classification.

4.1 Constructing a Correlation Coefficient Matrix of the 39 Objective Indices

For the 32 datasets, we obtained the rule sets using PART. This procedure is repeated 1000 times with bootstrap re-sampling for each dataset. As a representative value for each bootstrap iteration, the average for a rule set has been calculated. Then, we averaged the average values from 1000 times iterations.

With the average values for each dataset, we calculated correlation coefficients between each objective index.

4.2 Identifying Characteristic Relationships between Objective Indices Based on Correlation Coefficient Matrix Analysis

Based on the correlation coefficients, we identify characteristic relationship between each objective index. We defined the three characteristic relation ship as follows:

- Similar pair: two indices has strong positive correlation $r > 0.8$.
- Contradict pair: two indices has strong negative correlation $r < -0.8$.

– Independent index: a index has only weak correlations $-0.8 \leq r \leq 0.8$ for the other indices.

Table 2 shows similar pairs of objective indices on the correlation analysis. There are several groups having mutual correlations. The largest group, which has correlation to Cosine Similarity and F-Measure, includes 23 indices. Relative Risk and Odds Ratio make another group. χ^2-M1, χ^2-M4 and PSI also make different functional group. These pairs indicate distinct functional property for the rule sets.

Table 2. Similar pairs of objective indices on the correlation analysis

	Index Pair	Corr. Coefficient		Index Pair	Corr. Coefficient		Index Pair	Corr. Coefficient
Coverage	Precision	0.86	Leverage	Added Value	0.91	YulesY	Kappa	0.96
	Recall	0.81		Jaccard	0.97		KI	0.96
	Support	1.00		Certainty Factor	0.92		GOI	0.88
	Leverage	0.88		YulesY	0.92		Laplace Correction	0.88
	Added Value	0.82		Kappa	0.96		Gini Gain	0.90
	Jaccard	0.91		KI	0.96		YLI1	0.94
	Certainty Factor	0.84		GOI	0.89		YLI2	0.94
	KI	0.84		Laplace Correction	0.87		YZI	0.90
	BI	0.88		Gini Gain	0.97		Cosine Similarity	0.92
	GOI	0.86		J-Measure	0.84		F-Measure	0.91
	Laplace Correction	0.82		YLI1	0.81	Kappa	KI	0.93
	Gini Gain	0.83		YLI2	0.99		GOI	0.84
	J-Measure	0.90		YZI	0.95		Laplace Correction	0.80
	YLI2	0.82		Cosine Similarity	0.97		Gini Gain	0.95
	Cosine Similarity	0.91		F-Measure	0.97		YLI1	0.90
	F-Measure	0.92	Added Value	Jaccard	0.91		YLI2	0.99
Precision	Support	0.86		Certainty Factor	1.00		YZI	0.95
	Leverage	0.91		YulesQ	0.80		Cosine Similarity	0.94
	Added Value	0.99		YulesY	0.94		F-Measure	0.94
	Jaccard	0.89		Kappa	0.89	KI	GOI	0.96
	Certainty Factor	0.99		KI	0.99		Laplace Correction	0.93
	YulesY	0.83		BI	0.83		Gini Gain	0.91
	YulesQ	0.91		GOI	0.99		YLI1	0.87
	Kappa	0.85		Laplace Correction	0.93		YLI2	0.95
	BI	0.98		Gini Gain	0.85		YZI	0.89
	GOI	0.87		YLI1	0.86		Cosine Similarity	0.98
	Laplace Correction	0.98		YLI2	0.90		F-Measure	0.97
	Gini Gain	0.97		YZI	0.82	BI	GOI	0.90
	YLI1	0.83		Cosine Similarity	0.96		Laplace Correction	0.80
	YLI2	0.87		F-Measure	0.94		K-Measure	0.92
	Cosine Similarity	0.96	Relative Risk	Odds Ratio	0.80		Cosine Similarity	0.82
	F-Measure	0.94	Jaccard	Certainty Factor	0.91		F-Measure	0.80
Recall	Support	0.81	Certainty Factor	YulesQ	0.82	GOI	Laplace Correction	0.91
	Leverage	0.89		YulesY	0.93		Gini Gain	0.81
	Added Value	0.80		Kappa	0.88		YLI2	0.85
	Jaccard	0.95		KI	0.99		K-Measure	0.85
	YulesY	0.85		BI	0.84		Cosine Similarity	0.96
	Kappa	0.93		GOI	0.99		F-Measure	0.94
	KI	0.84		Laplace Correction	0.95	Laplace Correction	GiniGain	0.81
	GiniGain	0.88		Gini Gain	0.85		YLI2	0.83
	YLI2	0.91		YLI1	0.84		CosineSimilarity	0.90
	YZI	0.91		YLI2	0.89		F-Measure	0.87
	Cosine Similarity	0.91		YZI	0.81	ChiSquare-one	ChiSquare-four	0.96
	F-Measure	0.92		Cosine Similarity	0.96		PSI	0.89
Support	Leverage	0.88		F-Measure	0.95	ChiSquare-four	PSI	0.98
	AddedValue	0.82	YulesQ	BI	0.85	Gini Gain	J-Measure	0.82
	Jaccard	0.91		GOI	0.85		YLI2	0.98
	Certainty Factor	0.84		J-Measure	0.81		YZI	0.99
	KI	0.84		K-Measure	0.92		Cosine Similarity	0.92
	BI	0.88					F-Measure	0.92
	GOI	0.86				J-Measure	Cosine Similarity	0.83
	Laplace Correction	0.82					F-Measure	0.84
	Gini Gain	0.83				YLI1	YLI2	0.85
	J-Measure	0.90					YZI	0.80
	YLI2	0.82					Cosine Similarity	0.81
	Cosine Similarity	0.91				YLI2	YZI	0.97
	F-Measure	0.92					Cosine Similarity	0.95
							F-Measure	0.95
						YZI	Cosine Similarity	0.90
							F-Measure	0.91
						Cosine Similarity	F-Measure	1.00

As shown in Table 3, there are smaller number of discrepant pairs of objective indices on the correlation analysis. Accuracy and Prevalence has discrepant property each other. Likewise, BI and BC also indicate discrepant property because of negative correlation between them. BC shows discrepant property to several indices, which belong to the biggest group of similar pairs.

Table 3. Discrepant pairs of objective indices on the correlation analysis

Index Pair		Corr. Coefficient
Accuracy	Prevalance	−0.98
YulesQ	BC	−0.92
BI	GOI	−0.92
	BC	−0.85
K−Measure	BC	−1.00

Figure 2 shows scatter plots of representative pair of each relationship. Where $r = 1.00$ is not correctly 1. Also, $r = -1.00$ is not correctly -1.

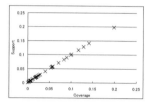 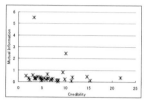 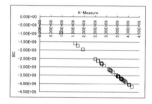

Fig. 2. Scatter plots of representative pairs. (Coverage vs. Support $(r = 1.00)$, Credibility vs. Mutual Information $(r = -0.10)$ and K-Measure vs. BC $(r = -1.00)$).

5 Discussion

With regarding to Table 2, we can say that the following objective indices indicate similar property: Coverage, Precision, Recall, Support, Leverage, Added Value, Jaccard, Certainty Factor, YulesQ, YulesY, Kappa, KI, BI, GOI, Laplace Correction, Gini Gain, J-Measure, YLI1, YLI2, YZI, K-Measure, Cosine Similarity, and F-Measure. The other groups also show similar functional property to the classification rule sets based on information gain ratio. Considering their definitions, although they have different theoretical backgrounds, their functional property is to represent correctness of rules. This indicates that these indices evaluate given rules optimistically.

On the other hand, looking at Table 3, BC indicates opposite functional property comparing with the former indices. Therefore, the result indicates that BC evaluate given rules not so optimistically. As for Accuracy and Prevalence, although Accuracy measures ratio of both of true positive and false negative for each rule, Prevalence only measures ratio of mentioned class value of each rule. It is reasonable to indicate discrepant property, because accurate rules have high Accuracy values irrespective of their mentioned class value.

As for the independent indices, GBI and Peculiarity suggested with the different theoretical background comparing with the other indices. Therefore, what they have different functional properties is reasonable. However, Corrective Strength, Credibility, Mutual Information and ϕ Coefficient indicate the different functional property comparing with the other indices which have the same theoretical backgrounds (**P**,**S** and **N**). These indices evaluate given rules from each different viewpoint.

6   Conclusion

In this paper, we described the method to analyze functional properties of objective rule evaluation indices.

We investigated functional properties of objective indices with 34 UCI datasets and their rule sets as an actual example. With regarding to the result, several groups are found as functional similarity groups in cross-sectional manner for their theoretical backgrounds.

In the future, we will investigate functional properties of objective indices to other kind of rule sets obtained from the other rule mining algorithms. At the same time, we will investigate other rank correlation coefficients and other correlations.

References

1. Hilderman, R.J., Hamilton, H.J.: Knowledge Discovery and Measure of Interest. Kluwer Academic Publishers, Dordrecht (2001)
2. Tan, P.N., Kumar, V., Srivastava, J.: Selecting the right interestingness measure for association patterns. In: Proceedings of International Conference on Knowledge Discovery and Data Mining KDD-2002, pp. 32–41 (2002)
3. Yao, Y.Y., Zhong, N.: An analysis of quantitative measures associated with rules. In: Zhong, N., Zhou, L. (eds.) PAKDD 1999. LNCS (LNAI), vol. 1574, pp. 479–488. Springer, Heidelberg (1999)
4. Freitas, A.A.: On rule interestingness measures. Knowledge-Based Systtems 12(5-6), 309–315 (1999)
5. Vaillant, B., Lenca, P., Lallich, S.: A clustering of interestingness measures. In: Proceedings of the Discovery Science 2004, pp. 290–297 (2004)
6. Huynh, X.H., Guillet, F., Briand, H.: A data analysis approach for evaluating the behavior of interestingness measures. In: Proceeding of the Discovery Science 2005, pp. 330–337 (2005)
7. Blanchard, J., Guillet, F., Gras, R., Briand, H.: Using information-theoretic measures to assess association rule interestingness. In: Proceedings of the fifth IEEE International Conference on Data Mining ICDM 2005, pp. 66–73. IEEE Computer Society, Los Alamitos (2005)
8. Ohsaki, M., Kitaguchi, S., Kume, S., Yokoi, H., Yamaguchi, T.: Evaluation of rule interestingness measures with a clinical dataset on hepatitis. In: Boulicaut, J.-F., Esposito, F., Giannotti, F., Pedreschi, D. (eds.) PKDD 2004. LNCS (LNAI), vol. 3202, pp. 362–373. Springer, Heidelberg (2004)
9. Explora: A Multipattern and Multistrategy Discovery Assistant. In: Fayyad, U.M., Piatetsky-Shapiro, G., Smyth, P., Uthurusamy, R. (eds.) Advances in Knowledge Discovery and Data Mining, pp. 249–271. AAAI/MIT Press, California (1996)
10. Ali, K., Manganaris, S., Srikant, R.: Partial classification using association rules. In: Proceedings of the International Conference on Knowledge Discovery and Data Mining KDD-1997, pp. 115–118 (1997)

11. Brin, S., Motwani, R., Ullman, J., Tsur, S.: Dynamic itemset counting and implication rules for market basket data. In: Proceedings of the ACM SIGMOD International Conference on Management of Data, pp. 255–264 (1997)
12. Rijsbergen, C.: Information retrieval, ch. 7 (1979),
    http://www.dcs.gla.ac.uk/Keith/Chapter.7/Ch.7.html
13. Gray, B., Orlowska, M.E.: CCAIIA: Clustering categorical attributes into interesting association rules. In: Wu, X., Kotagiri, R., Korb, K.B. (eds.) PAKDD 1998. LNCS, vol. 1394, pp. 132–143. Springer, Heidelberg (1998)
14. Hamilton, H.J., Shan, N., Ziarko, W.: Machine learning of credible classifications. In: Australian Conf. on Artificial Intelligence AI-1997, pp. 330–339 (1997)
15. Goodman, L.A., Kruskal, W.H.: Measures of association for cross classifications. Springer Series in Statistics, vol. 1. Springer, Heidelberg (1979)
16. Smyth, P., Goodman, R.M.: Rule induction using information theory. In: Piatetsky-Shapiro, G., Frawley, W.J. (eds.) Knowledge Discovery in Databases, pp. 159–176. AAAI/MIT Press (1991)
17. Piatetsky-Shapiro, G.: Discovery, analysis and presentation of strong rules. In: Piatetsky-Shapiro, G., Frawley, W.J. (eds.) Knowledge Discovery in Databases, pp. 229–248. AAAI/MIT Press (1991)
18. Gago, P., Bento, C.: A metric for selection of the most promising rules. In: European Conference on the Principles of Data Mining and Knowledge Discovery PKDD-1998, pp. 19–27 (1998)
19. Zhong, N., Yao, Y.Y., Ohshima, M.: Peculiarity oriented multi-database mining. IEEE Transactions on Knowledge and Data Engineering 15(4), 952–960 (2003)
20. Hettich, S., Blake, C.L., Merz, C.J.: UCI repository of machine learning databases, Department of Information and Computer Science. University of California, Irvine (1998), http://www.ics.uci.edu/~mlearn/MLRepository.html
21. Frank, E., Witten, I.H.: Generating accurate rule sets without global optimization. In: The Fifteenth International Conference on Machine Learning, pp. 144–151 (1998)
22. Witten, I.H., Frank, E.: Data Mining: Practical Machine Learning Tools and Techniques with Java Implementations. Morgan Kaufmann, San Francisco (2000)

Description Logic Based Consistency Checking upon Data Mining Metadata

Xiaodong Zhu, Zhiqiu Huang, and Guohua Shen

College of Information Science and Technology
Nanjing University of Aeronautics and Astronautics
Nanjing, 210016, P.R. China
{zhuxd,zqhuang,ghshen}@nuaa.edu.cn

Abstract. During the process of constructing data mining metadata, the evolution of data mining techniques, the different experiences and views of related organizations inevitably cause inconsistencies. However, current data mining metadata lacks precise semantic due to their description with natural language and graphs, so the automatic consistency checking upon them has not been resolved well. In this paper, a formal logic $DLR_{\mathcal{DM}}$ in the description logic family is proposed. Subsequently, a formal reasoning method based on $DLR_{\mathcal{DM}}$ is designed to automatically check the consistency of data mining metadata. With the description logic $DLR_{\mathcal{DM}}$, formalization upon the metamodel and metadata of data mining is analyzed in detail. The reasoning engine Racer is applied into the method to check the consistency upon the data mining metadata. Results on the RacerPro reasoning system indicate the method is encouraging.

Keywords: Automated reasoning, Data mining, Metadata, Consistency checking, Knowledge representation.

1 Introduction

Data mining is a process of knowledge discovery from a great deal of data. It has attracted much attention since the relational database theory was proposed in the last century. A lot of data mining algorithms were proposed, which involve symbolic statistics, classification, clustering, association rules and other aspects. More and more corporations brought their data mining products such as SPSS Clementine, Oracle Data Miner and IBM DB2 Intelligent Miner.

With the constantly development of data mining techniques, the standardization of data mining become a new focus problem in recent years [1]. Different corporations look forward to common data mining metadata for the sharing, exchanging, integration and standardization of their data mining products. The data mining metadata provides unified criteria, rules, models and frameworks for data mining. In recent years, some data mining oriented industry criteria are proposed, for example, CRISP-DM[1], PMML[2] and CWM [2]. The CRISP-DM

---

[1] CRoss Industry Standard Process for Data Mining. See *http://www.crisp-dm.org*
[2] Predictive Modeling Markup Language. See *http://www.dmg.org*

G. Wang et al. (Eds.): RSKT 2008, LNAI 5009, pp. 475–482, 2008.
© Springer-Verlag Berlin Heidelberg 2008

gives a process criterion of whole data mining life cycle. Now it becomes the standard method of data mining developing projects. However, it is described with natural language and didn't make exact criterion for data mining metadata. PMML is defined by the data mining group DMG, which aims to define a unified XML format that saves the contents of mining models. So it benefits the metadata exchanging of data mining products. However, it is hard to guarantee the consistence of data mining metadata with the evolution of data mining techniques.

Common warehouse metamodel, CWM for short, provides detailed and sufficient graph description for data warehouse and data mining [2]. It is developed by the OMG Group and focuses on business intelligence, which includes the definitions of OLAP and data mining metadata. The aim of CWM is to solve the integration and management problems of metadata, so different application programs can integrate each other under different conditions. In the CWM criteria, data mining metadata is defined in detail. However, different experiences of different corporations, different views when describing data and continual development of data mining techniques inevitably bring conflicts upon metadata. However, the description with natural language and graphs in CWM make it hard to automatically discover the conflict information through reasoning. However, this problem has not been resolved well up to now. In this research, we put forward a description logic based formal reasoning method for the consistency checking of data mining metadata.

2   Related Work

Zhao first proposed a formal logic DL_{id} in description logic family to formalize CWM based data warehouse metadata, which could automatically check the inconsistency in data warehouse metadata [3]. However, this logic has following deficiency. First, it lacks the description upon the whole CWM architecture such as meta-meta model MOF. Second, it has not described the dependency in CWM based metadata. So the logic easily lead to the missing of conflict information. At the same time, the formalization upon association relation can be further improved in the DL_{id}. Therefore, the DL_{id} has restriction when it formalizes the CWM based metadata. Hauch et al expatiated on the conflict problems in construction and integration of metadata and presented some strategies of resolving the conflicts [4]. Cali et al proposed methods of formalizing and reasoning UML class graphs, but they did not utilize the information in metamodel [5]. Zubcoff et al constructed a data mining metamodel for classification analysis using abundant semantic information supplied by CWM [6]. However, Up to now, little research was paid upon the consistency checking on data mining metadata. Based on Zhao's research, we expand the functions of the description logic DL_{id} and construct more complete formal logic DLR_{DM}, which is used to automatically check the inconsistency in data mining metadata.

3   Description Logic DLR$_{\mathcal{DM}}$

The idea of applying description logic into data mining metadata management derives from that metadata pattern can be expressed by description logic repository, so reasoning upon description logic can detect the inconsistency problems in data mining metadata. In this research, we extend the expressive ability of DLid in terms of dependency relation, association relation and description in meta-meta model. A novel formal logic DLR$_{\mathcal{DM}}$ that belongs to the description logic family is proposed. This logic is a subset of DLR which is proposed by Calvanese [7] and a superset of DLid proposed by Zhao [3]. In the DLR$_{\mathcal{DM}}$, basic elements include concept, role, individual and subsumption. From the standing point of symbolic logic, they are unitary predication, binary predication, constant and hierarchy relation respectively. Let A and P denote atomic concept and atomic role respectively, then the syntax rules of any concept C or role R are shown as follows.

$$C ::= \top_1 | A | \neg C | C_1 \sqcap C_2 | (\leq k[i/2]R),$$
$$R ::= \top_2 | P | (i/2 : C) | \neg R | R_1 \sqcap R_2.$$

Here, k is a nonnegative integer; i denotes the i-th element, and i is either 1 or 2; $\leq k[i/2]R$ denotes that the multiplicity constraint of the i-th element of the relation R corresponding to R; $i/2 : C$ denotes that the i-th associated concept of the role R is C, brief for $(i : C)$. Differing from many other description logic, we omit $\perp$, $\sqcup$, $\rightarrow$, $\geq k[i]R$, $\exists[i]R$ and $\forall[i]R$. In fact, $\perp \equiv \neg\top$, $C_1 \sqcup C_2 \equiv \neg(\neg C_1 \sqcap \neg C_2)$, $C_1 \rightarrow C_2 \equiv \neg C_1 \sqcup C_2$, $\geq k[i]R \equiv \neg(\leq (k-1)[i]R)$, $\exists[i]R \equiv \geq 1[i]R$ and $\forall[i]R \equiv \neg\exists[i]\neg R$.

The repository Σ of DLR$_{\mathcal{DM}}$ is constituted by TBox and ABox. The TBox is the set of axioms that describe the domain structure. The TBox contains axioms of describing concepts and axioms of declaring subsumption relations, denoted by $A \equiv C$ and $A \sqsubseteq C$ respectively. Besides, DLR$_{\mathcal{DM}}$ includes the identification constrain and dependency constrain. The syntax of identification constrain is $(id\ C\ [i_1]R_1, ..., [i_n]R_n)$, whose meaning is that if a concept C is the i_j-th $(i_j \in \{1, 2\})$ concept of the role R_j, then any two different instances of C induce two different instances of R_j, here $j \in 1, ..., n$. In the dependency constrain, the syntax is $(fd\ R\ i_1 \rightarrow i_2), i_1, i_2 \in \{1, 2\}$, whose meaning is that the value of the associated i_2-th concept upon the role R depend on the value of the associated i_1-th concept. More semantics of the DLR$_{\mathcal{DM}}$ are list as following Fig. 1.

An interpretation of the repository Σ of DLR$_{\mathcal{DM}}$ is denoted by $I = (\Delta^I, \cdot^I)$, which is constituted by an interpretation domain Δ^I and an interpretation function $\cdot^I$. This function maps every concept C into a subset C^I of the domain Δ^I, and every role R into a subset R^I of the Cartisian product $(\Delta^I)^2$.

The ABox of DLR$_{\mathcal{DM}}$ is the set of axioms that describe concrete status, which includes concept declaration and roles declaration. The concept declaration denotes that whether or not an object belongs to a concept, marked with $a :: C$. The role declaration denotes that whether or not two objects satisfy a role R, marked with $< a, b >:: R$.

$\top_1{}^I = \Delta^I$	$(C_1 \sqcap C_2)^I = C_1{}^I \cap C_2{}^I$	$(i/2:C)^I = \{t \in \top_2{}^I \mid t[i] \in C^I\}$
$A^I \subseteq \Delta^I$	$\top_2{}^I = (\Delta^I)^2$	$(\neg R)^I = \top_2{}^I \setminus R^I$
$(\neg C)^I = \Delta^I \setminus C^I$	$P^I \subseteq (\Delta^I)^2$	$(R_1 \sqcap R_2)^I = R_1{}^I \cap R_2{}^I$

$(\leq k[i/2]R)^I = \{a \in \Delta^I \mid \#\{t \in R^I \mid t[i]=a\} \leq k\}$ here # denotes cardinality of a set

$(\text{id } C\ [i_1]R_1,...,[i_n]R_n \mid i_j \in \{1,2\})^I = \{\text{for } \forall\ a,b \in C^I, \text{ and } \forall t_1, s_1 \in R_1{}^I,...,t_n, s_n \in R_n{}^I : \{a=t_1[i_1]=...=t_n[i_n]\}$
$\wedge \{b=s_1[i_1]=...=s_n[i_n]\} \wedge \{t_j[i] = s_j[i], \text{for } j \in \{1,...,n\}, \text{and for } i \neq i_j\} \rightarrow \{a=b\}. \}$

$(\text{fd } R\ i_1 \rightarrow i_2), i_1\ i_2 \in \{1,2\})^I = \{\text{for } \forall t, s \in R^I : \{t[i_1]=s[i_1],...,t[i_n] =s[i_n]\} \rightarrow \{a=b\}. \}$

Fig. 1. Semantics of DLR$_{\mathcal{DM}}$

4   Formalization of Data Mining Metamodel

Metadata is the data about data. The CWM based data mining metamodel is the metadata upon data mining systems and application data infrastructure. Fig. 2 shows a CWM based data mining metamodel. In what follows, we analyze the formalization of CWM based data mining metadata with the DLR$_{\mathcal{DM}}$.

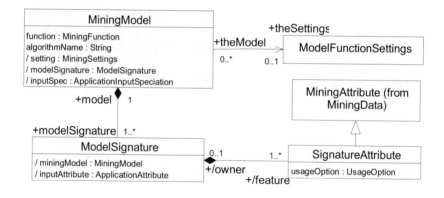

Fig. 2. Data Mining Metamodel

4.1   Meta Class

In the CWM based data mining models, metaclass is the description of a series of objects with same attributes, operations and semantics. A metaclass could be called class for short. For instance, the MiningModel class in the Fig. 2 is a meta-class in the data mining metamodel. Since concepts of DLR$_{\mathcal{DM}}$ and metaclasses of CWM are both used to describe objects, we naturally conceive the idea that describes the CWM based data mining metaclass with concepts of DLR$_{\mathcal{DM}}$.

A metaclass is graphically rendered as a rectangle divided into two parts. The first part contains the name of the metaclass; the second part contains the attributes of the metaclass, each denoted by a name with an associated class, which indicates the domain of the attribute values. Each "/" indicates that the type of the attribute is a metaclass already included in the metamodel. i.e., the metaclass to which the attribute belongs is associated with the metaclass that is the type of the attribute.

A CWM based data mining metamodel attribute a of type C' for a class C associates to each instance of C, so we think of an attribute a of type C' for a class C as a binary relation between instances of C and instances of C'. We capture such a binary relation by means of a role of $\text{DLR}_{\mathcal{DM}}$. To specify the type of the attribute we use the assertion: $C \sqsubseteq \forall[1](a \rightarrow (2 : C'))$.

Such an assertion specifies precisely that, for each instance c of the concept C, all objects related to c by a, are instances of C'. It also indicates that an attribute name is not necessarily unique in the whole metadata, and hence two different metaclasses could have the same attribute, possibly of different types.

4.2   Generalization and Inheritance

In CWM metamodel, one can use generalization between a parent class and a child class to specify that each instance of the child class is also an instance of the parent class. Hence the instances of the child class inherit the prosperities of the parent class, but typically they satisfy additional properties that do not hold for the parent class. Generalization is naturally supported in $\text{DLR}_{\mathcal{DM}}$. In CWM based data mining metamodel, the metaclass *Element* generalizes *ModelElement*, we can describe it by the $\text{DLR}_{\mathcal{DM}}$ assertion: $ModelElement \sqsubseteq Element$.

Inheritance between $\text{DLR}_{\mathcal{DM}}$ concepts works exactly as inheritance between CWM metaclasses. This is an obvious consequence of the semantics of " $\sqsubseteq$ " which is based on the subset notion. Indeed, in $\text{DLR}_{\mathcal{DM}}$, given an assertion $C_1 \sqsubseteq C_2$, every tuple in a role having C_2 as i-th argument type may have as i-th component an instance of C_1, which is in fact also an instance of C_2. As a consequence, in the formalization, each attribute of C_2, each aggregation association and each ordinary association involving C_2 are correctly inherited by C_1. Notice that the formalization in $\text{DLR}_{\mathcal{DM}}$ also captures directly multiple inheritances between metaclasses.

4.3   Association

Associations describe the relations of two classes. CWM based data mining associations includes ordinary associations, reflection associations and aggregation associations. Ordinary associations are described as follows, if instances of the metaclass C_1 have components that are instances of metaclass C_2 by association A, the multiplicity on C_1 is $m_1..m_2$, the multiplicity on C_2 is $n_1..n_2$, then A is formalized in $\text{DLR}_{\mathcal{DM}}$ by means of a role A, and the following assertions are added to the Tbox:

$A \sqsubseteq (1 : C_1) \sqcap (2 : C_2); C_1 \sqsubseteq (\geq n_1[1]A) \sqcap (\leq n_2[1]A); C_2 \sqsubseteq (\geq m_1[2]A) \sqcap (\leq m_2[2]A)$.

The second assertion specifies that for each instance of C_1, there can be at least n_1 and at most n_2 instances of C_2 related to it by role A. The reflection association is actually a special ordinary association. Its two association sides are the same element. The DLR$_{\mathcal{DM}}$ description upon reflection associations need the multiplicity of the two sides, however, the two sides is the same elements.

An aggregation association in data mining metamodel, graphically rendered as a black diamond shown in Fig. 2, is a binary relation between the instances of two metaclasses, denoting a part-whole relationship. Differing from ordinary association, the aggregation association has restrictions upon the multiplicity. That is, the multiplicity of the whole end must be 1 or 0..1. Following formulas show the restrictions:

$$A \sqsubseteq (1 : C_1) \sqcap (2 : C_2); C_1 \sqsubseteq (\geq n_1[1]A) \sqcap (\leq n_2[1]A); C_2 \sqsubseteq (\geq 0[2]A) \sqcap (\leq 1[2]A).$$

We add them into the TBox of DLR$_{\mathcal{DM}}$. In the Fig. 2, the aggregation association of the metaclass $ModelSignature$ and the metaclass $SignatureAttritute$ is formalized as follows: $A \sqsubseteq (1 : ModelSignature) \sqcap (2 : SignatureAttribute)$, $SignatureAttribute \sqsubseteq (\leq 1[2]A), ModelSignature \sqsubseteq (\geq 1[1]A)$.

4.4   Dependency

The dependency denotes a kind of relation of two elements. In the dependency relations, the change of the independent element affects the dependent element. For example, in CWM based data mining metamodel, Classification analysis package depends on the Supervised analysis package and the MiningCore package. And the Supervised analysis package depend on the MiningCore package. We naturally think of describing dependency relation in the data mining metamodel with function dependency assertion in DLR$_{\mathcal{DM}}$.

fd Supervised-Classification $Supervised \rightarrow Classification$
fd MiningCore-Supervised $MiningCore \rightarrow Supervised$
fd MiningCore-Classification $MiningCore \rightarrow Classification$

4.5   Formalization of MOF Meta-metamodel

Meta object framework MOF is a model driven distributed object management framework used for specifying, constructing, managing, exchanging metadata in an application system. It supports all kind of metadata. If necessary it can add new metadata and delete old metadata. MOF, located on the highest layer of the CWM based metadata architecture, is the meta-meta model of CWM based data mining metamodel. The core elements of MOF are class, object, attribute and operation. The core relations are association, aggregation, generalization, dependency. Same as the metamodel, the class, association, aggregation, generalization, inherence and dependency in the meta-meta model can be described with DLR$_{\mathcal{DM}}$.

4.6   Formalization of Metadata

In the CWM based metadata architecture, CWM metamodel is the instance of MOF meta-metamodel, so it is also called meta-metadata. The UML model is the instance of CWM metamodel, so it is also called metadata. The modeled data mining system and application data is the instance of UML model, they are ordinary data, **NOT** metadata. In the DLR$_{\mathcal{DM}}$, we describe CWM based metadata and meta-meta data with ABox. Here, we don't distinguish them. The instances of metamodel and meta-metamodel are both called metadata. If the metadata c is the instance of a metaclass C, then it is formalized with $c : C$ or $C(c)$. If the element a and b of metadata have the ordinary association relation A, then they are formalized with $< a, b >: A$.

5   Consistency Checking for Data Mining Metadata

Theorem 1. *DLR$_{\mathcal{DM}}$ is decidable, and it is exponent-time complete problem.*

Proof. The reasoning algorithm tableaux of Basic ALC language is exponent-time complete problem. DLR$_{\mathcal{DM}}$ can be mapped into a subset of ALC. Therefore, DLR$_{\mathcal{DM}}$ is decidable, and it is exponent-time complete problem [8].   ∎

Now description logic has equipped with good reasoning engineer LOOM and Racer. What's more, the new version RacerPro of Racer has a platform of creating repository [9]. Compared with LOOM, its reasoning algorithm tableaux is reliable and complete. So we construct a DLR$_{\mathcal{DM}}$ repository with RacerPro, then check the inconsistency with the reasoning mechanism of the RacerPro.

In the DLR$_{\mathcal{DM}}$ repository, the metaclasses Classification and Clustering are disjoint. Suppose that we should add new metaclass SVM, here SVM denotes support vector machine, which is a classification technique essentially and does not belong to the clustering. However, when adding dependencies of SVM upon Classification and Clustering, the process would lead to conflicts. For example, we add fd Classification-SVM *Classification* → *SVM* and fd Clustering-SVM *Clustering* → *SVM* into TBox. The function dependency symbol → is defined according to the semantic of DLR$_{\mathcal{DM}}$. Under the RacerPro query commands:
(In-knowledge-base data-mining-metadata)
;;=======T-box Reasoning=======
(tbox-conherent? data-mining-metadata)
;; Find all incoherent concept
We get following conflict information:
(TBOX-COHRENT? data-mining-metadata) $--$ > NIL. Concept(SVM) is incoherent in TBox data-mining- metadata. Error: ABox data-mining-metadata is incoherent.

Therefore, that SVM depends on both Classification metaclass and Clustering metaclass leads to the conflict. The method of resolving the conflict is deleting the dependency of SVM upon the Clustering metaclass in the data mining metadata. Correspondingly, in the DLR$_{\mathcal{DM}}$ repository, we delete this dependency, import the repository again and rerun the reasoning, the conflict is resolved. Besides, we can get some new conclusion from the results. For example, that SVM metaclass depends on Supervised metaclass due to that the Classification class depends on Supervised metaclass. Some further relation between metadata can be acquired.

6 Conclusion

During the process of constructing data mining metadata, it is usually hard to discover the conflicts in the metadata. However, the automatically discovering conflicts upon data mining metadata has not been resolved well up to now. The main contributing of this paper is proposing a description logic based consistency checking mechanism for data mining metadata. A formal logic $DLR_{\mathcal{DM}}$ in the description logic family is proposed to formalize the CWM based data mining metadata. We prove the validity of $DLR_{\mathcal{DM}}$ applied into consistency checking of data mining metadata. Finally we construct a data mining metadata repository with RacerPro. The reasoning results on the reasoning engineer Racer indicates that $DLR_{\mathcal{DM}}$ is correct and efficient when automatically discovering inconsistency information of data mining metadata.

This research is helpful to develop conflict checking system of data mining metadata, enhance the stability of data mining metadata criteria, and guarantee the reliability of data mining metadata integration as well as data mining products sharing. In the future, we will further investigate formal reasoning based on other data mining metadata such as PMML. We will also research how to automatically resolve conflicts based on inconsistency information discovered by $DLR_{\mathcal{DM}}$ upon data mining metadata.

References

1. KDD 2006 Workshop on Data Mining Standards, Services and Platforms (2006), http://www.ncdm.uic.edu/dm-ssp-06.htm
2. Object Management Group: Common Warehouse Metamodel Specification, Version 1.1 (2001), http://www.omg.org
3. Zhao, X., Huang, Z.: A Formal Framework for Reasoning on Metadata Based on CWM. In: Embley, D.W., Olivé, A., Ram, S. (eds.) ER 2006. LNCS, vol. 4215, pp. 371–384. Springer, Heidelberg (2006)
4. Hauch, R., Miller, A., Cardwell, R.: Information Intelligence: Metadata for Information Discovery, Access, and Integration. In: 2005 ACM SIGMOD International Conference on Management of Data, Baltimore, USA, pp. 793–798 (2005)
5. Cali, A., Calvanese, D., Giacomo, G.D.: A formal framework for reasoning on UML class diagrams. In: Hacid, M.-S., Raś, Z.W., Zighed, A.D.A., Kodratoff, Y. (eds.) ISMIS 2002. LNCS (LNAI), vol. 2366, pp. 503–513. Springer, Heidelberg (2002)
6. Zubcoff, J., Trujillo, J.: Conceptual Modeling for classification mining in Data Warehouses. In: Tjoa, A.M., Trujillo, J. (eds.) DaWaK 2006. LNCS, vol. 4081, pp. 566–575. Springer, Heidelberg (2006)
7. Calvanese, D., Giacomo, G.D., Lenzerini, M.: Identification Constraints and Functional Dependencies in Description Logics. In: 17th International Joint Conference on Artificial Intelligence (IJCAI 2001), Washington, pp. 155–160 (2001)
8. Horrocks, I., Sattler, U.: A Tableaux decision procedure for SHOIQ. In: 19th International Joint Conference on Artificial Intelligence (IJCAI 2005) (2005)
9. Haarslev, V., Moller, R., Wessel, M.: RacerPro Version 1.9 (2005), http://www.racer-systems.com

LRLW-LSI: An Improved Latent Semantic Indexing (LSI) Text Classifier

Wang Ding, Songnian Yu, Shanqing Yu, Wei Wei, and Qianfeng Wang

School of Computer Engineering and Science, Shanghai University,
200072 Shanghai, China

Abstract. The task of Text Classification (TC) is to automatically assign natural language texts with thematic categories from a predefined category set. And Latent Semantic Indexing (LSI) is a well known technique in Information Retrieval, especially in dealing with polysemy (one word can have different meanings) and synonymy (different words are used to describe the same concept), but it is not an optimal representation for text classification. It always drops the text classification performance when being applied to the whole training set (global LSI) because this completely unsupervised method ignores class discrimination while only concentrating on representation. Some local LSI methods have been proposed to improve the classification by utilizing class discrimination information. However, their performance improvements over original term vectors are still very limited. In this paper, we propose a new local Latent Semantic Indexing method called "Local Relevancy Ladder-Weighted LSI" to improve text classification. And separate matrix singular value decomposition (SVD) was used to reduce the dimension of the vector space on the transformed local region of each class. Experimental results show that our method is much better than global LSI and traditional local LSI methods on classification within a much smaller LSI dimension.

Keywords: Data mining, Text classification, Latent Semantic Indexing (LSI).

1 Introduction

Traditional text classification is based on explicit character, and the common method is to represent textual materials with space vectors using Vector Space Model (VSM) [3][4][5][6], finally, confirm the category of the test documents by comparing the degree of similarity.

With more and more textual information available on the internet, conceptual retrieval has become more important than word matching retrieval. Traditional information retrieval system such as VSM retrieves relevant documents by lexical matching with query. The drawback of VSM is that it cannot retrieve the conceptually relevant documents with respect to query, and the semantic information may lose during the process of VSM.

In order to overcome the drawbacks of VSM, we apply Latent Semantic Indexing (LSI), which is widely used as the information retrieval technique, in

G. Wang et al. (Eds.): RSKT 2008, LNAI 5009, pp. 483–490, 2008.

the proposed method. While LSI is applied to text classification, there are two common methods. The first one is called "Global LSI", which performs SVD directly on the entire training document collection to generate the new feature space. This method is completely unsupervised, that is, it pays no attention to the class label of the existing training data. It has no help to improve the discrimination power of document classes, so it always yields no better, sometimes even worse performance than original term vector on classification [3]. The other one is called "Local LSI", which performs a separate SVD on the local region of each topic. Compared with global LSI, this method utilizes the class information effectively, so it improves the performance of global LSI greatly. However, due to the same weighting problem, the improvements over original term vector are still very limited.

Typically all documents in the local region are equally considered in the SVD computation in local LSI. But intuitively, first, more relevant documents to the topic should contributes more to the local semantic space than those less-relevant ones; second, tiny less local relevant documents may be a little more global relevant. So based on these ideas, we propose a new local LSI method "Local Relevancy Ladder-Weighted LSI (LRLW-LSI)", which selects documents to the local region in a ladder way so that the local semantic space can be extracted more accurately considering both the local and global relevancy. Experimental results shown later prove this idea and it is found LRLW-LSI is much better than global LSI and ordinary local LSI methods on classification performance within a much smaller LSI dimension.

2   Prelminaries

2.1   Singular Value Decomposition

SVD is one of the most important matrix decomposition in numerical linear algebra. It has been applied in many fields such as image processing [1], neural networks [2] and others. In this paper, we use matrix singular value decomposition to reduce the dimension of the vector space and remove the influences of synonymy and polysemy.

Let A denote an matrix of real-valued. Without loss of generality we assume that $m \geq n$, and $rank(A)$ is r. The Singular value decomposition (SVD) of A is its factorization into a product of three matrices

$$A = USV^T \tag{1}$$

Where U is an orthogonal matrix, V an orthogonal matrix, and S an matrix and $U^T U = V^T V = I_n$, $S = diag(\sigma_1 \cdots \sigma_n)$ $\sigma_i > 0$ for $1 \leq i \leq r$ and $\sigma_j = 0$ for $j \geq r + 1$. The first r columns of the orthogonal matrix U and V define the orthonormal eigenvectors associated with the r nonzero eigenvalues of AA^T and $A^T A$, respectively. The columns of U and V are referred to as the left and right singular vectors, respectively, and the singular values of A are defined as the diagonal elements of S which are the nonnegative square roots of the n eigenvalues of AA^T [4].

The following two theorems illustrate how the SVD can reveal important information about the structure of a matrix.

Theorem 1: Let the SVD of A be given by Equation (1) and $\sigma_1 \geq \sigma_2 \geq \cdots \sigma_r > \sigma_{r+1} = \cdots = \sigma_n = 0$ let $R(A)$ and $N(A)$ denote the range and null space of A, respectively. Then, (1) Rank property: $rank(A) = r$, $N(A) \equiv span\{v_{r+1}, \cdots, v_n\}$ and $R(A) \equiv span\{u_1, \cdots, u_r\}$, where $U = [u_1 u_2 \cdots u_m]$ and $V = [v_1 v_2 \cdots v_n]$. (2) Dyadic decomposition:

$$A = \sum_{i=1}^{r} u_i \cdot \sigma_i \cdot v_i^T$$

(3) Norms: $\|A\|_F^2 = \sigma_1^2 + \cdots \sigma_2^2 \cdots + \sigma_r^2$ and $\|A\|_2^2 = \sigma_1$ Proof. See [4].

Theorem 2: Let the SVD of A be given by Equation (1) with and define

$$A_k = \sum_{i=1}^{k} u_i \cdot \sigma_i \cdot v_i^T \tag{2}$$

Then we can get:(Proof. See [5]).

$$\min_{rank(B)=k} \|A - B\|_F^2 = \|A - A_k\|_F^2 = \sigma_{k+1}^2 + \cdots + \sigma_p^2$$

In other words, A_k which is constructed from the k-largest singular triplets of A, is the closest rank-k matrix to A. In fact, A_k is the best approximation to A for any unitarily invariant norm. Hence,

$$\min_{rank(B)=k} \|A - B\|_2 = \|A - A_k\| = \sigma_{k+1} \tag{3}$$

2.2 Latent Semantic Indexing (LSI)

Firstly, we briefly describe the Vector space model (VSM). Vector space model of text document is put forward by Salton [6] and used in SMART system. In the vector space model, a document is represented by a vector of words. And a word-by-document matrix A used to represent a collection of documents, where each entry represents the occurrences of a word in a document, e.g., $A = (a_{wd})$ where a_{wd} is the weight of the word w in the document d.

In order to implement Latent Semantic Indexing [7], a matrix of terms by documents must be constructed. In this paper, we use Vector Space Model mentioned above to construct the original and rough matrix A for LSI. Since every word does not normally appear in each document, the matrix A is usually sparse. The matrix A is factored into product of three matrices using the singular value decomposition. The SVD derives the latent semantic structure model from the orthogonal matrices U and V containing left and right singular vectors of A respectively, and the diagonal matrix, S, of singular values of A. These matrices reflect a breakdown of the original relationships into linearly-independent vectors or factor values. The use of k factors or k-largest singular triplets is equivalent to approximation the original term-document matrix by A_k in Equation (2). The SVD captures most of the important underlying structure in the association of terms and documents, yet at the same time removes the noise or variability in word usage.

3 Proposed Algorithm

The most straightforward method of applying LSI for text classification is the global LSI method as discusses in section 2.2, which performs SVD directly on the entire training set and then testing documents transformed by simply projecting them onto the left singular matrix produced in the original decomposition. However, global LSI has many drawbacks which are discussed above. In order to overcome the drawbacks of Global LSI, we proposed a local LSI method. We name it "Local Relevancy Ladder-Weighted LSI (LRLW-LSI)".

In local LSI, each document in the training set is first assigned with a relevancy score related to a topic, and then the documents whose scores are larger than a predefined threshold value are selected to generate the local region. Then SVD is performed on the local region to produce a local semantic space. This process can be simply described as the jump curve in Figure 4.2. That is 0/1 weighting method is used to generate the local region where documents whose scores are larger than the predefined threshold value are weighted with 1 and others are weighted with 0. The 0/1 weighting method is a simple but crude way to generate local region. It assumes that the selected documents are equally important in the SVD computation. But intuitively, first, different documents should play different roles to the final feature space and it is expected that more relevant documents to the topic should contributes more to the local semantic space than those less-relevant ones; second, less local relevant documents may be more global relevant. So based on these ideas, we propose the local LSI method "Local Relevancy Ladder-Weighted LSI (LRLW-LSI)", which selects documents to the local region in a ladder way. In other words, LRLW-LSI gives same weight among a ladder-range and different weight located in different ladder-range to documents in the local region according to its relevance before performing SVD so that the local semantic space can be extracted more accurately considering both the local and global relevancy and more relevant documents can be introduced with higher weights, which make they do more contribution to SVD computation. Hence, the better local semantic space which results in better classification performance can be extracted to separate positive documents from negative documents. Ladder-Curve is described in Figure 1.

LRLW-LSI Algorithm For each class, assume an initial classifier C_0 has been trained using training documents in term vector representation and here we use SVM classifier. Then the training process of LRLW-LSI contains the following six steps.

(1) The initial classifier C_0 of topic c is used to assign initial relevancy score RS_0 to each training document.

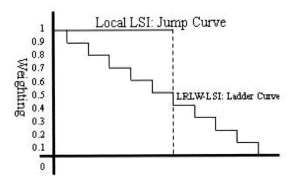

Fig. 1. Local Relevancy Ladder-Weighted LSI (LRLW-LSI)

(2) Each training document is first weighted according to equation (4). The weighting function is a Sigmoid function which has two parameters a and b. Then, assign the belonged average ladder weight according to the first raw weight. E.g. if the first raw weight is 0.91, then we will assign 0.95 for the new weight, which is the average of the top ladder.

(3) Top n documents are selected to generate the local term-by-document matrix of the topic c.

(4) The SVD is performed to generate the local semantic space.

(5) All other weighted training documents are folded into the new space.

(6) All training documents in local LSI vector are used to train a real classifier C of topic c.

$$\vec{t} = \vec{t} * f(rs_i), where f(rs_i) = \frac{1}{1 + e^{-a(rs_i+b)}} \qquad (4)$$

Then the testing process of LRLW-LSI contains the following three steps. When a testing document comes in,

(1) It is classified by the initial classifier C_0 to get its initial relevancy score.

(2) It is weighted according to the equation (4) and then folded into the local semantic space to get its local LSI vector.

(3) The local LSI vector generated in step 2 is finally used to be classified by the classifier RC to decide whether it is belongs to the topic or not.

4   Experiment Results

In this section, we evaluate Local Relevancy Ladder-Weighted LSI method. SVM light (http://svmlight.joachims.org) is chosen as the classification algorithm, SVDPAKC/sis (www.netlib.org/svdpack) is used to perform SVD and F-Measure is used to evaluate the classification results. Two common data sets are used, including Reuters-21578 and Industry Sector.

Before performing classification, a standard stop-word list is used to remove common stop words and stemming technology is used to convert variations of the same words into its base form. Then those terms that appear in less than 3 documents are removed. Finally tf*idf (with "ltc" option) is used to assign the weight of each term in each document.

4.1 Data Set

Two text collections, Reuters-215783 (www.daviddlewis.com/resources/test-collections) and Industry Sector4 (www-2.cs.cmu.edu/afs/cs.cmu.edu), are used in our experiment.

Reuters-21578 (Reuters) is the most widely used text collection for text classification. There are total 21578 documents and 135 categories in this corpus. In our experiments, we only chose the most frequent 25 topics and used "Lewis" split which results in 6314 training examples and 2451 testing examples.

Industry Sector (IS) is a collection of web pages belonging to companies from various economic sectors. There are 105 topics and total 9652 web pages in this dataset. A subset of the 14 categories whose size are bigger than 130 is selected for the experiments.

4.2 Experimental Results and Discussion

For local relevancy ladder-weighted LSI, we use SVM classifier as the initial classifier C_0 to generate each document's initial relevancy score. And the parameters a and b of Sigmoid function are initially set with 5.0 and 0.2. The number of ladder step is assigned 10.

Figure 2 and Figure 3 show the classification results on the data set, Reuters-21578 and Industry Sector. The lines of term vector are displayed only as the reference points in terms of performance comparison. From these figures, the following observations can be made:

First, compared to term vector, LRLW-LSI improves the both F1 performances greatly on both data. For example, using 20 dimensions on Reuters-21578, the micro-averaging F1 is improved by 1.1% and the macro-averaging F1 is improved by 3.7%; using 50 dimensions on Industry Sector, the micro-averaging F1 is improved by 7.2% and the macro-averaging F1 is improved by 9.8%.

Second, Figure 4 shows the run time of different LSI methods on a PC with Pentium IV 1.7GHz and 256M memory. The runtime includes both training procedure and testing procedure. As can be seen, term vector is the fastest and it needs only hundred seconds. Global LSI needs much more time than term vector due to the costly SVD computation on entire training set. Although SVD computation on local region is very fast, the overall computation on all topics is extremely high, so local LSI is not expected to be used in practice. Similar with local LSI, LRLW-LSI has to perform a separate SVD on local region of each topic, but such a low LSI dimension makes LRLW-LSI be extremely rapid. It needs only less than 3 times of runtime of term vector, so it can be widely used in practice.

Third, with the LSI dimension increases, the performances decrease slowly. But even in a relatively high dimension, the performances are still above the performances of term vector. Using 150 dimensions, for example, on Reuters-21578 the micro-averaging F1 is improved by 1.1% and the macro-averaging F1 is still improved by 1.2%; on Industry Sector, the micro-averaging F1 and the macro-averaging F1 are still improved by 3.7%. In this paper, we propose a Local Relevancy Ladder-Weighted LSI (LRLW-LSI) method to help improve the text classification performance. This method is developed from Local LSI, but

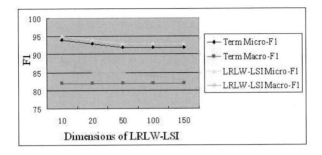

Fig. 2. Results on Reuters Figure

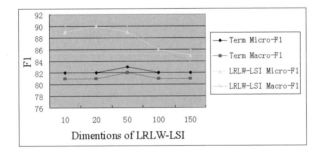

Fig. 3. Results on Industry Sector

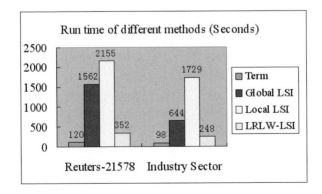

Fig. 4. Run time of different methods

different from Local LSI in that the documents in the local region are introduced using a ladder-descending curve so that more relevant documents to the topic are assigned higher weights and the global relevancy is also considered. Therefore, the local SVD can concentrate on modeling the semantic information that is actually most important for the classification task. The experimental results verify this idea and show that LRLW-LSI is quite effective.

Acknowledgements

This research was supported by the international cooperation project of Ministry of Science and Technology of PR China, grant No. CB 7-2-01, and by "SEC E-Institute: Shanghai High Institutions Grid" project.

References

1. Liu, J., Niu, X.M., Kong, W.H.: Image Watermarking Based on Singular Value Decomposition. In: Proceedings of the 2006 Intelligent Information Hiding and Multimedia Signal Processing, pp. 457–460 (2006)
2. Kanjilal, P.P., Dey, P.K., Banerjee, D.N.: Reduced-size neural networks through singular value decomposition and subset selection. Electronics Letters 29, 1516–1518 (1993)
3. Torkkola, K.: Linear Discriminant Analysis in Document Classification. In: Proceedings of the 2001 IEEE ICDM Workshop Text Mining (2001)
4. Golub, G., Loan, C.V.: Matrix Computations, 2nd edn. Johns-Hopkins, Baltimore (1989)
5. Golub, G., Reinsch, C.: Handbook for automatic computation: linear algebra. Springer, New York (1971)
6. Salton, G.: Introduction to modern information retrieval, Auckland. McGraw-Hill, New York (1983)
7. Deerwester, S., Dumais, S., Furnas, G., Landauer, T., Harshman, R.: Indexing by latent semantic analysis. Journal of the American Society for Information Science, 391–407 (1990)

Multi-granularity Classification Rule Discovery Using ERID

Seunghyun Im[1], Zbigniew W. Raś[2], and Li-Shiang Tsay[3]

[1] Department of Computer Science
University of Pittsburgh at Johnstown
Johnstown, PA 15904, USA
sim@pitt.edu
[2] Department of Computer Science
University of North Carolina
Charlotte, NC 28223, USA
ras@uncc.edu
[3] Department of Electronics, Computer and Information Technology
North Carolina A&T University
Greensboro, NC 27411, USA
ltsay@ncat.edu

Abstract. This paper introduces the use of *ERID* [1] algorithm for classification rule discovery at various levels of granularity. We use an incomplete information system and attribute value hierarchy to extract rules. The incomplete information system is capable of storing weighted attribute values and the domains of those attributes are organized using a hierarchical tree structure. The granularity of attribute values can be adjusted using the attribute value hierarchy. The result is then processed through *ERID*, which is designed to discover rules from partially incomplete information systems. The capability of handling incomplete data enables to build more specific and general classification rules.

Keywords: knowledge discovery, incomplete information system, attribute hierarchy, rough sets.

1 Introduction

Real world data are often collected from a wide variety of sources using different methods, which makes data aggregation and analysis difficult. Attribute hierarchy (a.k.a. concept hierarchy or attribute taxonomy) provides a solution to the problem by helping to organize such data into gradual levels of abstraction so that data may be accessed and analyzed at multiple levels of granularity. For example, suppose that users need an answer for flights departing around 10:00 AM on a day. Instead of searching for a flight departing at that particular time a more general definition of 10:00AM (e.g. morning) is drawn from the attribute value hierarchy to answer the query. The same principle can be used to build a knowledge discovery system that is able to discover rules at different levels of abstraction based on the user's interest; e.g. finding out the dependency between

G. Wang et al. (Eds.): RSKT 2008, LNAI 5009, pp. 491–499, 2008.

the flight schedule and ticket sales to increase the profit. Such system also helps to discover more precise rules from a distributed knowledge discovery system [6] [7]. For example, the granularity of data (or knowledge) at remote sites may be finer or coarser than that of local site. Then, data with similar semantics of common attributes but different granularity can be used jointly to discover hidden knowledge. By using the notion of semantic bridge [7] [8] and attribute hierarchy we will be able to adapt the discovered knowledge at remote sites for local use so that data with the similar meanings are processed as a single property.

Multiple attribute values can be used to describe a single object property when hierarchical attributes exist. Eye color {black} of a person (object) may be described by {dark black or light black} if finer granularity of description is required. The difficulty in providing a flexible ad-hoc classification rule discovery lies in how to deal with those multi-valued objects in the rule discovery process. In this paper, we describe and illustrate the use of *ERID* (Extracting Rules from Incomplete Decision systems) [1] to discover classification rules from an information system with hierarchical attributes. An incomplete information system is suitable for storing data with attribute value hierarchies because it enables to contain multiple weighted values as a single attribute value. *ERID* is particularly designed to extract rules from such partially incomplete information systems.

2   Incomplete Information System

We begin with the definition of an incomplete information system [8] which is a generalization of an information system given by [4]. In the conventional information system, the value of an object is either precisely known or not known at all. This implies that either a single value of an attribute is assigned to an object as its property or no value is assigned. However, it happens quite often that users do not have exact knowledge about some objects, which makes it difficult to determine a unique set of values describing them. The notion of an incomplete information system was introduced to model such cases.

By an information system S, we mean $S = (X, A, V)$ where $X = \{x_1, x_2, \ldots, x_n\}$ is a finite set of objects, $A = \{a_1, a_2, \ldots, a_m\}$ is a finite set of attributes, and $V = \{a_1(x_1), a_2(x_2), \ldots, a_m(x_n)\}$ is set of their values. In particular, we say that S is an incomplete information system of type λ if the following three conditions hold:

- $a(x)$ is defined for any $x \in X$, $a \in A$,
- $(\forall x \in X)(\forall a \in A)[(a(x) = \{(v_i, w_i) : 1 \leq i \leq k\}) \rightarrow \sum_{i=1}^{k} w_i = 1]$,
- $(\forall x \in X)(\forall a \in A)[(a(x) = \{(v_i, w_i) : 1 \leq i \leq k\}) \rightarrow (\forall i)(w_i \geq \lambda)]$.

The threshold value λ is used to determine the minimum confidence of an attribute value that can be assigned to S.

Example 1. An attribute value for an object in an incomplete information system can be {(dark black,0.5),(light black, 0.5)}, meaning that the object may be dark black or light black with equal weight.

Example 2. Suppose that $\{(v_1, 0.2), (v_2, 0.4), (v_3, 0.4)\}$ is being assigned as a single attribute value in S with $\lambda = 0.3$. Then, it is automatically converted to $\{(v_2, 0.5), (v_3, 0.5)\}$ because the minimum confidence of $v_1 < 0.3$, and sum of the weights of v_i is 1 by definition. Weights for v_2 and v_3 are calculated by the ratio of its weight and the total wight after eliminating v_1, that is $p_2 = \frac{0.4}{0.4+0.4}$, $p_3 = \frac{0.4}{0.4+0.4}$

The capability of assigning a set of weighted values makes changing the granularity of data simpler and more effective, which will be discussed in the next section.

3   Rule Discovery from Incomplete Hierarchical Information System

3.1   Granularity Adjustment in Incomplete Information System

An incomplete hierarchical information system [5] is an incomplete information system in which the domain of each attribute is defined as a tree-like structure, called attribute value hierarchy (*AVH*). A set of *AVH*s can be derived from domain ontology [3] or independently defined by a domain expert before data are collected.

For an incomplete information system $S_t = (X, A, V)$, we define a set of attribute value hierarchies $T = \{t_{a_1}, t_{a_2}, \cdots, t_{a_m}\}$, where t_{a_i} is the *AVH* for $a_i \in A$. For simplicity, we use the notation $t = \{(a_{[p]})\}$ to indicate the set of node values in an *AVH*, where a is an attribute name and p is the path from the root to the node. (e.g. $a_{[1,1]}$ is the value of the left most element at depth 2 in Figure 1). a_ε, where ε is the empty sequence, is equivalent to the attribute name.

We assume that two values $a_{[p]}$ and $a_{[p']}$ on the same path of t have the same semantic type, but different semantic granularity. Therefore, it is possible to replace one with another in S_t. We use the term *generalization* to denote a replacement of $a(x)$ with a value in coarser granularity, and *specification* to denote a replacement of $a(x)$ with a value in finer granularity. The following definitions are used to determine the value and weight of the replaced attribute value, $a'(x)$.

1. The sum of the weights for $a'(x) = 1$ (unchanged by the definition of an incomplete information system).
2. If $a_i(x) = \{(v, w)\}$ is generalized to level k, $a'(x) = \{(v', w)\}$, where v' is an ancestor node of v at depth k, and w is carried over.
3. If $a_i(x) = \{(v, w)\}$ is specified to level k, each v is replaced by one or more descendant nodes $\{v'_1, v'_2, ..., v'_n\}$, and w' for each v' is $\frac{1}{n}$, where n is the number of sibling nodes at depth k.
4. If $t_i \in T$ is not balanced, and the granularity of $a_i(x)$ is given to a level where no node exists, we choose the closest parent node.

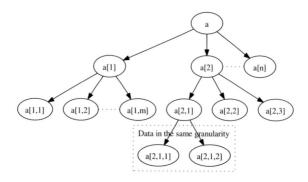

Fig. 1. An example of attribute value hierarchy

Example 3. If an attribute value $\{(a_{[2,1]}, 0.5), (a_{[2,2]}, 0.5)\}$ is generalized to one level higher in Figure 1, it becomes $(a_{[2]}, 1)$. For instance, dark black and light black are black for certain. If an attribute value $\{(a_{[1,1]}, 0.7), (a_{[2,1]}, 0.3)\}$ is generalized to one level up, new attribute value is $\{(a_{[1]}, 0.7), (a_{[2]}, 0.3)\}$.

Example 4. If $\{(a_{[2]}, 1)\}$ is specified to the values in depth 3, the attribute value in S changes to $\{(a_{[2,1,1]}, \frac{1}{6}), (a_{[2,1,2]}, \frac{1}{6}), (a_{[2,3]}, \frac{1}{3}), (a_{[2,3]}, \frac{1}{3})\}$. This replacement satisfies the definition of an incomplete information system because the sum of the weights is $1 = (\frac{1}{6} + \frac{1}{6} + \frac{1}{3} + \frac{1}{3})$.

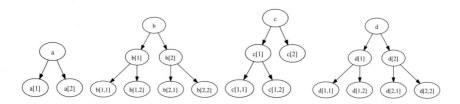

Fig. 2. Attribute value hierarchy for S_t

Now, let us consider the information system S_t shown in Table 1 and the corresponding *AVH* set T given in Figure 2. Suppose that a user is interested in discovering rules with granularity at $(a = 1, b = 2, c = 2, d = 1)$ where the number indicates the depth of the trees in (t_a, t_b, t_c, t_d), respectively. Then, S'_t (see Table 2) is the information system after granularity adjustment. For a and c, all attribute values are at the desired granularity so no replacement was performed. Note that $a(x_2)$ is not replaced with (a_1, a_2) because a null assigned to an object is interpreted as an unknown value. For b, values $b(x_4)$ and $b(x_5)$ remain unchanged. Other attribute values are replaced with the values in their child nodes, and their weights are also adjusted based on the definition described above. For d, all attribute values are generalized to the values in their parent nodes.

Table 1. Information System S_t

X	a	b	c	d
x_1	$(a_{[1]},\frac{1}{2})(a_{[2]},\frac{1}{2})$	$b_{[2]}$	$c_{[1,1]}$	$d_{[1,1]}$
x_2		$b_{[1]}$	$c_{[1,1]}$	$d_{[2,1]}$
x_3	$a_{[1]}$	$b_{[2]}$	$(c_{[1,1]},\frac{1}{2})(c_{[1,2]},\frac{1}{2})$	$d_{[2,2]}$
x_4	$a_{[1]}$	$b_{[1,1]}$	$c_{[1,1]}$	$d_{[1,1]}$
x_5	$a_{[1]}$	$b_{[1,1]}$		$d_{[1,2]}$
x_6	$a_{[2]}$	$b_{[2]}$	$c_{[1,1]}$	$d_{[2,1]}$

Table 2. Information System S'_t

X	a	b	c	d
x_1	$(a_{[1]},\frac{1}{2})(a_{[2]},\frac{1}{2})$	$(b_{[2,1]},\frac{1}{2})(b_{[2,2]},\frac{1}{2})$	$c_{[1,1]}$	$d_{[1]}$
x_2		$(b_{[1,1]},\frac{1}{2})(b_{[1,2]},\frac{1}{2})$	$c_{[1,1]}$	$d_{[2]}$
x_3	$a_{[1]}$	$(b_{[2,1]},\frac{1}{2})(b_{[2,2]},\frac{1}{2})$	$(c_{[1,1]},\frac{1}{2})(c_{[1,2]},\frac{1}{2})$	$d_{[2]}$
x_4	$a_{[1]}$	$(b_{[1,1]},1)$	$c_{[1,2]}$	$d_{[1]}$
x_5	$a_{[1]}$	$(b_{[1,1]},1)$		$d_{[1]}$
x_6	$a_{[2]}$	$(b_{[2,1]},\frac{1}{2})(b_{[2,2]},\frac{1}{2})$	$c_{[1,1]}$	$d_{[2]}$

3.2   ERID

Now, we run *ERID* algorithm [1] on S'_t to discover classification rules. *ERID* has some similarities with *LERS* [2]. It initially generates sets of weighted objects having only one-value properties. Then, some pairs of those sets are used for constructing classification rules if their support and confidence are higher than given threshold values. This process is recursively continued by moving to sets of weighted objects containing more properties. The distinctive feature of *ERID* is the ability to discover rules from incomplete information systems in which attribute values are incomplete or partially incomplete (multiple weighted values).

Assume that the minimum support and confidence are given as 1 and 0.75. The first step is to find the pessimistic interpretation in S of all attribute values in V, as shown below. The resulting sets are called granules. The granule $a_{[1]}{}^*$ associated with attribute value $a_{[1]}$ in S is the set of objects having property $a_{[1]}$ (e.g. objects $\{(x_1, \frac{1}{2}), (x_3, 1), (x_4, 1), (x_5, 1)\}$).

$$
\begin{aligned}
a_{[1]}{}^* &= \{(x_1, \tfrac{1}{2}), (x_3, 1), (x_4, 1), (x_5, 1)\} \\
a_{[2]}{}^* &= \{(x_1, \tfrac{1}{2}), (x_6, 1)\} \\
b_{[1,1]}{}^* &= \{(x_2, \tfrac{1}{2}), (x_4, 1), (x_5, 1)\} \\
b_{[1,2]}{}^* &= \{(x_2, \tfrac{1}{2})\} \\
b_{[2,1]}{}^* &= \{(x_1, \tfrac{1}{2}), (x_3, \tfrac{1}{2}), (x_6, \tfrac{1}{2})\} \\
b_{[2,2]}{}^* &= \{(x_1, \tfrac{1}{2}), (x_3, \tfrac{1}{2}), (x_6, \tfrac{1}{2})\} \\
c_{[1,1]}{}^* &= \{(x_1, 1), (x_2, 1), (x_3, \tfrac{1}{2}), (x_6, 1)\} \\
c_{[1,2]}{}^* &= \{(x_3, \tfrac{1}{2})(x_4, 1)\}
\end{aligned}
$$

$$d_{[1]}{}^* = \{(x_1, 1), (x_4, 1), (x_5, 1)\}$$
$$d_{[2]}{}^* = \{(x_2, 1), (x_3, 1), (x_6, 1)\}$$

Next, recursively, we check the relationship between classification attributes and the decision attribute from values in V [9]. For example, the relationship between $a_{[1]}{}^*$ and $d_{[1]}{}^*$, which is the set-theoretical inclusion between objects from $a_{[1]}{}^*$ and objects from $d_{[1]}{}^*$, depends on how high is the support of the corresponding rule $a_{[1]} \rightarrow d_{[1]}$. We calculate the support of the rule $a_{[1]} \rightarrow d_{[1]}$ by calculating the sum of the products of weights, $(\frac{1}{2} \cdot 0) + (1 \cdot 0) + (1 \cdot 1) + (1 \cdot 1)$. If support is below 1, the corresponding relationship does not hold and it is not considered in later steps (marked as "-"). Otherwise, the confidence of the rule is checked. If the confidence is greater than or equal the threshold value (0.75), the rule is approved and the relationship is marked as "+". When the confidence is below 0.75, the relationship remains unmarked to be considered in later steps as shown below.

$a_{[1]}{}^* \subseteq d_{[1]}{}^*$	$\sup = \frac{5}{2}$, conf = 0.72	+
$a_{[1]}{}^* \subseteq d_{[2]}{}^*$	$\sup = 1$, conf = 0.29	
$a_{[2]}{}^* \subseteq d_{[1]}{}^*$	$\sup = \frac{1}{2}$	−
$a_{[2]}{}^* \subseteq d_{[2]}{}^*$	$\sup = 1$, conf = 0.67	
$b_{[1,1]}{}^* \subseteq d_{[1]}{}^*$	$\sup = 2$, conf = 0.8	+
$b_{[1,1]}{}^* \subseteq d_{[2]}{}^*$	$\sup = \frac{1}{2}$	−
$b_{[2,1]}{}^* \subseteq d_{[1]}{}^*$	$\sup = \frac{1}{2}$	−
$b_{[2,1]}{}^* \subseteq d_{[2]}{}^*$	$\sup = 1$, conf = 0.67	
$b_{[2,2]}{}^* \subseteq d_{[1]}{}^*$	$\sup = \frac{1}{2}$	−
$b_{[2,2]}{}^* \subseteq d_{[2]}{}^*$	$\sup = 1$, conf = 0.67	
$c_{[1,1]}{}^* \subseteq d_{[1]}{}^*$	$\sup = 3$, conf = 0.67	
$c_{[1,1]}{}^* \subseteq d_{[2]}{}^*$	$\sup = \frac{1}{2}$	−
$c_{[1,2]}{}^* \subseteq d_{[1]}{}^*$	$\sup = 2$, conf = 0.8	+
$c_{[1,2]}{}^* \subseteq d_{[2]}{}^*$	$\sup = \frac{1}{2}$	−

This process is performed recursively to build terms containing $n + 1$ elements from previous terms with n elements using unmarked ones[1] (e.g. $a_{[1]}{}^*$ with $b_{[2,2]}{}^*$). The relationship between 2-element terms and the decision attribute is shown below.

$(a_{[1]} \cdot b_{[2,1]})^* \subseteq d_{[2]}{}^*$	$\sup = 1$, conf = 0.67,	
$(a_{[1]} \cdot b_{[2,2]})^* \subseteq d_{[2]}{}^*$	$\sup = 1$, conf = 0.67,	
$(a_{[1]} \cdot b_{[2,2]})^* \subseteq d_{[1]}{}^*$	$\sup = 0$,	+
$(a_{[1]} \cdot c_{[1,1]})^* \subseteq d_{[2]}{}^*$	$\sup = 1$, conf = 1,	-
$(b_{[1,1]} \cdot c_{[1,1]})^* \subseteq d_{[1]}{}^*$	$\sup = 2$, conf = 0.8,	+
$(b_{[2,1]} \cdot c_{[1,1]})^* \subseteq d_{[2]}{}^*$	$\sup = \frac{1}{2}$,	-

The process stops at this point because terms cannot be concatenated further. The relations marked with + symbol are the rules satisfying the the minimum support and confidence.

---

[1] The motivation and detailed description of the algorithm can be found in [1].

4   Implementation

The core algorithm (*ERID* and *AVH* program) was implemented in Matlab and the graphic interface was coded in Visual Basic.NET. The program was executed on a PC running Windows XP as shown in Figure 3. A sample table that contains 4884 objects with 12 selected attributes was extracted from the census bureau database of the *UCI* Knowledge Discovery in Databases Archive [10]. A set of *AVH*s with one to three levels of depth was built based on the description given by the data provider (see Table 3).

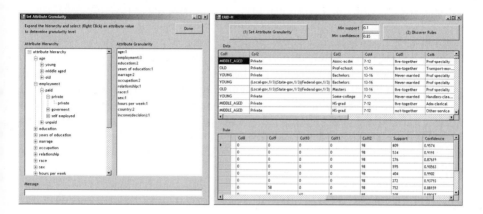

Fig. 3. A snapshot of the implementation

Table 3. Levels of attribute granularity for the test dataset

Attribute	# of Levels	Attribute	# of Levels
(1) age	3	(7) relationship	1
(2) employment	3	(8) race	1
(3) education	2	(9) gender	1
(4) years of education	2	(10) hours per week	2
(5) marriage	3	(11) country	2
(6) occupation	2	(12) income (decision)	1

Table 4. Number of rules at different attribute granularity

Exp #	Attribute Granularity	Min.Sup	Min.Conf	Num. of rules
1	2 2 2 2 2 2 1 1 1 1 1 1	10%	0.85%	6
2	3 3 2 2 3 2 1 1 1 2 2 1	10%	0.85%	5
3	1 1 1 1 1 1 1 1 1 1 1 1	10%	0.85%	8

To test the program for different attribute granularity (finer and coarser), we first converted the data set to medium levels of granularity and extracted rules as shown in the first experiment of Table 4, then, extracted rules using two different AVH granularity sets. For the sample data set, rule extraction using coarser attribute granularity generated a few more rules due to the generalized values in two attributes, age and hours per week.

5   Related Work

There have been many studies on the use of a concept hierarchy in data mining. Srikant and Agrawal [16] discussed methods for generating generalized association rules using a taxonomy. To improve the performance of the algorithm, higher levels of association rules were acquired by filtering and pre-computing ancestors of transaction items in the taxonomy. In [17], authors proposed a method that extracts classification rules at multiple levels of abstraction by incorporating ontological information into the rule discovery process. In [14], a method called rule based attribute oriented induction was introduced. In their method, generalization of data was performed along the direction provided by rules.

The authors in [18] discussed ontology-driven decision tree learning algorithm to learn generalized classification rules. The authors in [12] also studied methods for generating a multiple-level decision tree. They improved efficiency by removing some attributes if higher level concept did not exist and by applying the relevance analysis to the generalized data. In [11], feature hierarchies were used to discretize categorical values and to obtain compact bayesian network representation. In [13], structured attributes were used to group continuous data into discrete ranges. A method for constructing a decision tree using partially missing data was introduced in [15]. In their method, instances specified at different levels of precision were considered as partially specified instances.

The main difference between our proposed method and the methods previously used is that our method is able to extract rules at more specific levels as well as more abstract levels.

6   Conclusion

This paper described the use of $ERID$ for discovering multi-granularity classification rules from an incomplete information system having attribute value hierarchies. $ERID$ is especially suitable to extract rules from information systems containing multiple weighted values, and this feature enables to handle partial incompleteness of attribute values when the levels of abstraction are changed.

References

1. Dardzińska, A., Raś, Z.: Extracting Rules from Incomplete decision Systems: System Erid. In: Lin, T.Y., Ohsuga, S., Liau, C.J., Hu, X. (eds.) Foundations and Novel Approaches in Data Mining of Studies in Computational Intelligence, vol. 9, pp. 143–154. Springer, Heidelberg (2006)
2. Grzymala-Busse, J.: A New Version of the Rule Induction System LERS. Fundamenta Informaticae 31(1), 27–39 (1997)

3. Guarino, N., Giaretta, P.: Ontologies and Knowledge Bases, Towards A Terminological Clarification. In: Towards Very Large Knowledge Bases: Knowledge Building and Knowledge Sharing, pp. 25–32. IOS Press, Amsterdam (1995)

4. Pawlak, Z.: Rough Sets: Theoretical Aspects of Reasoning about Data. Kluwer Academic Publishers, Norwell (1992)

5. Raś, Z., Dardzińska, A.: KD-Chase Based Query Answering for Hierarchical Information Systems. In: Czaja, L. (ed.) CS&P 2005 Workshop in Ruciane-Nida, Warsaw University, vol. 2, pp. 445–454 (2005)

6. Collaborative Query Processing in DKS Controlled by Reducts. In: Rough Sets and Current Trends in Computing. LNCS, vol. 2475, pp. 189–196. Springer, Heidelberg (2002)

7. Ontology Based Distributed Autonomous Knowledge Systems. Information Systems 29(1), 47–58 (2004)

8. Joshi, S.: Query Approximate Answering System for An Incomplete DKBS. Fundamenta Informaticae 30(3/4), 313–324 (1997)

9. Skowron, A.: Rough Sets and Boolean Reasoning. In: Granular Computing: an Emerging Paradigm, pp. 95–124. Physica-Verlag (2001)

10. Blake, C.L., Hettich, S., Merz, C.J.: UCI Repository of Machine Learning Databases (1998)

11. DesJardins, M., Getoor, L., Koller, D.: Using Feature Hierarchies in Bayesian Network Learning. In: Choueiry, B.Y., Walsh, T. (eds.) SARA 2000. LNCS (LNAI), vol. 1864, pp. 260–270. Springer, Heidelberg (2000)

12. Kamber, M., Winstone, L., Gong, W., Cheng, S., Han, J.: Generalization and Decision Tree Induction: Efficient Classification in Data Mining. In: The 7th International Workshop on Research Issues in Data Engineering (RIDE 1997) High Performance Database Management for Large-Scale Applications, p. 111. IEEE Computer Society, Los Alamitos (1997)

13. Kaufman, K., Michalski, R.S.: A Rethod for Reasoning with Structured and Continuous Attributes in the INLEN-2 Multistrategy Knowledge Discovery System. In: The Second International Conference on Knowledge Discovery and Data Mining, pp. 232–237. AAAI Press, Menlo Park (1996)

14. Cheung, D., Fu, A., Han, J.: Knowledge Discovery in Databases: A Rule-based Attribute-oriented Approach. In: Raś, Z.W., Zemankova, M. (eds.) ISMIS 1994. LNCS, vol. 869, pp. 164–173. Springer, Heidelberg (1994)

15. Zhang, J., Honavar, V.: Learning Decision Tree Classifiers from Attribute Value Taxonomies and Partially Specified Data. In: The Twentieth International Conference on Machine Learning, pp. 880–887. AAAI Press, Menlo Park (2003)

16. Srikant, R., Agrawal, R.: Mining Generalized Association Rules. Future Generation Computer Systems 13(2-3), 161–180 (1997)

17. Taylor, M., Stoffel, K., Hendler, J.: Ontology-based Induction of High Level Classification Rules. In: SIGMOD 1997 Workshop on Research Issues in Data Mining and Knowledge Discovery, pp. 40–47. ACM Press, New York (1997)

18. Zhang, J., Silvescu, A., Honavar, V.: Ontology-driven Induction of Decision Trees at Multiple Levels of Abstraction. In: Koenig, S., Holte, R.C. (eds.) SARA 2002. LNCS (LNAI), vol. 2371, pp. 316–323. Springer, Heidelberg (2002)

Feature Selection on Chinese Text Classification Using Character N-Grams

Zhihua Wei[1,2], Duoqian Miao[1], Jean-Hugues Chauchat[2], and Caiming Zhong[1]

[1] Tongji University, Key laboratory "Embedded System and Service Computing"
Ministry of Education, Shanghai, Cao'an Road, China 201804
[2] Université Lumière Lyon 2, Laboratoire ERIC,
5 avenue Pierre Mendès-France, 69676 Bron Cedex, France

Abstract. In this paper, we perform Chinese text classification using n-gram text representation on TanCorp which is a new large corpus special for Chinese text classification more than 14,000 texts divided into 12 classes. We use different n-gram feature (1-, 2-grams or 1-, 2-, 3-grams) to represent documents. Different feature weights (absolute text frequency, relative text frequency, absolute n-gram frequency and relative n-gram frequency) are compared. The sparseness of "document by feature" matrices is analyzed in various cases. We use the C-SVC classifier which is the SVM algorithm designed for the multi-classification task. We perform our experiments in the TANAGRA platform. We found out that the feature selection methods based on n-gram frequency (absolute or relative) always give better results and produce denser matrices.

Keywords: Chinese text classification, N-gram, Feature selection.

1 Introduction

In recent years, much attention has been given to the Chinese text classification (TC) with the rapidly increasing quantity of web sources and electronic texts in Chinese. The great difference between Chinese TC and Latin languages TC lies in the text representation. Unlike most of the western languages, the Chinese words do not have a remarkable boundary. This means that the word segmentation is necessary before any other preprocessing. The use of a dictionary is necessary. The word sense disambiguation issue and the unknown word recognition problem limit the precision of word segmentation [1]. This makes Chinese representation using words, phrases, meanings, and concepts more difficult.

In this paper, we use a method independent of languages which represents texts with character n-grams. A character n-gram is a sequence of n consecutive characters. The set of n-grams (usually, n is set to 1, 2, 3 or 4) that can be generated for a given document is basically the result of moving a window of n characters along the text. The window is moved one character at a time. Then, the number of occurrences of each n-gram is counted [2]. There are several advantages of using n-grams in TC tasks [3]. One of them is that by using n-grams, we do not need to perform word segmentation. In addition, no dictionary

G. Wang et al. (Eds.): RSKT 2008, LNAI 5009, pp. 500–507, 2008.

or language specific techniques are needed. However, n-gram extraction on a large corpus will yield a large number of possible n-grams, but only some of them will have significant frequency values in vectors representing the texts and good discriminate power. Our contribution is twofold: first we present how to choose the value of n in using n-gram to represent Chinese texts; and second the most suitable kind of feature weight is proposed.

Usually, there are two steps in the construction of an automated text classification system. The first one is that the texts are being preprocessed into a representation more suitable for the learning algorithm that is applied afterwards. The second step regards the learning algorithm that is chosen. In this work we focus on the first step. There are various ways of representing a text such as by using word fragments, words, phrases, meanings, and concepts [4]. Different text representations have different dependence on the language used in the text.

The reminder of this paper is organized as follows. Section 2 presents the text representation forms in our work. Section 3 gives our feature choosing strategies. Section 4 introduces the experiment dataset and the experiment scenarios. Section 5 analyzes the experimental results and section 6 concludes.

2   Text Representation Using N-Grams Frequencies

We adopt the VSM (Vector Space Model), where each document is considered to be a vector in feature space. Thus, given a set of N documents, d_1, $d_2...d_N$, the table of "document by feature" is constructed such as that shown in table 1, where each document is represented by a core "w_{ij}". Generally, w_{ij} has two kinds of value:

i) w_{ij} = frequency of feature j in document i;
ii) w_{ij} = 0 or 1, w_{ij} = 1, if feature j appears in document i, otherwise, w_{ij}=0.

In our work, we choose the first form. In table 1, F_i is n-gram. Chinese texts representation by using n-grams is concerned in some researches. [3] regards that 2-grams are best features for Chinese texts. [5] gives some experimental results by using n-gram combination. In their papers, the best result is by using 1-, 2-, 3-, 4-grams, the second best is by using 1-,2-grams, the third best is by using 2-grams, the case of using 2-, 3-, 4-grams follow and the worst one is by using 1-grams.

In Chinese language, the most part of words are made of one character (for example, some frequently used nous) or two characters. Some proper names or scientific terms have more characters [1]. It seems that the combination of 1-, 2-, 3-, 4-grams even 5-gram, 6-gram will produce a better result. However, we should not extract too many n-grams since it will produce a very large set of candidate features for a corpus which includes more than 14,000 documents. As a result, we choose the combination of 1-, 2-grams. We also do some experiments by using 1-, 2-, 3-grams in order to include some proper names or unknown words.

Table 1. "Document by feature" vector table

D	F_1	F_2	...	F_j	...	F_M	Class	Status
d_1	w_{11}	w_{12}	...			w_{1M}	A	Learning
d_2	w_{21}	w_{22}	...			w_{2M}	B	Learning
...		...					...	Learning
d_i	w_{i1}	w_{i2}	...	w_{ij}		w_{iM}	C	...
...		...					...	Testing
d_N	w_{N1}	w_{N2}	...			w_{NM}	A	Testing

3   Feature Selection

Feature selection is a term space reduction method which attempts to select the more discriminative features from preprocessed documents in order to improve classification quality and reduce computational complexity. As many n-grams are extracted from Chinese texts, we perform two steps of feature selection. The first is reducing the number of features inter-class. The second is choosing the more discriminate features among all the classes in training set.

3.1   Some Definitions

In text classification, the text is usually represented as a vector of weighted features. The difference between various in text representations comes from the definition of "feature". This work explores four kinds of feature building methods with their variations.

In the training set, each text in corpus D belongs to one class c_i. Here, $c_i \in C$, $C = \{c_1, c_2...c_i...c_n\}$, C is the class set defined before classification.

- Absolute text frequency is noted as $Text\_freq_{ij}$,which is the number of texts which include n-gram j in class c_i;
- Relative text frequency is noted as $Text\_freq\_relative_{ij}$,which is got from $Text\_freq_{ij}/N_i$, here, N_i is the quantity of texts in class c_i in training set;
- Absolute n-gram frequency is noted as $Gram\_freq_{ij}$, which is the number of n-gram j in all texts in class c_i in training set;
- Relative n-gram frequency is noted as $Gram\_freq\_relative_{ij}$, which is got from $Gram\_freq_{ij}/N_i'$, here, N_i' is the total of occurrence of all n-grams in all texts in class c_i in training set.

3.2   Inter-class Feature Number Reduction

We extract all the 1-, 2-grams or 1-, 2-, 3-grams in all the texts of the corpus and divide the corpus into training set and testing set. In our work, 70% texts in each class are selected by random to constitute the learning set and the rest 30% are used for the testing set. The following inter-class feature number reduction algorithm is performed only on training set.

Algorithm 1

Begin

For $c_i \in C$, $C = \{c_1, c_2...c_i...c_n\}$,

$Term'_i = \emptyset$, $Term = \emptyset$;

For $n - gram_j \in Term_i$,

If $Text\_freq\_relative_{ij} > \alpha$, then $n - gram_j \in Term'_i$.

$Term = \{Term'_1, Term'_2...Term'_i...Term'_n\}$.

End.

Here, $Term_i$ include all the n-grams extracted from the texts in class c_i , $Term'_i$ include all the n-grams selected in class c_i and $Term$ is n-gram set in all classes selected by algorithm 1. We choose $\alpha = 0.02$ as threshold in order to keep as many as possible features in each class. After this selection, there are 7000 features in each class in average which are enough for text classification task. In the case of $Text\_freq\_relative_{ij} < 0.03$, there are only 4,000 features left in each class in average. It is not enough for the further steps.

3.3   Cross-Class Feature Selection

We construct "feature by class" matrix (noted as $Matrix_{cf}$) by algorithm 2 to select discriminative features.

Algorithm 2

Begin

For $c_i \in C$, $C = \{c_1, c_2...c_i...c_n\}$,

For $n - gram_j \in Term$,

If $n - gram_j \notin Term_i$, $\{O_{ij}\} = 0$

Else $\{O_{ij} = Text\_freq_{ij}$ or $Text\_freq\_relative_{ij}$ or

$Gram\_freq_{ij}$ or $Gram\_freq\_relative_{ij}\}$,

End.

In $Matrix_{cf}$, each feature "j" is assigned a numeric score based on its occurrence within the different document classes c_i. The choice of the scoring method in this work is the CHI-Square test. There are many other tests available as summarized in [6], but the CHI-Square is often cited as one of the best tests for the feature selection. It gives a similar result as Information Gain because it is numerically equivalent as shown by [7]. The score of n-gram "j" is:

$$\sum_i \frac{(O_{ij} - E_{ij})^2}{E_{ij}} \tag{1}$$

Where "i" is the class, "j" is the n-gram and O_{ij} is the observed value. E_{ij} represent the expectation value in the hypothesis of independence of classes and features:

$$E_{ij} = \frac{O_{i+} * O_{+j}}{O_{++}} \tag{2}$$

Here, we define four kinds of values on O_{ij} (as described in algorithm 2) in different experiment scenarios. According to the result of CHI-Square, we separately perform the classification using the 200, 500, 800, 1000, 2000... 5000 features.

4   Experiment

We adopt TanCorp-12 corpus, a collection of 14,150 texts in Chinese language, has been collected and processed by Songbo Tan [8]. It contains 12 categories (art, car, career, computer, economy, education, entertainment, estate, medical, region, science and sport). The biggest class contains 2865 texts (4.17M) and the smallest class contains 150 texts (0.49M).

In order to test the results given from different kinds of methods in feature selection, we set different experiment scenarios, as described in table 2. In the following section, we use a short name (e.g. Ex_1) and a long name (e.g. $1, 2 - gram\&ngram - re$) to describe each experiment scenario. The first part of the long name can be "1,2-gram" or "1,2,3-gram" and it notes the items extracted from texts as features. The second can be "text-re", "ngram-re", "ngram-ab" or "text-ab" notes the feature selection method cross-class.

Table 2. Experiment scenarios list

Experiment scenario	N-gram combination	Feature selection cross-class
Ex_1: 1,2-gram&ngram-re	1+2-gram	Relative n-gram frequency
Ex_2: 1,2-gram&text-re	1+2-gram	Relative text frequency
Ex_3: 1,2-gram&ngram-ab	1+2-gram	Absolute n-gram frequency
Ex_4: 1,2-gram&text-ab	1+2-gram	Absolute text frequency
Ex_5: 1,2,3-gram&ngram-ab	1+2+3-gram	Absolute n-gram frequency
Ex_6: 1,2,3-gram&text-re	1+2+3-gram	Relative text frequency

We use the C-SVC classifier which was introduced in LIBSVM [9]. It is the SVM algorithm designed for the multi-classification task. We use a linear kernel. Learning parameters are set to $gamma = 0$ and $penaltycost = 1$. We perform our experiments in the platform TANAGRA which is a free data mining software for academic and research purposes developed by Ricco Rakotomalala [10].

We use the F1 measure introduced by [11]. This measure combines recall and precision in the following way for bi-class case.

$$Recall = \frac{number\ of\ correct\ positive\ prediction}{number\ of\ positive\ examples} \quad (3)$$

$$Precision = \frac{number\ of\ correct\ positive\ prediction}{number\ of\ positive\ predictions} \quad (4)$$

$$F1 = \frac{2 * Recall * Precision}{Recall + Precision} \quad (5)$$

For more than 2 classes, the F1 scores are summarized over the different categories using the Micro-averages and Macro-averages of F1 scores.

1) Micro - F1 = average in documents and classes
2) Macro - F1 = average of within - category F1 values

5   Results and Discussions

5.1   Comparison of Macro-F1 and Micro-F1 in All Experiment Cases

Figure 1 show that Ex_3 and Ex_1 have the best performance, Ex_5 has the second best, Ex_2 and Ex_4 follow and the Ex_6 has the worst results. The first three best results are in the Ex_3, Ex_1 and Ex_5 which are all using n-gram frequency (relative or absolute) for feature selection. In the situation of absolute frequency (Ex_3) and relative frequency (Ex_1), the results are similar. The results indicate that using n-gram frequency for feature selection is better than using text frequency. Also the relative frequency does not give better results than the absolute frequency. We use 1-, 2-grams in Ex_1, Ex_2, Ex_3 and Ex_4 and use 1-, 2-, 3-grams in $Ex5$ and Ex_6. Fig.1 shows that the results produced by using 1-, 2-grams are little better than those produced by using 1-, 2-, 3-grams.

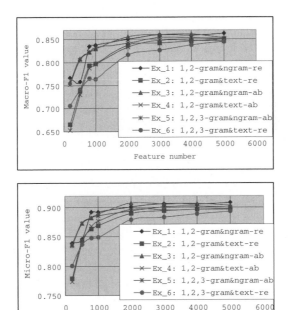

Fig. 1. Macro-F1 and Micro-F1 comparison on all experiment cases

Our experiments also indicates that the number of 2-grams and 3-grams increase with the increasing of feature number. In the case of more than 3000 features, the percentage of 1-grams, 2-grams and 3-grams do not change greatly.

Either in the case of using 1-, 2-, 3-grams or using 1-, 2-grams, 2-grams are always the most important features. There are 1108 3-grams in 5000 features and 206 3-grams are words. Most of them are new words, scientific terms, proper names, abbreviations and phrases which are very difficulties in Chinese word segmentation. In this regard, the method based on n-gram can solve the problem of unknown words recognition to some degree.

5.2   Sparseness Comparison

[12] shows that the computational time is more linked with the number of non-zero values in the cross-table (document by feature) than with its number of columns (features). Fig. 2 shows the non-zero value distribution in the "document by feature" matrix for six experiment cases. Ex_2 (1,2-gram&text-re) has about two times less non-zero cells than Ex_1 (1,2-gram&ngram-re), which indicates that it will produce less dense matrices after cross-class feature selection, so in this way the computation will be faster. Similarly, Ex_4 (1,2-gram&text-ab) has about two times less non-zero cells than Ex_3 (1,2-gram&ngram-ab). Ex_6 (1,2,3-gram&text-re) has two times less non-zero cells than Ex_5 (1,2,3-gram&ngram-ab). The matrices are denser when we use an absolute frequency than a relative frequency. For example, Ex_3 has more non-zero cells than Ex_1 and Ex_4 has more non-zero cells than Ex_2. The number of non-zero cells in the cases of using 1-, 2-, 3-grams is less than that of using 1-, 2-grams.

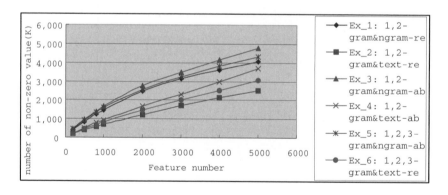

Fig. 2. Comparison of non-zero value in "text by feature" matrix on six cases

6   Conclusion

In this paper, we perform Chinese text categorization on a large corpus using n-gram text representation and different cross-class feature selection methods. Our experiments show that a combination of 1-, 2-grams is little better than that of 1-, 2-, 3-grams for Chinese text classification.

The feature selection methods based on n-gram frequency (absolute or relative) always give better results than those based on text frequency (absolute or relative). Relative frequency is not better than the absolute frequency. Methods based on n-gram frequency also produce denser "document by feature" matrices. Our further work are exploring more excellent methods for feature selection using 1-, 2-grams in Chinese text classification, for example, the methods based on rough set.

Acknowledgments. This paper is sponsored by the National Natural Science Foundation of China (No. 60475019 and No. 60775036) and the Research Fund for the Doctoral Program of Higher Education of China (No. 20060247039). Our colleagues, Ruizhi WANG and Anna Stavrianou gave many good advices for this paper. We really appreciate their helps.

References

1. Miao, D.Q., Wei, Z.H.: Chinese Language Understanding Algorithms and Applications. Tsinghua University Press (2007)
2. Radwan, J., Chauchat, J.-H.: Pourquoi les n-grammes permettent de classer des textes? Recherche de mots-clefs pertinents l'aide des n-grammes caractèristiques. In: JADT 2002: 6es Journées internationales d'Analyse statistique des Données Textuelles, pp. 381–390 (2002)
3. Alain, L., Halleb, M., Delprat, B.: Recherche d'information et cartographie dans des corpus textuels à partir des fréquences de n-grammes. In: Mellet, S. (ed.) 4èmes Journées Internationales d'Analyse statistique des Données Textuelles, Université de Nice - Sophia Antipolis, pp. 391–400 (1998)
4. Joachims, T.: Learning to Classify Text Using Support Vector Machines. University Dortmund (February 2001)
5. Zhou, S.G., et al.: A Chinese Document Categorization System Without Dictionary Support and Segmentation Processing. Journal of Computer Research and Development 38(7), 839–844 (2001)
6. Sebastiani, F.: Machine learning in automated text categorization. ACM Computing Surveys 34(1), 1–47 (2002)
7. Benzécri, J.-P., L'Analyse, D.: T1 = la Taxinomie. DUNOD, Paris (1973)
8. Tan, S.B., et al.: A novel refinement approach for text categorization. In: CIKM 2005, pp. 469–476 (2005)
9. Fan, R.-E., Chen, P.-H., Lin, C.-J.: Working set selection using second order information for training SVM. Journal of Machine Learning Research, 1889–1918 (2005)
10. Ricco, R.: TANAGRA: un logiciel gratuit pour l'enseignement et la recherché. In: EGC 2005, RNTI-E-32, pp. 697–702 (2005)
11. Van Rijsbergen, C.J.: Information Retrieval. Butterworths, London (1979)
12. Artur, Š, et al.: Detailed experiment with letter n-gram method on Croatian-English parallel corpus. In: EPIA 2007, Portuguese Conference on Artificial Intelligence (2007)

Minimum Spanning Tree Based Spatial Outlier Mining and Its Applications

Jiaxiang Lin[1], Dongyi Ye[2], Chongcheng Chen[3], and Miaoxian Gao[4]

[1] Key Lab of Spatial Data Mining and Information Sharing of Ministry of Education
Fuzhou University, Fujian, 350002, P.R. China
linjx2000@gmail.com
[2] College of Mathematics and Computer Science
Fuzhou University, Fujian, 350002, P.R. China
yiedy@fzu.edu.cn
[3] Key Lab of Spatial Data Mining and Information Sharing of Ministry of Education
Fuzhou University, Fujian, 350002, P.R. China
chencc@fzu.edu.cn
[4] Spatial Information Research Center of Fujian
Fuzhou University, Fujian, 350002, P.R. China
gaomiaoxian126@163.com

Abstract. Spatial outliers are spatial objects whose non-spatial attributes are significantly different from the values of their neighborhoods. Detection of spatial outliers will provide the user with meaningful, interesting and potential information. Usually, algorithms for outlier mining on traditional business-oriented datasets are no longer applicable to spatial datasets. A new algorithm based on MST clustering is proposed in this paper to identify spatial outliers. The algorithm organically integrates the approach of minimum spanning trees and the density-based mechanism for outlier mining. Basic spatial structure characteristics of spatial objects are maintained by Delaunay Triangles and MST clustering is achieved by cutting off several most inconsistent edges. It turns out that the algorithm can find true spatial outliers, and it doesnt require any parameter for the algorithm be specified firstly. Experiments on real application problems indicate that the proposed algorithm is feasible and effective for identifying outliers from the large-scale spatial datasets.

Keywords: Spatial outliers, Outlier mining, MST Clustering, D-TIN.

1 Introduction

With the rapid development of technologies for spatial data processing since 1990s, spatial data mining has become a hot field in data mining[1]. Nowadays, researches on spatial data mining are mainly concentrated on common rules of spatial distribution, spatial clustering, spatial association, spatial division, spatial evolvement, and so on[2]. They are concerned with the common rules and the prevalent characteristics, but not the minority and abnormal feature of spatial data. As a matter of fact, anomaly in spatial datasets may contain some

G. Wang et al. (Eds.): RSKT 2008, LNAI 5009, pp. 508–515, 2008.

surprising and useful information. Hence, how to detect the abnormal(spatial outliers) from spatial datasets and to explain the reason that causes the anomaly in specific application of geographic information system have become more and more interesting to the researchers. Spatial outlier mining is of great significance and can be used in many fields, such as the analysis of images gained from remote-sensing, weather forecast, ecology, location based services, and so on.

Algorithms based on clustering are important ways to detect anomalies in a dataset. The main idea is to define outliers as small clusters after clustering. *CLARANS, DBSCAN, BIRCH, STING, ROCK, OPTICS*[3,4] are some typical clustering algorithms that have the function of outlier detection. However, due to the diversity of clustering algorithms, it is difficult to choose a proper one for outlier mining. Furthermore, the main purpose of clustering is to discover the principal features of the dataset and outliers are often the by-products of clustering, but it hardly meets the real need. Therefore, when choosing a clustering algorithm for outlier mining, it is better to choose one whose mechanism is much more similar to the outlier mining's so as to greatly cut down the computation and reduce the complexity of the algorithm. In this paper, a new algorithm based on minimum spanning tree(MST) will be proposed to detect spatial outliers, and it will be applied to the analysis of soil chemical elements data.

2 Related Works

As an important direction of outlier detection, Shekhar[2] is the first to bring forward the definition of spatial outliers, that's "spatial outliers are the spatial objects whose non-spatial attributes are obviously different from the others in the neighborhood", and now it is widely adopted by the researchers. Due to the uniqueness of spatial data, if one wishes to find some meaningful information, both spatial attributes and thematic attributes should be carefully considered, such as the position, shape, geometric feature and the relationship of spatial objects[5]. Therefore, spatial outlier mining is usually much more complicated than outlier mining in relational databases.

As the deepening of research, different algorithms for spatial outlier mining were proposed. According to the mechanism adopted by the algorithm, they were generally divided into two categories, graphic approaches and multi-dimensional quantitative tests[5,6]. Graphic approaches highlight spatial outliers based on visualization of spatial data; typical methods include variogram clouds, pocket plots, scatterplot, and Moran scatterplot[7,8]. Quantitative methods provide tests upon data attributes to identify spatial outliers based on statistical techniques, *z-value*, *Iterative z-value, Iterative z-ratio*[9,10,11] are 3 representative one.

Targeted to the analysis of soil chemical elements data inspected by the project *Ecological Geochemical Survey of Fujian Coastal Economic Belt*, we drew our attention on the representation of spatial neighboring relationship and the preferences for the algorithm, Delaunay Triangle Net($D\text{-}TIN$) was constructed and method of density based was adopted, and finally, the goal of MST based spatial outlier mining is achieved.

3 MST-SOM Algorithm

3.1 Basic Ideas

Basic idea of *MST-SOM* algorithm rests on the full use of *MST*. Firstly, it extracts the spatial attributes and translates them into feature space, during the period, *D-TIN* is used to maintain the spatial structure characteristics; then the algorithm constructs *MST* in the feature space and performs density-based clustering[13] by cutting off the most inconsistent edges of *MST*; finally, imposes outlier detection on the small clusters left by clustering, and finds out potential spatial outliers.

3.2 Implementation

In the algorithm *MST-SOM*, dataset is denoted as a spanning tree, and thus the issue of multidimensional data clustering is translated into *MST* division. What we should do is to perform a local clustering in the sub-tree. Given the connected graph with weight $G(V, E)$, whose spanning tree T is also with weight. The sum of each edge's weight is called T's weight, it is denoted as $W(T) = \sum w(d_i, d_j)$, for every $(d_i, d_j) \in TE$, where TE denotes the set of edges and $w(d_i, d_j)$ denotes the weight of edge(d_i, d_j), the spanning tree with the smallest weight is called minimum spanning tree.

Similarity denotes the degree of resemblance between two models, and it is vital to choose a proper measurement of similarity in the clustering algorithm. In this paper, it is supposed that spatial attributes are under standardization, and Euclidean distance is used as the measurement of spatial object's similarity.

3.2.1 D-TIN Construction Based on MST

Suppose that $V = \{x_1, x_2, \ldots, x_n\}$ is a dataset with n scattered points in the planar R^2. If $G(V, E)$ is a graph, where V is the set of vertexes and E is the set of edges; then minimum spanning tree of G can be denoted as $EMST(V)$; $e_{ik} = (v_i, v_k)$ and $e_{ij} = (v_i, v_j)$ are two edges of G, and they are called a pair of relative edges, the angle between $v_i v_k$ and $v_i v_j$ is the angle of the relative edges.

D-TIN construction based on *MST* can be divided into three major steps[14]. First of all, obtain *MST* according to the spatial similarity between objects; then, construct triangle by inserting edge one by one until the last edge is added; finally, perform local optimization procedure on the triangles according to the principle of the largest minimum interior angle, until *D-TIN* is achieved. During the process of adding a new edge to $EMST(V)$, there are 2 rules to be abided, the first is never-intersected(there is not intersect between triangle), and the second is never-included(any one triangle will not include another). If the information of scatter points and triangles was stored in the data structure of *D-TIN* which is denoted as TML, and the relative edges in $EMST(V)$ or the current graph is denoted as $IEPL(V)$, then the procedure of *D-TIN* construction based on *MST* is shown as follows:

1. Specify each edge's weight of graph to be the similarity between vertexes;
2. Gain minimum spanning tree $EMST(V)$ out of the point set V, where method of Kruskal is adopted;
3. Gain the initial set of relative edges $IEPL(V)$ according to $EMST(V)$;
4. Sort $IEPL(V)$ in an ascending order according to the angle of the relative edges;
5. Initialize $TML = null$;
6. If there was a pair of edges left in the $IEPL(V)$, then select the pair of edge (e_{ik}, e_{ij}) whose angle is the smallest, and mark it with 1; if triangle $\triangle v_i v_j v_k$ composed by the vertexes of relative edge (e_{ik}, e_{ij}) satisfies the two principles above, then $\triangle v_i v_j v_k$ is added to TML; new edge $e_{jk} = (v_j, v_k)$ is added to $EMST(V)$; new edge $e_{jk} = (v_j, v_k)$ is inserted into the set of $IEPL(V)$ in an ascending order in term of the angle;
7. Perform local optimization procedure according to the principle of the largest minimum interior angle, and then gain D-TIN.

3.2.2   MST Clustering

Given a spatial dataset $D \subset R^d$ with n objects. Firstly, perform MST construction based on D-TIN gained in the former step.

Then, try to cut off the most M inconsistent edge that is obviously different from the edges adjacent to it so as to gain $M + 1$ sub-tree $T[m]$, where $m = 1, 2, \ldots, M + 1$. Consequently, the dataset is divided into $M + 1$ local areas, and similar objects are partitioned into a same cluster, which is well accordant to the principle of clustering[15]. Presently, there are many ways to define the most inconsistent edge, in the paper, the weightiest one or the lightest one is simply seen as the most inconsistent edge. As to the parameter M, there was little influence on the clustering result, if a small value was set, then the bigger clusters would be subdivided.

Finally, the algorithm utilizes the idea of MST and density based clustering algorithm[13], computes objects's minimum number $MinPts[m]$ and the area $T[m]$'s radius $Eps[m]$, based on which spatial data clustering is achieved. In this paper, parameter $MinPts[m]$ was set to sub-tree $T[m]$'s degree; and as to the value $Eps[m]$, it was usually gained according to the graph's k-nearest-distance. However, to get parameter $Eps[m]$, a large amount of computation is required; moreover, it has great impact on the result of clustering. If $Eps[m]$ was too small, then there would be a large amount of clusters; on the other side, the algorithm would be unable to wipe off the noises. In this paper, to achieve a proper value that is neither too big nor too small[11], according to the properties of MST, $Eps[m]$ is set to $2\frac{w}{N[m]}$, where $N[m]$ is the number of edges in sub-tree $T[m]$, and w is the total weight of $N[m]$ edges.

3.2.3   Spatial Outlier Mining Based on MST Clustering

Outliers usually fall on a sub-tree which is small, so when the algorithm finds outliers from the datasets, only the sub-trees that are small should be detected. As a result, time consumption on scanning of the whole dataset is reduced, and the efficiency of outlier mining is greatly improved. In the paper, a simple mechanism Nested Loop was used to detect outliers from the fragment left by clustering.

3.3 Performance

Given a spatial dataset $D \subset R^d$ with n objects, d denotes the number of spatial objects's attributes, then the average time complexity for translating spatial objects into a feature space is $O(nlogn)$[14]; time complexity for MST construction is $O(nlogn)$, if parallel computing was adopted, it would be improved to $O(nlog_{3/2}n)$; If there is r spatial objects in the largest sub-tree $T[m]$, then the average time complexity for MST clustering in the sub-tree is $O(rlogr)$, and the total time complexity to clustering in $M + 1$ sub-tree is equal to $O(nlogn)$; outlier mining based on clustering merely detect those spatial objects in small cluster, so the time complexity is approximate to a linear one $O(n)$. Taking all these steps together, the time complexity for $MST\text{-}SOM$ algorithm is $O(nlogn)$.

4   Application

$MST\text{-}SOM$ algorithm is implemented in java and used in the analysis of soil chemical elements data inspected by the project *Ecological Geochemical Survey of Fujian Coastal Economic Belt*. There are 53 attributes totally in the soil chemical elements data, including 2 spatial attribute known as longitude and latitude, and many thematic attributes, such as As, Ag, Al_2O_3, Au, B, Ba, Be, Bi, Br, CaO, Ce, Cl, Co, Cr, Cu, F, Fe_2O_3, Ga, Ge, Hg, I, K_2O, La, Li, MgO, Mn, Mo, Na_2O, Nb, Ni, N, P, Pb, Ph, Rb, S, Sb, Sc, Se, SiO_2, Sn, Sr, Th, Ti, U, V, W, Zn, Zr, Organic Carbon, sum of carbon.

In the real application on soil chemical elements data, spatial structure characteristics of the soil are maintained by $D\text{-}TIN$, and the algorithm $MST\text{-}SOM$ processes MST clustering on the basis of $D\text{-}TIN$, then it implements density based spatial outlier mining. According to the strategy adopted by $MST\text{-}SOM$ algorithm, the most 15 inconsistent edges were cut off, and the result of MST clustering was shown as Fig.1.

Several small clusters are gained after MST clustering, which were seen as the candidate. For the application on the soil chemical elements data analysis, candidate outliers are shown in Table1, among which p* denotes the *th spatial object in the datasets.

After a narrow examination on the candidate spatial outliers, object p75, p16, p49, p133, p148 and p227 are considered to be the real spatial outliers, for their outlier scores are much bigger than the other's, which are shown in Table2.

Table 1. Candidate Spatial Outliers Left after MST Clustering ($\times 10^5$,unit:°)

Point	Coordinate	Point	Coordinate	Point	Coordinate	Point	Coordinate
p16	(7.03, 27.73)	p26	(7.01, 27.63)	p49	(6.73, 27.57)	p217	(7.01, 27.71)
p75	(6.71, 27.33)	p104	(6.89, 27.73)	p55	(7.01, 27.69)	p227	(6.87, 27.77)
p118	(6.71, 27.57)	p131	(6.81, 27.55)	p109	(6.91, 27.71)	p199	(6.71, 27.31)
p148	(6.73, 27.45)	p184	(7.02, 27.62)	p133	(6.99, 27.55)	p245	(7.03, 27.63)

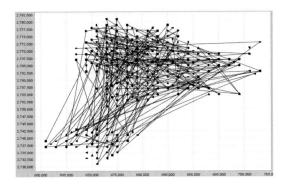

Fig. 1. Clusters Info of Soil Chemical Elements Data after MST Clustering

If combined with geographic layers, the distribution of spatial outliers in soil chemical elements data can be shown as Fig.2.

As a matter of fact, soil outliers are mainly caused by human damage and environmental contamination, so in this paper, much attention is paid on the analysis of the harmful elements of soil. Detail information of harmful soil chemical elements for the 6 spatial outliers was shown in Table3.

Through the comparison between the values inspected and the true of the harmful chemical elements, object p16 becomes an outlier by the reason that inspected values of chemical elements Cd, Cu, Pb, Zn, Cr, Ni are seriously abnormal. Similarly, object p49 becomes an outlier by the reason that inspected value of chemical elements As, Pb, Zn, Cr, Ni, Be are seriously abnormal; object p75 becomes an outlier by the reason that inspected values of chemical elements Hg, Pb, Cr, Tl are seriously abnormal; object p133 becomes an outlier by the reason that inspected values of chemical elements Zn, Cr are seriously abnormal, etc. To sum up, abnormity of harmful soil chemical elements data may be derived from exterior factor during the development of industry, such as

Table 2. Outlier Score of Candidate Spatial Outliers

Point	Score	Point	Score	Point	Score	Point	Score
p16	0.952(2ed)	p75	1.548(1st)	p131	0.41(7th)	p199	0.407(8th)
p26	0.116(16th)	p104	0.358(10th)	p133	0.73(4th)	p217	0.348(12th)
p49	0.824(3rd)	p109	0.333(13th)	p148	0.695(5th)	p227	0.498(6th)
p55	0.128(15th)	p118	0.269(14th)	p184	0.382(9th)	p245	0.357(11th)

Table 3. Harmful Elements Values of Spatial Outliers

Item	Cd	Hg	As	Cu	Pb	Zn	Cr	Ni	Sb	Be	Se	Tl	Ag
p16	0.4	0.04	2.31	31.8	20.5	121.7	41.9	44	0.39	1.44	0.31	0.59	0.13
p49	0.08	0.09	7.59	14.5	46.9	92.3	33.5	17.7	0.43	2.84	0.27	0.91	0.1
p75	0.05	0.26	2.59	11	37.5	46.3	12.2	5.6	0.37	1.55	0.16	0.57	0.12
p133	0.06	0.10	2.07	9.6	33.7	52.3	11.6	5.7	0.42	2.14	0.12	0.89	0.08
p148	0.09	0.04	6.82	24.7	42.1	90.8	34	17.5	0.46	2.05	0.15	0.74	0.13
p227	0.10	0.06	8.31	23.1	37	120.9	64.6	31.2	0.54	2.67	0.2	0.8	0.09
Avg	0.08	0.11	2.47	10.5	32.87	50.91	17.59	7.81	0.33	1.9	0.17	0.82	0.09

Fig. 2. Distribution of Spatial Outliers in Soil Chemical Elements Data

the stack of garbage; and it may also be derived from the internal factor during the shaping of soil, like the enrichment of chemical elements.

5   Conclusion

In this paper, a new algorithm for spatial outlier mining based on *MST* clustering is proposed and applied to harmful soil chemical elements detection. The algorithm not only retains the inherent advantages of *MST* clustering, but also maintains basic spatial structure characteristics of spatial objects in an effective way, and it proved to be suitable for a complex situation of unsupervised spatial outlier mining. Further work involves spatial data visualization and its application on data mining, multi-resource spatial data outlier mining, multi-scale spatio-temporal data outlier mining.

Acknowledgements

This research is sponsored by the National Natural Science Foundation of China *(No.60602052)* and supported by Fujian Provincial Science-Technology key program *(2005H086, 2005H028, 2006J0131)*.

References

1. Lu, W., Han, J., Oöi, B.C.: Discovery of General Knowledge in Large Spatial Databases. In: 1993 Far East Workshop on Geographic Information Systems (FEGIS 1993), Singapore, pp. 275–289 (1993)
2. Shekhar, S., Lu, C.T., Zhang, P.: A Unified Approach to Detecting Spatial Outliers. Geo-Informatica: An International Journal on Advances of Computer Science for Geographic Information System 7(2), 139–166 (2003)
3. Ng, R.T., Han, J.: Efficient clustering methods for spatial data mining. In: Proc. The 20th International Conference on Very Large Data Bases, Santiago, pp. 144–155 (1994)

4. Ester, M., Kriefel, H.P., Sander, J., et al.: A density-based algorithm for discovering clusters in large spatial databases with noise. In: Proc. The 2nd International Conference on Knowledge Discovery and Data Mining, pp. 226–231. AAAI Press, Portland (1996)

5. Lu, C.T., Chen, D.C., Kou, Y.F.: Algorithms for spatial outlier detection. In: Proceedings of the Third IEEE International Conference on Data Mining, pp. 597–600. Melbourne (2003)

6. Kou, Y., Lu, C.T., Chen, D.: Spatial weighted outlier detection. In: The 2006 SLAM Conference on Data Mining, Bethesda, Maryland, pp. 613–617 (2006)

7. Shekhar, S., Lu, C.T., Zhang, P.S.: Detecting graph-based spatial outliers: algorithms and applications (a summary of results). In: Proceedings of the seventh ACM SIGKDD international conference on Knowledge discovery and data mining, San Francisco, California, pp. 371–376 (2001)

8. Shekhar, S., Lu, C.T., Zhang, P.: Detecting graph-based spatial outlier. Intelligent Data Analysis: An International Journal 6(5), 451–468 (2002)

9. Lu, C.T., Chen, D.C., Kou, Y.F.: Detecting spatial outliers with Multiple Attributes. In: Proceeding of the 15th IEEE International Conference on tools with artificial intelligence, Sacramento, California, USA, pp. 122–128 (2003)

10. Wang, Z.Q., Wang, S.K., Hong, T., Wan, X.H.: A spatial outlier detection algorithm based multi-attributive correlation. In: Proceedings of the Third International Conference on Machine Learning and Cybernetics, Shanghai, China, pp. 1727–1732 (2004)

11. Wang, Z.Q., Li, J.H., Yu, H.Q., Chen, H.B.: Research of spatial outlier detection based on quantitative value of attributive correlation. In: Proceedings of the 6th World Congress on Intelligent Control and Automation, Dalian, China, pp. 5906–5910 (2006)

12. Ai, T.H.: Research on the supporting data model and methodology of city map database integration. Doctoral dissertation, Wuhan University (2000) (in Chinese)

13. Markus, M.B., Hans-Peter, K., Raymond, T.N., Jörg, S.: LOF: identifying density-based local outliers. In: Proceedings of the 2000 ACM SIGMOD international conference on Management of data, Dallas, Texas, USA, pp. 93–104 (2000)

14. Ma, X.H., Dong, J., Pan, Z.G., Shi, J.Y.: A Graph-Based Algorithm for Generating the Delaunay Triangulation of a Planar Point Set. China Journal of Image and Graphics 2(1), 7–11 (1997) (in Chinese)

15. Cui, G.Z., Cao, L.Z., Zhang, X.C., Wang, Y.F.: Research of Density-Based MST Clustering Algorithm. Computer Engineering and Applications 05, 155–158 (2006) (in Chinese)

A New Algorithm for High-Dimensional Outlier Detection Based on Constrained Particle Swarm Intelligence

Dongyi Ye and Zhaojiong Chen

College of Mathematics and Computer, Fuzhou University, Fuzhou 350002, China
{yiedy, chenzj}@fzu.edu.cn

Abstract. In this paper we present an algorithm for outlier detection in high-dimensional spaces based on constrained particle swarm optimization techniques. The concept of outliers is defined as sparsely populated patterns in lower dimensional subspaces. The search for best abnormally sparse subspaces is done by an innovative use of particle swarm optimization methods with a specifically designed particle coding and conversion strategy as well as some dimensionality-preserving updating techniques. Experimental results show that the proposed algorithm is feasible and effective for high-dimensional outlier detection problems.

1 Introduction

Outlier detection has now become a hot issue in the area of data mining with numerous applications. Most of the existing algorithms for outlier detection use concepts of proximity to define and detect outliers, such as the notion of distance-based or density-based outliers [1]-[6]. However, as explained in [7], these algorithms are not appropriate for high dimensional cases because the data are sparse and the notion of proximity fails to retain effectiveness. Actually, the sparsity of high dimensional data implies that every point is an almost equally good outlier from the perspective of proximity-based definitions [6][7]. Given the fact that most applications as mentioned above are high dimensional problems, it is of significance to reasonably define outliers in a high dimensional space and to design specific algorithms for their detection.

Recent research results indicate that mining outliers in lower-dimensional projections is feasible for high dimensional data. On one hand, it is in practice not qualitatively meaningful to detect outliers in full dimensional space. On the other hand, only the subsets of the attributes are affected in some applications, such as credit card fraud, and so on. Along this line, Aggarwal and Yu [7] presented a new way of defining outliers in a high dimensional space as well as a new technique for their detection. The point is to observe the density distributions of projections from the data that could be measured by the so called sparsity coefficients and a data point is considered an outlier, if it is located in some abnormally low density subspace. Hence, the outlier detection in this context boils down to finding those combinations of dimensions with most abnormally sparse data. This turns out to

G. Wang et al. (Eds.): RSKT 2008, LNAI 5009, pp. 516–523, 2008.

be a very difficult combinatorial optimization problem since the combinations of dimensions exponentially increase with increasing dimensionality and it is hard to examine all possible subsets of dimensions. Actually, a brute-force algorithm will become computationally untenable even for a problem with moderate dimensionality. To solve this problem, Aggarwal and Yu [7] developed an effective evolutionary computation based algorithm, denoted hereafter as Gen^o, by making an innovative use of genetic algorithms(GAs). The algorithm works no worse than the brute-force(or exhaustive search) algorithm in terms of solution quality but with much less computational effort. Chen and Ye [8] presented a modified version of Gen^o that may possibly find more outlier patterns but requires many extra computations.

The purpose of this paper is to develop another evolutionary computation based algorithm for high-dimensional outlier detection problems. The idea is to use particle swarm optimization(PSO) techniques in a specific and appropriate way. A PSO method, featured by intuitiveness, simplicity, easy implementation, and fast convergence, has turned out to be more competent than GAs on some NP-hard optimization problems [9]-[13]. Therefore, it is quite natural to come up with the idea of developing an effective PSO based outlier detection algorithm. As we are going to see, this attempt of exploration is worthwhile.

The paper is organized as follows. In section 2, we review some basic concepts and discuss some related work. In section 3, we describe some specific strategies for both particle coding and dimensionality preserving search. In section 4, an outlier detection algorithm based on constrained PSO techniques is presented. Experimental results are given in section 5 and section 6 concludes.

2   Some Basic Concepts and Related Work

An abnormal projection is one in which the density of the data is exceptionally lower than average. Aggarwal and Yu presented a definition of sparsity coefficient to measure the density of a lower k-dimensional projection [7]. The coefficient is a key to outlier detection and is hereby quoted below. Assume throughout the paper that there is an n-dimensional data set having a total of N points. Each attribute of the data is divided into Φ equi-depth ranges. Each range contains a fraction $f = 1/\Phi$ of total points. The reason of using equi-depth instead of equi-width has been well explained in [7]. If the data are uniformly distributed, then the number of points in a cube is expected to be $N \times f^k$.

Definition 1. *Let $n(D)$ be the number of points in a k-dimensional cube D. The sparsity coefficient $S(D)$ of the cube D is defined as follows:*

$$S(D) = \frac{n(D) - N \times f^k}{\sqrt{N \times f^k \times (1 - f^k)}} \tag{1}$$

Apparently, the number of points in k-dimensional cube D is lower than expected when $S(D)$ value is negative. The most negative $S(D)$ value means that the cube D has least points. Cubes that are empty are considered infeasible and their sparsity coefficients are set to the very high value of 10^6 from an implementation perspective. Now, the problem of detecting outliers in high-dimensional spaces in our context is to find the most sparse k-dimensional *nonempty* cubes in which the points are considered outliers.

It is important to note that unlike problems such as frequent itemset detection, the problem of finding the most sparse k-dimensional nonempty cubes in the data is of such nature that there are no upward-or downward closed properties in the set of dimensions that are unusually sparse [7]. It may be often the case that although particular regions may be densely populated on certain sets of dimensions, they can be very sparsely populated when such dimensions are combined together. In general, it is very hard or even impossible to predict the behavior of the data when two sets of dimensions are joined together. The best projections (with the most negative sparsity coefficients) are often created by an a priori unknown combination of dimensions, which can not be determined by looking at any lower-dimensional projection. Moreover, subsets of dimensions that are sparsely populated are scarce and finding these patterns is somewhat like finding a needle in a haystack since one is looking for them in an exponentially increasing space of all possible projections.

Aggarwal and Yu proposed an innovative use of genetic algorithms by introducing some specifically designed crossover and mutation operations [7]. In a run of their algorithm, the search for the best projections is carried out in lower-dimensional subspaces of a given dimensionality k. This dimensionality needs to be kept in mind while performing crossover and mutation operations. That is, the two children after recombination should also correspond to a k-dimensional projection. To achieve this goal, they introduced a specific individual coding strategy, an optimized crossover operation and a two-branch mutation operation with two mutation probabilities. For convenience of our later presentation and comparison, we briefly describe this coding method. As assumed previously, each attribute(dimension) of the data set is divided into Φ ranges. Thus, the value of each attribute can be any of the values 1 through Φ, or the value "*", which denotes a "don't care". Hence, there are a total of $\Phi+1$ values that any dimension can take on. This defines the individual coding policy. That is, an individual string consists of n genes, each gene taking any of the values 1 through Φ, or the value "*". For example, consider a 4-dimensional data set with $\Phi=10$. Then, individual "*2*7" corresponds to a 2-dimensional projection whose second and fourth dimensions are identified and the other dimensions are "don't care"s.

3   Particle Conversion and Updating Strategy

3.1   Basics of PSO Methods

The first PSO algorithm was introduced in 1995 by Kennedy and Eberhart [9] for continuous optimization problems and since then many improved versions of it have been presented [10]-[12]. It is a population-based optimization algorithm inspired by the social behavior of birds and, like other algorithms of its kind, it is initialized with a population of L possible solutions (called particles) randomly located in a d-dimensional solution space. A fitness function determines the quality of a particle's position. A particle at time step t has a position vector and a velocity vector. The algorithm iterates updating the trajectories of the swarm through the solution space on the basis of information about each particle's previous best performance and the best previous performance of its neighbors until a stopping criterion is met. More precisely, it updates the positions $x^i(t)$ and velocity $v^i(t)$ of particles according to the following equations:

$$v^i(t+1) = w(t)v^i(t) + c_1r_1(t)(pb^i(t) - x^i(t)) + c_2r_2(t)(gb(t) - x^i(t)), \qquad (2)$$

$$x^i(t+1) = x^i(t) + v^i(t+1), i = 1, \cdots, d, \qquad (3)$$

where $pb^i(t)$ is the previous best performance position of particle i and $gb(t)$ is the best previous performance position of the whole swarm; $w(t) = \frac{T-t}{T}$ is the inertia weight; c_1 and c_2 are the learning coefficients; $r_1(t), r_2(t)$ are uniformly randomized numbers in the interval $[0,1]$.

One of important issues in designing a PSO based algorithm is the design of a suitable particle coding strategy. As mentioned earlier, in each run, the search for sparsely populated cubes is conducted within k-dimensional subspaces. Therefore, updated particles should correspond to k-dimensional projections. However, as the mechanism of generating offspring particles is quite different from that in a genetic algorithm, the above mentioned coding strategy does not directly apply to PSO case. Actually, if we use the said coding for a particle, then it is easy to see that the update equation (3) will turn a particle into an infeasible solution or even a solution that does not make any sense. Hence, some adjustment is needed in the context of PSO. The idea is to take directly R^n as the space of particles incorporated with a pattern conversion step so as to establish a correspondence between a particle and a projection pattern.

3.2   Pattern Conversion Operation

The pattern conversion operates as follows: Let $x = (x_1, \cdots, x_n)$ be a particle, define its conversion pattern $cx = (cx_1, \cdots, cx_n)$ by $cx_i = \lfloor x_i \rfloor, i = 1, \cdots, n$, where $\lfloor\ \rfloor$ is the floor function. If for some i, $cx_i \leq 0$ or $cx_i > \Phi$, then set cx_i to $*$. Similarly, the positions with "$*$" are called "don't care" ones and the others are called "identified" ones of both x and cx. Let m be the number of "identified"s in cx, then cx corresponds to a m-dimensional cube in the data space. We denote this number m by $Dim(x)$. The sparsity coefficient of this cube can then be calculated by definition and is used to determine the fitness of the particle. For example, considering a 4-dimensional problem with $\Phi=10$, the conversion pattern of the particle $(12.3,1.27,3.13,-2.32)$ is $(*,1,3,*)$, which corresponds to a 2-dimensional projection.

3.3   ModifyPSO Operation

Since the search for abnormally sparse lower dimensional projections in a run should be conducted in subspaces of a given dimensionality, the traditional particle updating strategy needs to be modified. Given a particle $x = (x_1, \cdots, x_n)$

satisfying $Dim(x) = k$, let $x' = (x'_1, \cdots, x'_n)$ be its updated particle by Equation (3). Our goal is to modify x' so as to get a new particle with $Dim(x') = k$. We distinguish 3 cases.

Case 1: $Dim(x') = Dim(x) = k$. This means that after the pattern conversion, cx' still corresponds to a k-dimensional projection. Accept this updated particle as a new particle.

Case 2: $Dim(x') = k1 < k$. This means that values at some "identified" positions of x have gone beyond the range $[1, \Phi]$. Let $k2$ be the number of such positions. Then, $k - k1 \leq k2$. Randomly choose $k - k1$ positions from among $k2$ "identified" positions. For each chosen position j, modify x' as follows:

$$x'_j = \begin{cases} \Phi, \, x'_j > \Phi; \\ 1, \, x'_j < 1. \end{cases}$$

After such modification, the new particle x' satisfies $Dim(x') = k$.

Case 3: $Dim(x') = k1 > k$. If the sparsity coefficient of $k1$-dimensional cube cx' is positive, then randomly set $k1 - k$ "identified" positions of x' to value "*" to get a new particle x'; Otherwise, select through enumeration k positions from among $k1$ "identified" positions such that the resulting k-dimensional conversion pattern cube has the least sparsity coefficient. Although there are C_{k1}^k possibilities, the time cost is not expensive since $k1, k$ are usually small. Therefore, in either case, we will get a new particle x' with $Dim(x') = k$.

4   A Constrained PSO-Based Outlier Detection Algorithm

With the above mentioned pattern conversion and dimensionality preserving particle updating strategies, we are now in the position to outline our PSO-based outlier detection algorithm as depicted in Fig.1. Since the updating of particles is subject to the dimensionality preserving constraint, our algorithm can be regarded as a constrained PSO algorithm, denoted by CPSO.

The input parameters include the number m of the best solutions with most negative sparsity coefficients, a given lower dimensionality k. Aside from this, there are other parameters such as the range Φ, the size p of particle population, the maximum allowed iteration number T and velocity. They are to be set up depending on the data sets involved. In general, it is not an easy task to get a suitable choice of values of k and Φ. If we pick too big a Φ, that will result in many empty or sparsely populated subcubes; whereas if we pick too small a Φ, then there would be rare outlier patterns. A desirable value for Φ is such that there are sufficient numbers of intervals on each dimension that corresponds to a reasonable notion of locality. As for a reasonable choice of the values of k, [7] suggested using the relationship $k = \lfloor log_\Phi(N/s^2 + 1) \rfloor$ where $s = -3$. We shall also follow this suggestion when running our outlier detection algorithm. The algorithm will be terminated if either the m best solutions(denoted as *BestSet* in Fig.1) are found or the maximum allowed iteration number is reached.

The Constrained PSO based outlier detection algorithm
Algorithm CPSOOutlierSearch(Number: m, Dimensionality: k)
begin

> S=Initial population of p n-dimensional particles
> $BestSet=null$;
> **while** not(termination criterion) **do begin**
> S=PSOUpdate(S)(using Equations (2) (3));
> S=ModifyPSO(S);
> Perform PatternConversion(S);
> Update $BestSet$ to be the $m$solutions in $BestSet \bigcup$ PatternConversion(S) with
> most negative sparsity coefficients;
> **end**;
> O=Set of data points covered by $BestSet$;
> return($BestSet,O$);

end

Fig. 1. The constrained PSO based outlier detection algorithm

5 Experimental Results

In order to evaluate the performance of the proposed algorithm, we tested it on some real data sets obtained from the UCI machine learning repository. In order to make comparisons with the baseline algorithm Gen^0, we picked the same data sets as in [7] each only with continuously-valued attributes. The results are given in Table 1. Both algorithms were implemented on a 1.6-GHz PC running Windows XP with 256MB of main memory. Parameter settings were as follows. For all data sets, we picked $\Phi = 7$ and $m =20$ as in [7]. However, as the mutation probabilities $p1$ and $p2$ were not explicitly given in [7], we just set both of them to the value of 0.05 for Gen^0 algorithm. The size p of particle population was set to 30 and the maximum allowed number of iterations was set to 500. The learning coefficients were picked as $c_1 = c_2 = 2$. In addition, the data sets were cleaned in order to take care of categorical and missing attributes. For instance, in the housing data set, 13 of 14 attributes were selected, eliminating the single binary attribute.

In Table 1, we have reported the time cost(in second) as well as the average sparsity coefficients of the best 20 projections indicated under the column(quality). Note that we did not regain the results by using Gen^0 as reported in [7]. This could be partly because of our different choice of mutation probabilities(as indicated earlier) and the sizes of populations, and partly because of the differences in its implementations. It could also be because we only reported the results in a run of the algorithm that are not necessarily the best possible solutions. Nevertheless, the results in Table 1 show to some extent the performance of both algorithms. We note that in four of the six data sets, the average quality of the 20 best projections was the same(if truncated to the second decimal) using either Gen^0 or the CPSO algorithm. These cases are marked with a "*". For the other two data sets, our algorithm outperformed Gen^0 in one data

Table 1. Performance for different data sets

dataset	Gen^0(time)	CPSO(time)	Gen^0(quality)	CPSO(quality)
Breast cancer(14)	11	10	-3.28	-3.28(*)
Ionosphere(34)	87	80	-3.05	-3.05(*)
Segmentation(19)	13	11	-3.02	-3.02 (*)
Musk(160)	292	279	-2.63	-2.63(*)
Machine(8)	3	4	-3.25	-3.17
Housing(13)	24	21	-4.11	-4.18

set but was inferior to Gen^0 in the other data set. Thus, our algorithm works equally well as Gen^0 on these data sets. When it comes to the time cost, it can be observed that in most cases CPSO algorithm consumed less than Gen^0 algorithm. This advantage may be attributed to the fast local search ability of PSO methods. To sum up, the experimental results show that CPSO, a constrained PSO based algorithm may work equally well as or sometimes even better than the baseline GA-based algorithm Gen^0 in terms of computational efficiency and outlier detection quality.

6   Conclusion

In this paper, we discussed the applicability of particle swarm optimization techniques to the problem of detecting outliers in high dimensional spaces where the outliers are defined as abnormally sparse lower dimensional patterns. A specific dimensionality preserving updating strategy for particles was introduced to handle the subspace constraints. It turned out that PSO-based algorithms can also be used to effectively detect such outliers in high-dimensional spaces and can work equally well as or sometimes even better than the existing GA-based detection algorithms with suitably modified particle updating and search strategies.

Acknowledgement

This work was partly funded by Natural Science Foundation of China (No. 60602052), Provincial Natural Science Foundation of Fujian of China(No. A0610014) and High Tech. Research Project of Fujian of China(No. 2005H028).

References

1. Knorr, E.M., Ng, R.T.: Algorithms for Mining Distance-Based Outliers in Large Datasets. In: 24th International Conference on Very Large Databases, pp. 392–403. ACM Press, New York (1998)
2. Knorr, E.M., Ng, R.T.: Finding Intensional Knowledge of Distance-based Outliers. In: 25th International Conference on Very Large Data Bases, pp. 211–222. Morgan Kaufmann, Edinburgh (1999)

3. Ramaswamy, S., Rastogi, R., Kyuseok, S.: Efficient Algorithms for Mining Outliers From LargeData Sets. In: Proceedings of the ACM SIGMOD International Conference on Management of Data, pp. 427–438. ACM Press, Dallas (2000)
4. Knorr, E.M., Ng, R.T., Tukacov, V.: Distance-based Outliers: Algorithms and Applications. The VLDB Journal 8, 237–253 (2000)
5. Angiulli, F., Basta, S., Pizzuti, C.: Distance-Based Detection and Prediction of Outliers. IEEE Transactions on Knowlodge and Data Engineering 18, 145–160 (2006)
6. Beyer, K., Goldstein, J., Ramakrishnan, R., Shaft, U.: When Is Nearest Neighbors Meaningful? In: Beeri, C., Bruneman, P. (eds.) ICDT 1999. LNCS, vol. 1540, pp. 217–235. Springer, Heidelberg (1998)
7. Aggarwal, C., Yu, P.: An Effective and Efficient Algorithm for High-Dimensional Outlier Detection. The VLDB Journal 14, 211–221 (2005)
8. Chen, G.P., Ye, D.Y.: An Improved Evolutinary Algorithm Based Approach for Outlier Detection in High Dimensional Spaces. Journal of Communication and Computer 3, 5–8 (2006)
9. Kennedy, J., Eberhart, R.: Particle Swarm Optimization. In: IEEE International Conference on Neural Networks, pp. 1942–1948. IEEE Service Center, Piscataway (1995)
10. Shi, Y., Eberhart, R.: A Modified Particle Swarm Optimizer. In: IEEE International Conference on Evolutionary Computation, pp. 69–73. IEEE Press, Piscataway (1998)
11. Clerc, M.: Discrete Particle Swarm Optimization: New Optimization Techniques in Engineering. Springer, Heidelberg (2004)
12. Zeng, J.C.: Particle Swarm Optimization Algorithms. Science Press, Beijing (2004)
13. Ye, D.Y., Chen, Z.J., Liao, J.K.: A New Algorithm for Minimum Attribute Reduction Based on Binary Particle Swarm Optimization with Vaccination. In: Zhou, Z.-H., Li, H., Yang, Q. (eds.) PAKDD 2007. LNCS (LNAI), vol. 4426, pp. 1029–1036. Springer, Heidelberg (2007)

A New Decision Tree Construction Using the Cloud Transform and Rough Sets*

Jing Song[1,2], Tianrui Li[1], and Da Ruan[3,4]

[1] School of Information Science and Technology
Southwest Jiaotong University, Chengdu 610031, P.R. China
songj@s-ec.com, trli@swjtu.edu.cn
[2] Research Center for Secure Application in Networks and Communications
Southwest Jiaotong University, Chengdu 610031, P.R. China
[3] Belgian Nuclear Research Centre (SCK•CEN), 2400 Mol, Belgium
druan@sckcen.be
[4] Transportation Research Institute, Hasselt University, 3590 Diepenbeek, Belgium

Abstract. Many present methods for dealing with the continuous data and missing values in information systems for constructing decision tree do not perform well in practical applications. In this paper, a new algorithm, Decision Tree Construction based on the Cloud Transform and Rough Set Theory under Characteristic Relation (DTCCRSCR), is proposed for mining classification knowledge from the data set. The cloud transform is applied to discretize continuous data and the attribute whose weighted mean roughness under the characteristic relation is the smallest will be selected as the current splitting node. Experimental results show the decision trees constructed by DTCCRSCR tend to have a simpler structure, much higher classification accuracy and more understandable rules than C5.0 in most cases.

Keywords: Rough sets, Cloud transform, Decision trees, Weighted mean roughness, Characteristic relation.

1 Introduction

Decision trees are considered as one of the most popular data-mining techniques for knowledge discovery. It systematically analyzes information contained in a large amount of data source to extract valuable rules and relationships [1]. Many approaches for constructing decision trees have been presented. One of the representative methods is ID3 algorithm, which is based on the information theory and attempts to minimize the expected number of comparisons [2]. The basic idea of the induction algorithm is that the attribute which has a maximum gain value of information entropy will be chosen as the current splitting node. C4.5 [3] and C5.0 [4], based on ID3, allow the use of missing data, continuous data and

---

* This work is partially supported by NSFC (No.60074014), the Research Fund for the Doctoral Program of Higher Education (No.20060613007) and the Basic Science Foundation of Southwest Jiaotong University (No.2007B13).

G. Wang et al. (Eds.): RSKT 2008, LNAI 5009, pp. 524–531, 2008.

improved techniques for splitting. For example, when a decision tree is built by C4.5, continuous data are divided into ranges based on the attribute values, while missing data are simply ignored. To classify a record with a missing attribute value, the value for that item can be predicted based on the attribute values for other records [1]. However, the existing algorithms for dealing with the continuous data and missing values in information systems do not perform well in real applications.

The classical rough set theory (RST), proposed by Pawlak in 1982, is a mathematical tool to deal with vagueness and uncertainty and has been applied successfully in data mining [5,6,8,9]. The rough approximation-based algorithms which can be used to select splitting node in the construction of decision trees were discussed in [8,9]. However, these approaches are under the assumption that information systems are complete. To deal with incomplete data directly, an extension of classical rough sets, the characteristic relation-based rough sets, was proposed in [5]. This extension better reflects real conditions of incomplete information systems (IIS).

Cloud model that integrates the properties of fuzziness and randomness was proposed in [12] for realizing the uncertainty transition between qualitative concept and quantitative description. A method, called cloud transform, for discretization of the continuous data is introduced in [13] based on the cloud model, which is especially suitable for processing a large amount of data. Cloud transform can partition the domain of every continuous attribute into many concepts represented by cloud models.

To overcome the difficulty in handling missing data and continuous data for classification tasks, a novel mining algorithm, Decision Tree Construction based on Cloud Transform and RST under Characteristic Relation (DTCCRSCR), is proposed. The cloud transform is firstly applied to discretize continuous data and then the attribute whose weighted mean roughness under the characteristic relation is the smallest will be selected as the current splitting node. Experimental results show that the decision trees constructed by DTCCRSCR tend to have a simpler structure, higher classification accuracy and more understandable rules than C5.0 in general.

The material of the paper is organized as follows. In Section 2, basic concepts of characteristic relation-based rough sets and the cloud model as well as the cloud transform are introduced. The DTCCRSCR method for constructing decision trees in IIS is illustrated in Section 3. Experimental evaluation of the proposed method compared with C5.0 is shown in Section 4. Section 5 concludes the research work of this paper.

2   Preliminaries

2.1   Rough Set Theory under Characteristic Relation

The following basic concepts of rough sets under characteristic relation and their extensions will be used in this paper.

Definition 1. *[7] An information system is defined as a pair $< U, A >$ where U is a non-empty finite set of objects, $A = C \cup D$ is a non-empty finite set of attributes, C denotes the set of condition attributes and D denotes the set of decision attributes, $C \cap D = \emptyset$. Each attribute $a \in A$ is associated with a set V_a of its value, called the domain of a.*

Definition 2. *[7] $< U, A >$ is an IIS if there exists a in A and x in U that satisfy that the value $a(x)$ is missing. All the missing values are denoted by "?" or "∗", where the lost value is denoted by "?", "do not care" condition is denoted by "∗".*

In [7], Grzymala-Busse presented that the characteristic set and characteristic relation can be determined by using the idea of blocks of attribute-values pairs which is defined as follows.

Definition 3. *[7] Let b be an attribute and v be a value of b for some cases. If $t = (b, v)$ is an attribute-value pair, $v \neq ?$ and $\ast$, then a block of t, denoted $[t]$, is a set of all cases from U that attribute b have value v. If there exists a case x such that $v = b(x) = ?$, then the case x is not included in the block $[(b, v)]$ for any value v of attribute b. If there exists a case x such that $v = b(x) = \ast$, then the case x is included in the block $[(b, v)]$ for all value v of attribute b.*

Definition 4. *[7] Let $B \subseteq A$ be a subset of attributes. The characteristic set $I_B^C(x)$ is the intersection of blocks of attribute-value pairs (b, v) for all attributes b from B for which $b(x)$ is specified and $b(x) = v$.*

Definition 5. *[7] Let $B \subseteq A$ be a subset of attributes. The characteristic relation, denoted by C_B, is defined as: $(x, y) \in C_B \Leftrightarrow y \in I_B^C(x)$.*

The characteristic relation C_B is reflexive but not symmetric nor transitive. Obviously, it is a generalization of the indiscernibility, tolerance and similarity relations in information systems [7].

Definition 6. *[7] The lower and upper approximations of X with regard to B under the characteristic relation are $X_B^C = \cup \{ I_B^C(x) | x \in X, I_B^C(x) \subseteq X \}$, $X_C^B = \cup \{ I_B^C(x) | x \in X, I_B^C(x) \cap X \neq \emptyset \} = \cup \{ I_B^C(x) | x \in X \}$, respectively.*

Similarly, we can define the concept of the weighted mean roughness in the extension of RST under characteristic relation as that in [9].

Definition 7. *Let $< U, A >$ be an IIS. $X \subseteq U$, $B \subseteq A$, $u_B(X) = \frac{card(X_B^C)}{card(X_C^C)}$ is a precision of X with regard to B under the characteristic relation $(0 \leq u_B(X) \leq 1)$. The weighted mean roughness of X with regard to B is defined as:*

$$\beta(B) = 1 - (\sum_{j=1}^{m} \omega_j u_B(X_j)) \tag{1}$$

where j is the jth decision class of decision attributes, $j = 1, 2, ..., m$, m is the number of decision class; X_j is the jth set of decision class; ω_j, the percent of X_j in U, is defined as: $\omega_j = card(X_j)/card(U)$.

According to the definition of the weighted mean roughness under the characteristic relation, we know the value of $\beta(B)$ ranges from 0 to 1. When $\beta(B) = 0$, there is no uncertainty. When $\beta(B) = 1$, this means the set B leads to the greatest uncertain partition. As $\beta(B) \to 0$, the uncertainty decreases.

2.2   Cloud Model and Cloud Transform

Cloud model integrates the properties of fuzziness and randomness for realizing the uncertainty transition between qualitative concepts and quantitative descriptions. Let $U = \{u\}$ be the universe of discourse and T be a linguistic term associated with U. The membership degree of u in U to the linguistic term T, $C_T(u)$, is a random variable with a probability distribution. $C_T(u)$ takes values in $[0, 1]$. A cloud is a mapping from the universe of discourse U to the unit interval $[0, 1]$ [12]. It is described with only three digital characteristics, expected value Ex, entropy En and hyper entropy He. Ex is the position at the universe of discourse, corresponding to the center of gravity of the cloud. En is a measure of the fuzziness of the concept over the universe of discourse showing how many elements in the universe of discourse could be accepted to the linguistic term. He is a measure of the uncertainty of the entropy En. Close to the waist of the cloud, corresponding to the center of gravity, cloud drops are most dispersed, while at the top and bottom the focusing is much better [12].

Cloud transform is a method for discretization of continuous data based on the cloud model which is especially suitable for processing a huge data. The domain of every continuous attribute can be partitioned by the cloud transform into many concepts represented by cloud models. The number of cloud produced by the cloud transform is given by users. The cloud transform is expressed by the following formula:

$$g(x) \approx \sum_{i=1}^{m} (c_i * f_i(x)) \quad 0 < MAX(|g(x) - \sum_{i=1}^{m} (c_i * f_i(x))|) < \varepsilon \qquad (2)$$

where $g(x)$ is the data distribution function, $f_i(x)$ is the expected function of the cloud model, m is the number of cloud, ε is the error threshold. The course of the cloud transform is to find out the expected value, entropy and hyper entropy of every cloud [13].

3   A New Decision Tree Construction Using the Cloud Transform and RST

3.1   Decision Tree Construction Based on RST

Based on the definition of rough sets under characteristic relation, we develop the DTCRSCR algorithm which combines the characteristic relation-based RST for mining classification knowledge from IIS [17]. It firstly computes the weighted mean roughness of every condition attribute under the characteristic relation. Then, the attribute whose weighted mean roughness is the smallest will be selected as the splitting node. The algorithm is described as follows.

Data: Data set *sample* (all of values of attributes are discrete), the collection of condition attributes *attribute_list*.

Result: *decision_tree*.

Step1. With respect to *sample*, firstly compute the lower and upper approximations of every condition attribute with regard to every partition set of decision attributes. Then, calculate the weighted mean roughness of every condition attribute.

Step2. The attribute B whose weighted mean roughness under the characteristic relation-based rough sets is the smallest will be selected as the current splitting node.

Step3. For every value of the selected attributes B, we obtain a data set Q of corresponding branch by using test $B.value = v$.

Step4. For every branch Q, if it has not reached the leaf then call DTCRSCR($Q, attribute\_list \backslash \{B\}$).

Step5. Return.

Algorithm 1. The DTCRSCR Algorithm

3.2 Decision Tree Construction Using the Cloud Transform and RST

The proposed DTCRSCR algorithm cannot handle continuous data. Namely, the attribute domain must be divided into categories at the beginning. Then a new algorithm, Decision Tree Construction based on the Cloud Transform and RST under Characteristic Relation (DTCCRSCR), is proposed which integrates the cloud transform for dealing with continuous data (see Alg. 2).

Data: Data set *sample* (Not all attribute values are categorical data in *sample.*), the collection of condition attributes *attribute_list*.

Result: *decision_tree*.

Step1. The cloud transform algorithm is applied to discretize continuous attributes in *sample* and a new data set $sample^*$ which only contains categorical data will be gained.

Step2. Call the DTCRSCR algorithm for data set $sample^*$.

Step3. Return.

Algorithm 2. The DTCCRSCR Algorithm

4   Experimental Evaluation

Experiments are performed on an 864MHz Pentium Server with 512MB of memory, running windows XP server and SQL server 2000. Algorithms are coded in C#. Ten data sets, publicly available from the UC Irvine Machine Learning Database Repository [14], are chosen as benchmark datasets for the performance tests. The descriptions of experimental data are shown in Table 1.

Where E means the data set contains '?' or '*' and R means that we randomly replace some data with '?' or '*' in the data set.

Table 1. The Description Table of Experimental Data

Data set	Tuples	?	*	Continuous	Attribute(C/D)
monks-1_learn.tab	124	R	R	No	6/1
monks-1_test.tab	432	R	R	No	6/1
monks-2_learn_tab	169	R	R	No	6/1
monks-3_learn.tab	122	R	R	No	6/1
monks-3_test.tab	432	R	R	No	6/1
breast-cancer.tab	286	E	R	Yes	9/1
hayes-roth_learn.tab	132	R	R	Yes	5/1
flare1.tab	322	R	R	No	13/1
post-operative.tab	90	E	R	No	8/1
lymphography.tab	148	R	R	No	18/1

We first take experiments on the data set, breast-cancer.tab, to test whether classification rules produced by DTCCRSCR is more understandable than those of C5.0. Table 2 shows the episode of classification rules produced by C5.0.

Table 2. The episode of classification rules obtained by C5.0

ID of Rules	Content of Rules
7	node_caps=no, inv_nodes=3-5, menopause=premeno, age<=46, class=recurrence-events
18	node_caps=no, inv_nodes=0-2, irradiat=no, deg_malig=3, tumor_size<=23, class=no-recurrence-events
29	node_caps=no, inv_nodes=0-2, irradiat=yes, deg_malig=3, breast=right, age>49, class=recurrence-events

Obviously, C5.0 does not handle continuous data well. The continuous domain space is divided into some rectangular regions. It is uneasy to understand the rules with continuous attributes in practical applications. Table 3 shows the digital characteristics of five cloud models that represent concept "age". They are defined as the linguistic terms of low, middle-low, middle, middle-high and high, while the episode of classification rules obtained by DTCCRSCR is listed in Table 4. It can be observed that the decision trees constructed by DTCCRSCR tend to have more understandable rules than C5.0.

Table 5 shows the result of the performance evaluation of DTCCRSCR compared with C5.0 in constructing decision trees.

From Table 5, in most of data sets (7 out of 10 data sets), the decision trees (here the number of leaves and nodes of the whole tree are listed) constructed by DTCRSCR tend to have simpler structure and higher classification accuracy than C5.0. Only in the data set, "monks-1_test.tab", the decision trees constructed by DTCRSCR have a simpler structure than C5.0 and with the same classification accuracy as C5.0. In these two data sets, "monks-3_test.tab"

Table 3. The digital characteristics of the cloud model that represent a concept "age"

Linguistic term	Ex	En	He
low	38	2.99	0.24
middle-low	42	1.49	0.56
middle	48	5.97	0.97
middle—high	59	4.48	0.98
high	65	2.99	0.43

Table 4. The episode of classification rules obtained by DTCCRSCR

ID of Rules	Content of Rules
35	inv_nodes=9-11, age=middle_high, node_caps=yes, class=recurrence-events
23	inv_node =3-5, deg_malig=3, tumor_size=middle_big, node_caps=yes, irradiat=no, class=no-recurrence-events
15	inv_nodes=0-2, breast=right, breast_quad=left_low age=high, tumor_size=middle, deg_malig=2, class=no-recurrence-events

Table 5. A performance evaluation of DTCCRSCR compared with C5.0

Data sets	C5.0 (leaves/nodes)	C5.0 (accuracy)	DTCCRSCR (leaves/nodes)	DTCCRSCR (accuracy)
monks-1_learn.tab	50/59	91.87%	46/62	98.374%
monks-1_test.tab	220/371	50.926%	220/369	50.926%
monks-2_learn_tab	146/255	90.374%	124/218	94.083%
monks-3_learn.tab	88/203	83.097%	86/161	86.066%
monks-3_test.tab	22/27	100%	25/34	100%
breast-cancer.tab	228/248	96.503%	90/135	97.902%
hayes-roth_learn.tab	33/41	85.610%	29/40	96.212%
flare1.tab	64/132	84.290%	38/75	99.690%
post-operative.tab	61/104	90.412%	56/106	91.111%
lymphography.tab	47/138	98.649%	59/144	98.649%

and "lymphography.tab", the decision tree constructed by DTCRSCR has more complex structures than C5.0 but having the same classification accuracy.

5   Conclusions

Decision tree is one of the most significant classification methods in data mining. However, most decision tree algorithms cannot handle missing data and continuous data effectively. In this paper, a new algorithm, DTCCRSCR, based on the cloud transform and the characteristic relation-based rough sets for construction of decision trees demonstrates that DTCCRSCR performs better than C5.0 in most cases. But DTCCRSCR is quite time-consuming compared to C5.0 and is thus unsuitable for data mining tasks in large data sets. Our future research work is to study how to improve the DTCCRSCR algorithms.

References

1. Dunham, M.H.: Data Mining – Introductory and Advanced Topics. Prentice Hall, Englewood Cliffs (2002)
2. Quinlan, J.R.: Induction of decision tree. Machine Learning 11, 80–106 (1986)
3. Quinlan, J.R.: C4.5: Programs for machine learning. Morgan Kaufmann, San Francisco (1993)
4. http://www.rulequest.com/see5info.html
5. Grzymala-Busse, J.W., Siddhaye, S.: Rough set approaches to rule induction from incomplete data. In: Proc. of the Int. Conf. on IPMU in KBS, pp. 923–930 (2004)
6. Li, T., Ruan, D., Geert, W., Song, J., Xu, Y.: A rough sets based characteristic relation approach for dynamic attribute generalization in data mining. Knowledge-Based Systems 20(5), 485–494 (2007)
7. Grzymala-Busse, J.W.: Characteristic relations for incomplete data: A generalization of the indiscernibility relation. Transactions on Rough Sets IV. 58–68 (2005)
8. Wei, J., Huang, D., Wang, S., Ma, Z.: Rough Set Based Decision Tree. In: Proc. of the 4th World Congress on Intel. Cont. and Auto, pp. 426–430 (2002)
9. Jiang, Y., Li, Z., Qi, Z., Liu, Y.: New method for constructing decision tree based on rough set theory. Computer Application 24(8), 21–23 (2004)
10. Peters, J.F., Suraj, Z., Shan, S., Ramanna, S., Pedrycz, W., Pizzi, N.: Classification of meteorological volumetric radar data using rough set methods. Pattern Recognition Letters 24, 911–920 (2003)
11. http://www.ics.uci.edu/~mlearn/MLRepository.html
12. Li, D., Meng, H., Shi, X.: Cloud model and cloud model generator. Computer Research and Development 6, 15–20 (1995)
13. Fan, J., Li, D.: Mining classfictions knowledge based on cloud models. In: Zhong, N., Zhou, L. (eds.) PAKDD 1999. LNCS (LNAI), vol. 1574, pp. 26–28. Springer, Heidelberg (1999)
14. Song, J., Li, T., Wang, Y., Qi, J.: Decision Tree Construction based on rough set theory under Characteristic Relation. In: Li, T., Xu, Y., Ruan, D. (eds.) Proc. of the 2007 ISKE, pp. 788–792. Atlantis Press, Paris (2007)

Improved Unsupervised Anomaly Detection Algorithm

Na Luo[1,3], Fuyu Yuan[2], Wanli Zuo[1], Fengling He[1], and Zhiguo Zhou[3]

[1] College of Computer Science and Technology, Jilin University,
Changchun, China 130012
[2] Changchun Institute of Applied Chemstry, Chinese Academy of Sciences,
Changchun, China 130022
[3] Computer Sicence Department, Northeast Normal University,
ChangChun, China 130117

Abstract. In recent years, the network infrastructure has been improved constantly and the information techniques have been applied broadly. Because the misuse detection and anomaly detection methods both have individual benefits and drawbacks, this paper supports the point that combines these two methods to construct the whole intrusion detection system by data mining technique. In this paper, we focus on the improvement of the anomaly detection module in MINDS(Minnesota Intrusion Detection System). By analysis, we use the method of multi-dimension outlier point detection and adapt the connection score with dynamic weight to improve the performance of intrusion detection system. The improved unsupervised anomaly detection algorithm, also named IUADA, is non-linear, and reduces both the response time and the false alarm rate.

1 Introduction

The goal of intrusion detection is to discover intrusions in the computer or network by observing various network activities or attributes. Here, the intrusion refers to any set of actions that threaten the integrity, availability, or confidentiality of a network resource. However, intrusion detection is not yet a perfect technology, and gives the data mining technique the opportunity to make several important contributions in the field of intrusion detection. It has becoming a hot field and attracted some experts and researchers. The applications of data mining in IDS consist of two directions:

Research in misuse detection has focused mainly on detecting network intrusions using various classification algorithms. Models of misuse with data mining are created automatically, and can be more comprehensible and precise. But they still can not detect unknown type attacks, and labeling data instances may require a great deal of time and effort.

In supervised anomaly detection, given a set of normal data to train on, and given a new set of test data, the goal is to determine whether the test data is normal or anomalous. Recently, there have been several efforts in it, such as ADAM [1],

G. Wang et al. (Eds.): RSKT 2008, LNAI 5009, pp. 532–539, 2008.

PHAD(Packet Header Anomaly Detection) [2],etc. The major benefit of anomaly detection algorithms is to potentially detect the attacks of unknown type. The limitation of anomaly detection systems is a possible high false alarm rate.

In the paper, we introduced an effective detection techniques namely MINDS [3]. The basic features used in MINDS include source and destination IP addresses, source and destination ports, protocol, flags, number of bytes and number of packets, and the derived features include time-window and connection-windows based on the basic features. The data is fed into the MINDS anomaly detection module that uses an outlier detection algorithm to assign an anomaly score to each network connection. Because the score is computed by the derived features, MINDS has the drawbacks such as long response time and high false alarm rate.

This paper focus on the unsupervised anomaly detection technique. In Section 2, the outlier detection method is introduced, and takes the option of combining the misuse detection and anomaly detection, and MINDS is introduced in this section, too. We present an improved unsupervised anomaly detection algorithm (IUADA) in the following section. Then a set of experiments are taken in section 4, and the result show that our method has higher performance. Finally, we draw the conclusions.

2   Anomaly Detection Techniques

2.1   Outlier Point Detection

In anomaly detection system, the data points are mapped into a feature space, so we can determine what data points are outliers based on the position of the points in the feature space. We label points which are in sparse regions of the feature space as anomalous ones.

There are lots of outlier point detection algorithms and variant of them in [4]. Here, we introduce the Density Based Local Outliers approach.

The outlier factor of a data point is local in the sense that it measures the degree of being an outlier with respect to its neighborhood. For each data example, the density of the neighborhood is first computed. The LOF [5] of a specific data example represents the average of the ratios of the density of its neighbors. LOF requires the neighborhood around all data points be constructed. This involves calculating pairwise distances between all data points, which is an $O(n^2)$ process and makes it computationally infeasible for millions of data points. To address this problem, we sample a training set from the data and compare all data points to this small set, which reduces the complexity to $O(n * m)$, where n is the size of the data and m is the size of the sample.

2.2   Combination of Misuse Detection and Anomaly Detection

In section 1, we have shown that misuse detection and anomaly detection both have individual limitations: misuse detection systems can not detect unknown attack type and make network volatile easily, while anomaly detection systems have high false alarm. So the combination of these two methods would contain the individual benefits and avoid their drawbacks. There is an ideal solution for intrusion detection. First, the misuse detection module can use its repository to matching the attack connections of known type. Through this phase, system can filter out some data to improve processing speed. It is more important that the phase plays the role of data cleaning for following modules. Under the good condition, the anomaly detection module can get good results by anomaly detection algorithm such as outlier detection. Thus, the solution overcomes the individual drawbacks of both modules. The incorporative system improves the speed and accuracy, and can capture unknown intrusions. There are some systems proved this point in [1] and [3].

3   Improved Unsupervised Anomaly Detection Algorithm

3.1   The Metrics of Evaluating Detection Systems

There are lots of metrics in [3] and [6]. The most common metrics include true positive rate, false negative rate, false positive rate and true negative rate.

The metrics that were developed for evaluating network intrusions in our method usually correspond to detection rate (true positive rate) as well as false alarm rate (false positive rate).

Assume that a given network traffic in some time intervals, each connection is assigned a score value which represented as a vertical line in Figure 1. This score represents how anomalous the connection is. The dashed line represents the real attack curve that is zero for normal network connections and one for intrusive connections. The full line corresponds to the predicted attack curve, and for each connection it is equal to its assigned score. These two curves allow us to compute the error for every connection as the difference between the real connection and the assigned score to the connection[3].

3.2   Improved Unsupervised Anomaly Detection Algorithm (IUADA)

IUADA is based on MINDS, and we only focus on anomaly detection module which uses an outlier detection algorithm to assign an anomaly score to each

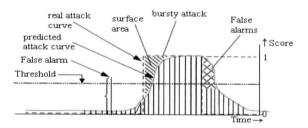

Fig. 1. Assigning scores in network intrusion detection

network connection. We select the algorithm of LOF approach which detects outlier points in MINDS to compare with our method.

In Figure 1, we observe the features of the burst attack. The form of the real attack curve is square-wave, while the predicted attack curve is a non-smooth curve. At the left border of attack curve in a burst, we can find that there is a large interval between the beginning of attack till the time when the first network connection has the score value higher than the threshold. Because the score in MINDS is computed by derived features partly, the score value will increase slowly at the beginning of the burst, and the value will decrease slowly in the end. According to the experiment result in MINDS[3], in proportion as the width of burst increase the response time will rise nearly. Similarly, at the right border of the real attack curve in a burst, the predicted attack curve falls slowly and causes false alarms. So we present an idea to improve the performance of unsupervised anomaly detection system by reduce the area between the predicted attack curve and the real attack curve.

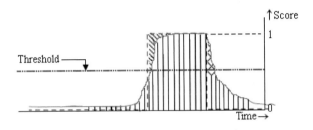

Fig. 2. Adapted scores in network intrusion detection

Thinking as follow, we increase the ascending speed by multiply the original connection score by a weight greater than 1 at the ascending phase, while multiplying the original connection score by a weight less than 1 can increase the descending speed at the descending phase. Fig.2 shows the adapted scores in network intrusion detection. We can find that the response time and false alarm rate are both reduced by comparing Figure 1 with Figure 2.

The attention focuses on how to construct the dynamic weight. The dynamic weight should have such characteristics:

(1) When the score value increases, the weight value should be greater than 1, while the weight value should be less than 1 when the score value decreases.

(2) To increase the increasing speed, the weight value should growth monotonously in ascending phase, while the weight value should decline monotonously to increase the decreasing speed in the descending phase.

(3) The above two characteristics have a potential question that the score value multiplied by weight may be more sensitive to noise during non-burst periods. For solving this problem, the weight should change dynamically. The change is small at the beginning, and then it enlarges with time elapse.

For these three reasons, we use the dynamic weight to get ideal result in the experiment. To avoid the acute fluctuation of connection score, we apply the Gauss weight curve and the Reversed Gauss weight curve. In the detection, we need retain all of the original score value of connections to compare. With time elapsed, we adapt the score value by changing the weight dynamically.

The functions of two curves are expressed as follow:

$$f(x) = 1 + \exp(-(\frac{x-a}{b})^2).$$ (1)

$$g(x) = 1 - \exp(-(\frac{x-a}{b})^2).$$ (2)

The gauss function is dynamic because the width of bursts are not same as each other. The parameter a in equality (1) and (2) increase in the detection process.

The algorithm IUADA is as Algorithm 1

Algorithm 1. IUADA

1: Repeat until the detection is over
2:    Get the current original score $OS2$ and compare it with previous one $OS2$;
3:      If $OS2 > OS1$ Then
4:        If $flag <> INCREASE$ Then
5:          increase phase: $weight = 1; flag = INCREASE; i = 1$;
6:        Else
7:          $weight = Gauss\_Increase(i++)$;
8:      If $OS2 < OS1$ Then
9:        If $flag <> DECREASE$ Then
10:          decrease phase: $weight = 1; flag = DECREASE; i = 1$;
11:        Else
12:          $weight = Gauss\_Decrease(i++)$;
13:      Adapted connection score $MS = OS2 * weight$;
14:      If $MS >= threshold$ Then
15:        The connection is intrusive;
16:      Else
17:        The connection is normal;
18:      $OS1 = OS2$;

---

where the function $Gauss\_Increase$ and $Gauss\_Decrease$ are used to compute the dynamic weight. Take the $Gauss\_Increase$ function for example, parameter i represents the detection schedule, the function adjust the curve function and compute the weight value.

We apply the LOF approach to detect outlier data points in actual experiment. To avoid the high time complexity, we sample a small data set namely referring set from all of the data points. Outlier in this small set, would affect accuracy of experiment. So we filter the outlier from the referring set before using LOF approach.

4   Experiments

We applied the improved unsupervised anomaly detection algorithm to 1998 DARPA Intrusion Detection Evaluation Data[6]. The DARPA' 98 data contains training data and test data. The training data consists of 7 weeks of network-based attacks inserted in the normal background data. Attacks in training data are labeled. The test data contained 2 weeks of network-based attacks and normal background data. 7 weeks of data resulted in about 5 million connection records. The data contains four main categories of attacks: Denial of Service (DoS), R2L, U2R and Probing.

In our experiment, we built the environment like MINDS, and the system consisted of three modules: feature abstraction, misuse detection and unsupervised anomaly detection. In the anomaly detection module, we adapted the score values of network connection using a dynamic weight algorithm. For comparing with MINDS, we sampled sequences of normal connection records in order to create the normal data set that had the same distribution as the original normal data set. Then we used the sample data set to train our anomaly detection, and examined how well the detection worked. We focus on response time and false alarm rate to evaluate the performance of our algorithm.

We used the TCP connections from 5 weeks of training data (499,467 connections), where we sampled 1% records that correspond to the normal connections. Also we considered a random sample of 1000 connection records that correspond to normal data in order to determine the false alarm rate.

The first 3 columns in Table 1 show connection bursts of attack types and their burst lengths., and the last 4 columns represent the experiment results. The response time (t) represents the first connection for which the score value is larger than the threshold. It is apparent from Table 1 that the method with smaller response time is improved unsupervised anomaly algorithm since it uses dynamic weight. The larger the number of burst length, the larger the difference between the LOF and IUADA response time. It is the result of dynamic weight working on connection scores.

We compared IUADA with the Nearest Neighbor(NN) and Mahalanbis Distance methods. For the NN method, the number of nearest neighbor is 1 and the threshold is 2%; for the Mahalanbis Distance method, the threshold is 2%, and for IUADA, the threshold is 60. Fig 3 show the roc results, the true alarm rate of IUADA is better than NN and Mahalanobis methods, except when the false alarm rate is 0.02.

Table 1 reports on another metric for evaluation of bursty attacks, namely false alarm area. The smaller the false alert area between the real and the predicted attack curve, the better the intrusion detection algorithm. The false alarm area (FA) in Table 1 was normalized, such that the total surface area was divided by the total number of connections from the corresponding attack burst. We can see the improved unsupervised anomaly algorithm is better than LOF approach from Table 1.

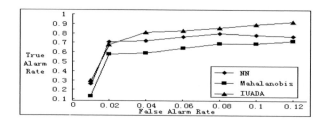

Fig. 3. ROC Curve

Table 1. The connection bursts and results in experiment

Burst position	Burst length	Attack type	LOF		IUADA	
			FA	t	FA	t
Week1,burst1	15	neptune(DoS)	0.03	1	0.03	1
Week2,burst1	50	guest(U2R)	0.22	1	0.15	1
Week2,burst2	102	portsweep(probe)	0.5	20	0.33	14
Week2,burst3	898	ipsweep(probe)	0.61	2	0.4	2
Week2,burst4	1000	back(DoS)	0.3	3	0.2	3
Week3,burst1	15	satan(probe)	0.89	0	0.58	0
Week3,burst2	137	portsweep(probe)	0.8	30	0.52	14
Week3,burst3	105	nmap(probe)	0.3	2	0.2	2
Week3,burst4	1874	nmap(probe)	0.33	13	0.22	9
Week3,burst5	5	imap(R2L)	0.14	2	0.1	2
Week3,burst6	17	warezmaster(U2R)	0.08	1	0.06	1
Week4,burst1	86	warezclient(U2R)	0.56	1	0.37	1
Week4,burst2	6104	satan(probe)	0.12	10	0.09	7
Week4,burst3	1322	pod(DoS)	0.34	1	0.23	1
Week4,burst4	297	portsweep(probe)	0.48	17	0.32	12
Week4,burst5	2304	portsweep(probe)	0.2	1	0.14	1
Week5,burst1	3067	satan(probe)	0.06	21	0.05	14
Week5,burst2	5	ffb(R2L)	0.86	0	0.56	0
Week5,burst3	1021	portsweep(probe)	0.49	8	0.32	6
Total	18424					

5   Conclusions

In anomaly detection, the task is to detect a lot of unknown network connections and recognize the outlier points from them, which are regarded as intrusive behaviors. This task is base the hypothesis that the number of normal behaviors is larger than that of anomalous behaviors, and anomalous behaviors are different enough from normal ones. If the number of anomalous behaviors is large, anomaly detection can not be effective. This paper emphasizes the combination of misuse detection and anomaly detection, which has great benefits. The system can filter some of intrusive behaviors by match-ing connections with known attack types, so the unsupervised anomaly detection can be work effectively.

Additionally, human experts can mining the attack model from the anomalous behaviors and reinforce the known model detection.

MINDS uses a suite of data mining techniques to automatically detect attacks against computer networks and systems. There are four modules in MINDS, the most important one is anomaly detection module. The data is fed into the MINDS anomaly detection module that uses an outlier detection algorithm to assign an anomaly score to each network connection to determine if they are actual attacks or other interesting behavior. But there are still the problems of long response time and high false alarm rate in anomaly detection. In this paper, we present the thinking to improve the performance of unsupervised anomaly detection. The algorithm reduces the response time and false alarm rate by using a dynamic weight to change the connection scores. Considering the DARPA 98' data, performed experiment demonstrates that our algorithm is more successful than anomaly detection module in MINDS.

Acknowledgement

This work was supported by Technological Development Projects of JiLin Province under Grant No.20070533 and the 2008's Natural Science Funds for Young Scholars of Northeast Normal University under Grant No. 20081004.

References

1. Barbara, D., Jajodia, S.: Applications of Data Mining in Computer Security, pp. 72–78. Kluwer Academic Publishers, Dordrecht (2002)
2. Mahoney, M.V., Chan, P.K.: Learning nonstationary models of normal network traffic for detecting novel attacks. In: 8th ACM SIGKDD International Conference on Knowledge Discovery and Data mining, pp. 376–385. ACM Press, New York (2002)
3. Ertoz, L., Eilertson, E., Lazarevic, A., Tan, P., Srivastava, J., Kumar, V., Dokas, P.: The MINDS - Minnesota Intrusion Detection System, Next Generation Data Mining, pp. 65–86. MIT/AAAI Press, New York (2004)
4. Han, J., Kamber, M.: Data Mining: Concepts and Techniques, pp. 146–165. Morgan Kaufmann, San Francisco (2000)
5. Breunig, M., Kriegel, H.P., Ng, R., Sander, J.: LOF: Identifying Density-Based Local Outliers. In: Proceedings of ACM SIGMOD Int. Conf. on Management of Data (SIGMOD 2000), Dallas, Texas, pp. 93–104 (2000)
6. Lippmann, R., et al.: Evaluating Intrusion Detection Systems: The 1998 DARPA Off-line Intrusion Detection Evaluation. In: Proceedings of the DARPA Information Survivability Conference and Exposition, pp. 12–26. IEEE Press, New York (2000)

Applications of Average Geodesic Distance in Manifold Learning

Xianhua Zeng

School of Computer Science, China West Normal University
Nanchong 637002, China
xianhuazeng2005@163.com

Abstract. Manifold learning has become a hot issue in the research fields of machine learning and data mining. Current manifold learning algorithms assume that the observed data set has the high density. But, how to evaluate the denseness of the high dimensional observed data set? This paper proposes an algorithm based on the average geodesic distance as the preprocessing step of manifold learning. Moreover, for a high dense data set evaluated, we further utilize the average geodesic distance to quantitatively analyze the mapping relationship between the high-dimensional manifold and the corresponding intrinsic low-dimensional manifold in the known ISOMAP algorithm. Finally, experimental results on two synthetic Swiss-roll data sets show that our method is feasible.

1 Introduction

In recent years we observe intensive research in manifold learning. The purpose is to map a set of data on a manifold in high-dimensional space to a lower-dimension space. Moreover, manifold learning has been associated with human visual perception [1]. It has important applications in areas such as pattern analysis, data mining, and multimedia data processing. Some known manifold learning algorithms have been proposed and fully researched [3][5]. ISOMAP is a promising method because of the use of geodesic distances [8]. Intuitively, geodesic distance between a pair of points on a manifold is the distance measured along the manifold. Owing to geodesic distance reflects the underlying geometry of data, data embedding using geodesic distance is expected to unfold highly twisted data manifolds. But, in ISOMAP method, geodesic distance estimation using a neighborhood graph constructed by using either K-NN or ε-neighbor approach has the following several problems [4][6]: a) Neither K-NN nor ε-neighbor approach guarantees the connectivity of the neighborhood graph. In particular, both approaches fail to build connected neighborhood graphs when the data spread across multiple clusters. Most applications would require that geodesic distance between every pair of data points is measurable. Using a disconnected neighborhood graph, however, we cannot measure the graph distance between every pair of data points. Consequently, the data cannot be embedded into a single low-dimensional coordinate system. b) How to choose a proper value of the parameter K or epsilon is a difficult problem. If it were chosen too small,

G. Wang et al. (Eds.): RSKT 2008, LNAI 5009, pp. 540–547, 2008.
© Springer-Verlag Berlin Heidelberg 2008

the neighborhood graph would be disconnected. If it were chosen too large, a so-called "short-circuit" problem [2] would occur and the constructed neighborhood graph would not reflect the nature of the data manifold. This short-circuit problem makes the constructed neighborhood graph sensitive to noise in the data. c) The constructed neighborhood graph contains many holes and looks like a Swiss cheese [7], which means that graph distances, especially the short ones, may not be good estimations of the corresponding geodesic distances.

These problems motioned above have a great relationship with the density of the observed data set. In practical applications of manifold learning, we assume that the set of data is high dense and smooth [2][3]. But, whether the assumption can be guaranteed or not, there isn't a quantitative measure at present. We proposed the limit of the average geodesic length between any two points as the measure criterion of the density of data points. With the increasing of data points, if the limit of the average geodesic length between any two points exists, the data manifold is a sufficiently high density and manifold learning algorithms output coordinate vectors in a lower-dimensional Euclidean space that can just represent the intrinsic geometry of the data. In addition, how to quantitatively analyze the mapping relationship between the intrinsic dimensions and the observation space still is a difficult problem [6]. The proportion of the average geodesic length between the intrinsic lower-dimensional space and the observation space was discussed. The proportion may be thought as the magnification factor from the observation space to the intrinsic low-dimensional space.

The rest of the paper is organized as follows. In Section 2, we discuss ISOMAP algorithm and the density assumption of data set. In Section 3, the measure criterion of the density of data points is proposed and analyzed. In Section 4, we briefly describe the opinion of using the average geodesic distance for quantitatively analyzing the embedding relationship. In section 5, an algorithm based on the average geodesic distance is put forward. Experimental results are reported in Section 6. Finally, we give some concluding remarks and future works in Section 7.

2   ISOMAP Algorithm

ISOMAP rests on an assumption. That is, as the number of data points increases, the graph distances provide increasingly better approximations to the intrinsic geodesic distances, and become arbitrarily accurate in the limit of infinite data which has a sufficiently high density. Under this condition, ISOMAP algorithm has three steps [2][3]. The first step determines which points are neighbors on data manifold, based on Euclidean distance $d_X(i, j)$ between pairs of points (i, j) in the input space X. There are two methods which connect each point to all points of its K-nearest neighbors or to all within some fixed radius ε. These neighborhood relations are represented as a weighted graph G over data points, with edges of weight $d_X(i, j)$ between neighboring points. The second step is to estimate the geodesic distances $d_M(i, j)$ between all pairs of points on the manifold M by computing their shortest path distances $d_G(i, j)$ in the graph G.

The final step applies classical MDS algorithm to the matrix of geodesic distances $D_G = \{d_X(i,j)\}$, computing an embedding of the data in a lower-dimensional space Y that best preserves the manifold's estimated intrinsic geometry. The simple illustration [4] of ISOMAP is shown in Fig. 1.

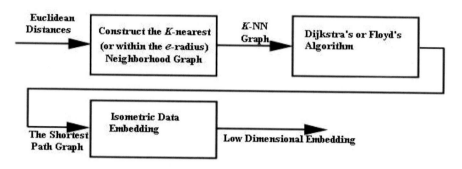

Fig. 1. Data embedding based on geodesic distances

For data points with a sufficiently high density, it can always choose a neighborhood size (K or ε) large enough that the graph will (with high probability) have a path not much longer than the true geodesic, but small enough to prevent edges from "short circuit" in the true geometry of manifold. In literature [3], more precisely, given arbitrarily small values of λ_1, λ_2 and μ, we can guarantee that, with probability $1 - \mu$ at least, estimates of the form $(1 - \lambda)d_M(i,j) \leq d_G(i,j) \leq (1 + \lambda)d_M(i,j)$ will hold uniformly over all pairs of data points (i,j). But, how to know that a data set has a sufficiently high density? We give a technique which can quantitatively resolve the problem.

3   The Average Geodesic Distance

Let $X_n = \{x_1, x_2, \cdots, x_n\}$ be a compact data set in smooth manifold M. A Riemann metric [9][10] is a mapping which associates each point $x \in M$ to an inner product $g(\bullet, \bullet)$ between tangent vectors at x. Then, Riemann manifold (M, g) is just a smooth manifold with a given Riemann metric g. As an example, when M is a sub-manifold of the Euclidean space R^d, the natural Riemann metric on M is just the usual dot product between tangent vectors. For any tangent vector v at point x, we can define its norm as $|v|_{g_x} = g_x(v, v)$. Using this norm, it is natural to define the length of a piecewise smooth curve $\Gamma(t)$ on manifold M as

$$L(\Gamma) = \int_0^s |\frac{d}{dt}\Gamma(t)|_{g_x} dt \qquad (1)$$

Then, the geodesic distance between points $x_1, x_2 \in M$ is the length of the shortest piecewise smooth curve between two points, that is:

$$d_g(x_1, x_2) = \inf_\Gamma \{L(\Gamma) : \Gamma(0) = x_1, \Gamma(s) = x_2\} \qquad (2)$$

For compact data set $X_n = \{x_1, x_2, \cdots, x_n\}$ on manifold M, we define the sum of the geodesic distances between any two points as:

$$L_n = \sum_{i<j} d_g(i,j) \tag{3}$$

Theorem 1. *Let M be a smooth compact manifold, and $X_n = \{x_1, x_2, \cdots, x_n\}$ is a high-dense data set on M, then,*

$$\lim_{n\to\infty} \frac{L_n}{n(n-1)} = \lim_{n\to\infty} \frac{\sum\limits_{i<j} \bar{d}_g(x_i, x_j)}{n(n-1)} = constant \tag{4}$$

where $\bar{d}_g(x_i, x_j)$ is the approximation of $d_g(x_i, x_j)$ on M.

Proof. Set $X_n = \{x_1, x_2, \cdots, x_n\}$ and $X_{n+1} = \{x_1, x_2, \cdots, x_n, x_{n+1}\}$ be two high dense data sets on manifold M which has boundary. Look for the nearest neighborhood of x_{n+1} in $\{x_1, x_2, \cdots, x_n\}$. If x_i is the nearest neighborhood, then $|x_i - x_{n+1}|$ is sufficiently small quantity because data set X_{n+1} is high dense. So,

$L_{n+1} = L_n + \sum\limits_{j=1, j\neq i} \bar{d}(x_i, x_j) + (n-1)\varepsilon_1 + \varepsilon_2,$

where $\varepsilon_1, \varepsilon_2$, are sufficiently small quantities. Then,

$\lim\limits_{n\to\infty} \left(\frac{L_{n+1}}{(n+1)n} - \frac{L_n}{n(n-1)} \right) = \lim\limits_{n\to\infty} \frac{(n-1)L_{n+1} - (n+1)L_n}{(n+1)n(n-1)}$

$= \lim\limits_{n\to\infty} \dfrac{(n-1)(L_n + \sum\limits_{j=1, j\neq i} d(x_i, x_j) + (n-1)\varepsilon_1 + \varepsilon_2) - (n+1)L_n}{(n+1)n(n-1)} = 0$

Corollary 1. *Set $X_n = \{x_1, x_2, \cdots, x_n\}$ and $Y_m = \{x_1, x_2, \cdots, x_m\}$ be any two high dense data sets on the same manifold M which has boundary. Then*

$$\lim_{\substack{n\to\infty \\ m\to\infty}} \left(\frac{L_n}{n(n-1)} - \frac{L_m}{m(m-1)} \right) = 0 \tag{5}$$

4   Analysis of the Embedding Relationship

In the section, we briefly describe the opinion of using the average geodesic distance for quantitatively analyzing the embedding relationship. In the field of manifold learning, how to quantitatively analyze the mapping relationship between the intrinsic low-dimensional space and the observation space still is a difficult problem. In literature [6], magnification factor is introduced and used to describing the proportion of the changing trend between observed data manifold and corresponding low-dimensional manifold. In some degree, it quantitatively describes the curved degree in high-dimensional space. Magnification factor is defined as the proportion of the "micro-arc" between the intrinsic lower-dimensional space and the observed space. But it is hard to present the micro-arc. In this paper, we use the proportion of the corresponding average geodesic distance between observed data manifold and corresponding low-dimensional manifold to describe the magnification factor. This is a kind of simple and advantageous technique.

5   Applications of Average Geodesic Distance

For high dense data set, any point has a best neighborhood point. When any point is deleted, the average geodesic distance is hardly influenced. In sparse data set, some points have large distances with their neighborhood points. When one of these points is deleted, the average geodesic distance will have larger change. So we propose a manifold learning algorithm based on average geodesic distance which can distinguish whether the observed data set is highly dense or not. To the data set of satisfying requirement, the algorithm can come out the low-dimension presentation of all data points and a new presentation of magnification factor for analyzing the embedding relationship. The detailed steps of our algorithm are described as follows:

Step 1)   Initialize K, ρ and set $i = 1$, $g[1 \cdots n] = 0$, (where K denotes the number of the nearest neighborhood points of any point. ρ denotes the compact degree boundary of the observed data set. i denotes the i-th data point. The element $g[i]$ of the array $g[1 \cdots n]$ stores the average geodesic distance after deleting the i-th point. n denotes the number of points in data set).

Step 2)   To every $i = 1, ..., n$, delete the i-th data point from the observed data set. Then execute the following a-b-c steps:
    a.   Construct k-nearest neighborhood graph G.
    b.   Compute shortest paths between any two points. Then, get the geodesic distance between them and the geodesic connected graph G.
    c.   Compute the average geodesic distance $g[i]$ of the graph G .

Step 3)   To the full data set, construct the geodesic connected graph G and compute the average geodesic distance f by a-b-c steps in step 2. If there exists i and $|f - g[i]| > \rho$, the data set isn't compact and exit the program. Else run step 4.

Step 4)   For the full data set, compute low-dimensional embeddings D by manifold learning algorithm and compute the average geodesic distance f_d of the low-dimensional coordinates.

Step 5)   Compute the magnification factor f_d / f .

6   Experiment and Analysis

The proposed approach has been applied to two synthetic Swiss-roll data sets containing 1000 and 500 data points which are respectively chose the fore part from Swiss_roll_data.mat at http://isomap.stanford.edu/[2][3], named as A1-DATA-SET and A2-DATA-SET. A1-DATA-SET containing 1000 points is analyzed by the algorithm motioned above. After deleting a point from A1-DATA-SET, the average geodesic distance is calculated. Repeating 1000 times in the second step of the algorithm, the errors between these average geodesic distances and that of A1-DATA-SET change in the range of 0.1, as showed in Table 1. When the given compact degree of data set is more than $\rho = 0.1$, A1-DATA-SET is considered as a high dense data set. Under the condition of $\rho = 0.1$,

A2-DATA-SET containing 500 points is tested and the result shows that A2-data-set is considered as the relative sparse, as showed in Table 2. Of course, when sampling from the whole manifold, any point may be interfere the degree of the average geodesic distance depending on the density of data set and the number of data points isn't an absolute factor.

ISOMAP is a promising method because of the use of geodesic distances. ISOMAP should not be applied to the sparse data set, such as A2-DATA-SET tested above. To A2-DATA-SET, the K-nearest neighborhood graphs show in Fig. 2. As $K = 2$, the K-nearest graph has disconnected components in Fig.2(a). The K-nearest graph appears the obvious 'short-cut' phenomenon when $K \geq 3$, as shown in Fig.2(b). That is, any K value can't guarantee that the K-nearest neighborhood graph is full connectivity which is not obvious "short-cut" phenomenon. The proposed algorithm has still been applied to A1-DATA-SET. The results show that it is a high dense data set when the given dense degree of data set is more than 0.01. In the case of $K = 2$ and 3, the K-nearest neighborhood graphs aren't the full connected graphs. Their largest connected components contain respectively 185 and 985 data points. The K-nearest neighborhood graph is full connected graph when $K \geq 4$. Especially, for the 4-nearest neighborhood graph, any path of two points is accessible, and it doesn't appear obvious "short-cut" phenomenon in Fig. 3(a). The two-dimensional embedding which presents

Table 1. The change of average geodesic distance after deleted a point from A1-DATA-SET ($K = 5$)

The i-th Deleting	The Average Geodesic Distance $g[i]$	The Error $|f - g[i]|$
...	...	...
501	22.7426	0.0637
502	22.6614	0.0175
503	22.7115	0.0326
504	22.6621	0.0168
505	22.6746	0.0043
...	...	...

Table 2. The change of average geodesic distance after deleted a point from A2-DATA-SET ($K = 5$)

The i-th Deleting	The Average Geodesic Distance $g[i]$	The Error $|f - g[i]|$
...	...	...
201	17.5094	0.0153
202	17.4791	0.0150
203	17.4860	0.0081
204	17.4666	0.0275
205	17.1675	0.3266
...	...	...

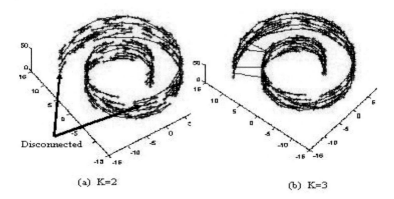

(a) K=2

(b) K=3

Fig. 2. The K-nearest neighborhood graph of A2-DATA-SET

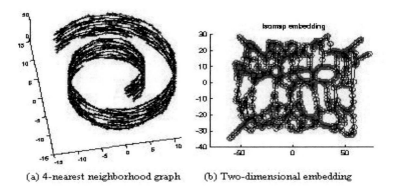

(a) 4-nearest neighborhood graph

(b) Two-dimensional embedding

Fig. 3. Experimental results of A1-DATA-SET

the geometric structure of A1-DATA-SET is showed in Fig. 3(b). In practical application, the value of K is chose as a minimize integer which can guarantees full access between any two points in K-nearest neighborhood graph. Finally, we calculate the average geodesic distance of original A1-DATA-SET and the corresponding distance of two-dimensional embedding. Their values are respectively 24.4536 and 19.7750. So, the magnification factor is equal to 0.8087. It shows that the scale of the embedding manifold is almost 80.87% while the relationship between the observed data points is preserved.

7   Conclusion

This paper proposes an algorithm based on the average geodesic distance as the preprocessing step of manifold learning. Moreover, for a high dense data set evaluated, we further utilize the average geodesic distance to quantitatively analyze the mapping relationship between the high dimensional manifold and the corresponding intrinsic low dimensional manifold in the known ISOMAP algorithm.

Finally, experimental results on two synthetic Swiss-roll data sets show that our method is feasible. In practical application such as pattern analysis, data mining and face recognition, we will generalize the novel algorithm and maybe further to expand the promising ISOMAP algorithm. In addition, the part technique may be used to pre-process the data set in LLE algorithm [5].

Acknowledgment

The authors would like to thank the anonymous reviewers for their comments and suggestions which greatly improved this paper. The work was supported by Scientific Research Fund of Sichuan Provincial Education Department of China (no. 07ZA121) and Research Fund of China West Normal University (no. 07A024).

References

1. Seung, H.S., Lee, D.: The Manifold Ways of Perception. Science 290, 2268–2269 (2000)
2. Balasubramanian, M., Schwartz, E.L., Tenenbaum, J.B., de Silva, V., Langford, J.C.: The Isomap Algorithm and Topological Stability. Science 295, 7a–7 (2002)
3. Tenenbaum, J.B., de Silva, V., Langford, J.C.: A global geometric framework for nonlinear dimensionality reduction. Science 290, 2319–2323 (2000)
4. Li, Y.: Building k Edge-disjoint Spanning Trees of Minimum Total Length for Isometric Data Embedding. IEEE Transactions on Pattern Analysis and Machine Intelligence 27, 1680–1683 (2005)
5. Roweis, S.T., Lawrance, K.S.: Nonlinear Dimensionality reduction by locally linear embedding. Science 290, 2323–2326 (2000)
6. He, L., Zhang, J.P., Zhou, Z.H.: Investigating Manifold Learning Algorithm Based on Magnification Factors and Principle Spread Directions. Chinese Journal of Computers 28, 2000–2009 (2005)
7. Lee, J.A., Lendasse, A., Verleysen, M.: Nonlinear Projection with Curvilinear Distances: Isomap versus Curvilinear Distance Analysis. Neurocomputing 57, 49–76 (2003)
8. Yeh, W.C., Lee, I.H., Wu, G., Wu, Y., Chang, E.Y.: Manifold Learning, A Promised Land or Work in Progress? In: IEEE International Conference on ICME 2005 (Multimedia and Expo), Amsterdam, pp. 1154–1157 (2005)
9. Costa, J.A., Hero, A.O.: Entropic Graphs for Manifold Learning. Signals. In: Systems and Computers, Conference Record of the Thirty-Seventh Asilomar Conference on n Signals, Systems and Computers. Asilomar Conference, pp. 316–320 (2003)
10. Costa, J.A., Hero, A.O.: Geodesic Entropic Graphs for Dimension and Entropy Estimation in Manifold Learning. IEEE Transactions on Acoustics, Speech, and Signal Processing 52, 2210–2221 (2004)

Fast Generation of Local Hasse Graphs for Learning from Structurally Connected Instances*

Wuming Pan

College of Computer Science, Sichuan University, Chengdu 610065, P.R. China
pan.wuming@gmail.com

Abstract. This paper introduces a variety of graph structures on data. The properties of the local graphs around an instance are studied. A fast approach for generating local graphs and classifying by structurally connected instances is then sketched out.

1 Introduction

In recent years, the use of graph representations has gained popularity in data mining and knowledge discovery[1][2]. These methods now have been successfully applied in many areas, such as graphical symbol recognition, character recognition, shape analysis, biometric person authentication by means of facial images and fingerprints, computer network monitoring, and Web document analysis [1][3][4].

Pan et al. [5] showed that a variety of graph structures can be easily generated from relational data. For real-valued data, these graphs are generated from order relations on each dimension. With these graphs, we can determine the "nearest neighbors" of an input vector by collecting directly connected instance of it, and then use the principle of structure similarity to classify input vectors. Once structures are generated from data, we can be free of some restrictions on original data. These restrictions include that data are relational or multi-relational, that data attributes are real value or categorical value, that there are domain knowledge about data or not, that there are missing values in data or not. In most data mining methods, these restrictions lead to different algorithms design.

But the computational complexity of the straightforward approach is very high in classifying new vectors. In this paper, we provide a faster algorithm. The rest of the paper is organized as follows. Section 2 introduces the definition of Hasse graph and its application in classification. Some properties of local Hasse graph are provided in Section 3. Section 4 presents a faster approach without the dimensional effect on computation to refine the intuitive approach. Section 5 concludes the research work of this paper.

---

* This work is partly supported by the Natural Science Foundation of Sichuan Province, China (05JY029-021-2).

G. Wang et al. (Eds.): RSKT 2008, LNAI 5009, pp. 548–555, 2008.

2   Hasse Graphs on Data

We consider the standard classification problem. Let $\langle \mathcal{D}, f, \mathcal{C} \rangle$ be the training data, where $\mathcal{D}$ is a collection of n-dimension real-valued vectors in $X_1 \times X_2 \times \ldots \times X_n$, $\mathcal{C}$ is the set of labels, and $f : \mathcal{D} \to \mathcal{C}$ is a function which assigns each vector in $\mathcal{D}$ a label in $\mathcal{C}$. And $\mathcal{C}$ contains m classes $C_j, j = 1, 2, \ldots, m$, where C_j is a nominal data. But there will be different instances d_1 and d_2 in $\mathcal{D}$ are the same vector $\langle x_1, x_2, \cdots, x_n \rangle$ in $X_1 \times X_2 \times \ldots \times X_n$, and $f(d_1)$ and $f(d_2)$ are equal or not equal in $\mathcal{C}$. To differentiate instances in $\mathcal{D}$, we add identity number to instances in $\mathcal{D}$ such that each instance is represented as $\langle k, x_1, x_2, \cdots, x_n \rangle$, where $k(=1, 2, \cdots, l)$ is the id of the instance, and l is the size of $\mathcal{D}$.

Now we can describe the process of generating structures from data. For each $X_i, i = 1, 2, \cdots, n$, we consider the ordered set $\mathbf{P}_i = \langle X_i, \geqslant \rangle$, where $\geqslant$ is the order relation on domain of X_i – the set of real numbers. The Cartesian product of two ordered sets $\mathbf{P} = \langle P, \geqslant \rangle$ and $\mathbf{P}' = \langle P', \geqslant \rangle$ is naturally an ordered set $\mathbf{P} \times \mathbf{P}' = \langle P \times P', \geqslant_{P \times P'} \rangle$, where $(P \times P')$ is the Cartesian product of P and P', and $(x_1, x_2) \geqslant_{P \times P'} (y_1, y_2)$ iff $x_1 \geqslant y_1$ and $x_2 \geqslant y_2$. Let $\mathbf{P}_i^* = \langle X_i, \leqslant \rangle$ be the dual ordered set of $\mathbf{P}_i = \langle X_i, \geqslant \rangle$, $\mathbf{P}_1$ and $\mathbf{P}_2$ be the order sets on X_i and X_2. Then there are four ordered sets, $\mathbf{P}_1 \times \mathbf{P}_2$, $\mathbf{P}_1^* \times \mathbf{P}_2$, $\mathbf{P}_1 \times \mathbf{P}_2^*$ and $\mathbf{P}_1^* \times \mathbf{P}_2^*$, on $X_1 \times X_2$. Similarly, let $\mathbf{P}_i^0 = \mathbf{P}_i = \langle X_i, \geqslant \rangle$, $\mathbf{P}_i^1 = \mathbf{P}_i^* = \langle X_i, \leqslant \rangle$, $i \in \{1, 2, \cdots, n\}$, there are 2^n order relations on $X_1 \times X_2 \times \ldots \times X_n$:

$$\mathbf{M}_j = \mathbf{P}_1^{j_1} \times \mathbf{P}_2^{j_2} \times \cdots \times \mathbf{P}_n^{j_n} = \langle X_1 \times X_2 \times \ldots \times X_n, \geqslant_j \rangle \tag{1}$$

where $j = 0, 1, 2, \cdots, 2^n - 1$, and for $i \in \{1, 2, \cdots, n\}$, j_i is either 0 or 1, counted from following equation:

$$j_i = \frac{(2^n + j) \bmod 2^i - (2^n + j) \bmod 2^{(i-1)}}{2^{(i-1)}} \tag{2}$$

This means that:

$$j = \sum_{i=1}^{n} j_i \cdot 2^{(i-1)} \tag{3}$$

and the binary representation of j is

$$\underbrace{j_n j_{n-1} \cdots j_2 j_1}_{n}$$

If we also write $\geqslant$ as $\geqslant^0$ and $\leqslant$ as $\geqslant^1$, then $(x_1, x_2, \cdots, x_n) \geqslant_j (y_1, y_2, \cdots, y_n)$ iff $x_i \geqslant^{j_i} y_i$, $i = 1, 2, \cdots, n$.

The order relation $\geqslant_j$ in $\mathbf{M}_j$ restricting on data set $\mathcal{D}$ may not lead to order relation on $\mathcal{D}$, for there may be two instances which have the same value – they are duplicate vectors but may be labeled differently. This breaks the anti-symmetric property of order relations. For $d_k \in \mathcal{D}$, $d_k = (k, x_1^k, x_2^k, \ldots, x_n^k)$, let $v_k = (x_1^k, x_2^k, \ldots, x_n^k)$, d_k is also written as (k, v_k).

Definition 1. *A graph* $\mathbf{H}_j^{\mathcal{D}} = \langle \mathcal{D}, E_j \rangle$ *is called the* Hasse graph *on the instances set* $\mathcal{D}$ *if and only if for any* $d_s, d_r \in \mathcal{D}$, $(d_s, d_r) \in E_j$, v_s *is a* **cover** *of* v_r *according to the order relation* $\geqslant_j$ *of ordered set* $\mathbf{M}_j$, *or* $v_s = v_r$.

If $(d_s, d_r) \in E_j$, we also write $d_s \succcurlyeq_j d_r$. The *local graph* of an instance $d \in \mathcal{D}$ is considered as the subgraph of Hasse graph $\mathbf{H}_j^{\mathcal{D}} = \langle \mathcal{D}, E_j \rangle$, in which only those instances connected with d is contained, and we define it as

$$\mathbf{Loc}_j^{\mathcal{D}}(d) = \langle loc_j(d), E_j(d) \rangle. \tag{4}$$

where

$$loc_j(d) = \{d_s \,|\, (d_s, d) \in E_j \text{ or } (d, d_s) \in E_j \},$$

$$E_j(d) = \{(d_s, d_r) \,|\, d_s = d \text{ or } d_r = d \}.$$

There are 2^n local structures of the query instance d in $\mathcal{D} \cup \{d\}$. Because of dual of the order relation, we have $loc_j(d) = loc_{2^n - 1 - j}(d)$.

Let $\mathcal{D}$ be the training data set, C_i, $i = 1, 2, \cdots, m$, be the class labels, and $\lambda_j(C_i, d)$ be the number of the instances in $loc_j(d) - \{d\}$ whose class label is C_i. The **support** of assigning the query instance d to class C_i, according to data set $\mathcal{D}$, is defined as

$$\mathbf{support}(C_i, d) = \frac{\sum_{j=0}^{2^n - 1} \lambda_j(C_i, d)}{\sum_{j=0}^{2^n - 1} |loc_j(d) - \{d\}|}. \tag{5}$$

The conceptually straightforward approach to predict the class label of d is to assign the query instance d the class label C^d such that **support**(C^d, d) is the maximum of all **support**(C_i, d), $i = 1, 2, \cdots, m$. We use $LGM(\mathcal{D}, d)$ to denote this predicted class label of d, then

$$LGM(\mathcal{D}, d) = \underset{C_i \in \{C_1, C_2, \dots, C_m\}}{\arg\max} \mathbf{support}(C_i, d) \tag{6}$$

3   The Properties of the Local Graphs

We can easily see that computational complexity of the above approach is very high in classification of new vectors. For a feature space of n dimensions, there are 2^n local graphs of the query instance d. Because of the dual of the structures, we need to consider 2^{n-1} local structures. The straight forward approach is to search 2^{n-1} local structures from Hasse graphs $\mathbf{H}_j^{\mathcal{D} \cup \{d\}}$, $j = 0, 1, 2, \cdots, 2^{n-1} - 1$. Hence, the computational complexity grows exponentially with the dimensionality of the feature space. To consider faster algorithms we study the properties of the local graphs first.

Definition 2. *For a query instance d, its* total local points set *is*

$$loc(d) = \bigcup_{0}^{2^n-1} loc_j(d) = \bigcup_{0}^{2^n-1} loc_j(d),$$

where $loc_j(d)$ is the set of vertices of the local graph around d in $\mathbf{H}_j^{\mathcal{D}\cup\{d\}}$.

Proposition 1. *Let $d' = (k', x_1', x_2', \ldots, x_n')$, $d = (k, x_1, x_2, \ldots, x_n)$ and*

$$J\left(d' - d\right) = \sum_{i=1}^{n} 2^{i-1} \mathrm{sgn}^* \left(x_i' - x_i\right), \tag{7}$$

where

$$\mathrm{sgn}^* \left(x\right) = \begin{cases} 0, x \geqslant 0 \\ 1, x < 0 \end{cases}.$$

We have $d' \geqslant_{J(d'-d)} d$.

Proof. For $\mathbf{M}_j = \mathbf{P}_1^{j_1} \times \mathbf{P}_2^{j_2} \times \cdots \times \mathbf{P}_n^{j_n}$, by Equation (3), we have

$$j = \sum_{i=1}^{n} j_i \cdot 2^{(i-1)}.$$

Comparing this equation with Equation (7), we have

$$\mathbf{M}_{J(d'-d)} = \mathbf{P}_1^{\mathrm{sgn}^*\left(x_1'-x_1\right)} \times \mathbf{P}_2^{\mathrm{sgn}^*\left(x_2'-x_2\right)} \times \cdots \times \mathbf{P}_n^{\mathrm{sgn}^*\left(x_n'-x_n\right)}.$$

Hence $d' \geqslant_{J(d'-d)} d$.

Theorem 1. (*Eliminable instances theorem*) *If $d' \notin loc_{J(d'-d)}(d)$, then for any j, $d' \notin loc_j(d)$. Conversely, if for some j, $d' \in loc_j(d)$, then $d' \in loc_{J(d'-d)}(d)$.*

Proof. If $d' \notin loc_{J(d'-d)}(d)$, then $d' \succcurlyeq_{J(d'-d)} d$ is not true. By Definition 1 of Hasse graphs, there must be a path from d' to d in Hasse graph $\mathbf{H}_{J(d'-d)}^{\mathcal{D}}$, that is, there are $d^1, d^2, \cdots, d^r$ such that

$$d' \succ_{J(d'-d)} d^1 \succ_{J(d'-d)} d^2 \succ_{J(d'-d)} \cdots \succ_{J(d'-d)} d^r \succ_{J(d'-d)} d.$$

Namely,

$$d^r \in loc_{J(d'-d)}(d),$$

and

$$d^r \neq d'.$$

Suppose for some j, $d' \in loc_j(d)$, then $d' \succ_j d$ or $d \succ_j d'$. If $d' \succ_j d$, let

$$j = \sum_{i=1}^{n} j_i \cdot 2^{(i-1)},$$

and

$$J\left(d'-d\right) = \sum_{i=1}^{n} j_i^a \cdot 2^{(i-1)}.$$

Let $\mathbf{M}_j = \mathbf{P}_1^{j_1} \times \mathbf{P}_2^{j_2} \times \cdots \times \mathbf{P}_n^{j_n}$, and

$$\mathbf{M}_{J(d'-d)} = \mathbf{P}_1^{j_1^a} \times \mathbf{P}_2^{j_2^a} \times \cdots \times \mathbf{P}_n^{j_n^a},$$

we have

$$x_i' \geqslant^{j_i^a} x_i^r \geqslant^{j_i^a} x_i$$

and

$$x_i' \geqslant^{j_i} x_i.$$

Then if for some i, $j_i \neq j_i^a$, there must be

$$x_i' = x_i^r = x_i.$$

Hence, for any i,

$$x_i' \geqslant^{j_i} x_i^r \geqslant^{j_i} x_i,$$

is also true. Because $d' \succcurlyeq_j d$, This says that $v' = v^r$ or $v^r = v$. Similarly, for any $s \in \{1, 2, \cdots, r\}$, we have that $v' = v^s$ or $v^r = v$. Finally, we obtain that $\{d', d^1, d^2, \cdots, d^r\} \subseteq loc_{J(d'-d)}(d)$, this is contrary to the condition $d' \notin loc_{J(d'-d)}(d)$. If $d \succcurlyeq_j d'$, it similarly leads to the contrary. Hence, the proposition holds.

Corollary 1. *Let*

$$r\left(d_s - d_r\right) = \sum_{i=1}^{n} \left(1 - |\mathrm{sgn}\left(x_i^s - x_i^r\right)|\right), \tag{8}$$

where

$$\mathrm{sgn}\left(x\right) = \begin{cases} 1, x > 0 \\ 0, x = 0 \\ -1, x < 0 \end{cases},$$

If for some $j^a \neq j^b$, $d_s \geqslant_{j^a} d_r$ and $d_s \geqslant_{j^b} d_r$ are both true, then $r\left(d_s - d_r\right) \neq 0$ hold.

Corollary 2. *Notation being as Proposition 1, if for some $i^\triangle$, $x_{i^\triangle}' = x_{i^\triangle}$, let*

$$J^\triangle\left(d'-d\right) = \left(\sum_{i=1,2,\cdots,n;\ i \neq i^\triangle} 2^{i-1} \mathrm{sgn}^*\left(x_i' - x_i\right) + 1\right) + 2^{i^\triangle - 1}. \tag{9}$$

Then $d' \geqslant_{J^\triangle(d'-d)} d$. Furthermore, $d' \geqslant_{J(d'-d)} d$, $d' \geqslant_{J^\triangle(d'-d)} d$, $d' \geqslant_{2^n - J(d'-d)+1} d$ and $d' \geqslant_{2^n - J^\triangle(d'-d)+1} d$ are all true.

Theorem 2. (Disjoint local graphs theorem) *Let*

$$d' = (k', x_1', x_2', \ldots, x_n') \in loc(d),$$

and

$$R(d' - d) = \begin{cases} r(d' - d) + 1, & r(d' - d) \neq n \\ n, & r(d' - d) = n \end{cases}$$

Then there are a set $\left\{ j^1, j^2, \cdots, j^{2^{R(d'-d)}} \right\} \subseteq \{0, 1, 2, \cdots, 2^n - 1\}$ *such that* $d' \in loc_{j^t}(d)$, $t = 1, 2, \cdots, 2^{R(d'-d)}$, *and for any* $j' \notin \left\{ j^1, j^2, \cdots, j^{2^{R(d'-d)}} \right\}$, $d' \notin loc_{j'}(d)$.

Proof. We know there must be some j such that $d' \in loc_j(d)$. Theorem 1 says that for any j, if $d' \in loc_j(d)$, then $d' \in loc_{J(d'-d)}(d)$. Suppose we have listed all $i_1^*, i_2^*, \cdots, i_s^*$ such that $\text{sgn}\left(x_{i_u^*}' - x_{i_u^*} \right) = 0$, $u = 1, 2, \cdots, s$. Thus by Corollary 2 we can **enumerate** all $j^1, j^2, \cdots, j^{2^{R(d'-d)}}$ such that $d' \in loc_{j^t}(d)$, $t = 1, 2, \cdots, 2^{R(d'-d)}$. Hence there are not any j' such that $j' \notin \left\{ j^1, j^2, \cdots, j^{2^{R(d'-d)}} \right\}$ and $d' \in loc_{j'}(d)$.

Theorem 3. (Support counting in total local structure points set) *Let* $\langle \mathcal{D}, f, \mathcal{C} \rangle$ *be the training data, and* $\mathcal{C}$ *contains* m *labels* $C_i \in \mathcal{C}$, $i = 1, 2, \ldots, m$; $r(d' - d)$ *is obtained by equation 8. We have*

$$\text{support}(C_i, d) = \frac{\sum\limits_{d' \in loc(d), d' \neq d, f(d') = C_i} 2^{R(d'-d)}}{\sum\limits_{d' \in loc(d), d' \neq d} 2^{R(d'-d)}}. \tag{10}$$

Proof. By Proposition 2, each $d' \in loc(d)$ will only appear in $2^{R(d'-d)}$ local structures. By Equation (5), Equation (10) holds.

4   Faster Generation of Local Hasse Graphs

In this section, we will present a faster algorithm for classification and do not need to search 2^{n-1} times on $\mathcal{D}$ when classifying a new instance. By virtues of above propositions, corollaries and theorems, we can reduce computational cost of the presented method through using following tactics:

1. Instead of searching $\mathbf{H}_1^{\mathcal{P}}$ to $\mathbf{H}_{2^n}^{\mathcal{D}}$, we only search local graphs $\mathbf{Loc}_{J(d'-d)}(d)$ of input instance d in Hasse graphs $\mathbf{H}_{J(d'-d)}^{\mathcal{D}}$ for all d' in $\mathcal{D}$. By the *eliminable instances theorem*, if for some j, $d' \in loc_j(d)$, then $d' \in loc_{J(d'-d)}(d)$. So we easily know

$$loc(d) = \bigcup_{d' \in \mathcal{D}} loc_{J(d'-d)}(d).$$

Table 1. The Training Data

$\mathcal{D}$	d_1	d_2	d_3	d_4	d_5	d_6	d_7	d_8	d_9	d_{10}	d_{11}
x value	0.2	−0.2	−2	1	0.5	−0.6	−0.7	−1	3	−2	−0.3
y value	2	3.5	2	1	0.5	0.2	0.2	−1	−2	−2.5	−3
label in $\mathcal{C}$	C_1	C_2	C_1	C_1	C_1	C_1	C_2	C_2	C_2	C_2	C_1

2. By the *eliminable instances theorem*, for processing each $\mathbf{H}^{\mathcal{D}}_{J(d'-d)}$, we eliminate all $d^* \notin loc_{J(d'-d)}(d)$ such that $d^* \geqslant_{J(d'-d)} d$ or $d \geqslant_{J(d'-d)} d^*$ since it is not included in $loc(d)$.

3. Once we get $loc(d)$, we use Equation (10) to evaluate **support** (C_i, d), $i = 1, 2, \ldots, m$.

Example 1. Assume the feature space is R^2 and the training data $\langle \mathcal{D}, f, \mathcal{C} \rangle$ is shown in Table 1, and the input point is $d(0, 0)$.

For this case, there are four Hasse graphs, and the input point $d(0, 0)$ has 4 local structures. Initially start with d_1. That is, we search $\mathbf{Loc}_{J(d_1-d)}(d)$ firstly. The local points of $(0, 0)$ are d_1, d_5, d_7, d_{12}. Because $d_4 \geqslant_1 d$, $d \geqslant_1 d_8$ and $d \geqslant_1 d_{10}$, then d_4, d_8 and d_{10} are not considered for another search.

After first search, except d_2, d_3, d_6 and d_9, other elements are not considered for another search as just mentioned. Now we search $\mathbf{Loc}_{J(d_2-d)}(d)$ from $\mathbf{H}^{\{d_2,d_3,d_6,d_9,d\}}_{J(d_2-d)}$. The local points of $(0, 0)$ is d_2, d_6, d_9. Because $d_3 \geqslant_2 d$, then d_3 is eliminated. Now d_2, d_3, d_6 and d_9 are deleted. Hence there are no instance to search. The process finishes.

According to Equation (10),

$$\mathbf{Support}(C_1, d) = \frac{\displaystyle\sum_{d' \in \{d_1,d_5,d_7,d_{12},d_2,d_6,d_9\}, f(d')=C_1} 2^{R(d'-d)}}{\displaystyle\sum_{d' \in \{d_1,d_5,d_7,d_{11},d_2,d_6,d_9\}} 2^{R(d'-d)}} = 4/7$$

and

$$\mathbf{Support}(C_2, d) = \frac{\displaystyle\sum_{d' \in \{d_1,d_5,d_7,d_{12},d_2,d_6,d_9\}, f(d')=C_2} 2^{R(d'-d)}}{\displaystyle\sum_{d' \in \{d_1,d_5,d_7,d_{11},d_2,d_6,d_9\}} 2^{R(d'-d)}} = 3/7$$

Since $\mathbf{Support}(C_1, d) > \mathbf{Support}(C_2, d)$, $d(0, 0)$ will be labeled as C_1.

Remark 1. In the most naive approach, k-nearest neighbor algorithm, we inspect each stored point in turn, calculate its Euclidean distance to the input instance d, retain the identity only of the current closest one. The complexity of calculating each distance is $O(n)$ ($l = |\mathcal{D}|$), and thus the complexity of this search is $O(lkn)$. In above algorithm we need to search at most l Hasse graphs, and each local graph contains at most l nodes, therefore the time complexity of our algorithm is $O\left(l^3 n\right)$. But the elements in $loc\,(d)$ is always fewer than l. Moreover, through eliminate some elements in searching process, we may only need search Hasse graphs less than l. Therefore, the complexity of proposed algorithm may match the naive k-nearest neighbor algorithm in solving real problems, though the complexity of our algorithm seems worse than the naive $k-$nearest neighbor algorithm.

5   Conclusions

In this paper, a variety of graph structures on data was introduced and the properties of the local graphs around an instance are studied. A fast approach for generating local graphs and classifying according to structurally connected instances is then sketched out. Our future work is to study how to preconstruct some data structures on training data beforehand and use the sampling technique to eliminate the complexity of the proposed algorithm. Another future work is to apply this approach to data mining tasks other than classification on real-valued data.

References

1. Cook, D., Holder, L. (eds.): Mining Graph Data. John Wiley & Sons, New Jersey (2006)
2. Neuhaus, M., Bunke, H.: Self-organizing Maps for Learning the Edit Costs in Graph Matching. IEEE Trans. on Systems, Man and Cybernetics Part B 35, 503–514 (2005)
3. Pawlak, Z.: Some Issues on Rough Sets. In: Peters, J.F., Skowron, A., Grzymała-Busse, J.W., Kostek, B.z., Świniarski, R.W., Szczuka, M. (eds.) Transactions on Rough Sets I. LNCS, vol. 3100, pp. 1–58. Springer, Heidelberg (2004)
4. Carpineto, C., Romano, G.: Concept Data Analysis Theory and Applications. John Wiley & Sons, Chichester (2004)
5. Pan, W., Li, T.: From Analogy Reasoning to Instances Based Learning. In: Da Ruan (ed.) Applied Artificial Intelligence (FLINS 2006), pp. 177–184. World Scientific Publishing, Singapore (2006)

Approximation and Prediction of Wages Based on Granular Neural Network

Milan Marček[1] and Dušan Marček[2]

[1] Faculty of Philosophy and Science, Silesian University, 746 01 0pava, Czech Republic &
MEDIS Nitra, Ltd., Pri Dobrotke 659/81, 949 01 Nitra-Dražovce, Slovak Republic
marcek@fria.utc.sk
[2] Faculty of Philosophy and Science, Silesian University, 746 01 0pava, Czech Republic &
Faculty of Management Science and Informatics, University of Zilina
010 26 Zilina, Slovak Republic
dusan.marcek@fpf.slu.cz, dusan.marcek@fri.uniza.sk

Abstract. This article offers a detailed computational algorithm used in that type of neural networks, extends their applications to fit and predict the data of wages time series, conducts experiments and indicates the gain of granular neural networks, specifically conducting experimentation using the classical (statistical) or econometric methods and conventional/soft RBF neural networks. Results are analysed and opportunities for future research are suggested.

Keywords: Probabilistic time-series models, Fuzzy system, Classic and soft RBF network, Cloud models, Granular computing.

1 Introduction

As mentioned in Liao et al. [5] and Zhang et al. [11] neural networks document competitive performance on a larger number of time series, indicating the use of increased computational power to automate NN forecasting on a scale suitable for automatic forecasting.

The scope of the paper is confined to some statistical forecasting methods and methods based on granular computing. According to Zadeh [10] granulation plays an essential role in human cognition and has a position of centrality in both granular computing and rough set theory. The exploitation of granular concept for forecasting purposes can be found in many works. In Yu et al. [9] a method is proposed based on information granulation model and the granular discretization method to form fuzzy rules from granular time series for a fuzzy forecasting system. In Yang et al. [8] methods of time series prediction based on cloud methods and on the different time (short term and long term granularities) were presented and described respectively. Lastly in Marcek et al. [6] a new approach of function estimation is shown for time series model of daily sales by means of a granular RBF neural network.

In comparison with [6], this paper extends the application of granular network to fit and predict the quarterly data of wages time series, gives new calculating algorithm for the specific granular network and compares obtained results with those obtained

G. Wang et al. (Eds.): RSKT 2008, LNAI 5009, pp. 556–563, 2008.

using statistical procedures. The organization of the paper involves 5 sections. Section 2 outlines necessary prerequisites for predictors based on the fuzzy system and RBF neural network approach. In Section 3, we briefly offer a granular extension of RBF neural networks. In Section 4 we give the complete algorithm for weights updating in the granular RBF network and for calculating of the output values and the statistical summary measures of a model´s forecast accuracy. In Section 5 concluding remarks and opportunities for future work are given.

2   Fuzzy and Neural Function Estimators

In function estimation of complex input-output systems, the fuzzy systems and neural networks estimate a function without requiring a mathematical description of how output functionally depends on the output. The most popular centroidal defuzzification technique uses all the information in the fuzzy distribution to compute the crisp y value as the centroid $\tilde{y}$. When the output membership functions are singletons, then, in the case of an $\Re^k \to \Re$ function, the formula for the centroid calculation is

$$\tilde{y} = \sum_{j=1}^{n} y_j \mu_j(x) / \sum_{j=1}^{n} \mu_j(x) \tag{1}$$

where y_j stands for the centre of gravidity of the jth output singleton, the notation μ is used for a membership function and n denotes the number of rules. Next, we will show, how to obtain fuzzy rules and how to determine the weights w_i for a fuzzy system using RBF networks.

In Fig 1 (a), the classic RBF neural network structure and in Fig. 1 (b) its soft or fuzzy logic version is shown. The output layer neuron is linear and has a scalar output given by

$$\hat{y}_t = G(\mathbf{x}_t, \mathbf{c}, \mathbf{v}) = \sum_{j=1}^{s} v_{j,t} \psi_2(\mathbf{x}_t, \mathbf{w}_j) = \sum_{j=1}^{s} v_j o_{j,t}, \qquad t = 1, 2, ..., N \tag{2}$$

where N is the size of data samples, s denotes the number of the hidden layer neurons, v_j are the trainable weights connecting the component of the output vector $\mathbf{o}$. The hidden layer neurons receive the Euclidian distances $(\|\mathbf{x} - \mathbf{c}_j\|)$ and compute the scalar values $o_{j,t}$ of the Gaussian function $\psi_2(\mathbf{x}_t, \mathbf{c}_j)$ that form the hidden layer output vector $\mathbf{o}_t$, where $\mathbf{x}_t$ is a k-dimensional neural input vector, $\mathbf{w}_j$ represents the hidden layer weights, ψ_2 are radial basis (Gaussian) activation functions. Note that for an RBF network, the hidden layer weights $\mathbf{w}_j$ represent the centres $\mathbf{c}_j$ of activation functions ψ_2.

A serious problem is how to determine the number of hidden layer (RBF) neurons. The most used selection method is to preprocess training (input) data by some clustering algorithm. After choosing the cluster centres, the shape parameters σ_j must be

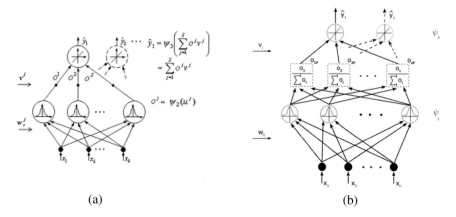

(a)   (b)

Fig. 1. Classic (a) and fuzzy logic (soft) (b) RBF neural network architecture

determined. These parameters express an overlapping measure of basis functions. For Gaussians, the standard deviations σ_j can be selected, i.e. $\sigma_j \sim \Delta c$, where Δc denotes the average distance among the centres.

As mentioned in [3, 6], if in RBF neural networks, the scalar output values $o_{j,t}$ from the hidden layer will be normalised, where the normalisation means that the sum of the outputs from the hidden layer is equal to 1, then the RBF network will compute the "normalised" output data set $\hat{y}_t$ as follows

$$\hat{y}_t = G(\mathbf{x}_t, \mathbf{c}, \mathbf{v}) = \sum_{j=1}^{s} v_{j,t} \frac{o_{j,t}}{\sum_{j=1}^{s} o_{j,t}} = \sum_{j=1}^{s} v_{j,t} \frac{\psi_2(x_t, c_j)}{\sum_{j=1}^{s} \psi_2(x_t, c_j)}, \, t = 1, 2, ..., N. \tag{3}$$

The graphical representation of the form of RBF neural networks which produce the output values according to the formula (3) is shown in Fig. 1 (b).

The frequently used learning technique uses clustering to find a set of centres which more accurately reflect the distribution of the data points. For example by using K-means clustering algorithm, the member of K centres must be decided in advances. After choosing the centres **w**, the standard deviations σ_j can be selected as $\sigma_j \sim \Delta c_j$ where c_j denotes the average distance among the centres w_j. To train the weights v_j, the first-order gradient procedure is used. These weights can be adapted by the error back-propagation algorithm. In this case, the weight update is particularly simple. If the estimated output for the single output neuron is $\hat{y}_t$, and the correct output should be y_t , then the error e_t is given by $e_t = y_t - \hat{y}_t$ and the learning rule has the form

$$v_{j,t} \leftarrow v_{j,t} + \eta \, o_{j,t} \, e_t, \qquad j = 1, 2, ..., s; \quad t = 1, 2, N \tag{4}$$

where the term η is a constant called the learning rate, $o_{j,t}$ is the normalised output signal from the hidden layer. Typically, the updating process is divided into epochs. Each epoch involves updating all the weights for all the examples.

3   Granular RBF Network

As we have already seen the goal of learning is to find a smooth mapping from $\mathbf{x}$ to $\mathbf{y}$ which captures the underlying systematic aspects of the data, without fitting noise or uncertainty on the data. If we now, in the soft RBF neural network, replace the standard Gaussian activation (membership) function of RBF neurons with functions based on the normal cloud concept [2], then the inherent uncertainty and randomness of the data are simultaneously included into estimation process. Then, in the case of soft RBF network, the Gaussian membership function $\psi_2(./.)$ in Eq. (3) has the form

$$\psi_2(x_t, c_j) = \exp\left[-(\mathbf{x}_t - E(\mathbf{x}_j)/2(En')^2\right] = \exp\left[-(\mathbf{x}_t - c_j)/2(En')^2\right] \tag{5}$$

where En' is a normally distributed random number with mean En (the entropy as the uncertainty measurement of the qualitative concept) and standard deviation He (the hyper entropy) which represents the uncertain degree of entropy, E is the expectation operator (for more details see [6]).

4   Experiment Design and Results

To illustrate the statistical or econometric modeling methodology, consider the quarterly wages time readings $\{y_t\}$ of the Slovak economy. The data were collected for the period January 1, 1991 to December 31, 2006 which provides total of 64 observations (displayed in Fig. 2 (a)). This time series shows an increasing trend with apparent periodic structure. To build a forecast model, we define the sample period for analysis $y_1, ..., y_{64}$, i.e. the training data set denoted as A, and the ex ante forecast period or the validation data set denoted as E.

Firstly, in the case of statistical modeling we shall present the model based on Box-Jenkins methodology [1]. Experimenting with this method, the following reasonable model formulation was found (B-J model, ARMA(1, 3) process)

$$y_t = -0.0016557 - 0.4567y_{t-1} + \varepsilon_t + 0.90516\varepsilon_{t-1} + 0.58768\varepsilon_{t-2} + 0.36497\varepsilon_{t-3};$$
$$MSE_A = 0.014, \ MSE_E = 0.048 \tag{6}$$

The forecast values of the ARMA(1, 3) model are seen in Fig. 2 (b). Secondly, the estimated appropriate econometric model based on the economic theory [7] has the following form

$$y_t = 0.239347 + 1.04044 \ y_{t-4} \qquad MSE_A = 0.0026, \ MSE_E = 0.0033 \tag{7}$$

where ε_t is the white noise disturbance term.

In the classic and soft (fuzzy logic) RBF neural network representation of model (7), the non-linear function was estimated according to the expressions (2) and (3). The fuzzy logic RBF neural network was extended towards estimation with (a priori known) noise levels of the entropy. We select, for practical reasons, that the noise level is a multiple, say 0.015 of entropy. Then, the non-linear input – output approximation function was estimated according to the formula (3) by substituting the

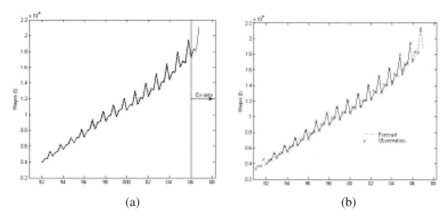

(a)   (b)

Fig. 2. Nominal average wages (a) (January 1991 - December 2006) and forecasts of wages data - ARMA(1,3) model (b)

Gaussian function $\psi_2(x_i, c_j)$ with Eq. (5). In Table 1, we give the achieved results of approximation ability in dependence on various number of RBF neurons. The mean square error (MSE$_A$) was used to measure the approximation ability.

The detailed computational algorithm for the MSE$_A$ values in Table 1, the weight update rule for the granular network and detailed computational algorithm used for this type of neural networks are shown in Appendix. The mean (centre) and standard deviation of clusters (RBF neurons) are computed using K-means algorithm. The accuracy of our one quarter forecasts is presented in Table 2.

Comparing both approaches, i.e. econometric model and model based on the RBF network we see that both approaches give approximately identical results. As shown in Table 1, models that generate the "best" MSE$_A$´s are soft RBF networks.

The accuracy of our one quarter forecasts is presented in Table 2. As can be seen, RBF networks have worse forecasting power than econometric model. We show that too many parameters results in overfitting, i.e. a curve fitted with to many parameters follows all the small fluctuations, but is poor for generalisation.

Table 1. Approximation results of various RBF´s networks related to the different number of clusters (RBF neurons)

Neural network architecture	Classic RBF network	Soft RBF network	Granular network
Number of RBF neurons	RBF network representation for model (7)		
		MSE$_A$	
3	0.202	0.079	0.018
5	0.103	0.018	0.022
10	0.097	0.002	0.007
15	0.031	0.009	0.009

Table 2. Ex-ante prediction results of various RBF´s networks related to the different number of clusters (RBF neurons)

Neural network architecture	Classic RBF network	Soft RBF network	Granular network
Number of RBF neurons	RBF network representation for model (7)		
	MSE$_E$		
3	2.484	0.380	0.018
5	2.111	0.299	0.008
10	1.688	1.878	0.123
15	1.799	0.180	0.262

5 Conclusion

To approximate the input-output function of a economic process, the RBF neural network approach was applied on the quarterly data of average nominal wages of the Slovak economy and compared with an approach based on the statistical procedures. For the sake of approximation abilities we evaluated 24 models. Two models were based on statistical (econometric) approach, and 24 models are based on the neural (fuzzy logic) methodology. Using the disposable data a very appropriate model is the soft RBF network with activation functions based on the granular concept. It is also interesting to note that the most computationally intensive models, the model based on the Box-Jenkins methodology, is newer considered "best". The econometric model (8) apparently ignores some of the example points, but best fits out-of training data set (forecasts).

In our view, this initial research step is only the beginning. Much of the cloud RBF neural networks, we still not understand. For instance, the empirical results suggest that the granular network is indeed capable of discovering the basic structures underlying a set of noisy data. But how precisely does this relate to its sensitiveness to expectation when the number of clusters is increased. We were expecting that the accuracy of ex post forecasts increases when the number of clusters (RBF neurons) increases. This is not truth for Gaussian RBF network with normal cloud concept and soft RBF one with 15 RBF neurons. Does it means overfiting? Can we find minimal number of RBF neurons (i.e. minimal model complexity) that is able to represent real deterministic structure? This is generally a very hard problem. We shall continue our research in this direction.

Acknowledgement

The authors would like to thank the anonymous reviewers for their insightful comments and helpful suggestions that have improved the paper greatly. This research has been supported by the grants GAČR 402/08/0022 and VEGA 1/0024/08.

References

1. Box, G.E.P., Jenkins, G.M.: Time Series Analysis, Forecasting and Control. Holden-Day, San Francisco. CA (1970)
2. Liu, C.Y., Li, D.Y., Du, Y., Han, X.: Normal Cloud Models and Their Interpretation. In: 11th IFSA World Congress, Beijing, pp. 1540–1543. Springer, Heidelberg (2005)
3. Kecman, V.: Learning and Soft Computing: Support Vector Machines, Neural Networks, and Fuzzy logic Models. The MIT Press, Massachusetts (2001)
4. Kosko, B.: Neural Networks and Fuzzy Systems a Dynamic Approach to Machine Intelligence. Prentice-Hall, Englewood Cliffs (1992)
5. Liao, K.P., Fildes, R.: The Accuracy of a Procedural Approach to Specifying feedforward Neural Networks for Forecasting. Computers & Operations Research 32, 2151–2169 (2005)
6. Marcek, M., Marcek, D.: RBF Neural Network Implementation of Fuzzy Systems: Application to Time Series Modeling. In: An, A., Stefanowski, J., Ramanna, S., Butz, C.J., Pedrycz, W., Wang, G. (eds.) RSFDGrC 2007. LNCS (LNAI), vol. 4482, pp. 500–507. Springer, Heidelberg (2007)
7. Palenik, V., Bors, L., Kvetan, V., Vokoun, J.: Costruction and Verification of Macroeconomic Model ISW97q3. Journal of Economics 46, 228–466 (1998)
8. Yang, F., Han, P., Li, D.: Uncertain Time Series Prediction Based on Different Time Granularity. In: 11th IFSA World Congress, Beijing, pp. 1524–1528. Springer, Heidelberg (2005)
9. Yu, F., Pedricz, W., Yuan, J.: Finding Fuzzy Rules from Ganular Time Series. In: 11th IFSA World Congress, Beijing, pp. 1731–1735. Springer, Heidelberg (2005)
10. Zadeh, L.A.: Granular Computing and Rough Sets Theory. In: JRS 2007. Infobright, York University Toronto, May 14-16, Keynote Talk (2007)
11. Zhang, G.P., Qi, M.: Neural Network Forecasting for Seasonal and Trend Time Series. European J. of Operational Research 160, 501–514 (2005)

Appendix

The algorithm for updating weights in the granular RBF network and for calculating statistical summary measures (MSE) from Table 1.

function MSE(*granular_RBF_network, examples, η, s, c_j, σ_j, Hej*) **returns**
   a network with the *MSE* value for Table 1
   inputs: *granular_RBF_network*, *MSE* a Gaussian soft RBF network with normal cloud concept
      examples, a set of *N* input/output observed data pairs: x, y
      η, the learning rate
      s, the number of clusters (RBF neurons)
      c_j, the center of the *j*-th cluster, $j = 1, 2, ..., s$
      σ_j, the standard deviation *j*-th cluster, $j = 1, 2, ..., s$
      He_j, the hyper entropy
   - Initialize weights: v_j, $j = 1, 2, ..., s$ leading to the output neuron.
   - Initialize the learning rate: η.
   - Initialize the input values: s, c_j, σ_j, He_j (see text for details)

repeat

    $MSE \leftarrow 0$

    for each *example* x **in** examples **do**

        /* Generate normally distributed random numbers He'_j with the means σ_j and the standard deviations He_j */

        $He'_j \leftarrow$ RUN-NORMAL-RANDOM-GENERATOR(σ_j, He_j), $j = 1, 2, ..., s$

        /* Calculate the outputs from the RBF neurons */

        $o_j \leftarrow \psi_2(x, c_j)$, $\quad j = 1, 2, ..., s$

        /* where ψ_2 is the Gaussian function: $\psi_2 = \exp(-(x - c_j)/2He'_j))$ */

        /* Calculate the normalized outputs $o_j^{(N)}$ */

$$o_j^{(N)} \leftarrow o_j / \sum_{j=1}^{s} o_j, \quad j = 1, 2, ..., s$$

        /* Calculate the output $\hat{y}$ from the output neuron */

$$\hat{y} \leftarrow \sum_{j=1}^{s} v_j o_j^{(N)}$$

        /* Calculate the output layer neuron's error e */

        $e \leftarrow y - \hat{y}$

        /* Update the output layer weight v_j */

$$v_j \leftarrow v_j + o_j^{(N)} e, \quad j = 1, 2, ..., s$$

    end

        /* Calculate the mean square error */

    for each *example* x, y **in** examples **do**

$$MSE = MSE + [y - v_j \sum_{j=1}^{s} \exp(-(x - c_j)/2He'_j)]^2$$

    end

  $MSE = MSE/N$

until *MSE* has converged

return *MSE, network*

Kernel Matching Reduction Algorithms for Classification

Jianwu Li and Xiaocheng Deng

Beijing Key Lab of Intelligent Information, School of Computer Science and
Technology, Beijing Institute of Technology, Beijing 100081, China
jianwuli@hotmail.com

Abstract. Inspired by kernel matching pursuit (KMP) and support vector machines (SVMs), we propose a novel classification algorithm: kernel matching reduction algorithm (KMRA). This method selects all training examples to construct a kernel-based functions dictionary. Then redundant functions are removed iteratively from the dictionary, according to their weights magnitudes, which are determined by linear support vector machines (SVMs). During the reduction process, the parameters of the functions in the dictionary can be adjusted dynamically. Similarities and differences between KMRA and several other machine learning algorithms are also addressed. Experimental results show KMRA can have sparser solutions than SVMs, and can still obtain comparable classification accuracies to SVMs.

Keywords: Kernel matching reduction algorithms, Kernel matching pursuit, Support vector machines, Radial basis function neural networks.

1 Introduction

Kernel-based pattern classification techniques have been widely used, especially during the past decade [1]. The SVM approach, proposed by Vapnik [2], is the representative one, which can achieve state-of-the-art performance for many classification problems, and uses only a fraction of the set of all training examples to form solutions. Although SVMs can get sparse expressions for classification, some techniques have been proposed to make solutions much sparser. In [3], Kernel matching pursuit (KMP) is introduced to learn a weighted sum of basis functions from a kernel-based dictionary, and can produce much sparser models than SVMs.

KMP appends functions to an initially empty basis sequentially, from a redundant dictionary of functions, to approximate a classification function by using a certain loss criterion. The basic matching pursuit algorithm, as well as its two refinements: back-fitting and pre-fitting, are described in [3]. To make KMP practical for large datasets, a stochastic version is proposed as an approximation of the original KMP [4].

Different from KMP and its variants as mentioned above, kernel matching reduction algorithms (KMRAs), are proposed to perform a reverse procedure

G. Wang et al. (Eds.): RSKT 2008, LNAI 5009, pp. 564–571, 2008.

in this paper. Firstly, all training examples are selected to construct a function dictionary. Then the function dictionary is reduced iteratively by linear support vector machines (SVMs). During the reduction process, the parameters of the functions in the dictionary can be adjusted dynamically.

The rest of this paper is organized as follows. KMRAs are described in section 2. Compared are similarities and differences between KMRAs and other machine learning algorithms, such as KMP, SVMs, hidden space SVMs (HSSVMs), and radial basis function neural networks (RBFNN). Experimental results are presented in Section 4, then some conclusions and further thoughts are given in the last section.

2   Kernel Matching Reduction Algorithms (KMRAs)

Inspired by KMP and SVMs, we propose kernel matching reduction algorithms. The detailed procedures are expatiated through the following 2.1-2.3.

2.1   Constructing a Kernel-Based Dictionary

For a binary classification problem, assume there exist l training examples, which form the training set $S = \{(x_1, y_1), (x_2, y_2), \ldots, (x_l, y_l)\}$, where $x_i \in R^d, y_i \in \{-1, +1\}$, and y_i represents the class label of the point $x_i, i = 1, 2, \ldots, l$.

Given a kernel function $K : R^d \times R^d \to R$, similar to KMP [3], we use kernel functions, centered on the training points, as our dictionary: $D = \{K(x, x_i)|i = 1, \ldots, l\}$. Throughout this paper, the Gaussian kernel function,

$$K(x, x_i) = exp(\frac{-\|x - x_i\|^2}{2\sigma_i^2}) \,, \tag{1}$$

is selected. The value of σ_i should be set to keep the influence of the local domain around x_i and prevent x_i from having a high activation for the field far from x_i. Therefore, we adopt the heuristic method for σ_i from the reference [5], which is used to design RBFNNs and can be represented as

$$\sigma_i = (\frac{1}{p}\sum_{j=1}^{p}\|x_i - \hat{x}_j\|^2)^{\frac{1}{2}} \,, \tag{2}$$

where $\hat{x}_j$ are p nearest neighbors of x_i. Such, the receptive width of each point is determined to cover a certain region in the sample space.

2.2   Reducing the Kernel-Based Dictionary by Linear SVMs

Using all the kernel functions from the kernel-based dictionary $D = \{K(x, x_i)|i = 1, \ldots, l\}$, we construct a mapping from original space to feature space. Any training example x_i in S is mapped to a corresponding point z_i in S', where $z_i = (K(x_i, x_1), K(x_i, x_2), \ldots, K(x_i, x_l))$. Thus, the training set $S = \{(x_1, y_1), (x_2, y_2), \ldots, (x_l, y_l)\}$ in original space is mapped to $S' = \{(z_1, y_1), (z_2, y_2), \ldots, (z_l, y_l)\}$ in feature space.

Subsequently, we design a linear decision function $g_l(z_t) = sign(f_l(z_t))$ in feature space, and

$$f_l(z_t) = b + \sum_{i=1}^{l} w_i z_{ti} \,, \qquad (3)$$

which corresponds to the nonlinear form in original space:

$$f_l(x_t) = b + \sum_{i=1}^{l} w_i K(x_t, x_i) \,, \qquad (4)$$

where $w = (w_1, w_2, \ldots, w_l)$ represents weights of every dimension in z. Then we can decide which kernel functions are important for classification, and which are not, according to their weights magnitudes $|w_i|$ in (3) or (4), where $|w_i|$ denotes the absolute value of w_i. Those redundant kernel functions, which have lowest weights magnitudes, can be deleted from the dictionary to reduce the model.

How do we create the decision function? If we use the usual least squares error criterion to find this function, it is not practical, since the number of training examples, at the beginning, is equal to, or near to, the dimension number of the feature space S', and we will confront the problem of the not-invertible matrix. In fact, support vector machines (SVMs), based on the structural risk minimization, are fit for solving supervised classification problems with high dimensions [2]. In [6], a linear SVM is utilized to select genes from broad patterns of gene expression data by recursive feature elimination. Moreover, feature scoring and selection, based on weights from linear classification models for document classification, is explored in [7], and experimental results show that feature selection using weights from linear SVMs yields better classification performance than other feature weighting methods. Illuminated by [6] and [7], we also adopt linear SVMs to find the classification function in (3) or (4) on S'.

The optimization objective of linear SVMs is to minimize

$$\tau(w, \xi) = \frac{1}{2}(w \bullet w) + C \sum_{i=1}^{l} \xi_i \,, \qquad (5)$$

subject to the constraints

$$y_i[(w_i \bullet z_i) + b] \geq 1 - \xi_i, \text{ and } \xi_i \geq 0, i = 1, 2, \cdots, l \,,$$

where C is a penalty factor, which can be determined by v-fold cross validation. We can solve (5) by using its dual representation to get the result (3) (see [2] for detailed procedures). w_i denotes the contribution of z_i to the classifier in (3), and the higher the value of $|w_i|$, the more contribution of z_i to the model. Consequently, we can rank z_i according to the values of $|w_i|$ $(i = 1, 2, \cdots, l)$ from large to small. We can also rank x_i by $|w_i|$, because x_i is the preimage of

z_i in the original space. The x_i with the smallest $|w_i|$ can be deleted from the dictionary D, and D can be reduced to D'. Then we can continue this procedure on the new dictionary D'. Thus, the process can be iteratively performed until a given stop criterion is satisfied.

Note that, each σ should be computed again on the new dictionary D', according to (2), after D is reduced to D' every time, such that the receptive widths of kernel functions in D' can always cover the whole sample space.

We can set a tolerant minimum accuracy δ for the training examples, as the termination criterion of this procedure. Thus, we expect to gain the simplest model under the condition of guaranteeing the satisfied classification accuracy for all training examples. In fact, this accords with the principles of minimum description length and Occam's Razor [8]. Therefore, this algorithm can be expected to have a good generalization ability. In addition, different from KMP which appends kernel functions to the last model gradually, this reduction strategy can expect to avoid local optima, just due to deleting redundant functions from the functions dictionary iteratively.

2.3   The Detailed Procedure of KMRAs

We present below the program of KMRAs.

Algorithm KMRA:

Step 1, Set the parameter p in (2), the cross validation fold number v for determining C in (5), and the required classification accuracy δ on the training examples.

Step 2, Input training examples $S = \{(x_1, y_1), (x_2, y_2), \ldots, (x_l, y_l)\}$.

Step 3, Compute each σ by the equation (2), and construct the kernel-based dictionary $D = \{K(x, x_i) | i = 1, \ldots, l\}$.

Step 4, Transform S to S' by the dictionary D.

Step 5, Determine C by v-fold cross validation.

Step 6, Train the linear SVM with the penalty factor C on S', and obtain the classification model, including w_i, $i = 1, 2, \ldots, l$.

Step 7, Rank x_i by their weights magnitudes $|w_i|$, $i = 1, 2, \ldots, l$.

Step 8, If the classification accuracy of this model for training data is higher than δ, delete from D the $K(x, x_i)$ which has the smallest $|w_i|$, then adjust each σ for new D by (2), and go to Step 4; Otherwise go to Step 9.

Step 9, Output the classification model, which satisfies the accuracy δ with the simplest structure.

The reduction step 8 can be generalized to remove more than one basis function per iteration for improving the training speed.

3   Comparing with Other Machine Learning Algorithms

Although KMRAs, KMP, SVMs, HSSVMs, and RBFNNs can all generate a similar decision function shape as the equation (4), KMRAs have distinct characteristics, in the essence, compared with several other algorithms.

3.1   Differences with KMP

Both KMRA and KMP build kernel-based dictionaries, but they adopt different ways to select basis functions for last solutions. KMP appends kernel functions iteratively to the classification model. By contrary, KMRAs reduce the size of the dictionary step by step, by deleting redundant kernel functions. Moreover, different from KMP, KMRAs utilize linear SVMs to find solutions in the feature space.

3.2   KMRA Versus SVM

The main difference between KMRA and SVM consists in the approaches of producing feature spaces. KMRAs create the feature space by a kernel-based dictionary, whereas SVMs by kernel functions. Kernel functions in SVMs must satisfy Mercers theorem [2], while KMRAs have no restrictions on kernel functions in the dictionary (no positive-definiteness constraint, could be asymmetrical, even could include several different kernel shapes, etc.). The comparison between KMRAs and SVMs is similar to that between KMP and SVM, as addressed in [3]. In fact, we select Gaussian kernel functions in this paper, which can have different kernel widths obtained by the equation (2), but those Gaussian kernel functions, for all support vectors of SVMs, have the same kernel width.

3.3   Linking with HSSVMs

Hidden space support vector machines (HSSVMs), proposed in [9], also map input patterns into a high-dimensional hidden space by a set of nonlinear functions, and then train linear SVMs in the hidden space. From this viewpoint of constructing feature spaces and performing linear SVMs, KMRAs are similar to HSSVMs. But we adopt an iterative procedure to eliminate redundant kernel functions, until obtaining a condense solution. So, KMRAs can be considered as an improved version of HSSVMs.

3.4   Relation with RBFNNs

Although RBFNNs also build feature spaces using usually Gaussian kernel functions, they create discrimination functions in the least square sense. However, KMRAs use linear SVMs, i.e. the idea of structural risk minimization, to find solutions. In a broad sense, we can think of KMRAs as a special model of RBFNNs with a new configuration design strategy.

4   Experiments

4.1   Description on Data Sets and Parameter Settings

Because SVM is a representative benchmark classification technique, we compare KMRAs with SVMs, on four datasets: Wisconsin Breast Cancer, Pima Indians

Diabetes, Heart, and Australian, in which the former two (Breast and Diabetes) are from the UCI machine learning databases [10], and the latter two (Heart and Australian) from the Statlog database [11]. In addition, we directly use the LIBSVM software package [12] for performing the normal SVM.

Throughout the experiments:
1. All training data and test data are normalized to $[-1, 1]$.
2. Two-thirds of examples are randomly selected as training examples, and the remaining one-third as test those.
3. Gaussian kernel functions are chosen for SVMs, in which the kernel width σ and the penalty parameter C are decided by ten-fold cross validation on the training set.
4. $p = 2$, in equation (2), is adopted, as suggested by [5].
5. $v = 5$, in Step 5 of algorithm KMRA, is set.
6. For any dataset, SVM is firstly trained, and then according to the classification accuracy of SVM, we determine the stop accuracy δ for KMRAs.

4.2   Experimental Results

We first illustrate the results from standard SVMs, including their parameters C and σ in Table 1, and support vector numbers #SVs, and the prediction accuracy in Table 2.

Table 1. Parameter settings of SVMs by 10-fold cross validation for 4 datasets

Dataset	Breast	Diabetes	Heart	Australian
C	8	2048	2048	3.125×10^{-2}
σ	0.125	1.221×10^{-4}	1.221×10^{-4}	0.125

We set the termination accuracy $\delta = 0.97, 0.8, 0.8$, and 0.9 in KMRAs for these four datasets respectively, according to the classification accuracies of SVMs in Table 2.

We perform KMRAs on these datasets, and record classification accuracies for test datasets per iteration with algorithms running. Then we also show the results in Fig. 1.

In Fig. 1, the accuracies of SVMs on test examples are expressed in the thick straight lines, and the thin curves represent the classification performance of KMRAs. The row axis denotes iteration times of KMRAs, that is to say, numbers of kernel functions in the dictionary decrease gradually from left to right.

For Diabetes and Australian, we can find the prediction accuracies of KMRAs are improved gradually with kernel functions in the dictionary reducing. At the beginning of KMRAs' runs, we can conclude that the overfittings happen. Before KMRAs end, the performance of KMRAs approaches to, even is superior to, that of SVMs.

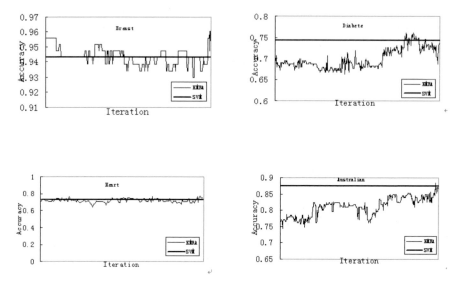

Fig. 1. Experimental results of KMRAs for every iteration on 4 datasets

Table 2. The last and the best experimental results from KMRAs, as well as the results from SVMs, for 4 datasets

	Dataset	Breast	Diabetes	Heart	Australian
The last	#SVs	11	32	8	52
KMRA	Accuracy	95.614%	73.8281%	75.5556%	86.5217%
The best	#SVs	12	109	11	59
KMRA	Accuracy	96.0526%	76.1719%	77.7778%	88.2609%
SVM	#SVs	38	284	55	373
	Accuracy	94.2982%	74.2188%	73.3333%	87.3913%

For Breast and Heart, from the beginning to the end, the curves of KMRAs fluctuate up and down around the accuracy lines of SVMs.

We further illustrate, in the Table 2, the numbers of kernel functions (i.e. #SVs), which appear in the last classification functions, as well as the corresponding prediction accuracies, when KMRAs terminate. Moreover, we record the best performance during the iterative process of KMRAs, and also list them in the Table 2.

From Table 2, compared with SVMs, KMRAs use much sparser support vectors, whereas they can obtain comparable results.

5   Conclusions

We propose KMRAs, which delete redundant kernel functions from a kernel-based dictionary, iteratively. Therefore, we expect KMRAs can avoid local optima, and can have a good generalization ability. Experimental results demonstrate that, compared with SVMs, KMRAs show comparable accuracies, but with typically much sparser representations. This means that KMRAs can have a fast classification speed for test examples than SVMs. In addition, analogous to SVMs, we can extend KMRAs to solve multi-classification problems, though we only consider the two-class situation in this paper.

We can also find, KMRAs gain sparser models at the expense of the long training time. Consequently, future work should attempt to explore how to reduce the training cost. Clustering techniques are used in [13], and approximation models are designed in [4], respectively, to facilitate the training of KMP. These can also enlighten us to further improve KMRAs. In conclusion, KMRAs provide a new problem solving approach for classification.

References

1. Shawe-Taylor, J., Cristianini, N.: Kernel Methods for Pattern Analysis. Cambridge University Press, Cambridge (2004)
2. Vapnik, V.: The Nature of Statistical Learning Theory. Springer, Berlin (1995)
3. Vincent, P., Bengio, Y.: Kernel Matching Pursuit. Mach. Learn. 48(1), 165–187 (2002)
4. Popovici, V., Bengio, S., Thiran, J.P.: Kernel Matching Pursuit for Large Datasets. Pattern Recogn. 38, 2385–2390 (2005)
5. Moody, T.J., Darken, C.J.: Fast Learning in Networks of Locally Tuned Processing Units. Neural Comput. 1, 151–160 (1989)
6. Guyon, I., Weston, J., Barnhill, S., Vapnik, V.: Gene Selection for Cancer Classification Using Support Vector Machines. Mach. learn. 46, 389–422 (2002)
7. Mladenić, D., Brank, J., Grobelnik, M., Milic-Frayling, N.: Feature Selection Using Linear Classifier Weights: Interaction with Classification Models. In: Proceedings of the 27th annual international ACM SIGIR conference on Research and development in information retrieval, pp. 234–241. ACM Press, New York (2004)
8. Mitchell, T.: Machine Learning. McGraw Hill, New York (1997)
9. Zhang, L., Zhou, W., Jiao, L.: Hidden Space Support Vector Machines. IEEE Trans. Neural Networ. 15(6), 1424–1434 (2004)
10. Blake, C., Keogh, E., Merz, C. J.: UCI Repository of Machine Learning Databases, http://www.ics.uci.edu/~mlearn/MLRepository.html
11. Michie, D., Spiegelhalter, D.J., Taylor, C.C.: Machine Learning, Neural and Statistical Classification, ftp://ncc.up.pt/pub/statlog
12. Chang, C.C., Lin, C.J.: LIBSVM: a Library for Support Vector Machines, http://www.csie.ntu.edu.tw/~cjlin/libSVM
13. Gou, S., Li, Q., Zhang, X.: A New Dictionary Learning Method for Kernel Matching Pursuit. In: Wang, L., Jiao, L., Shi, G., Li, X., Liu, J. (eds.) FSKD 2006. LNCS (LNAI), vol. 4223, pp. 776–779. Springer, Heidelberg (2006)

Text Categorization Based on Topic Model

Shibin Zhou, Kan Li, and Yushu Liu

School of Computer Science and Technology
Beijing Institute of Technology, Beijing 100081, P.R. China
{guoguos.zhou, likan, liuyushu}@bit.edu.cn

Abstract. In the text literature, many topic models were proposed to represent documents and words as topics or latent topics in order to process text effectively and accurately. In this paper, we propose LDACLM or Latent Dirichlet Allocation Category Language Model for text categorization and estimate parameters of models by variational inference. As a variant of Latent Dirichlet Allocation Model, LDACLM regard documents of category as Language Model and use variational parameters to estimate maximum a posteriori of terms. Experiments show LDACLM model to be effective for text categorization, outperforming standard Naive Bayes and Rocchio method for text categorization.

Keywords: Latent Dirichlet Allocation, Variational Inference, Category Language Model.

1 Introduction

In the text analysis, standard algorithms are unsatisfactory because terms often were supposed independent, which was recognized as "bag of words" model. However, the "bag of words" model offers a rather impoverished representation of the data because it ignores any relationships between the terms.

In the recent past, a new class of generative models called Topic Model has quickly become more popular in some text-related tasks. Topic Model supposes documents and corpus composed of mixture topics and then documents can be thought of "bag of topics". Thus, these models can handle the problem effectively about terms dependency. Topics can be view as a probability distribution over words, where the distribution implies semantic coherence. For example, a topic related to fruit would have high probabilities for the words "orange", "apple", and even "juicy". Wallach [10] demonstrated the "bag of topics" to surpass in performance to "bag of words" in unigram and bigram schemas.

There are many Topic Models proposed by researchers in the past such as Latent Semantic Analysis or LSA [3], the probabilistic Latent Semantic Indexing or pLSI [6], Latent Dirichlet allocation or LDA [1] and so on.

Latent Semantic Analysis (LSA) [3] is an approach that combines both term and document clustering. LSA usually takes a term-document matrix in the vector space representation as input, and uses a singular value decomposition of the input matrix to identify a linear subspace in the space of tf-idf features

G. Wang et al. (Eds.): RSKT 2008, LNAI 5009, pp. 572–579, 2008.

that captures most of the variance in the collection. Thus LSA can map text elements to a representation in the latent semantic space and can capture some aspects of basic linguistic notions such as synonymy and polysemy.

The probabilistic Latent Semantic Indexing (pLSI) model introduced by Hofmann [6], also known as the aspect model, was designed as a discrete counterpart of LSI or LSA to provide a better fit to text data and to overcome deficiencies of Latent Semantic Indexing (LSI). pLSI is a latent variable model that models each document as a mixture of topics. Although there are some problems with the generative semantics of pLSI, Hoffmann has shown some encouraging results in Information Retrieval.

One of these models, Latent Dirichlet Allocation (LDA) has quickly become one of the most popular probabilistic text modeling techniques in Information Retrieval. LDA has been shown to be effective in some text-related tasks. Processing fully generative semantics, LDA overcomes the drawbacks of previous topic models such as probabilistic Latent Semantic Indexing (pLSI) which is a MAP/ML estimated LDA model under a uniform Dirichlet distribution according to Girolami and Kaban discovery [4]. Latent Dirichlet allocation represents documents as mixtures over latent topics differentiated with pLSI, which each topic is characterized by a distribution over words. In [11], Wei and Croft shown the LDA-based document model had good performance in Information Retrieval. Moveover, Griffiths and Steyvers [5] apply LDA model to find scientific document topics.

Our goal in this paper is to address a variants of LDA and a extension of Language Model [9], which is a novel model for text categorization as we known. This generative model represents words set of each category with a mixture of topics assumed independent, as in state-of-the-art approaches like Latent Dirichlet Allocation [1], and extends these approaches to estimate maximum a posteriori of category language model parameters by assuming that variance parameters would be multinomial and dirichlet parameters of category language model.

In Section 2, we demonstrate our approaches on how to estimate parameters of models and classify documents. In section 3, we evaluate accuracy of our model on Reuters21578 and 20Newsgroups datesets. We conclude the paper with a summary, and a brief discussion of future work in section 4.

2   Latent Dirichlet Allocation Category Language Model

In this section we introduce our model that extends Latent Dirichlet Allocation and Language Model called Latent Dirichlet Allocation Category Language Model and manifest methods of inferring and estimating parameters.

2.1   Model Structure

Latent Dirichelt Allocation Category Language Model or LDACLM is a variant of LDA, which is used as classifier of text documents. Rather, LDA described in [1] used as dimension reducer in the discriminative framework of documents classification. The prominent feature of LDACLM is that the model assume each word would be a independent topic that we called word topic and assume extra topics other than word topics would be model the correlation among the words. As we known, this distinguish to LDA and also tradeoff between effective and time consuming. The following process similar to LDA generates documents in the LDACLM model.

- For each category language model or words set $\mathbf{w}$, pick multinomial distribution $p(\theta_\mathbf{w})$ from a symmetric Dirichlet distribution $p(\theta_\mathbf{w}|\alpha)$ with prior scalar parameter α which is identity to all category language models.
- Pick a topic $z \in \{1, 2, \ldots, K\}$ from a multinomial distribution $p(z|\theta_\mathbf{w})$ with parameter vector $\theta_\mathbf{w}$.
- Generate a word w_t from a multinomial distribution $p(w_t|z, \beta)$ with parameter vector β, where each parameter β_z in the vector β is related to specific z respectively.

2.2   Inference

The maximum likelihood of category language model $\mathbf{w}$ with model parameter vector β and model dirichlet parameter α may formulate as:

$$p(\mathbf{w}|\alpha, \beta) \propto \int \left(\prod_{k=1}^{K} \theta_k^{\alpha-1} \right) \left(\prod_{t=1}^{V} \left\{ \sum_{k=1}^{K} (\theta_k \beta_{k,t}) \right\}^{tf_{t,\mathbf{w}}} \right) \mathrm{d}\theta$$

Where words set $\mathbf{w}$ containing words form corpus $\mathcal{D}$ who has a vocabulary of size V and $tf_{t,\mathbf{w}}$ stores the number of occurrences of a word w_t in words set $\mathbf{w}$.

Similar to LDA [1], We develop a variational approximation [8] for LDACLM by defining an approximating family distribution $q(\theta, z|\mathbf{w}, \gamma, \phi)$, and choose the variational Dirichlet parameter vector γ and variational multinomial parameter vector ϕ which are different sets for each category language model to yield a tight approximation to the true posterior. Suppose the factorized variational parameters distribution is $q(\theta, z|\mathbf{w}, \gamma, \phi) = q(\theta|\mathbf{w}, \gamma) \prod_{t=1}^{V} q(z_t|\mathbf{w}, \phi_t)$ with variational Dirichlet parameter vector γ and variational multinomial parameter vector ϕ. Especially, for each category language model, there is a different set of Multinomial and Dirichlet variational parameter vectors. Thus, minimization of the KL divergence $D(q(\theta, z|\mathbf{w}, \gamma, \phi) \| p(\theta, z|\mathbf{w}, \alpha, \beta))$ we can derive approximation of $p(\theta, z|\mathbf{w}, \alpha, \beta)$.

So, we can take decreasing steps in the KL divergence and converge to optimizing parameter by an iterative fixed-point method, bounding the marginal likelihood of a document using Jensen's inequality [8].

$$\log p(\mathbf{w}|\alpha, \beta) \geq E_q \{\log p(\theta, z, \mathbf{w}|\alpha, \beta)\} - E_q \{\log q(\theta, z|\mathbf{w}, \gamma, \phi)\} \qquad (1)$$

Letting $\mathcal{L}(\gamma, \phi|\mathbf{w}, \alpha, \beta)$ denote the right-hand side of Eq.(1) and expand it, we have

$$
\mathcal{L} = \log\Gamma\left(\sum_{k=1}^{K}\alpha_k\right) - \sum_{k=1}^{K}\log\Gamma\left(\alpha_k\right) + \sum_{k=1}^{K}\left(\alpha_k - 1\right)\left(\Psi\left(\gamma_k\right) - \Psi\left(\sum_{j=1}^{K}\gamma_j\right)\right)
$$

$$
+ \sum_{t=1}^{V}\sum_{k=1}^{K}\phi_{t,k}\left(\Psi\left(\gamma_k\right) - \Psi\left(\sum_{j=1}^{K}\gamma_j\right)\right) + \sum_{t=1}^{V}\sum_{k=1}^{K}tf_{t,\mathbf{w}}\phi_{t,k}\log\beta_{t,k}
$$

$$
- \log\Gamma\left(\sum_{k=1}^{K}\gamma_k\right) + \sum_{k=1}^{K}\log\Gamma\left(\gamma_k\right) - \sum_{k=1}^{K}\left(\gamma_k - 1\right)\left(\Psi\left(\gamma_k\right) - \Psi\left(\sum_{j=1}^{K}\gamma_j\right)\right)
$$

$$
+ \sum_{t=1}^{V}\sum_{k=1}^{K}\phi_{t,k}\log\phi_{t,k} \tag{2}
$$

Where Γ is gamma function, Ψ is digamma function.

Firstly, we maximize Eq.(2) with respect to $\phi_{t,k}$, the probability that the word t was generated by latent topic z. This is a constrained maximization with constraint $\sum_{k=1}^{K}\phi_{t,k} = 1$. With $\beta_{t,k}$ reference to $p\left(w_t|z_t = k, \beta\right)$, we form the Lagrangian by isolating the terms which contain $\phi_{t,k}$ and adding the appropriate Lagrange multipliers, so we have

$$
\mathcal{L}^{\mathbf{w}}_{[\phi_{t,k}]} = \phi_{t,k}\left(\Psi\left(\gamma_k\right) - \Psi\left(\sum_{j=1}^{K}\gamma_j\right)\right)
$$

$$
+ tf_{t,\mathbf{w}}\phi_{t,k}\log\beta_{t,k} + \phi_{t,k}\log\phi_{t,k} + \lambda_t\left(\sum_{k=1}^{K}\phi_{t,k} - 1\right)
$$

Taking derivatives with respect to $\phi_{t,k}$ and setting the derivative to zero yields the maximized , we have

$$
\phi_{t,k} \propto \left(\beta_{t,k}\right)^{tf_{t,\mathbf{w}}} \exp\left(\Psi\left(\gamma_k\right) - \Psi\left(\sum_{j=1}^{K}\gamma_j\right)\right) \tag{3}
$$

Secondly, we maximize Eq.(2) with respect to γ_k, the kth component of the posterior Dirichlet parameter. Take the derivative with respect to γ_k and setting to zero yields a maximum:

$$
\gamma_k = \alpha_k + \sum_{t=1}^{V}\phi_{t,k} \tag{4}
$$

2.3   Estimating

Given a corpus of $\mathcal{D} = \{\mathbf{w}_1, \ldots, \mathbf{w}_M\}$ that $\mathbf{w}$ is a category language model, we use a variational EM algorithm (EM with a variational E Step) [1] to find the parameters and which maximize a lower bound on the log marginal likelihood:

$$\ell(\alpha, \beta) = \sum_{\mathbf{w} \in \mathcal{D}} \mathrm{log} p(\mathbf{w}|\alpha, \beta)$$

As we have described above, we can bound the log likelihood using

$$\mathrm{log} p(\mathbf{w}|\alpha, \beta) = \mathcal{L}(\gamma, \phi|\mathbf{w}, \alpha, \beta) + D(q(\theta, z|\mathbf{w}, \gamma, \phi)\|p(\theta, z|\mathbf{w}, \alpha, \beta)) \quad (5)$$

Which exhibits $\mathcal{L}(\gamma, \phi|\mathbf{w}, \alpha, \beta)$ as a lower bound because the KL term is positive. We now obtain a variational EM algorithm that repeats the following two steps until Eq.(5) converges:

- (E step) For each category language model, optimize values for the variational parameter vectors γ and ϕ, the update rules are Eq.(3) and Eq.(4).
- (M step) Maximize the resulting lower bound on the log likelihood with respect to the model parameter α and parameter vector β. We can do this by finding the maximum likelihood estimates with expected sufficient statistics computed in the E-step.

Firstly, we maximize Eq.(2) with respect to $\beta_{t,k}$. This is a constrained maximization with constraint $\sum_{t=1}^{V} \beta_{t,k} = 1$, so we form the Lagrangian by isolating the terms which contain $\beta_{t,k}$ and adding the appropriate Lagrange multipliers, so we have

$$\mathcal{L}_{[\beta_{t,k}]} = \sum_{\mathbf{w} \in \mathcal{D}} \sum_{t=1}^{V} \sum_{k=1}^{K} tf_{t,\mathbf{w}}\phi_{t,k} \mathrm{log}\beta_{t,k} + \sum_{k=1}^{K} \lambda_k \left(\sum_{t=1}^{V} \beta_{t,k} - 1 \right)$$

Taking derivatives with respect to $\beta_{t,k}$ and setting the derivative to zero yields the maximized $\beta_{t,k}$, we have

$$\beta_{t,k} \propto \sum_{\mathbf{w} \in \mathcal{D}} tf_{t,\mathbf{w}}\phi_{t,k}$$

Secondly, we maximize Eq.(2) with respect to α. Then, take first derivative and second derivative with respective to α (α is a scalar dirichlet parameter). So according Newton-Raphson formula, we can find the maximal α by iteration as following:

$$\alpha^{\mathrm{new}} = \alpha - \frac{M(\Psi(K\alpha) - K\Psi(\alpha)) + \sum_{\mathbf{w} \in \mathcal{D}} \sum_{k=1}^{K} \left\{ \Psi(\gamma_{k,\mathbf{w}}) - \Psi\left(\sum_{k=1}^{K} \gamma_{k,\mathbf{w}} \right) \right\}}{M \times K \times (\Psi'(K\alpha) - \Psi'(\alpha))}$$

where Ψ' is trigamma function.

2.4   Maximum a Posteriori of Multinomial Parameter

After model parameter α, model parameter vector β and variational parameter vector ϕ converged, we can fit the variational parameter vector γ as Eq.(4) description. Hereafter, to specific category language model $\mathbf{w}$, the maximum a posteriori of multinomial parameter in vector $\theta_{\mathbf{w}}$ can be computed approximately as

$$\theta_k^{\mathrm{MAP}} = \frac{\gamma_k}{\sum_{k=1}^{K} \gamma_k} \qquad k = \{1, 2, \dots K\}$$

Eventually, based on our model, we can derive maximum likelihood of document d generating by category language model $\mathbf{w}$ as following formula:

$$p(d) \propto \prod_{t \in d} \left\{ \sum_{k=1}^{K} \left(\theta_k^{\text{MAP}} \beta_{t,k} \right) \right\}^{tf_{t,d}}$$

3   Experiments and Results

We have conducted experiments on two real-world datasets, Reuters21578 and 20newsgroups, to evaluate the effectiveness of our proposed model for text categorization.

The Reuters21578 dataset contains documents collected from Reuters newswire articles are assigned to 135 categories. However, some categories are empty and thus there are only non-empty 118 categories, among which the 10 most frequent categories called R10 by Debole [2] contain about 75% of the documents as Table 1 show. There are several ways to split the documents into training and testing sets: 'ModLewis' split, 'ModApte' split, and 'ModHayes' split. The 'ModApte' train/test split is widely used in text classification research. We followed the ModApte split in which the 10 most frequent categories and the numbers of documents are used for training and testing.

Table 1. Number of Training and Test documents About R10

Category name	Num Train	Num test
earn	2877	1087
acq	1650	719
money-fx	538	179
grain	433	149
crude	389	189
trade	369	118
interest	347	131
wheat	212	71
ship	197	89
corn	182	56

The 20Newsgroups(20NG) dataset is a collection of approximately 20,000 documents that were collected from 20 different newsgroups. This collection consists of 19,974 non-empty documents distributed evenly across 20 newsgroups and we selected 19,946 non-empty documents which are all the same after feature selection . We use the newsgroups to form categories, and randomly select 70% of the documents to be used for training and the remaining 30% for testing.

On the "Gerneral Text Toolkit" developing by our laboratory, We have tried our proposed LDACLM with 100 topics modeling the relationship among words, NaiveBayes with Laplace smoothing, and Rocchio algorithm [7] with TF-IDF scheme to these datasets respectively. Furthermore, We apply to Information

Table 2. Experimental results on the 20NG dataset

	NaiveBayes	LDACLM	Rocchio
macro-averaging precision	0.809	0.824	0.736
macro-averaging recall	0.808	0.813	0.739
macro-averaging F1	0.808	0.818	0.738
micro-averaging accuracy	0.803	0.813	0.736

Table 3. Experimental results on the Reuters21578 R10

	NaiveBayes	LDACLM	Rocchio
macro-averaging precision	0.662	0.660	0.647
macro-averaging recall	0.616	0.714	0.661
macro-averaging F1	0.638	0.686	0.654
micro-averaging accuracy	0.804	0.840	0.787

Gain [12] feature selecting method to the documents of both 20NG and Reuters-21578 R10 datasets with threshold 0.055 to 20NG and 0.3 to Reuters. The results of macro-averged and micro-averaged to 20NG and Reuters datasets are shown in Tables 2 and 3 for LDACLM, NaiveBayes and Rocchio respectively.

Specially, All results are averaged across 5 random runs for 20NG datatset. According experimental results, LDACLM outperform NaiveBayes with Laplace smoothing and Rocchio algorithm.

4   Conclusion and Future Work

This paper proposed Latent Dirichlet Allocation Category Language Model, a novel model based on LDA model. We have presented variational inference approach, and parameters estimation method which is similar to LDA [1] in category language model. As Results on 20NG and Reuters21578 datasets shown above, LDACLM cannot significantly improve performance. In our opinion, we think that it was because the topics modeling the relationship among words is not abundant which constraint by computer memory. In the future work, we will try use topics by collection from Wordnet based on Gibbs sample, and this maybe create many topics which approximate words dependency than variational inference do.

Acknowledgement

We would like to thank the anonymous reviewers for their valuable comments and suggestions. We are grateful for Zhao Cao's helpful discussion and advice. Many thanks also give to Shidong Feng, Yingfan Gao, Jian Cao, Jinghua Bai, and Xu Zhang for their suggestions regarding this paper.

References

1. Blei, D., Ng, A., Jordan, M.: Latent Dirichlet allocation. Journal of Machine Learning Research 3, 993–1022 (2003)
2. Debole, F., Sebastiani, F.: An Analysis of the Relative Difficulty of Reuters-21578 Subsets. Journal of the American Society for Information Science and Technology 56(2), 584–596 (2004)
3. Deerwester, S., Dumais, S., Furnas, G., Landauer, T., Harshman, R.: Indexing by latent semantic analysis. Journal of the American Society for Information Science 41(6), 391–407 (1990)
4. Girolami, M., Kaban, A.: On an equivalence between PLSI and LDA. In: Proceedings of the 26th annual international ACM SIGIR conference on Research and development in informaion retrieval, Toronto, pp. 433–434 (2003)
5. Griffiths, T., Steyvers, M.: Finding scientific topics. Proceedings of the National Academy of Sciences 101, 5228–5235 (2004)
6. Hofmann, T.: Probabilistic Latent Semantic Indexing. In: Proceedings of the 22nd annual international ACM SIGIR conference on Research and development in information retrieval, Berkeley, California, pp. 50–57 (1999)
7. Joachims, T.: A Probabilistic Analysis of the Rocchio Algorithm with TFIDF for Text Categorization. In: Proceedings of the 14th International Conference on Machine Learning, Nashville, TN, USA (1997)
8. Jordan, M., Ghahramani, Z., Jaakkola, T., Saul, L.: An introduction to variational methods for graphical models. Machine Learning 37, 183–233 (1999)
9. Ponte, J., Croft, W.: A Language Modeling Approach to Information Retrieval. In: Proceedings of the 21st annual international ACM SIGIR conference on Research and development in information retrieval, Melbourne, Australia, pp. 275–281 (1998)
10. Wallach, H.: Topic modeling: beyond bag-of-words. In: Proceedings of the 23rd International Conference on Machine Learning (2006)
11. Wei, X., Croft, W.: LDA-Based Document Models for Ad-hoc Retrieval. In: Proceedings of the 29th Annual International ACM SIGIR Conference on Research and Development on Information Retrieval, Seattle, pp. 178–185 (2006)
12. Yang, Y., Pedersen, J.: A Comparative Study on Feature Selection in Text Categorization. In: Proceedings of the 14th International Conference on Machine Learning, Nashville, TN, USA, pp. 412–420 (1997)

A Time Weighted Neighbourhood Counting Similarity for Time Series Analysis

Hui Wang[1] and Zhiwei Lin[1,2]

[1] School of Computing and Mathematics
University of Ulster at Jordanstown
BT37 0QB, Northern Ireland, UK
{h.wang, z.lin}@ulster.ac.uk
[2] Key Lab of Network Security and Cryptology
School of Mathematics and Computer Science
Fujian Normal University, P.R. China

Abstract. Time series data abound and analysis of such data is challenging and potentially rewarding. One example is *financial time series analysis*. In time series analysis there is the issue of *time dependency*, that is, the state in the nearer past is more relevant to the current state than that in the more distant past. In this paper we study this issue by introducing time weighting into similarity measures, as similarity is one of the key notions in time series analysis methods.

We consider the generic neighbourhood counting similarity as it can be specialised for various forms of data by defining the notion of neighbourhood in a way that satisfies different requirements. We do so with a view to capturing time weights in time series. This results in a novel time weighted similarity for time series. A formula is also discovered for the similarity so that it can be computed efficiently.

Keywords: time series analysis, time dependency, time weighting, neighbourhood counting.

1 Introduction

A time series is a sequence of observations, representing the measurements of one or more variables at (usually) equal time intervals [1]. When more than one variable are involved we get a *multidimensional time series* (MTS). Examples of MTS include: stock price movements, volume of sales over time, daily temperature readings and ECG data [1].

Time series forecasting and analysis (TSFA) is challenging but potentially very rewarding, especially in financial TSFA. Most of the intelligent data analysis methods can be applied in principle, but the classical methods for TSFA are auto-regressive moving average model, exponential smoothening and spectral decomposition, etc [1,2]. More recently evolutionary computing is becoming increasingly popular for TSFA as well as financial modelling [3].

G. Wang et al. (Eds.): RSKT 2008, LNAI 5009, pp. 580–587, 2008.

In time series analysis there is the issue of *time dependency*, that is, the state in the nearer past is more relevant to the current state than that in the more distant past. In this paper we study this issue by introducing time weighting into similarity measures, as similarity is one of the key notions in time series analysis methods.

Euclidean distance is perhaps the most widely used measure of distance/ similarity in any distance related tasks, including TSFA. Other measures that are often used in TSFA include: longest common subsequences [4,5], dynamic time warping [6], normalization of sequences [7] and landmarks [8]. These measures are originally defined or extended for one dimensional time series, but need generalisation for use in multidimensional time series. Unfortunately time weighting is not considered in the above mentioned distance/similarity measures. The need for time weighting is mentioned in [9,10] but this problem has not been adequately addressed in the literature, to the best of our knowledge.

We start with the generic neighbourhood counting (NCM) similarity [11,12], which measures similarity of two data items by counting their common neighbourhoods. NCM needs specialised by defining the notion of neighbourhood in a way that carries or encodes the desired or intended characteristics. NCM has been specialised for multivariate data [12,11] as well as sequence data [13], resulting in novel and competitive similarity measures.

We define neighbourhood by taking into account both the time dimension and space dimension of multidimensional time series data, thus we specialise NCM as a time weighted similarity for multidimensional time series – *TWNCM*. We derive formulas by which the number of such neighbourhoods can be efficiently calculated, with a computational complexity in the same order as the Euclidean distance.

The rest of the paper is organised as follows. In the next section we review some of the well known similarity measures for time series. In Section 2 we present our time weighted similarity for time series in the neighbourhood counting framework. The final section concludes the paper.

2   Neighbourhood in the Space of Multidimensional Time Series

We consider a system that is characterised by a set of attributes $R = \{a_1, a_2, \cdots, a_n\}$, where the domain of attribute $a_i \in R$ is $dom(a_i)$. The state of the system evolves over time, but is observed and recorded at fixed intervals (e.g., every hour or day), indexed by $T = \{0, 1, 2, \cdots, \}$. This results in a multidimensional time series $(d(0), d(1), \cdots, d(t), \cdots)$, where $d(t)$ is the state at time t and is described by the values of the attributes valid for the system at time t, written as $d(t) = < v_1, v_2, \cdots, v_n >$. A multidimensional time series can be understood as a set of n time series aligned by time. Such an MTS reflects the underlying operational mechanism of the system.

An attribute can be in either categorical or numerical scale[1]. We assume that, if attribute a_i is categorical then $dom(a_i)$ is finite; if it is numerical then there is a lower bound and an upper bound, denoted by $\min(a_i)$ and $\max(a_i)$ respectively, and its domain is quantized and then mapped to the natural numbers.

At any time instant, the state of the system is a data point in the n dimensional space defined by $\Omega \overset{\text{def}}{=} \prod_{i=1}^{n} dom(a_i)$. The system can be observed over any length of time, giving rise to time series of any length. A time series of length k is a data point in the k-MTS space defined by $\Omega^k = \prod_{t=1}^{k} \Omega$, and a time series of any length is thus a data point in the MTS space $\Omega^{mts} = \sum_k \Omega^k$. We call a data point in an MTS space, an mts. The period of time when a time series is observed is called a $window$, and k is then the $window\ size$.

Consider an mts of length k, which consists of states over k consecutive time instants. We write such an mts by $d = (d(t), d(t-1), \cdots, d(t-k+1))$, where $d(i) \in \Omega$ and $\{t, t-1, \cdots, t-k+1\}$ are $time\ indexes$. Since Ω is defined by n attributes, $d(i)$ is a vector or simple tuple $< d(i)_1, d(i)_2, \cdots, d(i)_n >$ where $d(i)_j \in dom(a_j)$.

A premise for time series analysis based on historical data is that the state of system at time t is dependent on states at past times $t - i$ for $i = 1, 2, \cdots, k$. In this paper we consider the $time\ dependency\ relevance$ assumption that the degree of dependence is a function of how far a past state is to the current. More specifically $d(t)$ depends on $d(t-1), d(t-2), \cdots, d(t_k)$; but $d(t)$ depends more on $d(t-1)$ than on $d(t-2)$ and, in general, $d(t)$ depends more on $d(t-a)$ than on $d(t-b)$ when $a \le b$. In other words the influence of the past states on the current is weighted. The difficult question is: how to determine weighting?

In this section we present a similarity for multidimensional time series, which gives the past states varying weightings, the closer to the current state the higher the weighting. This similarity is derived from the generic NCM similarity [11] under a specialisation of the general notion of neighbourhood.

In the following sections we first of all discuss how neighbourhood should be defined from both the time and space aspects, which captures the intended time weighting. We then discuss how neighbourhoods are counted based on this definition, in order to provide formulas for the time weighted NCM similarity.

---

[1] Scale of measurement of an attribute (variable) describes how much information the values of the attribute contain. Different operations on values are meaningful for different scales. There are four recognised scales [14]: $nominal\ scale$, where values are names or labels and the only operations are 'equality' and 'inequality'; $ordinal\ scale$, where values represent the rank order (1st, 2nd, 3rd etc) of the objects measured and, in addition to 'equality' and 'inequality' operations, the 'greater than' and 'less than' operations are meaningful; $interval\ scale$, where intervals of values can be meaningfully compared and additional operations are 'addition' and 'subtraction'; and $ratio\ scale$, where ratios between arbitrary pairs of values are meaningful and additional operations are 'multiplication' and 'division'. Most physical quantities, such as mass, length or energy are measured on ratio scales. In data mining we usually consider two scales: categorical (nominal) and numerical (ratio). So all the above operations can be meaningfully applied.

2.1 Time Dependent Neighbourhood for Multidimensional Time Series

The notion of neighbourhood used in [11] is the same as that in the mathematical study of topology. Loosely speaking a neighbourhood is a region in a data space and a region of a data point x is a region that covers x.

We consider a simple mts of length 3, $\alpha = (d(t), d(t-1), d(t-2))$, where $d(i) \in \Omega$ is a simple tuple[2]. A shorter mts $\alpha' = (d(t), d(t-1))$ can be understood as $(d(t), d(t-1), ?)$, where ? means the position can be any value, which represents a set of mts of length 3. When ? is taken to be $d(t-2)$, $(d(t), d(t-1), ?)$ becomes $(d(t), d(t-1), d(t-2))$. Therefore $(d(t), d(t-1))$ is a neighbourhood of α. Similarly $(d(t))$ is also a neighbourhood of α. By *time dependency relevance* assumption, $d(t)$ depends the most on $d(t)$ itself, then on $d(t-1)$ and the least on $d(t-2)$ – such dependency needs to be weighted. To accommodate this weighted dependency, we need to break the symmetry of generating neighbourhoods from both ends of an mts. Therefore we do not regard $(d(t-1), d(t-2))$ and $(d(t-2))$ as neighbourhoods of α.

In general we consider an mts of length k, $\alpha = (d(t), d(t-1), d(t-2), \cdots, d(t-k+1))$. Any mts $(d(t), d(t-1), d(t-2), \cdots, d(t-k+m))$, for $1 \le m \le k$, is a neighbourhood of α. Such neighbourhoods are due to time.

2.2 Space Dependent Neighbourhood for Multidimensional Time Series

In addition to the time related aspect of neighbourhood, there is another aspect to consider. Any $d(t)$ in an mts is a data point in the multidimensional space Ω, therefore it has neighbourhoods as regions in Ω. Such neighbourhoods can be defined by distance/similarity in the usual sense or as hypertuples, as adopted in [11]. A hypertuple [15] is a vector $\langle s_1, s_2, \cdots, s_n \rangle$ where $s_i \subseteq dom(a_i)$ for $i = 1, 2, \cdots, n$. Such neighbourhoods are due to space.

To sum up, a neighbourhood is simply one set of neighbours satisfying the same set of conditions. A neighbourhood of an mts has two aspects: time and space. The combination of them characterises or defines all neighbourhoods of an mts. The generic NCM similarity seeks to find the number of all common neighbourhoods. Later in this paper we will discuss how many common neighbourhoods there is for a pair of mts.

2.3 An Example

Example 1. To illustrate the notions of MTS and neighbourhood, we consider a data space Ω which is defined by three attributes a_1, a_2, a_3. a_1 and a_2 are numerical with the same domain of $\{1, 2, 3, 4, 5\}$ and a_3 is categorical with a

---

[2] A *simple tuple* is a data tuple in the traditional sense, i.e., a vector of values. The adjective 'simple' is used in order to differentiate it from the notion of *hyper tuple* [15], which is a vector of sets.

Table 1. A toy example

ID	t			t-1			t-2		
	a_1	a_2	a_3	a_1	a_2	a_3	a_1	a_2	a_3
α_1	3	2	+	2	2	+	2	1	+
α_2	2	3	+	1	3	+	1	2	+
α_3	4	4	-	3	4	-	3	3	+
α_4	5	4	-	4	4	-	4	3	-
α_5	4	5	-	3	5	-	3	4	-

Table 2. All neighbourhoods of α_1 due to time

t			t-1			t-2		
a_1	a_2	a_3	a_1	a_2	a_3	a_1	a_2	a_3
3	2	+	2	2	+	2	1	+
3	2	+	2	2	+			
3	2	+						

Table 3. Some neighbourhoods of α_1 due to space

t			t-1			t-2		
a_1	a_2	a_3	a_1	a_2	a_3	a_1	a_2	a_3
$[3]$	$[2]$	$\{+\}$	$[2]$	$[2]$	$\{+\}$	$[2]$	$[1]$	$\{+\}$
$[3,5]$	$[2,4]$	$\{+,-\}$	$[2,3]$	$[2,5]$	$\{+\}$	$[2,4]$	$[1,3]$	$\{+,-\}$
$[1,3]$	$[1,3]$	$\{+\}$	$[1,3]$	$[1,5]$	$\{+,-\}$	$[1,4]$	$[1,4]$	$\{+\}$

domain of $\{+,-\}$. The time window under consideration is $k = 3$. Table 1 is a sample of 5 mts. Consider α_1 for an example, and we want to see what its neighbourhoods are. Table 2 lists all neighbourhoods of α_1 due to time, while Table 3 lists some neighbourhoods of α_1 due to space.

2.4 Number of Common Neighbourhoods

The neighbourhood counting similarity is simply the number of all common neighbourhoods of two data points. Now that we have defined neighbourhoods for multidimensional time series we need to see how neighbourhoods can be counted efficiently.

Space Aspect. Consider two mts $\alpha =< x(k-1), x(k-2), \cdots, x(0) >$ and $\beta =< y(k-1), y(k-2), \cdots, y(0) >$. According to [11], the *number of common neighbourhoods* of α and β at time t is

$$N(t)(\alpha, \beta) = \prod_{i=1}^{n} C(x(t)_i, y(t)_i), \qquad (1)$$

where

$$
C(x(t)_i, y(t)_i) = \begin{cases} (\max(a_i) - \max(\{x(t)_i, y(t)_i\}) + 1) \times (\min(\{x(t)_i, y(t)_i\}) \\ \quad - \min(a_i) + 1), \\ \quad \text{if } a_i \text{ is ordinal} \\ 2^{m_i - 1}, \quad \text{if } a_i \text{ is nominal and } x(t)_i = y(t)_i \\ 2^{m_i - 2}, \quad \text{if } a_i \text{ is nominal and } x(t)_i \neq y(t)_i \end{cases}
$$

$$(2)$$

where $m_i = |dom(a_i)|$ is the domain size of categorical attribute a_i, and $\max(a_i)$ and $\min(a_i)$ are the maximal and minimal values of numerical attribute a_i. For simplicity we will write $N(t)$ when α and β are obvious from the context.

When Time Aspect is Considered. Consider two simple mts of length 3, $\alpha = (x(t), x(t-1), x(t-2))$ and $\beta = (y(t), y(t-1), y(t-2))$. Suppose there are $N(t)$, $N(t-1)$ and $N(t-2)$ space-based common neighbourhoods for the three time instants respectively. Without the time aspect considered, there are $N(t) * N(t-1) * N(t-2)$ common neighbourhoods.

Consider one common neighbourhood $c(i)$ for each of the three time instants. When time aspect is considered, we have the following common neighbourhoods for α and β:

$$
(c(t), c(t-1), c(t-2))
$$
$$
(c(t), c(t-1))
$$
$$
(c(t))
$$

Correspondingly we have the following number of common neighbourhoods respectively:

$$
N(t) * N(t-1) * N(t-2)
$$
$$
N(t) * N(t-1)
$$
$$
N(t)
$$

Therefore the total number of common neighbourhoods is

$$
N(t) * N(t-1) * N(t-2) + N(t) * N(t-1) + N(t)
$$

In general we assume the length of α and β is k. The *total number of common neighbourhoods*, or the *time weighted NCM similarity* (TWNCM), is

$$
TWNCM(\alpha, \beta) = NCM_{mts}(\alpha, \beta) = \sum_{i=1}^{k} \prod_{j=1}^{i} N(t - j + 1) \tag{3}
$$

Note that $N(i)$ is the number of common neighbourhoods due to state at time i. This number is used a various number of times, thus giving varying weightings to states at different times. The closer a past state is to the current, the higher its weighting is. We therefore call the number of times $N(i)$ is used the *weighting index* of state i. More precisely, let k be the size of the time window, t be the time index of the current state, and i be the time index of a past state. Then the weighting index of i is: $k - t + i$. Clearly, when a past state is the current state, i.e., $i = t$, its weighting index is k; when a past state is the last one in the time window, i.e., $i = t - k + 1$, its weighting index is 1.

2.5   Example

To illustrate the TWNCM similarity we consider Table 1 again. Let $\beta = (< 2, 4, - >, < 1, 4, + >, < 1, 3, + >)$ be another mts. We want to calculate the TWNCM similarity between β and each mts in the table. Consider α_1 first. Following Eq.(1) we have

$$N(t)(\beta, \alpha_1) = 2 * 3 \times 2 * 2 \times 1 = 24$$
$$N(t - 1)(\beta, \alpha_1) = 1 * 4 \times 2 * 2 \times 2 = 32$$
$$N(t - 2)(\beta, \alpha_1) = 1 * 4 \times 1 * 3 \times 2 = 24$$

According to Eq.(3) we have

$$TWNCM(\beta, \alpha_1) = 24 * 32 * 24 + 24 * 32 + 24 = 19224$$

Similarly we have

$$TWNCM(\beta, \alpha_2) = 48 * 60 * 60 + 48 * 60 + 48 = 175728$$
$$TWNCM(\beta, \alpha_3) = 64 * 24 * 54 + 64 * 24 + 64 = 84544$$
$$TWNCM(\beta, \alpha_4) = 32 * 16 * 18 + 32 * 16 + 32 = 9760$$
$$TWNCM(\beta, \alpha_5) = 32 * 12 * 18 + 32 * 12 + 32 = 7328$$

3   Conclusion

Time dependency is a problem in time series forecast and analysis. The premise of this problem is that the current state of a system is dependent on the state of the same system in the past and, furthermore, the state of a system is dependent more on the nearer past than the more distance past.

In this paper we study this problem by introducing time weighting in the definition of a new similarity measure. We consider the generic NCM similarity as it can be specialised for various forms of data. We define the notion of neighbourhood so that both the space and time aspects are separately taken into account. This results in a novel concept of time weighted NCM similarity for time series, TWNCM, as well as a formula by which the similarity can be computed efficiently.

Acknowledgment

This work is partially supported by the open funding of Key Lab of Fujian Province University Network Security and Cryptology (NO. 07B003) and Fujian Provincial Department of Science and Technology(NO. 2007F5036).

References

1. Gunopulos, D., Das, G.: Time series similarity measures. In: Tutorial notes of the sixth ACM SIGKDD international conference on Knowledge discovery and data mining (2000)
2. Wang, H., Weigend, A.: Data mining for financial decision making: an editorial. Journal of Decision Support Systems 37(4) (2004)
3. Brabazon, A., O'Neill, M.: Biologically Inspired Algorithms for Financial Modelling. Springer, Heidelberg (2006)
4. Hirschberg, D.S.: Algorithms for the longest common subsequence problem. Journal of ACM 24(4), 664–675 (1977)
5. Agrawal, R., Lin, K.I., Sawhney, H., Shim, K.: Fast similarity search in the presence of noise, scaling, and translation in time-series databases. In: Proc. VLDB 1995 (1995)
6. Berndt, D., Clifford, J.: Using dynamic time warping to find patterns in time series. In: AAAI 1994 workshop on knowledge discovery in databases, pp. 229–248 (1994)
7. Goldin, D.Q., Kanellakis, P.C.: On similarity queries for time-series data: Constraint specification and implementation. In: Montanari, U., Rossi, F. (eds.) CP 1995. LNCS, vol. 976, pp. 137–153. Springer, Heidelberg (1995)
8. Perng, C.S., Wang, H., Zhang, S.R., Parker, D.S.: Landmarks: a new model for similarity-based pattern querying in time series databases. In: Proceedings of ICDE (2000)
9. Steingold, S., Wherry, R., Piatetsky-Shapiro, G.: Measuring real-time predictive models. In: Proceedings of the 2001 IEEE International Conference on Data Mining (ICDM01) (2001), Slides available at http://www.kdnuggests.com/gpspubs/ICDM-2001-RT-measuring-models/index.htm
10. Ding, Y., Li, X.: Time weight collaborative filtering. In: Proceedings of the 14th ACM international conference on Information and knowledge management, pp. 485–492. ACM Press, New York (2005)
11. Wang, H.: Nearest neighbors by neighborhood counting. IEEE Transactions on Pattern Analysis and Machine Intelligence 28(6), 942–953 (2006)
12. Wang, H., Dubitzky, W.: A flexible and robust similarity measure based on contextual probability. In: Proceedings of IJCAI 2005, pp. 27–32 (2005)
13. Wang, H.: All common subsequences. In: Proc. IJCAI 2007, pp. 635–640 (2007)
14. Stevens, S.S.: Mathematics, Measurement, and Psychophysics. In: Handbook of Experimental Psychology, Wiley, Chichester (1951)
15. Wang, H., Düntsch, I., Gediga, G., Skowron, A.: Hyperrelations in version space. International Journal of Approximate Reasoning 36(3), 223–241 (2004)

Reasoning Problems on Distributed Fuzzy Ontologies*

Bo Zhou[1], Jianjiang Lu[1], Yanhui Li[2], Yafei Zhang[1], and Dazhou Kang[2]

[1] Institute of Command Automation, PLA University of Science and Technology,
Nanjing 210007, China
[2] School of Computer Science and Engineering, Southeast University,
Nanjing 210096, China

Abstract. The distributed fuzzy ontologies play an important role in semantic web. Most current solutions are proposed respectively on one of these two aspects. To represent distributed fuzzy ontology, we combine fuzzy description logics and E-connections, and present distributed extended fuzzy description logic ($\mathcal{DEFSHOIN}$). This novel logical approach couples both fuzzy and distributed features within description logics. The main contribution of this paper is to propose a combined tableau to achieve reasoning within $\mathcal{DEFSHOIN}$.

Keywords: ontology; distributed; fuzzy; description logic.

1 Introduction

Description logics (DLs) [1] are widely used in the semantic web as a logic foundation for knowledge representation and reasoning. The most popular ontology language OWL uses DLs as their underlying formalism. The fuzzy ontologies play an important role in many web applications such as text mining and multimedia information system. To represent fuzzy ontologies, many fuzzy DLs have been presented. Straccia presented a fuzzy extension of typical $\mathcal{ALC}$ ($\mathcal{FALC}$), and gave a constraint propagation calculus for reasoning with empty TBoxes [2]. Some reasoning techniques for FDLs were discussed in [3,4]. Calegari developed a suited plug-in of the KAON Project in order to introduce fuzziness in an ontology, which is based on corresponding fuzzy ontology and Fuzzy-OWL notions [5]. However, to some degree, Straccias fuzzy framework only brings limited fuzzy expressive power in some complex fuzzy cases. To overcome its insufficiency, we pointed out a new family of extended fuzzy description logics (EFDLs), in which cut sets of fuzzy concepts and fuzzy roles are considered as the atomic concepts and atomic roles [6]. Some complexity results and reasoning techniques in EFDLs were proposed in [6,7], and in these papers, expressive advantages of our framework were detailed discussed.

For the number of independently developed ontologies increases rapidly in current semantic web, many research works on distributed description logics have been presented. Borgida proposed distributed description logics (DDLs), in which multiple description logic knowledge bases are combined by inter-ontology bridge rules [8]. E-connection is another popular technique for coupling different ontologies [9]. Cuenca

---

* This work was supported by the National High Technology Research and Development Program of China (No. 2007AA01Z126, 863 Program).

G. Wang et al. (Eds.): RSKT 2008, LNAI 5009, pp. 588–595, 2008.

integrated the E-connections formalism into OWL by defining links that stand for the inter-ontology relations [10]. He also pointed out that the extension is strictly more expressive than C-OWL. Parsia made some more expressive extension of links and gave tableau calculus for reasoning within such links [11].

The extensions of description logic in fuzzy and distributed aspects respectively have done a lot, however, only a few published works have considered on both of these two extensions. We combined EFDLs and E-connection, and got the distributed EFDL (DEFDL) [12]. However, we have not yet proposed the reasoning method for DEFDL.

In this paper, we combine the extended fuzzy description logic $\mathcal{EFSHOIN}$ and E-connections, thus getting Distributed $\mathcal{EFSHOIN}$ ($\mathcal{DEFSHOIN}$). The remainder of this paper is organized as follows: a brief introduction of $\mathcal{EFSHOIN}$ is given in section 2; by introducing cut links that couple different $\mathcal{EFSHOIN}$ knowledge bases (KBs), we propose a general definition of syntax, semantics in section 3; section 4 gives a combined tableau for $\mathcal{DEFSHOIN}$; finally section 5 concludes this paper.

2　Extended Fuzzy Description Logic

Let N_C and N_R be two disjoint and countable infinite sets of atomic fuzzy concepts (denoted B) and of atomic fuzzy roles (denoted R, S). For any $B \in N_C$, $R \in N_R$ and $0 < n \leq 1$, we call $B_{[n]}$ an atomic cut concept and $R_{[n]}$ an atomic cut role, where B and R are the prefixes of n, and n is the suffix of B and R. The semantics of atomic fuzzy concepts and roles, and their cut sets are defined in terms of a fuzzy interpretation $\mathcal{I} = \langle \Delta^{\mathcal{I}}, \cdot^{\mathcal{I}} \rangle$. The domain $\Delta^{\mathcal{I}}$ is a nonempty set and the interpretation function $\cdot^{\mathcal{I}}$ maps every fuzzy atomic concept B into a membership function $B^{\mathcal{I}} : \Delta^{\mathcal{I}} \to [0, 1]$, and every fuzzy atomic role R into a membership function $R^{\mathcal{I}} : \Delta^{\mathcal{I}} \times \Delta^{\mathcal{I}} \to [0, 1]$. Additionally, $\cdot^{\mathcal{I}}$ maps $B_{[n]}$ and $R_{[n]}$ into n-cuts of $B^{\mathcal{I}}$ and $R^{\mathcal{I}}$, which are subsets of $\Delta^{\mathcal{I}}$ and $\Delta^{\mathcal{I}} \times \Delta^{\mathcal{I}}$:

$$B_{[n]}^{\mathcal{I}} = \{d | B^{\mathcal{I}}(d) \geq n\} \tag{1}$$

$$R_{[n]}^{\mathcal{I}} = \{(d, d') | R^{\mathcal{I}}(d, d') \geq n\} \tag{2}$$

Obviously B and R are fuzzy sets w.r.t $\Delta^{\mathcal{I}}$ and $\Delta^{\mathcal{I}} \times \Delta^{\mathcal{I}}$, while their cuts $B_{[n]}$ and $R_{[n]}$ are actually crisp sets.

Starting with atomic cut concepts $B_{[n]}$ and atomic cut roles $R_{[n]}$, cut concept descriptions and cut role descriptions (cut concepts and cut roles, for short) can be inductively defined by applying concept and role constructors in classical description logics. The syntax and semantics of these constructors are summarized in Table 1.

An $\mathcal{EFSHOIN}$ knowledge base (KB for short) $\mathcal{K}(\mathcal{T}, \mathcal{R}, \mathcal{A})$ contains terminological axioms about cut concepts and roles in TBox $\mathcal{T}$ and RBox $\mathcal{R}$, and assertions about individuals in ABox $\mathcal{A}$.

A TBox $\mathcal{T}$ is a finite set of cut concept axioms of the form $C_{[n_1,...,n_k]} \sqsubseteq D_{[m_1,...,m_l]}$, where for any $1 \leq i \leq n$, $1 \leq j \leq l$, $n_i, m_j \in (0, 1]$. Any interpretation $\mathcal{I}$ satisfies $C_{[n_1,...,n_k]} \sqsubseteq D_{[m_1,...,m_l]}$ iff $C_{[n_1,...,n_k]}^{\mathcal{I}} \subseteq D_{[m_1,...,m_l]}^{\mathcal{I}}$. $\mathcal{I}$ is a model of TBox $\mathcal{T}$ iff $\mathcal{I}$ satisfies all axioms in $\mathcal{T}$.

An RBox $\mathcal{R}$ is a finite set of cut role axioms of the form $R_{[n]} \sqsubseteq S_{[m]}$, where $n, m \in (0, 1]$, $R_{[n]}$ and $S_{[m]}$ are cut roles. An interpretation $\mathcal{I}$ satisfies the above cut

Table 1. Concept and role constructors

Name	Syntax	Semantics
Top concept	$\top$	$\Delta^{\mathcal{I}}$
Bottom concept	$\bot$	$\emptyset$
Conjunction	$C_{[n_1,\ldots,n_h]} \sqcap D_{[n_{h+1},\ldots,n_k]}$	$C^{\mathcal{I}}_{[n_1,\ldots,n_h]} \cap D^{\mathcal{I}}_{[n_{h+1},\ldots,n_k]}$
Disjunction	$C_{[n_1,\ldots,n_h]} \sqcup D_{[n_{h+1},\ldots,n_k]}$	$C^{\mathcal{I}}_{[n_1,\ldots,n_h]} \cup D^{\mathcal{I}}_{[n_{h+1},\ldots,n_k]}$
Complement	$\neg C_{[n_1,\ldots,n_k]}$	$\Delta^{\mathcal{I}} - C^{\mathcal{I}}_{[n_1,\ldots,n_k]}$
Value restriction	$\forall R_{[n]}.C_{[n_1,\ldots,n_k]}$	$\{d | \forall d', (d,d') \in R^{\mathcal{I}}_{[n]} \to d' \in C^{\mathcal{I}}_{[n_1,\ldots,n_k]}\}$
Existential restriction	$\exists R_{[n]}.C_{[n_1,\ldots,n_k]}$	$\{d | \exists d', (d,d') \in R^{\mathcal{I}}_{[n]} \wedge d' \in C^{\mathcal{I}}_{[n_1,\ldots,n_k]}\}$
Unqualified number restriction	$\geq N R_{[n]}$ $\leq N R_{[n]}$	$\{d | \#\{d' | (d,d') \in R^{\mathcal{I}}_{[n]}\} \geq n\}$ $\{d | \#\{d' | (d,d') \in R^{\mathcal{I}}_{[n]}\} \leq n\}$
Inverse role	$\mathrm{Inv}(R_{[n]})$	$\{(d,d') | (d',d) \in R^{\mathcal{I}}_{[n]}\}$
Transitive role	$\mathrm{Trans}(R_{[n]})$	$R^{\mathcal{I}}_{[n]} = (R^{\mathcal{I}}_{[n]})^{+}$
Nominal	$\{o\}$	$\{o\}^{\mathcal{I}}(d) = 1$ iff $d = o^{\mathcal{I}}$

role axiom, iff $R^{\mathcal{I}}_{[n]} \subseteq S^{\mathcal{I}}_{[m]}$ holds. $\mathcal{I}$ is a model of TBox $\mathcal{R}$ iff $\mathcal{I}$ satisfies all axioms in $\mathcal{R}$. And we introduce $\sqsubseteq^*$ as the transitive-reflexive closure of $\sqsubseteq$ on $\mathcal{R} \cup \{\mathrm{Inv}(R_{[n]}) \sqsubseteq \mathrm{Inv}(S_{[m]}) | R_{[n]} \sqsubseteq S_{[m]} \in \mathcal{R}\}$.

An ABox $\mathcal{A}$ is a finite set of cut concept assertions of the form $a : C_{[n_1,\ldots,n_k]}$, cut role assertions of the form $(a,b) : R_{[n]}$ and individual assertions of the form $a \neq b$, where a and b are individuals and their interpretation $a^{\mathcal{I}}$ and $b^{\mathcal{I}}$ are elements in $\Delta^{\mathcal{I}}$. $mathcalI$ satisfies an assertion $a : C_{[n_1,\ldots,n_k]}$, $(a,b) : R_{[n]}$ or $a \neq b$, iff $a^{\mathcal{I}} \in C^{\mathcal{I}}_{[n_1,\ldots,n_k]}$, $(a^{\mathcal{I}}, b^{\mathcal{I}}) \in (R_{[n]})^{\mathcal{I}}$ or $a^{\mathcal{I}} \neq b^{\mathcal{I}}$. $\mathcal{I}$ satisfies an ABox $\mathcal{A}$ iff $\mathcal{I}$ satisfies any assertion in $\mathcal{A}$, such $\mathcal{I}$ is called a model of $\mathcal{A}$.

$\mathcal{I}$ satisfies an $\mathcal{EFSHOIN}$ KB $\mathcal{K}(\mathcal{T}, \mathcal{R}, \mathcal{A})$ (written $\mathcal{I} \models \mathcal{K}$) iff $\mathcal{I}$ is a model of $\mathcal{T}$, $\mathcal{R}$ and $\mathcal{A}$; such $\mathcal{I}$ is also called a model of $\mathcal{K}$.

3 Distributed Extended Fuzzy Description Logic

A Distributed $\mathcal{EFSHOIN}$ ($\mathcal{DEFSHOIN}$ for short) KB is a pair $\sum = (\mathcal{K}_S, L_S)$, where $\mathcal{K}_S$ is a set of $\mathcal{EFSHOIN}$ KBs: $\mathcal{K}_S = \{\mathcal{K}^i(\mathcal{T}^i, \mathcal{R}^i, \mathcal{A}^i) | 1 \leq i \leq s\}$, and L_S is a set of LBoxes that connect two KBs in KS: $L_S = \{L^{ij} | 1 \leq i, j \leq s$ and $i \neq j\}$.

In any LBoxes L^{ij}, we introduce fuzzy links(denoted E^{ij}, F^{ij}) and cut links (denoted $E^{ij}_{[n]}, F^{ij}_{[n]}$), and we propose cut link axioms $E^{ij}_{[n]} \sqsubseteq F^{ij}_{[m]}$ to describe constraints among cut links, where $0 < n, m \leq 1$. The semantics of fizzy links, cut links and cut link axioms are defined in terms of a link fuzzy interpretation $\mathcal{I}_{ij} = \langle \Delta^{\mathcal{I}_{ij}}, \cdot^{\mathcal{I}_{ij}} \rangle$. Let $\mathcal{I}_i$ and $\mathcal{I}_j$ be the fuzzy interpretations of $\mathcal{K}^i$ and $\mathcal{K}^j$, $\mathcal{I}_{ij}$ are constructed based on $\mathcal{I}_i$ and $\mathcal{I}_j$: $\Delta^{\mathcal{I}_{ij}} = \Delta^{\mathcal{I}_i} \times \Delta^{\mathcal{I}_j}$. And the interpretation function $\cdot^{\mathcal{I}_{ij}}$ maps E^{ij} and $E^{ij}_{[n]}$ as:

$$(E^{ij})^{\mathcal{I}_{ij}} : \Delta^{\mathcal{I}_i} \times \Delta^{\mathcal{I}_j} \to [0,1], (E^{ij}_{[n]})^{\mathcal{I}_{ij}} = \{x | (E^{ij})^{\mathcal{I}_{ij}}(x) \geq n\} \tag{3}$$

A link interpretation $\mathcal{I}_{ij}$ satisfies a cut link axiom $E^{ij}_{[n]} \sqsubseteq F^{ij}_{[m]}$, iff $(E^{ij}_{[n]})^{\mathcal{I}_{ij}} \subseteq (F_{[m]})^{\mathcal{I}_{ij}}$. $\mathcal{I}_{ij}$ satisfies L^{ij}, iff it satisfies any axiom in L^{ij}.

Additionally, for any cut concept $C^j_{[n_1,\ldots,n_k]}$ in $\mathcal{K}^j$ and any cut link $E^{ij}_{[n]}$ in L^{ij}, the following expressions $\forall E^{ij}_{[n]}.C^j_{[n_1,\ldots,n_k]}$ and $\exists E^{ij}_{[n]}.C^j_{[n_1,\ldots,n_k]}$ are also considered as cut concepts in $\mathcal{K}^i$.

A combined interpretation of a $\mathcal{DEFSHOIN}$ KB is also a pair $\mathcal{I} = \langle \{\mathcal{I}_i\}, \{\mathcal{I}_{ij}\} \rangle$, where $\mathcal{I}_i$ is an interpretation of $\mathcal{K}^i$ and correspondingly $\mathcal{I}_{ij}$ is an interpretation of L^{ij}. The interpretation of cut concepts, cut roles and cut links are inductively defined as:

1. for any cut concept $C^j_{[n_1,\ldots,n_k]}$ in $\mathcal{K}^j$, $(C^j_{[n_1,\ldots,n_k]})^{\mathcal{I}} = (C^j_{[n_1,\ldots,n_k]})^{\mathcal{I}_j}$;
2. for cut role $R^j_{[n]}$ in $\mathcal{K}^j$, $(R^j_{[n]})^{\mathcal{I}} = (R^j_{[n]})^{\mathcal{I}_j}$;
3. for any cut link $E^{ij}_{[n]}$ in L^{ij}, $(E^{ij}_{[n]})^{\mathcal{I}} = (E^{ij}_{[n]})^{\mathcal{I}_{ij}}$;
4. for any individual a^j in $\mathcal{K}^j$, $(a^j)^{\mathcal{I}} = (a^j)^{\mathcal{I}_j}$;
5. for $\forall E^{ij}_{[n]}.C^j_{[n_1,\ldots,n_k]}$ and $\exists E^{ij}_{[n]}.C^j_{[n_1,\ldots,n_k]}$, their interpretation are defined as:

$$(\forall E^{ij}_{[n]}.C^j_{[n_1,\ldots,n_k]})^{\mathcal{I}} = \{d | \forall d', (d,d') \in (E^{ij}_{[n]})^{\mathcal{I}_{ij}} \rightarrow d' \in (C^j_{[n_1,\ldots,n_k]})^{\mathcal{I}_j}\}$$
$$(\exists E^{ij}_{[n]}.C^j_{[n_1,\ldots,n_k]})^{\mathcal{I}} = \{d | \exists d', (d,d') \in (E^{ij}_{[n]})^{\mathcal{I}_{ij}} \text{ and } d' \in (C^j_{[n_1,\ldots,n_k]})^{\mathcal{I}_j}\}$$

A combined interpretation $\mathcal{I}$ is a combined model of $\sum = (\mathcal{K}_S, L_S)$, iff $\mathcal{I}$ satisfies any $\mathcal{K}^i$ in $\mathcal{K}_S$ and any L^{ij} in L_S.

4   Combined Tableau

The combined interpretation of $\mathcal{DEFSHOIN}$ contains a set of domains and interpretation functions, where the interpretation function describes the relations between elements in the domain and cut concepts, cut roles, cut links and individuals. In the following definition of combined tableau, the nonempty set S_i corresponds to each domain; L_i, M_i and N_i describe the relations between elements in the domain and cut concepts, cut roles and individuals respectively; M_{ij} describes the relations between cut links and elements in S_i and S_j.

Before expressing the tableau, here we introduce some notations. Assume that the cut concepts appearing in tableau algorithms are written in NNF. The set of sub-concepts of a concept is denoted as $sub(C)$. For a $\mathcal{DEFSHOIN}$ KB $\mathcal{K}^i$, we define $sub(\mathcal{K}^i)$ as the union of all $sub(C)$, for any concept C appearing in $\mathcal{K}^i$, and finally, define the set of relevant sub-concepts of $\mathcal{K}^i$ as follows:

$$cl(\mathcal{K}^i) = sub(\mathcal{K}^i) \bigcup \{nnf(\neg C_{[n_1,\ldots,n_k]}) | C_{[n_1,\ldots,n_k]} \in sub(\mathcal{K}^i)\}$$
$$\bigcup \{\forall S_{[m]}.C_{[n_1,\ldots,n_k]} | C_{[n_1,\ldots,n_k]}(nnf(\neg C_{[n_1,\ldots,n_k]})) \in sub(\mathcal{K}^i)\} \tag{4}$$

Definition 1. (Combined Tableau). *Let R^i and $N_I{}^i$ be the sets of cut roles and individuals appearing in $\mathcal{K}^i$, and E_{ij} be the set of cut links appearing in L_{ij}. A combined tableau* T *for $\mathcal{DEFSHOIN}$ KB $\Sigma = (\mathcal{K}_S, L_S)$ is a tuple:* T $= \langle \text{T}_i, \text{M}_{ij} \rangle$, T$_i = \langle S_i, L_i, M_i, N_i \rangle, i \neq j, 1 \leq i, j \leq s, i \neq j$, *where*

- S_i: a nonempty set of nodes, with $S_i \cap S_j = \emptyset$;
- L_i: $S_i \rightarrow 2^{cl(\mathcal{K}^i)}$, maps each node to a set of concepts;
- M_i: $R_i \rightarrow 2^{S_i \times S_i}$,maps each cut role in R^i to a set of pairs of nodes in S_i;
- N_i: $N_I^i \rightarrow S_i$,maps any individual in N_I^i into a corresponding node in S_i;
- M_{ij}: $E_{ij} \rightarrow 2^{S_i \times S_j}$ maps each cut link in E_{ij} to a set of pairs of nodes, where the first element of the pair is an element of S_i and the second an element of S_j.

In any T_i, for any $x, y \in S_i$, $a, b, o \in N_I^i$, $C^i_{[n_1,...,n_k]}$, $D^i_{[m_1,...,m_k]} \in cl(\mathcal{K}^i)$, and $R^i_{[n]}, S^i_{[m]} \in R^i$, the following conditions must hold:

1. There does not exist $\geq NR^i_{[n]}$ and $\leq (N-1)R^i_{[n]}$ in $L_i(x)$, where $0 \leq m \leq n \leq 1$;
2. There does not exist $B_{[n]}$ and $\neg B_{[m]}$ in $L_i(x)$, where $0 \leq m \leq n \leq 1$;
3. if $C^i_{[n_1,...,n_k]} \sqsubseteq D^i_{[m_1,...,m_l]} \in \mathcal{T}^i, \forall x \in S_i, \neg C^i_{[n_1,...,n_k]} \sqcup D^i_{[m_1,...,m_l]} \in L_i(x)$;
4. if $C^i_{[n_1,...,n_h]} \sqcap D^i_{[m_1,...,m_l]} \in L_i(x)$, $C^i_{[n_1,...,n_h]} \in L_i(x)$ and $D^i_{[m_1,...,m_l]} \in L_i(x)$;
5. if $C^i_{[n_1,...,n_h]} \sqcup D^i_{[m_1,...,m_l]} \in L_i(x)$, $C^i_{[n_1,...,n_h]} \in L_i(x)$ or $D^i_{[m_1,...,m_l]} \in L_i(x)$;
6. if $\forall R^i_{[n]}.C^i_{[n_1,...,n_k]} \in L_i(x)$ and $(x, y) \in M_i(R^i_{[n]})$, $C^i_{[n_1,...,n_k]} \in L_i(y)$;
7. if $\exists R^i_{[n]}.C^i_{[n_1,...,n_k]} \in L_i(x)$, there is some $y \in S_i$ such that $(x, y) \in M_i(R^i_{[n]})$ and $C^i_{[n_1,...,n_k]} \in L_i(y)$;
8. if $\forall S^i_{[m]}.C^i_{[n_1,...,n_k]} \in L_i(x)$ and $(x, y) \in M_i(R^i_{[n]})$ for some $R^i_{[n]} \sqsubseteq^* S^i_{[m]}$ with $\mathrm{Trans}(R^i_{[n]}) =$True, $\forall R^i_{[n]}.C^i_{[n_1,...,n_k]} \in L_i(y)$;
9. if $\geq NR^i_{[n]} \in L_i(x)$, $\#\{y|(x, y) \in M_i(R^i_{[n]})\} \geq N$;
10. if $\leq NR^i_{[n]} \in L_i(x)$, $\#\{y|(x, y) \in M_i(R^i_{[n]})\} \leq N$;
11. if $(x, y) \in M_i(R^i_{[n]})$, $(x, y) \in M_i(R^i_{[m]})$, where $0 < m \leq n \leq 1$;
12. if $(x, y) \in M_i(R^i_{[n]})$ and $R^i_{[n]} \sqsubseteq^* S^i_{[m]}$, $(x, y) \in M_i(S^i_{[m]})$;
13. $(x, y) \in M_i(R^i_{[n]})$ iff $(y, x) \in M_i(\mathrm{Inv}(R^i_{[n]}))$;
14. if $\{o\} \in L_i(x) \cap L_i(y)$ for some $o \in N_I^i$, $x = y$;
15. for each $o \in N_I^i$, there is some $x \in S_i$ with $\{o\} \in L_i(x)$;
16. if $a \in C^i_{[n_1,...,n_k]} \in \mathcal{A}^i$, $C^i_{[n_1,...,n_k]} \in L_i(N_i(a))$;
17. if $(a, b) \in R^i_{[n]} \in \mathcal{A}^i$, $(N_i(a), N_i(b)) \in M_i(R^i_{[n]})$;
18. if $a \neq b \in \mathcal{A}^i$, $N_i(a) \neq N_i(b)$;

Condition 1 ensures the tableau contains no clash. Condition 2 deals with TBoxes. Conditions 3-15 are necessary for the soundness of combined tableaux. Conditions 16-18 ensure the correctness of individual mapping function $N_i()$.

Additionally, we add some constraints to deal with cut links. For any $x \in S_i$, $w \in S_j$, $E^{ij}_{[n]}$, $F^{ij}_{[m]} \in E^{ij}$ and $C^j_{[n_1,...,n_k]} \in cl(K^j)$, the following conditions hold:

19. if $(x, w) \in M_{ij}(E^{ij}_{[n]})$ and $E^{ij}_{[n]} \sqsubseteq F^{ij}_{[m]} \in L^{ij}$, $(x, w) \in M_{ij}(F^{ij}_{[m]})$;
20. if $\forall E^{ij}_{[n]}.C^j_{[n_1,...,n_k]} \in L_i(x)$ and $(x, w) \in M_{ij}(E^{ij}_{[n]})$, $C^j_{[n_1,...,n_k]} \in L_j(w)$;
21. if $\exists E^{ij}_{[n]}.C^j_{[n_1,...,n_k]} \in L_i(x)$, there is some $w \in S_j$ such that $(x, w) \in M_i(E^{ij}_{[n]})$ and $C^j_{[n_1,...,n_k]} \in L_j(w)$;

Condition 19 guarantees that the combined tableau satisfies the restriction of LBoxes. Conditions 20-21 are distributed extensions of classical conditions to deal with forall and restrictions.

Theorem 1. *For any* $\mathcal{DEFSHOIN}$ *KB* $\sum = (\mathcal{K}_S = \{\mathcal{K}^i \langle \mathcal{T}^i, \mathcal{R}^i, \mathcal{A}^i \rangle\}, L_S = \{L^{ij}\})$, *has a combined model iff it has a combined tableau* T.

Proof. $\Rightarrow$) Let $\mathcal{I} = \langle \{\mathcal{I}_i\}, \{\mathcal{I}_{ij}\} \rangle$ be a combined model of $\sum$. We can create a combined tableau T $= \langle \mathrm{T}_i, \mathrm{M}_{ij} \rangle$, $\mathrm{T}_i = \langle S_i, L_i, M_i, N_i \rangle$ from $\mathcal{I}$ in the following steps:

S_i: $S_i = \Delta^{\mathcal{I}_i}$;
L_i: $\forall x \in S_i, L_i(x) = \{C^i_{[n_1,\dots,n_k]} | x \in (C^i_{[n_1,\dots,n_k]})^{\mathcal{I}_i}\}$;
M_i: $\forall R^i_{[n]} \in R_i, M_i(R^i_{[n]}) = \{(x,y) | (x,y) \in (R^i_{[n]})^{\mathcal{I}_i}\}$;
N_i: $\forall a \in N^i_I, N_i(a) = a^{\mathcal{I}_i}$;
M_{ij}: $\forall E^{ij}_{[n]} \in \mathrm{E}^{ij}, M_{ij}(E^{ij}_{[n]}) = \{(x,w) | (x,w) \in (E^{ij}_{[n]})^{\mathcal{I}_{ij}}, x \in S_i, w \in S_j\}$

Now, we prove T is a combined tableau for $\sum$, that means T satisfies the 21 restrictions:

3. if $C^i_{[n_1,\dots,n_k]} \sqsubseteq D^i_{[m_1,\dots,m_l]} \in \mathcal{T}^i$, for $\mathcal{I}_i$ satisfies $\mathcal{T}^i$, then $\Delta^{\mathcal{I}_i} = (\neg C^i_{[n_1,\dots,n_k]} \sqcup D^i_{[m_1,\dots,m_l]})^{\mathcal{I}_i}$. For any $x \in S_i$, $x \in S_i = \Delta^{\mathcal{I}_i} = (\neg C^i_{[n_1,\dots,n_k]} \sqcup D^i_{[m_1,\dots,m_l]})^{\mathcal{I}_i}$. Therefore $\neg C^i_{[n_1,\dots,n_k]} \sqcup D^i_{[m_1,\dots,m_l]} \in L_i(x)$;

4. if $C^i_{[n_1,\dots,n_h]} \sqcap D^i_{[m_1,\dots,m_l]} \in L_i(x)$, from the construction of L_i, $x \in (C^i_{[n_1,\dots,n_h]} \sqcap D^i_{[m_1,\dots,m_l]})^{\mathcal{I}_i}$. Then $x \in (C^i_{[n_1,\dots,n_h]})^{\mathcal{I}_i}$ and $x \in (D^i_{[m_1,\dots,m_l]})^{\mathcal{I}_i}$. Therefore $C^i_{[n_1,\dots,n_h]} \in L_i(x)$ and $D^i_{[m_1,\dots,m_l]} \in L_i(x)$;

14. if $\{o\} \in L_i(x) \cap L_i(y)$ for some $o \in N^i_I$, then $x = o^{\mathcal{I}_i} = y$;

15. for each $o \in N^i_I$, there must exist an element x such that $x = o^{\mathcal{I}_i}$. From the definition of L_i, $\{o\} \in L_i(x)$;

16. if $a \in C^i_{[n_1,\dots,n_k]} \in \mathcal{A}^i$, for $\mathcal{I}_i$ satisfies $\mathcal{A}^i$, $N_i(a) = a^{\mathcal{I}_i} \in (C^i_{[n_1,\dots,n_k]})^{\mathcal{I}_i}$. From the definition of L_i, $C^i_{[n_1,\dots,n_k]} \in L_i(N_i(a))$;

19. if $(x,w) \in M_{ij}(E^{ij}_{[n]})$ and $E^{ij}_{[n]} \sqsubseteq F^{ij}_{[m]} \in L^{ij}$, $(x,w) \in (E^{ij}_{[n]})^{\mathcal{I}_{ij}} \subseteq (F^{ij}_{[m]})^{\mathcal{I}_{ij}}$. We can get $(x,w) \in M_{ij}(F^{ij}_{[m]})$;

$\Leftarrow$) Let T $= \langle \mathrm{T}_i, \mathrm{M}_{ij} \rangle$ be a combined tableau for $\sum$, We can create a combined model $\mathcal{I} = \langle \{\mathcal{I}_i\}, \{\mathcal{I}_{ij}\} \rangle$ from T in the following steps:

1. the local domain $\Delta^{\mathcal{I}_i}$ is defined as: $\Delta^{\mathcal{I}_i} = S_i$
2. for any individual a that occurs in $\mathcal{A}^i$, $a^{\mathcal{I}_i} = N_i(a)$;
3. for any atomic cut concept $B^i_{[n]}$ and atomic cut role $R^i_{[n]}$ in $\mathcal{K}^i$,
$(B^i_{[n]})^{\mathcal{I}_i} = \{x | (B^i_{[n]}) \in L_i(x)\}$, $(R^i_{[n]})^{\mathcal{I}_i} = \{(x,y) | (x,y) \in M_i(R^i_{[n]})\}$;
4. for any cut link $E^{ij}_{[n]} \in \mathrm{E}^{ij}$, $(E^{ij}_{[n]})^{\mathcal{I}^{ij}} = \{(x,y) | (x,y) \in M_{ij}(E^{ij}_{[n]})\}$;
5. the interpretations of complex cut concepts in $\mathcal{K}^i$ are inductively defined based on 3 and 4.

Now we prove that $\mathcal{I}$ is a combined model of $\sum$.

1. According to tableau's conditions 4-15 and 20-21, it is easily verified that for any $C^i_{[n_1,...,n_k]}$, $C^i_{[n_1,...,n_k]} \in L_i(x)$, $x \in (C^i_{[n_1,...,n_k]})^{\mathcal{I}}$;

2. Consider $C^i_{[n_1,...,n_k]} \sqsubseteq D^i_{[m_1,...,m_l]} \in T^i$, we can get for any $x \in S_i$, $\neg C^i_{[n_1,...,n_k]} \sqcup D^i_{[m_1,...,m_l]} \in L_i(x)$. And from 1, $x \in (\neg C^i_{[n_1,...,n_k]} \sqcup D^i_{[m_1,...,m_l]})^{\mathcal{I}}$, then from the definition of $\Delta^{\mathcal{I}_i}$, $(\neg C^i_{[n_1,...,n_k]} \sqcup D^i_{[m_1,...,m_l]})^{\mathcal{I}} = \Delta^{\mathcal{I}_i}$, Therefore $\mathcal{I}$ satisfies $C^i_{[n_1,...,n_k]} \sqsubseteq D^i_{[m_1,...,m_l]}$;

3. we can follow the similar steps to prove $\mathcal{I}$ satisfies axioms in every RBox, ABox and LBox.

From theorem 1, an algorithm that constructs a combined tableau of can be considered as a decision procedure for the consistency of $\mathcal{DEFSHOIN}$. Unfortunately, there is no practical reasoning algorithm supporting reasoning within $\mathcal{DEFSHOIN}$. A calculus for satisfiability of concepts in $\mathcal{SHOIQ}$ [13] can be considered as a referenced solution. This algorithm exhibits a "pay as you go" behavior: if an input TBox, RBox and concept do not involve any one of inverse roles, number restrictions or nomimals, then the NN-rule will not be applied, and the corresponding non-deterministic guessing is avoided. For deciding the consistency of $\mathcal{DEFSHOIN}$ KBs, this algorithm needs improving by "replacing" classic concepts and roles with cut concepts and roles, and adding rules applied to cut links. Because the cut links used in current $\mathcal{DEFSHOIN}$ are not transitive, adding the related rules is not difficult. However, the technique "internalization" [14] used in $\mathcal{SHOIQ}$ is not available for $\mathcal{DEFSHOIN}$. No believable evidences guarantee that these translations can also be straightforwardly used in our fuzzy cases. Therefore, the present proposing $\mathcal{DEFSHOIN}$ doesn't include very complex concept axioms in TBoxes.

5   Conclusion

By development of the semantic web, fuzzy and distributed extensions of ontology languages attract more and more attention. This paper proposes a novel logical approach to couple fuzzy and distributed features. By integrating E-connection into $\mathcal{EFSHOIN}$, we point out $\mathcal{DEFSHOIN}$ and give a combined tableau for $\mathcal{DEFSHOIN}$ KBs. The main direction for future work involves the computational aspect. We are now addressing the issue to develop a calculus for reasoning within $\mathcal{DEFSHOIN}$. Another direction is in extending $\mathcal{DEFSHOIN}$ with more expressive forms. Currently, the cut links are not transitive, but the cut concept in one KB may have relations with another concept in other KBs through the transitivity of cut links. So we will extend the transitive closure of $\sqsubseteq$ on L_S to achieve this expression. Finally, We believe that $\mathcal{DEFSHOIN}$ is of great interest to the Semantic Web community, as it can be considered as a logical foundation to support reasoning with distributed fuzzy ontologies.

References

1. Baader, F., Sattler, U.: An overview of tableau algorithms for description logics. Studia Logica 69(1), 5–40 (2001)
2. Straccia, U.: A fuzzy description logic. In: Proceedings of AAAI 1998, 15th Conference of the American Association for Artificial Intelligence, Madison, Wisconsin, pp. 594–599 (1998)

3. Straccia, U.: Reasoning within fuzzy description logics. Journal of Artificial Intelligence Research 14, 137–166 (2001)

4. Li, Y.H., Xu, B.W., Lu, J.J., Kang, D.Z.: Discrete tableaus for fshi. In: Proceedings of 2006 International Workshop on Description Logics - DL 2006, The Lake District of the UK (2006)

5. Calegari, S., Ciucci, D.: Fuzzy ontology and fuzzy-owl in the kaon project. In: Proceedings of the 2007 Fuzzy Systems Conference, London, pp. 1–6 (2007)

6. Li, Y.H., Xu, B.W., Lu, J.J., Kang, D.Z.: On the computational complexity of the extended fuzzy description logic with numerical constraints. Journal of Software 17(5), 968–975 (2006)

7. Lu, J.J., Xu, B.W., Li, Y.H.: Extended fuzzy alcn and its tableau algorithm. In: Wang, L., Jin, Y. (eds.) FSKD 2005. LNCS (LNAI), vol. 3613, pp. 232–242. Springer, Heidelberg (2005)

8. Borgida, A., Serafini, L.: Distributed description logics: Assimilating information from peer sources. In: Spaccapietra, S., March, S., Aberer, K. (eds.) Journal on Data Semantics I. LNCS, vol. 2800, pp. 153–184. Springer, Heidelberg (2003)

9. Kutz, O., Lutz, C., Wolter, F.: E-connections of abstract description systems. Artificial Intelligence 156(1), 1–73 (2004)

10. Grau, B.C., Parsia, B., Sirin, E.: Working with multiple ontologies on the semantic web. In: Proceedings of the 3rd International Semantic Web Conference (2004)

11. Parsia, B., Cuenca Grau, B.: Generalized link properties for expressive e-connections of description logics. In: Proceedings of the Twentieth National Conference on Artificial Intelligence, Pittsburgh, pp. 657–662 (2005)

12. Li, Y.H., Lu, J.J., Zhou, B., Kang, D.Z.: A distributed and fuzzy extension of description logics. In: Gabrys, B., Howlett, R.J., Jain, L.C. (eds.) KES 2006. LNCS (LNAI), vol. 4251, pp. 655–662. Springer, Heidelberg (2006)

13. Horrocks, I., Sattler, U.: A tableaux decision procedure for shoiq. Journal of Automated Reasoning 39(3), 249–276 (2007)

14. Horrocks, I., Sattler, U., Tobies, S.: Practical reasoning for expressive description logics. In: Ganzinger, H., McAllester, D., Voronkov, A. (eds.) LPAR 1999. LNCS, vol. 1705, Springer, Heidelberg (1999)

An Approach for Kernel Selection Based on Data Distribution

Wenjian Wang[1], Jinling Guo[2], and Changqian Men[1]

[1] Institute of System Engineering, Faculty of Computer and Information Technology
Shanxi University, Taiyuan, P.R. China 030006
[2] Department of Information and Engineering, Business College of Shanxi University
Taiyuan, P.R. China 030031

Abstract. This paper presents a data based kernel selection approach, which utilizes the geometry distribution of data. Once the approximate distribution can be confirmed as a special one like circle, cirque, sphere cylinder, et al, some known kernel functions corresponding to the special distribution can then be used. Four datasets are used to verify the presented approach, and simulation results demonstrate the rationality and effectiveness of the presented approach.

1 Introduction

Support vector machine (SVM), developed by V. Vapnik [12-13], is a general powerful tool for solving classification, regression and time series prediction problems, et al [2,6,10-11]. SVM is a kernel based approach and the generalization performance of SVM is largely dependent on the used kernel function. In case when we have some prior knowledge of smoothness of the input data space, we can utilize them to choose a good kernel [3]. However, for most applications, prior knowledge about input data are very difficult to be obtained because the data often originated from unknown areas or the prior knowledge are uneasy to be extracted and formulated. Therefore, prior knowledge based approach for kernel selection is limited and constrained. Some alternative approaches can then be used for kernel selection in practical applications. At present, the common used methods include leave-one-out or cross validation [4], bootstrapping [1], estimation of VC dimension derived from statistical learning theory [8,4] and Bayesian learning [7] etc. Leave-one-out and bootstrapping are available and effective but with heavy computational burden because more training iterations are needed for each specific parameter, and they are not practical especially for large data sets. For the approach of VC dimension estimation, the kernel parameter is determined by minimizing the upper bound of VC dimension, while the radius of hypersphere enclosing the data in nonlinear feature space must be estimated. Because upper bound estimation of VC dimension is loose and empirical risk may not thought over, the obtained optimal parameter may be less accurate than leave-one-out. Bayesian methods interpret SVM as maximizing a posteriori solutions to inference problems with Gaussian Process priors. This probabilistic

G. Wang et al. (Eds.): RSKT 2008, LNAI 5009, pp. 596–603, 2008.

interpretation can provide intuitive guidelines for choosing a good SVM kernel, which is shown to define a Gaussian process prior over functions on the input space. This method relates kernel selection to prior assumptions about the kinds of learning problem at hand, i.e., it is actually the prior based approach.

Data dependent method is another way to select the optimal kernel, which is only related to the given training samples. Based on these empirical data, kernel and relative parameters can be optimized in advance or adjusted during learning but with limit iterations. It only needs less computational expense than data independent methods like leave-one-out. Therefore, data dependent method is general without additional prior assumption about input data and can be used to any area and any problem.

This paper presents a data based kernel selection strategy, which well utilizes the information embedded in data, i.e., approximate geometry distribution of data (for high dimension data, they need to be firstly projected to a low dimension like two or three dimension so as to determine their geometry distribution easily). If the data distribution is a special one like circle, cirque, sphere or cylinder et al, some known kernel functions corresponding to the special data distribution can be used. This paper provides an approach to determine some special distributions and appropriate kernels corresponding to these distributions. For an uninterpretable distribution data set, the most common used Gaussian or polynomial kernel may be a good choice.

2   The Proposed Approach

For a real world problem, all the usable information to guide kernel selection is usually originated from the given data, and the most direct clue may be the geometry distribution of the data. This paper focuses on utilizing data distribution to determine the most appropriate kernel function for a given problem. The approach judges the approximate geometry distribution firstly, and then it selects an appropriate kernel function based on the distribution. Because kernel selection is accomplished before SVM is trained, the computation cost can be largely decreased. Moreover, the features contained in data can be incarnated as well.

Since majority data are high-dimensional, the data need to be firstly projected to a lower dimension space so as to determine their distribution easily. Here, two common used methods, principle component analysis (PCA) [9] and multidimensional scaling (MDS) [5] are utilized. As we know, PCA can be applied in various areas but with many loss of information, so its performance may be at a discount especially for nonlinear high dimensional data. While, MDS takes the relations among the original data into account, so the loss of information embedding in the data is less after data are projected to a lower dimensional space. The experiments with both methods in next section also support the contention.

In the sequel, the proposed approach decides the distribution of the projected data. Four kinds of special data distribution, circle, cirque, sphere and cylinder, can be determined, and some transmutations of these distributions can be judged as well. Corresponding to these special distributions, polar kernel K_{polar}, sphere kernel K_{sphere} and cylinder kernel $K_{cylinder}$ can be regarded as the optimal kernel, respectively. For those uninterpretable data, Gaussian and polynomial kernels can be adopted alternatively.

The main idea of the proposed approach is concluded as follow:

> *Step 1: Decide whether the given dataset needs to be projected to a lower dimensional space. For a given dataset G, if the dimension of data vector is higher than three, the dataset G will be projected to a two or three dimensional space by PCA or MDS. The new dataset is denoted as G'.*
> *Step 2: Determine whether the approximate distribution of G' is the special one, i.e., circle, cirque, sphere or cylinder.*
> *Step 3: Select an appropriate kernel for the given dataset from a candidate kernel class $K(\Sigma)$ (Theoretically, $K(\Sigma)$ can be finite or infinite, and it is often finite for a practical application):*
> $K(\Sigma) = \{K_{gau}, K_{pol}, K_{spl}, K_{nn}, K_{polar}, K_{sphere}, K_{clinder}, \cdots\}$
> *Where,*
> *Gaussian kernel: $K_{gau}(\sigma) = \exp(-\|x - y\|^2/2\sigma^2)$;*
> *Polynomial kernel: $K_{pol}(d) = (\langle x, y \rangle + c)^d$;*
> *Spline kernel: $K_{spl}(\tau_i, \rho) = \sum_{\tau=0}^{\rho=l} x^\tau y^\tau + \sum_{s=1}^{N}(x - \tau_s)_+^\rho (y - \tau_s)_+^\rho$;*
> *Neural network kernel: $K_{nn}(\rho, \theta) = \tanh(\rho x^T y + \theta)$;*
> *Polar kernel: $K_{polar}(\alpha) = \alpha \tan(x_2/x_1) \cdot \alpha \tan(y_2/y_1) + \|x\|\|y\|$;*
> *Sphere kernel:*
> $K_{sphere}(\alpha) = \alpha \cos(x_3/\|x\|) \cdot \alpha \cos(y_3/\|x\|) + \alpha \tan(x_2/x_1) \cdot \alpha \tan(y_2/y_1) + \|x\|\|y\|$;
> *Cylinder kernel:*
> $K_{clinder}(\alpha) = x_3 y_3 + \alpha \tan(x_2/x_1) \cdot \alpha \tan(y_2/y_1) + \sqrt{(x_1^2 + x_2^2)(y_1^2 + y_2^2)}$;
> *If the distribution approximates to circle or cirque, the polar kernel K_{pol} will be selected. If it approximates to sphere or cylinder, the sphere kernel K_{sphere} or cylinder kernel $K_{clinder}$ will be selected, respectively. While, when the distribution can not be determined, the Gaussian kernel K_{gau} and the polynomial kernel K_{pol} may be common adopted.*

The proposed approach takes the approximate distribution of dataset into account, and regards it as a guide to select an appropriate kernel for the given data. In so doing, the generalization performance of SVM may be improved.

3   Simulation Results

To evaluate the proposed approach, four datasets are adopted. In the experiments, the classification error, the number of support vectors and CPU time are discussed for different kernels. For each dataset, Gaussian and/or polynomial kernels are used as comparison conferences.

3.1   Dataset D1

In the experiments, 96 data are generated randomly including 30 training data for each class and others take as testing data. Fig 1 shows the respective

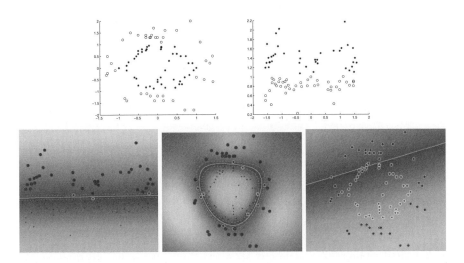

Fig. 1. Classification results on dataset D1 with approximate circle distribution. (The first one is the distribution of D1, the second one is the distribution of D1 in the feature space with polar kernel. The last three are the classification results with polar, Gaussian and polynomial kernels.)

distributions of dataset D1 in the original and feature space, classification results with different kernels.

Table 1 compares the classification error, the number of support vectors and CPU time by three different kernels, i.e., polar, Gaussian and polynomial kernels. From Fig. 1 and Table 1, it can be observed that the polar kernel is the optimal for dataset D1.

Table 1. Comparisons of training and testing results on D1 by three kernels

Kernel	error	#SVs	CPU time(second)
Polar	0	3	1.5
Gaussian	0	12	2.5
Polynomial	27	56	36

3.2 Dataset D2

In the experiments, 120 data are generated randomly including 40 training data for each class and others take as testing data. Fig 2 shows the respective distributions of dataset D2 in the original and feature space, classification results with different kernels.

Table 2 compares the classification error, the number of support vectors and CPU time by polar, Gaussian and polynomial kernels. From Fig. 2 and Table 2, it can be observed that the polar kernel is the optimal for dataset D2.

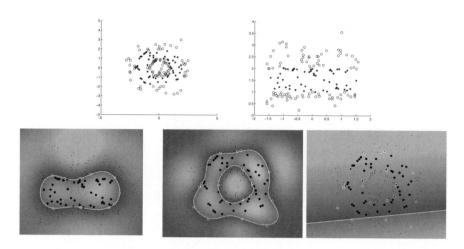

Fig. 2. Classification results on dataset D2 with approximate cirque distribution. (The first one is the distribution of D2, the second one is the distribution of D2 in the feature space with polar kernel. The last three are the classification results with polar, Gaussian and polynomial kernels.)

Table 2. Comparisons of training and testing results on D2 by three kernels

Kernel	error	#SVs	CPU time(second)
Polar	0	7	1.4
Gaussian	2	17	3.8
Polynomial	56	78	50

3.3   Dataset D3

In the experiments, 100 data are generated randomly including 30 training data for each class and others take as testing data. Fig 3 shows the respective distributions of dataset D3 in the original and feature space, classification results with different kernels.

Table 3 compares the classification error, the number of support vectors and CPU time by sphere, Gaussian and polynomial kernels. From Fig. 3 and Table 3, it can be observed that the sphere kernel is the optimal for dataset D3.

Table 3. Comparisons of training and testing results on D3 by three kernels

Kernel	error	#SVs	CPU time(second)
sphere	0	4	1.5
Gaussian	3	15	2.9
Polynomial	25	34	39

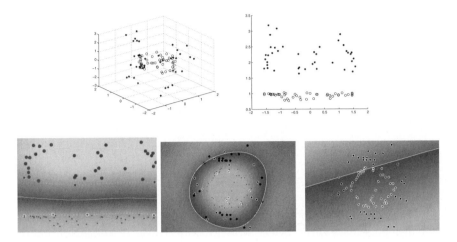

Fig. 3. Classification results on dataset D3 with approximate sphere distribution. (The first one is the distribution of D3, the second one is the distribution of D3 in the feature space with sphere kernel. The last three are the classification results with sphere, Gaussian and polynomial kernels.)

3.4 Dataset D4

In the experiments, 100 data are generated randomly including 30 training data for each class and others take as testing data. Fig 4 shows the respective distributions of dataset D4 in the original and feature space, classification results with different kernels.

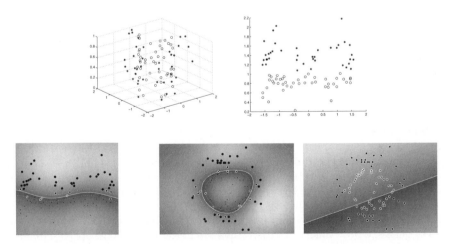

Fig. 4. Classification results on dataset D4 with approximate circle distribution. (The first one is the distribution of D4, the second one is the distribution of D4 in the feature space with cylinder kernel. The last three are the classification results with cylinder, Gaussian and polynomial kernels.)

Table 4 compares the classification error, the number of support vectors and CPU time by cylinder, Gaussian and polynomial kernels. From Fig. 4 and Table 4, it can be observed that the cylinder kernel is the optimal for dataset D4.

Table 4. Comparisons of training and testing results on D4 by three kernels

Kernel	error	#SVs	CPU time(second)
cylinder	0	4	1.5
Gaussian	2	15	2.9
Polynomial	25	34	39

4   Conclusion

This paper presents a kernel select strategy, which takes data distribution into account. Some special distributions and corresponding appropriate kernels are provided as well. The experiment results on four datasets support the rationality and effectiveness of the proposed approach. Although the data distributions of practical problems may be diverse, some new algorithms for verifying data distribution can be easily appended. Therefore, the proposed approach has practicability and extensity, and it can be regarded as an assistant tool for kernel selection.

Acknowledgement

The work described in this paper was partially supported by the National Natural Science Foundation of China (No. 60673095, 70471003), Hi-Tech R&D (863) Program (No. 2007AA01Z165), Program for New Century Excellent Talents in University (NCET), Project for Young Learned Leader, Program for Science and Technology Development in University (No. 200611001), and Program for Selective Science and Technology Development Foundation for Returned Overseas of Shanxi Province.

References

1. Anguita, D., Boni, A., Ridella, S.: Evaluating the generalization ability of support vector machines through the bootstrap. Neural Processing Letters 11, 51–58 (2000)
2. Burges, C.J.C.: A tutorial on support vector machines for pattern recognition. Data Mining and Knowledge Discovery 2(2), 121–167 (1998)
3. Burges, C.J.C.: Geometry and invariance in kernel based method. In: Scholkopf, B., Burges, C.J.C., Smola, A. (eds.) Advances in Kernel Methods, pp. 89–116. MIT Press, Cambridge (1999)
4. Chapelle, O., Vapnik, V.: Model selection for support vector machines. In: Smola, A., Leen, T., Muller, K. (eds.) Advances in Neural Information Processing Systems 12, MIT Press, Cambridge (2000)
5. Cox, T., et al.: Multidimensional Scaling. Chapman and Hall, London (1994)

6. Drucker, H., Burges, C.J.C., Kaufman, L., Smola, A., Vapnik, V.: Support vector regression machines. In: Mozer, M., Jordan, M., Petsche, J. (eds.) Advances in Neural Information Processing Systems 9, pp. 155–161. MIT Press, Cambridge (1997)
7. Gao, J.B., Gunn, S.R., Harris, C.J.: A probabilitic framework for SVM regression and error bar estimation. Machine Learning 46, 71–89 (2002)
8. Gunn, S.: Support vector machines classification and regression. ISIS Technical Report, Image Speech & Intelligent Systems Group, University of Southampton (1998)
9. Jabri, M.: Robust Principle Component Analysis. IEEE Transaction on Neural Network 1, 289–298 (2000)
10. Mattera, D., Haykin, S.: Support vector machines for dynamic reconstruction of a chaotic system. In: Scholkopf, B., Burges, C.J.C., Smola, A. (eds.) Advances in Kernel Methods-Support Vector Learning, pp. 211–242. MIT Press, Cambridge (1999)
11. Müller, K.R., Smola, A., Ratsch, G., Scholkopf, B., Kohlmorgen, J., Vapnik, V.: Predicting time series with support vector machines. In: Gerstner, W., Hasler, M., Germond, A., Nicoud, J.-D. (eds.) ICANN 1997. LNCS, vol. 1327, pp. 999–1004. Springer, Heidelberg (1997)
12. Vapnik, V.: The Nature of Statistical Learning Theory. Springer, New York (1995)
13. Vapnik, V.: Statistical Learning Theory. Wiley, New York (1998)

Phylogenetic Trees Dissimilarity Measure Based on Strict Frequent Splits Set and Its Application for Clustering*

Jakub Koperwas and Krzysztof Walczak

Institute of Computer Science, Warsaw University of Technology, Nowowiejska 15/19, 00-665 Warsaw, Poland
J.Koperwas@elka.pw.edu.pl,K.Walczak@ii.pw.edu.pl

Abstract. This paper focuses on clustering of leaf-labelled trees. As opposed to many other approaches from literature it is suitable not only for trees on the same leafset, but also for trees where the leafset varies. A new dissimilarity measure, constructed on the frequent subsplit term is used as the fundament of clustering technique. The clustering algorithm is designed to maximize the clustering quality measure. The computational time saving improvements are used. The initial results on phylogenetic and duplication trees are presented.

Keywords: Clustering, Leaf-labelled trees, Phylogenetic trees, Frequent subsplits.

1 Introduction

Comparing phylogenetic trees and clustering of phylogenetic trees are one of important issues in phylogenetic analysis. One of the important problems of existing approaches is that they are mainly applicable for tress on the same leafset. Measuring similarity or distance between phylogenetic trees is a key step for approaches like tree clustering. Many distance measures were proposed, however the Robinson Foulds distance is especially often used, because of its simplicity and intuitive interpretation. In this paper we propose a new dissimilarity measure for leaf-labelled trees. Unlike the Robinson Foulds distance, it is also suitable for trees where the leafset varies (trees on a free leafset). Moreover, it produces even more intuitive results than R-F for the trees on the same leafset and it is also easy to interpret. The measure is based on the strict maximal split set, which is a technique for extracting common information in trees where the leafset varies. It represents this information as maximal subsets of taxa connected with the same relations. In the next part we extend the quality-maximization approach presented in [1] in such way it is applicable for trees, where the leafset varies, and does not apriori discard any taxa like in [2].We also propose a time-saving improvement of proposed algorithm. This paper is closed by the initial results of clustering of phylogenetic and duplication trees.

* The research has been partially supported by grant No 3 T11C 002 29 received from Polish Ministry of Education and Science.

G. Wang et al. (Eds.): RSKT 2008, LNAI 5009, pp. 604–611, 2008.
© Springer-Verlag Berlin Heidelberg 2008

2   Basic Notions

Leaf-labelled trees are very often represented as a set of splits [3] . Split (or Bipartition) $A|B$ (of a tree T with leafset L), corresponding to an edge e is a pair of leafsets A and B, which originated in splitting tree T into two disconnected trees whilst removing an edge e from a tree T, $A \cup B = L$. In this paper we will refer to the leafset of a given split s as $L(s)$.

Definition 1 (Restricted Split Equality(z-equality)). *Splits s_1 and s_2 are restrictedly equal on the leafset z, if those two splits after removing leaves not in z are equal:*

$$s_1 =^z s_2 \iff s_1^z = s_2^z.[2]$$
(1)

For example, splits $abcd|efxy$ and $abcd|efwz$ are z-equal on leafset $\{abcdef\}$.

Definition 2 (Subsplit and supersplit). *Split s_1 is a subsplit of s_2 , and s_2 is a supersplit of s_1 , iff s_1 is restrictedly equal to s_2 on the leafset of s_1 , and leafset of s_1 is a subset of the leafset of s_2:*

$$s_2 \subseteq s_1 \iff (s_1 =^z s_2) \wedge z = L(s_1) \wedge (L(s_1) \subseteq L(s_2)),$$
(2)

it can also be presented alternatively:

$$s_2(A|B) \subseteq s_1(C|D) \iff (A \subseteq D \wedge B \subseteq C) \vee (A \subseteq C \wedge B \subseteq D).$$
(3)

For example $abcd|ef$ is a subsplit of $abcd|efxy$.

Various distance measures were proposed for leaf-labelled trees, among others Robinson-Foulds [4], (the most popular), Jaccard [5], nearest neighbour interchange (NNI), bisection and reconnection (TBR), subtree prune and regraft (SPR) [6]. The distances are suitable for only leaf-labelled trees on the same leafset, and the last three are NP-hard. As the R-F distance is the most popular and easy to interpret we will correspond to this measure in our paper.

R-F Distance - The most popular distance measure for the trees on the same leafset is R-F distance. R-F distance [4] between two trees T_1 and T_2 with set of splits S_1 and S_2 respectively is defined as follows:

$$d_{R-F}(T_1, T_2) = |S_1 \cup S_2| - |S_1 \cap S_2|.$$
(4)

Frequent subsplit s with support minsup in a profile of trees is a split that is a subsplit of at least one split in at least minsup of trees. The minsup parameter is called minimal support. It may be an absolute value which denotes the minimal number of trees the split is supposed to be found in (as a subsplit). It can also be given as the relative value, given as a minimal percentage of tree the split is supposed to be found in.

Maximal Strict Frequent Split Set (SFS) - a set that contains maximal frequent subsplits with minsup=100%. In other words, it contains subsplits s that occur in all input trees and there is no other frequent subsplit r that occurs

in all input trees that is also a supersplit of s. More formally, the SFS can be
represented as follows:

$$SFS(T_1,\ldots,T_n) = \{s : (\forall_{i\in\{1\ldots n\}} s\widetilde{\in}T_i) \wedge (\sim \exists_r(\forall_{i\in\{1\ldots n\}} r\widetilde{\in}T_i) \wedge s \subset r)\}, \quad (5)$$

where

$$s\widetilde{\in}T \iff (\exists_{r\in T} s \subseteq r). \quad (6)$$

The SFS has a very good interpretation, as it returns different, maximal sets
of taxa that are connected with the same relations in the input trees. It can
be visualized as the profile of trees or the one tree based on the intersection of
leafsets. More details about SFS and it's interpretation can be found in [7].

For example from the Figure 1:
T_1:$a|bcd, b|acd, c|abd, d|abc, ab|cd$
T_2:$a|bcdxywz, b|acdxywz, c|abdxywz, d|abcxywz, abxy|cdwz, x|abcdywz,$
$y|abcdxwz, w|abcdxyz, z|abcdxyw$
SFS: $a|bcd, b|acd, c|abd, d|abc, ab|cd$ (tree T_3 is the visualisation of the SFS).

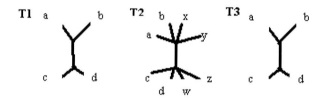

Fig. 1. Two trees on different leafset, together with the visualisation of their SFS

A more difficult example, where SFS produce more then one tree, is presented
on the Figure 2.
T_1: $cd|abefghi, bcd|aefghi, abcd|efghi, hi|abcdefg, ghi|abcdef,$
$fghi|abcde, a|bcdefghi, b|acdefghi, c|abdefghi, d|abcefghi, e|abcdfghi,$
$f|abcdeghi, g|abcedfhi, h|abcdefgi, i|abcdefgh$
T_2: $bc|adefghj, abc|defghj, abcd|efghj, hj|abcdefg, ghj|abcdef,$
$fghj|abcde, a|bcdefghj, b|acdefghj, c|abdefghj, d|abcefghj, e|abcdfghj,$
$f|abcdeghj, g|abcedfhj, h|abcdefgj, j|abcdefgh$
SFS of the input trees is as follows:$abcd|efgh, gh|abcdef, fgh|abcde,$
$bc|aefgh, h|abcdefg, a|bcdefgh, b|acdefgh, c|abdefgh, d|abcefgh, e|abcdfgh,$
$f|abcdegh, g|abcedfh,$

3   Clustering of Trees Using Frequent Splitset Approach

Various clustering techniques suitable for leaf-labelled trees were presented. Authors of paper [8] examined a few algorithms like agglomerative clustering or
k-mean-like algorithms for a partitioning set of phylogenetic trees with respect
to R-F distance. This approach was limited to leaf-labelled trees. The more

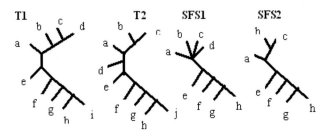

Fig. 2. Two input trees (T_1 and T_2) and trees built on their SFS (SFS_1 and SFS_2)

general, quality optimization approach was presented in [1]. In [2] we extended the techniques in the way it could be used for leaf-labelled trees on the free leafset. The methods, although simple and efficient, were only a partial problem solution for trees with a different leafset. Here we provide a clustering solution not only suitable for trees with a free leafset, but also providing better results for trees with the same leafset. Our approach aims to divide the trees into clusters in such a way, that it maximizes the quality of clustering which is dependent on the size of a frequent splitset of particular clusters. This way, we maximize common knowledge extractable from a cluster, which is usually the aim of clustering.

3.1   Dissimilarity Measure

At first, we would like to define a new dissimilarity measure between two trees (or splitsets) with the help of the frequent itemset term. Here we present a measure that is not only applicable for trees with different leafsets but also gives more intuitive results for trees with the same leafset:

$$d(T_1, T_2) = 1 - \frac{|SFS|}{|S_1 \widetilde{\cup} S_2|}. \qquad (7)$$

Where SFS is a strict-frequent splitset and $S_1 \widetilde{\cup} S_2$ is the modified sum of both splitsets, making that, if for splits $s_1 \in S_1, s_2 \in S_2$, s_1 is a supersplit of s_2 only supersplit (s_1) is included in result. Formally it can be represented as follows:

$$S_1 \widetilde{\cup} S_2 = \{s : (s \widetilde{\in} S_1 \vee s \widetilde{\in} S_2) \wedge (\sim \exists_r ((r \widetilde{\in} S_1 \vee r \widetilde{\in} S_2) \wedge s \subset r))\}. \qquad (8)$$

Such a measure determines dissimilarity on the basis of how many subsplits they share in common. Let us compare this measure to the most popular: R-F distance. Consider the example from the Figure 3:

$SFS(T_1, T_2) = trivial(5) + ab|cd$, $SFS(T_1, T_3) = trivial$, $SFS(T_2, T_3) = trivial$

$dRF(T_1, T_2) = 4$, $d(T_1, T_2) = 1 - 6/9 = 4/9$

$dRF(T_2, T_3) = 4$, $d(T_2, T_3) = 1 - 5/9 = 5/9$

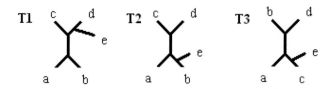

Fig. 3. Three different trees on the same leafset

It is clear that R-F distance states that T_1 and T_2 are the same dissimilar as T_2 and T_3 whilst our measure states differently, which is an intuitive result as both T_1 and T_2 share common non-trivial subsplit $ab|cd$. For trees on the different leafset, the R-F distance does not work at all whilst our measure does.

As for example from the Figure 1: $d(T_1, T_2) = 1 - 5/9 = 4/9$.
As for example from the Figure 2: $d(T_1, T_2) = 1 - 12/30 = 18/30 = 3/5$.

3.2 Quality Measure

For assessing the clustering quality, we use the method which bases on the cluster information loss, i.e. the amount of information that is lost while removing a cluster of trees with any chosen representative tree:

$$IG = \frac{\Delta I_{C_0} - \Delta I_C}{\Delta I_{C_0}}, \tag{9}$$

where ΔI_C is the sum of information loss for each cluster ΔI_{C_k}, given with expression:

$$\Delta I_{C_k} = \sum_{C(i)=k} d(T_i, T_R). \tag{10}$$

ΔI_{C_0} is the information loss of one-cluster clustering and T_R is a representative tree or split-set of given cluster, here we use SFS. For comprehensive explanation of motivations and characteristics of such quality measure see [1] [2]. In this approach we take a strict frequent splitset as a representative set and the dissimilarity measure presented in previous section.

3.3 Clustering of Leaf Labelled Trees, Maximizing Quality Measure for Strict Frequent Splitset as a Representative Set

Here we present the algorithm that attempts to maximize the proposed quality measure. It is based on agg-inf proposition given in [1]. As the base of algorithm the hierarchical agglomerative clustering is chosen, however as a merging condition we choose such two clusters to merge, so that they minimize the information loss after the merging. The following properties hold:

Property 1

$$SFS(S_1, SFS(S_2, S_3)) = SFS(S_1, S_2, S_3). \tag{11}$$

Proof. At first we will show that

$$s \in SFS(S_1, S_2, S_3) \Rightarrow s \in SFS(S_1, SFS(S_2, S_3)), \tag{12}$$

which is equivalent to:

$$s \in SFS(S_1, S_2, S_3) \Rightarrow \\ s\widetilde{\in}SFS(S_1, SFS(S_2, S_3)) \wedge \sim \exists_r(r\widetilde{\in}SFS(S_1, SFS(S_2, S_3)) \wedge s \subset r). \tag{13}$$

a) $s \in SFS(S_1, S_2, S_3) \Rightarrow s\widetilde{\in}SFS(S_1, SFS(S_2, S_3))$
is true according to SFS definition.

b) $s \in SFS(S_1, S_2, S_3) \Rightarrow \sim \exists_r(r\widetilde{\in}SFS(S_1, SFS(S_2, S_3)) \wedge s \subset r)$, assume
that it is not true then:

$$\exists_r(r\widetilde{\in}SFS(S_1, SFS(S_2, S_3)) \wedge s \subset r) \Rightarrow r\widetilde{\in}S_1 \wedge r\widetilde{\in}SFS(S_2, S_3) \\ \Rightarrow r\widetilde{\in}S_1 \wedge r\widetilde{\in}S_2 \wedge r\widetilde{\in}S_3 \Rightarrow s \notin SFS(S_1, S_2, S_3), \tag{14}$$

which is a contradiction because according to the assumption $s\widetilde{\in}SFS(S_1, S_2, S_3)$.
Now we will show that:

$$s \in SFS(S_1, SFS(S_2, S_3)) \Rightarrow s \in SFS(S_1, S_2, S_3), \tag{15}$$

which is equivalent to:

$$s \in SFS(S_1, SFS(S_2, S_3)) \\ \Rightarrow s\widetilde{\in}SFS(S_1, S_2, S_3) \wedge \sim \exists_r(r\widetilde{\in}S_1 \wedge r\widetilde{\in}S_2 \wedge r\widetilde{\in}S_3 \wedge s \subset r). \tag{16}$$

a) $s \in SFS(S_1, SFS(S_2, S_3)) \Rightarrow s\widetilde{\in}SFS(S_1, S_2, S_3)$ is true according to SFS
definition.

b) $s \in SFS(S_1, SFS(S_2, S_3)) \Rightarrow \sim \exists_r(r\widetilde{\in}S_1 \wedge r\widetilde{\in}S_2 \wedge r\widetilde{\in}S_3 \wedge s \subset r)$,
assume that it is not true then:

$$\exists_r(r\widetilde{\in}S_1 r \wedge \widetilde{\in}S_2 r \wedge \widetilde{\in}S_3 \wedge s \subset r) \\ \Rightarrow r\widetilde{\in}SFS(S_1, SFS(S_2, S_3)) \wedge s \subset r \Rightarrow s \notin SFS(S_1, SFS(S_2, S_3)), \tag{17}$$

which is a contradiction because according to the assumption $s \in SFS(S_1, SFS(S_2, S_3))$

The implication: $r\widetilde{\in}S_1 r \wedge \widetilde{\in}S_2 r \wedge \widetilde{\in}S_3 \wedge s \subset r \Rightarrow r\widetilde{\in}SFS(S_1, SFS(S_2, S_3))$,
can be explained as follows:

$$r\widetilde{\in}S_2 \wedge r\widetilde{\in}S_3 \Rightarrow r\widetilde{\in}SFS(S_2, S_3) \\ r\widetilde{\in}SFS(S_2, S_3) \wedge r\widetilde{\in}S_1 \Rightarrow r\widetilde{\in}SFS(S_1, SFS(S_2, S_3)). \tag{18}$$

Property 2. For any splitset X:

$$S\widetilde{\cup}SFS(S, X) = S. \tag{19}$$

Proof. According to SFS definition, $\forall_s(s \in SFS(S, X) \Rightarrow s\widetilde{\in}S \Rightarrow \exists_{r \in S}s \subset r)$
Now according to the modified sum definition:

$$S\widetilde{\cup}SFS(S, X) = \{s : (s\widetilde{\in}S \wedge s\widetilde{\in}SFS(S, X)) \wedge (\sim \exists_r(r\widetilde{\in}S \wedge r\widetilde{\in}SFS(S, X) \wedge s \subset r))\}, \tag{20}$$

From the definition of SFS it follows: $s\widetilde{\in}SFS(S, X) \Rightarrow s\widetilde{\in}S$, therefore

$$S\widetilde{\cup}SFS(S, X) = \{s : s\widetilde{\in}S \wedge (\sim \exists_r(r\widetilde{\in}S \wedge r\widetilde{\in}SFS(S, X) \wedge s \subset r))\} = S. \tag{21}$$

We take the information loss minimization criteria as the merging condition:

$$\Delta I' - \Delta I = \sum_{C(i)=z} d(T_i, SFS_z) - \sum_{C(i)=x} d(T_i, SFS_x) - \sum_{C(i)=y} d(T_i, SFS_y), \tag{22}$$

due to the properties 1 and 2 we can transform this to:

$$\Delta I' - \Delta I = \sum_{C(i)=x}(1 - \frac{|SFS_x| - |SFS(SFS_y, SFS_x)|}{|S_i|}) \\ + \sum_{C(i)=y}(1 - \frac{|SFS_y| - |SFS(SFS_y, SFS_x)|}{|S_i|}). \tag{23}$$

Where $C(i) = x$ selects trees assigned to cluster x, $C(i) = y$ selects trees assigned to cluster y, and $C(i) = z$ selects trees assigned to cluster z, which originated from merging cluster x and cluster y. Such a condition can be effectively counted as it does not require counting all possible information losses. The merging decision is based only on the frequent splitsets and the size of the splitset of trees assigned to clusters, which significantly reduces time required for computation.

4   Results

Here we present the results of clustering of leaf-labelled trees with the proposed approach(agg-inf-fs). We choose the phylogenetic trees datasets (Camp and the Caesal), which contain leaf-labelled trees on the same leafset. We compare it to the classical agg-inf approach as proposed in [1], by counting the quality measure based on a frequent subsplit approach. The other dataset contains duplication trees which are leaf-labelled trees on a free leafset. Here we compare the result to the z-restricted version of agg-inf [2], on the common leafset. The results are presented in Table 1.

Table 1. Value of IG parameter for clustering leaf-labelled trees with agg-inf and agg-inf-fs approach

k	Camp		Caesal		Dupl-trees	
	Agg-inf	Agg-inf-fs	Agg-inf	Agg-inf-fs	Agg-inf	Agg-inf-fs
5	0.14	0.19	0.58	0.58	0.09	0.18
7	0.16	0.21	0.57	0.57	0.10	0.19
8	0.18	0.23	0.54	0.54	0.12	0.21
9	0.20	0.25	0.53	0.53	0.13	0.22

The results presented above show that the proposed algorithm, based on frequent splitset approach can give at least not worse (Caesal) and up to 31% (Camp) and 56% (Dupl-trees) better results then the classical split-based approach. This proves that the proposed methods are reasonable.

5   Discussion

In this paper we have presented a new method of clustering leaf-labelled trees where the leafset varies, which is a novel approach. The presented method is also suitable for trees with the same leafset, and for such it provides better results than other known methods. Presented algorithm is based on the dissimilarity measure derived from frequent subsplitset which is a novel method of extracting common knowledge from trees. The presented results prove that proposed method is reasonable. Future work is needed to further improve the performance. The studies on possible metrics derived form frequent subplitset are also challenging.

References

1. Koperwas, J., Walczak, K.: Clustering of leaf-labelled trees. In: Beliczynski, B., Dzielinski, A., Iwanowski, M., Ribeiro, B. (eds.) ICANNGA 2007. LNCS, vol. 4431, pp. 702–710. Springer, Heidelberg (2007)
2. Koperwas, J., Walczak, K.: Clustering of leaf labeled-trees on free leafset. In: Kryszkiewicz, M., Peters, J.F., Rybinski, H., Skowron, A. (eds.) RSEISP 2007. LNCS (LNAI), vol. 4585, pp. 736–745. Springer, Heidelberg (2007)
3. Bryant, D.: Building trees, hunting for trees, and comparing trees. Theory And Method. In: Phylogenetic Analysis. Ph.D Thesis University of Canterbury (1997)
4. Robinson, D., Foulds, L.: Comparison of phylogenetic trees. Math. Biosciences. 53, 131–147 (1981)
5. Jaccard, P.: Bulletin del la Société Vaudoisedes Sciences Naturelles 37, 241–272 (1901)
6. Bryant, D.: The Splits in the Neighborhood of a Tree. Ann. Combinatorics 8 (2004)
7. Koperwas, J., Walczak, K.: Frequent subsplit representation of leaf-labelled trees. In: EvoBIO 2008. LNCS, vol. 4973, pp. 95–105. Springer, Heidelberg (2008)
8. Stockham, C., Wang, L.S., Warnow, T.: Statistically based postprocessing of phylogenetic analysis by clustering. Bionformatics 18, 285–293 (2002)
9. Finden, C.R., Gordon, A.D.: Obtaining common pruned trees. Journal of Classification 2, 255–276 (1985)
10. Bryant, D.: A classification of consensus methods for phylogenetics, In Bioconsensus. DIMACS Series in Discrete Mathematics and Theoretical Computer Science, vol. 61, pp. 163–184. AMS Press, New York (2002)
11. Amenta, N., Klingner, J.: Case study: Visualizing sets of evolutionary trees. In: 8th IEEE Symposium on Information Visualization, pp. 71–74 (2002)
12. Agrawal, R., Srikant, R.: Fast algorithms for mining association rules. In: Proc. of the 20th Int'l. In: Conference on Very Large Databases (VLDB 1994), Santiago, Chile, pp. 487–499 (1994)

A Multicriteria Model Applied in the Diagnosis of Alzheimer's Disease

Ana Karoline Araujo de Castro, Placido Rogerio Pinheiro,
and Mirian Caliope Dantas Pinheiro

Master Degree in Applied Computer Sciences
University of Fortaleza
Av. Washington Soares, 1321 - Bloco J sala 30,
CEP: 60811-905, Fortaleza, Ceara, Brazil
akcastro@gmail.com, placido@unifor, caliope@unifor.br

Abstract. This study considers the construction of a multicriteria model to assist in the diagnosis of Alzheimer's disease. This disease is considered to be one of the most frequent of the dementias and it is responsible for about 50% of the cases. Due to this fact and the therapeutical limitations in the most advanced stage of the disease, the diagnosis made of the correct way is extremely important and it can provide better life conditions to patients and their families. For the application of the model two scenarios: 1) The battery of standardized assessments developed by The Consortium to Establish a Registry for Alzheimer's disease (CERAD) and 2) The questionnaire with 120 questions that was used in a study realized in the city of Sao Jose dos Campos, SP, Brazil were used.

Keywords: Diagnosis, Alzheimer's, Multicriteria, MACBETH, HIVIEW.

1 Introduction

Demographic studies in developed and developing countries have showed a progressive and significant increase in the elderly population in the last years [9].

Alzheimer's disease is the most frequent cause of dementia and is responsible (alone or in association with other diseases) for 50% of the cases in western countries [9]. According to [8], despite its high incidence, doctors fail to detect dementia in 21 to 72% of their patients.

Considering the few alternative therapies and greater effectiveness of treatments after diagnosis, identifying the cases that are high-risk for becoming dementia take on capital importance [8].

The main focus of this work is to develop a multicriteria model for aiding in decision making for the diagnosis of Alzheimer's disease.

In this work, the modeling and evaluation processes have been conducted with the aid of a medical expert and bibliographic sources. Data sets can be used during the validation of the model, but achieving these data sets can be difficult. An example of a data set is the battery of standardized assessments

G. Wang et al. (Eds.): RSKT 2008, LNAI 5009, pp. 612–619, 2008.

developed by the Consortium to Establish a Registry for Alzheimer's disease (CERAD). Another example is the questionnaire with 120 questions that was used in a study realized in the city of Sao Jose dos Campos, SP, Brazil. We have sought to discover which questions are most relevant for the diagnosis of Alzheimer's disease by using these data sets.

We have provided a ranking with the classification of these questions. This ranking is composed of the construction of judgment matrixes and constructing value scales for each Fundamental Point of View already defined. The construction of cardinal value scales was implemented through MACBETH multicriteria methodology. At the end of this study, a comparison was made of the results obtained in the application of the model in both data sets.

2 Diagnosis of Alzheimer's Disease

Diagnosis of Alzheimer's disease is carried out in several steps. Initially, syndromic diagnosis is defined, which informs whether the patient presents the diagnostic criteria for dementia. After Dementia is confirmed, etiological diagnosis follows, which informs which disease is causing the dementia. In this case, we are looking for Alzheimer's disease [8] and [9].

3 Definition of the Multicriteria Model

In studies developed by [3] and [4] the application of the multicriteria model for aiding in diagnosis of Alzheimer's disease was presented. A substantial reading on MCDA methods can be found in [1], where the authors address the definitions and the problems that are involved in the decision making process.

In the present study, we sought to develop a multicriteria model that helps in the decision making related to the diagnosis of Alzheimer's disease. This model consists of several steps that will be described below.

Initially were defined the problem value tree corresponding to the Fundamental point of view (FPV) that are used in evaluation of the diagnosis. The evaluation process is composed of the construction of judgment matrixes and constructing value scales for each FPV already defined. The construction of cardinal value scales will be implemented through the MACBETH methodology developed by [1]. We used the MCDA tool to help in the construction of matrixes: M-MACBETH (**http : //www.m − macbeth.com**).

From the family of FPVs it is possible to evaluate the attractiveness of the options for each interest. Although the definition of the describers of impact is a difficult task, its decisiveness contributes for a good formation of judgments and a justified and transparent evaluation.

3.1 Describers

An FPV is operational in the moment that has a set of levels of associated impacts (describers). These impacts are defined for Nj, that can be ordered in decreasing form according to the describers [1].

In this step of construction of the describers, the decisions were made during the meetings with the health professional involved in the process.

Each FPV was operationalized in such a way as to evaluate the influence of the questions evaluated in the elderly patients that correspond to each criterion during the definition of the diagnosis of Alzheimer's disease.

For the evaluation of each FPV, the possible states were defined. Each FPV has a different quantity of states. These states were defined according to the exams or questions involved for each describer.

Its important remember that the describers has a structure of complete pre-order, otherwise, a superior level is always preferable a least level.

3.2 Analysis of Impacts

In this step, the analysis of impacts is carried out, according to each FPV: (i) the lowest and highest values of the impacts; and (ii) the relevant aspects of the distribution of the impacts in each one.

In this work, for each describer, the same values were considered to get the value function for each FPV. Therefore, scores higher than 60, obtained through the judgments matrixes were considered risk describers during the evaluation of diagnosis, in other words, the elderly person that has a great number of answers considered right in the definition of the diagnosis, becomes part of the group of people with a great probability of developing Alzheimer's Disease. This perception was defined by the health professional.

3.3 Evaluation

After the definition of the FPVs family and the construction of the describers, the next step is the construction of the cardinal value scales for each FPV. The evaluations of the judgments matrixes were made according to the opinion of the decision maker, the health area professional.

After evaluating the alternatives of all the FPVs individually, an evaluation of the FPVs in one matrix only was carried out. For this, a judgment matrix was created in which the decision maker's orders are defined according to the preference of the decision maker. The decision maker defined the order based on what he judged to be more important in deciding on a diagnosis.

3.4 Results

In this step we show the final result of the model, the contribution of the criterion for the diagnosis of Alzheimer's disease. We can see the describer values for each criteria. With these values we can conclude that the questions that are part of these describers should be preferentially applied during the definition of diagnosis of Alzheimer's disease.

3.5   Evaluation Results

In this step the value analyses of the sensitivity and the dominance of potential actions in the process of evaluation to discover which questions are most relevant for diagnosis of Alzheimer's disease by using data sets. HIVIEW software [6] was used to carry out these analyses. It is one of the instruments used in the decision support processes, essentially for evaluation of models obtained through Multi-criteria Methodologies for decision support, in virtue of the fact that it is used as a function of additive aggregation, yet it is compatible with the procedures developed in this study.

4   Application of the Multicriteria Model

The application of the model in multicriteria defined in this study was realized through the use of two scenarios that we describe in subsections 4.1 and 4.2. We gave a brief explanation in the two applications of the model. More information about the steps of the scenarios is shown in [5].

4.1   Scenario 1

The original mandate of the Consortium to Establish a Registry for Alzheimer's Disease (CERAD) in 1986 was to develop a battery of standardized assessments for the evaluation of cases with Alzheimer's disease who were enrolled in NIA-sponsored Alzheimer's Disease Centers (ADCs) or in other dementia research programs [7]. Despite the growing interest in clinical investigations of this illness at that time, uniform guidelines were lacking as to diagnostic criteria, testing procedures, and staging of severity. This lack of consistency in diagnosis and classification created confusion in interpreting various research findings. CERAD was designed to create uniformity in enrollment criteria and methods of assessment in clinical studies of Alzheimer's Disease and to pool information collected from sites joining the Consortium.

CERAD developed the following standardized instruments to assess the various manifestations of Alzheimer's disease: Clinical Neuropsychology, Neuropathology, Behavior Rating Scale for Dementia, Family History Interviews and Assessment of Service Needs.

According to the definition of the model, we first defined the problem value tree in figure 1. Next, we defined the describers. Each FPV has a different quantity of states. The FPV1 has 3 states. The FPV2 has 5 states. The FPV3 has 3 states. The FPV4 has 3 states. The FPV5 has 3 states. The FPV6 has 3 states. The FPV7 has 13 states. Next, the evaluation of the FPVs was made through the construction of the judgment matrixes.

After the evaluation we can see the final result of the model, the contribution of the criteria for the diagnosis of Alzheimer's.

Analyzing the FPV1 (History), we can see that only one describer achieved a value above that which was defined in the impact analysis. It is describer N03 with a value of 87.50. In FPV2 (Gross Examination), two describers achieved a

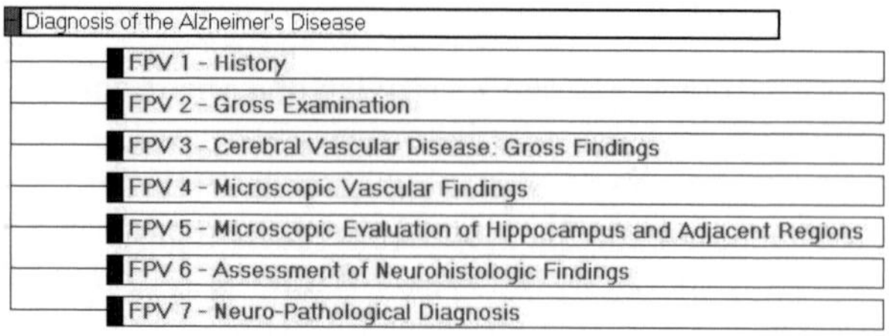

Diagnosis of the Alzheimer's Disease
FPV 1 - History
FPV 2 - Gross Examination
FPV 3 - Cerebral Vascular Disease: Gross Findings
FPV 4 - Microscopic Vascular Findings
FPV 5 - Microscopic Evaluation of Hippocampus and Adjacent Regions
FPV 6 - Assessment of Neurohistologic Findings
FPV 7 - Neuro-Pathological Diagnosis

Fig. 1. Problem value tree

value above that which was defined in the impact analysis. They are describers N04 and N05 with values of 69.23 and 92.31 respectively.

In FPV3 (Cerebral Vascular Disease: Gross Findings), only one describer achieved the defined minimum value. It is describer N03 with a value of 88.89. In FPV4 (Microscopic Vascular Findings), only one describer achieved the defined minimum value. It is describer N03 with a value of 88.89.

In FPV5 (Microscopic of Evaluation of Hippocampus and Adjacent Regions), only one describer achieved the minimum value. Describer N05 achieved a value of 92.86.

In FPV6 (Assessment of Neuro-histologic Findings) only one describer achieved the minimum value. Its value is 87.50. In FPV7 (Neuro-pathological Diagnoses), there were 10 describers which achieved the minimum value in impact analysis. They were describers N04 to N13, with values of 70.13, 80.52, 90.26, 91.56, 92.86, 94.16, 95.45, 96.75, 98.05 and 99.35.

In the last step of the application of the model we made the global evaluation of the actions when they are confronted with the seven FPVs. It was found that FPV1 to FPV7 have a total participation of 20%, 14%, 18%, 11%, 1%, 18% and 19%. Thus, action N03 proved to be potentially better with 85 points and the action N01 proved to be potentially least.

4.2   Scenario 2

In scenario 2 we used information obtained from the study realized in 2005 with 235 elderly people in the city of Sao Jose dos Campos, SP, Brazil. In this study a questionnaire with 120 questions that supplied demographic-social data, analyzed the subjective perception of the elderly, their mental and physical health (cognitive and emotional aspects), day-to-day independence, in addition to family and social support and the use of services [2].

According to the data collected with the application of the questionnaire we sought to carry out a study that showed among the people interviewed which have the possibility of being diagnosed with Alzheimer's disease. It's important to consider that at no time were questions answered by anyone that had already been diagnosed with the disease.

According to the definition of the model, we first defined the problem value tree in figure 2. Next, we defined the describers. Each FPV has a different

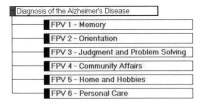

Fig. 2. Problem value tree

quantity of states. The FPV1 has 4 states. The FPV2 has 6 states. The FPV3 has 7 states. The FPV4 has 8 states. The FPV5 has 6 states. The FPV6 has 4 states.

After the evaluation we can see the final result of the model, the contribution of the criteria for the diagnosis of Alzheimer's.

Analyzing the FPV1 (Memory), we can see that two describers achieved a value above that which was defined in the impact analysis. It is describer N03 with a value of 77.78 and N4 with a value of 94.44. In FPV2 (Orientation), three describers achieved a value above that which was defined in the impact analysis. They are describers N04 with a value of 68.75, N05 with a value of 87.50 and N06 with values of 93.75.

In FPV3 (Judgment and Problem Solving), three describers achieved a value above that which was defined in the impact analysis. They were describers N05 to N07, with values of 69.57, 78.26 and 95.65. In FPV4 (Community Affairs), there were six describers which achieved the minimum value in impact analysis. They were describers N03 to N08, with values of 60.98, 80.49, 90.24, 92.68, 95.12 and 97.56.

In FPV5 (Home and Hobbies), five describers achieved a value above that which was defined in the impact analysis. They are describers N02 to N06, with values of 60.00, 80.00, 85.00, 90.00 and 95.00.

In FPV6 (Personal Care), two describers achieved the minimum value. It is describer N03 with a value of 72.73 and N04 with a value of 90.91.

In the last step of the application of the model we made the global evaluation of the actions when they are confronted with the seven FPVs. It was found that FPV1 to FPV6 have a total participation of 23%, 14%, 18%, 11%, 1% and 18%. Thus, action N04 proved to be potentially better with 84 points and the action N01 proved to be potentially least.

5 Comparation of the Results

Analyzing the results obtained with the application of the model in the two scenarios we can make several comparations. In scenario 1 we have the FPV7 (Neuro-pathological Diagnoses) as the criterion with the largest number of describers which obtained the minimum values required by the model. This result shows the importance of the definition of the diagnosis of Alzheimer's. Many studies show the importance of the definition of the diagnosis of dementia that should be carried out before the definition of the diagnosis of Alzheimer's disease [10], because many diseases can be confused with dementias and as a consequence, be confused with Alzheimer's disease. This is merely to underline the importance of this criterion for the solution of this problem. In scenario 2 we have FPV4 (Community Affairs) as the criterion with the largest number of describers which obtained the minimum values required by the model. This criterion has the characteristic of social integration with and without family members and also the execution of the daily activities. This criterion also contain factors related to health (problems of hearing, difficulty to speak and so on.), and mental health (satisfaction with life, sense of loneliness, and headaches, are some examples). Analyzing the characteristics of these two criteria, we see that despite being in different scenarios, they have similarities, as both the FPV7 in scenario 1 and FPV4 in scenario 2 are related primarily, the possible mental health problems that may occur with the patient and that may cause confusion in the definition of the diagnosis of Alzheimer's disease. The first result was obtained from the judgment of matrixes through the M-MACBETH software.

6   Conclusion

The diagnosis of Alzheimer's Disease is made up of many steps. The first step is to discover if the patient has dementia and then the patient is assessed to see if he has Alzheimer's.

Due to these limitations, this study sought to find the best way possible in the decision making process of defining this diagnosis. By using two data sets during the validation of the model: 1) The battery of standardized assessments developed by Consortium to Establish a Registry for Alzheimer's disease (CERAD) and 2) The questionnaire with 120 questions that was used in the study realized in the city of Sao Jose dos Campos, SP, Brazil. We attempted to select the main questions involved in diagnosis of Alzheimer's.

The MACBETH multicriteria method was used to aid in decision making. The criteria were defined according to the CERAD areas of assessment. In the questionnaire applied in the Brazilian study we defined the criteria based on to demographic-social data, analyzed the subjective perception of the elderly, their mental and physical health.

The questions that make up the battery of assessments and in the questionnaire were defined as the describers of the problem. With this information, the judgment matrixes were constructed using M-MACBETH software.

After evaluating the matrixes, a ranking was obtained showing all the questions, from most important to least important with respect to the diagnosis of Alzheimer's. At the end of this study, a comparison was made of the results obtained in the application of the model in both data sets.

An extension of the model is in the submission process [5] using Bayesian Networks. We defined the mapping of the questions of CERAD and in the questionnaire applied in the Brazilian study and through these mappings we can observe the questions of the biggest decisive impact for the diagnosis. In addition, with this Bayesian Network, we structured a multicriteria model that shows the questions of the biggest decisive impact for the diagnosis. Were showing the percentage of elderly people that presented the symptoms of Alzheimers disease.

As a future project, this model can be extended with the inclusion of new criteria or new models which can be developed using other data sets [2].

Acknowledgments. The authors thank the Consortium to Establish a Registry for Alzheimer's Disease (CERAD), Amanda Cecilia Simoes da Silva and Mischel Carmem Neyra Belderrain (Instituto Tecnologico de Aeronautica) for the divulgation of the data utilized in this case study. Ana Karoline Araujo de Castro is thankful to FUNCAP for the support she has received for this project.

References

1. Bana & Costa, C.A., Corte, J.M.D., Vansnick, J.C. Macbeth, LSE-OR Working Paper (2003)
2. Belderrain, M.C.N., Silva, A.C.S., Vianna, P.V.C.: Functional capacity of the elderly in city of Sao Jose dos Campos - Sao Paulo - Brazil. In: 21st European Conference on Operational Research - EURO XXI, Iceland. EURO XXI Proceedings (2006)
3. Castro, A.K.A., Pinheiro, P.R., Pinheiro, M.C.D.: Applying a Decision Making Model in the Early Diagnosis of Alzheimer's Disease. In: Yao, J., Lingras, P., Wu, W.-Z., Szczuka, M., Cercone, N.J., Ślęzak, D. (eds.) RSKT 2007. LNCS (LNAI), vol. 4481, pp. 149–156. Springer, Heidelberg (2007)
4. Castro, A.K.A., Pinheiro, P.R., Pinheiro, M.C.D.: A Multicriteria Model Applied in the Early Diagnosis of Alzheimer's Disease: A Bayesian Approach. In: International Conference on Operational Research for Development (ICORDVI), Brazil (2007)
5. Castro, A.K.A., Pinheiro, P.R., Pinheiro, M.C.D.: Applying Bayesian Networks and a Multicriteria Model in the Diagnosis of Alzheimer's Disease. BMC Medical Informatics and Decision Making (2008) (paper has been submitted), ISSN: 1472-6947
6. Keysalis: Hiview for Windows. Krysalis, London (1995)
7. Morris, J.C., Heyman, A., Mohs, R.C., et al.: The Consortium to Establish a Registry for Alzheimer's Disease (CERAD): Part 1. Clinical and Neuropsychological Assessment of Alzheimer's Disease. Neurology 39, 1159–1165 (1989)
8. Petersen, R.C., Smith, G.E., Waring, S.C., et al.: Mild Cognitive Impairment: Clinical Characterization and Outcome. Archives Neurology 56, 303–308 (1999)
9. Porto, C.S., Fichman, H.C., Caramelli, P., Bahia, V.S., Nitrini, R.: Brazilian Version of the Mattis Dementia Rating Scale Diagnosis of Mild Dementia in Alzheimer.s Disease. Arq Neuropsiquiatr. 61(2-B), 339–345 (2003)
10. Takada, L.T., Caramelli, P., Radanovic, M., et al.: Prevalence of Potentially Reversible Dementias in Adementia Outpatient Clinic of a Tertiary University Affiliated Hospital in Brazil. Arq Neuropsiquiatr 61, 881–890 (2003)

A New Complexity Metric Based on Cognitive Informatics

Sanjay Misra and Ibrahim Akman

Department of Computer Engineering, Atilim University, Ankara, Turkey
smisra@atilim.edu.tr, akman@atilim.edu.tr

Abstract. In this paper, a new complexity metric based on cognitive informatics is proposed for object oriented(OO) code. This is the single metric, which covers cognitive complexity of the OO system, method complexity and complexity due to inheritance together. The proposed metric was evaluated against Weyuker set of measurement principles. It was found that seven Weyuker properties are satisfied by this measure.

Keywords: Software metrics, object oriented programming, cognitive complexity, method complexity, inheritance, Weyuker properties.

1 Introduction

Cognitive Informatics (CI) [11] is an interdisciplinary area that tackles the problems related to modern informatics, computation, software engineering, artificial intelligence and cognitive science. The cognitive complexity measures the human effort needed to perform a task or the difficulty in developing software and is based on CI. Many software complexity measures based on cognitive informatics have been proposed in the last few years [5], [7], [8], and [12]. However, they all belong to procedural languages.

OO approach is characterized by its classes and objects, which are defined in terms of attributes (data) and operations (methods). Understandability of a code is mainly based on the operations of the code, which is closely related to its cognitive process and hence to cognitive complexity. Numbers of researchers have proposed variety of metrics for OO software development [2], [3], [4], [6]. These metrics generally do not involve the cognitive characteristics. On the other hand, the metrics on operation level like cyclomatic number, line of code etc., do not capture the features of object-oriented systems like polymorphism and inheritance.

In this study, we propose a cognitive complexity metric in terms of cognitive weights. It suggests that the complexity is not only due to operations in methods, but also due to an important feature of OO Programs (OOP): Inheritance. The proposed metric also includes the complexity caused by message calls between classes and, therefore, gives valuable information for coupling.

In the next section the proposal of the new complexity measure is introduced. Evaluation of the proposed metric through Weyuker properties is given in section 3. Conclusions constitute the last section.

G. Wang et al. (Eds.): RSKT 2008, LNAI 5009, pp. 620–627, 2008.

2   Complexity Metric for Object Oriented Programming

The cognitive weights of software are defined as the extent of difficulty or relative time and effort to comprehend the given software, modeled by a number of BCS [9], [12]. They measure the complexity of logical structures of the software. These logical structures reside in the method (code) of OOP and are classified as sequence, branch, iteration and call, whose weights are one, two, three and two, respectively [12].

The proposed measure first calculates the weight of individual method in a class by associating a number (weight) with each member function (method), and then it simply adds all the weights of all methods. This gives the complexity (weight) of a single class. There are two cases for calculating the whole complexity of the entire system depending on the architecture:

1. if the classes are in the same level then their weights are added.
2. if they are subclasses or children of their parent classes then their weights are multiplied (Inheritance).

If there are m level of depth in the OO code and level j has n classes then the class complexity (CC) of the system is given by,

$$\text{Class Complexity} = \text{CC} = \prod_{j=1}^{m} \left[\sum_{k=1}^{n} W_{c_{jk}} \right] \tag{1}$$

where W_c is the weight of the concerned class. The weight of a single operation is given by

$$W_e = \sum_{j=1}^{q} \left[\prod_{k=1}^{m} \sum_{i=1}^{n} W_c(j,k,l) \right] \tag{2}$$

where total cognitive weight of a software component W_c is defined as the sum of cognitive weights of its q linear blocks composed in individual BCS's. Each block consists of m layers of nested BCS's and each layer has n linear BCS. A higher weight indicates a higher level of effort required to understand the software and reduced maintainability. The Class Complexity Unit (CCU) of a class is defined as the cognitive weight of the simplest software component (having a single class which includes single method and also the method include only a linear structure). This corresponds to sequential structure in BCS and hence its cognitive weight is taken as 1.

The proposed complexity metric given by equation 1, is demonstrated with the programming example whose class diagram is shown in figure 1. This program (appendix 1) processes a person of a university system.

In this example, the main class **Person** has two subclasses **Employee** and **Student**. The class **Employee** is also a class with two subclasses **Faculty** and **Administrative**. As given in the following equation CC value of **Person** is 8 CCU since it has six methods. Similarly, the CC values of **Employee**, **Student**,

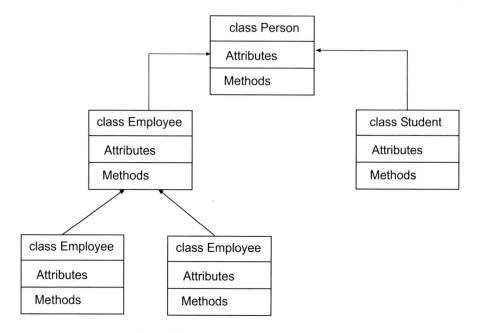

Fig. 1. Class diagram of program example

Faculty, and `Administrative` classes are found as 3, 6, 2 and 3 CCU respectively.

$$CC = \sum W_c = W_{p1} + W_{p2} + W_{p3} + W_{p4} + W_{p5} + W_{p6} = 1+1+3+1+1+1 = 8CCU$$

It can be seen from figure 1 that `Faculty` and `Administrative` classes are on the same level and they inherit their properties from `Employee`. Therefore, the CC of `Faculty` and `Administrative` is $2+3$ and CC of `Employee` is $(2+3)*3$. Similarly, `Employee` and `Student` are subclasses of `Person` and inherit their properties from `Person`. Therefore, total complexity of the system is:

CC = CC of class Person $*$ ((CC of class Employee $*$ (CC of class Faculty + CC of class Administrative) + CC of class Student)
$= 8 * (3 * (3+2) + 6) = 168CCU$

It is important to note that, in the case of a message passing between two classes, the complexity of the method is the sum of the weight of the called method and the weight due to that call (i.e. two). This means the proposed approach includes the complexity of the class due to messages and hence provides some indication of level of coupling. In other words, if the number of messages between the classes/objects increases, the overall complexity increases. This, clearly, indicates that a high complexity value represents high coupling between classes.

3   Proposed Metric Against Weyuker Criteria

Weyuker [13] suggested nine properties to provide a formal approach to determine the effectiveness of various software complexity measures. Number of researchers evaluated the object-oriented metrics by using these properties [1], [4], [10]. These properties were also employed to evaluate the proposed measure by using the example given in Appendix 1 when applicable.

Property 1. $(\exists P)\ (\exists Q)\ (|P| = |Q|)$, where P and Q are the two different classes. This property states that a measure should not rank all classes as equally complex. For the proposed measure, numbers of classes exist whose weights may vary. Therefore, the proposed measure is satisfied by this property. For example, CC values of **Student** and **Employee** are 6 and 3 respectively.

Property 2. Let c be a non-negative number, and then there are only finite number of classes and programs of complexity c.

   This property states that there are only a finite number of classes of the same weight. A possible largest number can be assumed, without harm, to be the upper bound for the number of BCSs in a program since the total complexity is defined as the sum of cognitive weights of all the methods (operations) in a program. Therefore, for a given number of BCSs, there are only finitely many programs having that many BCSs. Hence, the proposed complexity metric does hold this property.

Property 3. There are distinct classes P and Q such that $|P| = |Q|$.

   This property states that there are multiple classes of the same complexity. The proposed measure is satisfied by this property since total weights for a class may be the same for many different classes. For example, the class complexity for both of the objects **Administrative** and **Employee** is 3.

Property 4. $(\exists P)\ (\exists Q)\ (P \equiv Q\ \&\ |P| \neq |Q|)$.

   This property states that even if two classes have the same functionality, they may differ in terms of details of implementation. Since the cognitive weights depend on the internal architecture, then the cognitive weights for two classes with the same output may be different. Therefore, the proposed measure is satisfied.

Property 5. $(\forall P)\ (\forall Q)\ (|P| \leq |P; Q|\ \&\ |Q| \leq |P; Q|)$, where, $P + Q$ are the concatenation of P and Q.

   This property states that if the combined class is constructed from class P and class Q, the value of the class complexity for the combined class is larger than the value of the class complexity for class P or class Q. For the proposed measure, the class complexity is given by the associated cognitive weight, which is an integer, then

$$(\forall P)(\forall Q)(P \leq P + Q) \text{ and } (Q \leq P + Q).$$

   This property and Weyuker property 5 are analogous. This implies that the proposed measure is satisfied by property 5. For example, **Student**,

Administrative, Faculty, Employee, and Person are the classes whose CC values are 6,3,2,3 and 8 respectively, which all are less than CC value (i.e.168) of entire code.

Property 6

$$\mathbf{a} : (\exists P)(\exists Q)(\exists R)(|P| = |Q|). \text{ and } |P; R| \neq |Q; R|)$$
$$\mathbf{b} : (\exists P)(\exists Q)(\exists R)(|P| = |Q|). \text{ and } |R; P| \neq |R; Q|)$$

This property states that if a new class is appended to two classes which have the same class complexity, the class complexities of two new combined classes are different or the interaction between P and R can be different than the interaction between Q and R resulting in different complexity values for P+ R and Q + R. The cognitive weights of methods are fixed. Therefore, joining program R with P and Q adds same amount of complexity. This means property 6 is not satisfied by the proposed measure.

Property 7. There are program bodies P and Q such that Q is formed by permuting the order of the statements of P and $(|P| \neq |Q|)$

The intent is to ensure that metric values change as a result of permutation of classes. In any class, changing the order in which methods or attributes are declared does not affect the order in which they are executed. Thus, the proposed measure does not satisfy this property.

Property 8. If P is renaming of Q, then $|P| = |Q|$.

This property requires that when the name of a class changes it will not affect the complexity of the class. It means, even if the member function or member data name in the class changes, the class complexity should remain unchanged. For the proposed complexity measure, there is no effect in complexity by renaming, so this property is also satisfied.

Property 9. $(\exists P)(\exists Q)(|P| + |Q|) < (|P; Q|)$.

This property states that the class complexity of a new class combined from two classes is greater than the sum of two individual class complexities. In general, two classes can have a finite number of identical methods with some cognitive weights. A combination of two classes would result in one class version of the identical methods becoming redundant. Therefore, the complexity of the combined classes in terms of cognitive weight reduces. Therefore, the proposed complexity measure is satisfied by this property. For example, the sum of individual class complexities of Student, Administrative, Faculty, Employee, and Person is 22 $(|P| + |Q| = 22)$, which is less than the CC value of their combined form (i.e. $|P; Q| = 168$).

4   Conclusion

In this paper, a new complexity measure for object-oriented programming is presented. It calculates the cognitive complexity due to inheritance for OO systems. It also considers internal architecture of the methods and message passing between the classes. It is language independent since it uses cognitive weights, which are the same in all programming languages. It is also robust because it compasses all the major parameters that have a bearing on the difficulty of comprehension. The proposed metric is validated by evaluating it against nine Weyuker properties. It satisfies seven properties out of nine. Therefore, this metric established itself as a good and comprehensive measure for OOP.

Acknowledgement

We gratefully acknowledge the effort of Dr. Murat Koyuncu, Mr. Ziya Karakaya and Mr. Mustafa Kahraman for preparing the manuscript in proper format.

References

1. Aggrwal, K.K., Singh, Y., Kaur, A., Melhotra, R.: Software Design Metrics for Object Oriented Software. Journal of Object Technology 6(1), 121–138 (2006)
2. Briand, L., Wust, J.: Modeling Development Effort in Object Oriented System using Design Properties. IEEE Transactions on Software Engineering 27(11), 963–986 (2001)
3. Costagliola, G., Tortora, G.: Class points: An Approach for the Size Estimation of Object-Oriented Systems. IEEE Transactions on Software Engineering 31, 152–174 (2005)
4. Chidamber, S.R., Kermer, C.F.: A Metric Suite for Object Oriented Design. IEEE Transactions on Software Engineering 6, 476–493 (1994)
5. Klemola, T., Rilling, J.: A Cognitive Complexity Metric based on Category learning. In: Second IEEE International Conference on Cognitive Informatics (ICCI 2003), pp. 103–108 (2003)
6. Lorenz, M., Kidd, J.: Object-Oriented Software Metrics. Prentice-Hall, Englewood Cliffs (1994)
7. Misra, A.K., Kushwaha, D.S.: Robustness Analysis of Cognitive Information Complexity Measure using Weyuker Properties. ACM SIGSOFT SEN 31(1), 1–6 (2006)
8. Misra, S.: Modified Cognitive Complexity Measure. In: Levi, A., Savaş, E., Yenigün, H., Balcısoy, S., Saygın, Y. (eds.) ISCIS 2006. LNCS, vol. 4263, pp. 1050–1059. Springer, Heidelberg (2006)
9. Misra, S., Misra, A.K.: Evaluating Cognitive Comlexity Measure with Weyuker Properties. In: Second IEEE International Conference on Cognitive Informatics (ICCI 2004), pp. 103–108 (2004)
10. Sharma, N., Joshi, P., Joshi, R.K.: Applicability of Weyuker Property 9 to Object Oriented Metrics. IEEE Transactions Software Engineering 32(3), 209–211 (2006)
11. Wang, Y.: On Cognitive Informatics. In: Second IEEE International Conference on Cognitive Informatics (ICCI 2002), pp. 34–42 (2002)
12. Wang, Y., Shao, J.: A New Measure of Software Complexity based on Cognitive Weights. Can. J. Electrical and Computer Engineering, 69–74 (2003)
13. Weyuker, E.: Evaluating Software Complexity Measures. IEEE Transactions on Software Engineering 14, 1357–1365 (1988)

A   Appendix

```cpp
#include <iostream>
#include <string>
using namespace std;
/*Person Class is base class. Student and Employee both
  inherits Person Class
 ******** PERSON CLASS ********** */
class Person{
    string name; int age; char sex;
public:
    Person(string="" ,int=0, char='\0');     // WP1=1
    Person(const Person &person);             // WP2=1
    void print()const;                        // WP3=WP31+WP32=2+1=3
    string getName(){return name;}            // WP4=1
    int getAge(){return age;}                 // Wp5=1
    char getSex(){return sex;}                // Wp6=1
};
//Person-default constructor
Person :: Person(string in, int ia, char is){
    name = in; age = ia; sex = is;}
Person :: Person(const Person &p){name=p.name; age=p.age; sex=p.sex;}
void Person :: print()const{
    cout<<"Name\t : "<<name<<"\nAge\t : "<<age<<'\n' ;  //Wp31=1
    if (sex=='F')                                       //Wp32=2
        cout<<"Sex\t : Female" <<'\n' ;
    else
        cout<<"Sex\t : Female" <<'\n' ;
}
/* ******** STUDENT CLASS ********** */
class Student: public Person{int sid; float gpa;
public:
    Student(const Person &p,int student_id,float igpa): Person(p)
{        sid = student_id; gpa = igpa; }      //WS1=1
    void print()const;                        //WS2=Ws21+Ws22*Ws23=1+2*2=5
};
void Student :: print()const{
    Person :: print();
    cout<<"S.ID\t:"<<sid<<"\nGPA\t:"<<gpa<<endl; //Ws21=1
    if (gpa>=2.0)                                //Ws22=2
        cout<<" Student is successful"<<endl;
    else {  if (gpa>=1.7)                        //Ws23=2
        cout<<"Student must improve GPA"<<endl;
    else  cout<<"Student must repeat" <<endl; }
}
/* ******** EMPLOYEE CLASS ********** */
```

```
class Employee: public Person{float salary;
public:
    Employee(const Person &p,float sal):Person(p),salary(sal){}  //WE1=1
    Employee(const Employee &employee):Person(employee)
    {salary=employee.salary;}                       //WE2=1
    void print()const;                              //WE3=1
};
void Employee::print() const{
    Person::print();
    cout<<"salary: "<<salary<<endl;}
/* ******** FACULTY CLASS ********** */
class Faculty: public Employee{
    string branch;                          //Physics,Math,, etc .
public:
    Faculty(const Employee &e, string b):Employee(e),branch(b){} //WF1=1
        void print()const;                  //WF2=1
 };
/* ******** ADMINISTRATIVE CLASS ********** */
class Administrative: public Employee{
    string duty;                            //Secretary, Accountant
public:Administrative(const Employee &e,string d="\0"):Employee(e)
{   duty=d; }                               //WA1=1
    void print()const;                      //WA2=1
    void sendMessage(string msg,Faculty&fac){  //WA3=1
        cout<<"The incoming message :"<<msg<<"\nMessage to ";
        cout<<fac.getName(); }
};
/* ******************** MAIN ******************** */
int main(void)
{   Person * per[3];
    per[0]=new Person ("Aysegul",27,'f'); per[1]=new Person ("Remzi",23,'m');
    per[2]=new Person ("Ali",30,'m'); Person person1("fatmagul",27,'f');
/* Calculate average age */
    int sum=0, average=0, i;
    for (i=0; i<3;i++) sum=per[i]->getAge();
    average=sum/i;
    cout <<"Average personnel age : "<<average<<endl;
    Employee employee1(* per[0],1000); employee1.print();
    Student student1(* per[1],9299,3.5); student1.print();
    Employee employee2(*per[0],2000);
    AdministrativeadmEmployee(employee1,"Secretary  ");
    Faculty facEmployee(employee2,"Computer");
    admEmployee.sendMessage("Today there is a seminarat
    your university. You are invited",facEmployee);
}
```

Combined Analysis of Resting-State fMRI and DTI Data Reveals Abnormal Development of Function-Structure in Early-Onset Schizophrenia

Ming Ke[1,2], Xing Huang[2], Hui Shen[2], Zongtan Zhou[2], Xiaogang Chen[3], and Dewen Hu[2,4,*]

[1] Lanzhou University of Technology, Lanzhou 730050, P.R. China
[2] College of Mechatronics and Automation, National University of Defense Technology, Changsha 410073, P.R. China
[3] The Second Xiangya Hospital of Center South University, Changsha 410073, P.R. China
[4] National Key Laboratory of Cognitive Neuroscience and Learning, Beijing Normal University, Beijing 100875, P.R. China
dwhu@nudt.edu.cn

Abstract. A combined analysis by integrating the resting-state functional magnetic resonance imaging (fMRI) and diffusion tensor imaging (DTI) is presented in this paper to investigate the structure-function relationship in early-onset schizophrenia. After fractional anisotropy in white matter was assessed, functional connectivity was analyzed with regions of reference which was defined according to the regions showed lower fractional anisotropy values. We found significant decreased functional connectivity in early-onset schizophrenia correlated with the regions of anisotropy changes. Our results demonstrated that the white matter lesions may disrupt the anatomical connectivity and inflect the functional connectivity between the frontal regions and other brain regions.

Keywords: Functional connectivity, Anatomical connectivity, Resting fMRI, Fractional anisotropy, Early-onset schizophrenia.

1 Introduction

Neuroimaging techniques such as functional magnetic resonance imaging (fMRI) and diffusion tensor imaging (DTI) provided more advantages and details about brain abnormalities in schizophrenia research [1]. In the previous studies, fMRI was utilized to determine the brain function correlated with behavior by identifying the brain regions that become "active" during the performance of specific tasks. Recently, resting-state fMRI attracted more attentions. Some investigators reported that activity of distributed regions were spontaneously increasing and decreasing in functional-anatomic networks during resting state. And they thought the resting activity patterns would reflect functional activity as important as the activity evoked by tasks [2, 3]. Furthermore, self-generated mental

---

* Corresponding author.

G. Wang et al. (Eds.): RSKT 2008, LNAI 5009, pp. 628–635, 2008.

activity at rest is thought to be possibly relevant to the neuropathologic mechanisms, which cause schizophrenia [4].

DTI, which is a non-invasive technique derived from diffusion magnetic resonance imaging, has been increasingly applied to study the pathways of anatomical connectivity *in vivo* [5]. To describe the anisotropy of diffusion, *Basser* proposed the commonly used indices that included relative anisotropy (RA), fractional anisotropy (FA), and volume ratio (VR) [6]. FA is a measure of the fraction of the magnitude of the tensor that can be ascribed to the anisotropic diffusion. It provides information about myelination, density, coherence, and integrity of fibers. Alteration in diffusion anisotropy may be as the result of various disease processes or abnormal development [7].

Studies have used the combining measure of the task-related fMRI and DTI to investigate the function-structure relationships in healthy subjects and patients with brain diseases [8, 9, 10]. Toosy et al. found the mean FA in the optic radiations by tractography algorithm were correlated with fMRI meature in visual cortex activity [8]. Another study from Baird et al. explored the reaction time related to cortical activity and the integrity of the fibers [9]. Some reports also revealed the correlation between white matter abnormalities and brain activation in schizophrenia with the working memory task [10]. These previous investigations indicated the dynamic development of function and structure and the relevance of the combining method.

Early-onset schizophrenia is defined as schizophrenia with onset of psychosis before the age of 18 years and has shown a lot of the same neurobiological abnormalities observed in adult-onset schizophrenia [11]. The review from *Nicolson* indicated that early-onset schizophrenia was clinically and biologically continuous with adult-onset schizophrenia [12]. However, some studies of early-onset schizophrenia have reported that patients with early-onset schizophrenia tended to be more severe form of the disorder than adult patients [11, 13]. The underlying pathology of early-onset schizophrenia remains little known.

The purpose of the present study is to investigate that the changes in white matter structure are related to altered patterns of functional connectivity in gray matter during resting state in patients with early-onset schizophrenia by combining fMRI and DTI.

2   Methods and Materials

2.1   Subjects

The study included 24 patients with early-onset schizophrenia with the DSM-IV psychiatric diagnoses (9 girls, 15 boys; ages 14-18 years, mean=15.80, SD=1.10). All the patients were recruited from outpatient departments and inpatient units at the department of Psychiatry, Second Xiangya Hospital of Central South University of China, between October 2006 and March 2007. Thirty-first normal control subjects were recruited from advertisements and community centers (18 girls, 13 boys; ages 14-18 years, mean=16.26, SD=1.00). All the participants were right-handedness. During the scanning, subjects were instructed simply to keep their eyes closed, remain awake and perform no specific cognitive exercise.

2.2   Data Acquisition

Images were acquired on a 1.5T GE Signa System (GE Signa, Milwaukee, Wisconsin, USA). Subject's head was fixed by using foam pads with a standard birdcage head coil. Functional MRI images were collected by using a gradient-echo EPI sequence (TR=2000ms, TE=40ms, FOV=24cm, FA=90°, matrix=64×64, slice thickness=5mm, gap=1mm, a volume=20 axial levels, 180 volumes). Then diffusion tensor images were acquired by a diffusion weighted single-shot echo planar imaging sequence with the following parameters: TR=12000ms, TE=105ms, FOV=24cm, NEX=5, matrix=128×128×30.

2.3   Diffusion Tensor Image Analysis

The raw diffusion tensor imaging (DTI) data were realigned for head motion correction. Then the diffusion tensor matrix was calculated according to the Stejskal and Tanner equation [14], and the matrix were diagonalized for obtaining eigenvalues λ_1, λ_2, λ_3 and eigenvectors ε_1, ε_2, ε_3. FA maps separately for each subject were computed in a voxel-based way using the following formula:

$$FA = \sqrt{\frac{3[(\lambda_1 - \overline{\lambda})^2 + (\lambda_2 - \overline{\lambda})^2 + (\lambda_3 - \overline{\lambda})^2]}{2(\lambda_1^2 + \lambda_2^2 + \lambda_3^2)}} \qquad \overline{\lambda} = \frac{\lambda_1 + \lambda_2 + \lambda_3}{3} \qquad (1)$$

We proposed a method called quasi-OVBM to normalize FA images on the basis of optimized voxel-based morphometry (OVBM) [15], where the raw non-diffusion-weighted image ($b0$ image) of each subject was used as T1 structure image. The procedure contained two steps. One step was to create the user-template as follows: (a) The raw $b0$ images of the subjects were segmented into gray matter, white matter, and cerebrospinal fluid. (b) White matter images were normalized by using the standard white matter template. Then the transformation matrix was applied to normalize the raw $b0$ image. (c) The normalized $b0$ images were segmented and the segmented images were smoothed with an 8mm full-width half-maximum(FWHM) Gaussian kernel. Thus the user-template had been created. The other was to normalize the FA images by using 12-parameter affine transformation. The optimum 12-parameter affine transformation was obtained by normalizing the white matter images to the user white matter template, which resulted from the repeated segmentation of the raw $b0$ images. Further, each FA image was smoothed with a Gaussian kernel of 8mm FWHM.

In statistical analysis, a two-sample t-test framework was used to contrast the FA maps. The statistical significance was evaluated between patients with early-onset schizophrenia and normal healthy subjects. FA reductions were tested in patients across the entire volume (at $p < 0.005$, uncorrected).

2.4   Functional Connectivity Analysis

The Corresponding functional MRI data sets were first preprocessed by the statistical parametric mapping software package (SPM2, Wellcome Department of Cognitive Neurology, London, UK). The volumes were realigned to the first volume by registering and reslicing for head motion correction. All the subjects in this study had less than $1mm$ translation in x, y, or z axis and $1°$ of rotation in each axis. Then the volumes were normalized to the standard EPI template and spatially smoothed with a $8mm$ FWHM Gaussian kernel.

In DTI analysis, two clusters that showed FA reductions in early-onset schizophrenia patients related to normal subjects were medial frontal gyrus and subgyral of right frontal region. We defined respectively these two regions of gray matter as the two regions of reference by using the software developed in the Functional MRI Laboratory at the Wake Forest University School of Medicine [16]. Then, two separate correlation analyses were performed. Correlations between the reference time course and time courses in other regions of the brain were computed by using Pearson's correlation coefficients method:

$$cc = \frac{\sum (r - \bar{r})(R - \bar{R})}{\sqrt{\sum (r - \bar{r})^2}\sqrt{\sum (R - \bar{R})^2}} \qquad (2)$$

where R is the reference time course, and r denotes one of the time courses of other voxels. $\bar{R}$ and $\bar{r}$ are the time average of R and r respectively. Correlation maps were converted to z values using Fisher's z-transform for improving the normality of correlation coefficients: $z = (1/2) \times \log_e [(1 + cc)/(1 - cc)]$.

Differences between the schizophrenia group and the normal group were examined by two-sample t-tests at each voxel on the corrected z values. Voxels that t value greater than 3.27 ($p <$0.001, uncorrected) and a spatially contiguous cluster size of 10 voxels or greater were considered to be significant.

3   Results

3.1   Comparison of Mean Fractional Anisotropy between Patients and Controls

Mean fractional anisotropy between two groups were compared, and the results showed significantly lower FA values in the right frontal lobe of the patients. Two cluster were included the right medial frontal gyrus {peak coordinate(MNI) [x=10, y=56, z=-16], t level=3.57}, and the subgyral of right frontal area {peak coordinate(MNI) [x=12, y=44, z=-16], t level=3.44}.

3.2   Comparison of Functional Connectivity between Patients and Controls

Two seed regions and their associated correlation maps were compared between early-onset schizophrenia patients and healthy subjects. Results revealed that

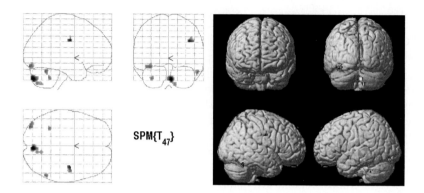

Fig. 1. SPM{t} of right medial frontal gyrus maps statistically significant differences in brain functional connection between patients with early-onset patients and healthy controls during resting state

Table 1. Reduced brain regions of patients with early-onset schizophrenia relative to normal subjects in right medial frontal gyrus maps

Anatonmical structure	BA area	MNI coordinates			t level
		x	y	z	
Precentral gyrus	BA6(R)	40	-8	32	4.48
Inferior occipital gyrus	BA18(L)	-36	-88	-16	3.95
Cerebellum, posterior lobe	L	-10	-54	-50	3.77
Cerebellum, posterior lobe	R	56	-56	-26	3.91
Cerebellum, posterior lobe	R	12	-74	-48	3.77

Note. (L), left hemisphere; (R), right hemisphere.

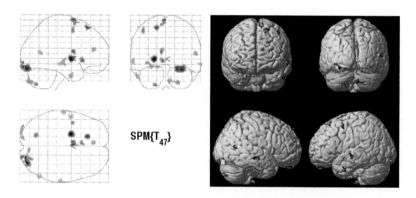

Fig. 2. SPM{t} of right subgyral maps statistically significant differences in brain functional connection between patients with early-onset patients and healthy controls during resting state

Table 2. Reduced brain regions of patients with early-onset schizophrenia relative to normal subjects in right subgyral maps

Anatonmical structure	BA area	MNI coordinates			t level
		x	y	z	
Inferior occipital gyrus	BA17/BA18(R)	30	-88	-20	5.30
Inferior occipital gyrus	BA17/BA18(L)	-38	-94	-8	4.37
Superior frontal gyrus	BA6(R)	-4	4	74	4.69
Middle frontal gyrus	BA6(L)	-44	2	54	4.05
Middle temporal gyrus	BA21/BA39(R)	68	-14	-10	3.80
Middle temporal gyrus	BA21(L)	-64	-16	-6	3.69
Inferior temporal gyrus	BA20(L)	-54	-2	-34	3.66
Anterior cingulate	BA24/BA32(L)	-6	20	-4	4.34
Cerebellum, posterior lobe	L	-20	-70	-50	4.27
Cerebellum, posterior lobe	R	24	-80	-44	3.98

Note. (L), left hemisphere; (R), right hemisphere.

there are decreased functional connections in patients compared with controls, respectively. Fig. 1 and Table 1 showed the abnormal regions of functional connection in right medial frontal gyrus maps between groups. These regions were mostly distributed over frontal lobe, occipital lobe and the cerebellum. Significant reduced connection in the right subgyral maps was shown in Fig. 2 and Table 2, which displayed the abnormal regions in frontal lobe, occipital lobe, temporal lobe, limbic lobe and the cerebellum.

4   Discussion

In the present study, we compared the significance of the mean FA and found the significant reduced FA in two white matter regions of the right frontal lobe. Then, the correlation in low-frequency fluctuations at rest between the two corresponding regions of reference in gray matter and other brain regions were analyzed for exploring the differences of functional connectivity between patients and controls. Unlike the former study combining DTI and fMRI on schizophrenia [10], we analyzed the resting-state functional connectivity patterns of the early-onset schizophrenia from a functional integration perspective, and selected the ROI based on the results from DTI. To our knowledge, no functional connectivity analysis with resting-state fMRI and DTI measures had been combined in the assessment of schizophrenia. This is the first study that DTI and resting-state fMRI measures have been combined to explore the early-onset schizophrenia. Three major contributions were generated from this study.

First, comparison of FA maps between early-onset schizophrenia patients and matched controls revealed reduced FA distributed in the frontal areas by using the quasi-OVBM analysis. This method would offer advantages, such as independence of user bias, in comparison of anatomically variable structures in voxel-based way. Two brain regions of the right frontal lobe [5, 17], which are medial frontal gyrus and subgyral in right frontal lobe, showed significantly lower FA values in the patients than in the controls. Reduced FA implied the differences of fibre architecture or the differences in myelinization of the fibres [10]. Our results indicated disruption of the white matter integrity in frontal areas of the patients and suggested that the white matter deficit might affect anatomical connectivity in a network.

Second, several regions showed relatively decreased functional connectivity with the two regions of reference in early-onset schizophrenia patients compared with that of normal controls. A number of studies have applied resting-state fMRI to explore the functional connectivity in normal subjects as well as in patients with various diseases [4, 18]. And Spontaneous low-frequency fluctuations observed in resting state have been attributed to spontaneous neural activity [18]. Our results are consistent with the opinion of dysfunction of the frontal-cerebellar network, frontal-temporal network, and frontal-occipital network in schizophrenia patients demonstrated in the previous studies [19]. Connection abnormalities during resting state found in early-onset patients as similar as in adult patients in this study suggest that early brain developmental disturbance in early-onset schizophrenia patients may led to the premonition of adult disorder [13].

Third, anatomical connectivity and functional connectivity are two fundamental issues in neuroscience research. We explored the relationship between them in early-onset schizophrenia. The findings in this study revealed that reduced FA in two frontal clusters was related with the abnormal brain functional connectivity between frontal region and other separate brain regions. Utilized fMRI and DTI to make population level inferences could analyze, both macroscopically and quantitatively, the relationship between anatomical connectivity and cortex function [8]. Our results demonstrated that altered functional patterns were relative to specific white matter changes in early-onset schizophrenia. It suggests structural deficit might affect connectivity within the network and functional activity in remote areas [10].

In summary, our analyses not only demonstrate the relationship of changes in white matter and alterations in functional connection, but also enhance the understanding of the mechanism of underlying pathology in early-onset schizophrenia. Further studies should investigate the longitudinal studies of patients through their disease courses and predict the severe course of illness.

Acknowledgement. This work is supported by National Basic Research Program of China (2003CB716104), Natural Science Foundation of China (60771062, 60575044).

References

1. Niznikiewicz, M.A., Kubicki, M., Shenton, M.E.: Recent structural and functional imaging findings in schizophrenia. Current Opinion in Psychiatry 16, 123–147 (2003)
2. Raichle, M.E., Mintun, M.A.: Brain work and brain imaging. Annu. Rev. Neurosci. 29, 449–476 (2006)

3. Buckner, R.L., Vincent, J.L.: Unrest at rest: the importance of default activity and spontaneous network correlations. NeuroImage 37, 1091–1096 (2007)
4. Malaspina, D., Harkavy-Friedman, J., Corcoran, C., Mujica-Parodi, L., Printz, D., Gorman, J.M., Heertum, R.V.: Resting neural activity distinguishes subgroups of schizophrenia patients. Biol. Psychiatry 56, 931–937 (2004)
5. Kanaan, R.A.A., Kim, J., Kaufmann, W.E., Pearlson, G.D., Barker, G.J., McGuire, P.K.: Diffusion tensor imaging in schizophrenia. Biol. Psychiatry 58, 921–929 (2005)
6. Basser, P.J., Pierpaoli, C.: Microstructural and physiological features of tissues elucidated by quantitative-diffusion-tensor MRI. J. Magn. Reson. B. 111, 209–219 (1996)
7. Beaulieu, C.: The basis of anisotropic water diffusion in the nervous system—a technical review. NMR Biomed 15, 435–455 (2002)
8. Toosy, A.T., Ciccarelli, O., Parker, G.J.M., Wheeler-Kingshott, C.A.M., Miller, D.H., Thompson, A.J.: Characterizing function-structure relationships in the human visual system with functional MRI and diffusion tensor imaging. NeuroImage 21, 1452–1463 (2004)
9. Baird, A.A., Colvin, M.K., VanHorn, J.D., Inati, S., Gazzaniga, M.S.: Functional connectivity: Integrating behavioral, diffusion tensor imaging, and functional magnetic resonance imaging data sets. J. Cognitive Neuroscience. 17, 687–693 (2005)
10. Schlösser, R.G.M., Nenadic, I., Wagner, G., Güllmar, D., von Consbruch, K., Köhler, S., Schultz, C.C., Koch, K., Fitzek, C., Matthews, P.M., Reichenbach, J.R., Sauer, H.: White matter abnormalities and brain activation in schizophrenia: A combined DTI and fMRI study. Schizophrenia Research 89, 1–11 (2007)
11. Rhinewine, J.P., Lencz, T., Thaden, E.P., Cervellione, K.L., Burdick, K.E., Henderson, I., Bhaskar, S., Keehlisen, L., Kane, J., Kohn, N., Fisch, G.S., Bilder, R.M., Kumra, S.: Neurocognitive profile in adolescents with early-onset schizophrenia: clinical correlates. Biol. Psychiatry 58, 705–712 (2005)
12. Nicolson, R., Rapoport, J.L.: Childhood-onset schizophrenia: rare but worth Studying. Biol. Psychiatry 46, 1418–1428 (1999)
13. White, T., Kendi, A.T.K., Lehéricy, S., Kendi, M., Karatekin, C., Guimaraes, A., Davenport, N., Schulz, S.C., Lim, K.O.: Disruption of hippocampal connectivity in children and adolescents with schizophrenia– A voxel-based diffusion tensor imaging study. Schizophrenia Research 90, 302–307 (2007)
14. Stejskal, E.O., Tanner, J.E.: Spin diffusion measurements: spin echoes in the presence of a time-dependent field gradient. J. Chem. Phys. 42, 288–292 (1965)
15. Good, C.D., Johnsrude, I.S., Ashburner, J., Henson, R.N., Friston, K.J., Frackowiak, R.S.: A voxel-based morphometric study of ageing in 465 normal adult human brains. NeuroImage 14, 21–36 (2001)
16. Maldjian, J.A., Laurienti, P.J., Kraft, R.A., Burdette, J.H.: An automated method for neuroanatomic and cytoarchitectonic atlas-based interrogation of fMRI data sets. NeuroImage 19, 1233–1239 (2003)
17. Kubicki, M., Westin, C., McCariley, R.W., Shenton, M.E.: The application of DTI to investigate white matter abnormalities in schizophrenia. Ann. N.Y. Acad. Sci. 1064, 134–148 (2005)
18. Biswal, B., Yetkin, F.Z., Haughton, V.M., Hyde, J.S.: Functional connectivity in the motor cortex of resting human brain using echo-planar MRI. Magn. Reson. Med. 34, 537–541 (1995)
19. Fallon, J.H., Opole, I.O., Potkin, S.G.: The neuroanatomy of schizophrenia: circuitry and neurotransmitter systems. Clinical Neuroscience Research 3, 77–107 (2003)

Qualitative Mapping Model of Pattern Recognition*

Jiali Feng

College of Information Engineering, Shanghai Maritime University, Pudong Ave 1550,
200135, Shanghai, China
jlfeng@shmtu.edu.cn

Abstract. A basic principle of pattern recognition is presented in this paper,
by which a qualitative mapping model of pattern recognition can be induced,
and an example of pattern recognition based on qualitative mapping model is
discussed.

Keywords: Qualitative mapping; Pattern recognition.

1 Introduction

In recently years, many methods for recognizing some different kinds of patterns have
been developed by computer scientists and artificial intelligent experts, but some of
basic and very important problems in pattern recognition, such as "why does a computer can recognize a pattern?" or "What is the principle of pattern recognition?" have
not been solved yet.

In [1, 2, 3], an intelligent fusing model, called the qualitative mapping, in which
some methods such as Expert System, Artificial Neural Network and Support vector
can be fused and unified together, has been proposed, and a logic circuit, called Attribute Network Computing can be constructed by the relationship between the input
and the output of qualitative mapping. Some application of qualitative mapping model
and its implementation of attribute network computing has been given. But it is not
clear yet that why does a pattern can be recognized by the qualitative mapping? In
other words, what is the basic principle of pattern recognition using qualitative mapping and attribute computing network? A mathematical discussion about these problems will be given in this paper.

2 Qualitative Mapping Unit and Attribute Computing Network

Definition 1 Let ai(u), i=1,2, be an attbirute of object u, $x_i \in X_i$ a quatity of attribute
ai(u), pij(u)$\in$Pu jth quality of ai(u), j=1,2,3, [α_{ij},β_{ij}]$\subseteq X_i$ qualitative criterion
of pij(u), Γ={{[α_{ij},β_{ij}]}} collection of qualitative criterion, and satisfying
that:[α_{ij},β_{ij}]$\cap$[α_{il},β_{ik}]= $\varnothing$, k=1,2,3, k$\neq$j, and $X_i = \bigcup_{j=1}^{3}[\alpha_{ij},\beta_{ij}]$. Let $a(u) = \bigwedge_{i=1}^{2} a_i(u)$ be the

---

* This paper was supported in part by the Project of Science and Technology development of
Shanghai Education Committee under Grant 06FZ005 and Shanghai Maritime University
under Grant XL0101-2.

G. Wang et al. (Eds.): RSKT 2008, LNAI 5009, pp. 636–643, 2008.

conjunction attribute of attributes $a_1(u)$ and $a_2(u)$, $x=(x_1,x_2) \in X= X_1 \times X_2 \subseteq R^2$ quantity of attributr of $a(u)$. Let $(\alpha_{i_k j_l}, \beta_{i_k j_l}]$ be the qualtative criterion of the j_l-th property $p_{i_k j_l}(u)$ of attrbibutes $a_{i_k}(u)$, $i_k \in \{1,2\}$, $j_l \in \{1,2,3\}$, let $(\alpha_{v(i_k j_l, i_t j_s)}, \beta_{v(i_k j_l, i_t j_s)}] = (\alpha_{i_k, j_l}, \beta_{i_k, j_l}] \times (\alpha_{i_t, j_s}, \beta_{i_t, j_s}]$, here, $i_t \in \{1,2\}, j_s \in \{1,2,3\}$, $i_k \neq i_t$, $j_l \neq j_s$, $(i_k j_l, i_t j_s)$ be a combination of subscripts $i_k j_l$ and sand $i_t j$, $v=v(i_k j_l, i_t j_s)$ the order number of the combination of $(i_k j_l, i_t j_s)$. Because for each i_k and i_t, there are 3 different choices j_l, the tatol number of diffrent combination is $3 \times 3=3^2=9$, and $v(i_k j_l, i_t j_s) \in \{1,...,9=3^2\}$. Let $(\alpha_{i_k j_l, i_t j_s}, \beta_{i_k j_l, i_t j_s}] = (\alpha_{v(i_k j_l, i_t j_s)}, \beta_{v(i_k j_l, i_t j_s)}]$, and $p_{i_k j_l, i_t j_s}(u)=p_{v(i_k j_l, i_t j_s)}(u)=p_{i_k j_l}(u) \wedge p_{i_t j_s}(u)$, then it is a conjunction property whose criterion is $(\alpha_{i_k j_l, i_t j_s}, \beta_{i_k j_l, i_t j_s}]$, $\Gamma^2=\{[\alpha_v, \beta_v]_3^2\}$ is the collection of all criteria latticers $[\alpha_v, \beta_v]_3^2$, let

$$([\alpha_v, \beta_v]_3^2) = \begin{pmatrix} [\alpha_{11,23}, \beta_{11,23}] & (\alpha_{12,23}, \beta_{12,23}] & (\alpha_{13,23}, \beta_{13,23}] \\ [\alpha_{11,22}, \beta_{11,22}] & (\alpha_{12,22}, \beta_{12,22}] & (\alpha_{13,22}, \beta_{13,22}] \\ [\alpha_{11,21}, \beta_{11,21}] & (\alpha_{12,21}, \beta_{12,21}] & (\alpha_{13,21}, \beta_{13,21}] \end{pmatrix}$$ be the grid of $9=3^2$

2-dimension matrixs $[\alpha_v, \beta_v]_3^2$, $\tau: X \times \Gamma^2 \rightarrow \{0,1\}$, for any $x \in X$, if there is $[\alpha_v, \beta_v] \in \Gamma^2$ and the property $p_v(u) \in P_u$ whose criterion is $[\alpha_v, \beta_v]_3^2$, such that

$$T\left((x_1, x_2), \begin{pmatrix} [\alpha_{11,23}, \beta_{11,23}] & (\alpha_{12,23}, \beta_{12,23}] & (\alpha_{13,23}, \beta_{13,23}] \\ [\alpha_{11,22}, \beta_{11,22}] & (\alpha_{12,22}, \beta_{12,22}] & (\alpha_{13,22}, \beta_{13,22}] \\ [\alpha_{11,21}, \beta_{11,21}] & (\alpha_{12,21}, \beta_{12,21}] & (\alpha_{13,21}, \beta_{13,21}] \end{pmatrix} \right)$$

$$= \mathop{\vee}_{j_l, j_s=1(j_l \neq j_s)}^{3} \mathop{\wedge}_{i_k, i_t=1(i_k \neq i_t)}^{2} \{(x_1, x_2) \in [\alpha_{i_k j_l, i_t j_s}, \beta_{i_k j_l, i_t j_s}]\}$$

$$= \mathop{\vee}_{j_l, j_s=1(j_l \neq j_s)}^{3} \mathop{\wedge}_{i_k, i_t=1(i_k \neq i_t)}^{2} \tau_{i_k j_l, i_t j_s}(x_1, x_2)\}$$

(1)

Here, $\tau_{v(i_k j_l, i_t j_s)}(x) = \begin{cases} 1 & iff & x \in [\alpha_v, \beta_v]_3^2 \\ 0 & iff & x \notin [\alpha_v, \beta_v]_3^2 \end{cases}$ (2)

Then called (1) is the qualitative mapping that judging wherter the object u with vector x propose the property $p_v(u)$, and (2) is a qualitative mapping whose qualitative criterion is $[\alpha_v, \beta_v]_3^2$, or a factor mapping of mapping (1), and also noted it as $\tau_p(x, [\alpha_v, \beta_v])$.

The input-output of qualitative mapping (1) whose criterion is the 3×3 grid, and its corresponding Electro-Circuit Unit of Attribute Computing Network[1,2] are respectively shown in Fig.1.

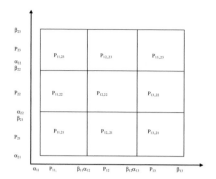 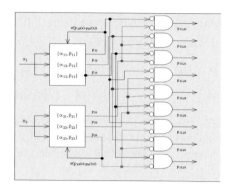

Fig. 1. The relation between input-output of qualitative mapping and its Electro-Circuit Unit

3   A Mathematical Principle of Pattern Recognition

In general, since the recognition of some patterns that varies with time t or one variable x, such as Electrocardiograph, stocks and etc, can be considered as the recognition of graph of one variable function y=f(x), it is a very basic and important problem, whether the recognition method and model of the graph of function y=f(x) could be found out or not.

Furthermore, If some complex patterns that vary with two or more then two variables could be discomposed into a series of simple functions, so that their recognition can be discomposed into a combination of some recognitions for one variable function y=f(x), then the recognition method or model of the graph of function y=f(x) will be a foundation for pattern recognition.

First of all, in order to conveniently, Let us give some relevant definitions.

Definition 2. Let X,Y be two set, if for each $x \in X$, there is a rule f and an element $y \in Y$, such that y=f(t), then the rule f is a function from set X to set Y, noted by f:X→Y, X is called domain of function f, {y|y=f(x),x∈X}⊆Y is called co-domain of f.

A function f is called an one to one corresponding function from domain [a,b]⊆X to co-domain [c,d]⊆Y, f:X→Y, if there two $x_1,x_2 \in [a,b] \subseteq X$, such that $y=f(x_1)=f(x_2)$, then $x_1=x_2$.

Let $x_j \in [a,b]$, j=1,...,m, be m+1 points which is selected by a computer program, $y'(x_j)$ the computing value of function y=f(x) at point x_j. Let $\{(x_j,y'(x_j))\}$ be the set of the pair of variables x_j and their computing values $y'(x_j)$, $P(\{(x_j,y'(x_j))\})$ the pattern constructed by set $\{(x_j,y'(x_j))\}$ in the 2 dimension coordinate system X-Y. Because the number of memory for any computer is finite, but the number of $x \in [a,b]$ are infinite, not only there exist infinite $x(\neq x_j) \in [a,b]$ that never been selected by the program, so that their function value f(x) could not be computed by computer, but also at least there exist q points $\{x_k\} \subseteq \{x_j\}$, $k=j_1..., j_q$, whose computing values $y'(x_k)$ equals not their function values $f(x_k)$, i.e,. $y'(x_k) \neq f(x_k)$. In the case, the pattern $P(\{(x_k,y'(x_k))\})$ is not the graph of function y=f(x), denoted by $P(\{x_k,y'(x_k)\}) \neq P(f(x_k))$, as show in fig.1.

In despite of $y'(x_k) \neq f(x_k)$, we always consider $y'(x_k)$ as to be $f(x_k)$, and take $P(\{(x_j,y'(x_j))\})$ as the graph of function f(x). Why we can do so? We are doing what is

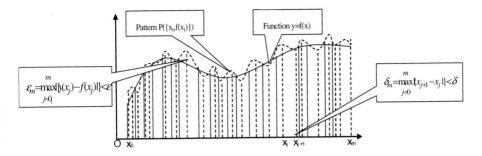

Fig. 2. The competition between pattern of computing values of one variable function y=f(x) and graph of y=f(x)

the principle? Obviously, If the principle could be found out, then a model of pattern recognition based on the principle could be established.

From the program design, we show that the step long $\Delta x_j = x_{j+1} - x_j$ must be designed, before the computing of values of function y=f(x). If the step Δx_j is too long to fine the result of function, then it must be shorted. Second, an error θ that stops the operation of computer must be selected by designer, so that if $|y^{(n)}(x_j) - y^{(n-1)}(x_j)| < \theta$, after the n-th computing, and we get $|y'(x_j) - f(x_j)| < \varepsilon$, here $\varepsilon > 0$ is a arbitrary small position number, then let $y'(x_j) = y^{(n)}(x_j)$, and the machine stop.

From the above description, the basic principle or theorem about why the computing value $y'(x_j)$ equal to the function value $f(x_j)$: $y'(x_j) = f(x_j)$? And the pattern $P(\{x_j, y'(x_j)\})$ can be considered to be the graph of f(x), can be shown as following

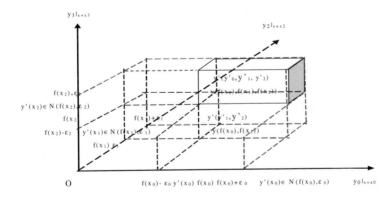

Fig. 3. The qualitative criterion $N(f(x),\varepsilon) = N(f(x_0),\varepsilon_0) \times \ldots \times N(f(x_m),\varepsilon_m)$ constructed by the projection $x=x_j$ to be an axes $y_j|_{x=xj}$

Basic Principle or Theorem. For two given arbitrary small position number $\delta > 0$ and $\varepsilon > 0$, and all j, j=0,...m, if there exist $\delta_m = \max_{j=0}^{m} |x_{j+1} - x_j|\}$ and

$\varepsilon_m = \max\limits_{j=0}^{m}\{|\, y'(x_j) - f(x_j)|\}$, such that when $|x_{j+1}\text{-}x_j|<\delta_m<\delta$ and $|y'(x_j)\text{-}f(x_j)|<\varepsilon_m <\varepsilon$, we

get the following limit

$$\lim_{m\to\infty} y'(x_0,\cdots x_m) = (y'(x_0),\cdots, y'(x_m)) = (f(x_0),\cdots, f(x_m)) = y(x) \qquad (3)$$

Proof. Let $\{\, y_j \mid_{x=x_j}\}$ be a m+1 dimension coordinate system as shown in Fig.3, its

axis $y_j \mid_{x=x_j}$ respectively is the line $x=x_j$ from the 2 dimension coordinate system

X-Y as shown in Fig.2. Let $N(f(x_j),\varepsilon_j)$ on the axis $y_j \mid_{x=x_j}$ be the neighborhood

whose center is $f(x_j)$, the radial is ε_j, and it come from the interval $[f(x_j)\text{-}\varepsilon_j, f(x_j)\text{-}\varepsilon_j]$ on
the line $x=x_j$ in X-Y, then a m+1dimension hypercube $N(f(x_j),\varepsilon) = N(f(x_0),\varepsilon_0)\times...\times$
$N(f(x_m),\varepsilon_m))$ is got as show in Fig.3.

Then, a qualitative mapping in new system $\{\, y_j \mid_{x=x_j}\}$ can be given as follow

$$
\begin{aligned}
&\tau((y'(x_0),\cdots y'(x_m)), N[(f(x_0),\cdots, f(x_m)), (\varepsilon_0,\cdots \varepsilon_m)] \\
&= (y'(x_1),\cdots y'(x_m)) \underset{?}{\in} N[(f(x_0),\cdots, f(x_m)), (\varepsilon_0,\cdots \varepsilon_m)] \\
&= \begin{cases} 1 & y'(x_0)\in N[(f(x_0),\varepsilon_0)] \wedge \cdots \wedge y'(x_m)\in N[(f(x_m),\varepsilon_m)] \\ 0 & \text{othrewise} \end{cases}
\end{aligned}
\qquad (4)
$$

If let $m = \dfrac{b-a}{\delta_m}$, or $\delta_m = \dfrac{b-a}{m}$, and $|x_1\text{-}x_0|=...=|x_m\text{-}x_{m-1}|=\delta_m$, then the interval
or domain [a,b] could be divided into m subintervals, so that $[a,b]=[x_0,x_1]\cup(x_1,x_2]$
$\cup...\cup(x_{m-1},x_m]$. When $m\to\infty$, since $\delta_m\to0$, the number of element of set $\{x_j\}$ and the
computing values of function $\{y'(x_j)\}$ will go to (countable) infinite. If let the number
of memory of computer also be infinite, and let the point $\{x_j\}$ run all over the rational
number in [a,b], then, expect the function value at the irrational number $\{(x_s\}$
$\{y'(x_s)\}$, the set $\{(x_j,y'(x_j))\}$ will be over all rational number point $\{x_j\}$.

On the other hand, since set $\{x_j\}$ run over all rational number, [a,b] will be divided
into countable infinite subintervals, such that $[a,b]=[x_0,x_1]\cup(x_1,x_2]\cup...\cup (x_{m-1},x_m]...$,
whose divided points are all rational number $\{x_j\}$. In the case, similarly, the coordinate system $\{\, y_j \mid_{x=x_j}\}$ will be became an infinite dimension Hilbert space, and the

qualitative criterion $N(f(x),\varepsilon) = N(f(x_0),\varepsilon_0)\times...\times N(f(x_m),\varepsilon_m))...$, will also became an
infinite dimension hypercube.

According to the program, when the n-th computing value $|y^{(n)}(x_j)\text{-} y^{(n-1)}(x_j)|<\theta$,
then let $y'(x_j)=y^{(n)}(x_j)$, and we get $|y'(x_j)\text{-}f(x_j)|<\varepsilon$. Therefore, when the machine stop,
in despite of the qualitative criterion of qualitative mapping is a finite hypercube, but
the different between $y'(x_j)$ and $f(x_j)$, $y'(x_j)\text{-}f(x_j)$ is smaller then any small given position number ε. So we get the following limit

$$\lim_{m\to\infty} y'(x_0,\cdots x_m) = (y'(x_0),\cdots, y'(x_m)) = (f(x_0),\cdots, f(x_m)) = y(x) \qquad (5)$$

From (3) we can get the following

$$\lim_{m\to\infty} P(\{(x_j, y'(x_j)\}) = ((x_0, y'(x_0)), \cdots, (x_m, y'(x_m)) = P(f(x)) \qquad (6)$$

Then the proof is got.

It is shown, from the proof, that because a new coordinate system whose axes $y_j \mid_{x=x_j}$ come from the line $x=x_j$ in X-Y coordinate system, not only a m+1 dimension hypercube $N(f(x),\varepsilon) = N(f(x_0),\varepsilon_0)\times\ldots\times N(f(x_m),\varepsilon_m))$ can be constructed, but also a qualitative mapping model (4) for pattern recognition can be given.

4   A Example of Pattern Recognition Using Qualitative Mapping

Let's discuss the applications of qualitative mapping model of pattern cognition by a example of Electrocardiograph detection.

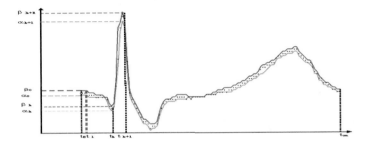

Fig. 4. The Top and Down of sampling t_j of EGR $[\alpha_j,\beta_j]$

Let ECG_u be the Electrocardiograph of u, since it could be considered as a function from interval $[t_0,t_m]$ to current set Y $y:[t_0,t_m]\to Y$, for any $t\in[t_0,t_m]$, there is a $y_u\in Y$, such that $t\to y_u(t)$, the coordinate of any point of ECG_u is $(t,y_u(t))$.

Let $t=t_j\in[t_0,t_m]$,j=0,...,m, be a serial of samplings, $y_u(t_0,..., t_m)=(y_u(t_0),...,y_u(t_m))$ a m+1dimansion vector, $(y_u(t_j)$ the j-th values of function $y=y_u(t_j)$.In 2 dimension coordinate system T-Y, let $P(\{(t_j,y_u(t_j))\})=((t_0,y_u(t_0)),..., (t_m,y_u(t_m))$ be the pattern constructed by the set $\{(t_j,y_u(t_j))\}$ whose a component are respectively m+1 points of $ECG_u=y_u(t)$, as shown in Fig.4.

From the basic principle or theorem, we know that when m go to infinite, the vector $y_u(t_0,..., t_m)$ will be trend to $y_u(t)$, i.e,. $y_u(t_0,..., t_m)=(y_u(t_0),...,y_u(t_m))\approx y_u(t)$, and the pattern $P(\{(t_j,y_u(t_j))\})$ will approximately equal to the electrocardiograph ECG_u, then we get $ECG_u\approx P(\{(t_j,y_u(t_j))\})=((t_0,y_u(t_0)),..., (t_m,y_u(t_m))$.This is mean that each point of ECG_u can be considered as a pair of t and $y_u(t)$, $(t,y_u(t))$.

Let $E=\{ECG_i,i=1,...,n\}$ be a set of all normal ECG_i, $Top(ECG_i)=Max\{ECG_i\}$ and $Down(ECG_i)=Min\{ECG_i\}$ respectively the upper limit of n $\{ECG_i\}$ and the lower limit one, and $N(ECG_i)$ the neighborhood which boundaries are respectively $Top(ECG_i)$ (red line in Fig.4), $Down(ECG_i)$(green line), $t=t_0$ and $t=t_m$. If let (t_j,α_j) and

(t_j,β_j) be respectively Top(ECG$_i$) and Down(ECG$_i$) of current value at the time t=t$_j$, then $[\alpha_j,\beta_j]$ is the qualitative criterion judging whether the value ECG$_u$(t$_j$) of ECG$_u$ is normal or not. And we get a qualitative mapping as following

$$\tau(ECG_u(t_j),[\alpha_j,\beta_j]) = ECG_u(t_j) \underset{?}{\in} [\alpha_j,\beta_j] = \begin{cases} 1 & ECG_u(t_j) \in [\alpha_j,\beta_j] \\ 0 & ECG_u(t_j) \notin [\alpha_j,\beta_j] \end{cases} \quad (7)$$

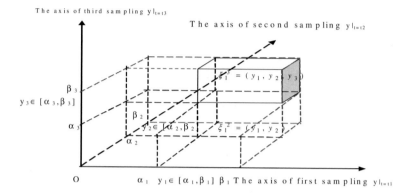

Fig. 5. The qualitative Criterion $[\alpha,\beta]=[\alpha_1,\beta_1]\times\ldots\times[\alpha_m,\beta_m]$ in m dimension coordinate System

Let $\{y_j \mid_{t=t_j}\}$ be the new coordinate System or the Hilbert space, whose axis respectively sampling y$\mid_{t=tj}$, as show In fig.5, because the hypercube $[\alpha,\beta]=[\alpha_1,\beta_1]\times\ldots\times[\alpha_m,\beta_m]$ is the qualitative criterion judging whether the vector $((t_0,y_u(t_0)),\ldots,(t_m,y_u(t_m))$ belongs to the Normal neighborhood N(ECG$_i$) or not, when m go to infinite, we get the qualitative mapping to detect whether the ECG$_u$ is normal or not as follow

$$\tau(ECG_u(t),[\alpha,\beta]) = ECG_u(t) \underset{?}{\in} [\alpha,\beta] = \begin{cases} 1 & ECG_u(t) \in [\alpha,\beta] \\ 0 & ECG_u(t) \notin [\alpha,\beta] \end{cases} \quad (8)$$

The following is an example of application of qualitative mapping model of pattern recognition in ECG detection.

Example A training algorithm and the recognition algorithm based on qualitative mapping model for detecting cardiograph are proposed by Xuguang Liu. First of all, taking 600 normal cardiograms as training examples, 1000 amplitudes A$_j$(car$_i$) for eeach cardiograph car$_i$, i=1,…, 600, j=1,…1000, are samplinged, let α_j=min{A$_j$(car$_i$)}, noteed the down threshold of 600 normal cardiograph, and β_j=max{A$_j$(car$_i$)}, noted the top threshold of 600 normal cardiograph, then a strip between two red lines which is discribed respectively by 1000 qualitative criteion $[\alpha_j,\beta_j]$, as shown in fig 4.

Second, as shown in fig 5, a qualitative criterion, the 1000 dimension parallelepiped $[\alpha,\beta]=[\alpha_1,\beta_1]\times\ldots\times[\alpha_{1000},\beta_{1000}]$ for detecting of the normal cardiograph is created by the transfomation from criterion $[\alpha_j,\beta_j]$ in the sampling space into the 1000 dimension Space.Thereupon, a normal cardiograph that sandwiching in the strip

Fig. 6. One point of deviant cardiograph breaks through the strip of the cardiograph

of qualitative criterion be transfomed as a point in the 1000 dimension parallelepiped $[\alpha,\beta]=[\alpha_1,\beta_1]\times\ldots\times[\alpha_{1000}, \beta_{1000}]$, As shown in fig 5. But one poin of a deviant cardiograph breaks through the strip of qualtative criterion, as shown in fig 6. And the fig7 shows that the deviant cardiograph is identified as abnormality by the qualitative mapping model.

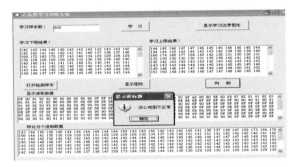

Fig. 7. The result of detection of deviant cardiograph using qualitative mapping model

Acknowledgements

I would like to thank you Prof. Li Tianrui and Prof. Wang Guoying for their concern.

References

1. Feng, J.: Qualitative Mapping Orthogonal System Induced by Subdivision Transformation of Qualitative Criterion and Biomimetic Pattern Recognition. Chinese J. Elec. Special Issue on Bio. Pat. Rec. 15(4), 850–856 (2006)
2. Feng, J.: Attribute Computing Network Induced by Linear Transformation and Granular Transformtion of Qualitative Mapping. In: 2007 IEEE International Conference on Granular Computing, IEEE Computer Society, pp. 83–88. IEEE Press, Los Alamitos (2007)

FLPI: An Optimal Algorithm for Document Indexing

Jian-Wen Tao[1,2], Qi-Fu Yao[1], and Jie-Yu Zhao[2]

[1] Department of Information Engineer
Zhejiang Business Technology Institute, Ningbo, P.R. China
{tjw, yqf}@zjbti.net.cn
[2] College of Information Science and Engineering, Ningbo University, Ningbo, P.R. China
zjy@nbu.edu.cn

Abstract. LPI is not efficient in time and memory which makes it difficult to be applied to very large data set. In this paper, we propose a optimal algorithm called FLPI which decomposes the LPI problem as a graph embedding problem plus a regularized least squares problem. Such modification avoids eigen decomposition of dense matrices and can significantly reduce both time and memory cost in computation. Moreover, with a specifically designed graph in supervised situation, LPI only needs to solve the regularized least squares problem which is a further saving of time and memory. Real and synthetic data experimental results show that FLPI obtains similar or better results comparing to LPI and it is significantly faster.

Keywords: Locality preserving indexing (LPI), Latent semantic indexing (LSI), Document indexing, Dimensionality reduction.

1 Introduction

Recently, Locality Preserving Indexing (LPI) is proposed to discover the discriminant structure of the document space. It has shown that it can have more discriminative power than LSI[1]. However, the computational complexity of LPI is very expensive because it involves eigen-decompositions of two dense matrices. It is almost infeasible to apply LPI on very large data set.

Given a set of documents $\{x_i\}_{i=1}^{m} \subset R^n$, which can be represented as a term-document matrix $X = [x_1, x_2, \ldots, x_m]$. Suppose the rank of X is r, Latent Semantic Indexing(LSI) decompose the X by using SVD as follows:

$$X = U\Sigma V^T , \tag{1}$$

where $\Sigma = \mathrm{diag}(\sigma 1, \ldots, \sigma r)$ and $\sigma_1 \geq \sigma_2 \geq \ldots \geq \sigma_r$ are the singular values of X, U = $[u_1, \ldots, u_r]$ and u_i's are called left singular vectors, V = $[v_1, \ldots, v_r]$ and v_i's are called right singular vectors.

Given a similarity matrix W, LPI can be obtained by solving the following minimization problem:

$$a^* = \underset{a^T XDX^T a=1}{arg\ min} \sum_{i=1}^{m}\sum_{j=1}^{m} (a^T x_i - a^T x_j)^2 W_{ij} = \underset{a^T XDX^T a=1}{arg\ min}\ a^T XLX^T a , \tag{2}$$

G. Wang et al. (Eds.): RSKT 2008, LNAI 5009, pp. 644–651, 2008.

where D is a diagonal matrix whose entries are column sums of W ($D_{ii} = \sum_j W_{ji}$) and $L = D-W$ is the graph Laplacian [2]. LPI constructs the similarity matrix W as:

$$W_{ij} = \begin{cases} \dfrac{x_i^T x_j}{\| x_i \| \| x_j \|}, if\ x_i \in N_p(x_j)or\ x_j \in N_p(x_i), \\ 0, otherwise. \end{cases} \tag{3}$$

where $Np(xi)$ is the set of p nearest neighbors of *xi*. Thus, the objective function in LPI incurs a heavy penalty if neighboring points x_i and x_j are mapped far apart [5]. The basis functions of LPI are the eigenvectors associated with the smallest eigen values of the following generalized eigen-problem:

$$XL X^T a = \lambda XD X^T a.$$

Since we have $L = D-W$, it is easy to check that the minimization problem in Eqn. (1) is equivalent to the following maximization problem:

$$a^* = \underset{a^T XD_X T a=1}{arg\ min}\ a^T XW X^T a, \tag{4}$$

and the optimal *a*'s are also the maximum eigenvectors of eigen-problem:

$$XW X^T a = \lambda XD X^T a, \tag{5}$$

which in some cases can provide a more numerically stable solution [3]. To get a stable solution of the above eigen-problem, the matrix XDX^T is required to be non-singular [3]. When XDX^T is singular, SVD can be used to solve this problem [1].

Suppose we have the SVD decomposition of X shown in Eqn.(1). Let $\tilde{x} = U^T X = \sum V^T$ and $b = U^T a$, we have

$$a^T XD X^T a = a^T U \textstyle\sum V^T DV \sum U^T a = b^T \tilde{X}D \tilde{X}^T b .$$

and

$$a^T XL X^T a = a^T U \textstyle\sum V^T LV \sum U^T a = b^T \tilde{X}L \tilde{X}^T b .$$

Now, the objective function of LPI in (4) can be rewritten as:

$$b^* = \underset{b^T \tilde{X}D \tilde{X}^T b=1}{arg\ min}\ b^T \tilde{X}W \tilde{X}^T b .$$

and the optimal *b*'s are the maximum eigenvectors of eigen-problem:

$$\tilde{X}W \tilde{X}^T b = \lambda \tilde{X}D \tilde{X}^T b , \tag{6}$$

It is easy to check that $\tilde{x} D \tilde{x}^T$ is nonsingular and the above eigen-problem can be stably solved. After we get b^*, the a^* can be obtained by solving a set of linear equations systems $U^T a = b^*$.

In this paper, we propose a new algorithm called FLPI. FLPI decomposes the LPI problem as a graph embedding problem plus a regularized least squares problem. Such modification avoids eigen decomposition of dense matrices and can significantly reduce both time and memory cost in computation. Moreover, with a specifically designed graph in supervised situation, the graph embedding problem in FLPI becomes trivial and FLPI only needs to solve the regularized least squares problem which is a further saving of time and memory.

2   Algorithm for FLPI

2.1   FLPI Algorithm Design

In order to solve the eigen-problem in Eqn. (5) efficiently, we use the following theorem:

Theorem 1. Let y be the eigenvector of eigen-problem

$$W_y = \lambda D_y , \tag{7}$$

with eigen value λ. If $X^T a = y$, then a is the eigenvector of eigen-problem in Eqn. (5) with the same eigenvalue λ.

Proof. We have $Wy = \lambda Dy$. At the left side of Eqn. (5), replace $X^T a$ by y, we have

$$XW X^T a = X W_y = X\lambda D_y = \lambda X D_y = \lambda XD X^T a .$$

Thus, a is the eigenvector of eigen-problem Eqn. (5) with the same eigen value λ. $\square$

Theorem 1 shows that instead of solving the eigen-problem in Eqn.(5), the LPI basis functions can be acquired through two steps:

1. Solve the eigen-problem in Eqn. (7) to get y.
2. Find a which satisfies $X^T a = y$. A possible way is to find a which can best fit the equation in the least squares sense:

$$a = \arg\min_{a} \sum_{i=1}^{m} (a^T x_i - y_i)^2 , \tag{8}$$

where y_i is the i-th element of y.

We may have infinite many solutions for the linear equations system $X^T a = y$ (the system is underdetermined).The most popular way to solve this problem is to impose a penalty on the norm of a [4]:

$$a = \arg\min_{a} \sum_{i=1}^{m} (a^T x_i - y_i)^2 + \alpha \|a\|^2 , \tag{9}$$

The $\alpha \geq 0$ is a parameter to control the amounts of shrinkage.

Given a set of documents $\{ x_i \}_{i=1}^{m} \subset R^n$, the algorithmic procedure of FLPI is stated as below:

Step 1. Adjacency graph construction: Let G denote a graph with m vertices, each vertex represents a document. Let W be a symmetric $m \times m$ matrix with Wij having the weight of the edge joining vertices i and j.

$$W_{ij} = \begin{cases} \dfrac{x_i^T x_j}{\| x_i \| \| x_j \|}, if \ x_i \in N_p(x_j) or \ x_j \in N_p(x_i), \\ 0, otherwise. \end{cases} \tag{10}$$

where $Np(xi)$ is the set of p nearest neighbors of xi.

Step 2. Eigen decomposition: Solve the eigen-problem

$$W_y = \lambda D_y, \tag{11}$$

where D is a diagonal matrix whose entries are column (or row, since W is symmetric) sums of W, $D_{ii} = \sum_j W_{ji}$. Let $\{y_0, y_1, \dots, y_d\}$ be the $d + 1$ eigenvectors with respect to the $d + 1$ maximum eigen values $\lambda_0 \geq \lambda_1 \geq \dots \geq \lambda_d$. It is easy to check that $\lambda_0 = 1$ and y_0 is a vector of all 1 [2].

Step 3. Regularized least squares: Find d vectors $a_1, \dots, a_d \in R^n$, where a_j ($j = 1, \dots, c - 1$) is the solution of regularized least squares problem:

$$a_j = \arg \min_a \sum_{i=1}^m (a^T x_i - y_i^j)^2 + \alpha \|a\|^2, \tag{12}$$

where y_i^j is the i-th element of y_j.

Step 4. FLPI Embedding: Let $A = [a_1, \dots, a_d]$, the embedding is as follows: $x \rightarrow z = A^T x$, where z is a d-dimensional representation of the document x and A is the transformation matrix.

2.2 FLPI in Supervised Situation

Suppose the m documents $\{x_i\}_{i=1}^m$ belong to c classes. Let m_j be the number of documents in the j-th class, $\sum_{j=1}^c m_j = m$. With label information available, a natural way to define graph W can be:

$$W_{ij} = \begin{cases} 1, if \ x_i and x_j share \ the \ same \ label \\ 0, otherwise \end{cases}, \tag{13}$$

To simplify our exposition, we assume that the documents $\{x_i\}_{i=1}^m$ are ordered according to their labels. It is easy to check that the matrix W defined in Eqn. (13) has a block-diagonal structure

$$W = \begin{bmatrix} W^{(1)} & 0 & \dots & 0 \\ 0 & W^{(2)} & \dots & 0 \\ \cdot & \cdot & \cdot & \cdot \\ \cdot & \cdot & \cdot & \cdot \\ \cdot & \cdot & \cdot & \cdot \\ 0 & 0 & \dots & W^{(c)} \end{bmatrix}, \tag{14}$$

where $W^{(j)}$ is a $m_j \times m_j$ matrix with all the elements equal to 1. We also have the D as the diagonal matrix. Thus, the eigen values and eigenvectors of $Wy = \lambda Dy$ are the union of the eigen values and eigenvectors of its blocks (the latter padded appropriately with zeros) [3]: $W^{(j)} y^{(j)} = \lambda D^{(j)} y^{(j)}$.

It is straightforward to show that the above eigen-problem has the eigenvector $e^{(j)} \in R^{mj}$ associated with the eigen value 1, where $e^{(j)} = [1, 1, \ldots , 1]^T$ [2]. Also $Rank(W^{(j)}) = 1$, there is only one non-zero eigen value of $W^{(j)}$. Thus there are exactly c eigenvectors of eigen-problem $Wy = \lambda Dy$. They are

$$y_j = [\underbrace{0,\ldots,0}_{\Sigma_{i=1}^{j-1} m_i},\underbrace{1,\ldots,1}_{m_j},\underbrace{0,\ldots,0}_{\Sigma_{i=j+1}^{c} m_i}]^T , \tag{15}$$

These eigenvectors correspond to the same eigen value 1. Since 1 is a repeated eigen value, we could just pick any other c orthogonal vectors in the space spanned by $\{yj\}$ in Eqn. (15), and define them to be our c eigenvectors [3]. The vector of all ones is naturally in the spanned space. This vector is useless since the corresponding projective function will embed all the documents to the same point. In reality, we can pick the vector of all ones as our first eigenvector and use Gram-Schmidt process to get the remaining $c-1$ orthogonal eigenvectors. The vector of all ones can then be removed.

The above analysis shows that with the W defined as in Eqn.(13) in supervised case, the first two steps of FLPI become trivial. We can directly get the y's which is a significant saving of both time and memory for FLPI computation. It makes FLPI applicable for very large scale supervised learning tasks.

3   Experiments Results and Analysis

3.1   Unsupervised Experiments

The following six methods are compared in the experiment:

- K-means on original term-document matrix, which is treated as our baseline (denoted as **Baseline**)
- K-means after LSI (denoted as **LSI**)
- K-means after LPI (denoted as **LPI**)
- K-means after FLPI (denoted as **FLPI**)
- Clustering using Probabilistic Latent Semantic Indexing (denoted as **PLSI**) [6].
- Nonnegative Matrix Factorization-based clustering (denoted as **NMF-NCW**[9]).

In this experiment, we use the same graph for LPI and FLPI and the parameter p (number of nearest neighbors) was set to 7. The parameter α in FLPI was set to 0.1.

All these algorithms are tested on the TDT2 corpus[1]. In this experiment, those documents appearing in two or more categories were removed, and only the largest 30 categories were kept, thus leaving us with 9,394 documents in total.

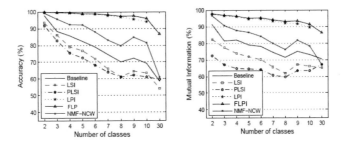

Fig. 1. Performance comparisons on clustering

Two metrics, the accuracy (AC) and the normalized mutual information metric (MI) are used to measure the clustering performance [1], [9].

Fig. 1 shows the average accuracy and average mutual information of the six algorithms. Both LPI and FLPI achieve significant improvements over other four algorithms. Fig. 2 shows the processing time of the six algorithms. Consider both accuracy and efficiency, FLPI is obviously the best among the six compared algorithms for document clustering.

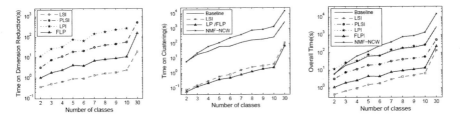

Fig. 2. Processing time on TDT2

3.2 Supervised Experiments

The following three classifiers are used in the experiment:

- k Nearest Neighbor (kNN). The only parameter in kNN is the number of nearest neighbors k.
- Support Vector Machine (SVM) (http://www.csie.ntu.edu.tw/~cjlin/libsvm). There is a parameter C to control the trade-off between large margin and the training error.
- Nearest Centroid (NC). There is no parameter in this method.

All the three classifiers are performed in original document space (Baseline) as well as LSI (PLSI, LPI and FLPI) subspace. The dimension of the LSI (PLSI and LPI) subspace is the number of categories $c(= 20)$ and the dimension of the FLPI subspace is c-1(= 19). The value of parameter α in FLPI is also set to 0.1. The parameter k in kNN and C in SVM are tuned to achieve the best Baseline performance.

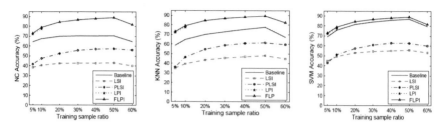

Fig. 3. Performance comparisons on classification

We use 20Newsgroups(bydate version) as a data set [7]. The original split is separated in time, with 11,314 (60%) training documents and 7,532 (40%) testing documents. The classification results of the three classifiers on five document representation methods are listed in Fig. 3. Table 1 shows the dimensionality reduction time of the four algorithms.

Table 1. Computational time

Train Set Size	LSI	PLSI	LPI	FLPI
5%	3.84	2.33	2.97	9.560
10%	4.08	86.05	73.83	12.01
20%	5.43	131.7	– *	16.37
30%	6.35	173.4	– *	20.97
40%	7.51	214.5	– *	26.22
50%	9.02	254.3	– *	32.38
60% (Orig. Split)	9.75	294.1	– *	39.52

*LPI can not be applied due to the memory limit

4   Conclusions

We have proposed a novel algorithm for document indexing and representation based on LPI called FLPI which avoids the expensive computation. FLPI can be computed by a sparse matrix eigen-decomposition followed with a regularized least squares. Moreover, with a graph designed specifically in supervised case, the eigen-decomposition becomes trivial and FLPI only needs to solve a set of regularized least squares problems. Such property makes FLPI can be efficiently computed even for a large scale data set.

Acknowledgments

This work was supported by National Natural Science Foundation of China (NSFC60273094) and Natural Science Foundation of Ningbo City (2006A610012). Thank the anonymous reviewers for their valuable remarks.

References

1. Cai, D., He, X., Han, J.: Document Clustering using Locality Preserving Indexing. IEEE Transactions on Knowledge and Data Engineering 17(12), 1624–1637 (2005)
2. Chung, F.R.K.: Spectral Graph Theory. volume 92 of Regional Conference Series in Mathematics. AMS (1997)
3. Golub, G.H., Loan, C.F.V.: Matrix Computations. Johns Hopkins University Press (1996)
4. Hastie, T., Tibshirani, R., Friedman, J.: The Elements of Statistical Learning: Data Mining, Inference, and Prediction. Springer, New York (2001)
5. He, X., Cai, D., Liu, H., Ma, W.-Y.: Locality Preserving Indexing for Document Representation. In: 2004 Int. Conf. on Research and Development in Information Retrieval, Sheffield, UK, pp. 96–103 (2004)
6. Hofmann, T.: Probabilistic Latent Semantic Indexing. In: 1999 Int. Conf. on Research and Development in Information Retrieval, Berkeley, CA, pp. 50–57 (1999)
7. Lang, K.: Newsweeder: Learning to Filter Netnews. In: The Twelfth International Conference on Machine Learning, pp. 331–339 (1995)
8. Penrose, R.: A Generalized Inverse for Matrices. The Cambridge Philosophical Society 51, 406–413 (1955)
9. Xu, W., Liu, X., Gong, Y.: Document Clustering Based on Non-negative Matrix Factorization. In: 2003 Int. Conf. on Research and Development in Information Retrieval, Toronto, Canada, pp. 267–273 (2003)

Question Answering System Based on Ontology and Semantic Web

Qinglin Guo and Ming Zhang

Department of Computer Science and Technology
Peking University, Beijing 100871, China
qlguo88@sohu.com, mz@163.com

Abstract. Semantic web and ontology are the key technologies of Question Answering system. Ontology is becoming the pivotal methodology to represent domain-specific conceptual knowledge in order to promote the semantic capability of a QA system. In this paper we present a QA system in which the domain knowledge is represented by means of Ontology. In addition, a Chinese Natural Language human-machine interface is implemented mainly through a NL parser in this system. An initial evaluation result shows the feasibility to build such a semantic QA system based on Ontology, the effectivity of personalized semantic QA, the extensibility of ontology and knowledge base, and the possibility of self-produced knowledge based on semantic relations in the ontology.And experiments do prove that it is feasible to use the method to develop a QA System, which is valuable for further study in more depth.

Keywords: WWW, Ontology, Semantic Web, Question Answering.

1   Introduction

Semantic web technologies bring new benefits to knowledge-based Question Answering system. Especially, Ontology is becoming the pivotal methodology to represent domain-specific conceptual knowledge in order to promote the semantic capability of a QA system.Specific research in the areas of QA has been advanced in the past couple of years particularly by TREC-QA [1]. The QA competitions focus on open-domain systems that can potentially answer any generic question.in contrast, a QA system working on a specific technical domain can make use of the specific domain-dependent terminology to recognize the true meaning included in a segment of natural language text. So we realize that the terminology plays a pivotal role in a technical domain such as Java programming. A great deal of work has been done representing domain-specific concepts and the terminology by means of Ontology [2]. Recent research advancements on Knowledge Representation with Semantic Web and Ontology have proved that this methodology is able to promote the semantic capability of a QA system.

The Semantic Web is a Web that includes documents, or portions of documents, describing explicit relationships between things and containing semantic information intended for automated processing by our machines. It operates on

G. Wang et al. (Eds.): RSKT 2008, LNAI 5009, pp. 652–659, 2008.

the principle of shared data. When you define what a particular type of data is, you can link it to other bits of data and say "that's the same", or some other relation. Although it gets more complicated than this, which is basically what the Semantic Web is all about, sharing data through ontologies, and processing it logically. Trust is also important, as the trust of a certain source is fully in the hands of the user. Although the Semantic Web is a Web of data, it is intended primarily for humans; it would use machine processing and databases to take away some of the burdens we currently face so that we can concentrate on the more important things that we can use the Web for.For example, recent research in information processing has focused on health care consumers [3]. These users often experience frustration while seeking online information, due to their lack of understanding of medical concepts and unfamiliarity with effective search strategies. We are exploring the use of semantic relationships as a way of addressing these issues. Semantic information can guide the lay health consumer by suggesting concepts not overtly expressed in an initial query. We present an analysis of semantic relationships that were manually extracted from questions asked by health consumers as well as answers provided by physicians. Our work concentrates on samples from Ask-the-Doctor Web sites. The Semantic Network from the Unified Medical Language System (UMLS) [4] served as a source for semantic relationship types and this inventory was modified as we gained experience with relationship types identified in the texts.A semantic relationship associates two concepts expressed in text and conveys a meaning connecting those concepts. A large variety of such relationships have been identified in several disciplines, including linguistics, philosophy, computer science, and information science. Some researchers have organized hierarchies of semantic relationships into meaningful but not formal structures.

2 Semantic Web and Agent-Based Semantic Web Services Query

Making the Web more meaningful and open to manipulation by software applications is the objective of the Semantic Web initiative. Knowledge representation and logical inference techniques form the backbone. Annotations expressing meaning help software agents to obtain semantic information about documents [5]. For annotations to be meaningful for both creator and user of annotations, a shared understanding of precisely defined annotations is required. Ontologies – the key to a semantic Web – express terminologies and semantic properties and create shared understanding. Web ontologies can be defined in DAML+OIL – an ontology language based on XML and RDF/RDF Schema.Some effort has already been made to exploit Semantic Web and ontology technology for the software engineering domain [6]. DAML-S is a DAML+OIL ontology for describing properties and capabilities of Web services, which shows the potential of this technology for software engineering. Formality in the Semantic Web framework facilitates machine understanding and automated reasoning. DAML+OIL is equivalent to a very expressive description logic [7]. This fruitful connection provides well-defined semantics and reasoning systems. In the conventional Web Services approach exemplified by WSDL or even by DAML Services, the communicative intent of a message is not separated from the application domain. This is at odds with the convention from the multi-agent systems world, where there is a clear separation between the intent of a message, which is expressed using an agent communication language. This separation between intent and domain is beneficial because it reduces the brittleness of a system. If the character of the application domain changes, then only that component which deals with the domain-specific information need change; the agent communication language component remains unchanged.

When the service in the QA example is invoked, the value of the input parameter should be an instance of the class restriction that is given as the input parameter types in both the profile and the process descriptions. For the various query performatives, this input parameter contains the query expression that would be contained in the message content in a conventional agent-based system. However, there is as yet no standard query language for RDF, DAML+OIL or OWL, although there are several under development, including DAML Rules [8,9].As an example, the domain ontology that we have designed for this application is centred on events and reports of events. We have taken the approach that communication in the system will be about these events and reports, so the queries can be expressed using the anonymous resource technique by specifying the properties that the report must possess. It should be noted, however, that we did not specifically design the ontology in this report to circumvent the expressive limitations of our chosen query language, but rather that the query language was chosen because it was appropriate for use with the domain ontology that we had already designed.

3   The Stochastic Syntax-Parse Model Named LSF of Knowledge-Information in QAS

Local environment information is regarded as an important means to WSD in sentence structure all along [10]. But in some lingual models, which are assigned by probability on the basis of rules traditionally, the probability of grammar-producing model is only decided by non-terminal, while is independent of glossarial example in analyzing tree. This quality of non-vocabulary makes lingual phenomena description inadequate for probability model. Therefore, QAS adopts the stochastic syntax-parse model named LSF.

Here, we describe a sort of basic probability depending model. It is named lexical semantic frame (LSF for short [11]) in order to be put easily. LSF is supposed as a result of character string $s = w_i \ldots w_j$, $\mathrm{SR}(R, h, w_i)$ denotes that w_i among LSF relies on the word h through semantic relation, thus we can write down the function $\mathrm{SR}(i) = \mathrm{SR}(R, h, w_i)$. Analyzing semantic probability $p(\mathrm{SR}(i)|h, w_i)$ among words is on the basis of this model. The model supposes that there exists high conjunction between depending relation R and Hyponym node, the contradiction of data sparsely is less. So we can give LSF the analyzing probability from $w_i \ldots w_j$. Unlike rules probability model, the probability model parameter based on vocabulary association is usually gained from supervised training as well as using tagged corpus. In fact, The reasons that we use both the words in corpus and their Hyponym POS information to estimate P(LSF $|w_i \ldots w_j)$ are:

(1) Vocabulary information plays a vital role on QA system.

(2) Considering the limit to corpus scale, words repetition has little probability in sentence analysis, we must deal with statistic result smoothly [12]. Vocabulary information is needed to "magnify" to reduce the degree of data sparseness with the help of Hyponym part of speech. But the close word class such as preposition or adverb uses statistic information of words. We may use parameter smoothing technology. In analysis course, dynamic scheming pruning process and probability computing process are similar to rules probability model. If the analysis of the two parts in one cell case having the same attribute structure, then the analysis result of the part which has lower probability will be cast aside and will not participate in the following analyzing-combining process.

Supposing that we inputting a sentence in QAS: "She eats pizza without anchovies", now we have:

$$P(T_1){=}P(AGT|eat, she)P(OBJ|eat, pizza)P(MOD \mid pizza, anchovies) \quad (1)$$

$$P(T_2){=} P(AGT|eat, she)P(OBJ|eat, pizza)P(MOD|eat, anchovies) \quad (2)$$

Supposing that we can gain the correlative model parameter through corpus statistics such as Table 1, then:

$P(T_1){=}0.0025{\times}0.002{\times}0.003{=}1.5{\times}10^{-6} P(T_2){=}0.0025{\times}0.002{\times}0.0001{=}5{\times}10^{-8}$

T_1 may be chosen to be the right result according to this. If we convert "anchovies" to "hesitation", then $P(T_1){=}5{\times}10^{-8}$, $P(T_2){=}4{\times}10^{-7}$. We find that language model may also help us to choose sound analysis result with the change of words in sentence. This is just about its merit.

Table 1. Interrelated model parameters

PFUNC(X)	Value
P(AGT|eat, she)	0.0025
P(OBJ |eat, pizza)	0.002
P(MOD|pizza, anchovies)	0.003
P(MOD|eat, anchovies)	0.0001
P(MOD|pizza, hesitation)	0.0001
P(MOD|eat, hesitation)	0.0008

4   Explaining Answers from the Semantic Web

Semantic Web aims to enable applications to generate portable and distributed justifications for any answer they produce. Users need to decide when to trust answers before they can use those answers with confidence. We believe that the key to trust understands. Explanations of knowledge provenance and derivation history can be used to provide that understanding [13]. In one simple case, Users may need to inspect information contained in the deductive proof trace that was used to derive implicit information before they trust the system answer. Some users will decide to trust the deductions if they know what reasoner was used to deduce answers and what data sources were used in the proof. Other users may need additional information including how an answer was deduced before they will decide to trust the answer. Users may also obtain information from hybrid and distributed systems and they may need help integrating answers and solutions. Inference Web addresses the issues of knowledge provenance with its registry infrastructure called Semantic Web Ontology [14]. It also addresses the issues concerned with inspecting proofs and explanations with its browser. It addresses the issues of explanations with its language axioms and rewrite rules.

In order to present the findings, the analyst may need to defend the conclusions by exposing the reasoning path used along with the source of the information. In order for the analyst to reuse the previous work, s/he will also need to decide if the source information and assumptions used previously are still valid. Inference Web includes a new explanation dialogue component that was motivated by usage observations. The goal is to present a simple format that is a typical abstraction of useful information supporting a conclusion. The current instantiation provides a presentation of the question and answer, the ground facts on which the answer depended, and an abstraction of the metal information about those facts. There is also a follow-up action option that allows users to browse the proof or explanation, obtain the assumptions that were used, get more information about the sources; provide input to the system, etc.

5 Implement of QAS

Our Automatic Question Answer System includes three models: question's semantic comprehension model based on Ontology and Semantic Web, FAQ-based question similarity match model, document warehouse-base automatic answer fetching model. The question's semantic comprehension model combines many natural language processing techniques, including Ontology and Semantic Web, Segmentation and Part-Of-Speech Tagging, the confirmation of the question type, the extarction of keywords and extending, the confirmation of the knowledge unit, Through these works, the intention of the user is held, which greatly helped the last work of this system. The FAQ-based question similarity match model is implemented by semantic sentence similarity computation, which is improved by our system, this model can answer frequently-asked question fast and concisely. The document warehouse-base automatic answer fetching model firstly deal with the document warehouse beforehand and construct inversed index, then use high efficient information retrieval model to search in the base and return some relevant documents, lastly, we use answer extraction technique to get the answer from these relevant documents and present it to users. For the question that cannot be answered by FAQ base, this model can automatically return exact answer fast. The document repository pre-processing module including Web pages crawlering, HTML format filtering, segmentation and Tagging etc. we receive a term-document matrix by computer the word frequency. This matrix is then analyzed to derive our particular latent semantic structure model for later document retrieval and passage retrieval. Question analysis module is important to QA system. Given a question, the system generates a number of weighted rewrite strings. And then, transform the query into a vector by those weighted rewrite strings. In this module, lay emphasis on question classification. Systems classifies a query into the predefined classes based on the type of answer it is looking for, and then use the question types to identify a candidate answer within the retrieved sentences. Answer extraction module including: document retrieval, passage retrieval and answer matching. System provides a varying method to calculate weight and sort the answer by the weight. Finally, the answer been restricted within 50 words long and returned to user.

The QAS focuses on the key techniques of pattern knowledge based question answering [15]. We design and implement the question answering system and take part in the evaluation of Text Retrieval Conference. We also apply the pattern matching technique to a new related research area Reading Comprehension, and a satisfied result is acquired. The key task to implement the pattern matching technique is to construct a perfect pattern knowledge base. We put forward a novel question classification hierarchy that is based on answer type and question pattern. It retains the semantic and structured information of questions. We make use of the questions on FAQ base as our training and test data. The answer patterns to different question types are studied and evaluated automatically. We have implemented pattern learning to questions with complex structure. It is more effective and reliable to extract the correct answer with answer patterns containing multiple question terms. For higher precision, we give semantic restriction to candidate answers that are extracted by answer patterns. We adopt generalization strategy to answer patterns using named entity information. It makes the answer patterns have better extending ability; the constituent elements of answer pattern contain both morphological and semantic information with better robustness. We evaluate all the answer patterns by the concept of Confidence and Support, which are borrowed from data mining. Answer patterns with higher confidence lead to choose the answer with greater reliability. Table 2 is the experimental results of QAS.

Table 2. Experimental results of QAS

number of questions	Answer correctly	Answer mistakenly	no responsion	Accuracy	recall
2000	1641	198	161	82.05	91.95

6   Conclusions

An initial evaluation is performed on our QA system, focusing on 4 aspects: the feasibility to build such a semantic QA system based on not traditional natural language text but Ontology, the effectivity of personalized semantic QA, the extensibility of ontology and knowledge base, and the possibility of self-produced knowledge based on semantic relations in the ontology. The test set includes 100 questions sampled from a set of questions asked by the students in a one-semester programming lesson, excluding the questions about reading a segment of program, writing a small program to finish a function and so on, which is beyond the ability of a QA system. At the same time, all these 100 questions are ensured within the covering scale. For the scale of the initial evaluation, we don't distinguish the situations between no answer and a false answer. These two situations are regarded as the same - no answer.

The initial evaluation result shows the feasibility of building a semantic QA system based on Ontology and Semantic Web. The personalized answering based on a user model benefits to focusing the user's more attentions on fresh learning material. A user can get the direct answers about some questions based on semantic QA, which shows the effectivity of the system. In no answer situation, the system takes a big proportion, which shows the good extensibility, for the answer can be easily supplied into the knowledge ontology without conflicting with the semantic relations defined in the ontology. At last, the system takes a small proportion, in which the ontology needs to be expanded and ontology consistency must be ensured. How to prove the possibility of self-produced knowledge based on semantic relations in the ontology? A simple example is that the property require is a transitive property, so if the fact that document A requires document B and document B requires document C is stated in the knowledge ontology, a new document relation, document A requires document C, would be self-produced based on the system inference. Afterwards, the update of inter-dependency between documents would bring new answer for a question. And experiments do prove that it is feasible to use the method based on Ontology and Semantic Web to develop a Question Answering System, which is valuable for further study in more depth.

Acknowledgments

We would like to acknowledge the support from the National Natural Science Foundation of China (90412010, 70572090), the National High Technology Research and Development Program (863 Program in china: 2004AA1Z2450), HP Labs China under "On line course organization", NSCF Grant #60573166.

References

1. Voorhees, E.M.: The TREC Question Answering Track. Natural Language Engineering 17, 361–378 (2006)
2. Laura, A., Thomas, C.: Semantic representation of consumer questions and physician answers. Int. J. of Meth. Inform. 11, 513–529 (2006)
3. Lindberg, D., Humphrey, B., McCray, C.: The Unified Medical Language System. Int. J. Meth. Inform. Med. 32, 281–289 (2003)

4. McCray, C., Hole, W.: The scope and structure of the first version of the UMLS Semantic Network. Int. J. Annu. Symp. Comput. 16, 126–130 (2006)
5. W3C Semantic Web Activity (2002), http://www.w3.org/sw
6. Paolucci, M., Kawamura, K., Sycara, K.: Semantic Matching of Web Services Capabilities. In: Cruz, I., Decker, S., Allemang, D., Preist, C., Schwabe, D., Mika, P., Uschold, M., Aroyo, L.M. (eds.) ISWC 2006. LNCS, vol. 4273, p. 279. Springer, Heidelberg (2006)
7. Baader, F., McGuiness, D., Schneider, P.P.: The Description Logic Handbook. Cambridge University Publishers, Baader, England (2003)
8. Decker, S.: DAML Rules—An RDF Query, Inference and Transformation Language (2007),
   http://wwwdb.stanford.edu/stefan/daml/2007/07/03/rules/damlrules.ps
9. Paolucci, M., Kawamura, T., Payne, T.R.: Semantic matching of Web services capabilities. In: Horrocks, I., Hendler, J. (eds.) ISWC 2002. LNCS, vol. 2342, pp. 206–211. Springer, Heidelberg (2002)
10. Collins, M.A.: New Statistical Parser Based on Bigram Lexical De-pendencies. In: Proceedings ACL 2006 - 44th Annual Meeting of the ACL, Jose, America, pp. 184–191 (2006)
11. Terje, B., John, A.: Natural language analysis for semantic document modeling. Int. J. of Data and Know. Eng. 28, 45–62 (2005)
12. Shichao, A., Mnhammed, J.: RMining Multiple Data Sources: Local Pattern Analysis. Int. J. of Data Mining and Knowledge Discovery 12, 121–125 (2006)
13. McGuinness, D.L.: Trusting answers on the web. New Directions in Question Answering, Berlin, Germany (2005)
14. Lambrix, P.: Evaluation of ontology development tools for bioinformatics. Int. J. of Bioinformatics 19, 1564–1571 (2005)
15. Eilbeck, K.: The sequence ontology: a tool for the unification of genome annotations. Int. J. Genome Biol. 6, 44–49 (2005)

Group Assessment of Web Source/Information Quality Based on WebQM and Fuzzy Logic*

Yan Zhu

School of Information Science and Technology
Southwest Jiaotong University
Chengdu 610031, P.R. China
yzhu@home.swjtu.edu.cn

Abstract. Web sources are open, dynamic, and autonomous. They contain a great deal of incomplete, imprecise, and unqualified information. These issues result in unacceptable Web source quality. Evaluating and selecting high quality Web source/information is a key for the success of Web-based applications. In this paper, Web quality is modeled by using a Web quality model, WebQM. Fuzzy TOPSIS (FTOPSIS) is applied to evaluate and screen Web sources for advanced Web applications, such as data warehousing, OLAP, and data mining. In addition, an expert-average group evaluation strategy is combined with FTOPSIS to obtain more objective and more precise results. To illustrate our evaluation process, an example is discussed.

Keywords: Web source/information quality assessment, WebQM, fuzzy TOP-SIS, group evaluation.

1 Introduction

Web has already become the richest information resource of the world. The development of Web-based applications more and more depends on the high quality of Web information, but Web source/information quality is not optimistic. Web sources are open, dynamic, and autonomous. They contain a great deal of incomplete, imprecise, and unqualified information. These issues result in unacceptable Web source quality. Evaluating and selecting high quality Web source/information is a key for a successful Web-based application. The objective of our work is to evaluate Web source/information quality and select the high quality Web data for advanced applications, such as, Web data warehousing, data mining, and Web-based decision making.

Web quality evaluation is a MCDM (multi-criteria decision making) problem. In classical MCDM [1], the performance rating values of each alternative and the weight of each criterion are certain and precise. Web information may be incomplete and uncertain due to its particular features. Evaluating Web quality using crisp values is difficult. In addition, the criteria weights assigned directly by evaluators can not be estimated precisely. Fuzzy logic is a kind of solutions for dealing with uncertainty and inaccuracy

* This work is supported by the Natural Science Foundation of China (Grant No. 60573165) and the Development Foundation of Southwest Jiaotong University (No. 2007A14).

G. Wang et al. (Eds.): RSKT 2008, LNAI 5009, pp. 660–667, 2008.

of real world problems. Therefore, the fuzzy logic and MCDM techniques should be combined for improving the Web quality evaluation.

Another issue we must consider is subjective biases in the Web quality assessment. The criteria weights and the quality rating values of each Web source in terms of the criteria are two key quality parameters. Because they are mostly assigned by evaluators directly, subjective biases may be produced due to the wrong or inaccurate rating of the evaluators. To solve this problem, we apply a group assessment solution, furthermore we improve the group evaluation process by combining expert- and average-decision making strategies.

The main contributions of this paper are:

- A comprehensive and feasible Web quality assessment system based on WebQM is constructed;
- Fuzzy logic and MCDM approaches are combined for reducing the uncertainty and inaccuracy of Web quality data;
- Expert- and average-group evaluation solutions are studied and combined with FM-CDM to reduce the subjective influence to the evaluation process.

Web source quality assessment is crucial for all successful Web-based applications, however, researches in this area is very insufficient so far. The related work can be roughly divided into two groups. One group [2,3,4,5] focuses on the utilization and modification of group fuzzy MCDM approaches for conventional products, but it doesn't take Web sources as assessment candidates into account. Another group deals with the challenge of Web quality [6,7], but has not given a complete Web quality evaluation system yet.

Different from the work of the above two groups, the Web quality features and MCDM approaches are intensively analyzed in our work, and a comprehensive Web quality assessment system based on WebQM is constructed in this paper. In this system, several techniques are studied and combined not only for evaluating Web quality, but also for tackling the particular issues in the assessment and for improving the reliability of the evaluation.

2 The Web Quality Model and Group Evaluation Techniques

Web quality model (WebQM) introduced in our former work [8,9] defines the static and dynamic features of Web quality as 3 quality dimensions: Web source quality (WebSQ), Web information quality (WebIQ), and application specific quality (WebAQ). Each dimension can be specified by a set of subdimensions, which are the assessment criteria. WebQM is the basis of the Web quality evaluation system and covers several key aspects of Web quality evaluation. The details of the 12 criteria can be found in [8]. Figure 1 illustrates the structure of WebQM.

Group evaluation techniques from Group Decision Support Systems (GDSS) [10] are the techniques of arriving at a judgment based on the opinions of a decision team. The main advantage is to obtain a sound solution that may not be gained by an individual alone or to make a decision based on information as unbiased as possible. There are 7 methods in GDSS. Each method has its strengths and weaknesses. For example,

Fig. 1. The Structure of WebQM

group average decision is a method that each group member is asked his/her opinion separately and the judgments are averaged to obtain a consistent result. The advantages are to filter out the extreme opinions, remove typical errors, and integrate the viewpoints of all group members. However, the opinions of the experienced members are treated the same as those of their inexperienced coworkers. As matter of fact, the experts can understand the goal and approaches better and can make decisions efficiently. Therefore, the experienced persons are selected as the group members in this work. Their opinions are averaged to the group rating scores, which as initial values are integrated into a synthetical quality score for each Web source.

3 Fuzzy Evaluation of Web Quality

3.1 The Web Quality Evaluation System

TOPSIS (Technique for Order Preference by Similarity to Ideal Solution)[11] is a classic approach of MCDM. In TOPSIS, two extreme points, the positive-ideal solution (PIS) and the negative-ideal solution (NIS) are built, PIS represents a virtual alternative with a set of possibly best performance scores in terms of each criterion, while NIS is a virtual alternative with a set of worst scores. The performance of each alternative is evaluated and compared with PIS and NIS. If the alternative is closest to PIS and farthest to NIS, it is the best choice.

FTOPSIS extends TOPSIS with fuzzy logic, where the criteria weights and the quality rating scores of an alternative in terms of each criterion are specified using linguistic terms. These linguistic terms are then transformed to the fuzzy numbers by means of a triangular fuzzy membership function in this paper.

In the Web quality evaluation system, Web sources are first preselected by their relevance to the application domain, then the Web source candidates are input to the evaluation process. In the process, FTOPSIS is applied to obtain final integrated quality scores, based on which the Web sources are ranked. The top Web sources are the qualified selections for the advanced applications. The Web quality evaluation system is shown in Figure 2.

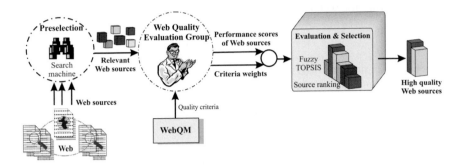

Fig. 2. The Fuzzy Evaluation System of Web Source/Information Quality

3.2 FTOPSIS with Group Evaluation Technique

Assume that there are M alternatives and N criteria in a quality fuzzy evaluation problem. An evaluation group consists of P equal-weighted experts. Each group member defines the criteria weights and the quality rating scores of M alternatives, which construct a fuzzy weight vector $\tilde{W}^k$ and a fuzzy evaluation matrix $\tilde{X}^k$, separately. The data of all group members is integrated into corresponding comprehensive group results, $\tilde{W}$ and $\tilde{X}$, by using the following formulas.

$$\tilde{W} = \frac{1}{P}(\tilde{w}_j^1 \oplus \tilde{w}_j^2 \cdots \oplus \tilde{w}_j^p)_N \tag{1}$$

$$\tilde{X} = \frac{1}{P}(\tilde{x}_{ij}^1 \oplus \tilde{x}_{ij}^2 \cdots \oplus \tilde{x}_{ij}^p)_{M \times N} \tag{2}$$

where $\tilde{w}_j^k$ is the weight of the jth criterion assigned by k^{th} group member, $\tilde{x}_{ij}^k$ denotes the rating value of the ith alternative in terms of the jth criterion by k^{th} decision maker, and $\oplus$ is the fuzzy arithmetic addition. $i = 1, \cdots, M, j = 1, \cdots, N$.

The remaining evaluation steps are:

1. **Normalizing the fuzzy weight vector $\tilde{W}$.**
   The normalized weight vector $\tilde{V}$ is obtained as follows:

$$\tilde{V} = [\tilde{v}_j]_N = [\tilde{w}_j(\div)\sum_j(\tilde{w}_j)] \tag{3}$$

where $\tilde{w}_j$ is the weight of jth criterion in $\tilde{W}$, $\sum$ denotes the fuzzy arithmetic addition, $(\div)$ is fuzzy arithmetic division, and $0 \leq \tilde{v}_j \leq 1$.

2. **Normalizing the fuzzy decision matrix $\tilde{X}$:**
   The normalized decision matrix $\tilde{Y}$ is calculated as follows:

$$\tilde{Y} = [\tilde{y}_{ij}]_{M \times N} = \left(\frac{\tilde{x}_{ij1}}{\tilde{x}_{ij3}^+}, \frac{\tilde{x}_{ij2}}{\tilde{x}_{ij2}^+}, \frac{\tilde{x}_{ij3}}{\tilde{x}_{ij1}^+}\right), \quad \tilde{x}_{ijk}^+ = max_j(\tilde{x}_{ijk}) \tag{4}$$

where $0 \leq \tilde{y}_{ij} \leq 1$.

3. Constructing the weighted normalized fuzzy decision matrix $\tilde{R}$

$$\tilde{R} = \tilde{V} \bigotimes \tilde{Y} = [\tilde{r}_{ij}]_{M \times N} \tag{5}$$

where $\bigotimes$ denotes the fuzzy arithmetic multiplication.

4. Determining PIS ($\tilde{S}^+$) and NIS ($\tilde{S}^-$)

$$\tilde{S}^+ = max(\tilde{r}_{ij}), \quad \tilde{S}^- = min(\tilde{r}_{ij}) \tag{6}$$

The most common PIS and NIS are $\tilde{S}^+ = (1, 1, 1), \tilde{S}^- = (0, 0, 0)$

5. Finding the Euclidean distance of each alternative to PIS and NIS

The Euclidean distances of an alternative to $\tilde{S}^+$ and $\tilde{S}^-$ represent the similarities between this alternative and PIS or NIS. According to [2,12], the distance between two positive triangular fuzzy numbers, $\tilde{a} = (a_1, a_2, a_3)$ and $\tilde{b} = (b_1, b_2, b_3)$, is calculated as:

$$d(\tilde{a}, \tilde{b}) = \sqrt{\frac{(a_1 - b_1)^2 + (a_2 - b_2)^2 + (a_3 - b_3)^2}{3}} \tag{7}$$

The Euclidean distances of M alternatives are calculated as:

$$D_i^+ = \sum_{j=1}^{N}(d(\tilde{r}_{ij}, \tilde{s}_j^+)), \quad D_i^- = \sum_{j=1}^{N}(d(\tilde{r}_{ij}, \tilde{s}_j^-)), \tag{8}$$

where $i = 1, 2, \cdots, M$.

6. Calculating the relative closeness of each alternative to PIS

$$C_i = \frac{D_i^-}{D_i^+ + D_i^-}, \quad 0 \leq C_i \leq 1, \quad i = 1, 2, 3, ..., M \tag{9}$$

The Web sources are ranked according to the value of C. If an alternative itself is PIS, $C = 1$; if an alternative itself is NIS, $C = 0$. The synthetical rates of all alternatives fall into the interval [0,1]. The larger the relative closeness value (C) is, the closer a Web source to the ideal solution and the farther to the negative solution is, and the higher the quality of this Web source is.

4   An Evaluation Example

In the example, a 3-person-group will evaluate 4 Web sources about discount computer book information. The Web sources are identified as A,B,C, and D. Two criteria per dimension, total 6 criteria, are selected for assessment to simplify the calculation. The triangular fuzzy number representation of the linguistic terms used in this example is shown in Figure 3. The criteria weights assigned by all group members using the fuzzy values are shown in Table 1, data in this table will be computed as an integrated weight vector using Eq. 1. Each group member rates the quality performance of 4 Web sources in terms of 6 criteria, separately. Table 2 is one of the evaluation matrices.

Applying Eq. 2 - Eq. 5, we obtain the group-synthetical weighted normalized fuzzy decision matrix in Table 3. The value of PIS and NIS is (1,1,1) and (0,0,0). Applying

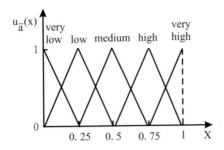

Fig. 3. The Triangular Fuzzy Values of the Lingustic Evaluation Scale

Table 1. Fuzzy Weights of the Quality Criteria by the Group Members

Criteria	Fuzzy weights and their linguistic terms		
	group member 1	group member 2	group member 3
availability	high (0.5, 0.75, 1)	medium (0.25, 0.5, 0.75)	high (0.5, 0.75, 1)
accessibility	medium (0.25, 0.5, 0.75)	high (0.5, 0.75, 1)	medium (0.25, 0.5, 0.75)
correctness	high (0.5, 0.75, 1)	medium (0.25, 0.5, 0.75)	medium (0.25, 0.5, 0.75)
completeness	medium (0.25, 0.5, 0.75)	high (0.5, 0.75, 1)	low (0.0, 0.25, 0.5)
relevance	very high (0.75, 1, 1)	high (0.5, 0.75, 1)	high (0.5, 0.75, 1)
presentation	medium (0.25, 0.5, 0.75)	very high (0.75, 1, 1)	medium (0.25, 0.5, 0.75)

Table 2. The Rating Scores of 4 Web Sources in Terms of Criteria by Group Member 1

sources criteria	A	B	C	D
availability	(0.25,0.5,0.75)	(0.5,0.75,1)	(0.25,0.5,0.75)	(0.25,0.5,0.75)
accessibility	(0.5,0.75,1)	(0,0.25,0.5)	(0.5,0.75,1)	(0.25,0.5,0.75)
correctness	(0.25,0.5,0.75)	(0.25,0.5,0.75)	(0.5,0.75,1)	(0,0.25,0.5)
completeness	(0.75,1,1)	(0.5,0.75,1)	(0.25,0.5,0.75)	(0,0.25,0.5)
relevance	(0,0.25,0.5)	(0.75,1,1)	(0.25,0.5,0.75)	(0.5,0.75,1)
presentation	(0.5,0.75,1)	(0.5,0.75,1)	(0,0.25,0.5)	(0,0.25,0.5)

Table 3. The Group-Synthetical Weighted Normalized Fuzzy Decision Matrix

sources criteria	A	B	C	D
availability	(0.03,0.12,0.48)	(0.04,0.13,0.52)	(0.02,0.12,0.59)	(0.02,0.12,0.59)
accessibility	(0.03,0.10,0.44)	(0,0.04,0.24)	(0.03,0.15,0.71)	(0.02,0.10,0.54)
correctness	(0.02,0.08,0.36)	(0.02,0.08,0.36)	(0.03,0.14,0.65)	(0,0.05,0.36)
completeness	(0.04,0.13,0.43)	(0.02,0.10,0.43)	(0.02,0.10,0.54)	(0.02,0.12,0.54)
relevance	(0,0.05,0.29)	(0.08,0.22,0.57)	(0.03,0.14,0.64)	(0.06,0.22,0.86)
presentation	(0.04,0.13,0.48)	(0.04,0.13,0.48)	(0,0.06,0.36)	(0.01,0.10,0.48)

Eq. 6 - Eq. 9 to Table 3, the results are: $D^+ = (5.03, 4.95, 4.79, 4.85)$ and $D^- = (1.47, 1.56, 2.06, 1.98)$. The relative closeness of each Web source is: $C_A=0.23$, $C_B=0.24$, $C_C=0.30$, $C_D=0.29$. The quality rank of 4 Web sources are C, D, B, A. Thus, C and D are of high quality.

The result shows that the synthetical quality of C and D is quite proximate and better than A and B, while A and B are also similar on their quality results. Such a result corresponds to the real state of the 4 Web sources. Both C and D are global computer book shops. They have approximate business goal and scope and mainly provide the discount computer books. While A and B are two computer book stores appertaining to a computer organization or a publisher, discount computer book selling is not their major business. Their source quality related to discount computer books is not as good as that of C and D.

5   Conclusion and Future Work

Web source quality is a key for the success of all Web-based applications. To this end, this paper discusses a fuzzy evaluation system of Web source/information quality and the key assessment approaches. The fuzzy multi-criteria group evaluation approach is detailed, where the expert- and average-group evaluation solutions are combined for improving evaluation performance. FTOPSIS algorithm is analyzed and carried out for integrating into the comprehensive quality scores of Web sources based on the group decision. The approach in this paper is easy understood and feasible. The evaluation results are reasonable. A prototype of the Web source quality evaluation has been implemented.

As future work, the sensitivity of evaluations, the suitability of different group decision models, as well as their combination will be analyzed. Different fuzzy MCDM approaches will be studied and compared in the system to obtain better assessment performance.

References

1. Jackson Jr., H.V.: A Structured Approach for Classifying and Prioritizing Product Requirements. PhD thesis, North Carolina State University (1999)
2. Chen, C.T.: Extensions of the TOPSIS for Group Decision Making under Fuzzy Environment. Fuzzy Sets and Systems 114, 1–9 (2000)
3. Chen, M.-F., Tzeng, G.-H., Ding, C.-G.: Fuzzy MCDM Approach to Select Service Provider. In: Proceedings of the IEEE International Conference on Fuzzy Systems, pp. 572–577. IEEE Press, New York (2003)
4. Lu, J., Zhang, G., Wu, F.: Web-based Multi-Criteria Group Decision Support System with Linguistic Term Processing Function. IEEE Intelligent Informatics Bulletin 5(1), 35–43 (2005)
5. Saghafian, S., Hejazi, A.R.: Multi-Criteria Group Decision Making Using a Modified Fuzzy TOPSIS Procedure. In: Proceedings of CIMCA-IAWTIC 2005, pp. 215–221. IEEE Press, Washington (2005)
6. Mich, L., Franch, M., Gaio, L.: Evaluating and Designing Web Site Quality. IEEE MultiMedia 10(1), 34–43 (2003)

7. Olsina, L., Rossi, G.: Measuring Web Application Quality with WebQEM. IEEE MultiMedia 9(4), 20–29 (2002)
8. Zhu, Y.: Integrating External Data from Web Sources into a Data Warehouse for OLAP and Decision Making. PhD thesis, Technical University of Darmstadt, Germany. Shaker Verlag, Aachen (2004)
9. Zhu, Y., Buchmann, A.P.: Evaluating and Selecting Web Sources as External Information Resources of a Data Warehouse. In: Proceedings of the 3rd International Conference on Web Information Systems Engineering, pp. 149–160. IEEE Press, Washington (2002)
10. Johnson, D.W., Johnson, F.P.: Joining together: Group Theory and Group Skills, 9th edn. Allyn & Bacon, Boston (2005)
11. Hwang, C.L., Yoon, K.: Multiple Attribute Decision Making. Lecture Notes in Economics and Mathematical Systems, vol. 186. Springer, Heidelberg (1981)
12. Fenton, N., Wang, W.: Risk and Confidence Analysis for Fuzzy Multicriteria Decision Making. Knowledge-Based Systems 19, 430–437 (2006)

Visolink:
A User-Centric Social Relationship Mining

Lisa Fan and Botang Li

Department of Computer Science, University of Regina
Regina, Saskatchewan S4S 0A2 Canada
{fan,li269}@cs.uregina.ca

Abstract. With the popularity of Web 2.0 websites, online social networking has thriven rapidly over the last few years. Lots of research attention have been attracted to the large-scale social network extraction and analysis. However, these studies are mostly beneficial to sociologists and researchers in the area of social community studies, but rarely useful to individual users. In this paper, we present a "friends ranking" system - visoLink which is a personal social network analysis service based on user's reading and writing interest. In order to provide a better understanding to user's personal network, a weighted personal social representation and visualization are proposed. Our system prototype shows a much more user friendly design on personal networks than the classical node-edge distance based network visualization.

Keywords: Web mining, Social network, User centric.

1   Introduction

Writing blogs, sharing photos and videos are the most popular user behaviors on the Web. In the past two years, Web 2.0 brought lots of user participation onto Internet, especially in the area of social networking. Millions of users are contributing contents including texts, pictures and videos to the social network sites. These huge amounts of contents and user activity patterns on the Web become a great source for social network analysis and Web data mining. Recently, researchers from computer science and sociology have been attracted to computational social networking study [2] [4] [5].

With the number of participants in online social networks increasing dramatically, for managing social relationships online, a common feature from the current online social networking sites is to provide users a linear "Friend List". The problem with this list is that while the number of contacts increases, users hardly find out the most important friends in the list. One proposed solution from Anthony Dekker is to define the distance function between network entities based on the frequency of the communications of the user with other friends [1]. However, traditional daily communications is hard to be captured and recorded without a mechanism.

G. Wang et al. (Eds.): RSKT 2008, LNAI 5009, pp. 668–675, 2008.

Blog-based social networking sites are content intensive. Most of the content reflects author's opinions and interests. From the computer science perspective, it contains much less noise data to mine user's interest. Our research motivation is to employ the latest Web Mining techniques to provide users a better way to manage their online social relationships. The proposed framework ranks user's friends based on their online reading and writing interest. In our system prototype, visoLink also provides a user friendly graphical interface to present personal network.

2 Related Work

Social network analysis mainly analyzes the relationships between people or groups of people within the social networks. Generally, a social network is computationally represented by a node-edge undirected graph. Most of the study in social network analysis use binary relationship representation. In [1], conceptual distance is considered in the social network analysis. The edge distance between every two entities in the social network, represents the closeness between two entities in the network. The link value is simply obtained by times of communication between two entities from daily life. For example, the value is assigned to 1.0, if the communication occurs every day; 0.6, if occurs once per week. It can be easily seen that the frequency of daily life communication is hard to be captured without a mechanism.

Because of the popularity of blog, interest similarity measure between bloggers has attracted researchers' attentions. [6] proposed an author-topic model to compute the similarity between authors over topics distributed on documents of their writings. Most of recent research works just focus on this kind of Web content analysis aspect using content mining techniques, but not on user's online activities pattern. The Web Mining technology opens the opportunity to mine relationships among users on the Web [7]. Times of online communications can be simply found from server log file. [2] evaluated the author-topic model and proposed their two-step method which combines probabilistic topics similarity in first step and finer content similarity measure in second step. The second step measuring considers the temporal factor of published post entries, since people's interest could be changed while time passes. The second step measure demonstrates the improvement by considering the time intervals related to author's interest. However, all of these methods are only based on author's writing interest. There are still lots of users surfing on the Web only being readers rather than writers. How to analyze user's reading interest? Web usage mining technique provides a possibility to find the solution. Web Usage Mining techniques are used to analyze user's behavior on a Website [7] [8] [14]. The study from [8] shows a proposed approach combining content and usage together to measure the similarity of behaviors between two visitors. In [10], authors introduce a model to find patterns between visitors in order to build an effective recommender system. Nevertheless, those studies are only classifying users based on their behaviors, but not their real interest.

3   The Proposed User-Centric Personal Network

In order to start our social network analysis, the proposed personal network is defined as follows:

- Each actor has his or her own network which is represented as a weighted graph $G = (V, E, W)$. In this network, a centric user represents the root node of the graph.
- Vertices V represent the friends of the centric user in the social network. The interest of each centric user is reflected by all the related content, including his or her own blog entries, and also other blog entries he or she browsed or read.
- Edges E represent the relationships between different users in the network.
- W denotes the weight of a relationship $Rel(i, j) = W_{ij}$, $Rel(i, j)$ denotes the relationship between *user i* and *user j*. W_{ij} indicates the closeness between two users.

According to our review study, there is nearly no previous research providing a mechanism to weight users social relationships. As a result, our study only focuses on personal network. Firstly, personal network is much less complex than the entire network. Secondly, personal network analysis is designed to be more user-oriented. Additionally, our proposed network design also considers that one relationship could have different values based on different centric-user. In other words, $Rel(i, j) \neq Rel(j, i)$. The importance of the relationship is different from each actor in the network.

4   User Interest Mining

In order to weight different relationships for centric user, two basic principles for interest mining are needed to design. First one is: if two share more similar interest, these two contacts should consider to have a closer relationship. The second principle: More times one spending or more frequently visiting the other one's website indicates that the later one's site owner or site content is more interesting and important. Thus, based on these two principles, our task here is converted to user interest similarity measure.

4.1   Writing Content Analysis

Writing content analysis concentrates on mining centric-user's self-generated content. Blog content mining has been studied in some recent research works [2] [3] [4] [5]. One of the two main approaches in the previous works is to utilize topic distribution model based on probabilistic theory. Another method uses the statistical term frequency content-based approach which is mainly used in the area of information retrieval.

Each blog entry from blog websites may contain several topics. All the text corpus from each user is viewed as a combination of different topics. Each topic occurring in a content corpus produces a probability value. With the help of entropy-based technology, such as KL-divergence, probabilities on the topics shared by two writers is able to be obtained. Topic model for learning the interest of authors from text corpus was introduced in [6] [8], and Rosen-Zvi proposed Author-Topic model to extend the basic LDA model [6]. Both of these two methods need to learn the parameters in estimation approach. In our study, the topic probability distributions are directly obtained from tags (keywords) distribution, since tags are inserted by authors themselves. Similar to the approach in [6], the similarity measure between user i and j is shown in Equation 1,

$$D(i,j) = \sum_{t=1}^{T}[\theta_{it} \log \frac{\theta_{it}}{\theta_{jt}} + \theta_{jt} \log \frac{\theta_{jt}}{\theta_{it}}], \tag{1}$$

where T denotes the set of topics, and θ_{it} denotes the probability of topic t from user i. This method applies KL-divergence to compute the similarity between user i and j.

The term-frequency model is well studied in the area of text document classification. After stop-word removal, spamming and low frequency terms removal, the terms in the text occurring more frequently contribute more importance to the whole document. According to [2], in its second stage of similarity computation, temporal factors are considered to affect the similarity. For example, the topics of two different pieces of content are very similar, but the interest similarity value is still low if the time interval between two published dates is large.

According to [2], the similarity function is defined in Equation 2, where $entry_k$ denotes a blog entry from the entry set E_{it} of user i, $|m(k) - m(l)|$ denotes the month difference of published date between $entry_k$ and $entry_l$. Additionally, in Equation 2, λ takes the value "1", if it is set to consider time difference; otherwise, it takes "0". In order to take average similarity value from all the entry content, the sum of similarity values are divided by the numbers of total entries from user i and j which denote as n_i and n_j.

$$Sim(i,j) = \frac{\sum_{k \in E_i} \sum_{l \in E_j} S(entry_k, entry_l) \cdot e^{-\lambda|m(k)-m(l)|}}{n_i \cdot n_j} \tag{2}$$

4.2   Reading Interest Analysis

Measuring user interest based on blog entry content, however, only considers user's writing content on the Web. Although large number of Web users are contributing contents, the majority of the Web users are still readers. Based on this reality, detecting reading interest of users is highly necessary.

Web log analysis is to study the access patterns of user's online activities. In the context of social networking, the browsing history of user i on j's website indicates user j's content is interested to user i. Therefore, if user i stays on page p longer than a threshold time length l, where p is not in E_i. E_i denotes the pages of user i's personal website. It can be concluded that user i is interested in the content of page p.

In the first stage of Web usage analysis, the raw data for usage analysis is extracted from the Web server log files. Since no user identities in Web Server log files which recorded IP address as client identification, problem encounters when multiple users logon using a same machine. Fortunately, In social networking websites, users log in and start their online social life with their own account. In our project, the logging history is extracted from application level, HTTP sessions. Once one logs in, the application would create a session for each user. Privacy issue may arise, if users do not want their browsing history being manipulated. As a result, in order to handle this situation, our proposed framework consider that browsing history is denied to be processed. A set of visited pages from browsing history for *user i* is denoted as R_i. R_i could be an empty set, if history data is denied to be processed.

4.3   Our Proposed Framework Combining Reading and Writing Interest

Two set of pages are defined in our proposed framework. One is a set of pages of which are centric-user generated content. The second set of pages is from content which the centric user has read. Based on these two sets of content, the system tries to analyze the content not only what users write, but also what users read. It attempts to address the problem that some users prefer reading other's content rather than writing his/her own blog content, which is a very common phenomenon on the Web.

The main task is to measure the similarity between centric-user i and a friend j. Due to the privacy issue needs to be considered, the whole measuring process is divided into five stages as follows:

- The similarity S_1 between user i and j based on their writings is computed using the Equation 3. The content data in this phase is from blog entries of user i and j. The result is multiplied by the weight factor β_0.
- Since users log data from both i and j is collected, the similarity S_2 between the content of i's writing and j's reading is able to be computed. The similarity result is multiplied by a weight factor β_1.
- Same to the process in phrase two, the similarity S_3 between the content of i's reading and j's writing is computed. The result is multiplied by a weight factor β_1.
- Similarly, the similarity S_4 between the content of i's reading and j's reading is computed. The result is multiplied by a weight factor β_2.
- Finally, we sum up S_1, S_2, S_3 and S_4 and then multiplies it with another weight factor α. *alpha* is a factor that considers how often user i visits j's website. If i visits j's website. User j means more important to user i.

$$S_1 = Sim(W_i, W_j) \cdot \beta_0, \tag{3}$$

$$S_2 = Sim(W_i, R_j) \cdot \beta_1, \tag{4}$$

$$S_3 = Sim(R_i, W_j) \cdot \beta_1, \tag{5}$$

$$S_4 = Sim(R_i, R_j) \cdot \beta_2, \tag{6}$$

$$Similarity(i, j) = (S_1 + S_2 + S_3 + S_4) \cdot \alpha, \tag{7}$$

where $Sim()$ function is content similarity measure function from Equation 2, weight factors $\beta_0 > \beta_1 > \beta_2$, W_i denotes the writing content from user i. R_i denotes the reading content of user i, and W_j does not belong to R_i.

If user i denies the application to process log data, S_3 will take value 0. Similarly, if $user\ j$ denies, S_2 takes 0. The values of weight factors β_0, β_1 and β_2 are defined as follows: $\beta_0 > \beta_1 > \beta_2$, because writing interest has more impact on reflecting personal interest than reading which could occur arbitrarily. α is the weight factor that indicates how often user i visits j's website.

In section 4.1, in equation 1, the content analysis model is introduced. By replacing $Sim(i, j)$ in equation 3 with equation 1, the similarity value between two users i and j is able to be obtained. After applying equation 3 to each relationship between each friend and centric-user, the values of ranking criteria for the friend list are generated. As a result, the system is able to rank the friend list based on the common sharing interest.

5   System Prototype Implementation

In order to evaluate our ranking method, the system prototype, namely visolink, has been under development. This prototype system provides the similar services as the current online social networking sites, such as blog service, photo sharing and friendship management. Experimental data is collected when users are using the site. For example, topic probabilities are extracted from the user's blog post tagging annotation. User's reading behaviors are extracted from the server Web logs. As shown in Figure 1, the personal interest are mainly represented by his or her writing content of his blog-based personal website, such as blog posts, photo titles, descriptions and comments on the other's website.

The final goal of the system is to present the ranking of social relationships. Actually showing the order of the ranking is more important than the actual ranking scores. As a result, system prototype visolink provides an enhanced view of friends ranking. Based on our principle system design concept, it is useful to show the order of online social relationship ranking, instead of show meaningless individual ranking score.

As shown in Figure 2, the personal social network of centric-user "Anson" is generated from an automatic graph drawing algorithm. The main contact "Anson", is placed into the center of the graph. Unlike the classical graph drawing using length of edges representing the distance between two entities, visoLink visualizes the network by using vector-based graphical technique which allows those less important nodes being smaller and more transparent. This kind of representation of the network with criteria of clearness and node size is much better for users to judge which nodes are more important, rather than letting users to measure the distance or length between nodes by using their eyes. We

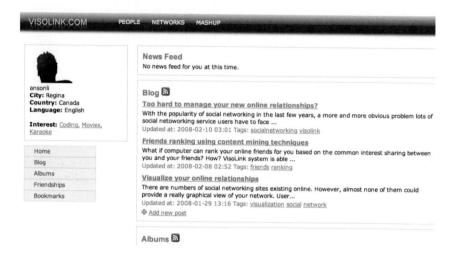

Fig. 1. A screenshot from a user's blog-based personal website of system prototype visoLink

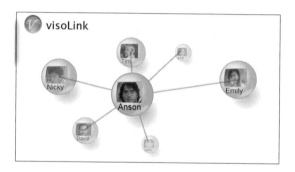

Fig. 2. A screenshot of our proposed visualization of personal network ranking result

design our visualization component to provide users a better understanding on their own personal networks. Most important contacts should be emphasized, and others that have low similarity values should be ignored. A "fake" 3D view of personal network is generated to end user as shown in Figure 2.

visoLink includes personal network friends ranking and recommendation. In the current phase, we have proposed a framework to generate ranking automatically. The prototype website has started to collect experimental user data.

6   Conclusions and Future Work

In this paper, an approach combining content and usage analysis for user interest mining of online social networks has been proposed. It measures user's interests based on both users' writing and reading interests. This similarity measure between online users provides a fundamental support for personal social network visualization and the personalized recommendation.

The existing dataset online available for our system to perform experiment is hard to be found. Because both blog content and application logging data are needed. In the next phase of the project, we will perform evaluation experiments to examine the accuracy and effect of the ranking method from our own site visolink.com. A recommendation system based on online social relationship ranking will be explored in the future.

References

1. Dekker, A.: Conceptual Distance in Social Network Analysis. Journal of Social Structure 6(3) (2005)
2. Shen, D., Sun, J., Yang, Q., Chen, Z.: Latent Friend Mining from Blog Data. In: 6th International Conference on Data Mining, Hong Kong, China, pp. 552–561 (2006)
3. Takama, Y., Matsumura, A., Kajinami, T.: Interactive Visualization of News Distribution in Blog Space. In: 2006 IEEE/WIC/ACM international conference on Web Intelligence and Intelligent Agent Technology, pp. 413–416. IEEE Press, Hong Kong, China (2006)
4. Markrehchi, M., Kamel, M.S.: Learning Social Networks from Web Documents Using Support Vector Classifier. In: 2006 IEEE/WIC/ACM International Conference on Web Intelligence, pp. 88–94. IEEE Press, Hong Kong, China (2006)
5. Spertus, E., Sahami, M., Buyukkokten, O.: Evaluating Similarity Measures: A Large-Scale Study in the Orkut Social Network. In: 11th ACM SIGKDD international conference on Knowledge discovery in data mining, Chicago, U.S.A., pp. 678–684 (2005)
6. Rosen-Zvi, M., Griffiths, T., Steyvers, M., Smyth, P.: The author-topic model for authors and documents. In: 20th conference on Uncertainty in artificial intelligence, Arlington, Virginia, U.S.A., pp. 487–494 (2004)
7. Liu, B.: Web Data Mining: Exploring Hyperlinks. Contents and Usage Data. Springer, Heidelberg (2006)
8. Blei, D.M., Ng, A.Y., Jordan, M.I.: Latent Dirichlet Allocation. J. Mach. Learn. Res. 3, 993–1022 (2003)
9. Murata, T., Saito, K.: Extracting User's interests from Web Log Data. In: 2006 IEEE/WIC/ACM International Conference on Web Intelligence, Hong Kong, China, pp. 343–346 (2006)
10. Mobasher, B., Dai, H., Luo, T., Sun, Y., Zhu, J.: Integrating Web Usage and Content Mining for More Effective Personalization. In: Int'l Conf. on E-Commerce and Web Technologies, ECWeb 2000, UK, pp. 165–176 (2000)

Rough Multi-category Decision Theoretic Framework

Pawan Lingras[1], Min Chen[1,2], and Duoqian Miao[2]

[1] Department of Mathematics & Computing Science, Saint Mary's University,
Halifax, Nova Scotia, B3H 3C3, Canada
[2] School of Electronics and Information Engineering, Tongji University, Shanghai
201804, P.R. China
pawan.lingras@smu.ca

Abstract. Decision theoretic framework has been helpful in providing
a better understanding of classification models. In particular, decision
theoretic interpretations of different types of the binary rough set classification model have led to the refinement of these models. This study
extends the decision theoretic rough set model to supervised and unsupervised multi-category problems. The proposed framework can be used
to study the multi-classification and clustering problems within the context of rough set theory.

Keywords: Rough sets, Web usage mining, Rough approximation,
k-means cluster algorithm.

1 Introduction

Probabilistic extensions have played a major role in the development of rough
set theory since its inception. Recently, Yao [10] explained a list of probabilistic
models under the decision theoretic framework. The models included in the
overview were: rough set-based probabilistic classification [7], 0.5 probabilistic
rough set model [4], decision-theoretic rough set models [8,9], variable precision
rough set models [11], rough membership functions [4], parameterized rough set
models [5], and Bayesian rough set models [6]. The study of such a variety of
models under a common framework also helps understand the similarities and
differences between the models. Such a comparison can help in choosing the right
model for the application on hand. It can also help in creating a new model that
combines desirable features of two or more models. Finally, it can also lead to a
unified model that can be moulded to a given application requirement. Yao [10]
described how the decision theoretic framework exposed additional issues in
probabilistic rough set models.

Rough set theory - like many other classification techniques - was originally
developed for binary classification. That is, an object either belongs to a given
class or does not. Many classification techniques are not easily extendible to a
multi-class problem. The objective of a multi-class problem is to assign an object
to any of the k possible classes. Whenever a technique cannot be easily extended

G. Wang et al. (Eds.): RSKT 2008, LNAI 5009, pp. 676–683, 2008.

to the multi-class problem, researchers have generally chosen two approaches, namely one-versus-one or one-versus-rest [1].

This paper describes how rough set theory does not need to use either the one-versus-one or one-versus-rest technique for extending the binary classification. The framework described in this paper uses the term *category* instead of class to emphasize the fact that it can be used in supervised and unsupervised learning. Conventionally, the classification techniques refer to only supervised learning. When the objects are categorized without the help of a supervisor, the categories are usually called clusters. The proposed multi-category framework is applicable to both classification and clustering problems.

The paper further extends the binary decision theoretic rough set framework for a multi-category problem. The extended framework is shown to reduce to Yao's binary classification approach when the number of categories is equal to two. Moreover, the framework is also shown to be applicable to rough clustering techniques. Finally, it is shown that the decision theoretic crisp categorization is a special case of the rough set based approach. The paper concludes with a discussion on the implications of introducing decision theoretic framework in further theoretical development, especially in the rough clustering area.

2 Literature Review

Due to space limitations, we assume familiarity with the rough set theory [5].

2.1 The Bayesian Decision Procedure

The Bayesian decision procedure deals with making decision with minimum risk based on observed evidence. Let $\Omega = \{\omega_1, \ldots, \omega_s\}$ be a finite set of s states, and let $A = \{a_1, \ldots, a_m\}$ be a finite set of possible m actions. Let $P(\omega_j|\mathbf{x})$ be the conditional probability of an object x being in state ω_j given that the object is described by $\mathbf{x}$. Let $\lambda(a_i|\omega_j)$ denoted the loss, or cost for taking action a_i when the state is ω_j. For an object x with description $\mathbf{x}$, suppose action a_i is taken. Since $P(\omega_j|\mathbf{x})$ is the probability that the true state is ω_j given $\mathbf{x}$, the expected loss associated with taking action a_i is given by:

$$R(a_i|\mathbf{x}) = \sum_{j=1}^{s} \lambda(a_i|\omega_j)P(\omega_j|\mathbf{x}) \tag{1}$$

The quantity $R(a_i|\mathbf{x})$ is also called the conditional risk.

Given a description $\mathbf{x}$, a decision rule is a function $\tau(\mathbf{x})$ that specifies which action to take. That is, for every $\mathbf{x}$, $\tau(\mathbf{x})$ takes one of the actions, $a_1, \ldots, a_m$. The overall risk $\mathbf{R}$ is the expected loss associated with a given decision rule, defined by:

$$\mathbf{R} = \sum_{\mathbf{x}} R(\tau(\mathbf{x})|\mathbf{x})P(\mathbf{x}) \tag{2}$$

If the action $\tau(\mathbf{x})$ is chosen so that $R(\tau(\mathbf{x})|\mathbf{x})$ is as small as possible for every object $\mathbf{x}$. For every $\mathbf{x}$, compute the conditional risk $R(a_i|\mathbf{x})$ for $i = 1, \ldots, m$

defined by equation (1) and select the action for which the conditional risk is minimum. If more than one action minimizes $R(a_i|\mathbf{x})$, a tie-breaking criterion can be used.

Yao proposed probabilistic rough set approximations in [10], which applies the Bayesian decision procedure for the construction of probabilistic approximations. The classification of objects according to approximation operators in rough set theory can be easily fitted into the Bayesian decision-theoretic framework. Let $\Omega = \{A, A^c\}$ denote the set of states indicating that an object is in A and not in A, respectively. Let $A = \{a_1, a_2, a_3\}$ be the set of actions, where a_1, a_2 and a_3 represent the three actions in classifying an object, deciding $POS(A)$, deciding $NEG(A)$, and deciding $BND(A)$, respectively. The probabilities $P(A|[x])$ and $P(A^c|[x])$ are the probabilities that an object in the equivalence class $[x]$ belongs to A and A^c, respectively. The expected loss $R(a_i|[x])$ associated with taking the individual actions can be expressed as:

$$R(a_1|[x]) = \lambda_{11} P(A|[x]) + \lambda_{12} P(A^c|[x]), \qquad (3)$$
$$R(a_2|[x]) = \lambda_{21} P(A|[x]) + \lambda_{22} P(A^c|[x]), \qquad (4)$$
$$R(a_3|[x]) = \lambda_{31} P(A|[x]) + \lambda_{32} P(A^c|[x]), \qquad (5)$$

where $\lambda_{i1} = \lambda(a_i|A)$, $\lambda_{i2} = \lambda(a_i|A^c)$, and $i = 1, 2, 3$. The Bayesian decision procedure leads to the following minimum-risk decision rules:

If $R(a_1|[x]) \leq R(a_2|[x])$ and $R(a_1|[x]) \leq R(a_3|[x])$, decide $POS(A)$;
If $R(a_2|[x]) \leq R(a_1|[x])$ and $R(a_2|[x]) \leq R(a_3|[x])$, decide $NEG(A)$;
If $R(a_3|[x]) \leq R(a_1|[x])$ and $R(a_3|[x]) \leq R(a_2|[x])$, decide $BND(A)$.

Tie-breaking criteria should be added so that each object is classified into only one region. Since $P(A|[x]) + P(A^c|[x]) = 1$, the rules to classify any object in $[x]$ can be simplified based on the probability $P(A|[x])$ and the loss function λ_{ij} ($i = 1, 2, 3$;$j = 1, 2$).

Based on the general decision-theoretic rough set model, it is possible to construct specific models by considering various classes of loss functions. In fact, many existing models can be explicitly derived from the general model. For example, the 0.5 probabilistic model can be derived when the loss function is defined as follows:

$$\lambda_{12} = \lambda_{21} = 1, \qquad \lambda_{31} = \lambda_{32} = 0.5, \qquad \lambda_{11} = \lambda_{22} = 0. \qquad (6)$$

A unit cost is incurred if an object in A^c is classified into the positive region or an object in A is classified into the negative region; half of a unit cost is incurred if any object is classified into the boundary region. The 0.5 model corresponds to the application of the simple majority rule.

3   Extension to the Multi-category Problem

Many classification techniques are originally designed for binary classification. Examples include Decision trees, Perceptrons, and Support Vector Machines. These techniques tend to classify objects into two classes such as the positive or negative regions in rough set theory. Some of these techniques have natural extensions for multi-class problems. Others use either the one-versus-one or one-versus-rest technique [1]. Let $C = \{c_1, \ldots, c_k\}$ be a set of categories. We will use the terms category, classes, and clusters interchangeably whenever it is appropriate in the context. In the one-versus-one approach, a binary classification model is created for every pair of classes (c_i, c_j). The training of such a model uses only the subset of those objects, which were classified as either c_i or c_j. It can be easily seen that there will be a total of $k \times (k-1)$ such models. Assuming uniform distribution, there will be $\frac{n}{k}$ objects belonging to each class, where n is the size of the complete training set. While it would require significant computational effort to train $k \times (k-1)$ models, on an average each model will have only $\frac{2 \times n}{k}$ objects. The one-versus-rest technique, on the other hand, creates a binary model for each class c_i by classifying objects as either belonging to c_i or not belonging to c_i. There are only k such models. However, the training set for each model is the same size as the complete training set, i.e. n. Moreover, the training set is biased towards objects not belonging to the class. For example, for any given class c_i there will be $\frac{n}{k}$ objects belonging to c_i and $\frac{(k-1) \times n}{k}$ objects not belonging to c_i. Therefore, the chances of a classification model erring towards predicting that an object does not belong to c_i are higher. As a result, studies have shown that the one-versus-one approach tends to be more accurate than the one-versus-rest approach. However, one-versus-one multi-classification creates a large number of models and works with a small amount of training data for each model. Smaller training data can lead to over-fitting and may explain the relative accuracy of the one-versus-one approach.

Given the inadequacies of both one-versus-one and one-versus-rest models, a classification technique that has a natural multi-class extension is more desirable. Rough set theory has such a natural extension. In this section, the multi-class extension of rough set is described. It should be noted that many implementation of rough set theory use similar philosophy for multi-classification. This section provides a formal framework that can be used with both supervised and unsupervised rough categories. We will start with formal definitions for the proposed framework.

Objects: Let $X = \{x_1, \ldots, x_n\}$ be a finite set of objects.

Categories: Let $C = \{c_1, \ldots, c_k\}$ be a finite set of k states given that C is the set of categories and each category is represented by a vector c_i ($1 \leq i \leq k$). Furthermore, let C partition the set of objects X.

Object and category similarity: For every object, x_l, we define a non-empty set T_l of all the categories that are similar to x_l. Clearly, $T_l \subseteq C$. We will use $x_l \to T_l$ to denote the fact that object x_l is similar to all the elements of set T_l. Let us further stipulate that object x_l can be similar to one and only one T_l. The definition of the similarity will depend on a given application. Later on we will see an example of how to calculate similarity using probability distribution.

Upper and lower approximations: If an object x_l is assigned to a set T_l, then the object belongs to the upper approximations of all categories $c_i \in T_l$. If $| T_l | = 1$, then x_l belongs to the lower approximation of the only $c_i \in T_l$. Please note that when $| T_l | = 1$, $\{c_i\} = T_l$. Therefore, upper $(\overline{apr})$ and lower $(\underline{apr})$ approximation of each category c_i can be defined as follows:

$$\overline{apr}(c_i) = \{x_l | x_l \to T_l, c_i \in T_l\}, \tag{7}$$

$$\underline{apr}(c_i) = \{x_l | x_l \to T_l, \{c_i\} = T_l\}. \tag{8}$$

Since we do not define upper and lower approximations of all the subsets of X, we cannot test all the properties of rough set theory. However, it can be easily shown that the resulting upper and lower approximations in fact follow important rough set theoretic properties given the fact that C is a partition of X specified by Lingras and West [2].

- An object can be part of at most one lower approximation   (P1)
- $x_l \in \underline{apr}(c_i) \Rightarrow x_l \in \overline{apr}(c_i)$ (P2)
- An object x_l is not part of any lower approximation   (P3)

$$\Updownarrow$$

x_l belongs to two or more upper approximations.

4   Loss Functions for Multi-category Problem

Following Yao [10], we define a set of states and actions to describe the decision theoretic framework for multi-category rough sets.

States: The states are essentially the set of categories $C = \{c_1, \ldots, c_k\}$.

An object is said to be in one of the categories. However, due to lack of information we are unable to specify the exact state of the object. Therefore, our actions are defined as follows.

Actions: Let $B = \{B_1, \ldots, B_s\} = 2^C - \{\emptyset\}$ be a family of non-empty subsets of C, where $s = 2^k - 1$. We will define a set of actions $b = \{b_1, \ldots, b_s\}$ corresponding to set B, where b_j represents the action in assigning an object x_l to the set B_j.

Note that some of the sets B_j's will be the same as the set T_l's defined in previous sections. The reason we choose to use a different notation is to emphasize the fact that we do not specify any similarity between x_l and B_j as we do in case of x_l and T_l. Note that there will be a total of n T_l's, one for each object, and they may not be distinctly different from each other. That is, two objects may be similar to the same subset of C. On the other hand, there will be exactly $s = 2^k - 1$ distinct B_j's.

Now we are ready to write the Bayesian decision procedure for our multi-category rough sets as follows.

Let $\lambda_{x_l}(b_j|c_i)$ denote the loss, or cost, for taking action b_j when an object belongs to c_i. Let $P(c_i|x_l)$ be the conditional probability of an object x_l being in state c_i. Therefore, the expected loss $R(b_j|\boldsymbol{x}_l)$ associated with taking action b_j for an object $\boldsymbol{x}_l$ is given by:

$$R(b_j|\boldsymbol{x}_l) = \sum_{i=1}^{k} \lambda_{\boldsymbol{x}_l}(b_j|c_i)P(c_i|\boldsymbol{x}_l) \tag{9}$$

For an object $\boldsymbol{x}_l$, if $R(b_j|\boldsymbol{x}_l) \leq R(b_h|\boldsymbol{x}_l)$, $\forall\, h = 1, \ldots, s$, then decide b_j.

We generalize the loss function for the 0.5 probabilistic model [3] given by Yao [10] as follows:

$$\lambda_{\boldsymbol{x}_l}(b_j|c_i) = \frac{|b_j - T_l|}{|b_j|} \qquad if \quad c_i \in b_j \;;$$

$$\lambda_{\boldsymbol{x}_l}(b_j|c_i) = \frac{|b_j - \emptyset|}{|b_j|} \qquad if \quad c_i \notin b_j \;. \tag{10}$$

When c_i belongs to b_j, the loss for taking action b_j corresponds to the fraction of b_j that is not related $\boldsymbol{x}_l$. Otherwise, the loss for taking action b_j will have the maximum value of 1.

It can be easily seen that when k is equal to 2, $C = \{c_1, c_2\}$. Therefore, $B = \{\{c_1\}, \{c_2\}, \{c_1, c_2\}\}$. Without loss of generality, we can designate c_1 to be the positive class, c_2 to be the negative class, and $\{c_1, c_2\}$ to be the boundary region. Then one can easily verify that $\lambda_{\boldsymbol{x}_l}(\{c_1\}|c_1) = 0$, $\lambda_{\boldsymbol{x}_l}(\{c_2\}|c_1) = 1$, and $\lambda_{\boldsymbol{x}_l}(\{c_1, c_2\}|c_1) = \frac{1}{2}$, which corresponds to the loss function described by Yao [10] for the 0.5 probabilistic model [3].

Let us illustrate the proposed rough multi-category expected loss function with the following example.

Example 1. Let $C = \{c_1, c_2, c_3, c_4\}$ and $B = 2^C - \{\emptyset\}$ ($|B| = 2^4 - 1 = 15$). For an object $\boldsymbol{x}_l$, let $\{P(c_1|\boldsymbol{x}_l), P(c_2|\boldsymbol{x}_l), P(c_3|\boldsymbol{x}_l), P(c_4|\boldsymbol{x}_l)\} = \{0.15, 0.2, 0.25, 0.4\}$. We will define the set T_l such that $\boldsymbol{x}_l \to T_l$ as: $T_l = \{c_h|P(c_h|\boldsymbol{x}_l) > 0.2\} = \{c_3, c_4\}$. The expected loss associated with taking action b_j is shown in Table 1. The values of the expected loss seem quite reasonable. The lowest value is obtained for the set $T_l = \{c_3, c_4\}$. It is highest for the sets that do not contain either c_3 or c_4. Since the probability of $P(c_4) > P(c_3)$, the sets containing c_4 tend to have lower loss than those containing c_3.

Example 2. One can also obtain a crisp categorization from the proposed formulation by stipulating that all the T_l's in our formulation are singleton sets. We can demonstrate this by using the same probability function, but changing the criteria for defining the set T_l such that $\boldsymbol{x}_l \to T_l$ as: $T_l = \{c_h\}$ such that $P(c_h|\boldsymbol{x}_l)$ is maximum. If more than one such c_h have the same (maximum) value, we arbitrarily choose the first c_h. This ensures that T_l is a singleton set. In our example, with $\{P(c_1|\boldsymbol{x}_l), P(c_2|\boldsymbol{x}_l), P(c_3|\boldsymbol{x}_l), P(c_4|\boldsymbol{x}_l)\} = \{0.15, 0.2, 0.25, 0.4\}$, $T_l = \{c_4\}$. The resulting expected loss function in this example is shown in Table 2.

Table 1. Expected loss for all the actions from Example 1

The expected loss $R(b_j|\boldsymbol{x}_l)$	Action
0.35	$\{c_3, c_4\}$
0.433	$\{c_2, c_3, c_4\}$
0.467	$\{c_1, c_3, c_4\}$
0.5	$\{c_1, c_2, c_3, c_4\}$
0.6	$\{c_4\}$
0.7	$\{c_2, c_4\}$
0.725	$\{c_1, c_4\}$
0.75	$\{c_3\}, \{c_1, c_2, c_4\}$
0.775	$\{c_2, c_3\}$
0.8	$\{c_1, c_3\}, \{c_1, c_2, c_3\}$
1	$\{c_1\}, \{c_2\}, \{c_1, c_2\}$

Table 2. Expected loss for all the actions from Example 2

The expected loss $R(b_j|\boldsymbol{x}_l)$	Action
0.6	$\{c_4\}$
0.75	$\{c_3\}$
0.8	$\{c_2\}$
0.85	$\{c_1\}$

5   Concluding Remarks

This paper describes an extension of the Bayesian decision procedure described by Yao [10] for multi-category rough sets. The proposal is a natural extension of the conventional binary rough set classification. Unlike some other classification techniques such as Perceptrons and Support Vector Machines, it is not necessary to create a multiple binary classifiers using either the one-versus-one or one-versus-rest approaches. This is a significant advantage of rough set theory as both one-versus-one and one-versus-rest approaches can be difficult to implement in practice. The one-versus-one approach can lead to large number of binary classifiers, which may overfit the training data. On the other hand, the one-versus-rest approach tends to have lower classification accuracy.

In addition to extending the Bayesian decision process from binary rough set classifiers to rough set multi-classifiers, the approach can easily be applied to unsupervised rough set classifiers. The definition of probability used in this paper is abstract as opposed to the frequency based values used in various probabilistic rough set models, including the unified framework proposed by Yao [10]. By changing the definition of the probability one can easily adopt the Bayesian decision process to rough set based clustering. Such an adoption can be useful in further theoretical development in rough clustering. Results of such development will be reported in future publications.

Acknowledgments. The authors would like to thank China Scholarship Council and NSERC Canada for their financial support.

References

1. Lingras, P., Butz, C.J.: Rough set based 1-v-1 and 1-v-r approaches to support vector machine multi-classification. Information Sciences 177, 3298–3782 (2007)
2. Lingras, P., West, C.: Interval set clustering of web users with rough k-means. Journal of Intelligent Information System 23, 5–16 (2004)
3. Pawlak, Z., Wong, S.K.M., Ziarko, W.: Rough sets: probabilistic versus deterministic approach. International Journal of Man-Machine Studies 29, 81–95 (1988)
4. Pawlak, Z., Skowron, A.: Rough membership functions. In: Yager, R.R., Fedrizzi, M., Kacprzyk, J. (eds.) Advances in the Dempster-Shafer Theory of Evidence, pp. 251–271. John Wiley and Sons, New York (1994)
5. Pawlak, Z., Skowron, A.: Rough sets: some extensions. Information Sciences 177, 28–40 (2007)
6. Slezak, D.: Rough sets and Bayes factor. In: Peters, J.F., Skowron, A. (eds.) Transactions on Rough Sets III. LNCS, vol. 3400, pp. 202–229. Springer, Heidelberg (2005)
7. Wong, S.K.M., Ziarko, W.: Comparison of the probabilistic approximate classification and the fuzzy set model. Fuzzy Sets and Systems 21, 357–362 (1987)
8. Yao, Y.Y., Wong, S.K.M.: A decision theoretic framework for approximating concepts. International Journal of Man-machine Studies 37, 793–809 (1992)
9. Yao, Y.Y., Wong, S.K.M., Lingras, P.: A decision-theoretic rough set model. In: Ras, Z.W., Zemankova, M., Emrich, M.L. (eds.) Methodologies for Intelligent Systems, vol. 5, pp. 17–24. North-Holland, New York (1990)
10. Yao, Y.Y.: Decision-theoretic rough set models. In: Yao, J., Lingras, P., Wu, W.-Z., Szczuka, M., Cercone, N.J., Ślęzak, D. (eds.) RSKT 2007. LNCS (LNAI), vol. 4481, pp. 1–12. Springer, Heidelberg (2007)
11. Ziarko, W.: Variable precision rough set model. Journal of Computer and System Sciences 46, 39–59 (1993)

Dynamic Features Based Driver Fatigue Detection

Xiao Fan, Baocai Yin, and Yanfeng Sun

Beijing Key Laboratory of Multimedia and Intelligent Software, College of Computer Science and Technology, Beijing University of Technology, Beijing 100022, China
wolonghongni@emails.bjut.edu.cn, ybc@bjut.edu.cn, yfsun@bjut.edu.cn

Abstract. Driver fatigue is an important reason for traffic accidents. To account for the temporal aspect of human fatigue, we propose a novel method based on dynamic features to detect fatigue from image sequences. First, global features are extracted from each image and concatenated into dynamic features. Then each feature is coded by the means of training samples, and weak classifiers are constructed on histograms of the coded features. Finally AdaBoost is applied to select the most critical features and establish a strong classifier for fatigue detection. The proposed method is validated under real-life fatigue conditions. The test data includes 600 image sequences with illumination and pose variations from thirty people's videos. Experiment results show the validity of the proposed method and the average recognition rate is 95.00% which is much better than the baselines.

Keywords: Computer vision, human fatigue, PCA, AdaBoost.

1 Introduction

Driver fatigue is an important reason for traffic accidents. In China, driver fatigue resulted in 3056 deaths in vehicular accidents in 2004, and caused 925 deaths in highway accidents which amounted to about 14.8%. Many computer vision based approaches have been proposed for fatigue detection.

The frequency and time of eye closed all increase when driver fatigue. Much attention is paid to eye features for fatigue detection. Based on the study by the Federal Highway Administration, percentage of eyelid closure (PERCLOS) [1] has been found to be the most reliable and valid measure of a person's alertness level among many drowsiness detection measures. Rangben Wang et al. [2] used Gabor wavelet to extract texture features of drivers' eyes, and used Neural Network classifier to identify drivers' fatigue behavior. Wenhui Dong et al. [3] decided whether the driver was fatigue by detecting the distance of eyelids. Fei Wang et al. [4] combined gray scale projection, edge detection with Prewitt operator and complexity function to judge whether the driver had his eyes closed. Yawning is also an important character of fatigue. Mouth features are extracted to detect fatigue in [5,6]. Rongben Wang et al. [5] took the mouth region's geometric features to make up an eigenvector as the input of a BP ANN, and they

G. Wang et al. (Eds.): RSKT 2008, LNAI 5009, pp. 684–691, 2008.

acquired the BP ANN output of three different mouth states that represent normal, yawning or talking state respectively. Tiesheng Wang *et al.* [6] represented the openness of the mouth by the ratio of mouth height to width, and detected yawning if the ratio was above 0.5 in more than 20 frames.

Most of these methods are spatial approaches, and they do not model the dynamics of fatigue and therefore do not utilize all of the information available in image sequences. In facial expression recognition, according to psychologists, analyzing an image sequence produces more accurate and robust facial expression recognition, and facial motion is fundamental to facial expression recognition. Fatigue is a cognitive state that is developed over time. It is our belief that dynamic features which capture the temporal pattern should be the optimal features to describe fatigue just as facial expression recognition[7].

To account for the temporal aspect of human fatigue, Qiang Ji *et al.* [8] introduced a probabilistic framework based on dynamic Bayesian networks for modeling and real-time inferring human fatigue by integrating information from various sensory data and certain relevant contextual information spatially and temporally, leading to a more robust and accurate fatigue modeling and inference. However, in summary, there is little research in extracting dynamic features for fatigue detection. Achieving high accuracy in fatigue detection is still a challenge due to the complexity and variation of facial dynamics.

In this paper, to account for the temporal characteristic of human fatigue, we propose a novel dynamic feature from image sequences. The framework of the proposed approach is illustrated in Fig. 1. First, each image in the input face image sequence is processed by face detection, geometric normalization and cropping. Then, PCA coefficients are extracted by PCA projection from each image and concatenated into a feature sequence as the dynamic features. Inspired by [7], The dynamic features are thresholded into binary codes and the histograms of the coded dynamic features are computed. Finally, weak classifiers are constructed on the histogram features and AdaBoost is applied to select the most discriminate features and build a strong classifier for fatigue detection.

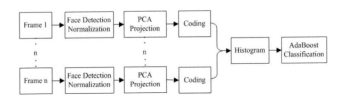

Fig. 1. Framework of the Proposed Approach

The paper is organized as follows. Section 2 introduces dynamic facial feature extraction. Dynamic Feature coding and Histogramming are showed in section 3. In section 4, feature selection and classifier learning are presented. Finally, experiments with analysis and conclusions are presented in section 5 and 6.

2   Dynamic Facial Feature Extraction

We extract facial features based on PCA projection from image sequences, and combine them into a sequence as the dynamic facial feature for fatigue detection.

2.1   Preprocessing

Before we extract PCA coefficients from face areas, original face images are preprocessed so that they are aligned in a predefined way. Face images in our experiments are gray scale and normalized into the size of 64*64.

2.2   Dynamic Facial Features

Feature extraction using PCA entails to represent an image in a low dimensional space, and receives considerable attention in the computer vision area. To obtain the dynamic feature, we first extract PCA coefficients from each image in the sequence, and then the same PCA coefficients in the consecutive images are combined into a sequence as the dynamic features (Fig. 2).

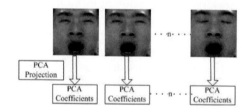

Fig. 2. Dynamic Facial Feature Extraction

Given one image sequence with n images, we label each image with I_i, where i is the index of the image. We set G as the PCA coefficient set which includes all the PCA coefficients in one face image. We label each PCA coefficient in image I_i with $g_{i,j}$, where j is the index of the PCA coefficient in the coefficient set G. Based on each PCA coefficient $g_{i,j}$, we obtain a dynamic feature unit s_j as $\{g_{0,j}, g_{1,j}, \ldots, g_{n-1,j}\}$. The temporal variation of driver fatigue can be effectively described by all the dynamic feature units.

3   Histograms of the Coded Dynamic Features

3.1   Dynamic Feature Coding

A dynamic feature unit is composed of a sequence of PCA coefficients. Thus, each dynamic unit is a feature vector. Considering weak learner construction for AdaBoost learning with one feature is much easier than with a feature vector, we further threshold each dynamic feature unit into a binary code sequence.

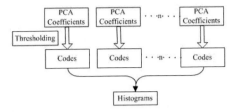

Fig. 3. Histograms of the Coded Dynamic Features

Thresholding is based on the means of dynamic features. First, we extract features from the training samples, and then the means for fatigue state and normal state can be computed. Based on the means, we can map each feature $g_{i,j}$ to $\{1, 0\}$ codes by the following formula.

$$C_{i,j} = \begin{cases} 1, & \text{if } D_{n,j} - D_{f,j} \geq 0, \\ 0, & \text{if } D_{n,j} - D_{f,j} < 0. \end{cases} \tag{1}$$

Where $D_{f,j} = |g_{i,j} - m_{f,j}|$, $D_{n,j} = |g_{i,j} - m_{n,j}|$ and $m_{f,j}$ is the means of features in fatigue state and $m_{n,j}$ is the means in normal state.

3.2 Histograms of the Coded Dynamic Features

We can map a dynamic feature unit s_j to a code sequence S_j based on Eq. 1.

$$S_j = \{\, C_{0,j}, C_{1,j}, ..., C_{n-1,j} \} \tag{2}$$

Histograms of the codes are computed over each code sequence, each histogram bin being the number of occurrences of the corresponding code in one code sequence. Finally, all the histograms estimated from all the code sequences are concatenated into a single histogram sequence to represent the given face image sequence (Fig. 3).

A histogram of a code sequence S_j can be defined as

$$H_{c,j} = \sum_{i=0}^{n-1} I\{C_{i,j} = c\}, c = 0, 1 \tag{3}$$

Where n is the number of images in one sequence and

$$I\{A\} = \begin{cases} 1, & A \text{ is true,} \\ 0, & A \text{ is false.} \end{cases} \tag{4}$$

There are 2 bins for one code sequence. A histogram of a code sequence is considered as a set of 2 individual features. When one bin is known, the other can be computed easily in the proposed approach. Therefore, we only select the bin for code 1. The resulting histograms on each code sequence are combined yielding the histogram for an image sequence. Assuming a face image is projected on m eigenfaces, the number of total histogram features will be m. In our case, there are 600 eigenfaces, and therefore, there are 600 histogram features for one image sequence. The histogram to descript an image sequence can be defined as

$$H = \{ H_{1,0}, H_{1,1}, ..., H_{1,m-1} \}, m = 600 \tag{5}$$

In this histogram, we effectively have a description of a face sequence on spatial and temporal levels.

4   Feature Selection and Classifier Learning

4.1   Weak Classifiers

Weak classifiers are the basis of AdaBoost. In our case, they are decision trees (Fig. 4) based on the histogram features. Decision tree is a tree graph, with leaves representing the classification results and nodes representing some predicates. Branches of the tree are marked with true or false.

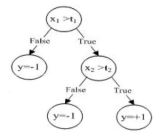

Fig. 4. Decision Tree with Two Splits

4.2   AdaBoost Learning

AdaBoost method[9] provides a simple yet effective stage-wise learning approach for feature selection and nonlinear classification at the same time. Due to its good generalization capability, fast performance and low implementation complexity, AdaBoost has achieved great success in face detection and other applications[7]. In the proposed approach, AdaBoost is used both to select a small set of features and train the classifier.

The final hypothesis of AdaBoost algorithm is

$$H(x) = \begin{cases} 1, & \text{if } \sum_{t=1}^{T} \alpha_t h_t(x) \geq \frac{1}{2} \sum_{t=1}^{T} \alpha_t, \\ 0, & \text{otherwise.} \end{cases} \tag{6}$$

The strong classifier is a linear combination of the T weak classifiers. The AdaBoost learning procedure is aimed to derive α_t and $h_t(x)$.

5   Experiments and Analysis

To validate the proposed approach, we conducted experiments on a fatigue face database that we built.

5.1   Data Set

There is no public database to test methods of fatigue detection. The test sets in a lot of researches are often very small with only several subjects. To test the fatigue detection methods, we built a fatigue face database. We used web cameras to catch videos of about forty persons. The videos of each person last several hours which are caught indoors without directions to the subjects.

Fig. 5. Examples of (a) Normal and (b) Fatigue Image Sequences in the Database

Totally, we got about 50GB videos in AVI format compressed by MPEG-4. We selected thirty subjects' face fatigue videos from the original ones. Then we extracted the fatigue image sequences from the videos and made up the fatigue face database. There are 600 image sequences of ten female subjects and twenty male subjects. Each subject has 20 image sequences (10 normal and 10 fatigue). There are 5 images in each image sequence, and each image in the sequences is gray scale with a resolution of 320*240 (Fig. 5). Randomly, we select 300 sequences of 15 persons for the gallery set and the other 300 sequences of 15 persons for the probe set.

5.2   Baselines

We present two statistical learning methods, one based on LDA (PCA+LDA classifier) and one on HMM (PCA+HMM classifier) as baselines. PCA+HMM employs HMM to classify the sequence of PCA coefficients extracted from a face image sequence. PCA+LDA employs LDA to classify the PCA coefficients of face images. Two experiments were made on the LDA based classifier. In the first one, LDA is used to classify the PCA coefficients from a single image. In the other, PCA coefficients from an image sequence are combined into a feature vector and classified by LDA.

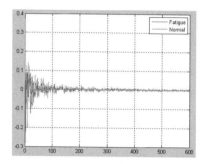

Fig. 6. Means of PCA Coefficients

Table 1. Confusion Matrix for Fatigue Detection

Method		Normal	Fatigue
PCA+HMM	Normal	88.00%	12.00%
	Fatigue	31.33%	68.67%
PCA+LDA	Normal	86.40%	13.60%
(Single)	Fatigue	21.60%	78.40%
PCA+LDA	Normal	83.33%	16.67%
(Sequence)	Fatigue	23.33%	76.67%
Proposed	Normal	95.33%	4.67%
Approach	Fatigue	5.33%	94.67%

5.3 Results and Analysis

From the training samples, we get the means of the PCA coefficients for the normal state and the fatigue state. Based on the means, we code the coefficient sequences. The means are showed in Fig. 6.

The recognition results of each classifier are listed in table 1. The classifiers include PCA+HMM, PCA+LDA (single), PCA+LDA (sequence) and the proposed approach. It becomes obvious that the proposed method enjoys a significant performance advantage on the data set. The average correct rates are 78.33%, 82.40%, 80.00% and 95.00% respectively. Our approach is much better than the other methods. PCA+LDA (single) is slightly better than PCA+LDA (sequence) in performance. This may be account for the not enough training samples. LDA is specifically optimized for discriminability but is susceptible to overfitting the training data.

6   Conclusions

We propose a novel dynamic feature to account for the spatial and temporal aspects of human fatigue in image sequences. A statistical learning algorithm is used to extract the most discriminative features and construct a strong classifier. The proposed approach is tested in a real-life fatigue environment with

30 human subjects of different poses and illuminations. The experiment results show that the proposed approach can achieve a much better performance than the baselines. In addition, this method can be easily extended to video based facial expression recognition. Future efforts will be focused on how to improve the dynamic feature extraction and get better performance. A hybrid of dynamic features and static features is the topic of our future research.

Acknowledgements

This research is sponsored by National Natural Science Foundation of China (No.60533030), Beijing Natural Science Foundation (No.4061001), and PHR (IHLB).

References

1. Dinges, D.F., Grace, R.: Perclos: A Valid Psychophysiological Measure of Alertness as Assessed by Psychomotor Vigilance. U.S. Dept. Transportation, Federal Highway Admin., Washington, DC, Tech. Rep. Publication No. FHWA-MCRT-98-006 (1998)
2. Wang, R., Guo, K., Shi, S., Chu, J.: A Monitoring Method of Driver Fatigue Behavior Based on Machine Vision. In: IEEE Proceedings on Intelligent Vehicles Symposium, pp. 110–113 (2003)
3. Dong, W., Wu, X.: Fatigue Detection Based on the Distance of Eyelid. In: IEEE International Workshop on VLSI Design and Video Technology, pp. 365–368 (2005)
4. Wang, F., Qin, H.: A FPGA Based Driver Drowsiness Detecting System. In: IEEE International Conference on Vehicular Electronics and Safety, pp. 358–363 (2005)
5. Wang, R., Guo, L., Tong, B., Jin, L.: Monitoring Mouth Movement for Driver Fatigue or Distraction with One Camera. In: The 7th International IEEE Conference on Intelligent Transportation Systems, pp. 314–319 (2004)
6. Wang, T., Shi, P.: Yawning Detection for Determining Driver Drowsiness. In: IEEE International Workshop on VLSI Design and Video Technology, pp. 373–376 (2005)
7. Yang, P., Liu, Q., Metaxas, D.N.: Boosting Coded Dynamic Features for Facial Action Units and Facial Expression Recognition. In: IEEE Conference on Computer Vision and Pattern Recognition, pp. 1–6 (2007)
8. Ji, Q., Lan, P., Looney, C.: A Probabilistic Framework for Modeling and Real-time Monitoring Human Fatigue. IEEE Transactions on Systems, Man and Cybernetics, Part A 36(5), 862–875 (2006)
9. Freund, Y., Schapire, R.E.: A Decision-theoretic Generalization of On-line Learning and an Application to Boosting. Journal of Computer and System Sciences 55(1), 119–139 (1997)

An Improved Method of Formula Structural Analysis

Xuedong Tian[1], Fei Wang[1], and Xiaoyu Liu[2]

[1] Faculty of Mathematics and Computer, Hebei University, 071002 Baoding, China
txdinfo@sina.com
[2] Library, Hebei Universyty, 071002 Baoding, China
wangfei-0984@163.com

Abstract. Structural analysis is an important step in mathematical formula recognition system. In this paper, an approach of structural analysis based on baseline strategy is proposed in which two strategies are employed to improve the robustness of the system. Firstly, a converse-matching algorithm is employed to solve the problem in traditional baseline method. Secondly, the feedback mechanism is built to correct the errors coming from the recognition result. The experiments show that the approach can obtain favorable results.

Keywords: Formula recognition, Structural analysis, Baseline, Converse Matching.

1 Introduction

With the high-speed of science and technology at full speed in our country, the literature of science and technology to which a large amount of mathematical formulas are applied in more and more fields. Current OCR (Optical Character Recognition) system shows high accuracy in recognizing the characters in printed documents, but has no way to handle the mathematical formulas among them. Therefore, it is necessary to develop the method of mathematical formula recognition to extend the application fields of traditional OCR technique.

This paper is focused on the structural analysis and comprehension of printed mathematical formulas. The problem has attracted the attention of several earlier workers. Anderson [1] adopted a purely top-down approach for parsing mathematical expressions. Chan and Yeung [2] designed on-line system to recognize mathematical expressions by using of structure and the syntax method. Zanibbi [3] etc. used a transformable technique called tree transform, where the information was represented as an attributed tree. This method consisted of three stages. In the first stage, an original Baseline Structure Tree (BST) was constructed to mainly describe the two-dimensional arrangement of input symbols. In the second stage, the Lexical pass produced a Lexed BST from the initial BST by grouping tokens comprised of multiple input symbols. At last, Lexed BST is translated into an Operator Tree, which describes the order and scope of operations in the input expression.

But the above-mentioned method is confined to some certain special type of mathematical formula only. In order to resolve the structural analysis problem based on baseline method, this paper proposes a converse matching method with syntax

G. Wang et al. (Eds.): RSKT 2008, LNAI 5009, pp. 692–699, 2008.

rules and carries out a post-processing on the mathematical formula analysis. The experiment has indicated this method that can improve the accuracy of the structural analysis effectively.

This paper is organized as follows. In Section 2, the concepts of baseline are introduced and some shortcomings of the baseline method are discussed. In section 3, the proposed method is described in detail. The test result is presented in the final Section.

2   Overview of Structural Analysis

Usually, a formula recognition system can be divided into two steps: symbol recognition and structural analysis. Structural analysis stage is to parse the logical and spatial relationships of the symbols according to the codes and their correlative information obtained from the character recognition module, moreover, to represent the relationship by a formula grammar tree. A structural analysis method based on baseline consists of three phases: (1) To seek the dominant baseline of the formula; (2) To find the other nested baseline according to the control rules and get the nested relationship of each symbol; (3) To describe the information of parsing by a grammar tree, in which a node [4] represents a symbol as shown in Figure 1. According to the symbol operating ranges [3] of mathematical formula, namely symbol dominance, seven kinds of mathematical relations are taken into account: *up, superscript, right, subscript, down, inclusion* and *left up*.

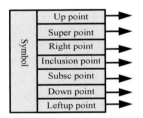

Fig. 1. The Node of Grammar Tree

2.1   Baseline

A baseline in mathematical formula is a linear horizontal arrangement of symbols, intended to be perceived as adjacent [4]. Generally, a formula includes the dominant baseline and the nested baseline, the former is a baseline in which the symbols are not nested relative to any other symbols in the mathematical formula. In this paper, we defined the dominant baseline is that baseline of the leftmost symbol of the mathematical formula which lies within a region. The latter is to have deflected some symbol on vertical direction or the baseline in which some symbols are surrounded by other symbols. Sometimes, the symbol located in nested baseline are controlled

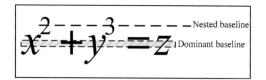

Fig. 2. Dominant Baseline and Nested Baseline

direct or indirect by other character. The instance containing two baselines in formula $x^2 + y^3 = z$, is shown in Figure 2. One baseline contains the symbols ("x", "y", "=", "z") and the other contains ("2", "3"). And the symbol "x" is a nested relationship with the symbol "2", which lies in nested baseline.

2.2   The Problem Suspended in the Baseline Method

In a mathematical formula, characters and symbols can be spatially arranged as a complex two-dimensional structure, possibly of different character and symbol sizes [5]. However, sometimes mathematical formulas are prone to ambiguities [6] especially when they are not typeset or scanned properly. Hence, there exist a lot of shortcomings when using the baseline method. Some of them are summarized and briefly discussed:

1) When building a baseline, baseline method often relies on the use of thresholds. In practice, threshold values cannot be chosen to work well on all of possible symbols.
2) In mathematical formula, the spatial relationship is determined based on the relative positions of symbols. It involves all the associated attributes of a symbol, especially the center of the symbol, which often refers to the typographical center. Sometimes, different typographical center of symbol will lead to the wrong relative positions of symbols.
3) A few special mathematical symbols, such as "-"(fraction), "Σ", "$\int$", "$\prod$", etc. which dominate their neighboring sub-formula. These symbols usually are called "structural symbol" [7]. Sometimes operand overlaps the confine of structural symbol, hence, the method of baseline can not be able to accurately allocate the dominant baseline for this type of mathematical formula. (As shown in figure 4).
4) In mathematical formula the type of spatial operators between symbols is determined based on the relative positions of symbols. It will affect the analysis result if the images were inclined. For example, as shown the first formula in Figure 3, symbol "x" in the last part of the longer formula should have a superscript "2", but digits "2" deviated the baseline which should belong to it self.
5) Some nodes of the tree may be missing in the list of objects obtained from the previous phase due to the existence of spatial operators in mathematical formulas. Typical failure example is shown in Figure 3 of the lower formula.

$$(x^3 + 3x^2 - 5x + 6) - (2x^2{}_3 - 5x + 1) - (x + x - 2)$$

$$\lim_{n \to \infty} a\overline{n} = 1$$

Fig. 3. The Wrong Results after Structural Analysis

$$\sum_{k=n+1}^{n+p} | \, u_n(x) \, | \qquad \iint_{x^2 + y^2 = z} dxdy$$

Fig. 4. Operand Overlap the Confine of Symbol

3   The Method of Post-processing

The massive experiment indicates structural analysis based on baseline strategy is able to achieve quite satisfactory results for the one-dimension formula. Moreover, it has the dissatisfactory with the nested formula and the complex formula. Against the above problem, this article proposed the system of post-processing using the reversion matching and the semantics regular method. So far, to the best of our knowledge, papers on error detection and correction in mathematical formula recognition are relatively rare [8]. In most cases, it is still at detection level [9]. In this paper, we will correct some simple errors with semantics regular.

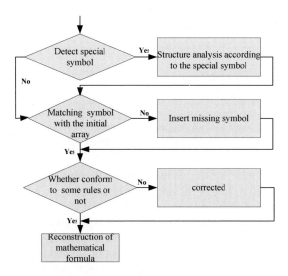

Fig. 5. System Flow Diagram for Post-processing

3.1   The Design of Post-processing System

The major steps in the system of post-processing are as follows. In the first step, a grammar tree has been constructed for each mathematical formula when in structural analysis step. This step of post-processing system will detect the formula which have structural symbol. If the analysis result is not accepted by the following rules, the result of the analysis step is corrected according to the rules. In the second step, we called the symbol matching, all nodes of the grammar tree by the analysis step are matched with the initial array which transmission is from symbol recognition step. In the third step, we revised some which did not conform to the semantic of mathematical formula according to following rule. The system block diagram is shown in Figure 5.

3.2   Concrete Step

Step 1. To search symbol of the grammar tree and detect whether it contains the structural symbol or not, such as "-"(fraction), "$\sum$", "$\int$", "$\prod$". Its algorithm is as follows:

```
if (Ssymbol.up==True && Ssymbol.down==True) then break;

if (Ssymbol.up==False && Ssymbol.down==True) or
(Ssymbol.up==True && Ssymbol.down==False)then
center=(Ssymbol.cx,Ssymbol.cy);

Return BuildBS( ).
```

1: Search Structural Symbol Alg.

Ssymbol.up , *Ssymbol.down* is the up point and down point of the structural symbol respectively. The structural symbol center coordinate is ($Ssymbol.cx, Ssymbol.cy$) . BuildBS (center) expresses build baseline which is take *center* in the regions

Step 2. Symbol/Character matching

Some symbols/characters center coordinate which *(centry_x,centry_y)* locates neither in dominant baseline scope nor in the nested baseline scope for different typesetting. It is very easy to leak such symbol which we have discussed above. We use the method of converse matching to search the missing symbol and redistribute these symbols.

1)   Matching the symbol of the grammar tree with the symbol in the initial array which is arranged according to *min_x,* and if matching is successful, this symbol will be deleted from the initial array.

2)   We detect the initial array whether it is an empty until searching the grammar tree is finished. We will assign the rest of symbol according to the *nearest neighbor rule* (NNR) if the initial symbol is not an empty. Here we make use of a Euclidean distance that uses two-dimensional statistics to fix the symbol position result. $S_j(centriodX, centriodY)$ is the center coordinate of missing

symbol, Let $S = (s_1, s_2, \cdots s_n)$ be the set of symbol in initial array. The NN rule is defined as follows:

$$d_{\min}(s_i, s_j) = \sum_{i=1}^{n} \sqrt{[centerx(s_i) - centerx(s_j)]^2 - [centery(s_i) - centery(s_j)]^2} \ (i \neq j) \ . \quad (1)$$

Step3. According to following rules, we correct the situation where it did not conform to the semantic relations of mathematical symbol by searching the point of symbol in grammar tree respectively. However, the following tiny rules, show the basic corrected idea of the revision step.

Rule1. There are no symbols in the superscript or subscript position of a numerical, left bracket, and so on. Suppose the set of $S_1, S_2, \cdots S_n$ is in the formula symbol, Let $E_1 E_2, \cdots E_n$ is the symbol set which has not the superscript or subscript, the match (S_i, E_j) is a function of two symbols matching, and correct (S_i) carries on error correction of the symbol S_i

if $\text{match}(S_i, E_j) == \text{True then correct}(S_i)$

Rule2. It is impossible that a symbol has the subsc point and the down point simultaneously. Suppose the set of $S_1, S_2, \cdots S_n$ is in the formula symbol, "*subsc*" is the point of subscript, "*down*" expresses the point of down of the symbol S_i. The function of correct (S_i) carries on error correction of the symbol S_i.

if $S_i.subsc == \text{True}\&\& \ S_i.down == \text{True}$

then $\text{correct}(S_i)$

4   Experimental Results and Conclusion

In our experiments, we used more than 20 kinds of documents which are scanned at a resolution of 600dpi to test our method and the method described was implemented in the VC++ workstation. The experiment was focused on the ability of the post-processing on detecting the failures at the analysis step. The experimental results are shown in Table 1 and Table 2.

Figures 6-8 show several results of the proposed method. In each of those figures, analysis results without post-processing (left) and their analysis results with post-processing (right) are shown. The correction results are represented as formula synthesized by applying LATEX complier.

Table 1. The results without post-processing

Formula type	The number of formulas	The number of formulas analysis	Accuracy of analysis
Normal formula	2567	2189	85%
Special formula	1967	1561	79%

Table 2. The results with post-processing

Formula type	The number of formulas	The number of formulas analysis	Accuracy of analysis
Normal formula	2567	2243	87%
Special formula	1967	1590	81%

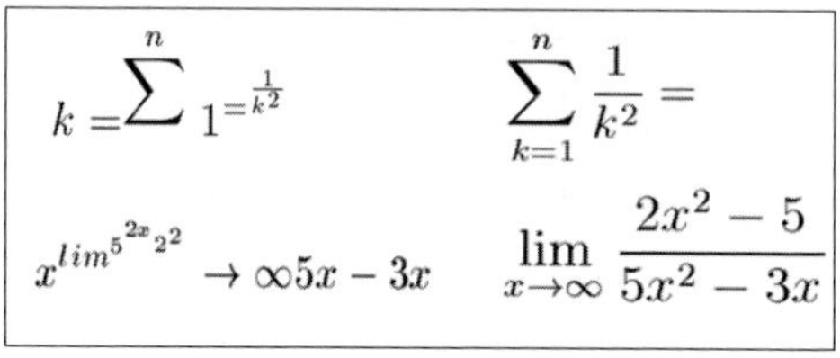

$$k = \sum_{}^{n} 1 = \frac{1}{k^2} \qquad \sum_{k=1}^{n} \frac{1}{k^2} =$$

$$x^{lim^{5^{2x}2^2}} \to \infty 5x - 3x \qquad \lim_{x \to \infty} \frac{2x^2 - 5}{5x^2 - 3x}$$

(a)Wrong result   (b)Right result

Fig. 6. Formula whose baseline errors were corrected by the post-processing

Figure 6 shows the results of the failure of dominant baseline at the structural analysis step. In the above example, dominant baseline was considered as the baseline which was decided by the leftmost symbol "k".

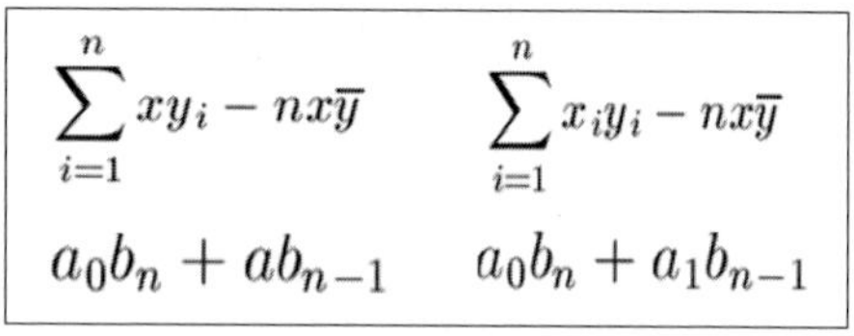

$$\sum_{i=1}^{n} xy_i - nx\bar{y} \qquad \sum_{i=1}^{n} x_i y_i - nx\bar{y}$$

$$a_0 b_n + a b_{n-1} \qquad a_0 b_n + a_1 b_{n-1}$$

(a) Lost character   (b)Find lost character

Fig. 7. Formula of missing character were corrected by the post-processing

Figure 7 shows that how the post-processing step could detect the loss character of the mathematical formula. In the first example of this figure, the subscript "i" of character "x" of the mathematical symbol ("$\sum$") is lost.

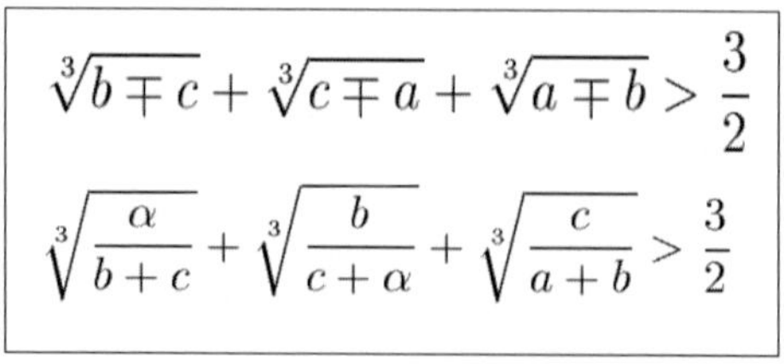

$$\sqrt[3]{b \mp c} + \sqrt[3]{c \mp a} + \sqrt[3]{a \mp b} > \frac{3}{2}$$

$$\sqrt[3]{\frac{\alpha}{b+c}} + \sqrt[3]{\frac{b}{c+\alpha}} + \sqrt[3]{\frac{c}{a+b}} > \frac{3}{2}$$

Fig. 8. Operand overlaps the confine of fraction

In Figure 8, the upper example is failure at the analysis step because denominator surpass the area of fraction. The lower examples that operand overlaps the area of fraction were successfully corrected by the post-processing.

In the future, our research will include: (1) finding new methods to choose proper thresholds that make a dominant baseline robust in mathematical formula, (2) adding syntax to analysis of the mathematical formula.

Acknowledgments. This work is supported by National Natural Science Foundation of China (60772073), Researching and Developing Project of Science and Technology of Hebei Province (06213598), and Scientific Research Project of the Education Department of Hebei Province (ZH2007104). The authors would like to thank other members of the lab for technical discussion.

References

1. Anderson, R.: Two-dimensional Mathematical Notation. In: Syntactic Pattern ecognition Applications, pp. 147–177. Springer, New York (1997)
2. Chan, K.F., Yeung, D.Y.: An Efficient Syntactic Approach to Structural Analysis of On-line Handwritten Mathematical Expressions. Pattern Recognition 33, 375–384 (2000)
3. Zanibbi, R., Blostein, D., Cordy, J.R.: Recognizing Mathematical Expressions Using Tree Transformation. IEEE Trans. Pattern Anal Mach Intel. 24(11), 1–12 (2000)
4. Tian, X.D., Li, N., Xu, L.J.: Research on Structural Analysis of Mathematical Expressions in Printed Documents. Computer Engineering 32(23), 202–204 (2006)
5. Chen, H.B., Wang, Q., Xu, X.R., Zhang, C.Y.: A Survey of Mathematical Expression Auto-recognition. Guangxi Sciences 11(1), 20–26 (2004)
6. Chen, Y., Shimizu, T., Yamauchi, K.: Ambiguous Problem Investigation in Off-line Mathematical Expression Understanding. In: SMC 2000 IEEE International Conference, vol. 4, pp. 291–2922 (2000)
7. Inoue, K., Miyazaki, R., Suzuki, M.: Optical Recognition of Printed Mathematical Documents. In: ATCM 1998, pp. 280–289 (1998)
8. Chan, K.F., Yeung, D.Y.: Error Detection, Error Correction and Performance Evaluation in Online Mathematical Expression Recognition. Pattern Recognition 34, 1671–1684 (2001)
9. Toyota, S., Uchida, S., Suzuki, M.: Structural Analysis of Mathematical Formulae with Verification Based on Formula Description Grammar. In: The 7th International Workshop DAS, pp. 153–163 (2006)

Fuzzy Inspection of Fabric Defects Based on Particle Swarm Optimization (PSO)

Juan Liu and Hui Ju

Department of Control Engineering, Chengdu University of Information Technology,
Chengdu 610225, China
liuxiaojuan1123@126.com

Abstract. A new approach for inspection of fabric defects based on
Principal Component Analysis (PCA) and Fuzzy C-Mean Clustering
(FCM) Based on Particle Swarm Optimization (PSO) is proposed. First,
the PCA is used to reduce the dimension of the original image and computation complexity. The dimension-reduced image features, which can
best describe the original image without unnecessary data, are recognized
by FCM based on PSO next. The recognition is carried out by the merits
of the overall optimizing and higher convergent speed of PSO combined
with FCM algorithm, which makes the algorithm have a strong overall searching capacity and avoids the local minimum problems of FCM.
At the same time, it reduce the degree of sensitivity of FCM that depends on the initialization values. The results show that the method is
more effective than the traditional one with BP neural networks based
on wavelet[1,2].

Keywords: Fabric Defects Inspection; Principal Component Analysis
(PCA); Particle Swarm Optimization (PSO); Fuzzy C-Mean Clustering
(FCM).

1   Introduction

Fabric defects inspection is a technology of fabric quality inspection, which use
image sensor to get the defect fabric images and consider them as a template
to recognize the defect fabric. Though the production efficiency of actual textile
increases tremendously, fabric defects are still inspected with manual work. This
method has many disadvantages, such as low speed, high mistakes and leaks,
wrong result due to the inspector's subjective consciousness, and so on. On
account of this, the technology of fabric defects inspection is a urgent need for
improvement in weave industry.

Since the 70s of the 20th century, many domestic and foreign experts have
done some researches about fabric defects inspection and educed some effective
methods. However, it is on the stating stage, and most of them are reported
as algorithm, patent and sample machine. Until now, adult fabric defects auto-
inspection system begin to enter into market[3]. With the wider and wider use
of image sensor, adoption of image information for fabric defects inspection becomes one of the head methods. The image method has two species: one is

G. Wang et al. (Eds.): RSKT 2008, LNAI 5009, pp. 700–706, 2008.

processed in time domain directly, like eigenvalue picking-up from spatial texture features. Another is in frequency domain just as Fourier transform, Gabor filter or wavelet transform[4]. In our paper, a new approach for fabric defects inspection based on Principal Component Analysis (PCA) and Fuzzy C-Mean Clustering (FCM) Based on Particle Swarm Optimization (PSO) is presented.

2   Principal Component Analysis

Principal Component Analysis is a common, simple and effective linear analysis method, which is proposed by Turkm and Pentland in the 90s of the 20th century[5]. It transforms the images by Karhunen-Loeve method based on their statistical speciality, then considers feature matrix that is composed of eigenvector of covariance matrix of random variable X in original space as transform matrix. Lastly it transforms the vectors of original space, changing the high-dimension and complex vectors into low-dimension and simple ones in feature space.

The transform has follow trait: the new transformed components are orthogonal or irrelevant; part new components are used to show the minimum of mean square error of original component; the transformed vectors, whose energy is more concentrated, are more stable.

Transforming the original space of fabric image by Karhunen-Loeve method is to construct the covariance matrix of the image's data sets, transform the matix orthogonally to get its eigenvectors, and order them according to the eigenvalue's size. Among these eigenvectors, everyone show a vector with different numbers in fabric images. They show a feature gather, all of which express a fabric image together. This method is in common use in face recognition, and a new test in fabric defects inspection.

The material algorithm is as follows: suppose that there are N training fabric images, and the dimension of everyone is $P \times Q$. Then, they can be figured as $L_1, L_2, L_3, \ldots, L_N$, while their mean image f and the distances d_i between every image and mean image is:

$$f = \frac{1}{N} \sum_{i=1}^{N} L_i \tag{1}$$

$$d_i = L_i - f \tag{2}$$

If matrix $A = \{d_1, d_2, d_3, \ldots, d_N\}$, we must extract the eigenvector of matrix AA^T to obtain the dimension-reduced image describe feature. At the same time, AA^T is a $(P \times Q) \times (P \times Q)$ high-dimension matrix that need a tremendous computation complexity. We can consider the Karhunen-Loeve transform make the energy concentrated correspondingly, and the eigenvalues related to the eigenvectors have follow speciality: $\lambda_m \leq \lambda_{m-1} \leq \ldots \lambda_2 \leq \lambda_1$.

Hence, get the eigenvalues related to the first m features to transform the vectors in original space, and obtain the minimum of the transformed mean square error. A lesser m eigenvectors is enough to recognize.

We firstly get the eigenvalues λ_j and the eigenvectors u'_j to obtain the first m eigenvectors u_j of AA^T:

$$u_j = Au'_j\lambda_j^{-\frac{1}{2}} \qquad (j = 1, 2, 3 \cdots, m) \qquad (3)$$

In above formula, u'_j is the eigenvectors of AA^T, while λ_j is the eigenvalues of AA^T. And u_j is the eigenvectors we need. Then, reduce the dimension of the training image to pick up features. A projective matrix composed of u_j can be showed as $W = \{u_1, u_2, u_3, \ldots, u_N\}$. All the dimension-reduced vectors of fabric images are as follows:

$$g_i = W^T d_i \qquad (i = 1, 2, 3, \cdots, N) \qquad (4)$$

With PCA, we can extract the eigenvectors to obtain the dimension-reduced image features from the original image for the next recognition by FCM based on PSO, which will reduce the computation complexity to improve the speed.

3   Fuzzy C-Mean Clustering (FCM) Algorithm Based on Particle Swarm Optimization (PSO)

In our paper, we consider the Fuzzy C-Mean Clustering as a measurement method of comparability. Confirm the comparabilities among the eigenvectors to recognize the kinds of fabric image.

3.1   Particle Swarm Optimization (PSO)

Particle Swarm Optimization (PSO) , which is an evolution calculate technology, is proposed by Kenney and Eberhart in 1995[6]. For the optimization problem, every particle in PSO is a possible solution. In the swarm the best position every particle has passed in the iterative process is the best solution found by the particle itself. The same, the best position that all the particles have passed is the best solution found by all. The former is called individual extremum, and the latter is overall extremum. Every particle refreshes itself according to the two extremum above continuously to produce a new particle. In this process, the whole swarm searches the solution domain roundly.

The position of the ith particle in the swarm is x_i, whose individual extremum is $pBest_i$. If v_i shows speed and $gBest$ shows overall extremum, a particle i will refresh its speed and position like this:

$$v_i(t+1) = \omega v_i(t) + c_1 r_1(t)(pBest_i(t) - x_i(t)) + c_2 r_2(t)(gBest(t) - x_i(t)) \quad (5)$$

$$x_i(t+1) = x_i(t) + v_i(t+1) \qquad (6)$$

In the formula, r_1 and r_2 are random numbers in the domain of (0, 1), and the three weights ω, c_1, c_2 decide the ability of searching space of the particle. Firstly, the inertia weight ω is to keep the particle moving, which reacts on the capability of overall balance and part searching. Secondly, the acceleration coefficient c_1, as the cognitive part, brings the cognitive capacity and strong overall searching ability to the particles, avoiding that the search fell in one of the minimums. Lastly, another acceleration coefficient c_2 describes information share among the particles as the society part.

As is known to all, the speed value v_i of every particle is limited to be less than a maximum $v_{max}(v_{max} > 0)$, which is a adjustable parameter. With time passed, the inertia weight ω could be reduced linearly.

The refresh formula of individual extremum and overall extremum of every particle is as follows:

$$pBest_i(t+1) = \begin{cases} x_i(t+1) & x_i(t+1) \geq pBest_i(t) \\ pBest_i(t) & x_i(t+1) < pBest_i(t) \end{cases} \tag{7}$$

$$gBest(t+1) = \max(pBest_i(t+1)) \tag{8}$$

3.2 Fuzzy C-Mean Clustering (FCM)

FCM is presented by Bezdek in 1981[7]. Suppose that finite the sample space $X = \{x_1, x_2, x_3, \ldots, x_n\}$, while x_i is a vector with arbitrary dimensions. Divide the samples into C clusters, the gather of whose center is $V = \{v_1, v_2, v_3, \ldots, v_C\}$. Define a criterion function J_m for a fuzzy C-mean cluster like this:

$$J_m(\tilde{U}, v) = \sum_{k=1}^{n} \sum_{i=1}^{C} (u_{ik})^m (d_{ik})^2 \tag{9}$$

$$d_{ik} = d(x_k - v_i) = \left[\sum_{j=1}^{m} (x_{kj} - v_{ij})^2 \right]^{\frac{1}{2}} \tag{10}$$

In the formula, u_{ik} is the Membership value of the kth data point in the ith cluster, whereas d_{ik} is Euclid Distance between the ith cluster's center and the kth data sets. Besides, m is a weighting parameter and in a limit area as $m \in [1, \infty]$. This parameter controls the size of ambiguity in the clustering process. The value of function J_m is variational. Thus the minimal one is correspond to the best clustering.

The Center Coordinate of every cluster can be calculated as follows:

$$v_{ij} = \frac{\sum_{k=1}^{n} u_{ik}^m x_{kj}}{\sum_{k=1}^{n} u_{ik}^m} \qquad (j = 1, 2, 3, \cdots, m) \tag{11}$$

And the new formula of membership matrix is:

$$
u_{ik}^{(r+1)} = \left[\sum_{j=1}^{C} \left(\frac{d_{ik}^{(r)}}{d_{jk}^{(r)}} \right)^{2/(m-1)} \right]^{-1}
\tag{12}
$$

FCM is to calculate the clustering solution that make the criterion function J_m minimal.

3.3   FCM Algorithm Based on PSO

Though every iteration process of the FCM follows the better solution, this method that is based on gradient descent algorithm is a local search algorithm essentially and inclined to fall in one of the minimums. This situation is more obvious with increasing clustering samples. What is more, the best solution calculated by FCM depends a lot on initial values, such as the initial clustering affects the final solution deeply. Therefore, it is necessary for FCM to test initial solution many times in order to obtain better one, which is wasting time and low efficient. But the initial value of PSO that based on the swarm operation is any possible solution on the uniform distributed solution space. It has powerful ability of overall searching, and is not apt to fall in one of the minimums. Its convergence speed is also very fast. Combine the two algorithms will possibly produce a better one.

FCM is finally concluded a process of minimizing the criterion function. To process this problem, PSO has strong advantages. Thus, we prefer to use PSO instead of the iteration process in FCM.

The core algorithm of FCM is to make sure the clustering center. So PSO is used to optimize the clustering center., which is considered as the particle swarm. The fitness function is:

$$
f(x) = \frac{1}{J(\tilde{U}, v) + 1}
\tag{13}
$$

The better the clustering solution is, the smaller the criterion function $J(\tilde{U}, v)$ is. Then the fitness function $f(x)$ is higher.

Because every particle shows a choice of a kind of clustering center, the high value of every fitness explain the good solution, while the low explain the bad. So estimate the fitness function is the way to estimate the efficiency of clustering solution. Choose the individual extremum and the overall extremum according to the fitness function $f(x)$ that corresponds to the clustering center and stop iterating on the same condition with FCM.

The detailed step of the algorithm is as follows:

1) Choose the number of the clustering center $C(2 \leq C \leq n)$ and parameter m. Initialize the membership matrix $\tilde{U}^{(0)}$, and define a termination error $\epsilon(\epsilon > 0)$. Suppose the iterating step is $t = 0$ and maximal step is `iter_max`.

2) Calculate the C clustering centers according to formula (11), and consider these centers as the particle swarm $x_i(t)$. Then refresh every particle's speed v_i.

3) Calculate the fitness of every particle according to fitness function, and refresh individual extremum $pBest_i$ and the overall extremum $gBest$ according to every fitness and formula (7) and (8). Then refresh every particle's speed v_i to produce a new particle $x_i(t + 1)$. Finally, calculate the membership matrix $\tilde{U}^{(t+1)}$ correspond to new particles according to formula (12). Thus, the criterion function J_m comes out.

4) If $||\tilde{U}^{(t+1)} - \tilde{U}^{(t)}|| \leq \epsilon$ or $J_m = \min J_m$, stop iterating. Else, let $t = t+1$ and return to step (2). When $t \geq iter\_max$, end the algorithm.

4   Experimental Simulation

Experiment adopt DVT Lengend510 intelligent image sensor to collect fabric image, whose size is 640×480. Four classes of images are collected, including normal, weft-lacking, warp-lacking and oil stain that exhibited in Fig. 1–4. Every class can have more than one defect, like Fig 5. In our experiment, 5 images of every class are considered as training images, and 50 images are testing samples.

Fig. 1. Normal   **Fig. 2.** Weft-lacking **Fig. 3.** Warp-lacking   **Fig. 4.** Oil stain

The experimental steps are as follows:

1) Choose training images from collected ones.
2) Normalize the images and transform them with PCA to get their features.
3) Cluster the features with the algorithm of FCM based on PSO to obtain the best sulotion. We can call it Detection-Template.
4) Input the testing images,transform it with PCA, and calculate the membership value from its feature description to every cluster. Then, we can judge the kind of the testing image.

This method is programmed with Matlab 7.0. It reduces the dimension of image to 20×20. The experimental result proves that the algorithm of FCM

Fig. 5. warp-lacking2

Table 1. Result compare of different recognition methods

Method	recognition rate	Training time(s)	Recognition mean time(s)
FCM	84%	140.156	4.844
PSO+FCM	94%	137.157	4.672
PCA+FCM	90%	116.141	3.375
PCA+PSO+FCM	98%	105.500	2.343

based on PSO presented in this paper is efficient. It produces higher recognition rate and searching speed than others, which is shown in Table 1.

To obtain higher recognition rate, we can add the numbers of the training image to get more precise template. Before we detect on line, the Detection-Template must be trained firstly. So, adding the training images just adds the training time. It will not reduce the recognition time to affect detection on line.

5   Conclusion

This method can not only be used in the four classes above, but also in any other class. In this case, what we should do is to add some training images of the class you want to recognize into the training processing(Experimental step 1–3). Therefore, it's still available in many other fields of recognition. Comparing with the traditional method[1,2], the algorithm presented in this paper reduces the dimension of inputting images and computation complexity, improving the inspection speed and making it easier to inspect on line. It also enhances the searching ability, reducing the possibility of falling in one of the minimums. Finally, it get a better clustering solution, increasing recognition rate.

References

1. Qu, P.G., Liu, S.G., Zh, H.: The Application of Wavelet Analysis in Fabric Defects Inspection. Basic Sciences Journal of Textile Universities 17(04), 364–367 (2004)
2. Xu, X.F., Duan, H., Wei, J.M.: Fabric Defect Detection with Two-Dimensional Wavelet Transform and BP Neural Network. Journal of Zhejiang Institute of Science and Technology 21(01), 15–19 (2004)
3. Chan, C.H., Grantham, P.K.H.: Fabric defect detection by fourier analysis. IEEE Transactions on Industry Application 36(5), 1267–1276 (2000)
4. Li, L.Q., Huang, X.B.: Woven Fabric Defect Detection with Features Based on Adaptive Wavelets. Journal of Donghua University(Natural Science Edition) 27(04), 82–87 (2001)
5. Turkm, P.A.: Eigenfaces for Recognition. Journal of Cognitive Neuroscience 3(1), 71–86 (1991)
6. Kennedy, J., Eberhart, R.: Particle Swarm Optimization. In: IEEE International Conference on Neural Networks, pp. 1942–1948. IEEE Service Center, Piscataway (1995)
7. Bezdek, J.C.: Pattern Recognition with Fuzzy Objective Function Algorithms. Plenum Press, New York (1981)

A Hybrid Feature Selection Approach Based on the Bayesian Network Classifier and Rough Sets

Li Pan[1], Hong Zheng[2], and Li Li[2]

[1] School of Remote Sensing and Information Engineering Wuhan University
129 Luoyu Road, Wuhan, Hubei 430079, P.R.China
panli@whu.edu.cn
[2] School of Electronic Information Wuhan University
129 Luoyu Road, Wuhan, Hubei 430079, P.R.China
zh@eis.whu.edu.cn

Abstract. The paper proposes a hybrid feature selection approach based on Rough sets and Bayesian network classifiers. In the approach, the classification result of a Bayesian network is used as the criterion for the optimal feature subset selection. The Bayesian network classifier used in the paper is a kind of naive Bayesian classifier. It is employed to implement classification by learning the samples consisting of a set of texture features. In order to simplify feature reduction using Rough Sets, a discrete method based on C-means clustering method is also presented. The proposed approach is applied to extract residential areas from panchromatic SPOT5 images. Experiment results show that the proposed method not only improves classification quality but also reduces computational cost.

Keywords: Rough Sets, Feature Selection, Naive Bayesian Network Classifier.

1 Introduction

Feature selection is the problem of choosing a small subset of features that is necessary and sufficient to describe target concept. The importance of feature selection is due to the potential for speeding up the processes of both concept learning reducing the cost of classification, and improving the quality of classification. Feature selection has been the key research issue in many fields such as pattern recognition, image understanding and machine learning. In general, the existing methods can be classified into two categories: open-loop or filter method and close-loop or classifier feedback method. The open-loop method does not consider the effect of selected features on a whole processing algorithm performance. The classifier feedback method is a kind of feature selection method which uses a classification rate as the criterion for a feature subset selection. The classifier feedback method has more advantages than the open-loop ones due to directly improving the classification results. The feature selection approach proposed in this paper is an instance of classifier feedback approach. It takes the

G. Wang et al. (Eds.): RSKT 2008, LNAI 5009, pp. 707–714, 2008.

classification results of naive Bayesian network classifier as the criterion function for feature subset selection using Rough Sets.

The existing methods for feature selection based on Rough Sets have many types. For example, ROSETT is a software systems based on searching for short reductions or reduction approximations [1,2]. The other methods are based on genetic algorithm with the fitness function measuring the quality of the selected reduction approximation [3,4]. However, the classification quality for these methods is still not good enough. In this paper, Bayesian network classifier is introduced as the classifier. Since Bayesian network classifiers can combine multiple source features, they have become an attractive approach to solve the problem of image classification. In this paper, we take use of a simple case of Bayesian network called naive Bayesian classifier for learning the samples and inferring about unknown regions. The samples used here are a set of texture features extracted from a large number of remote sensing sample images. Each texture feature vector is labeled by the C-means clustering method. The decision table used in Rough Sets is made up of feature values and related categories of samples. According to the classification rate of naive Bayesian network classifier, useless features or condition attributes are eliminated. At last, the optimal feature subset is found and used to extract interested objects from remote sensing images.

This paper is organized as follows: Section 2 presents the algorithm for the feature selection based on Rough Sets. Section 3 describes the classification method using a naive Bayesian network classifier. Section 4 shows experimental results and conclusions are given in Section 5.

2 The Feature Selection Based on the Rough Sets

Feature selection is a process of finding a subset of features from the original set of features according to the given criterion. In this paper, the classification rate of naive Bayesian network classifier is defined as the criterion of the feature selection based on Rough sets.

For an image with size $p \times q$, it is constituted with M-feature patterns. Let all M-features of a pattern generate a whole original feature set:

$$T_{total} = \{t_1, t_2, \dots t_m\}$$

The feature selection is a process to search for a subset

$$T_{sub} = \{t_1, t_2, \dots t_n\} \, (T_{sub} \subseteq T_{total}, n < m)$$

under a given criterion, which guarantees to obtain better classification results. Generally, an image can be classified into different parts according to color, size, etc. Hence, assume that there is a family of indicernibility relationship $I = \{I_1, I_2, \dots I_D\}$ over the universe U, which is equal to the whole original feature set $T_{total} = \{t_1, t_2, \dots t_m\}$. In Rough Sets, if minimal subset I_{small} of I can determine knowledge about the universe, $\cap I_{small} = \cap I$ will be called a reduction of I, where $\cap I_{small}$ is equal to a feature subset $T_{sub} = \{t_1, t_2, \dots t_n\}$. So the process of feature selection is to find minimal feature . In this paper, N samples are selected firstly, where the M features of each sample are calculated and converted into the discrete values by C-Means clustering method. The decision table consists of feature values and categories of selected samples. In the traditional Rough Sets, the process of reducing features mainly depends on the compatibility of the decision table. In other words, if there is a conflict of the rules in the decision table (DT) when one attribute or feature is deleted, the decision table is not compatible and the attribute should not be eliminated. Otherwise, the decision table is compatible and the attribute may be eliminated. Therefore, it is very important how to define the compatibility of the decision table .We take the classification results a Bayesian network classifier as the criterion of the decision table. If the correct classification ratea using the feature subset $T_{sub} = \{t_1, t_2, \dots t_n\}$ is equal to or greater than a given threshold, the decision table is compatible and the selected subset features $T_{sub} = \{t_1, t_2, \dots t_n\}$ is effective. As a result, the optimal subset has minimum features and the best classification results. Table 1 shows an original decision table, where $\{SAM_1, SAM_2 \dots SAM_5\}$ denote five samples, $\{CON_1, CON_2 \dots CON_9\}$ denote nine features or condition attributes and DES is the sample categories or the decision attribute. Reduced decision table also shown in Table 1 only includes two condition attributes, and $\{CON_1, CON_2\}$ is the optimal feature subset. The feature selection algorithm is described as follows:

Algorithm1: Feature selection algorithm using Rough Sets.
Given: $DT =< U, C \cup D, V, f >$

where DT is a decision table and U is the universe that is a family of indicernibility relationship of the all features. C is a set of condition attributes, which expresses a set of texture features. D is a set of decision attributes, which represents a set of sample categories. $V = \cup_{q \in C \cup D} V_q$, V represents a set of attributes which denote a set of sample texture feature values. $f : U \times (C \cup D) \rightarrow V$ is a decision function, which describes the relationship of texture feature values.

Assume $DX \in U \times C \rightarrow V$ is a rule of a decision table based on condition attribute C, in which the rule is made of texture features of samples. $DX \in U \times C \rightarrow V$ is a rule of a decision table based on decision attribute D, which not only includes the texture features of the samples, but also the categories of samples.

Step1: Eliminate one feature or attribute $r(r \subseteq C)$, if the classification rate of the Bayesian network classifier using the rest feature subset $T_{sub} = \{t_1, t_2, \dots t_i\}$ and $DX(i)$ is equal to greater than before, the decision table is compatible. Otherwise, the decision table is not compatible. For a compatible decision table, continue to eliminate another feature or attribute. For a non-compatible decision table, the previous deleted feature or attribute should be reserved firstly, then continue to eliminate another feature or attribute. Repeat above operations until all attributes or features are scanned.

Step2: Output final feature subset and decision table $DT_{SUB} =< U, C' \cup D, V', f' >$ where $C' \subseteq C$ $V' \subseteq V$ $f' \subseteq f$. The feature subset is the optimum feature subset.

Table 1. An original decision table(left) and its reduced decision table(right)

NO	CON_1	CON_2	....	CON_9	DES	NO	CON1	CON2	DES
SAM_1	$Value_{11}$	$Value_{12}$	...	$Value_{19}$	$Class_1$	SAM1	$Value_{11}$	$Value_{12}$	$Class_1$
SAM_2	$Value_{21}$	$Value_{22}$	...	$Value_{29}$	$Class_2$	SAM2	$Value_{21}$	$Value_{22}$	$Class_2$
SAM_3	$Value_{31}$	$Value_{32}$	...	$Value_{39}$	$Class_3$	SAM3	$Value_{31}$	$Value_{32}$	$Class_3$
SAM_4	$Value_{41}$	$Value_{42}$	...	$Value_{49}$	$Class_4$	SAM4	$Value_{41}$	$Value_{42}$	$Class_4$
SAM_5	$Value_{51}$	$Value_{52}$	...	$Value_{59}$	$Class_5$	SAM5	$Value_{51}$	$Value_{52}$	$Class_5$

3   The Bayesian Networks Approach for Image Classification

3.1   The Overview of Bayesian Networks Classifiers

A Bayesian network (BN) represents a joint probability distribution over a set of discrete and stochastic variables [7,8,9,10]. It is to be considered as a probabilistic white-box model consisting of a qualitative part specifying the conditional dependencies between the variables and a quantitative part specifying the conditional probabilities of the data set variables. Formally, a Bayesian network consists of two parts $B =< G, W >$.The first part G is a direct cyclic graph consisting of nodes and arcs. The nodes are the variables $X_1...X_n$ in the data set and the arcs indicate direct dependencies between the variables. The graph G encodes the independence relationships in the domain under investigation. The second part of the network, W, represents the conditional probability distributions. It contains a parameter $\theta_{x_i}/w_{x_i} = p(w_j/x_i)$ for each possible value x_i of X_i , given each combination of the direct parent variables of X_i, w_j denotes the set of direct parents of X_i ,in G. The network W then represents the following joint probability distribution:

$$P_B(X_1,...,X_n) = \prod_{i=1}^{n} P_B(X_i/w_{x_i}) = \prod_{i=1}^{n}\theta_{X_i}/W_{x_i} \qquad (1)$$

The first task of learning a Bayesian network is to find the structure G of the network. Once the network structure G is known, the parameters W need to be estimated. In general, these two estimation tasks are performed separately. In this paper, the empirical frequencies from training data D is used to estimate these parameters:

$$\theta_{x_i/w_{x_i}} = \widehat{p}_D(x_i/w_{x_i}) \qquad (2)$$

The simplifying assumption behind a Bayesian network classifier assumes that the variables are independently given the class label. In this paper, the naive Bayesian network classifier is used for image classification task. The global structure of the classifier, includes two classes $\{W_1, W_2\}$ and some textures features $\{x_1, x_2, x_3, x_4,..., x_n\}$. W_1 and W_2 represent residential area class and non- residential area class respectively. These features are independent each other and the conditional probability is calculated by

$$P(x_1,\dots,x_n/w_j) = \prod_{i=1}^{n} P(x_i/w_j) \qquad (3)$$

According to the Bayesian theory, a posterior probability for a set of feature $\{x_1, x_2, \dots, x_n\}$ from a pixel in an image is obtained by:

$$P(W_k|x_1, x_2,\dots,x_n) = P(W)\prod_{i=1}^{n}\frac{P(x_i/W)}{P(x_i)} k = 1, 2 \qquad (4)$$

Based on the values of the posterior probabilities, a pixel can be classified as a residential area class or a non-residential area class.

3.2 Image Classification Using Naive Bayesian Network Classifier

The classification procedure of naive Bayesian network classifier consists of two steps. The first step is to learn all parameters of the Bayesian classifier from the training data, which include the positive and negative samples. Learning provides a list of prior probabilities and a list of conditional probabilities. The second step is to test unknown images. According to the conditional probabilities of samples, the posterior probabilities of tested images are calculated. Then, a threshold of posterior probability is given to segment interested objects and background. We adopt discrete variables in the Bayesian model where continuous features are converted into discrete attribute values using an unsupervised clustering stage based on the C-Means algorithm. The whole procedure of algorithm is listed as follows:

Algorithm2: The classification algorithm using naive Bayesian network classifier.

Given: The collected positive negative samples, and tested images.

Step1. Training samples.

• The texture features of all samples are calculated. In this paper, twenty texture features from co-occurrence matrix and four texture features from four Law's energy templates are collected.

• The continuous texture features are converted into discrete values label using the C-Means algorithm, where the label of clusters is set for each feature.

• The frequencies of different labels for each feature are counted and calculated as

$$CON = \frac{Frequn}{Nub} \qquad (5)$$

where CON denotes the conditional probabilities of each features, $Frequm$ and Nub denotes the frequency of different labels and theamount of samples respectively.

- According to formula 3, calculate a list of conditional probabilities $P(x_1, x_2 \cdots x_i/w_j)$.

Step2. Tested images classification

- For each pixel of tested images, a 50×50 neighbor window is selected to calculate its texture feature. Then, discretize these texture features values using the C-Means clustering algorithm.
- According to the label of each pixel in tested images, the conditional probability is computed.
- The posterior probability p of each pixel in tested images is obtained by formula 4.
- Based on calculated posterior probability of a pixel, a threshold of posterior probabilities is given to classify the pixel as a residential area pixel or a non-residential area pixel.

4   Experiment Results

We test proposed approach on panchromatic SPOT5 to extract residential areas. Firstly, we collect positive and negative training samples from different resolution and scale panchromatic remote sensing images. The naive Bayesian classifier allows subjective definitions to be described in terms of easily computable objective attributes, which are based on texture, shape, spectral values, etc. In this paper, twenty texture features from gray level Co-occurrence matrix and four texture energy features from law's templates are used to describe residential areas and non-residential areas, which represent rough texture objects and non-rough texture objects respectively. The total number of samples is 500,inwhich300 samples are used to train Bayesian network classifier and 200 samples are used as tested images for features selection. Each sample has the size of 50×50 pixels. These samples are divided into two classes. One belongs to residential areas (positive samples), the others belong to non-residential areas (negative samples).

For each sample, twenty four texture features are calculated and converted into discrete values using C-means clustering method. Based on original texture features, the first decision table or feature list is built, then the compatibility of the decision table is calculated using naive Bayesian network classifier when a condition attribute or a feature is deleted in first decision table. According to the Algorithm 2, the first step is to learn positive and negative samples and obtain all parameters of Bayesian network classifier. Secondly, tested samples are classified and the classification rate is calculated. If the correct classification rate is greater than given threshold , it is proved that the eliminated condition attribute or feature is unnecessary and should be deleted .Otherwise the condition attribute or the feature should be reserved. A reduced decision table is obtained once a condition attribute is deleted. Based on the reduced decision table, Algorithm1 is used to find the optimal feature subset until all features is scanned. Fig 1 and Fig 2 show two image segmentations examples of panchromatic SPOT5 using the optimal feature subset and original feature set. The number of clusters in C-means clustering method is empirically set to 6. The posterior probability threshold for the segmentation process is set to 0.5. Fig 1 shows one original

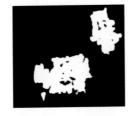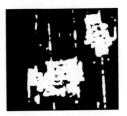

Fig. 1. Segmentation results using the optimal feature subset and the original feature set

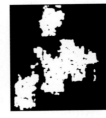

Fig. 2. Another segmentation results using the optimal feature subset and the original feature set

Table 2. The comparison results of proposed method and GA

Methods	Learning time(s)	Number of optimum features	Classification rate
GA	150	5	90.2%
Proposed method	15	5	91%

SPOT5 image including residential areas. The residential area extraction results using the optimal feature subset and original feature set are also shown in Fig 1. The other original SPOT5 image is shown in Fig 2. The residential area extraction results using the optimal feature subset and the original feature set are also shown in Fig 2. It can be seen that the correct classification rates using the optimal subset feature are more than 90%, while the correct classification rates based on original feature sets are less than 85%. This is because some disadvantage features in the original texture feature set influence classification results. In addition, the classification time using optimal subset feature and original set feature is 3 seconds and 5 seconds respectively. From our experiment results, it can be concluded that proposed approach not only improve the classification quality, but also reduce the computational cost of the classification.

In order to evaluate the performance of proposed method, genetic algorithm (GA), which is one of popular methods for the feature selection, is employed to select optimum features using same training samples. The comparison results with proposed method are listed in Table 2. From Table 2, it can been seen that both methods are able to find optimum features and reach good classification rate. However, the learning speed of proposed method is 10 times faster than that of GA.

5 Conclusions

The main study of this paper is to investigate how to optimize features using a Bayesian network classifier and Rough Sets. In this paper, a hybrid approach for feature selection is presented to reduce feature amount. In the approach, the criterion function of feature selection is defined by the correct classification rate from naive Bayesian network classifier. The naive Bayesian classifier is used for texture image classification by learning the positive and negative samples. Rough Sets is used to implement feature reduction with the correct classification rate from naive Bayesian network classifier. Experiment results on SPOT5 images show that the proposed method is feasibility and practical.

Acknowledgement

The work was supported by the National Science Foundation of China under grant No. 40571102 and the National Support Plan for Science and Technology of China under grant No. 2006BAB10B01.

References

1. Swiniarski, R.W.: Rough sets as a front end of neural networks texture classifiers. Neurocomputing 85–102 (2001)
2. Swiniarski, R.W., Skowron, A.: Rough set methods in feature selection and recognition. Pattern Recognition Letters 24, 833–849 (2003)
3. Yang, J., Honavar, V.: Feature subset selection using a genetic algorithm. Journal of Intelligent Information Systems 16, 215–228 (2001)
4. Pan, L., Nahavanid, S., Zheng, H.: The Rough set feature selection for tree recognition in color image using genetic algorithm. In: Proceedings of the 2nd International Conference on Computational Intelligence, Robotics and Autonomous Systems, pp. 15–18 (2003)
5. Vafaie, H., Jong, K.D.: Robust feature selection algorithms. In: Proceedings of International Conference on Tools with AI, Boston, Massachusetts, pp. 356–363 (2001)
6. Zheng, H., Zhang, J.X., Nahavandi, S.: Learning detect to texture objects by artificial immune approaches. Future Generation Computer Systems 20, 1197–1208 (2002)
7. Huang, Y., Luck, K., Zhang, C.: Texture classification by multi-model feature integration using Bayesian networks. Pattern Recognition Letters 24, 393–401 (2003)
8. Aksoy, S., Koperski, K.: Learning Bayesian classifiers for scene classification with a visual grammar. IEEE Trans. Geosci. Remote. Sens. 43, 581–589 (2005)
9. Paola, J.D., Schowengerdt, R.A.: The effect of neural network structure on a multi-spectral use/land-cover classification. Photogrammetry Engineering and Remote Sensing 63, 535–544 (1997)
10. Mitra, S.K., Lee, T.W., Goldbaum, M.: A Bayesian network based sequential inference for diagnosis of diseases from retinal images. Pattern Recognition Letters 26, 459–470 (2005)

An Application of Rough Set Concepts to Workflow Management

Georg Peters[1], Roger Tagg[2], and Richard Weber[3]

[1] University of Applied Sciences - München
Department of Computer Science and Mathematics
80335 Munich, Germany
georg.peters@hm.edu
[2] University of South Australia
School of Computer and Information Science
Mawson Lakes, SA 5095, Australia
roger.tagg@unisa.edu.au
[3] Universidad de Chile
Departamento de Ingeniería Industrial
República 701, Santiago, Chile
rweber@dii.uchile.cl

Abstract. Over the last decade, workflow management has become a significant tool in the effort of organizations to improve the efficiency of their processes. However, the scope for its adoption has been constrained by the variability, uncertainty and impreciseness that is inherent in business process execution. A number of attempts have been made by the workflow community to address this weakness, but no clear winner has emerged. In this paper, we consider the potential for applying Rough Set theory. One widely used modelling language for workflows is that offered by Petri Nets. In any process, the focal points of uncertainty are when decisions have to be taken. In the Petri Net model, this is represented by resolving, for each transition, whether it should fire or not. In particular at each OR-split (conditional branch) in a process instance, one has to decide the route to be taken by the tokens in the Petri Net. In our paper, therefore, we point out the potential of rough sets to resolve cases where contradictory information is available - or where information is missing. We introduce the concepts of rough places, rough tokens and rough transitions, and show how they can be utilized for workflow management.

Keywords: Uncertainty, Soft Computing, Rough Sets, Petri Nets, Workflow Management.

1 Introduction

On the one hand a central challenge in real life problems is how to deal with uncertainty and vague information. Therefore several concepts to describe uncertainty and vagueness have been suggested. The most established and oldest is probability theory which goes back to the 17th and 18th century when it was

G. Wang et al. (Eds.): RSKT 2008, LNAI 5009, pp. 715–722, 2008.

introduced by Bernoulli, Laplace, Pascal and others. In 1965 probability theory was joined by fuzzy set theory [1] and some 25 years ago Pawlak [2] suggested rough sets as tool to deal with uncertain information. And finally, for a few years Zadeh [3] has been promoting a general theory of uncertainty that provides a holistic framework to describe any kind of uncertainty.

On the other hand workflow management [4] has become well accepted approach to help to improve efficiency in companies. Unlike many semi-formal modelling languages, Petri Nets [5,6] are regarded as a precise and fully mathematically founded method to design and manage the primary *control flow* perspective of workflow systems. In Petri Nets the focal points of uncertainty are when decisions have to be taken. In general this is at each transition where one has to decide whether it should fire or not. In particular at each OR-split a decision has to be taken to specify the route in the Petri Net. So the decisions at OR-splits are of particular interest and importance in the context of our paper.

The objective of this paper, therefore, is to utilize rough set theory to effectively deal with uncertain and vague situations at OR-splits in Petri Nets. In particular we will suggest the concepts of rough places, rough tokens and rough transitions and show their potential use for the management of workflow systems.

The remainder of the paper is organized as follows. In the next section we give a short introduction to rough set theory. In Section 3 we use the concept of rough sets to make incomplete visible in Petri Nets. Section 4 discusses how these ideas might be applied to workflow systems. The paper ends with a short conclusion in Section 5.

2   Fundamentals of Rough Sets

Basic Properties of Rough Sets. Rough sets were introduced by Pawlak [2,7] in 1982. Since then they have gained increasing importance. Today they can be considered as important concept within the framework of soft computing. The fundamental idea of rough set theory is that there are two kinds of objects. While some objects are clearly distinguishable from each other some objects are indiscernible. The indiscernibility of the objects is normally caused by missing or incomplete information. To deal with such situations, Pawlak suggested the idea of describing a set by two approximations: a lower and an upper approximation of the set. While an object in a lower approximation of a set surely belongs to this set, an object in an upper approximation only *may* belong to the corresponding set.

Rough Decision Tables. In the context of our article the application of rough set theory to decision tables is of special importance. Consider the following example [8] dealing with a decision table of eight patients showing different symptoms (Table 1). Four of the patients are well (decision {Flu=no}) while the remaining four patients suffer from flu (decision {Flu=yes}).

The pair of patients #4 and #5 on the one hand and the pair of patients #6 and #8 on the other hand share the same symptoms {high, yes, yes} and

Table 1. Patient's Decision Table

#	Temperature	Headache	Nausea	Decision: Flu
1	high	yes	no	yes
2	very_high	yes	yes	yes
3	high	no	no	no
4	high	yes	yes	yes
5	high	yes	yes	no
6	normal	yes	no	no
7	normal	no	yes	no
8	normal	yes	no	yes

{normal, yes, no} respectively. However, the diagnosis differs, so the decision table does not lead to an unique result. Let us consider patients #4 and #5. While patient #4 suffers from flu patient #5 is well although the patients are indiscernible with respect to their symptoms. Therefore, in the terms of rough set theory, these patients belong to the upper approximations of both the sets {Flu=yes} and {Flu=no}. The same applies to the pair of patients #6 and #8. The diagnoses of the remaining patients do not cause the same problems as described above. Their symptoms lead to a clear diagnosis. While patients #1 and #2 are ill, #3 and #7 are well. So #1 and #2 belong to the lower approximation of the set {Flu=yes} while the patients #3 and #7 are members of the lower approximation of the set {Flu=no}. The implication for the diagnoses of new patients is straightforward; new patients with symptoms equal to #1, #2, #3 and #7 can be treated immediately while patients who have the symptoms {high, yes, yes} and {normal, yes, no} need to have some more detailed physical examination.

Rough Petri Nets. The potential of rough set theory to Petri Nets has already been investigated by J.F. Peters et al. who suggested rough Petri Nets [9,10,11,12]. Basically in rough Petri Nets transitions function as rough gates. J.F. Peters et al. applied this idea to, for example, sensor and filter models.

3   Making Incomplete Information Visible in Petri Nets

3.1   Some Notational Remarks

In Petri Net theory places are passive elements - containers storing tokens. In contrast to that transitions are active in the sense that only transitions can change the state of the net when they "fire": this means they consume tokens from their input places and produce tokens for their output places. Therefore transitions are the only elements in a Petri Net that have the capability to make decisions.

At an OR-split the Petri Net diverges and the transitions decide on the further path the case takes. The OR-split can be modelled in an explicit or implicit form [4]. In the implicit form the decision and the action are located in the same transition. In contrast to that the explicit OR-split separates the decision from the following action: the decision and the following action are modelled in two separate transitions.

In the following sections we - of course - do not question this concept. However, we graphically mark input places and/or tokens on input places to indicate whether one or more corresponding transitions can fire or not. The decision rules still remain in the only active elements of the Petri Net, the transitions.

3.2   Rough Places and Tokens

We propose the application of rough sets to OR-split in Petri Nets. Therefore let us consider the example given in the previous section again. The rules derived out of the decision table can be designed as a simple Petri Net consisting mainly of an OR-split. The patients are symbolized by tokens (see Figure 1 - for simplicity we only show the patients #1 and #2).

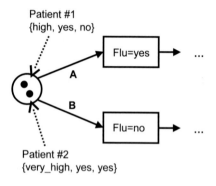

Fig. 1. Diagnosis Decision Tree as Part of a Petri Net

First Case: Rough Places. Obviously the decision rule at this place is insufficient to deal with all tokens. So some tokens get stuck on the input place of the OR-split. To indicate this we say that the place belongs to the upper approximations of both sets {Flu=yes} and {Flu=yes}. We indicate this by a "dashed circle" place notation as depicted in Figure 2.

Second Case: Rough Tokens. Now consider a patient phoning a General Practitioner (GP). The patients reports that she/he suffers from headache and nausea, However, she/he has not been able to check her/his temperature before phoning the GP. Formally the information provided can be described as: {?, yes, yes}. Since information is missing the GP cannot continue his treatment. In such a case we assign the token to the upper approximation. To graphically distinguish between tokens (patients) belonging to a lower or upper approximations we suggest their representation as shown in Figure 2, namely a "hollow" token for those in the upper approximation.

Relationship between Rough Places and Rough Tokens. The main difference between the rough places and rough tokens is related to who is responsible when a token gets stuck at a OR-split. In the first case discussed above the token carries all the required information. However, the firing rules at the OR-split are insufficient to take a decision. Therefore the responsibility is at the OR-split. Please note, the definition of a rough place depends on a token. Therefore rough places are no structural properties of Petri Nets.

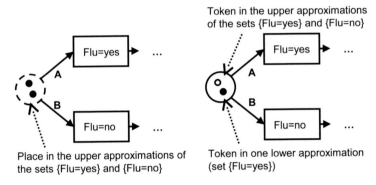

Fig. 2. Rough Places and Tokens

In the second case the token cannot provide the requested information. Therefore the token is accountable for its inability to proceed further, so it can be regarded as token in an upper approximation.

3.3 Rough Transitions

A token can only proceed when both the token as well as the place the token is assigned to belong to lower approximations. In this case the decision rule at the OR-split has sufficient information and a transition is enabled to fire.

However, when a token belongs to an upper approximation and/or the place belongs to an upper approximation then the token gets stuck. It is not defined which of the transitions may fire. This leads to the concept of rough transitions.

Let us define the following decision sets {fire=yes} and {fire=no}. Transitions which will surely fire belong to the lower approximation of the set {fire=yes} while transitions that surely won't fire belong to the lower approximation of the set {fire=no}. The remaining transitions belong to the upper approximations of both sets {fire=yes} and {fire=no}.

As an example, consider the Petri Net given in Figure 3[1]. Black solid-lined transitions will surely fire. Therefore they belong to the lower approximation

---

[1] Since the effects on the capability of making a decision are the same for rough tokens and places (in both cases a token cannot proceed) we will, for simplicity, only display rough places in the example.

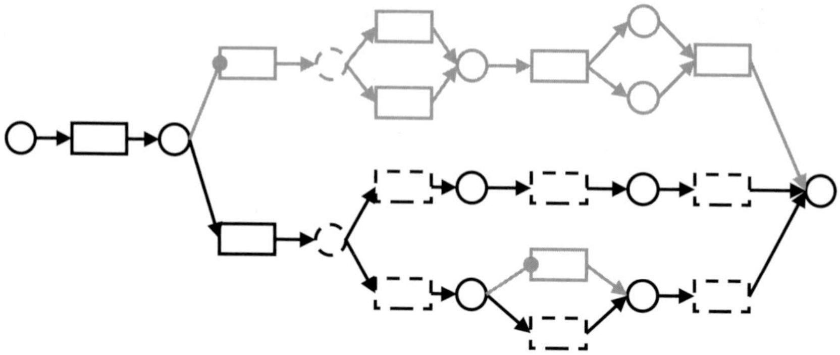

Fig. 3. Rough Transitions, an Example Process

of the set {fire=yes}. The grey transition surely won't fire[2], consequently they belong to the lower approximation of the set {fire=no}. The status of remaining dashed transition is unclear: they may or may not fire. So they belong to both upper approximations of the sets {fire=yes} and {fire=no}.

Please note, that rough transitions, like rough places, depend on the cases (tokens). Therefore they are no structural properties of Petri Nets.

4   Applications to Workflow Systems and Related Work

Potential Applications to Workflow Systems. There are several possible areas of application for the proposed method to workflow systems:

(1) Rough places: incomplete decision rule. The appearance of incomplete decision rules (*rough places*) may have two reasons. First, incomplete decision rule, indicated by the appearance of places in upper approximations of both sets {fire=yes} and {fire=no}, can be interpreted as a poorly designed workflow system. The system has to be improved to run properly without any further interruption. Second, a decision rule can intentionally be designed incomplete. Then, for example, the normal cases would pass the decision gate (OR-split) undisturbed. The exceptions would intentionally be "caught" in the upper approximation of a place and presented to the end user for further special treatment.

(2) Rough tokens: incomplete case information. A possible area of application of the proposed method is to provide early warning of potential delays within a workflow system that could be caused by incomplete information in certain business cases. The aim would be to get the workflow system to alert the end user when a choice is waiting on more information. If only the immediate decision is considered, the next transition will be held up. If the complete process including all potential downstream activities is considered, the alert is a warning that further down the track, a transition may be held up. Ideally, the workflow system

---

[2] The selected path is indicated by a normal arrowhead while the path that is not selected is indicated by a dot in Figure 3.

should monitor the arrival of the required extra data, so that transitions can be automatically enabled without user intervention. This may well involve facilities to set up software agents that an talk to the applications that manage this data. If, however, it can be seen in advance that certain combinations of case attributes mean that a choice cannot be resolved, the workflow template should probably be altered to allow for a "don't know" branch. The process owner would need to define how long cases can be left in this state, and what should happen to them when time runs out.

(3) Rough transitions: incomplete path information (resource management). Resource management is a crucial task in any company. The concept of *rough transitions* supports to more efficiently manage resources in the following way. As depicted in Figure 3 there are three categories of transition[3], (1) transition that will be performed surely, (2) transitions that will not be performed, (3) transition that may be performed. While in the first case resources have to be allocated to the transitions, in the second case any allocated resources can be released. Uncertainty is reduced to the third case in which it is unclear whether resources are needed to perform the transitions or not.

Related Work. In the workflow management literature, there has been some movement towards addressing the problems of handling the variability of individual process instances or 'cases', but rough set concepts have not generally been considered to date. Examples are *case handling* proposed by van der Aalst et al. [14] and *mixed-initiative management* (i.e. bringing in temporal and resource perspectives), as proposed by Rubinstein et al. [15]. Although the objective (efficiently handling process cases) is much the same as for introducing rough set theory, the emphasis in these other approaches is more on addressing the "non control flow" perspectives such as data and resource availability for performing tasks and for completing cases in a timely manner.

5 Conclusion

In this paper we have presented the potential use of rough sets for the management of missing or incomplete information in Petri Nets. We have suggested the concepts of rough places, rough tokens and rough transitions.

The main purpose is to utilize rough set theory to make incomplete information visible in order to deal with such a situation efficiently. In the first case, i.e. too much information, the rough concept of reducts can be applied. In the later case we identified three different occurrences: (1) incomplete decision rules, (2) incomplete case information and (3) incomplete path information.

The central advantage of the application of rough sets is that we can draw from a rich theoretical concept to efficiently manage such situations. Our future

---

[3] In Petri Nets transitions are only regarded as active entities in the sense that they can change the state of the net. So, generally, business activities can be mapped to transitions and places as well. However, in our context we follow the conventions of leading business process notations, like the EPC [13], where business activities can only be assigned to active entities ("functions").

research will concentrate on a more formal incorporation of these concepts into Petri Net theory and workflow management.

References

1. Zadeh, L.: Fuzzy sets. Information and Control 8, 338–353 (1965)
2. Pawlak, Z.: Rough sets. International Journal of Computer and Information Science 11, 341–356 (1982)
3. Zadeh, L.: Toward a generalized theory of uncertainty (GTU) an outline. Information Sciences 172, 1–40 (2005)
4. van der Aalst, W., Hee, K.: Workflow Management - Models, Methods, and Systems. MIT Press, Cambridge (2002)
5. Petri, C.: Kommunikation mit Automaten. Schriften IIM 2, University of Bonn, Institut für Instrumentelle Mathematik, Bonn (1962)
6. Murata, T.: Petri nets: Properties, analysis and applications. Proceedings of the IEEE 77, 541–580 (1989)
7. Pawlak, Z.: Rough Sets: Theoretical Aspects of Reasoning about Data. Kluwer Academic Publishers, Dordrecht (1992)
8. Grzymala-Busse, J.: Introduction to rough set theory and applications. In: Negoita, M.G., Howlett, R.J., Jain, L.C. (eds.) KES 2004. LNCS (LNAI), vol. 3213, Springer, Heidelberg (2004)
9. Peters, J., Skowron, A., Suraj, Z., Ramanna, S., Paryzek, A.: Modeling real-time decision-making systems with rough fuzzy petri nets. In: Proceed. EUFIT98 - 6th European Congress on Intelligent Techniques & Soft Computing, Aachen, Germany, pp. 985–989 (1998)
10. Peters, J., Skowron, A., Suraj, Z., Ramanna, S.: Guarded transitions in rough petri nets. In: Proceedings EUFIT 99 - 7th European Congress on Intelligent Systems & Soft Computing, abstract, p. 171 paper on CD. Aachen, Germany (1999)
11. Peters, J., Skowron, A., Suray, Z., Ramanna, S.: Sensor and filter models with rough petri nets. In: Burkhard, H., Czaja, L., Skowron, A., Starke, P. (eds.) Proceedings of the Workshop on Concurrency, Specification and Programming, Humboldt-University, Berlin, Germany, pp. 203–211 (2000)
12. Peters, J., Ramanna, S., Suraj, Z., Borkowski, M.: Rough Neurons: Petri Net Models and Applications. In: Pal, S., Polkowski, L., Skowron, A. (eds.) Rough-Neuro Computing, pp. 472–491. Springer, Berlin (2003)
13. Scheer, A.: ARIS - Business Process Modeling. Springer, Berlin (2000)
14. van der Aalst, W., Weske, M., Grünbauer, D.: Case handling: A new paradigm for business process support. Data and Knowledge Engineering 53, 129–162 (2005)
15. Rubinstein, Z., Corkill, D.: Mixed-initiative management of dynamic business processes. In: Proceedings 2003 IEEE International Workshop on Soft Computing in Industrial Applications, Binghamton, New York, USA, pp. 39–44 (2003)

Optimization on OLSR Protocol for Lower Routing Overhead

Yong Xue, Hong Jiang, and Hui Hu

Southwest University of Science and Technology,
Mianyang 621010, Sichuan, P.R. China
belat@163.com

Abstract. The optimized link state routing (OLSR) designed by the
IETF's mobile ad hoc networks (MANET) working group is one of the
four base routing protocols for ad hoc networks. This protocol is a table
driven, proactive protocol. It is particularly suitable for large and dense
mobile networks with less latency. However, high routing overhead is
a dominant disadvantage as comparing with reactive protocols. In this
paper, an optimizing scheme on OLSR by reducing the average size of
HELLO messages, as well as the size and the amount number of TC
messages is proposed. After analyzing and computing the overhead of
the optimized OLSR protocol, which is implemented and simulated on
NS-2 in different scenarios, the simulation results indicate that its routing
overhead is reduced; meanwhile, the latency and the average end-to-end
delay are still maintained in a low level without any change.

1 Introduction

Mobile ad hoc networks are infrastructure-less networks where mobile nodes
communicate wirelessly and the network topology changes constantly. The nodes
organize themselves to route packets in a multi-hop fashion from a source to a
destination. Reactive and proactive routing protocols have been proposed in the
literature. On the one hand, proactive protocols find and maintain paths to every
destination before they are actually required, which creates additional overhead.
On the other hand, reactive protocols find paths only when they are actually
required, without creating additional overhead [1]. Moreover, there are other
protocols, such as, hybrid protocols, combining the benefits of both protocols
by working proactive in the local neighborhood of a node and reactive for nodes
further away [2].

OLSR is a proactive protocol and has higher overhead than reactive protocols
and most other proactive protocols, although it has advantages in large and dense
network with less latency and average end-to-end delay, which characteristic
is shown by figures in section 2. Studying the routing protocols overhead is
necessary, especially in large and dense ad hoc networks, while many theoretical
analysis on overhead and performance have been done for OLSR.

In this paper, we present an approach specifically designed to minimize routing overhead of the OLSR, it differs from existing literature on the study performance of the OLSR in which some performance of latency and delay were

G. Wang et al. (Eds.): RSKT 2008, LNAI 5009, pp. 723–730, 2008.

partial sacrificed for minimizing routing overhead. Fish Eye approach is adapted to large networks in which the amount number of TC (Topology Control) message is reduced by defining three zones encircled one node with deferent radius and broadcasting frequency, but the number of invalid routes increases such that more time must be spent on updating topology message list and found new routes [3]. Hierarchical OLSR need select some node to build up backbone subnet supporting point-to-point wireless [4].

In this paper, some novel approaches are presented for enhancing the performance of OLSR which mainly comprises three ways to optimize OLSR protocol. At first, the size of HELLO message is shorted by comparing the link-states of neighbor set with the current ones so that only changed links and MPRs (Multi Point Relays) are transmitted. Second, the first TC message is no longer generated by node N, but is calculated and generated by its MPRs, so the amount number of TC message is decreased. Last, in the original protocol a node broadcasts link-state information between itself and its MPR selectors, in which TC message has redundant information when two nodes are selected as MPR by each other. The one link-state information is broadcasted by TC message twice times, so we select one of those nodes to advertise it.

For the experiments, the latest release of NS-2 (NS-2.29) is used. NS-2 is a discrete event simulator widely used in the networking research community. In general, the NS-2 installation will include all software extensions. It contains a detailed model of the physical and link layer behavior of a wireless network based on the 802.11 specifications and allows arbitrary movement of nodes within a network area. The new scheme presented in this paper is implemented and simulated on NS-2, several performance metrics were measured by varying the maximum speed of mobile hosts, routing overhead, latency and average end-to-end delay, etc., those performance metrics are analyzed and compared between original OLSR protocol and optimized one.

This paper is organized as follows. In Section 2, we give a brief description of OLSR's main operations. In Section 3, we illustrate the new optimization scheme for reducing the overhead of OLSR protocol, and how to compute the overhead of OLSR. The performance of optimized protocol is validated in Section 4 by confrontation with simulation results. Finally, we conclude in Section 5.

2   The OLSR Protocol

The protocol is an optimization of the classical link state algorithm tailored to the requirements of a mobile wireless LAN. In the protocol, MPRs are selected nodes which forward broadcast messages during the flooding process. This technique substantially reduces the message overhead as compared to a classical flooding mechanism, where every node retransmits each message when it receives the first copy of the message. In OLSR, link-state information is generated only by nodes elected as MPRs. Thus, a second optimization is achieved by minimizing the number of control messages flooded in the network. As a third optimization, an MPR node may chose to report only links between itself and its

MPR selectors. Hence, as contrary to the classic link state algorithm, partial link state information is distributed in the network. This information is then used for route calculation [5]. The two main OLSR functionalities, Neighbor Discovery and Topology Dissemination, are now detailed as follows.

2.1 Neighbor Discovery

A node must perform link sensing on each interface, in order to detect links between the interface and neighbor interfaces. Furthermore, a node must advertise its entire symmetric 1-top neighborhood on each interface in order to perform neighbor detection. Hence, for a give interface, a HELLO message will contain a list of links on that interface (with associated link types), as well as a list of the entire neighborhood. In principle, a HELLO message serves three independent tasks: Link sensing, Neighbor detection, MPR selection signaling.

Three tasks are all based on periodic information exchange within neighborhood nodes, and serve the common purpose of "local topology discovery". A HELLO message is therefore generated based on the information stored in the Local link Set, the Neighbor Set and the MPR Set form the local link information base.

The major improvement on OLSR in our work is focused on optimizing the format of HELLO message and the tactics of operating mode, so the more knowledge should be introduced about HELLO message format in detail. The proposed format of a HELLO message is shown in RFC 3626 (omitting packet, IP and UDP headers).

The data-portion of the general packet format with the "Message Type" set to HELLO_MESSAGE. Reserved field must be set to "0000000000000" to be in compliance with this specification. HTime field specifies the HELLO emission interval used by the node. Willingness field specifies the willingness of a node to carry and forward traffic for other nodes. Link Code field specifies information about the link between the interface of the sender and the following list of neighbor, the analysis and the improvement will be show more detailed in section 3. Link Message Size counted in bytes and measured from the beginning of the "Link Code" field and until the next "Link Code" field. Neighbor Interface Address is the address of an interface of a neighbor node.

2.2 Topology Dissemination

Each node of the network maintains topological information about the network obtained by means of TC messages. The nodes which were selected as a MPR by some of the neighbour nodes broadcast the TC message at every "TC interval". The TC message originated from one node to declare the set of nodes which having been selected as MPR. The TC messages are flooded to all network nodes and take advantage of MPRs to reduce the number of retransmissions. To optimize flooding, the OLSR forwarding rule is used: Any node forwards a broadcast message only if it is received for the first time from anode having been selected as MPR.

Thus, a node is reachable either directly or via its MPRs. The neighbor information and the topology information are refreshed periodically, and they enable
each node to compute the routes to all known destinations. These routes are
computed with Dijkstra's shortest path algorithm. Hence, they are optimal as
concerns the number of hops.

3   Optimization on OLSR Protocol

The optimization schemes presented in this paper mainly comprise three aspects
by reducing the size of HELLO message, the amount number and the average
size of TC message.

3.1   A Novel Operating Tactics about HELLO Messages

In HELLO interval a list of the entire neighborhood is transmitted periodically,
because the HELLO message contains a lot of link-states and generated in high
frequency, which results very high overhead, especially, in high density of nodes
scenarios. For reducing the size of HELLO messages we present a novel "neighbor
tuples" by which it is not necessary for HELLO messages to transmit the entire
neighborhood in every HELLO interval.

In OLSR a node records a set of "neighbor tuples" (N_neighbor_main_addr,
N_status, N_willingness), describing neighbors. N_neighbor_main_addr is the
main address of a neighbor, N_status specifies if the node is NOT_SYM or SYM.
N_willingness in an integer between 0 and 7, and specifies the node's willingness to carry traffic on behalf of other nodes. The new "neighbor tuples" adds
a field named N_modified which is a signal for indicating whether the link-state
is modified between the two periods.

Based on the mended neighbor tuples the generating and processing of HELLO
message can be designed as below operations:

HELLO Message Generation. A node N broadcasts its link-states once per
"HELLO interval", but not the entire neighborhood, just only include the links
and neighbor nodes information which have modified in a "HELLO interval" by
checking the field of N_modified in the neighbor tuples.

HELLO Message Processing. When a node received a HELLO message from
its neighbor node, at first, search the field of Original Address of the packet to
find whether the node of sender has been added in the neighbor set. If the
information of sender do not exist in the table, then attaches a item and fills
the every field with neighbor information, which means the node is a new joined
neighbor node, and transmits the entire neighborhood messages to the node of
sender. Otherwise, for an older neighbor it is not necessary to transmit the entire
Neighbor set, but just the information of node which link-state or attribute of
MPR has changed.

When any link-state between the node N and its neighbors is changed in a
slot of broadcasting the HELLO messages, the corresponding N_modified field of

"neighbor tuples" would be modified to one, which means the link-state should be updated and advertised to its neighbors. As soon as the changed link-states are broadcasted, the corresponding N_modified field would be reseted to zero.

3.2   Reduction of the Amount Number of TC Messages

An advertised link set is put in a TC message and transmitted to all neighbors of a node, only the nodes which were selector as a MPR by the sender node are responsible for forwarding control traffic. In actually, the entire information of link-states of one node and its MPR Selector has been gotten by MPRs through HELLO message excepting which link-states are between neighbor node and its MPR Selector, if we can indicate it by the Link Code field of HELLO message, then the first TC messages which broadcasts by a node to its all neighbors can be abolished , in result, the MPRs replace their MPR Selectors to disseminate the TC messages by mining the related information form HELLO messages, because the refreshing frequency of HELLO messages is higher than TC messages at least twice times, which can ensure that the link-state information is up to date.

For achieving the goal, we add a new Neighbor Types to Link Code field of HELLO message, which named SEL_NEIGH for indicating whether the sender is a MPR selected by its neighbor node. Then every MPR can build up the TC messages including the link-states between its MPR selectors and themselves MPR selectors, of course, the Original Address of TC messages must be the MPR selectors'.

3.3   Reduced on the Size of TC Messages

Through the "link types" a node can get whether it and its neighbor nodes are mutual MPR, if two node select its neighbor node as MPR each other, when the TC messages is building just one node which is chose to disseminate the link-state between themselves. The choice of node can simply be decided by comparing their address, although the effect on the whole overhead of routing protocol is small, but it has not any side effect also.

3.4   The Overhead Computation of Optimized OLSR Protocol

At first, we compute the overhead of original OLSR protocol, define some input parameters which characterize the ad hoc network configuration and can be seen as specifications.

The OLSR protocol configuration and scenario parameters include:

1. $hdr_{Hello}, hdr_{TC}, hdr_{msg}, hdr_{pck}$: size of OLSR message header and packet header.
2. f_H, f_{TC}: frequency of sending HELLO and TC.
3. N: total number of nodes.
4. M: total number of MPR nodes.
5. S: average number of retransmissions per TC message, including the first transmission by the originator.

6. n, m: average number of neighbor per node and MPR selectors per MPR, respectively.

The overhead of OLSR is defined as the average bandwidth, in bytes/s, the bandwidth can be decomposed to, on one hand, the bandwidth used for sending and receiving HELLO messages: O_{sendH}, O_{recvH}; on the other hand, the bandwidth used for sending and receiving TC messages: O_{sendTC}, O_{recvTC}. Let sz_H, sz_{TC} and respectively be the average HELLO and TC packet sizes. As each OLSR packet only has one OLSR message:

$$O_{sendH} = f_H \cdot sz_H$$
$$O_{sendTC} = O_{recvTC} = f_{TC} \cdot S \cdot sz_{TC}$$
$$O_{recvH} = f_H \cdot n \cdot sz_H$$

Thus, the overhead of the OLSR protocol is equal to:

$$O_{OLSR} = f_H (N + 1) sz_H + 2 \cdot f_{TC} \cdot S \cdot sz_{TC}$$

Let $sz_{addr}, sz_{linkcode}$, respectively be the address size and link code size. The link code is either symmetric, asymmetric, multipoint relay or lost. When the simulated network comes to a stable state, it should only have two kinds of link type (in the optimized protocol it adds to three kinds): symmetric and MPR links [3]. Thus the average number of link codes advertised in each HELLO messages is 2.

In OLSR protocol minimizes the overhead of TC message by using Message Grouping, through this means many TC messages are grouped into one single OLSR packet. Therefore, only one packet header is needed for many TC messages instead of one packet header for each TC message. The packet header and IP header are less but number of TC messages exchanged is unchanged, thus the overhead of due to HELLO and TC messages is:

$$O_{TC} = 2 \cdot f_{TC} \cdot S \cdot (m \cdot sz_{addr} + hdr_{TC} + hdr_{msg}) + 2 \cdot f'_{TC} \cdot S \cdot (hdr_{pck} + hdr_{IP})$$
$$O_H = f_H (n + 1) (n \cdot sz_{addr} + 2 \cdot sz_{linkcode} + hdr_H + hdr_{msg} + hdr_{pck} + hdr_{IP})$$

f'_{TC} is a coefficient related to the number of TC message received by one node in period of $1/f_{TC}$, which is smaller than f_{TC}.

In the optimized OLSR protocol the first TC message is constructed and broadcasted by the MPR instead of MPR selector, so the S in the overhead of O_{TC} decreases to $S - 1$, and the items of $n \cdot sz_{addr}$ of $2 \cdot sz_{linkcode}$ will be removed in a relatively stabile network scenario, but when the node move constantly it maybe reach to $3 \cdot sz_{linkcode}$.

Now we can give the reduced overhead of the optimized protocol in a relatively stabile network scenario by putting the values of size of different packet headers into above equations:

O_{TC} reduce $2 \cdot f_{TC} \cdot (4m + 16)$ bytes for one node in every period of $1/f_{TC}$.
O_H reduce $f_H \cdot (n + 1) (4m + 16)$ bytes for one node in every period of $1/f_{TC}$.

Mostly, f_H would be chose by 2 seconds, and f_{TC} by 5, thus, the gain brought by O_H is always several times than O_{TC}'s by a simply computation.

4   Simulation Results

Every scenario, static and mobile, was generated using the node-movement generator setdest, provided with NS-2. In order to allow for fair performance comparisons, the same scenarios were always utilized when evaluating each strategy. Each experimental stage is described next. We compute the overhead generated by OLSR as the number of bytes per second sent or received by an OLSR node at the IP level.

Simulation parameters of ship nodes are adopted as follows:

Network Type: 802.11, Transmission range: 250 m, Field Size: 1500 × 1500, Node Type: Static, HELLO message rate: Every 2 seconds, TC message rate: Every 5 seconds, Throughput: 2Mbps.

The density of nodes should have a significant influence on the routing protocols performance. In general, low density may cause the network to be frequently disconnected and high density increases the contention, resulting in a low per-node throughput. In simulations, the number of nodes per simulation area is increased from 100 to 180 nodes with the rest of the simulation parameters remain unchanged. The goal is to study how the optimized protocol improves the overhead comparing to the original protocol in different node densities.

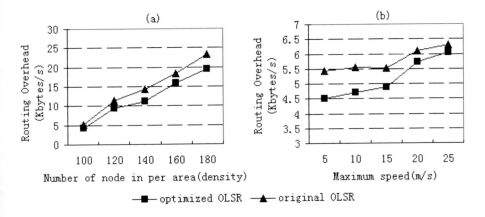

Fig. 1. Results of comparing routing overhead of two protocols

Fig.1.(a) shows the result of simulation, as a whole, the average routing overhead of the optimized protocol decrease to 83% approximately according to the original protocol, which verified the validity of the optimization approach for OLSR protocol. From the another Fig.1.(b), it can be seen that result is not very good comparing with the last figure's in mobile scenario, especially, when the node moving with speed of 25 m/s, the overhead of new protocol just decreases 4% approximately, obviously, the cause is the average size of HELLO messages increasing as a result of the neighborhood changing more frequently. In the whole process of simulation the average end-to-end delay and latency have not almost any different between two protocols.

5 Conclusion

In this paper, We have presented an optimizing approach to reduce the overhead through rebuilding the format and operating tactic of HELLO and TC messages. In the most case, it can decrease the routing overhead about 17% , and some extreme case, maybe less than that greatly. But we just studied an especial application of OLSR in which have not considered any redundancy, in fact OLSR protocol designed many schemes which have various information redundancy, such as number of MPRs and more link-states information encapsulated into TC message [6], in that case, the overhead problem is become more prominent. Through the reference [3] we can know in a large network with high density, the proportion of O_H in whole overhead is become preponderant, because in our optimization the gain brought by O_H is larger than O_{TC} greatly, the optimized OLSR protocol must more suitable to large network.

Although so many works have been done for reducing the overhead of OLSR protocol, Comparing to the other routing protocol, which still cost much width for routing; maybe using more advanced technology to construct backbone net in ad hoc network is a feasible approach.

This paper is supported by Defense Basic Research project under grant A3120060264 and SWSUT project under grant 06zx3106

References

1. Lye, P.C., McEachen, J.C.: A Comparison of Optimized Link State Routing with Traditional Routing Protocols in Marine Wireless Ad-hoc and Sensor Networks. In: 40th Annual Hawaii International Conference on System Sciences, pp. 293b–293b. IEEE Press, New York (2007)
2. Costa-Requena, J., Vadar, T., Kantola, R., Beijar, N.: AODV-OLSR scalable ad hoc routing proposal. In: Wireless Pervasive Computing, 2006 1st International Symposium, pp. 5–15. IEEE Press, New York (2006)
3. Nguyen, D., Minet, P.: Scalability of the OLSR Protocol with the Fish Eye Extension.Networking. In: 6th International Conference on Networking, pp. 88–95. IEEE Press, New York (2006)
4. Ge, Y., Lamont, L., Villasenor, L.: Hierarchical OLSR - A Scalable Proactive Routing Protocol for Heterogeneous Ad Hoc Networks. In: Wireless And Mobile Computing, Networking And Communications, vol. 3, pp. 17–23. IEEE Computer Society Press, Los Alamitos (2005)
5. Optimized Link State Routing Protocol, http://www.ietf.org/rfc/rfc3626.txt
6. Villanueva-Peña, P.E., Kunz, T., Dhakal, P.: Extending Network Knowledge: Making OLSR a Quality of Service Conducive Protocol. In: 2006 international conference on Wireless communications and mobile computing, pp. 103–108. ACM Press, New York (2006)

Password-Based Tripartite Key Exchange Protocol with Forward Secrecy

Guomin Li, Dake He, and Wei Guo

Laboratory of Information Security and National Computing Grid.
Southwest Jiaotong University, Chengdu, 610031, PR of China
li_gm95@{yahoo,163}.com, dkhe_scce@home.swjtu.edu.cn, gw1082@sina.com

Abstract. A tripartite authenticated key agreement protocol is designed for three entities to communicate securely over an open network particularly with a shared key. Password-authenticated key exchange (PAKE) allows the participants to share a session key using a human memorable password only. In this paper, A password-based authenticated tripartite key exchange protocol(3-PAKE) is presented in the standard model. The security of the protocol is reduced to the Decisional Bilinear Diffie-Hellman (DBDH) problem, and the protocol provides not only the properties of forward secrecy, but also resistance against known key attacks. The proposed protocol is more efficient than the similar protocols in terms of both communication and computation.

Keywords: Password-based tripartite authenticated key exchange, Forward secrecy, Known key attacks.

1 Introduction

A key agreement protocol is defined as a mechanism in which a shared secret key, often known as session key, is derived by two or more protocol entities as a function of information contributed by each of these parties such that no single entity can predetermine the resulting value. This secret key, usually established over a public network, can then be used to create a confidential or integrity-protected communication channel among the entities. In general, a key agreement protocol is called authenticated if the protocol is able to ensure that the session key is known only to the intended entities in a protocol run. Without authentication, a key agreement protocol would turn out to be insecure as an adversary can easily intrude the scheme by using the man-in-the-middle attack as well as other cryptographic attacks. The situation where three or more parties share a secret key is known as conference keying. The three-party (or tripartite) case is of most practical importance because it is the most common size for electronic conferences et al.

Joux[1] has initiated the development of one-round pairing-based tripartite Diffie-Hellman key agreement protocol in 2000. However, Shim[2] has pointed out that Joux's protocol does not provide authentication and therefore, it cannot resist the man-in-the-middle attack. Some researchers have further investigated the

G. Wang et al. (Eds.): RSKT 2008, LNAI 5009, pp. 731–738, 2008.
© Springer-Verlag Berlin Heidelberg 2008

scheme and proposed four tripartite authenticated key agreement protocols[6], which provided authentication using ideas from MTI[3] and MQV[5]. They used certificates of the parties to bind a party's identity with his static keys. The authenticity of the static keys provided by the signature of CA assures that only the parties who possess the static keys are able to obtain the session key. However, since the participants involved in the protocol should verify the certificate of the parties, a huge amount of computing time and storage is needed. In [4], Nalla et al. proposed authenticated tripartite ID-based key agreement protocols. The security of the protocol is discussed under the possible attacks. However, their protocol is not secure as they have claimed. Chen[9] showed the flaw of the protocol. Zhang et al.[14] designed an ID-based one round authenticated tripartite key agreement protocol and provided heuristic security analysis.

In practice, one finds several favors of key exchange protocols, each with its own benefits and drawbacks. Among the most popular ones is the 3-party Kerberos authentication system[10]. Another one is the 2-party SIGMA protocol[11] used as the basis for the signature-based modes of the Internet Key Exchange (IKE) protocol. Yet another favor of key exchange protocols which has received significant attention recently are those based on passwords. Compared to the protocols ID-based or PKI/CA-based, to the password-based protocols, a human is only required to remember a low entropy password shared between the participants. In fact, password-based schemes are suitable for implementation in many scenarios, especially those where no device is capable of securely storing high-entropy long term secret key, specially, password-based key agreement protocol has been extensively studied in the last few years [7,8,12,13]. Recently, Wen et al.[16] proposed a provably secure three-party password-based authenticated key exchange (three-PAKE) protocol using Weil pairings, unfortunately, Chien[18] point out the protocol can not resist to impersonation attack. In this paper, we present a password-based tripartite key agreement protocol using pairings, it seem that in the standard model. It allows three parties to negotiate a common session key via a shared password over an adversary controlled channel.

2   Preliminaries

In this section, we review some cryptographic assumptions that will be used throughout the paper.

2.1   Pseudorandom Function

Let $F : Keys(F) \times D \to R$ be a family of functions, and $f : D \to R$ a random function. A is an algorithm that takes an oracle access to a function and returns a bit. We consider two experiments:

$$Exp_{F,A}^{prf-1} : \{K \xleftarrow{R} Keys(F), d \leftarrow A^{F_K(\cdot)}, return(d)\}$$
$$Exp_{F,A}^{prf-0} : \{g \xleftarrow{R} Rand^{D \to R}, d \leftarrow A^{f(\cdot)}, return(d)\}$$

The advantage of an adversary A is defined as follows:

$$Adv_{F,A}^{prf}(\kappa, t, q) = |Pr[Exp_{F,A}^{prf-1} = 1] - Pr[Exp_{F,A}^{prf-0} = 1]| \tag{1}$$

where A is any adversary with time complexity t making at most q oracle queries. The scheme F is a secure pseudo random function family if the advantage of any adversary A with time complexity polynomial in κ is negligible.

2.2   The Bilinear Maps and Assumption

Let G_1 and G_2 be finite cyclic groups of prime order q. We assume that the discrete logarithm problems (DLP) in both G_1 and G_2 are hard to solve, and an efficient bilinear map $\hat{e} : G_1 \times G_1 \to G_2$.

Decisional Bilinear Diffie-Hellman (DBDH) problem. The decisional BDH problem is to distinguish between tuples of the form $(g_1, g_1^a, g_1^b, g_1^c, \hat{e}(g_1, g_1)^{abc})$ and $(g_1, g_1^a, g_1^b, g_1^c, \hat{e}(g_1, g_1)^d)$ for random $g_1 \in G_1$, and $a, b, c, d \in Z_q^*$. An algorithm A is said to solve the DBDH problem with an advantage of ϵ if

$$| Pr[A(g_1, g_1^a, g_1^b, g_1^c, \hat{e}(g_1, g_1)^{abc}) = 0] - Pr[A(g_1, g_1^a, g_1^b, g_1^c, \hat{e}(g_1, g_1)^d) = 0] | \geq \epsilon \tag{2}$$

Definition 1. We say that the DBDH assumption holds in G_1 if no polynomial time algorithm has advantage at least ϵ in solving the DBDH problem in G_1.

2.3   Security Notions

The model described in this section is based on model in [15]. Assume that the network is a broadcast network and a malicious adversary may intercept the broadcast messages and substitute his own messages for some of them. Assume that the users in set $U = \{U_1, U_2, U_3\}$ that shares a password pw uniformly distributed in a password space of size PW will negotiate a session key using the key exchange protocol. An instance of U_i is represented by an oracle Π_i^s, for any $s \in N$. Let sid_i^s be the concatenation of all (broadcast) messages that oracle Π_i^s has sent and received. Let a partner identifier pid_s^i for instance Π_i^s be a set of the identities of the users with whom Π_i^s intends to establish a session key, pid_s^i includes U_i itself. The oracles Π_i^s and Π_j^t are partnered if:

- $\Pi_i^s = \Pi_j^t$ and $sid_i^s = sid_j^t$.
  An attacker can make following queries.
- Execute(U): This query models passive attacks, where the adversary gets the instances of honest executions of a protocol by U.
- Send(Π_i^s, m): This query is used to send a message m to Π_i^s and get the response from Π_i^s. The number of on-line dictionary attacks can be bounded by the number of Send queries.
- Reveal(Π_i^s): This query models the adversarys ability to obtain session keys (known-key attacks). If a session key $sk_{\Pi_i^s}$ has previously been constructed by Π_i^s , it is returned to the adversary.
- Corrupt(U_i): This query models the adversarys ability to obtain long-term keys of parties (forward secrecy). We restrict that on Corrupt(U_i) the adversary only can get the password pw, but cannot obtain any internal data of U_i.

- Test(Π_i^s): This query is used to define the advantage of the adversary. This query is allowed only once by the adversary A, and only to fresh oracles, which is defined later. On this query a simulator flips a coin b. If b is 1, then the session $sk_{\Pi_i^s}$ is returned. Otherwise a string randomly drawn from a session key distribution is returned.

PAKE Security. Consider a game between an adversary A and a set of oracles. A asks the above queries to the oracles in order to defeat the security of a protocol P, and receives the responses. At some point during the game a Test query is asked to a fresh oracle, and the adversary may continue to make other queries. Finally the adversary outputs its guess b' for the bit b used by the Test oracle, and terminates. We define CG to be an event that A correctly guesses the bit b. The advantage of adversary A must be measured in terms of the security parameter k and is defined as follows:

$$Adv_{P,A}(k,t) = 2 \cdot Pr[CG] - 1.$$

where A is any adversary with time complexity t which is polynomial in k.

Freshness. An oracle Π_i^s is fresh if the following conditions hold:
- Π_i^s has computed a session key $sk \neq NULL$ and neither Π_i^s nor Π_j^t have been asked for a Reveal query, where Π_i^s nor Π_j^t are partnered.
- No Corrupt query has been made by the adversary since the beginning of the game.

Definition 2. A protocol P is a secure password-authenticated key exchange protocol if the following two properties are satisfied:

- Validity: if all oracles in a session are partnered, the session keys of all oracles are same.
- Key secrecy: $Adv_{P,A}(k,t)$ is bounded by $q_{se}/PW + \epsilon(k)$, where $\epsilon(k)$ is negligible, q_{se} is the number of Send queries, and PW is the size of the password space.

(1) A protocol P is a secure PAKE protocol if validity and key secrecy are satisfied when no Reveal and Corrupt queries are allowed.

(2) A protocol P is a secure PAKE-KK protocol if validity and key secrecy are satisfied when no Corrupt query is allowed.

(3) A protocol P is a secure PAKE-FS protocol if validity and key secrecy are satisfied when no Reveal query is allowed.

(4) A protocol P is a secure PAKE-KK-FS protocol if validity and key secrecy are satisfied.

3   Password-Based Tripartite Key Exchange Protocol (3-PAKE)

In this section, we propose a 3-PAKE protocol which achieves forward secrecy, and is secure against known-key attacks. 3-PAKE is designed without using the random oracle model and its security is proved under the DBDH assumption.

Public information: Let $U_i, U_j, U_l (1 \leq i, j, l \leq 3, j \neq l \neq i)$ be the identities of 3 users, $U_{ID} = U_1 \| U_2 \| U_3$, $U = \{U_1, U_2, U_3\}$, and let H be a cryptographic hash function with target collision resistant from $\{0, 1\}^* \rightarrow Z_p$, we throughout assume that U_i is the protocol initiator. Prior to the protocol execution, each user U_i computes $g = H(pw \| U_{ID} \| x) mod p$ and $x > 0$ is the smallest integer that makes g a generator of a multiplicative subgroup G_1 of order q in $GF(p)^*$. F is a pseudo random function family.

Stage 1 Message transfer.

User U_i chooses a random number $s_i (0 \leq s_i \leq q - 1)$, computes $D_i = g^{s_i}$. U_i broadcasts $D_i \| U_i$.

Stage 2 Key exchange and key confirmation.

After receiving D_j and D_l, Each user U_i computes $K_i = \hat{e}(D_j, D_l)^{s_i}$, U_i then broadcasts its key confirmation message C_i:

$C_i = F_{K_i}(sid \| U_i)$, where $sid = U_1 \| D_1 \| U_2 \| D_2 \| U_3 \| D_3$.

Stage 3 Key Computation.

After receiving C_j, C_l, user U_i checks whether the following equations holds:

$C_j \stackrel{?}{=} F_{K_i}(sid \| U_j)$ and $C_l \stackrel{?}{=} F_{K_i}(sid \| U_l)$. If the checks succeed, U_i computes its session key as: $sk_i = F_{K_i}(U_{ID} \| sid)$. Otherwise, U_i terminates the protocol execution as a failure.

4   Analysis of the Proposed Protocol

4.1   Security Analysis

We now present that the proposed protocol is secure against known-key attacks, and provides forward secrecy.

Theorem 1. Let F is a secure pseudo random function family. Then *3-PAKE* is a secure *3-PAKE-KK-FS* protocol under the DBDH assumption. Concretely,

$$Adv_P^{3PKF}(k, t, q_{ex}, q_{se}) \leq \frac{3 \cdot (q_{ex} + q_{se})^2}{2q} + Adv_F^{prf}(t) + \frac{q_{se}}{PW} + N_s \cdot Adv_{\hat{e}, G_1, G_2}^{DBDH}(t) \tag{3}$$

where *3PKF* denotes *3-PAKE-KK-FS*, t is the maximum total game time including an adversarys running time, and an adversary makes q_{ex} Execute queries and q_{se} Send queries. N_s is the upper bound of the number of sessions that an adversary makes, and PW is the size of the password space.

Proof. Consider an adversary A attacking the 3-PAKE in the sense of forward secrecy and security against known-key attacks. In this proof, we prove that the best strategy A can take is to eliminate one password from the password dictionary per initiated session. An adversary may get information about a particular session key if a collision appears on the transcripts (for the same set of users) during the experiment; i.e., there exists a user $U_i \in U$ and $t, s(t \neq s)$ such that the transcript used by instance Π_i^s is equal to the transcript used by instance Π_i^t. The other cases allow us to solve the DBDH problem and break a pseudo randomness of a pseudo random function family with probability related to the adversary's success probability. We now proceed with a more formal proof.

Case 1. Let Col be the event that a transcript is used twice by a particular user. The advantage with the event Col is bounded by the birthday paradox:

$$Adv_P^{3PKF}(k, t, q_{ex}, q_{se}) = 2Pr[CG \wedge Col] - 1 \leq 2Pr[Col] \leq \frac{3 \cdot (q_{ex} + q_{se})^2}{q} \quad (4)$$

where q is the size of the group G_1.

Case 2. The advantage without the event Col is from the following two cases: (Case2.1) For the Test oracle Π_i^s, all parties in pid_i^s have a partner oracle. (Case2.2) For the Test oracle Π_i^s, there exists at least one party $U_j (j \neq i \wedge U_j \in pid_i^s)$ such that U_j does not have a partner oracle. For $i \in \{1, 2\}$, let $Adv_P^{3PKF-C2.i}(k, t, q_{ex}, q_{se})$ be the advantage of an adversary from Case$i$. Then we have

$$Adv_P^{3PKF}(k, t, q_{ex}, q_{se}) = Adv_P^{3PKF-Col}(t) + Adv_P^{3PKF-\overline{Col}}(t)$$
$$= Adv_P^{3PKF-Col}(t) + Adv_P^{3PKF-C2.1}(t) + Adv_P^{3PKF-C2.2}(t) \quad (5)$$

Case 2.1. If the advantage of an adversary is from Case2.1, the password of the parties may be revealed by Corrupt queries. Although Corrupt queries are allowed by the definition of freshness, for the Test oracle Π_i^s, all instances in pid_i^s are executed by Execute queries. This case can be seen that there is no the password in the protocol, and thus we may ignore Corrupt queries, the protocol is the same as Joux's[1] tripartite key exchange from bilinear. Therefore, computing the upper bound of the advantage from Case2.1 is equal to that of Joux's protocol and hence we have:

$$Adv_P^{3PKF-C2.1}(t) \leq N_s \cdot Adv_{P-Joux}^{KE}(k, t, q_{ex}) \leq N_s \cdot Adv_{\hat{e}, G_1, G_2}^{DBDH}(t) \quad (6)$$

Case 2.2. To compute the upper bound of the advantage from Case2.2, we assume an adversary A gets the advantage from Case2.2. In this case, the password of the parties is not revealed by freshness conditions. Informally, there are only two ways an adversary can get information about a particular session key: either the adversary successfully breaks the authentication and key confirm parts of the protocol by guessing attacking; or correctly guesses the bit b involved in the Test query. let $Adv_P^{3PKF-C2.2}(k, t, q_{ex}, q_{se})$ be the advantage of an adversary from Case2.2. The attack game is simulated by the challenger as follows:

Initialize. Given a security parameter, k, the challenger generates a password $pw \in PW$, computes $g = H(pw \parallel U_{ID} \parallel x) mod p$, $U = \{U_1, U_2, U_3\}$, and the public system parameters:

$$param = \{U, U_{ID}, q, \hat{e}, G_1, G_2, H, F\}.$$

Challenge. The attacker A runs the protocol on the input of $param$. At some point, A terminates by outputting a guessed password pw^*. During its execution, A can make the following kinds of queries: Execute(U), Send(Π_i^s, m) and Reveal(Π_i^s). The challenger makes corresponding answer.

Without loss of generality, suppose A replaces D_i sent to $\Pi_j^s (1 \le j \le 3, j \ne i)$ with D_i^*, where $D_i^* = g_*^{s_i^*} = (H(pw^* \parallel U_{ID} \parallel x^*))^{s_i^*}$ is computed based on a guessed password pw^*, and $x^* (x^* > 0)$ is the smallest integer that makes g_* a generator of a multiplicative subgroup G_1 of order q in $GF(p)^*$, and s_i^* is randomly chosen by A.

Then A computes $K_i^* = \hat{e}(D_j, D_l)^{s_i^*} = \hat{e}(g^{s_j}, g^{s_l})^{s_i^*} = \hat{e}(g, g)^{s_i^* \cdot s_l \cdot s_j}$. Similarly, the user $U_j(\text{resp.} U_l)$ computes $K_j^* = \hat{e}(D_i^*, D_l)^{s_j} = \hat{e}((H(pw^* \parallel U_{ID} \parallel x^*))^{s_i^*}, g^{s_l})^{s_j} = \hat{e}(g_*, g)^{s_i^* \cdot s_l \cdot s_j}$ (resp. $K_l^* = \hat{e}(D_i^*, D_j)^{s_l} = \hat{e}((H(pw^* \parallel U_{ID} \parallel x^*))^{s_i^*}, g^{s_j})^{s_l} = \hat{e}(g_*, g)^{s_i^* \cdot s_j \cdot s_l}$). However, this is only way the adversary A can get information about K_j^* (resp. K_l^*) from C_j^* (resp. C_l^*), namely, A test the guessed password by checking whether $C_j = F_{K_j^*}(sid \parallel U_j)$ and $C_l = F_{K_l^*}(sid \parallel U_l)$ hold, where $sid = U_i \parallel D_i^* \parallel U_j \parallel D_j \parallel U_l \parallel D_l$. Therefore, in this case the advantage that A can obtain is:

$$Adv_P^{3PKF-C2.2}(t) \le Adv_F^{prf}(t) + \frac{q_{se}}{PW} \tag{7}$$

From Equations(4) to (7) lead to Equation(3).

4.2 Computational Overhead and Bandwidth

Our proposed protocol is similar to Ma et al.'s[17], in which the three users share a common password without a trusted server. Now we give a comparison with Ma et al.'s from the bandwidth and computational overhead of the protocols.

Table 1. Protocol Comparison

Protocol	Band-Width	Computation-Overhead
Ma et al[17]	$3\|E\|$	$5PA + 5EX + 7HA$
Our protocol	$\|E\| + \|ID\| + \|HA\|$	$PA + 2EX + 5HA$

Let Band-Width denote bandwidth per user, Computation-Overhead denote computation overhead per user, E denote a element of G_1, ID denote a user's identity, PA denote pairing, EX denote modular exponentiations, and HA hashing. The comparison of our protocol and Ma et al.'s is illustrated in Table 1. We can see that our protocol is more efficient than Ma et al's in terms of bandwidth and computational overhead.

5 Conclusions

A provably-secure protocol using ideal functions may be insecure if the ideal functions are implemented by the real-world functions. Thus a protocol without using ideal functions in proving its security is more desirable. In this paper, without using any ideal function, we design a password-based secure tripartite key agreement protocol that is suitable for the user who has no place to store the high-entropy long-term secret key or has not support from public key infrastructure.

References

1. Joux, A.: A on round protocol for tripartite Diffie-Hellman. In: Bosma, W. (ed.) ANTS 2000. LNCS, vol. 1838, pp. 385–394. Springer, Heidelberg (2000)
2. Shim, K.: Efficient One-round Tripartite Authenticated Key Agreement Protocol from Weil Pairing. Electronics Letters 39, 208–209 (2003)
3. Matsumoto, T., Takashima, Y., Imai, H.: On seeking smart public-key distribution systems. Transactions of IEICE of Japan E69, 99–106 (1986)
4. Nalla, D., Reddy, K.C.: ID-based tripartite authenticated key agreement protocols from pairings, http://eprint.iacr.org/2003/004
5. Law, L., Menezes, A., Qu, M., Solinas, J., Vanstone, S.: An efficient protocol for authenticated key agreement (1998), http://citeseer.nj.nec.com/law98efficient
6. Al-Riyami, S., Paterson, K.G.: Tripartite authenticated key agreement protocols from pairings. In: Paterson, K.G. (ed.) Cryptography and Coding 2003. LNCS, vol. 2898, pp. 332–359. Springer, Heidelberg (2003)
7. Lin, C., Sun, H., Hwang, T.: Three-party encrypted key exchange: Attacks and a solution. ACM SIGOPS Operating Systems Review 34(4), 12–20 (2000)
8. Gennaro, R., Lindell, Y.: A framework for password-based authenticated key exchange. In: Biham, E. (ed.) EUROCRYPT 2003. LNCS, vol. 2656, pp. 524–543. Springer, Heidelberg (2003)
9. Chen, Z.: Security analysis on Nalla-Reddys IDbased tripartite authenticated key agreement protocol, http://eprint.iacr.org/2003/103
10. Steiner, J.G., Neuman, B.C., Schiller, J.L.: Kerberos: An authentication service for open networks. In: Proceedings of the USENIX Winter Conference, Dallas, TX, USA, pp. 191–202 (1988)
11. Krawczyk, H.: SIGMA: The "SIGn-and-MAc" approach to authenticated Diffie-Hellman and its use in the IKE protocols. In: Boneh, D. (ed.) CRYPTO 2003. LNCS, vol. 2729, pp. 400–425. Springer, Heidelberg (2003)
12. Goldreich, O., Lindell, Y.: Session-key generation using human passwords only. In: Kilian, J. (ed.) CRYPTO 2001. LNCS, vol. 2139, pp. 408–432. Springer, Heidelberg (2001)
13. Abdalla, M., Fouque, P.A., Pointcheval, D.: Passwordbased authenticated key exchange in the three-party setting. In: Vaudenay, S. (ed.) PKC 2005. LNCS, vol. 3386, pp. 65–84. Springer, Heidelberg (2005)
14. Zhang, F., Liu, S., Kim, K.: ID-based one round authenticated tripartite key agreement protocol with pairings, http://eprint.iacr.org/2002/122
15. Tang, Q., Choo, K.R.: Secure Password-Based Authenticated Group Key Agreement for Data-Sharing Peer-to-Peer Networks. In: Zhou, J., Yung, M., Bao, F. (eds.) ACNS 2006. LNCS, vol. 3989, pp. 162–177. Springer, Heidelberg (2006)
16. Wen, H.A., Lee, T.F., Hwang, T.: Provably secure three party password-based authenticated key exchange protocol using Weil pairing. IEE Proc-Commun. 152(2), 138–143 (2005)
17. Ma, C., Ao, J., Li, J.: Provable password-based tripartite key agreement protocol, http://eprint.iacr.org/2007/184.pdf
18. Chien, H.-Y.: Comments on a Provably Secure Three-Party Password-Based Authenticated Key Exchange Protocol Using Weil Pairings (2006), http://eprint.iacr.org/2006/013.pdf

Credit Rating Method with Heterogeneous Information

Dan Meng[1] and Yang Xu[2]

[1] School of Economics Information engineering, Southwestern University of Finance
and Economics, Chengdu, Sichuan, China, 610074
mengd_t@swufe.edu.cn
[2] Department of Applied Mathematics, Southwest Jiaotong University, Chengdu,
Sichuan, China, 610074
yangxu@home.swjtu.edu.cn

Abstract. Corporate credit rating is a very important issue in finance
field. A lot of methods such as neural networks, genetic algorithm and
support vector machine have been proposed to solve this problem. The
credit rating is a complex problem which includes some determinate criteria and other uncertain criteria associating with human judgement
which may be vague or linguistic. Therefore, it includes both quantitative value and qualitative value in credit rating. Furthermore, even for
the same kind of determinate or uncertain criteria, or in other words, for
the same quantitative or qualitative criteria, the assessment domain and
scale are also diverse. Some traditional methods transform all the evaluation domain and scale to a uniform one. Accordingly, it may lead to
the loss of information so much as the final total departure of the assessment result. A method dealing with heterogeneous information proposed
by F. Herrera and L. Martinez et al. is a good solution for this problem
which includes various assessment domain and scale. Based on the above,
we take the corporate credit rating process as a multi-criteria evaluation
problem with heterogeneous information in this paper. And we propose a
corporate credit rating method based on multi-criteria evaluation model
with heterogeneous information on 2-tuple fuzzy linguistic model. And
we give a case study of an auto-manufacture corporate credit rating. The
case study shows that the method is feasible for corporate credit rating.

Keywords: Credit Rating, Numerical Information, Linguistic Information, Heterogeneous Information.

1 Introduction

A credit rating assesses the credit worthiness of an individual, corporation, or
even a country. Earlier credit ratings are calculated from financial history and
current assets and liabilities. A credit rating tells a lender or investor the probability of the subject being able to pay back a loan.It usually includes personal
credit ratings, corporate credit ratings and sovereign credit ratings. In this paper,
we will focus on corporate credit rating.

G. Wang et al. (Eds.): RSKT 2008, LNAI 5009, pp. 739–746, 2008.

Credit-risk evaluation decisions are important for the financial institutions and investors involved due to the high level of risk associated with wrong decisions. Credit rating can provide an decision support tools to manage credit risk and reduce loan losses for involved financial institution or investors. Because of the importance of credit rating, a lot of methods such as neural networks [1,25,20,21], genetic algorithm [19,11], fuzzy set [22,24,23] , support vector machine [10,16] and hybrid mining approaches in [10,13,12]have been proposed to solve this problem.

The corporate credit rating is a complex problem. With the development of finance market and credit rating system and methods, the credit rating involves not only financial conditions, but also management measure and competitions of the evaluated enterprises. Besides the determinate numerical indexes such as profitability ratios and efficiency ratios etc. financial indexes, there are other factors related to management measure such as administrator's experience which are usually linguistic and vague need to be dealt with in corporate credit rating.

We will deal with the credit rating problem from a different point of view in this paper. In our proposed method, we recognize the corporate credit rating as a multi-criteria evaluation problem with heterogeneous information.

2   Problem Formulation

The credit rating system and criteria may be different for various evaluated object and purpose. CAMELS rating system is an international bank-rating system with which bank supervisory authorities rate institutions according to the following six factors: C - Capital adequacy, A - Asset quality, M - Management quality ,E - Earnings, L - Liquidity, S - Sensitivity to Market Risk. However, the credit rating system for nonfinancial corporation is different from "CAMELS" rating system. But both CAMELS and other credit rating system include the heterogeneous information. We will illustrate an Chinese industry corporation credit rating model as an example in this section. Based on experts' opinion, we acquire the following credit rating criteria system and representation form of every criterion for industry corporation credit rating.

1. Financial Conditions
   - Profitability ratios
     - Operating margin (Percentage)
     - ROA, return on assets (Percentage)
     - ROE, return on equity (Percentage)
   - Efficiency ratios
     - Inventory turnover (time)
     - Receivable turnover (time)
     - Assets turnover (time)
   - Liquidity ratios
     - Quick ratio (Percentage)
     - Current ratio (Percentage)

- Leverage ratios
  - Times-interest-earned (times)
  - Total debt to assets (percentage)
  - Debt to equity ratio (Percentage)
2. Management Measure
  - administrator's management experiences (Fuzzy or Linguistic value)
  - stockholders structure type (Fuzzy or Linguistic value)
  - average sale growth rate during the last three years (Percentage)
  - conditions of capital increment during the last three years (Percentage)
3. Characteristics and Perspectives of the products and competitions
  - equipment and technologies (Fuzzy or Linguistic value)
  - product marketability (Fuzzy or Linguistic value)
  - economic conditions of the industry in the next year (Fuzzy or Linguistic value)

The assessment value which is percentage, time and times can be given in numerical value. However, other evaluation value which is related to human judgement is fuzzy or linguistic nature. Therefore, it is necessary to introduce some "appropriate" tool to deal with the vague and linguistic information relating to human judgement in credit rating.

Fuzzy linguistic variable and its corresponding approaches is one of the vague or imprecise information processing methods. The fuzzy linguistic variable and its related approach is presented by L. A. Zadeh in 1975 in [28]. After L. A. Zadeh's work in 1975, a lot of fuzzy linguistic approaches have been proposed and applied with very good results to different problems, such as, "education [15]", "information retrieval [2]", and "decision-making system [3,4,5],[6,7,9] [26,27]" etc. In addition, fuzzy approaches also can be used to deal with the vagueness or linguistic nature in credit rating problem in [22,24,14,23].

In fact, besides the existence of numerical evaluation value such as percentage, time and times, the fuzzy or linguistic value also exists in the above industry corporation credit rating system.

3   Credit Rating Method with Heterogeneous Information

A method which can deal with heterogeneous information in credit rating will be proposed in this section based on based on 2-tuple fuzzy linguistic model and the method proposed in [9,17],.

The following set $T = \{s_k | k = 1, 2, \cdots, 7\} = \{s_0 :$ None (abbr. to N), $s_1 :$ Very Low (abbr. to VL), $s_2 :$ Low (abbr. to L), $s_3 :$ Medium (abbr. to M), $s_4 :$ High (abbr. to H), $s_5 :$ Very High (abbr. to VH), $s_6 :$ Perfect (abbr. to P) $\}$ is used to represent the credit grades. We take triangular fuzzy set to denote s_k, $s_k = (a_k, b_k, c_k)$, where a_k denotes the left limit of the definition domain of the triangular membership function, c_k denotes the right limit of the definition domain of the triangular membership function, b_k denotes the value in which the membership value is 1. Its corresponding semantic is describe as in figure 1.

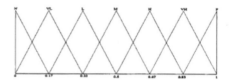

Fig. 1. Semantic of Set of Linguistic Fuzzy Sets

The credit rating method includes two main steps: transformation and aggregation. To prevent loss of information during value transformation, we will first introduce the following definitions to complete the transformation procedure.

Definition 1. *[18] The numerical values includes percentage, time and times which is absolute value and based on different scale. Actually, the credit evaluation value involved in aggregation should be a relative one based on industry background. We define the following transformation function as follows in order to prevent the loss of information. Let max be industry maximal value, min be industry minimal value, ave be industry average value, x is the evaluation value, define a function f which transform the absolute numerical value nv to a relative value $f(nv) \in [0,1]$ as follows:*

$$f(x) = \begin{cases} 1 - \dfrac{max - x}{max}, & x > avg, \\ 0.5, & x = avg, \\ \dfrac{x - min}{min}, & x < avg, \end{cases} \tag{1}$$

Definition 2. *[18] Let $T = \{s_k | k = 1, 2, \cdots, 7\} = \{s_0 : None (abbr. to N), s_1 : Very Low (abbr. to VL), s_2 : Low (abbr. to L), s_3 : Medium (abbr. to M), s_4 : High (abbr. to H), s_5 : Very High (abbr. to VH), s_6 : Perfect (abbr. to P) \}$ be credit grades set, define transformation function τ from numerical value in [0, 1] into fuzzy sets in T as follows:*

$\tau : [0,1] \to F(T),$
$\tau(v) = \{(s_0, \gamma_0), (s_1, \gamma_1), \cdots, (s_g, \gamma_g)\}, s_i \in T, r_i \in [0,1]$

$$\gamma_i = \mu_{s_i}(v) = \begin{cases} 0, & if\ v \notin Support(\mu_{s_i}(v)), \\ \dfrac{v - a_i}{b_i - a_i}, & if\ a_i < v < b_i, \\ 1, if\ v = b_i, \\ \dfrac{c_i - v}{c_i - b_i}, & if\ b_i < v < c_i \end{cases} \tag{2}$$

For different evaluated criteria in credit rating, the linguistic term set may be different from the credit grades set T, we define the following function ψ to transform all linguistic term into a unifying set T.

Definition 3. *[18] Let S be a evaluation linguistic term set, T be credit grades set and $lv \in S$ be a linguistic term provided by experts. We define the transformation function ψ from S to T as follows:*

$$\psi : S \to T :$$
$$\psi(lv) = \{(s_0, \gamma_0), (s_1, \gamma_1), \cdots, (s_7, \gamma_7)\}, \tag{3}$$
$$\gamma_i = max_y min\{\mu_{lv}(y), \mu s_k(y)\},$$

Definition 4. *[17] Transform the fuzzy sets in S_s into the 2-tuples in S_s:*

$$\mathcal{X} : F(S_s) \to S_s \times [-0.5, 0.5),$$
$$\mathcal{X}(\tau(v)) = \mathcal{X}(\{(s_j, \gamma_j), j = 0, 1, \cdots, g\}) = \triangle(\sum_{j=0}^{g} j \cdot \gamma_j / \sum_{j=0}^{g} \gamma_j) \tag{4}$$

Concretely, the transformation procedure is as follows:

- For numerical value x,
  1. We transform every numerical evaluation value nv to a relative value $f(nv)$, where $f(nv) \in [0, 1]$ as in definition 1;
  2. Transform the relative value $f(nv)$ into fuzzy sets $\tau(f(nv))$ in T as in definition 2;
  3. Transform fuzzy sets in T to $\mathcal{X}(\tau(f(nv)))$linguistic 2-tuples in T as in definition 4.
- For linguistic value v,
  1. Transform the linguistic value lv into fuzzy sets in T as $\psi(lv)$ as in definition 3 ;
  2. Transform fuzzy sets in T to linguistic 2-tuples in T as $\mathcal{X}\psi(lv)$ as in definition 4.

The transformed value can be aggregated as in definition 5.

Definition 5. *[8] Let $x = \{(r_1, \alpha_1), (r_2, \alpha_2), \cdots, (r_n, \alpha_n)\}$ be a set of 2-tuples and $W = (\omega_1, \omega_2, \cdots, \omega_n)$ be their associated vector. Then the weighted aggregation operator on 2-tuple is defined as:*

$$x \circ W = \triangle(\frac{\sum_{i=1}^{n} \triangle^{-1}(r_i, \alpha_i) \cdot \omega_i}{\sum_{i=1}^{n} \omega_i}) = \triangle(\frac{\sum_{i=1}^{n} \beta_i \cdot \omega_i}{\sum_{i=1}^{n} \omega_i}) \tag{5}$$

4 Case Study

In this section, we will choose two auto-manufacture enterprises "XX" and "YY" as the case study. All the assessment value is given in table 1. Most data is based on http://www.cnlist.com and http://mcin.macrochina.com.cn. Some linguistic evaluation value is the consensus of involved experts' evaluation.

By using the above proposed method, the credit rating result for XX and YY is (High, 0.7), (Medium, 0.8)respectively.

Table 1. Corporate Credit Rating Basic Data Set of Two Corporate

Year		2005	2005
Corporation		XX	YY
Profitability ratios	Operating margin(percentage)	19.5372	13.1393
	ROA, return on assets(percentage)	1.42	3.8434
	ROE,return on equity(percentage)	3.52	19.45
Efficiency ratios	Receivable turnover(time)	42.6851	38.6153
	Inventory turnover(time)	4.453	9.2347
	Assets turnover(time)	1.1497	2.8768
Liquidity ratios	Current ratio(times)	1.1172	1.0541
	Quick ratio(times)	0.6977	0.6695
Leverage ratios	Total debt to assets(percentage)	55.0177	71.9188
	Debt to equity ratio(percentage)	36.092	13.7952
	Times-interest-earned(times)	-26.7096	21.7646
administrator's management experiences		pretty good	pretty good
stockholders structure type		pretty good	good
average sale growth rate during the last three years(percentage)		25.94	70.28
conditions of capital increment during the last three years(percentage)		47.24	9.69
equipment and technology		very good	very good
product marketability		pretty good	pretty good
economic condition of the industry in the next year		rapid increase	slow increase

The value of time-interest earned of XX is negative because interest revenue is more than interest expense during 2005.

Table 2. Linguistic Term Set for Linguistic Evaluated Criterion

Evaluated Criterion	Linguistic Term Set
administrator's management experience	{very good, good, pretty good, medium, bad, pretty bad, very bad}
equipment and technology	{very good, good, pretty good, medium, bad, pretty bad, very bad}
product marketability	{very good, good, pretty good, medium, bad, pretty bad, very bad}
economic condition of the industry in the next year	{rapid increase, slow increase, fixed, slow decrease, rapid decrease}
stockholders structure type	{good, pretty good, bad}

5   Conclusions and Future Works

In this paper, a credit rating method with heterogeneous information is proposed.We perform an industry corporation credit rating case study based on the proposed method. and the case study shows that the method is feasible. It should be pointed out that we focus on proposing a credit rating method dealing with heterogeneous information other than establishing a general industry corporation credit rating system in this paper. Further result on constructing a general corporation credit rating system based on multi-criteria evaluation framework will be given in our future research.

Acknowledgements

This work was supported by the Scientific Research Fund of Southwestern University of Finance and Economics and the National Natural Science Fund of P.R.China (Grant No. 60474022) and the specialized Research Fund for the Doctoral Program of Higher Education of China under (Grant No. 20060613007).

References

1. Baesens, B., Setiono, R., Mues, C., Vanthienen, J.: Using neural network rule extraction and decision tables for credit-risk evaluation. Management Science 49(3), 312–329 (2003)
2. Bordogna, G., Passi, G.: A Fuzzy Linguistic Approach Generalizing Boolean Information Retrieval: A Model and Its Evaluation. J. Amer. Soc. Inform. Sci. 44, 126–132 (1993)
3. Herrera, F., Verdegay, J.L.: Linguistic Assessments in Group Decision. In: Proc. 1st European Cong. Fuzzy and Intelligent Technologies, Aachen, Germany, pp. 941–948 (1993)
4. Herrera, F., Herrera-Viedma, E., Verdegay, J.L.: A Sequential Selection Process in Group Decision Making with Linguistic Assessment. Information Science 85, 223–239 (1995)
5. Herrera, F., Verdegay, J.L.: A Linguistic Decision Process in Group Decision Making. Group Decision Negotiation 5, 165–176 (1996)
6. Herrera, F., Herrera-Viedma, E., Verdegay, J.L.: Direct approach processes in group decision making using linguistic OWA operators. Fuzzy Sets Syst. 79, 175–190 (1996)
7. Herrera, F., Herrera-Viedma, E.: Linguistic decision analysis: steps for solving decision problems under linguistic information. Fuzzy Sets Syst. 115, 67–82 (2000)
8. Herrera, F., Martinez, L.: A 2-tuple Fuzzy Linguistic Representation Model for Computing with Words. IEEE Transaction on Fuzzy Systems 8(6), 746–752 (2000)
9. Herrera, F., Martiez, L., Sanchez, P.J.: Managing non-homogeneous information in group decision making. European Journal of Operational Research 166, 115–132 (2005)
10. Huang, C.L., Chen, M.C., Wang, C.J.: Credit Scoring with a Data Mining Approach based on Support Vector Machines. Expert Systems with Applications (2006), doi:10.1016/j.eswa.2006.07.007
11. Huang, J.J., Tzeng, G.H., Ong, C.S.: Two-stage genetic programming (2SGP) for the credit scoring model. Applied Mathematics and Computation 174, 1039–1053 (2006)
12. Huang, Z., Chena, H., Hsua, C.J., Chenb, W.H., Wus, S.: Credit Rating Analysis with Support Vector Machines and Neural Networks: a Market Comparative Study. Decision Support Systems 37, 543–558 (2004)

13. Hsieh, N.C.: Hybrid mining approach in the design of credit scoring models. Expert Systems with Applications 28, 655–665 (2005)
14. Jiao, Y., Syau, Y.R., Lee, E.S.: Modelling Credit Rating by Fuzzy Adaptive Network. Mathematical and Computer Modelling (2006), doi:10.1016/j/mcm.2005.11.016
15. Law, C.K.: Using Fuzzy Numbers in Educational Grading System. Fuzzy Sets and Systems 83, 261–271 (1996)
16. Lee, Y.C.: Application of support vector machines to corporate credit rating prediction. Expert Systems with Applications 33, 67–74 (2007)
17. Martinez, L., Liu, J., Ruan, D., Liu, J.B.: Dealing With Heterogeneous Information In Engineering Evaluation Processes. Information Sciences (2006), doi: 10.106/j.ins.2006.07.005
18. Meng, D., Xu, Y.: SMEs Credit Rating Method with Heterogeneous Information: a Chinese Case. In: The 2007 International Conference on Intelligent Systems and Knowledge Engineering,
19. Ong, C.S., Huang, J.J., Tzeng, G.H.: Building credit scoring models using genetic programming. Expert Systems with Applications 29, 41–47 (2005)
20. Piramuthu, S.: Financial Credit Risk Evaluation with Neural and Neuro-Fuzzy Systems. European Journal of Operational Research 112, 310–321 (1999)
21. Piramuthr, S.: On preprocessing data for financial credit risk evaluation. Expert Systems with Applications 30, 489–497 (2006)
22. Rommelfanger, H.J.: Fuzzy Logic Based Systems For Checking The Credit Solvency of Small Business Firms. In: RIbeiro, R.A., Zimmermann, H.J., Yager, R.R., Kacprzk, J. (eds.) Soft Computing in Financial Engineering, pp. 388–401. Physica-Verlag, Heidelberg (1999)
23. Syau, Y.R., Hsieh, H.T., Lee, E.S.: Fuzzy Numbers in the Credit Rating of Enterprise Financial Condition. Review of Quantitative Finance and Accounting 17, 351–360 (2001)
24. Weber, R.: Application of Fuzzy Logic For Credit Worthiness Evaluation. In: Riberio, R.A., Zimmermann, H.J., Yager, R.R., Kacprzk, J. (eds.) Soft Computing in Financial Engineering, pp. 388–401. Physica Verlag, Heidelberg (1999)
25. West, D.: Neural network credit scoring models. Computers and Operations Research 27, 1131–1152 (2000)
26. Yager, R.R.: An Approach to Ordinal Decision Making. International Journal of Approximate Reasoning 12, 237–261 (1995)
27. Yager, R.R.: A New Methodology for Ordinal Multi-objective Decisions Based on Fuzzy Sets. In: Dubios, D., Prade, H., Yager, R.R. (eds.) Fuzzy Sets for Intel-ligent Systems, Morgan Kaufmann Publishers, San Maeto (1995)
28. Zadeh, L.A.: The Concept of a Linguistic Variable and Its Applications to Approximate Reasoning. Part I, Information Sciences 8, 199–249 (1975); Part II, Information Sciences. 8, 301–357 (1975); Part III, Information Sciences. 9, 43–80 (1975)

The K Shortest Transit Paths Choosing Algorithm in Stochastic Transit Network

Liyuan Zhao[1], Yiliang Xiong[2], and Hong Sun[3]

[1] College of Traffic and Transportation
Southwest Jiaotong University
Chengdu, 610031, P.R. China
xio_y@163.com

[2] College of Traffic and Transportation
Southwest Jiaotong University
Chengdu, 610031, P.R. China
wlxiong@gmail.com

[3] School of Air Transportation Management
Civil Aviation Flight University of China
Guanghan, 610031, P.R. China
hanksun@263.net

Abstract. The public transit route choosing problem is the key technology of public transit passenger information system. Considering travel time variety caused by uncertainty traffic congestion condition, firstly this paper designs the least transfer times algorithm and the K shortest transit paths algorithm in the stochastic transit network. On the basis of travel psychology analysis, transfer times, travel time and cost of each transit path plan are taken into account. By changing link travel time reliability, the algorithms generate different K shortest transit path plans under different traffic conditions. Computational experiments demonstrate the efficiency of the model and algorithm in stochastic transit network.

Keywords: K-shortest path, stochastic transit network, time reliability, least transfer times.

1 Introduction

Public transit system is an important part of city transportation system. As its high efficient utilization of resources, vigorously developing public transit and bus priority become an inevitable choice to ease the worsening traffic congestion status. Public transit paths choosing problem has been widely studied during the past decades [1,2,3,4]. All these studies are based on the average travel time value.

However, a lot of random factors change the state of traffic network, such as traffic accidents, weather conditions, road maintenance and even traffic jams. These random factors will lead directly to the variety of travel time. The reliability of travel time has become an important factor when traveler makes transit

G. Wang et al. (Eds.): RSKT 2008, LNAI 5009, pp. 747–754, 2008.

path choice. The low reliability of transit travel time has become a common phenomenon in many cities. This is also the main reason that makes bus less competitive.

To give a more reasonable description of the transit paths choice, transit link travel time is thought to follow normal distribution according to Bell and Iida's study [5]. The concept of travel time reliability is introduced. To predict travel time in the stochastic network, the expected reliability value is presented according to road traffic congestion status and different personal preference. During traffic peak time, the transit travel time is longer, and we set bigger reliability to predict, while in the low peak time, a smaller reliability value is set. There are always three factors impacting traveler make transit path choice: transfer times, travel time and cost. The research on passenger psychology (see [6]) shows that transfer times is the first considered factor, followed by cost and time.

This paper is organized as follows. First, the impacts on transit path choice are formulated including transit travel time in stochastic network and the least transfer times matrix. The improved Dijkstra algorithm is then introduced to compute the least transfer times. The K shortest transit path algorithm (see [7,8,9]) is designed based on the stochastic travel time. A numerical stochastic network example is provided. Conclusion follows in the last section.

2   Impacts on Transit Route Choosing

A public transit network is composed by some nodes (bus stop), the links connecting two nodes and bus lines. Define a public transit network as G, $G = \{N, E, R\}$, where $N = \{1 \leq i \leq n\}$ denotes the set of all nodes, and n is the number of nodes; the origin node and the destination node is O, D respectively. $E = \{1 \leq e \leq m\}$ is the set of all transit links, and m is the number of links; $R = \{1 \leq r \leq u\}$ is the set of all bus lines, and u is the number of links.

In this paper, we consider three factors: transfer times, travel time and cost. Transfer times are the most important factor. If there are some nonstop paths between the origin O and destination D, we choose the path with the least travel time and cost. If there is no nonstop path existing, then consider one time transfer paths. The method of determining travel time in stochastic network and transfer times is discussed follow.

2.1   Time Determination in Stochastic Network

First let's see the deficiency of the Previous Average Travel Time. The transit link travel time is thought to follow normal distribution.

The previous method, determining optimal route according to the average travel time, does not meet our need. For example, in Figure 1, the mean time of route 1 is less than route 2, so route 1 will be chosen in previous average time method. However, route 1 with bigger standard deviation, which means the travel time will fluctuate largely for frequent traffic congestion phenomenon.

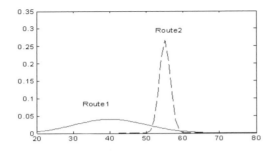

Fig. 1. Transit Path Normal Distribution Comparison

On the contrary, the travel time on route 2 has less fluctuation. In the traffic peak period, it would happen that route 1 takes 60 minutes for serious congestion while route 2 just needs 50 minutes. In this condition, route 2 will be a better choice. To solve this uncertainty circumstance, we adopt reliability theory to describe.

Transit travel time contains two parts: link travel time and transfer time. Transfer time generates when there is no nonstop bus route existing between origin and destination. In stochastic network, we suppose both follow normal distribution. T_e: $N(t_e,(\sigma_e)^2)$, T_c: $N(t_c,(\sigma_c)^2)$ where t is the mean and σ is the standard deviation.

Definition 1. Reliability of Link Travel Time

The *reliability of link travel time* can be defined as (see [10]): considering the uncertainty and randomness of the travel time, the probability of the link travel time to be less than some given time t_0: $\rho = P(T_e \leq t_0)$, $\rho \in [0,1]$.

$t_{e\rho}$: travel time on link e under reliability ρ, $P(T_e \leq t_{e\rho}) = \rho$. Set $\phi_e(x) = P(T_e \leq x)$ be distribution function of T_e, then $t_{e\rho} = \phi_e^{-1}(\rho)$. So we can calculate $t_{e\rho}$ if know T_e and ρ.

$t_{ij\rho}^r$: The estimated travel time from node i to j on bus line r, which is the sum of travel time $t_{e\rho}$ of passed links.

$t_{c\rho}$: The transfer times under reliability ρ.

During traffic peak time period, travel time is generally long, high reliability value will be appropriate, $\rho > 0.5$. In the low peak time, we set $\rho < 0.5$. If users belong to risk taking categories, they will tend to choose path with lower travel time, less reliability value will be set. The risk avoiding categories will tend to choose the route with higher reliability will be adopted.

2.2   The Least Transfer Times

Definition 2. The Adjacency Matrix

We denote $\delta = \{\delta_{ri}|r \in R, i \in N\}$ as the *adjacency matrix* between bus line r and node i. If the bus line r passes node i, then $\delta_{ri} = 1$, otherwise 0.

Definition 3. Nonstop Matrix

Nonstop matrix is defined as $A = \{a_{ij}|i,j \in N\}$, if there exist nonstop bus line passing from node i to j , then $a_{ij} = 1$; otherwise $a_{ij} = +\infty$, $a_{ii} = 0$. We can get a_{ij} by the correlation matrix: if $\max_r \delta_{ri} \cdot \delta_{rj} = 1$, then $a_{ij} = 1$; and $a_{ij} = +\infty$ when $\max_r \delta_{ri} \cdot \delta_{rj} = 0$.

Definition 4. Least Transfer times

We define the *least transfer times* matrix as: $W = \{w_{ij}|i,j \in N\}$, where w_{ij} is the least transfer times from node i to j . w_{iD} denotes the least transfer times from node i to D . If node i can't reach node j in the transit network, then $w_{ij} = +\infty$.

3   K Shortest Transit Paths Algorithm

The K shortest transit paths algorithm contains two parts. The first is the least transfer times matrix algorithm, which provides a heuristic value to the second algorithm, the K shortest transit paths search algorithm.

3.1   The Least Transfer Times Algorithm

In transit shortest algorithm, the least transfer times has been widely researched (see [11,12,13]). In this paper, we put forward an improved Dijkstra algorithm according to the nonstop matrix to calculate the least transfer times from all nodes i to the destination D: w_{iD}. The algorithm 1 is shown as follow.

Step1: $\forall i \in N$, construct the correlation matrix δ and nonstop matrix A, initialize $w_{iD} = +\infty$, $w_{DD} = 0$, put all notes in queue Q.

Step2: If Q is not empty, then do $\{y\} \leftarrow \{i|w_{iD} = \min w_{jD}, j \in Q\}$, $Q \leftarrow Q - \{y\}$; otherwise, the algorithm ends.

Step3: For all nodes $i \in N$, y and w_{iD} is finite, if $w_{iD} > w_{yD} + a_{iy}$, then $w_{iD} = w_{yD} + a_{iy}$, if not, go to Step2.

3.2   The K Shortest Transit Paths Search Algorithm

The past transit path search algorithms are mostly the shortest path algorithm (see [14,15]). In the shortest path problem there is a single label assigned to each node j, while in the K shortest paths problem we may have K labels for each node which can record K paths. The label of node j is formed by five K-tuple: $\pi_j, \sigma_j, \xi_j, \eta_j$ and θ_j. While $\pi_j^k, \sigma_j^k, \xi_j^k, \eta_j^k$ and θ_j^k is the kth respective component. π_j^k and σ_j^k, respectively denotes the travel time and transfer times of a path from original node O to node j; ξ_j^k and θ_j^k means the node i before j in that path and its position in ξ_i , respectively. η_j^k is the bus line connecting node j and the previous node i.

To improve the efficiency of search algorithm, we set a maximum tolerant transfer times Y from the origin O to the destination D, such that $\sigma_j^k + w_{jD} \leq Y$. It means that the sum of the transfer times from original node O to node j and the least transfer times from node j to destination D must be less or equal to the maximum tolerant transfer times. If this condition isn't satisfied, there is no need to search any further. In this way, unnecessary calculations are effectively avoided. The general form of the K shortest transit paths search algorithm is given in Alg.1. The value of w_{iD} can generate in Algorithm 1. Q is a first-in-first-out queue.

Data: $t^r_{ij\rho}, t_{c\rho}, w_{iD}$;
$\{\pi^1_i, ..., \pi^K_i\} \leftarrow \{+\infty, ..., +\infty\}, \forall i \in N$;
$\xi^1_O, \eta^1_O \leftarrow null$;
$\pi^1_O, \sigma^1_O \leftarrow 0, Q \leftarrow \{O\}$
while $Q \neq \emptyset$ **do**
   $i \leftarrow$ the top node of queue Q;
   $Q \leftarrow Q - \{i\}$;
   for *bus liner that passes node i* **do**
      for *node j that follows node i on the bus liner* **do**
         for $k \in \{1, ..., K\}$ *such that π^k_i is unused and finite* **do**
            $l \leftarrow$ order of $\max\{\pi^1_j, ..., \pi^K_j\}$ in $\{\pi^1_j, ..., \pi^K_j\}$
            while $\eta^l_j \neq r$, $\sigma^k_i + w_{iD} \leq Y$, *and* $\pi^k_i + t^n_{ij} + t_c < \pi^l_j$ **do**
               $\pi^l_j \leftarrow \pi^k_i + t^r_{ij\rho} + t_{c\rho}$;
               $\sigma^l_j \leftarrow \sigma^k_i + 1$;
               $\xi^l_j \leftarrow i$;
               $\eta^l_j \leftarrow r$;
               $\theta^l_j \leftarrow k$;
               while $j \notin Q$ **do**
                  push node j into queue Q;
               end
            end
         end
      end
   end
end
The k shortest paths can be built with label ξ_j, η_j and θ_j by recursive function.
1. The K Shortest Transit Paths Search Algorithm

4   Computational Experiments

Figure 2 is an instance of transit network, including 15 nodes , 22 links and 6 one-way bus lines, R1, R2...R6. The origin and the destination are node1, 15, respectively.The travel time on links in stochastic network show in Table 1, and the transfer time follows:$N(5,3)$.

As the principle of setting reliability, we give three reliability value$\rho=0.5, 0.2, 0.9$ to see the different choices in the three traffic condition, respectively: the general traffic condition, the low peak time traffic, and the serious traffic congestion condition. The results of the algorithm 1 and 2 are shown as the following to different reliability.

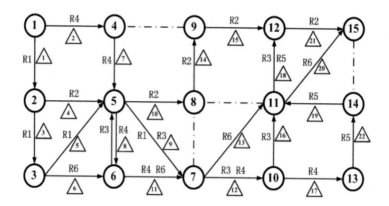

Fig. 2. Transit Network Example

Table 1. Link Travel Time Normal Distribution

Link	1, 2	3	4, 10, 14, 15	5, 6, 19	7, 9
Time	$N(10, 1)$	$N(10, 0.5)$	$N(8, 4)$	$N(9, 0.8)$	$N(8, 1)$
Link	8	11, 12	13, 18, 17, 21, 22	16	20
Time	$N(7, 1)$	$N(7, 0.8)$	$N(8, 0.8)$	$N(9, 1.2)$	$N(6, 0.8)$

Table 2. The K Shortest Paths under General Traffic Condition

Rank	Time	Cost	Transfer	Paths
1, 2	50	3	2	$1, R4, 5, R1, 7, R6, 151, R4, 5, R3, 7, R6, 15$
3,4	51	2	1	1,R4,7(6),R6,15;
5	55	2	1	1,R1,2,R2,15

Status 1. The general traffic condition, we set ρ=0.5.
In Table 2, Rank means the ranking of paths' travel time, and Transfer means the transfer times of paths. The path, taking [1,R4,7(6),R6,15] for example, means starting from node 1 to node 7 or 6 by bus line R4, then transfers for bus line R6 to the destination node 15.

The least transfer times is 1, which means nonstop bus line doesn't exist from node 1 to 15. If the transfer times (according to passenger's psychology analysis) take priority over other factors, then the best route is : [1,R4,7(6),R6,15]. If we prefer shorter travel time, then [1,R4,5,R1,7,R6,15] and [1,R4,5,R3,7,R6,15] will be chosen.

Status 2. Traffic low peak time period, we setρ=0.2.
The best route in the low peak time is [1,R1,2,R2,15], while the best path [1,R4,7(6),R6,15] in Status 1, which takes more time here, ranks 11th (and 12th). The influence of traffic condition is obvious here.

Table 3. The K Shortest Paths under Traffic Low Peak Condition

Rank	Time	Cost	Transfer	Paths
1	37.5	2	1	1,$R1$,2,$R2$,15
2	40	2	1	1,$R4$,5,$R2$,15
3,4	40.4	4	3	1,$R1$,2,$R2$,12,$R5$($R3$),11,$R6$,15
5	41	4	3	1,$R1$,2,$R2$,5,$R1$,7,$R6$,1
11,12	43.9	2	1	1,$R4$,7(6),$R6$,15

Table 4. The K Shortest Paths under Serious Traffic Congestion Condition

Rank	Time	Cost	Transfer	Paths
1,2	61.8	2	1	1,$R4$,7(6),$R6$,15;
3,4	63.3	3	2	1,$R4$,5,$R1$($R3$),7,$R6$,15
5	64.9	2	1	1,$R1$,3,$R6$,15
6	66.1	2	1	1,$R1$,7,$R6$,15
15	81.7	2	1	1,$R1$,2,$R2$,15

Status 3. Traffic peak time period, with serious traffic congestion, we setρ=0.9. In traffic peak time, paths [1,$R4$,7(6),$R6$,15] are the best choices being composed of links (2,7,8,11,13,20) with small standard deviation (see Table 1). There is little fluctuation on travel time. It means that in traffic peak time, paths with less time-variety are more likely to be chosen. Nevertheless, the best path [1,$R1$,2,$R2$,15] in Status 2 contains links (1,4,10,14,15,21) with big standard deviation, the travel time is really long in the traffic congestion condition and it ranks the 15th here.

The demonstration shows that in the traffic peak time, the path with low fluctuation on travel time is more likely to be chosen. But during the low peak time, people will choose the path with wide time-variety for less travel time.

5   Conclusion

The result of the demonstration shows that the K shortest transit paths search algorithm based on the stochastic transit network can effectively provide best choice under different traffic condition by setting different reliability value to link travel time. The running time and the results of the least transfer times algorithm and the K shortest transit paths choosing algorithm are satisfactory. In this paper, we propose the method of setting reliability value by the understanding of the traffic status. However, it's so hard to hold the entire road information that there must be some errors. How to set a right reliability being closer to practical situation should be further studied.

The factors of transit path choosing include travel time, cost and transfer times. In fact, there are a lot of factors working on the passengers' decision for complicated transit network, such as the bus departure frequency, transfer walking distance, the feasible walking range and individual preference. These facts should be taken into consideration to meet the needs of different users.

Acknowledgement

This paper is supported by the National Natural Science Foundation of China (60776820).

References

1. Weiner, E.: Urban Transportation Planning in The United States. Praeger Publishers (1987)
2. Ma, L., Liu, X.B.: The Shortest Path of Public Transport Net and Related Optimal Taking Bus Problem. Mathematics in Practice and Theory 34, 38–43 (2004)
3. Li, Y.Z., Fang, Y.M.: Design and Realization of Urban Public Traffic Inquiry System. Surveying and Mapping of Geology and Mineral Resources 22, 3–5 (2006)
4. Fu, D.: Algorithm of least transfer in traffic system and its implementation. Journal of Huaqiao University 22, 346–350 (2001)
5. Bell, M.G.H., Iida, Y.: Transportation Network Analysis. John Willey & Sons, Chichester (1997)
6. Yang, X.M., Wang, W.: Gis-Based public transit passenger route choice model. Journal of Southeast University 6, 59–62 (2000)
7. Fox, B.L.: Calculating Kth shortest paths. Canadian Journal of Operational Research 11, 66–70 (1973)
8. Eppstein, D.: Finding the K Shortest Paths. SIAM Journal on Computing 28, 652–673 (1999)
9. He, S.X., Fan, B.Q.: Improved optimal path searching algorithm in transit network. Journal of Shanghai for Science and Technology 28, 63–66 (2006)
10. Zhao, L.Y., Du, W.: The Fuzzy Lattice Order Decision-making Method for Travel Mode Choice Based on the Travel Time Reliability. In: Proceedings of the Second International Conference on Intelligent Systems and Knowledge Engineering, Chengdu, pp. 695–700 (2007)
11. Abkowitzm, J., Tozzi, J., Driscoll, M.K.: Operational Feasibility of Timed Transfer in Transit Systems. Journal of Transportation Engineering 113, 168–177 (1987)
12. Peter, K., Theo, M.: Optimized Transfer Opportunities in Public Transport. Transportation Science 29, 101–105 (1995)
13. Lo, H.K., Yip, C.W., Wan, K.H.: Modeling Transfer and Non-linear Fare Structure in Multi-modal Network. Transportation Research Part B 37, 149–170 (2003)
14. Su, A.H., Shi, F.Z.: Optimal Route Choice of Public Traffic Network Based on Shortest Path Searching. Journal of Engineering Graphics 4, 54–58 (2005)
15. Lozano, A., Storchi, G.: Shortest Viable Path Algorithm in Multimodal Network. Transportation Reasearch Part A 35, 225–241 (2001)

Rule Extraction Method in Incomplete Decision Table for Fault Diagnosis Based on Discernibility Matrix Primitive

Wentao Huang, Weijie Wang, and Xuezeng Zhao

School of Mechatronics Engineering, Harbin Institute of Technology
Harbin, 150001, P.R. China
{hwt,greatwang,zhaoxz}@hit.edu.cn

Abstract. Compared with extracting rules from complete data, it is more difficult to obtain rules from incomplete data for fault diagnosis. In this paper, based on the rough set theory, a method is proposed to directly extract optimal generalized decision rules from incomplete a decision table for fault diagnosis (IDTFD). The discernibility matrix primitive is defined and characterized to simplify the computing process. A definition of object-oriented discernibility matrix in IDTFD is also proposed. Using these concepts, an object-oriented discernibility function is constructed. With the basic equivalent forms in proposition logic such as distribution laws, absorption laws, a method is proposed to compute the minimal object-oriented reductions and to extract the optimal generalized decision rules in IDTFD. The proposed method is applied in fault diagnosis of operational states of an electric system. The effectiveness of this method is shown in our experiments.

Keywords: Fault diagnosis, Incomplete decision table, Discernibility matrix primitive, Rule extraction.

1 Introduction

Incompleteness, an important characteristic of fault in complex system, is a key research area in the study of the intelligent fault diagnosis theory and method [1]. Human experts mostly make decisions based on incomplete data. Therefore, intelligent fault diagnosis should be able to diagnose and infer in the same situation.

In fault diagnosis, a method is needed so that even when the data is incomplete, diagnosis decision rule can still be extracted for inference, decision-making and providing maximum probable solution. Doubtlessly, the research of rule extraction from incomplete data has practical value.

By using the Rough Set (RS) theory proposed by Pawlak [2] in 1982, the imprecise, inconsistent, incomplete and uncertain information can be effectively analyzed and handled. The hidden knowledge and latent rules can then be discovered. Therefore, this theory has been widely applied in machine learning, data mining, artificial intelligence etc.[3].

G. Wang et al. (Eds.): RSKT 2008, LNAI 5009, pp. 755–762, 2008.
© Springer-Verlag Berlin Heidelberg 2008

The research on incompleteness plays a key role in the study of rough set. Many researchers have studied incomplete data in knowledge acquisition. Chmielewski [4] processed data by changing incomplete data set into complete one through removing objects with null values or replacing null values with the most common values. Kryszkiewicz [5,6] expanded the equivalence relations in the traditional rough set theory and proposed similarity relation, which acquired rules in incomplete data set without the necessity of obtaining the null attributes-value. Leung [7] put forward a knowledge discovering method which combined the maximal consistent block technology and rough set method in the rule acquisition from incomplete data. Yin [8] described the object relationship in the incomplete information system through controlling non-symmetric similarity relations so that the categorization of the object generated was more reasonable. Leung [9] have proposed the concept of similarity classes in incomplete information system, which was used in mining certain and association rules in incomplete decision tables, and have also given the new quantitative measures.

For the incomplete data in fault diagnosis, Li [10] replaced each attribute of the example having incomplete attribute value with all the possible values that appeared in other observations with the same concept as the example. However, for data set with too many incomplete data, adoption of the method may drastically increase the scale. To solve this problem, based on the similarity relations forwarded by Kryszkiewicz, this paper proposes a rough set method which can extract optimal generalized diagnostic decision rule directly without changing the scale of the table.

2 Incomplete Decision Table for Fault Diagnosis

Definition 1. [11] *A incomplete decision table for fault diagnosis* $(IDTFD)$ *can be denoted by a triplet* $IDTFD =< U, M \cup \{d\}, f >$, *in which,*
(1) U *is finite nonempty set, called fault object domain;*
(2) M, *the finite nonempty set, is called fault symptom attributes set, where* d *is another attribute distinguished from the attributes in set* M, *which is called fault decision attribute. Then,* $d \notin M$. $V_M = \cup_{a \in M} V_a$ *is the set of attribute-value, where* V_a *is the domain value of symptom* a.
(3) *For every* $a \in M$, *there is a mapping* $f_a : U \rightarrow V_a$.
(4) *For at least one attribute* $a \in M$, V_a *contains null value* $(*)$, *and* $* \notin V_d$.

In Definition 1, the null value only occurs in the value domain of symptom attribute, while the value domain of decision attribute contains no null value. This paper considers the example in which the decision attribute-value contains null value as mistakes, which are then deleted.

In the following section, we will take the central database of electric power control center [11]. Its IDTFD is shown in Table 1, in which each row represents one type of operational points usually pre-designated by experts (operators or engineers), while the last column is the value of generalized decision function calculated from data object.

Table 1. The IDTFD and its generalized decision function

U	k	c_1	c_2	c_3	c_4	c_5	c_6	c_7	c_8	d	$\partial_M(x_i)$
x_1	2	M	M	L	N	H	N	1	1	S	$\{S\}$
x_2	1	$*$	M	$*$	N	H	$*$	1	1	S	$\{S, U2\}$
x_3	3	M	L	L	$*$	H	$*$	$*$	1	S	$\{S, U1\}$
x_4	2	$*$	$*$	M	N	N	N	0	1	S	$\{S\}$
x_5	1	M	$*$	L	$*$	H	N	1	1	S	$\{S, U1\}$
x_6	1	$*$	M	$*$	N	$*$	$*$	1	1	S	$\{S, U2\}$
x_7	3	L	$*$	L	$*$	$*$	H	$*$	1	$U2$	$\{S, U1, U2\}$
x_8	2	L	$*$	M	H	L	$*$	0	1	$U2$	$\{U1, U2\}$
x_9	1	$*$	M	M	$*$	L	H	$*$	1	$U2$	$\{S, U1, U2\}$
x_{10}	1	L	M	$*$	H	N	$*$	0	1	$U2$	$\{U1, U2\}$
x_{11}	1	H	$*$	M	L	H	L	$*$	1	$U2$	$\{U2\}$
x_{12}	4	$*$	M	$*$	H	$*$	H	$*$	1	$U1$	$\{U1, U2\}$
x_{13}	2	L	H	M	$*$	N	H	0	1	$U1$	$\{U1\}$
x_{14}	1	M	$*$	$*$	L	$*$	$*$	1	1	$U1$	$\{S, U1, U2\}$

Fault symptom attribute set $M = \{c_1, c_2, c_3, c_4, c_5, c_6, c_7, c_8\}$, where c_1, c_2, c_3 are the ratios of real power flows in transmission lines according to its nominal capacity, $V_{c_i} = \{L, M, H\}$, where $i = 1, 2, 3$, L, M, and H mean that the ratios are lower than 40%, larger than 40% but lower than 70%, and larger than 70%, respectively. c_4, c_5 and c_6 represent the normalized values of actual voltage in transmission line, $V_{c_i} = \{L, N, H\}$, $i = 4, 5, 6$, L, N, and H are the normalized value of actual voltage in transmission line, which are lower than 0.85, larger than 0.85 but lower than 1.05 and larger than 1.05, respectively. c_7 and c_8 are the states of circuit-breaker, $V_{c_i} = \{0, 1\}$, 0 means "on ", while 1 means "off ", $i = 7, 8$. The decision attributes are d, $V_d = \{S, U1, U2\}$, which correspond to the three operational states: safe (S), unsafe level 1 $(U1)$, and unsafe level 2 $(U2)$. k signifies the number of the points at the same state.

Obviously, in IDTFD, the indiscernibility relation in the traditional rough set is not valid anymore. Therefore, the similarity relation [5] is adopted in this paper to describe the object relation in fault state domain U. In IDTFD , let $B \subseteq M$, then, the similarity relation in U is denoted as: $S(B) = \{(x, y) \in U \times U \mid \forall a \in B, f_a(x) = f_a(y) \text{ or } f_a(x) = * \text{ or } f_a(y) = *\}$.

Let $S_B(x)$ be object set $\{y \in U \mid (x, y) \in S(B)\}$. $S_B(x)$ describes all the objects similar to x in domain U for symptom attribute set B. Let $U/S(B) = \{S_B(x) \mid x \in U\}$, which represents all the similarity classes of objects classified from domain U according to the symptom attribute B.

3   Optimal Generalized Diagnosis Decision Rules

In actual instance of fault diagnosis, the raw data acquired from equipment are, to some extent, usually inconsistent. In order to deal with the inconsistency, the generalized decision rule proposed by Kryszkiewicz,M. [6] is adopted to describe the inconsistency.In IDTFD , let $B \subseteq M$, for any $x \in U$, then the function that satisfies the following function,$\partial_B(x) = \{f_d(y) \mid y \in S_B(x)\}$,is the generalized decision function of IDTFD.

$\partial_B(x)$ defines the usable information, where object x is possibly classified in the decision class. If $card(\partial_M(x))=1$, then, x can be definitely classified in the decision class which includes only one decision value, $card(\cdot)$ denotes the cardinality of set. In Table 1, the generalized decision function of every object is shown in the last column.

The knowledge hidden in IDTFD can be discovered and expressed in the form of decision rules: $t \to s$, where $t = \wedge(a,v), a \in B \subseteq M, v \in V_a \setminus \{*\}, s = \vee(d,w), w \in V_d$. Henceforth, t and s stand for the condition and decision part of the rule $t \to s$, respectively. $\wedge$ and $\vee$ are the conjunction and disjunction in the logic calculation. In addition, the attribute-value pair (a,v) satisfies object set, then, $\{x_i \in U \mid f_a(x_i) = v\}$, which is denoted as $\| (a,v) \|$. Let $\| t \|$ be the object set that satisfies the condition part t and $\| s \|$ be the object set that satisfies the decision part s of the decision rule.

In IDTFD, $x \in U$, it is denoted that object x supports a fault diagnosis decision rule $t \to s$, if and only if x simultaneously satisfies the condition part t and the decision part s.

In IDTFD , the decision rule $t \to s$ is denoted as generalized , if and only if $\{x \in U \mid S_B(x) \cap \| t \| \neq \emptyset\} \subseteq \| s \|$,where B are all the symptom attributes that occur in the condition part t of decision rule.

In IDTFD, if a decision rule $t \to s$ is the generalized decision rule, then, the decision rule is regarded as consistent. In addition, if the conjunction of the decision rule obtained from IDTFD is generalized, then, IDTFD is called consistent, or else, it is inconsistent. The consistency of IDTFD means that diagnosis decisions can be made based on the condition provided by IDTFD without contradiction.

In IDTFD, a fault diagnosis rule $t \to s$ is called optimal generalized rule. If and only if it is generalized, then, the rules, formed by any proper subset of conjunction and disjunction occurring in t or s, are not generalized. IDTFD is consistent, if and only if for any $x \in U$, $card(\partial_M(x))=1$.

The example x_i in Table 1 and the corresponding row can all be denoted as decision rules. Then, it can be inferred that decision rule in Table 1 can be further reduced by deleting the attribute-value of condition part to form the optimal generalized decision rule.

4 Discernibility Matrix Primitive

In the fault diagnosis based on rough set, usually, the original decision table undergoes the symptom reduction and value reduction process to extract the concise diagnosis rule. In order to improve the efficiency of decision rule acquisition, this paper combines the symptom reduction and value reduction, and proposed an object-oriented reduction method. The method is used to extract the optimal generalized diagnosis decision rule from IDTFD.

Definition 2. *In IDTFD* $=< U, M \cup \{d\}, f >$, *the subset of symptom attributes* $B \subseteq M$ *is the reduction of the object-oriented* $x \in U$, *if and only if*

$$\partial_B(x) = \partial_M(x) \quad \text{and} \quad \forall B' \subset B, \partial_{B'}(x) \neq \partial_M(x). \tag{1}$$

In IDTFD, some objects contain unknown symptom attribute-values, so the classical equivalence relation is expanded to similarity relation, and the discernibility matrix for complete system needs to be expanded too. In the following part, the concept of discernibility matrix primitive in IDTFD is given first.

Definition 3. **[11]** *In IDTFD* $=< U, M \cup \{d\}, f >$, *where* $(x_i, x_j) \in U \times U$. $\beta_M(x_i, x_j)$, *the discernibility matrix primitive is denoted as*

$$\beta_M(x_i, x_j) = \begin{cases} \{a \in M \mid f_a(x_i) \neq *, f_a(x_j) \neq *, f_a(x_i) \neq f_a(x_j)\}, f_d(x_j) \notin \partial_M(x_i) \\ \emptyset, \qquad\qquad\qquad\qquad\qquad\qquad\qquad\qquad\quad f_d(x_j) \in \partial_M(x_i) \end{cases} \tag{2}$$

From the Definition 3, it can be seen that the discernibility matrix primitives may not be symmetrical, i.e., $\beta_M(x_i, x_j)$ and $\beta_M(x_j, x_i)$ are not always equal. The following properties give the equivalent condition of the discernibility matrix primitive [12].

Property 1. In IDTFD $=< U, M \cup \{d\}, f >$, $(x_i, x_j) \in U \times U$. $\beta_M(x_i, x_j)$ is the discernibility matrix primitive. If $\partial_M(x_i) \cap \partial_M(x_j) = \emptyset$, then $\beta_M(x_i, x_j) = \beta_M(x_j, x_i)$.

Property 2. In IDTFD $=< U, M \cup \{d\}, f >$, $(x_i, x_j) \in U \times U$. $\beta_M(x_i, x_j)$ is the discernibility matrix primitive. If $\partial_M(x_i) = \partial_M(x_j)$, then $\beta_M(x_i, x_j) = \beta_M(x_j, x_i) = \emptyset$.

Property 3. In IDTFD $=< U, M \cup \{d\}, f >$, $(x_i, x_j) \in U \times U$. $\beta_M(x_i, x_j)$ is the discernibility matrix primitive. If $\partial_M(x_j) \subset \partial_M(x_i)$, then $\beta_M(x_i, x_j) = \emptyset$.

From Properties 1, 2 and 3, the calculation process of discernibility matrix primitive can be simplified as follows: for two objects whose intersection of the generalized decision function value sets are null, since discernibility matrix primitives $\beta_M(x_i, x_j)$ and $\beta_M(x_j, x_i)$ are equal, only one of them needs to be calculated. When the two generalized decision function value sets are completely equal, the discernibility matrix primitives are null. For discernibility matrix primitives, if the generalized decision value set of column elements is the proper subset of the generalized decision value set of row element, then, the discernibility matrix primitive is null.

5   The Extraction of Optimal Generalized Decision Rule

According to Definition 3, the following part defines the object-oriented discernibility matrix in IDTFD, which then can be used to construct the object-oriented discernibility function so as to obtain the object-oriented reduction in the IDTFD.

Table 2. The object-oriented discernibility matrix of examples

$x_i \setminus x_j$	x_1	x_2	x_3	$\cdots$	x_{13}	x_{14}
x_1	$\emptyset$	$\emptyset$	$\emptyset$	$\cdots$	$c_1, c_2, c_3, c_5, c_6, c_7$	c_4
x_2	$\emptyset$	$\emptyset$	$\emptyset$	$\cdots$	c_2, c_5, c_6	c_4
x_3	$\emptyset$	$\emptyset$	$\emptyset$	$\cdots$	$\emptyset$	$\emptyset$
$\vdots$	$\vdots$	$\vdots$	$\vdots$	$\vdots$ $\vdots$	$\vdots$	$\vdots$
x_{12}	c_4, c_6	c_4	c_2	$\cdots$	$\emptyset$	$\emptyset$
x_{13}	$c_1, c_2, c_3, c_5, c_6, c_7$	c_2, c_5, c_7	c_1, c_2, c_3, c_5	$\cdots$	$\emptyset$	$\emptyset$
x_{14}	$\emptyset$	$\emptyset$	$\emptyset$	$\cdots$	$\emptyset$	$\emptyset$

Definition 4. *In IDTFD $=< U, M \cup \{d\}, f >$, card$(U) = n$, let $\beta_M(x_i, x_j)$ be the discernibility matrix primitive, where $x_i, x_j \in U$. The $n \times n$ matrix $[\beta_M(x_i, x_j)]_{n \times n}$ formed by discernibility matrix primitives is the object-oriented discernibility matrix, where $i, j = 1, 2, \ldots, n$. $\triangle(x_i) = \wedge \vee \beta_M(x_i, x_j), i, j = 1, 2, \ldots, n$, which is seen as the object-oriented discernibility function of x_i. If $\beta_M(x_i, x_j) = \emptyset$, then, let $\vee \beta_M(x_i, x_j) = 1$, or else, $\vee \beta_M(x_i, x_j)$ is the disjunctive variable corresponding to the symptom attributes included in $\beta_M(x_i, x_j)$.*

According to Definition 4, the object-oriented discernibility matrix corresponding to Table 1 is obtained and shown in Table 2.

In IDTFD, the object-oriented discernibility function object $\triangle(x_i)$, of the object x_i, is a conjunctive normal form. When it is converted to its equivalent disjunctive normal form, conjunctive normal forms of the disjunctive normal form define all the object-oriented reduction of object x_i. According to the definition of minimal reduction, it can be easily known that the conjunctive normal forms with the least symptom attributes are the minimal object-oriented reductions of x_i, denoted as $R_{min}(x_i)$. The proved calculation for the acquisition of minimal reduction is a problem of NP-hard. Therefore, combined with the basic equivalent forms of proposition logics, this paper proposes a method to calculate the minimal object-oriented reduction. The method is given below.

(1) The discernibility function $\triangle(x_i)$ of object-oriented x_i is denoted as the conjunctive normal form of several disjunctive normal forms.

(2) For every disjunctive normal form, starting from the one with the least symptom attributes, the absorption law of proposition logic is used to get rid of the superset of the disjunctive normal form.

(3) The number of occurrence of every symptom attribute is obtained. Then, the symptom with the highest number of occurrence (at least twice) is extracted as the common factor. The distribution law in the proposition logic is used to realize equivalent conversion.

(4) Repeat processes (2) and (3) until the absorption law cannot be applied. After the conversion, the conjunctive normal forms with the least symptom attributes in the disjunctive normal forms are the object-oriented reduction of object x_i.

Table 3. The optimal generalized diagnosis rule set of examples

Number	Optimal generalized diagnosis rule	Support degree	Support object
r_1	$(c_4, N) \wedge (c_6, N) \rightarrow (d, S)$	4	x_1, x_4
r_2	$(c_2, M) \wedge (c_4, N) \rightarrow (d, S) \vee (d, U2)$	4	x_1, x_2, x_6
r_3	$(c_4, N) \wedge (c_5, H) \rightarrow (d, S) \vee (d, U2)$	3	x_1, x_2
r_4	$(c_4, N) \wedge (c_7, 1) \rightarrow (d, S) \vee (d, U2)$	4	x_1, x_2, x_6
r_5	$(c_1, M) \wedge (c_2, L) \rightarrow (d, S) \vee (d, U1)$	3	x_3
$\vdots$	$\vdots$	$\vdots$	$\vdots$
r_{18}	$(c_2, M) \wedge (c_4, H) \wedge (c_6, H) \rightarrow (d, U1) \vee (d, U2)$	4	x_{12}
r_{19}	$(c_2, H) \wedge (c_3, M) \wedge (c_5, N) \wedge (c_6, H) \rightarrow (d, U1)$	2	x_{13}

After the object-oriented reduction of all the objects are obtained, the generalized fault diagnosis decision rule for IDTFD can be determined consequently. Then, the decision part and the condition part of the generalized diagnosis decision rule supported by every object in IDTFD are combined. The support degree (the number of examples supported by the rule) and the support object of the decision rule are taken as the evaluation criteria [11] for the optimal generalized diagnosis decision rule set supported by IDTFD, as shown in Table 3.

The diagnosis decision rule of Table 3 is the concise summary and direct demonstration in Table 1. It can provide decision support in the follow-up recognition of security state of the operation point for operators in the control center of electric power system. The rules provide a sound base for the follow-up construction of fault diagnosis knowledge base.

6 Conclusion

In this paper, the fault diagnosis from incomplete data was analyzed. The fault data provided by similarity relation in IDTFD was utilized and the concepts of optimal generalized decision rule in IDTFD were proposed. The consistency of the fault diagnosis was also studied. Some concepts such as the discernibility matrix primitive, object-oriented discernibility matrix and discernibility function were proposed. The two processes of symptom reduction and value reduction in the existing rough-set-based fault diagnosis method were combined. Along with the basic equivalent forms in the proposition logics, minimal object-oriented reduction method was proposed, which was used to directly obtain concise optimal generalized diagnosis decision rule in IDTFD. The fault decision rule contained incomplete data which made it convenient to infer or make decisions facing new fault state. This paper provided the application steps of diagnosis method based on a real example of the security status of operation point in an electric power system. The proved effectiveness showed that the method could also be used in the fault diagnosis in engineering field such as in mechanical equipments.

Acknowledgments

This paper was supported by the China Postdoctoral Science Foundation (No.20070410888).

References

1. Yang, S.Z., Ding, H., Shi, T.L., Zheng, X.J.: Diagnosis Reasoning Based on Knowledge. Tsinghua University Press, Beijing (1993)
2. Pawlak, Z.: Rough sets. Int. J. Comp. Inform. Sci. 11, 341–356 (1982)
3. Pawlak, Z., Grzymala-Busse, J., Slowinski, R., Ziarko, W.: Rough sets. Commun. ACM 38, 89–95 (1995)
4. Chmielewski, M.R., Grzymala-Busse, J.W., Peterson, N.W., Than, S.: The rule induction system LERS - a version for personal computers. Found. Comput. Decision Sci. 18, 181–212 (1993)
5. Kryszkiewicz, M.: Rough set approach to incomplete information systems. Inf. Sci. 112, 39–49 (1998)
6. Kryszkiewicz, M.: Rules in incomplete information systems. Inf. Sci. 113, 271–292 (1999)
7. Leung, Y., Li, D.Y.: Maximal consistent block technique for rule acquisition in incomplete information systems. Inf. Sci. 153, 85–106 (2003)
8. Yin, X.R., Jia, X.Y., Shang, L.: A new extension model of rough sets under incomplete information. In: Wang, G.-Y., Peters, J.F., Skowron, A., Yao, Y. (eds.) RSKT 2006. LNCS (LNAI), vol. 4062, pp. 141–146. Springer, Heidelberg (2006)
9. Leung, Y., Wu, W.Z., Zhang, W.X.: Knowledge acquisition in incomplete information systems: A rough set approach. Eur. J. Oper. Res. 168, 164–180 (2006)
10. Li, J.R., Khoo, L.P., Tor, S.B.: RMINE: A rough set based data mining prototype for the reasoning of incomplete data in condition-based fault diagnosis. J. Intell. Manuf. 17, 163–176 (2006)
11. Huang, W.T., Wang, W.J., Zhao, X.Z., Dai, L.Z.: Extracting optimal generalized decision rules for fault diagnosis from incomplete data based on rough set. Automation of Electric Power Systems 29, 49–54 (2005)
12. Huang, W.T.: A Study on Rule Extraction and Uncertainty Measure of Fault Diagnosis Based on Rough Set Theory. Ph.D thesis, Harbin Institute of Technology, Harbin (2004)

Author Index

Printing: Mercedes-Druck, Berlin
Binding: Stein+Lehmann, Berlin

Lecture Notes in Artificial Intelligence (LNAI)

Vol. 4766: N. Maudet, S. Parsons, I. Rahwan (Eds.), Argumentation in Multi-Agent Systems. XII, 211 pages. 2007.

Vol. 4760: E. Rome, J. Hertzberg, G. Dorffner (Eds.), Towards Affordance-Based Robot Control. IX, 211 pages. 2008.

Vol. 4755: V. Corruble, M. Takeda, E. Suzuki (Eds.), Discovery Science. XI, 298 pages. 2007.

Vol. 4754: M. Hutter, R.A. Servedio, E. Takimoto (Eds.), Algorithmic Learning Theory. XI, 403 pages. 2007.

Vol. 4737: B. Berendt, A. Hotho, D. Mladenic, G. Semeraro (Eds.), From Web to Social Web: Discovering and Deploying User and Content Profiles. XI, 161 pages. 2007.

Vol. 4733: R. Basili, M.T. Pazienza (Eds.), AI*IA 2007: Artificial Intelligence and Human-Oriented Computing. XVII, 858 pages. 2007.

Vol. 4724: K. Mellouli (Ed.), Symbolic and Quantitative Approaches to Reasoning with Uncertainty. XV, 914 pages. 2007.

Vol. 4722: C. Pelachaud, J.-C. Martin, E. André, G. Chollet, K. Karpouzis, D. Pelé (Eds.), Intelligent Virtual Agents. XV, 425 pages. 2007.

Vol. 4720: B. Konev, F. Wolter (Eds.), Frontiers of Combining Systems. X, 283 pages. 2007.

Vol. 4702: J.N. Kok, J. Koronacki, R. Lopez de Mantaras, S. Matwin, D. Mladenič, A. Skowron (Eds.), Knowledge Discovery in Databases: PKDD 2007. XXIV, 640 pages. 2007.

Vol. 4701: J.N. Kok, J. Koronacki, R. Lopez de Mantaras, S. Matwin, D. Mladenič, A. Skowron (Eds.), Machine Learning: ECML 2007. XXII, 809 pages. 2007.

Vol. 4696: H.-D. Burkhard, G. Lindemann, R. Verbrugge, L.Z. Varga (Eds.), Multi-Agent Systems and Applications V. XIII, 350 pages. 2007.

Vol. 4694: B. Apolloni, R.J. Howlett, L. Jain (Eds.), Knowledge-Based Intelligent Information and Engineering Systems, Part III. XXIX, 1126 pages. 2007.

Vol. 4693: B. Apolloni, R.J. Howlett, L. Jain (Eds.), Knowledge-Based Intelligent Information and Engineering Systems, Part II. XXXII, 1380 pages. 2007.

Vol. 4692: B. Apolloni, R.J. Howlett, L. Jain (Eds.), Knowledge-Based Intelligent Information and Engineering Systems, Part I. LV, 882 pages. 2007.

Vol. 4687: P. Petta, J.P. Müller, M. Klusch, M. Georgeff (Eds.), Multiagent System Technologies. X, 207 pages. 2007.

Vol. 4682: D.-S. Huang, L. Heutte, M. Loog (Eds.), Advanced Intelligent Computing Theories and Applications. XXVII, 1373 pages. 2007.

Vol. 4676: M. Klusch, K.V. Hindriks, M.P. Papazoglou, L. Sterling (Eds.), Cooperative Information Agents XI. XI, 361 pages. 2007.

Vol. 4667: J. Hertzberg, M. Beetz, R. Englert (Eds.), KI 2007: Advances in Artificial Intelligence. IX, 516 pages. 2007.

Vol. 4660: S. Džeroski, L. Todorovski (Eds.), Computational Discovery of Scientific Knowledge. X, 327 pages. 2007.

Vol. 4659: V. Mařík, V. Vyatkin, A.W. Colombo (Eds.), Holonic and Multi-Agent Systems for Manufacturing. VIII, 456 pages. 2007.

Vol. 4651: F. Azevedo, P. Barahona, F. Fages, F. Rossi (Eds.), Recent Advances in Constraints. VIII, 185 pages. 2007.

Vol. 4648: F. Almeida e Costa, L.M. Rocha, E. Costa, I. Harvey, A. Coutinho (Eds.), Advances in Artificial Life. XVIII, 1215 pages. 2007.

Vol. 4635: B. Kokinov, D.C. Richardson, T.R. Roth-Berghofer, L. Vieu (Eds.), Modeling and Using Context. XIV, 574 pages. 2007.

Vol. 4632: R. Alhajj, H. Gao, X. Li, J. Li, O.R. Zaïane (Eds.), Advanced Data Mining and Applications. XV, 634 pages. 2007.

Vol. 4629: V. Matoušek, P. Mautner (Eds.), Text, Speech and Dialogue. XVII, 663 pages. 2007.

Vol. 4626: R.O. Weber, M.M. Richter (Eds.), Case-Based Reasoning Research and Development. XIII, 534 pages. 2007.

Vol. 4617: V. Torra, Y. Narukawa, Y. Yoshida (Eds.), Modeling Decisions for Artificial Intelligence. XII, 502 pages. 2007.

Vol. 4612: I. Miguel, W. Ruml (Eds.), Abstraction, Reformulation, and Approximation. XI, 418 pages. 2007.

Vol. 4604: U. Priss, S. Polovina, R. Hill (Eds.), Conceptual Structures: Knowledge Architectures for Smart Applications. XII, 514 pages. 2007.

Vol. 4603: F. Pfenning (Ed.), Automated Deduction – CADE-21. XII, 522 pages. 2007.

Vol. 4597: P. Perner (Ed.), Advances in Data Mining. XI, 353 pages. 2007.

Vol. 4594: R. Bellazzi, A. Abu-Hanna, J. Hunter (Eds.), Artificial Intelligence in Medicine. XVI, 509 pages. 2007.

Vol. 4585: M. Kryszkiewicz, J.F. Peters, H. Rybinski, A. Skowron (Eds.), Rough Sets and Intelligent Systems Paradigms. XIX, 836 pages. 2007.

Vol. 4578: F. Masulli, S. Mitra, G. Pasi (Eds.), Applications of Fuzzy Sets Theory. XVIII, 693 pages. 2007.

Vol. 4573: M. Kauers, M. Kerber, R. Miner, W. Windsteiger (Eds.), Towards Mechanized Mathematical Assistants. XIII, 407 pages. 2007.

Vol. 4571: P. Perner (Ed.), Machine Learning and Data Mining in Pattern Recognition. XIV, 913 pages. 2007.

Vol. 4570: H.G. Okuno, M. Ali (Eds.), New Trends in Applied Artificial Intelligence. XXI, 1194 pages. 2007.

Vol. 4565: D.D. Schmorrow, L.M. Reeves (Eds.), Foundations of Augmented Cognition. XIX, 450 pages. 2007.

Vol. 4562: D. Harris (Ed.), Engineering Psychology and Cognitive Ergonomics. XXIII, 879 pages. 2007.

Vol. 4548: N. Olivetti (Ed.), Automated Reasoning with Analytic Tableaux and Related Methods. X, 245 pages. 2007.

Vol. 4539: N.H. Bshouty, C. Gentile (Eds.), Learning Theory. XII, 634 pages. 2007.